SAXON Math™

HOMESCHOOL

8/7

with Prealgebra

Stephen Hake
John Saxon

Saxon Publishers gratefully acknowledges the contributions of the following individuals in the completion of this project:

Authors: Stephen Hake, John Saxon

Editorial: Chris Braun, Brooke Butner, Brian E. Rice

Editorial Support Services: Christopher Davey, Jay Allman, Jenifer Sparks, Shelley Turner, Jean Van Vleck, Darlene Terry

Production: Alicia Britt, Karen Hammond, Donna Jarrel, Brenda Lopez, Adriana Maxwell, Cristi D. Whiddon

Project Management: Angela Johnson, Becky Cavnar

Printed in the United States of America

ISBN: 1-59141-320-6

Manufacturing Code: 2 3 4 5 6 7 8 062 13 12 11 10 09 08 07 06

CONTENTS

LETTER FROM AUTHOR STEPHEN HAKE

Dear Student,

We study mathematics because of its importance to our lives. Our study schedule, our trip to the store, the preparation of our meals, and many of the games we play involve mathematics. You will find that the word problems in this book are often drawn from everyday experiences.

As you grow into adulthood, mathematics will become even more important. In fact, your future in the adult world may depend on the mathematics you have learned. This book was written to help you learn mathematics and to learn it well. For this to happen, you must use the book properly. As you work through the pages, you will see that similar problems are presented over and over again. ***Solving each problem day after day is the secret to success.***

Your book is made up of daily lessons and investigations. Each lesson has four parts. The first part is a Warm-Up that includes practice of basic facts and mental math. These exercises improve your speed, accuracy, and ability to do math "in your head." The Warm-Up also includes a problem-solving exercise to familiarize you with strategies for solving complicated problems. The second part of the lesson is the New Concept. This section introduces a new mathematical concept and presents examples that use the concept. In the next section, the Lesson Practice, you have a chance to solve problems involving the new concept. The problems are lettered a, b, c, and so on. The final part of the lesson is the Mixed Practice. This problem set reviews previously taught concepts and prepares you for concepts that will be taught in later lessons. Solving these problems helps you remember skills and concepts for a long time.

Investigations are variations of the daily lesson that often involve activities. Investigations contain their own set of questions instead of a problem set.

Remember, solve every problem in every practice set, every problem set, and every investigation. Do not skip problems. With honest effort, you will experience success and true learning that will stay with you and serve you well in the future.

Stephen Hake
Temple City, California

PREFACE

Dear Parent-Teacher,

Congratulations on your decision to use *Saxon Math 8/7—Homeschool*! Proven results, including higher test scores, have made Saxon Math™ the hands-down favorite for homeschoolers. Only Saxon helps you teach the way your child learns best—step-by-step. With Saxon, each new skill builds on those already taught, daily reviews of earlier material increase understanding, and frequent, cumulative assessments ensure that your child masters each skill before new ones are added. The result? More confidence, more willingness to learn, more success!

SAXON PHILOSOPHY

The unique structure of Saxon Math™ promotes student success through the sound, proven educational practices of *incremental development* and *continual review.* Consider how most other mathematics programs are structured: content is organized into topical chapters, and topics are developed rapidly to prepare students for end-of-chapter tests. Once a chapter is completed, the topic changes, and often practice on the topic ends as well. Many students struggle to absorb the large blocks of content and often forget the content after practice on it ends. Chapter organization might be good for reference, but it is not the best organization for learning. Incremental development and continual review are structural designs that improve student learning.

Incremental development

With incremental development, topics are developed in small steps spread over time. One facet of a concept is taught and practiced before the next facet is introduced. Both facets are then practiced together until it is time for the third to be introduced. Instead of being organized in chapters that rapidly develop a topic and then move on to the next strand, Saxon Math™ is organized in a series of lessons that gradually develop concepts. This approach gives students the time to develop a deeper understanding of concepts and how to apply them.

Continual review

Through continual review, previously presented concepts are practiced frequently and extensively throughout the year. Saxon's cumulative daily practice strengthens students' grasp of concepts and improves their long-term retention of concepts.

John Saxon often said, "Mathematics is not difficult. Mathematics is just different, and time is the elixir that turns things different into things familiar." This program provides the time and experiences students need to learn the skills and concepts necessary for success in mathematics, whether those skills are applied in quantitative disciplines or in the mathematical demands of everyday life.

PROGRAM COMPONENTS

Saxon Math 8/7—Homeschool consists of three components: 1) textbook, 2) Tests and Worksheets, and 3) Solutions Manual. **Before using the program, please ensure that you have each component.**

Textbook

The *Saxon Math 8/7—Homeschool* textbook is divided into 120 lessons and 12 investigations. The textbook also contains an appendix topic that can be presented at the teacher's discretion, supplemental practice problems for remediation, an illustrated glossary, and a comprehensive index.

Tests and Worksheets

The *Saxon Math 8/7—Homeschool Tests and Worksheets* booklet provides all the worksheets and tests needed by one student to complete the program. It also contains the following recording forms for students to show their work and for parents to track student progress:

- Recording Form A: Facts Practice
- Recording Form B: Lesson Worksheet
- Recording Form C: Mixed Practice Solutions
- Recording Form D: Scorecard
- Recording Form E: Test Solutions

Directions for using the recording forms are provided in the Program Overview (below), as well as in the introduction to the Tests and Worksheets booklet.

Note: The recording forms are blackline masters that should be photocopied, as they may be used more than once.

Solutions Manual

The *Saxon Math 8/7—Homeschool Solutions Manual* contains step-by-step solutions to all textbook and test exercises.

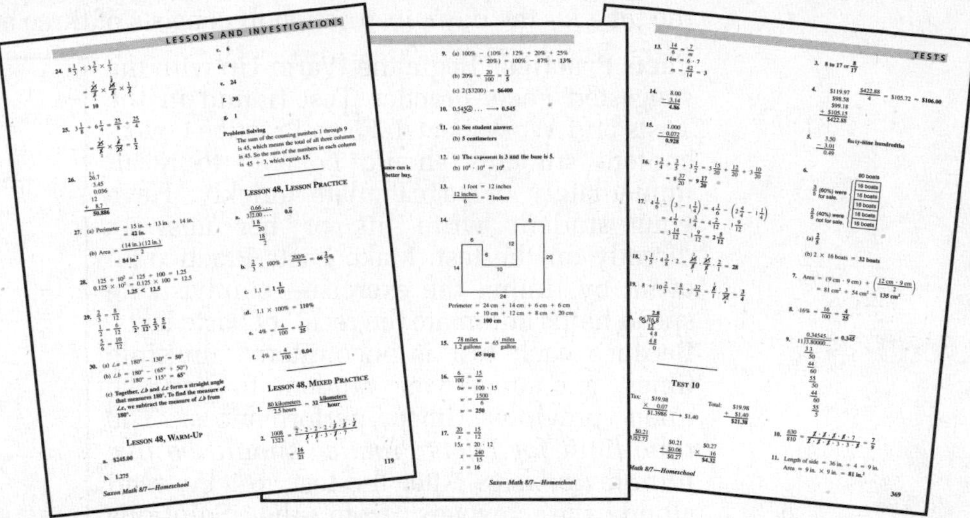

PROGRAM OVERVIEW

Saxon Math 8/7—Homeschool contains three types of math "sessions": lessons, investigations, and tests. Concepts are introduced and reviewed in a carefully planned sequence. **It is therefore crucial to complete all the lessons and investigations in *Saxon Math 8/7—Homeschool* in the given order.** If lessons are skipped or presented out of sequence, students will encounter problems on the tests and in the problem sets that they might not be equipped to solve.

By completing one lesson, investigation, or test per day, you can finish the entire program in thirty-one or thirty-two weeks. However, faster or slower paces may be appropriate, depending on students' individual learning styles.

Lessons

Each of the program's 120 lessons is divided into four sections: Warm-Up, New Concept(s), Lesson Practice, and Mixed Practice. Below we show a lesson from the textbook.

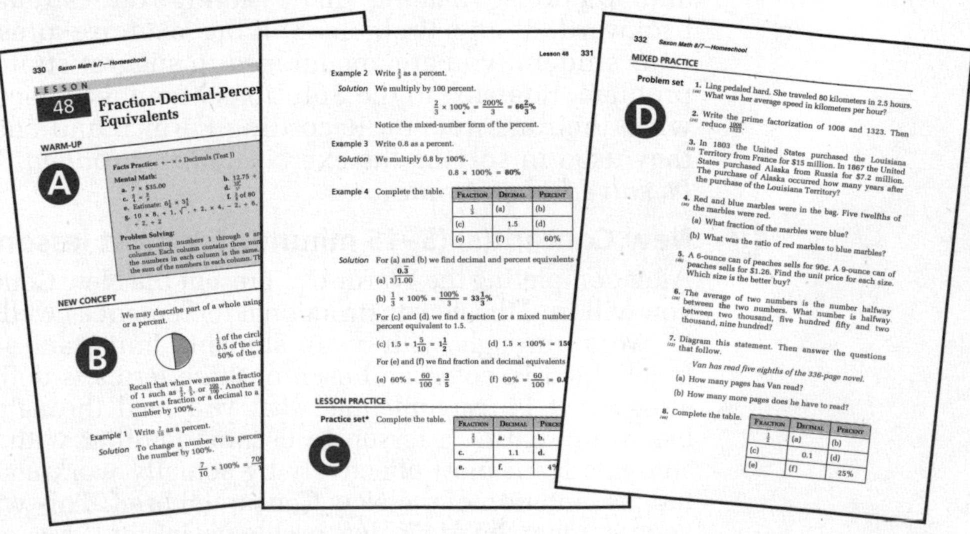

Ⓐ Warm-Up (10–15 minutes)

The Warm-Up promotes mental math and problem-solving skills and sets the tone for the day's instruction. It consists of three activities:

Facts Practice: Begin the Warm-Up with the suggested Facts Practice Test (found in the Tests and Worksheets). Facts Practice covers content students should be able to recall immediately or to calculate quickly. Have your student write his or her answers directly on the test. Make Facts Practice an event by timing the exercise—emphasizing speed helps automate the recall of basic facts. Because each test is encountered multiple times, encourage your student to improve upon previous timed performances. *The time limit for Facts Practice should be five minutes or less.* After the test, quickly read aloud the answers from the Solutions Manual as your student checks his or her answers. If desired, Facts Practice scores and times can be tracked on Recording Form A (from the Tests and Worksheets) or in a math notebook. The time invested in Facts Practice is repaid in students' ability to work more quickly.

FACTS PRACTICE TEST

J — + − × ÷ Decimals
For use with Lesson

Name ______
Time ______

Simplify these expressions.

0.8 + 0.4 =	0.8 × 0.4 =	0.8 ÷ 0.4 =
1.2 − 0.4 =	1.2 × 0.4 =	1.2 ÷ 0.4 =
1.2 + 0.04 =	1.2 × 0.04 =	1.2 ÷ 0.04 =
1.2 + 4 =	1.2 × 4 =	1.2 ÷ 4 =
6 − 0.3 =	6 × 0.3 =	6 ÷ 0.3 =
0.3 + 6 =	0.3 × 6 =	0.3 ÷ 6 =
0.01 − 0.01 =	0.01 × 0.01 =	0.01 ÷ 0.01 =

Saxon Math 8/7—Homeschool 100

Mental Math: Follow Facts Practice with Mental Math. Read the problems aloud while your student follows along. Have your student perform the calculations mentally and write the answers on a copy of Recording Form B or on blank paper. (*Note:* Students should **not** use pencil and paper to perform the calculations.) Mental math ability pays lifelong benefits and improves markedly with practice. *Complete the Mental Math activity in two to three minutes.* Mental Math answers are provided in the Solutions Manual.

Problem Solving: Finish the Warm-Up with the daily Problem Solving exercise. Problem Solving promotes critical-thinking skills and offers opportunities for students to use such strategies as drawing diagrams and pictures, making lists, acting out situations, and working backward. If the Problem Solving exercise presents difficulties for your student, you are encouraged to suggest strategies for tackling the problem, referring to the Solutions Manual as necessary. Students may write their answers on Recording Form B and check off the strategies they used in solving the exercise. *Most Problem Solving exercises can be solved in a few minutes.*

Ⓑ New Concept(s) (5–15 minutes for most lessons)

After completing the Warm-Up, present the New Concept(s). In this section you will find the new instructional increment as well as example problems to work through with your student. Important vocabulary terms are highlighted in color, and each of these terms is defined in the textbook's glossary. It is recommended that you read through the New Concept(s) before presenting a lesson to become familiar with the content. Because students learn most effectively by actually working math problems, keep the presentation of the New Concept(s) brief. This will maximize the time your student has to solve problems in the Lesson Practice and Mixed Practice problem sets (which are described in the next sections).

Some lessons involve activities that require the use of household items. Refer to page xxi for a list of necessary materials and the lessons in which they are used. Certain lessons also call for students to use Activity Sheets (see example at right). Activity Sheets are referenced in the textbook and can be found in the Tests and Worksheets booklet.

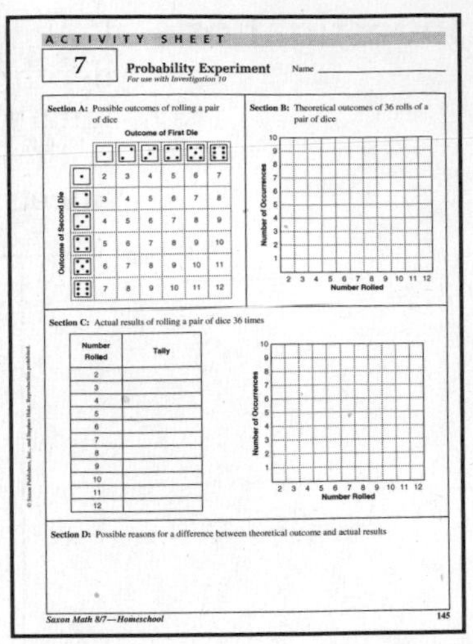

C Lesson Practice (5–10 minutes)

The Lesson Practice provides practice on the New Concept. Closely monitor student work on the Lesson Practice problems, providing immediate feedback as appropriate. Have your student solve **all** the problems in the Lesson Practice before proceeding to the next section of the lesson. Answers may be written on Recording Form B or on blank paper. If your student has difficulty with the Lesson Practice, you may wish to reteach the relevant examples in the New Concept section in order to identify the particular aspect of the concept that is causing problems.

Some Lesson Practice sets are marked with an asterisk (see example at right). An asterisk indicates that additional problems on the lesson's concept appear in the Supplemental Practice section of the textbook's appendix. Supplemental Practice problems are intended for remediation. Assign your student these problems only if he or she has difficulty with a concept several lessons after it is presented.

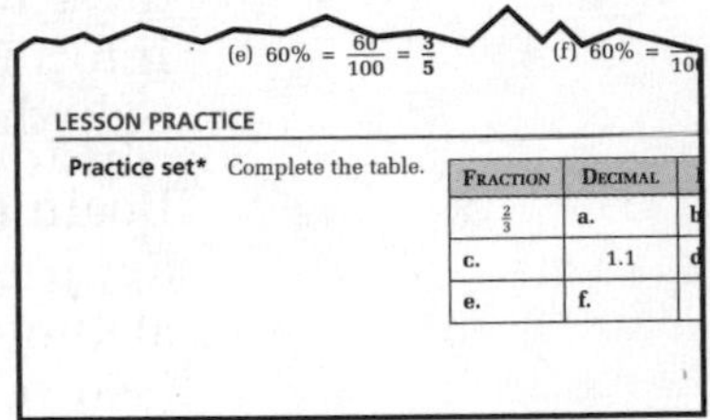

D Mixed Practice (20–40 minutes)

The Mixed Practice is the fourth and most important component of the lesson. This section contains twenty-five to thirty problems that prepare students for upcoming lessons, allow them to work with several strands of mathematics concurrently, and provide them with the distributed practice that promotes long-term retention of concepts. **Have your student work independently on the Mixed Practice, and ensure that no problem is skipped.** Students may show their work on a copy of Recording Form C or on blank paper.

> If your student encounters difficulty with Mixed Practice problems, have him or her refer to the Lesson Reference Numbers that appear in parentheses below each problem number. Lesson Reference Numbers indicate which lessons explain concepts relevant to the problems they label. Because many problems involve multiple concepts, more than one reference number might be given for a problem.

At the end of the math period, check your student's work, referring to the Solutions Manual as necessary. If there are incorrect answers, help your student identify which solution steps led to the errors. Then have the student rework the problems to achieve the correct answers. If desired, track the completion of your student's daily assignments on Recording Form D.

Investigations

Following every tenth lesson is an investigation. Investigations are in-depth treatments of concepts that often involve activities. Because of the length of investigations, no Warm-Up or Mixed Practice is included. As with lessons, investigations might call for students to use Activity Sheets, which can be found in the Tests and Worksheets booklet.

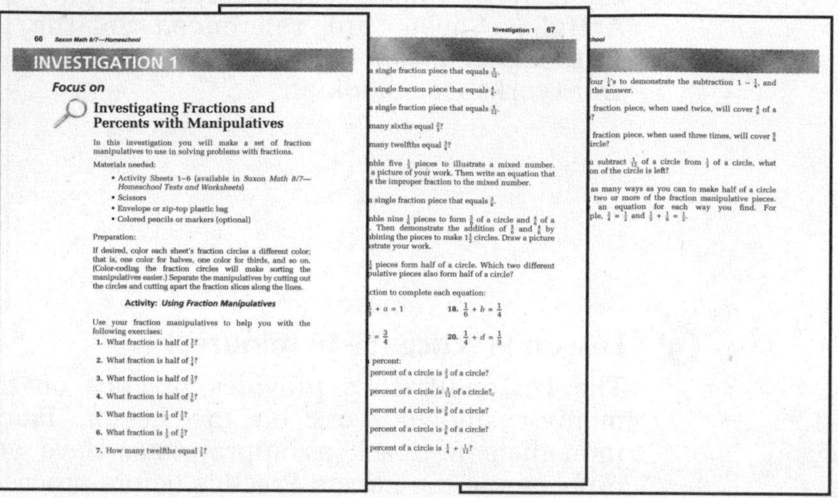

66 *Saxon Math 8/7—Homeschool*

INVESTIGATION 1

Focus on

Investigating Fractions and Percents with Manipulatives

In this investigation you will make a set of fraction manipulatives to use in solving problems with fractions.

Materials needed:

- Activity Sheets 1–6 (available in *Saxon Math 8/7—Homeschool Tests and Worksheets*)
- Scissors
- Envelope or zip-top plastic bag
- Colored pencils or markers (optional)

Preparation:

If desired, color each sheet's fraction circles a different color; that is, one color for halves, one color for thirds, and so on. (Color-coding the fraction circles will make sorting the manipulatives easier.) Separate the manipulatives by cutting out the circles and cutting apart the fraction slices along the lines.

Activity: ***Using Fraction Manipulatives***

Use your fraction manipulatives to help you with the following exercises:

1. What fraction is half of $\frac{1}{2}$?
2. What fraction is half of $\frac{1}{4}$?
3. What fraction is half of $\frac{1}{3}$?
4. What fraction is half of $\frac{1}{6}$?
5. What fraction is $\frac{1}{3}$ of $\frac{1}{2}$?
6. What fraction is $\frac{1}{3}$ of $\frac{1}{4}$?
7. How many twelfths equal $\frac{1}{2}$?

Tests

Twenty-three cumulative tests are provided in *Saxon Math 8/7—Homeschool.* The problems on the tests are similar to those in the textbook, and the tests are scheduled so that students have about five days to practice concepts before being assessed on them. For detailed information regarding when to give each test, refer to the Testing Schedule in the Tests and Worksheets booklet.

Begin each test day with Facts Practice. The appropriate Facts Practice Test is specified at the top of the scheduled cumulative test (see example at right). After the Facts Practice, administer the cumulative test. Have your student show his or her work and record his or her answers on a copy of Recording Form E. The textbook should not be used during the test.

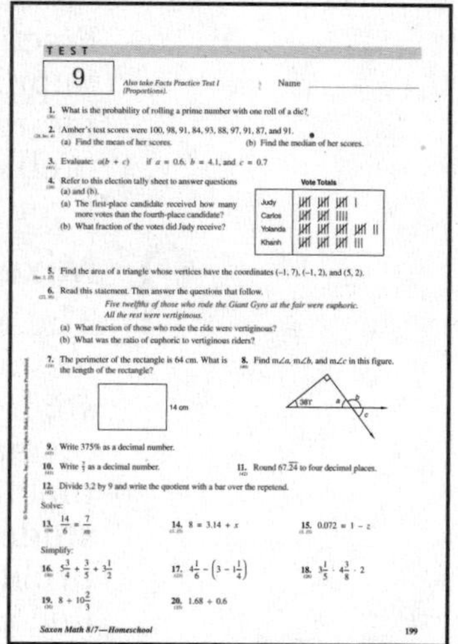

TEST 9 *Also take Facts Practice Test I (Proportions).* Name ________

1. What is the probability of rolling a prime number with one roll of a die?
2. Amber's test scores were 100, 98, 91, 84, 93, 88, 97, 91, 87, and 91.
   (a) Find the mean of her scores. (b) Find the median of her scores.
3. Evaluate: $a(b + c)$ if $a = 0.6$, $b = 4.1$, and $c = 0.7$
4. Refer to this election tally sheet to answer questions (a) and (b).
   (a) The first-place candidate received how many more votes than the fourth-place candidate?
   (b) What fraction of the votes did Judy receive?

Vote Totals	
Judy	卌 卌 卌 I
Carlos	卌 卌 IIII
Yolanda	卌 卌 卌 卌 II
Khanh	卌 卌 卌 III

5. Find the area of a triangle whose vertices have the coordinates (−1, 7), (−1, 2), and (5, 2).
6. Read this statement. Then answer the questions that follow.
   Five twelfths of those who rode the Giant Gyro at the fair were euphoric. All the rest were vertiginous.
   (a) What fraction of those who rode the ride were vertiginous?
   (b) What was the ratio of euphoric to vertiginous riders?
7. The perimeter of the rectangle is 64 cm. What is the length of the rectangle?
8. Find m∠a, m∠b, and m∠c in this figure.
9. Write 375% as a decimal number.
10. Write [illegible] as a decimal number.
11. Round $67.\overline{24}$ to four decimal places.
12. Divide 3.2 by 9 and write the quotient with a bar over the repetend.

Solve:

13. $\frac{14}{6} = \frac{7}{m}$
14. $8 = 3.14 + x$
15. $0.072 = 1 - z$

Simplify:

16. $5\frac{3}{4} + \frac{3}{5} + 3\frac{1}{2}$
17. $4\frac{1}{6} - \left(3 - 1\frac{1}{4}\right)$
18. $3\frac{1}{5} \cdot 4\frac{3}{8} \cdot 2$
19. $8 \div 10\frac{2}{3}$
20. $1.68 \div 0.6$

Saxon Math 8/7—Homeschool 199

After the test, compare each student answer to the one given in the Solutions Manual. Note any incorrect answers, and review the test with your student. Determine whether errors were caused by computational mistakes or conceptual misunderstandings. If necessary, stress to your student that computational errors can be prevented by writing neatly and by checking the work. If he or she misunderstands a concept, be sure to address the misunderstanding promptly. Work through textbook examples that demonstrate the concept (identify the appropriate lesson by referring to the Lesson Reference Numbers on the test), and assign additional practice problems for the student to solve. (Check for additional practice problems in the textbook's appendix.)

HOW TO GET HELP

If you need help implementing your homeschool program, you can call our Parent Support Line at (405) 217-1717, and our veteran teachers will counsel you on how to set up and teach Saxon Homeschool Math. If you and your student need help with a specific math problem, please e-mail our tutors at mathhelp@saxonhomeschool.com, and they will respond promptly.

We encourage you to visit the Saxon Homeschool Web site, www.saxonhomeschool.com, for descriptions of Saxon's math and phonics programs and for downloadable documents such as our homeschool catalog, placement tests, and state-standards correlations. The Web site also provides online math and phonics activities for your student.

We wish you success and enjoyment in the coming year, and please remember to contact us with any questions or comments!

LIST OF MATERIALS

The following materials are used throughout *Saxon Math 8/7—Homeschool.* We suggest you acquire these materials before beginning the program.

- inch/centimeter ruler

  (*Note:* a ruler that shows both customary and metric scales is preferred. However, separate customary and metric rulers are acceptable.)
- scientific calculator
- protractor
- graph paper (grid paper)
- compass (for drawing circles)
- scissors

Certain lessons and investigations contain activities that call for additional materials. Refer to the following list before beginning the specified lessons/investigations.

Investigation 1

- envelope or zip-top plastic bag (optional)
- colored pencils or markers (optional)

Lesson 61

- two pairs of plastic straws (The straws within a pair must be the same length. The two pairs may be different lengths.)
- thread or lightweight string
- paperclip (optional)

Lesson 66

- tape measure (preferably metric)
- circular objects

Lesson 70

- yardstick

Lesson 89

- length of string
- chalk
- masking tape (optional)

Lesson 97

- yardstick, ruler, and/or tape measure

Investigation 10

- pair of dot cubes

Investigation 11

- tape

Lesson 112

- two full length unsharpened pencils (or other straightedges)

Investigation 12

- envelope or zip-top plastic bag

LESSON

1 Arithmetic with Whole Numbers and Money • Variables and Evaluation

WARM-UP[†]

Facts Practice: 64 Multiplication Facts (Test A)

Mental Math: A score is 20. Two score and 4 is 44. How many is

a. 3 score **b.** 4 score **c.** 4 score and 7

d. Half a dozen **e.** 2 dozen **f.** 4 dozen

g. Start with a score. Add a dozen; divide by 4; add 2; then divide by 2. What is the answer?

Problem Solving:

What are the next three numbers in this pattern?

1, 3, 6, 10, 15, ...

NEW CONCEPTS

Arithmetic with whole numbers and money

The numbers we say when we count are called **counting numbers** or **natural numbers.** We can show the set of counting numbers this way:

{1, 2, 3, 4, 5, ...}

The three dots, called an *ellipsis,* mean that the list is infinite (goes on without end). The symbols { } are called *braces.* One use of braces is to designate a set. Including zero with the set of counting numbers forms the set of **whole numbers.**

{0, 1, 2, 3, 4, ...}

The set of whole numbers does not include any numbers less than zero, between 0 and 1, or between any **consecutive** counting numbers.

The four fundamental **operations of arithmetic** are addition, subtraction, multiplication, and division. In this lesson we will review the operations of arithmetic with whole numbers and with money. Amounts of money are sometimes indicated with a dollar sign ($) or with a cent sign (¢), but not both. We can show 50 cents either of these two ways:

$0.50 or 50¢

[†]For instructions on how to use the Warm-up activities, please consult the preface.

Occasionally we will see a dollar sign or cent sign used incorrectly.

This sign is incorrect because it uses a **decimal point** with a cent sign. This incorrect sign literally means that soft drinks cost not half a dollar but half a cent! Take care to express amounts of money in the proper form when performing arithmetic with money.

Numbers that are added are called **addends,** and the result of their addition is the **sum.**

addend + addend = sum

Example 1 Add:

(a) 36 + 472 + 3614

(b) \$1.45 + \$6 + 8¢

Solution (a) We align the digits in the ones place and add in columns. Looking for combinations of digits that total 10 may speed the work.

$$\begin{array}{r} {}^{111} \\ 36 \\ 472 \\ +\ 3614 \\ \hline \mathbf{4122} \end{array}$$

(b) We write each amount of money with a dollar sign and two places to the right of the decimal point. We align the decimal points and add.

$$\begin{array}{r} {}^{1} \\ \$1.45 \\ \$6.00 \\ +\ \$0.08 \\ \hline \mathbf{\$7.53} \end{array}$$

In subtraction the **subtrahend** is taken from the **minuend.** The result is the **difference.**

minuend − subtrahend = difference

Example 2 Subtract:

(a) 5207 − 948

(b) \$5 − 25¢

Solution (a) We align the digits in the ones place. We must follow the correct order of subtraction by writing the minuend (first number) above the subtrahend (second number).

$$\begin{array}{r} {}^{4\ \overset{1}{1}\ 9\ 1} \\ \not{5}\,\not{2}\,\not{0}\,7 \\ -\ \ 9\,4\,8 \\ \hline \mathbf{4\,2\,5\,9} \end{array}$$

(b) We write each amount in dollar form. We align decimal points and subtract.

$$\begin{array}{r} \overset{4\,9\,1}{\$\,\cancel{5}.\cancel{0}\,0} \\ -\ \$\,0.2\,5 \\ \hline \mathbf{\$\,4.7\,5} \end{array}$$

Numbers that are multiplied are called **factors.** The result of their multiplication is the **product.**

factor × factor = product

We can indicate the multiplication of two factors with a times sign, with a center dot, or by writing the factors next to each other with no sign between them.

$4 \times 5 \qquad 4 \cdot 5 \qquad 4(5) \qquad ab$

The parentheses in 4(5) clarify that 5 is a quantity separate from 4 and that the two digits do not represent the number 45. The expression *ab* means "*a* times *b*."

Example 3 Multiply:

(a) 164 · 23

(b) $4.68 × 20

(c) 5(29¢)

Solution (a) We usually write the number with the most digits on top. We first multiply by the 3 of 23. Then we multiply by the 20 of 23. We add the partial products to find the final product.

$$\begin{array}{r} 164 \\ \times\ \ 23 \\ \hline 492 \\ 328 \\ \hline \mathbf{3772} \end{array}$$

(b) We can let the zero in 20 "hang out" to the right. We write 0 below the line and then multiply by the 2 of 20. We write the product with a dollar sign and two decimal places.

$$\begin{array}{r} \$4.68 \\ \times\ \ \ \ \ 20 \\ \hline \mathbf{\$93.60} \end{array}$$

(c) We can multiply 29¢ by 5 or write 29¢ as $0.29 first. Since the product is greater than $1, we use a dollar sign to write the answer.

$$\begin{array}{r} 29¢ \\ \times\ \ 5 \\ \hline 145¢ \end{array} = \mathbf{\$1.45}$$

In division the **dividend** is divided by the **divisor.** The result is the **quotient.** We can indicate division with a division sign (÷), a division box ($\overline{)}$), or a division bar (–).

dividend ÷ divisor = quotient

$$\text{divisor}\overline{)\text{dividend}}^{\text{quotient}} \qquad \frac{\text{dividend}}{\text{divisor}} = \text{quotient}$$

Example 4 Divide:

(a) $1234 \div 56$

(b) $\frac{\$12.60}{5}$

Solution (a) In this division there is a remainder. Other methods for dealing with a remainder will be considered later.

```
     22 R 2
56)1234
   112
   ---
    114
    112
    ---
      2
```

(b) We write the quotient with a dollar sign. The decimal point in the quotient is directly above the decimal point in the dividend.

```
  $2.52
5)$12.60
   10
   --
    2 6
    2 5
    ---
      10
      10
      --
       0
```

Variables and evaluation

In mathematics, letters are often used to represent numbers—in formulas and expressions, for example. The letters are called **variables** because their values are not constant; rather, they vary. We **evaluate** an expression by calculating its value when the variables are assigned specific numbers.

Example 5 Evaluate each expression for $x = 10$ and $y = 5$:

(a) $x + y$ (b) $x - y$

(c) xy (d) $\frac{x}{y}$

Solution We substitute 10 for x and 5 for y in each expression. Then we perform the calculation.

(a) $10 + 5 = \mathbf{15}$ (b) $10 - 5 = \mathbf{5}$

(c) $10 \cdot 5 = \mathbf{50}$ (d) $\frac{10}{5} = \mathbf{2}$

LESSON PRACTICE

Practice set

a. This sign is incorrect. Show two ways to correct the sign.

b. Name a whole number that is not a counting number.

c. When the product of 4 and 4 is divided by the sum of 4 and 4, what is the quotient?

Simplify by adding, subtracting, multiplying, or dividing as indicated:

d. \$1.75 + 60¢ + \$3 **e.** \$2 − 47¢

f. 5(65¢) **g.** $250 \cdot 24$

h. \$24.00 ÷ 5 **i.** $\frac{234}{18}$

Evaluate each expression for $a = 20$ and $b = 4$:

j. $a + b$ **k.** $a - b$

l. ab **m.** $\frac{a}{b}$

MIXED PRACTICE

Problem set

1. When the sum of 5 and 6 is subtracted from the product of 5 and 6, what is the difference?

2. If the subtrahend is 9 and the difference is 8, what is the minuend?

3. If the divisor is 4 and the quotient is 8, what is the dividend?

4. When the product of 6 and 6 is divided by the sum of 6 and 6, what is the quotient?

5. Name the four fundamental operations of arithmetic.

6. Evaluate each expression for $n = 12$ and $m = 4$:

(a) $n + m$ (b) $n - m$

(c) nm (d) $\frac{n}{m}$

Simplify by adding, subtracting, multiplying, or dividing, as indicated:

7. $\begin{array}{r} \$43.74 \\ - \$16.59 \\ \hline \end{array}$

8. $\begin{array}{r} 64 \\ \times\ 37 \\ \hline \end{array}$

9. $\begin{array}{r} 7 \\ 8 \\ 4 \\ 6 \\ 9 \\ 3 \\ 5 \\ +\ 7 \\ \hline \end{array}$

10. 364 + 52 + 867 + 9

11. 4000 − 3625

12. (316)(18)

13. $43.60 ÷ 20

14. 300 · 40

15. 8 · 12 · 0

16. 3708 ÷ 12

17. 365 × 20

18. $25\overline{)767}$

19. 30(40)

20. $10 − $2.34

21. 4017 − 3952

22. $2.50 × 80

23. 20($2.50)

24. $\frac{560}{14}$

25. $\frac{\$10.00}{8}$

26. What is another name for *counting numbers?*

27. Write 25 cents twice, once with a dollar sign and once with a cent sign.

28. Which counting numbers are also whole numbers?

29. What is the name for the answer to a division problem?

30. Here we use a plus sign and an equal sign to show the relationship of addends and their sum:

addend + addend = sum

Use a minus sign, an equal sign, and the words *difference, subtrahend,* and *minuend* to show the relationship between the numbers in subtraction.

LESSON

2 Properties of Operations • Sequences

WARM-UP

Facts Practice: 64 Multiplication Facts (Test A)

Mental Math:

a. 2 score and 8 **b.** $1\frac{1}{2}$ dozen **c.** Half of 100

d. $400 + 500$ **e.** $9000 - 3000$ **f.** 20×30

g. Start with a dozen. Divide by 2; multiply by 4; add 1; divide by 5; then subtract 5. What is the answer?

Problem Solving:

When we add 1, 2, 3, and 4, the sum is 10. What is the sum when we add 1, 2, 3, 4, 5, 6, 7, 8, 9, and 10? (Try pairing numbers to make equal sums.)

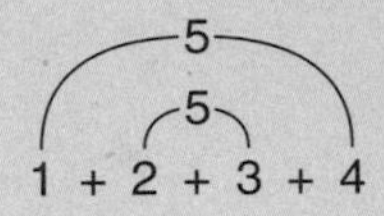

NEW CONCEPTS

Properties of operations

Addition and subtraction are **inverse operations.** We can "undo" an addition by subtracting one addend from the sum.

$$2 + 3 = 5 \qquad 5 - 3 = 2$$

Together, the numbers 2, 3, and 5 form an addition-subtraction **fact family.** With these numbers we can write two addition facts and two subtraction facts.

$$2 + 3 = 5 \qquad 5 - 3 = 2$$

$$3 + 2 = 5 \qquad 5 - 2 = 3$$

We see that both $2 + 3$ and $3 + 2$ equal 5. Changing the order of addends does not change the sum. This characteristic of addition is known as the **commutative property of addition** and is often stated in equation form using variables.

$$a + b = b + a$$

Since changing the order of numbers in subtraction may change the result, subtraction is not commutative.

Addition is commutative.

$$2 + 3 = 3 + 2$$

Subtraction is not commutative.

$$5 - 3 \neq 3 - 5$$

($\neq$ means "is not equal to")

The **identity property of addition** states that when zero is added to a given number, the sum is equal to the given number.

$$a + 0 = a$$

Thus, zero is the **additive identity.**

Multiplication and division are also inverse operations. Dividing a product by one of its factors "undoes" the multiplication.

$$4 \times 5 = 20 \qquad 20 \div 5 = 4$$

Together, the numbers 4, 5, and 20 form a multiplication-division fact family that can be arranged into two multiplication facts and two division facts.

$$4 \times 5 = 20 \qquad 20 \div 5 = 4$$

$$5 \times 4 = 20 \qquad 20 \div 4 = 5$$

Changing the order of the factors does not change the product. This characteristic of multiplication is known as the **commutative property of multiplication.**

$$a \times b = b \times a$$

Since changing the order of division may change the quotient, division is not commutative.

Multiplication is commutative.

$$4 \times 5 = 5 \times 4$$

Division is not commutative.

$$20 \div 5 \neq 5 \div 20$$

The **identity property of multiplication** states that when a given number is multiplied by 1, the result equals the given number. Thus, 1 is the **multiplicative identity.**

$$a \times 1 = a$$

The **property of zero for multiplication** states that when a number is multiplied by zero, the product is zero.

$$a \times 0 = 0$$

The operations of arithmetic are **binary,** which means that we only work with two numbers in one step. If we wish to add

$$2 + 3 + 4$$

we can add two of the numbers and then add the other number. The parentheses around $2 + 3$ in the expression below show that $2 + 3$ should be treated as a single quantity. Therefore, 2 and 3 should be added, and then 4 should be added to the sum.

$(2 + 3) + 4$	add 2 and 3 first
$5 + 4$	then add 5 and 4
9	sum

In the expression below, the parentheses indicate that 3 and 4 are to be added first.

$2 + (3 + 4)$	add 3 and 4 first
$2 + 7$	then add 2 and 7
9	sum

Notice that the sum is the same whichever way we group the addends.

$$(2 + 3) + 4 = 2 + (3 + 4)$$

The **associative property of addition** states that the grouping of addends does not change the sum. It is often stated as an equation using variables.

$$(a + b) + c = a + (b + c)$$

There is a similar property for multiplication. The **associative property of multiplication** states that the grouping of factors does not change the product.

$$(a \times b) \times c = a \times (b \times c)$$

The grouping of numbers in subtraction and division does affect the result, as we see in the following expressions. Thus, there is no associative property of subtraction, and there is no associative property of division.

$$(8 - 4) - 2 \neq 8 - (4 - 2)$$

$$(8 \div 4) \div 2 \neq 8 \div (4 \div 2)$$

Example 1 Name each property illustrated:

(a) $5 \cdot 3 = 3 \cdot 5$

(b) $(3 + 4) + 5 = 3 + (4 + 5)$

(c) $6 + 0 = 6$

(d) $6 \cdot 0 = 0$

Solution (a) **Commutative property of multiplication**

(b) **Associative property of addition**

(c) **Identity property of addition**

(d) **Property of zero for multiplication**

Example 2 Which property can we use to find each missing number?

(a) $8 + ? = 8$

(b) $1 \times ? = 9$

(c) $10 \times ? = 0$

Solution (a) **Identity property of addition**

(b) **Identity property of multiplication**

(c) **Property of zero for multiplication**

Sequences

A **sequence** is an ordered list of **terms** that follows a certain pattern or rule. A list of the whole numbers is an example of a sequence.

0, 1, 2, 3, 4, ...

If we wish to list the **even** or **odd** whole numbers, we could write the following sequences:

EVEN WHOLE NUMBERS	ODD WHOLE NUMBERS
0, 2, 4, 6, 8, ...	1, 3, 5, 7, 9, ...

A rule for both of these sequences is to add two to a term in the sequence to find the next term. However, the lists are different because the starting numbers of the sequences are different. To continue a sequence, we study the sequence to understand its pattern or rule; then we apply the rule to find additional terms in the sequence.

Example 3 The first four terms of a sequence are shown below. Find the next three terms in the sequence.

1, 4, 9, 16, ...

Solution We will describe two solutions. First we see that the terms increase in size by a larger amount as we move to the right in the sequence.

$$\overset{+3}{\frown}\ \overset{+5}{\frown}\ \overset{+7}{\frown}$$
$$1,\ 4,\ 9,\ 16, \ldots$$

The increase itself forms a sequence we may recognize: 3, 5, 7, 9, 11, ... We will continue the sequence by adding successively larger odd numbers.

$$+3 \quad +5 \quad +7 \quad +9 \quad +11 \quad +13$$
$$1,\ 4,\ 9,\ 16,\ 25,\ 36,\ 49, \ldots$$

We find that the next three numbers in the sequence are **25, 36,** and **49.**

Another solution to the problem is to recognize the sequence as a list of **perfect squares.** When we multiply a counting number by itself, the product is a perfect square.

$$1 \cdot 1 = 1 \qquad 2 \cdot 2 = 4 \qquad 3 \cdot 3 = 9 \qquad 4 \cdot 4 = 16$$

This relationship may be clearer if we compare each term to its position in the sequence.

Position (n)	1	2	3	4
Term (k)	1	4	9	16

Notice that each term is the square of its position in the sequence. In the table the variable n refers to the position of the term in the sequence, and k represents the value of the term, as is customary. The relationship between n and k can be represented by the following equation:

$$k = n \cdot n$$

To find the values of the fifth, sixth, and seventh terms, we may substitute 5, 6, and 7 for n.

Fifth term	Sixth term	Seventh term
$k = 5 \cdot 5$	$k = 6 \cdot 6$	$k = 7 \cdot 7$
$= 25$	$= 36$	$= 49$

So the next three terms are **25, 36,** and **49.**

Example 4 The rule of a certain sequence is $k = 2n$. Find the first four terms of the sequence.

Solution We substitute 1, 2, 3, and 4 for n to find the first four terms.

First term	Second term	Third term	Fourth term
$k = 2(1)$	$k = 2(2)$	$k = 2(3)$	$k = 2(4)$
$= 2$	$= 4$	$= 6$	$= 8$

The first four terms of the sequence are **2, 4, 6,** and **8.**

LESSON PRACTICE

Practice set

a. Which number is known as the additive identity? Which number is the multiplicative identity?

b. Which operation is the inverse of multiplication?

c. Use the letters x, y, and z to write an equation that illustrates the associative property of addition. Then write an example using counting numbers of your choosing.

d. Name the property we can use to find the missing number in this equation:

$$5 \times ? = 8 \times 5$$

Add, subtract, multiply, or divide as indicated to simplify each expression. Remember to work within the parentheses first.

e. $(5 + 4) + 3$

f. $5 + (4 + 3)$

g. $(10 - 5) - 3$

h. $10 - (5 - 3)$

i. $(6 \cdot 2) \cdot 5$

j. $6 \cdot (2 \cdot 5)$

k. $(12 \div 6) \div 2$

l. $12 \div (6 \div 2)$

m. Find the next three terms of this sequence:

1, 4, 9, 16, 25, 36, 49, ...

n. Use words to describe the rule of the following sequence. Then find the next three terms.

1, 2, 4, 8, ...

o. The rule of a certain sequence is $k = (2n) - 1$. Find the first four terms of the sequence.

MIXED PRACTICE

Problem set

[†]**1.** *(1)* When the product of 2 and 3 is subtracted from the sum of 4 and 5, what is the difference?

2. *(1)* Write 4 cents twice, once with a dollar sign and once with a cent sign.

3. *(1)* The sign shown is incorrect. Show two ways to correct the sign.

4. *(2)* Which operation of arithmetic is the inverse of addition?

5. *(1)* If the dividend is 60 and the divisor is 4, what is the quotient?

6. *(2)* For the fact family 3, 4, and 7, we can write two addition facts and two subtraction facts.

$3 + 4 = 7 \qquad 7 - 4 = 3$

$4 + 3 = 7 \qquad 7 - 3 = 4$

For the fact family 3, 5, and 15, write two multiplication facts and two division facts.

7. *(2)* Use words to describe the rule of the following sequence. Then find the next two terms.

1, 10, 100, ...

[†]The italicized numbers within parentheses underneath each problem number are called *lesson reference numbers.* These numbers refer to the lesson(s) in which the major concept of that particular problem is introduced. If additional assistance is needed, refer to the discussion, examples, or practice problems of that lesson.

Simplify:

8. (1) $\begin{array}{r} \$20.00 \\ -\ \$14.79 \\ \hline \end{array}$

9. (1) $\begin{array}{r} \$1.54 \\ \times\quad 7 \\ \hline \end{array}$

10. (1) $\dfrac{\$30.00}{8}$

11. (1) $4.36 + 75¢ + $12 + 6¢

12. (2) $10.00 − ($4.89 + 74¢)

13. (1) $\begin{array}{r} 8 \\ 5 \\ 4 \\ 6 \\ 5 \\ 4 \\ 3 \\ 7 \\ 2 \\ 4 \\ 1 \\ +\ 8 \\ \hline \end{array}$

14. (1) 3105 ÷ 15

15. (1) $40\overline{)1630}$

16. (2) 81 ÷ (9 ÷ 3)

17. (2) (81 ÷ 9) ÷ 3

18. (1) (10)($3.75)

19. (2) 3167 − (450 − 78)

20. (2) (3167 − 450) − 78

21. (1) $20.00 ÷ 16

22. (1) 70 · 800

23. (1) 3714 + 268 + 47 + 9

24. (1) 5 · 4 · 3 · 2 · 1

25. (2) $20 − ($1.47 + $8)

26. (1) 30 × 45¢

27. (2) Which property can we use to find each missing number?

(a) $10x = 0$ (b) $10y = 10$

28. (1) Evaluate each expression for $x = 18$ and $y = 3$:

(a) $x - y$ (b) xy

(c) $\dfrac{x}{y}$ (d) $x + y$

29. (2) Why is zero called the additive identity?

30. (1) Here we show the relationship between factors and their product.

factor × factor = product

Use a division sign, an equal sign, and the words *dividend, quotient,* and *divisor* to show the relationship between numbers in division.

LESSON

3 Missing Numbers in Addition, Subtraction, Multiplication, and Division

WARM-UP

Facts Practice: 64 Multiplication Facts (Test A)

Mental Math:

a. 3 score and 6 **b.** $2\frac{1}{2}$ dozen **c.** Half of 1000

d. 1200 + 300 **e.** 750 − 500 **f.** 30 × 30

g. Start with the number of minutes in an hour. Divide by 2; subtract 5; double that number; subtract 1; then divide by 7. What is the answer?

Problem Solving:

In one section of a theater there are twelve rows of seats. In the first row there are 6 seats, in the second row there are 9 seats, and in the third row there are 12 seats. If the pattern continues, how many seats are in the twelfth row?

NEW CONCEPT

An **equation** is a statement that two quantities are equal. Here we show two equations:

$$3 + 4 = 7 \qquad 5 + a = 9$$

The equation on the right contains a variable. In this lesson we will practice finding the value of variables in addition, subtraction, multiplication, and division equations.

Missing numbers in addition

Sometimes we encounter addition equations in which the sum is missing. Sometimes we encounter addition equations in which an addend is missing. We can use a letter to represent a missing number. The letter may be uppercase or lowercase.

MISSING SUM	MISSING ADDEND	MISSING ADDEND
$2 + 3 = N$	$2 + a = 5$	$b + 3 = 5$

If we know two of the three numbers, we can find the missing number. We can find a missing addend by subtracting the known addend from the sum. If there are more than two addends, we subtract all the known addends from the sum. For example, to find n in the equation

$$3 + 4 + n + 7 + 8 = 40$$

we subtract 3, 4, 7, and 8 from 40. To do this, we can add the known addends and then subtract their sum from 40.

Example 1 Find the missing number in each equation:

(a) $n + 53 = 75$ (b) $26 + a = 61$

(c) $3 + 4 + n + 7 + 8 = 40$

Solution In both (a) and (b) we can find each missing addend by subtracting the known addend from the sum. Then we check.

(a) Subtract. Try it.

$$\begin{array}{r} 75 \\ -\ 53 \\ \hline 22 \end{array} \qquad \begin{array}{r} 22 \\ +\ 53 \\ \hline 75 \end{array} \text{ check}$$

So the missing number in (a) is **22.**

(b) Subtract. Try it.

$$\begin{array}{r} 61 \\ -\ 26 \\ \hline 35 \end{array} \qquad \begin{array}{r} 26 \\ +\ 35 \\ \hline 61 \end{array} \text{ check}$$

So the missing number in (b) is **35.**

(c) We add the known addends.

$$3 + 4 + 7 + 8 = 22$$

Then we subtract their sum, 22, from 40.

$$40 - 22 = \mathbf{18}$$

We use the answer in the original equation for a check.

$$3 + 4 + 18 + 7 + 8 = 40 \quad \text{check}$$

Missing numbers in subtraction

There are three numbers in a subtraction equation. If one of the three numbers is missing, we can find the missing number.

Missing Minuend	Missing Subtrahend	Missing Difference
$n - 3 = 2$	$5 - x = 2$	$5 - 3 = m$

To find a missing minuend, we add the other two numbers. To find a missing subtrahend or difference, we subtract.

Example 2 Find the missing number in each equation:

(a) $p - 24 = 17$ (b) $32 - x = 14$

Solution (a) To find the minuend in a subtraction equation, we add the other two numbers. We find that the missing number in (a) is **41.**

$$\begin{array}{r} 17 \\ +\ 24 \\ \hline 41 \end{array} \qquad \begin{array}{r} 41 \\ -\ 24 \\ \hline 17 \end{array}\ \text{check}$$

Add. Try it.

(b) To find a subtrahend, we subtract the difference from the minuend. So the missing number in (b) is **18.**

Subtract. Try it.

$$\begin{array}{r} 32 \\ -\ 14 \\ \hline 18 \end{array} \qquad \begin{array}{r} 32 \\ -\ 18 \\ \hline 14 \end{array}\ \text{check}$$

Missing numbers in multiplication A multiplication equation is composed of factors and a product. If any one of the numbers is missing, we can figure out what it is.

MISSING PRODUCT	MISSING FACTOR	MISSING FACTOR
$3 \cdot 2 = p$	$3f = 6$	$r \times 2 = 6$

To find a missing product, we multiply the factors. To find a missing factor, we divide the product by the known factor(s).

Example 3 Find the missing number in each equation:

(a) $12n = 168$ (b) $7k = 105$ (c) $2 \cdot 3a = 30$

Solution In both (a) and (b) the missing number is one of the two factors. Notice that $7k$ means "7 times k." We can find a missing factor by dividing the product by the known factor.

(a) Divide. Try it.

$$12\overline{)168} = 14 \qquad \begin{array}{r} 168 \\ -\,12 \\ \hline 48 \\ -\,48 \\ \hline 0 \end{array} \qquad \begin{array}{r} 12 \\ \times\ 14 \\ \hline 48 \\ 12\ \ \\ \hline 168 \end{array}\ \text{check}$$

So the missing number in (a) is **14.**

(b) Divide. Try it.

$$7\overline{)105} = 15 \qquad \begin{array}{r} 105 \\ -\,7\ \ \\ \hline 35 \\ -\,35 \\ \hline 0 \end{array} \qquad \begin{array}{r} 15 \\ \times\ 7 \\ \hline 105 \end{array}\ \text{check}$$

So the missing number in (b) is **15.**

(c) In this equation there are three factors: 2, 3, and a. One way to find a is to divide the product, 30, by one of the known factors and then divide that result by the other known factor.

$$\frac{30}{2} = 15 \qquad \frac{15}{3} = \mathbf{5}$$

Another way to find a is to divide 30 by the product of the known factors. Since the product of 2 and 3 is 6, we divide 30 by 6.

$$\frac{30}{6} = \mathbf{5}$$

We check our answer, 5, in the original equation.

$$2 \cdot 3(5) = 30 \quad \text{check}$$

Missing numbers in division If we know two of the three numbers in a division equation, we can figure out the missing number.

MISSING QUOTIENT	MISSING DIVISOR	MISSING DIVIDEND
$\frac{24}{3} = n$	$\frac{24}{m} = 8$	$\frac{p}{3} = 8$

To find a missing quotient, we simply *divide* the dividend by the divisor. To find a missing divisor, we *divide* the dividend by the quotient. To find a missing dividend, we *multiply* the quotient by the divisor.

Example 4 Find the missing number in each equation:

(a) $\frac{a}{3} = 15$ (b) $\frac{64}{b} = 4$

Solution (a) To find a missing dividend, we multiply the quotient and divisor.

$$3 \times 15 = \mathbf{45} \quad \text{try it} \qquad 45 \div 3 = 15 \quad \text{check}$$

(b) To find a missing divisor, divide the dividend by the quotient.

$$4\overline{)64}\text{ with quotient } \mathbf{16} \quad \text{try it} \qquad \frac{64}{16} = 4 \quad \text{check}$$

LESSON PRACTICE

Practice set*† Find the missing number in each equation:

a. $a + 12 = 31$ **b.** $b - 24 = 15$ **c.** $15c = 180$

d. $\frac{r}{8} = 12$ **e.** $14e = 420$

f. $26 + f = 43$ **g.** $51 - g = 20$

h. $\frac{364}{h} = 7$ **i.** $3 \cdot 4n = 24$

j. $3 + 6 + m + 12 + 5 = 30$

MIXED PRACTICE

Problem set

1. (1) When the product of 4 and 4 is divided by the sum of 4 and 4, what is the quotient?

2. (1, 3) If you know the subtrahend and the difference, how can you find the minuend? Write a complete sentence to answer the question.

3. (2) Which property of addition is stated by this equation?

$$(a + b) + c = a + (b + c)$$

4. (3) If one addend is 7 and the sum is 21, what is the other addend?

5. (2) Use the numbers 3 and 4 to illustrate the commutative property of multiplication. Use a center dot to indicate multiplication.

6. (2) The rule of a certain sequence is $k = 3n$. Find the first four terms of the sequence.

Find each missing number:

7. (3) $x + 83 = 112$ **8.** (3) $96 - r = 27$ **9.** (3) $7k = 119$

10. (3) $127 + z = 300$ **11.** (3) $m - 137 = 731$

12. (3) $25n = 400$ **13.** (3) $\frac{625}{w} = 25$ **14.** (3) $\frac{x}{60} = 700$

---

†The asterisk after "Practice set" indicates that additional practice problems intended for remediation are available in the appendix.

15. Evaluate each expression for $a = 20$ and $b = 5$:
(1)

(a) $\frac{a}{b}$ (b) $a - b$

(c) ab (d) $a + b$

Simplify:

16. $96 \div (16 \div 2)$
(2)

17. $(96 \div 16) \div 2$
(2)

18. $16.47 + $15 + 63¢
(1)

19. $50.00 − ($6.48 + $31.75)
(2)

20. $\begin{array}{r} 47 \\ \times\ 39 \\ \hline \end{array}$
(1)

21. $\begin{array}{r} \$8.79 \\ \times\ \quad 80 \\ \hline \end{array}$
(1)

22. $\frac{4740}{30}$
(1)

23. $1100 - (374 - 87)$
(2)

24. $(1100 - 374) - 87$
(2)

25. $4736 + 271 + 9 + 88$
(1)

26. $30{,}145 - 4299$
(1)

27. $35\overline{)2104}$
(1)

28. $\frac{\$40.00}{32}$
(1)

29. $\begin{array}{r} \$0.48 \\ \times\ \quad 40 \\ \hline \end{array}$
(1)

30. Why is 1 called the multiplicative identity?
(2)

LESSON

4 Number Line

WARM-UP

Facts Practice: 64 Multiplication Facts (Test A)

Mental Math:

a. Five score b. Ten dozen c. Half of 500

d. 350 + 400 e. 50 × 50 f. 400 ÷ 10

g. Start with the number of feet in a yard. Multiply by 12; divide by 6; add 4; double that number; add 5; double that number; then double that number. What is the answer?

Problem Solving:

Simon held a dot cube so that he could see the dots on three of the faces. Simon said he could see 7 dots. How many dots could he not see?

NEW CONCEPT

A **number line** can be used to help us arrange numbers in order. Each number corresponds to a unique point on the number line. The zero point of a number line is called the **origin.** The numbers to the right of the origin are called **positive numbers,** and they are all **greater than zero.** Every positive number has an **opposite** that is the same distance to the left of the origin. The numbers to the left of the origin are called **negative numbers.** The negative numbers are all **less than zero.** Zero is neither positive nor negative.

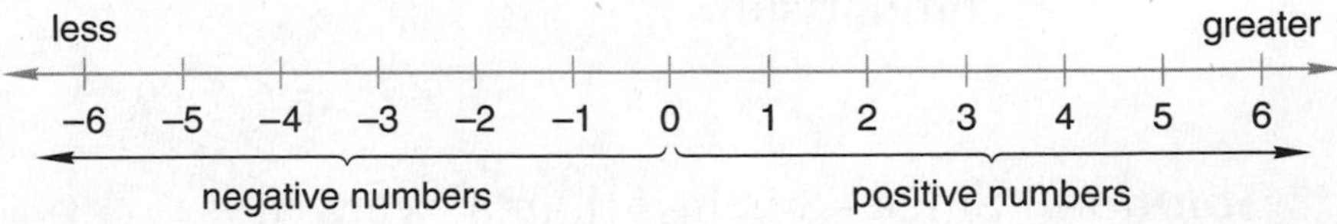

On this number line the **tick marks** indicate the location of **integers.** Integers include all of the counting numbers as well as their opposites—their negatives—and the number zero. Integers do not include fractions or any other numbers between consecutive tick marks on the number line.

INTEGERS

{..., –3, –2, –1, 0, 1, 2, 3, ...}

The ellipses to the left and the right indicate that the number of negative and positive integers is infinite. Notice that the negative numbers are written with a negative sign. For –5 we say "negative five." Positive numbers may be written with or without a positive sign. Both +5 and 5 are positive and equal to each other.

As we move to the right on a number line, the numbers become greater and greater. As we move to the left on a number line, the numbers become less and less. A number is greater than another number if it is farther to the right on a number line.

We **compare** two numbers by determining whether one is greater than the other or whether the numbers are equal. We place a **comparison symbol** between two numbers to show the comparison. The comparison symbols are the equal sign (=) and the greater than/less than symbols (> or <). We write these symbols so that the smaller end (the point) points to the number that is less. Below we show three comparisons.

$-5 < 4$	$3 + 2 = 5$	$5 > -6$
–5 is less than 4	3 plus 2 equals 5	5 is greater than –6

Example 1 Arrange these numbers in order from least to greatest:

$$0, 1, -2$$

Solution We arrange the numbers in the order in which they appear on a number line.

$$\mathbf{-2, 0, 1}$$

Example 2 Rewrite the expression below by replacing the circle with the correct comparison symbol. Then use words to write the comparison.

$$-5 \bigcirc 3$$

Solution Since –5 is less than 3, we write

$$\mathbf{-5 < 3}$$

Negative five is less than three.

We can use a number line to help us add and subtract. We will use arrows to show addition and subtraction. To add, we let the arrow point to the right. To subtract, we let the arrow point to the left.

Example 3 Show this addition problem on a number line: 3 + 2

Solution First we draw a number line. Next we start at the origin (at zero) and draw an arrow 3 units long that points to the right. From this arrowhead we draw a second arrow 2 units long that points to the right.

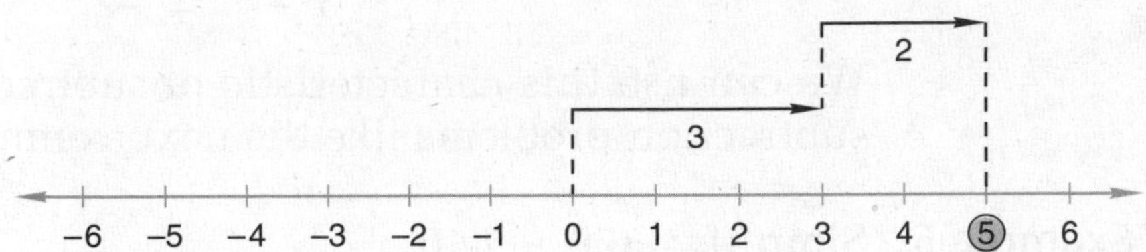

The second arrow ends at 5. We circle the 5 to show that 3 + 2 = **5.**

Example 4 Show this subtraction problem on a number line: 5 – 3

Solution We draw a number line. Then, starting at the origin, we draw an arrow 5 units long that points to the right. To subtract, we draw a second arrow 3 units long that points to the left. Remember to draw the second arrow from the first arrowhead.

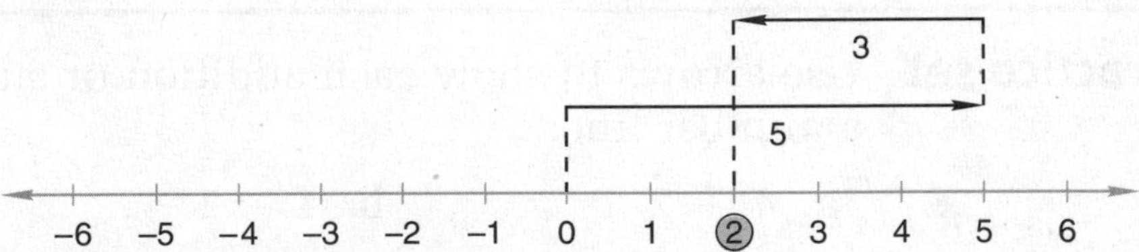

The second arrow ends at 2. This shows that 5 – 3 = **2.**

Example 5 Show this subtraction problem on a number line: 3 – 5

Solution We take the numbers in the order given. We always begin at the origin. Starting from the origin, we draw an arrow 3 units long that points to the right. From this arrowhead we draw a second arrow 5 units long that points to the left. The second arrow ends to the left of zero, which illustrates that the result of the subtraction is a negative number.

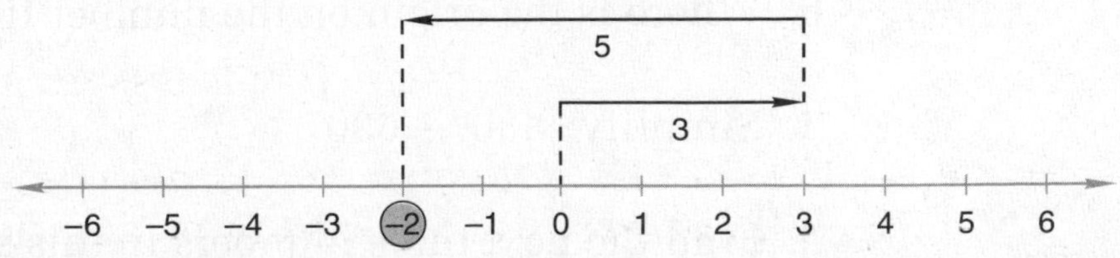

The second arrow ends at –2. This shows that 3 – 5 = **–2.**

Together, examples 4 and 5 show graphically that subtraction is not commutative; that is, the order of the numbers affects the outcome in subtraction. In fact, notice that reversing the order of subtraction results in opposite differences.

$$5 - 3 = 2$$

$$3 - 5 = -2$$

We can use this characteristic of subtraction to help us with subtraction problems like the next example.

Example 6 Simplify: $376 - 840$

Solution We see that the result will be negative. We reverse the order of the numbers to perform the subtraction.

$$\begin{array}{r} 840 \\ -\,376 \\ \hline 464 \end{array}$$

The answer to the original problem is the opposite of 464, which is **–464.**

LESSON PRACTICE

Practice set Use arrows to show each addition or subtraction problem on a number line:

a. $4 + 2$ **b.** $4 - 2$ **c.** $2 - 4$

d. Arrange these numbers in order from least to greatest:

0, –1, –2, –3

e. Use digits and other symbols to write "The sum of 2 and 3 is less than the product of 2 and 3."

Replace each circle with the proper comparison symbol:

f. $3 - 4 \bigcirc 4 - 3$ **g.** $2 \cdot 2 \bigcirc 2 + 2$

h. Where is the origin on the number line?

i. Simplify: $436 - 630$

j. Find the next three numbers in this sequence:

..., 3, 2, 1, 0, –1, ...

MIXED PRACTICE

Problem set

1. (1) What is the difference when the sum of 5 and 4 is subtracted from the product of 3 and 3?

2. (1, 3) If the minuend is 27 and the difference is 9, what is the subtrahend?

3. (4) What is the name for numbers that are greater than zero?

4. (1) Evaluate each expression for $n = 6$ and $m = 24$:

(a) $m - n$

(b) $n - m$

(c) $\frac{m}{n}$

(d) mn

5. (4) Use digits and other symbols to write "The product of 5 and 2 is greater than the sum of 5 and 2."

6. (4) Arrange these numbers in order from least to greatest:

$$-2, 1, 0, -1$$

7. (4) Replace each circle with the proper comparison symbol:

(a) $3 \cdot 4 \bigcirc 2(6)$

(b) $-3 \bigcirc -2$

(c) $3 - 5 \bigcirc 5 - 3$

(d) $xy \bigcirc yx$

8. (3) If you know the divisor and the quotient, how can you find the dividend? Write a complete sentence to answer the question.

9. (4) Show this subtraction problem on a number line: $2 - 3$

Find each missing number:

10. (3) $12x = 12$

11. (3) $4 + 8 + n + 6 = 30$

12. (3) $z - 123 = 654$

13. (3) $1000 - m = 101$

14. (3) $p + \$1.45 = \$4.95$

15. (3) $32k = 224$

16. (3) $\frac{r}{8} = 24$

Simplify:

17. $3.67 + 14¢ + $52.75
(1)

18. $100.00 − $36.49
(1)

19. 48(36¢)
(1)

20. 5 · 6 · 7
(1)

21. 9900 ÷ 18
(1)

22. 30(20)(40)
(1)

23. (130 − 57) + 9
(2)

24. 1987 − 2014
(1)

25. $68.60 ÷ 7
(1)

26. 46¢ + 64¢
(1)

27. $\frac{4640}{80}$
(1)

28. $\begin{array}{r} \$3.75 \\ \times \quad 30 \\ \hline \end{array}$
(1)

29. Use the numbers 2, 3, and 6 to illustrate the associative property of multiplication.
(2)

30. Use the numbers 10, 20, and 30 to write two addition facts and two subtraction facts.
(2)

LESSON

5 Place Value Through Hundred Trillions • Reading and Writing Whole Numbers

WARM-UP

Facts Practice: 64 Multiplication Facts (Test A)

Mental Math:

a. Half a score **b.** Twelve dozen **c.** Ten hundreds

d. 475 − 200 **e.** 25 × 20 **f.** 5000 ÷ 10

g. Start with the number of years in a century. Subtract 1; divide by 9; add 1; multiply by 3; subtract 1; divide by 5; multiply by 4; add 2; then find half of that number. What is the answer?

Problem Solving:

Copy this problem and fill in the missing digits:

```
  75_0
− _60_
  4_13
```

NEW CONCEPTS

Place value through hundred trillions In our number system the value of a digit depends upon its position within a number. The value of each position is its **place value.** The chart below shows place values from the ones place to the hundred-trillions place.

Whole Number Place Values

hundred trillions	ten trillions	trillions		hundred billions	ten billions	billions		hundred millions	ten millions	millions		hundred thousands	ten thousands	thousands		hundreds	tens	ones	decimal point
_	_	_	,	_	_	_	,	_	_	_	,	_	_	_	,	_	_	_	.

Example 1 (a) Which digit is in the trillions place in the number 32,567,890,000,000?

(b) In 12,457,697,380,000, what is the place value of the digit 4?

Solution (a) The digit in the trillions place is **2.**

(b) The place value of the digit 4 is **hundred billions.**

We write a number in **expanded notation** by writing each nonzero digit times its place value. For example, we write 5280 in expanded notation this way:

$$(5 \times 1000) + (2 \times 100) + (8 \times 10)$$

Example 2 Write 25,000 in expanded notation.

Solution $\mathbf{(2 \times 10{,}000) + (5 \times 1000)}$

Reading and writing whole numbers

Whole numbers with more than three digits may be written with commas to make the numbers easier to read. Commas help us read large numbers by separating the trillions, billions, millions, and thousands places. We need only to read the three-digit number in front of each comma and then say either "trillion," "billion," "million," or "thousand" when we reach the comma.

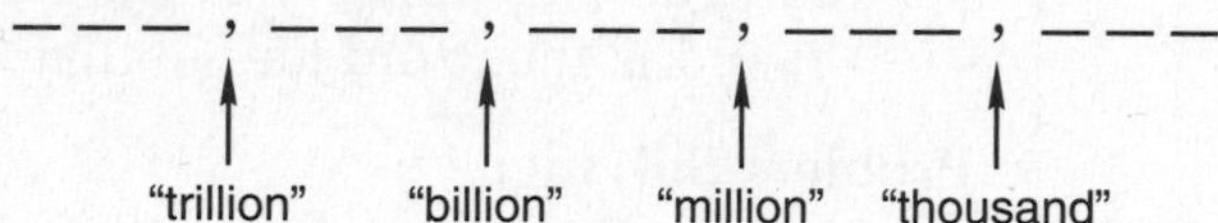

We will use the following guidelines when writing numbers as words:

1. Put commas after the words *trillion, billion, million,* and *thousand.*
2. Hyphenate numbers between 20 and 100 that do not end in zero. For example, 52, 76, and 95 are written "fifty-two," "seventy-six," and "ninety-five."

Example 3 Use words to write 1,380,000,050,200.

Solution **One trillion, three hundred eighty billion, fifty thousand, two hundred.**

Note: Since there are no millions, we do not read the millions comma.

Example 4 Use words to write 3406521.

Solution We start on the right and insert a comma every third place as we move to the left:

3,406,521

Three million, four hundred six thousand, five hundred twenty-one.

Note: We do not write "... five hundred *and* twenty-one." We never include "and" when saying or writing whole numbers.

Example 5 Use digits to write twenty trillion, five hundred ten million.

Solution It may be helpful to first draw a "skeleton" of the number. We see that the number is more than one trillion, so we draw the following skeleton, using abbreviations for "trillion," "billion," "million," and "thousand":

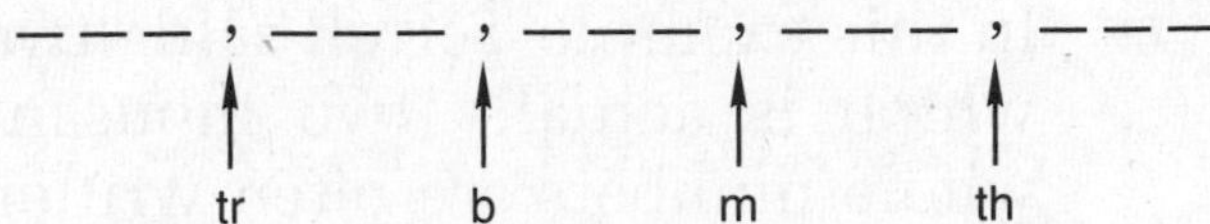

Now we will read the number until we reach a comma and then pause to write what we have read. We read "twenty trillion," so we write 20 before the trillions comma.

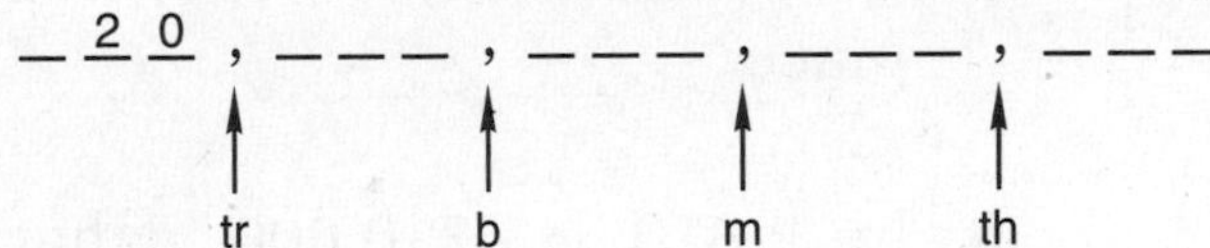

Next we read "five hundred ten million." We write 510 before the millions comma.

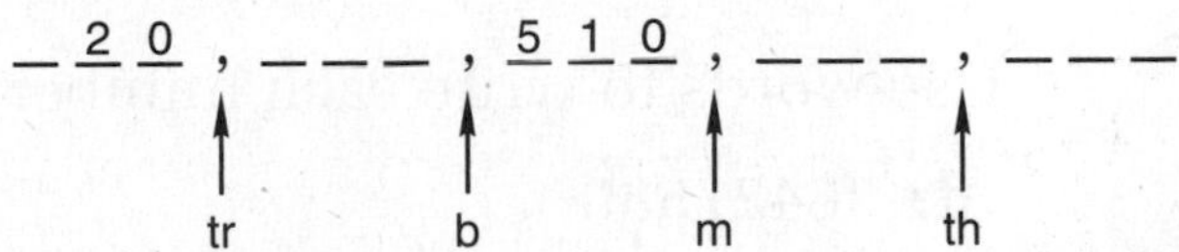

Since there are no billions, we write zeros in the three places before the billions comma.

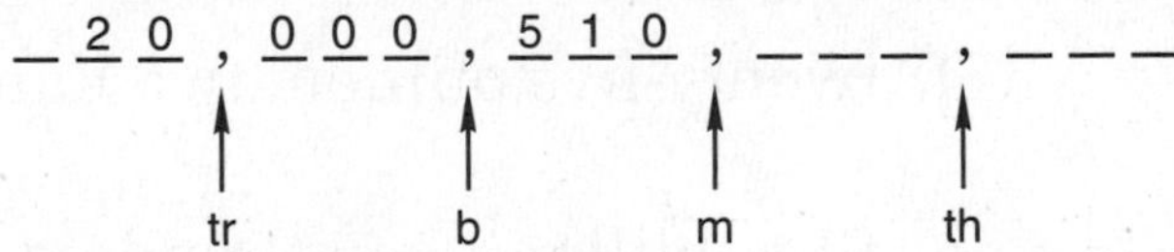

To hold place values, we write zeros in the remaining places. Now we omit the dashes and write the number.

20,000,510,000,000

Large numbers that end with many zeros are often named using a combination of digits and words, such as $3 billion for $3,000,000,000.

Example 6 Use only digits and commas to write 25 million.

Solution **25,000,000**

Example 7 Terrell said he drove twenty-four hundred miles on his trip. Use digits to write that number.

Solution Counting up by hundreds, some people say "eight hundred," "nine hundred," "ten hundred," "eleven hundred," and so on for 800, 900, 1000, 1100, ...

In this example Terrell said "twenty-four hundred" for 2400, which is actually two thousand, four hundred. Four-digit whole numbers are often written without commas, so either of these forms is correct: **2400** or **2,400.**

LESSON PRACTICE

Practice set

a. In 217,534,896,000,000, which digit is in the ten-billions place?

b. In 9,876,543,210,000, what is the place value of the digit 6?

c. Write 2500 in expanded notation.

Use words to write each number:

d. 36427580

e. 40302010

Use digits to write each number:

f. twenty-five million, two hundred six thousand, forty

g. fifty billion, four hundred two million, one hundred thousand

h. $15 billion

MIXED PRACTICE

Problem set

1. (5) What is the sum of six hundred seven and two thousand, three hundred ninety-three?

2. (4, 5) Use digits and other symbols to write "One hundred one thousand is greater than one thousand, one hundred."

3. Use words to write 50,574,006.
(5)

4. Which digit is in the trillions place in the number 12,345,678,900,000?
(5)

5. Use digits to write two hundred fifty million, five thousand, seventy.
(5)

6. Replace the circle with the proper comparison symbol. Then write the comparison as a complete sentence, using words to write the numbers.
(4)

$$-12 \bigcirc -15$$

7. Arrange these numbers in order from least to greatest:
(4)

$$-1, 4, -7, 0, 5, 7$$

8. Describe how to show this subtraction problem on a number line: $5 - 4$
(4)

9. How many units is it from negative 5 to positive 2 on the number line?
(4)

Find each missing number:

10. $2 \cdot 3 \cdot 5 \cdot n = 960$
(3)

11. $a - 1367 = 2500$
(3)

12. $b + 5 + 17 = 50$
(3)

13. $\$25.00 - k = \$18.70$
(3)

14. $6400 + d = 10{,}000$
(3)

15. $\frac{144}{f} = 8$
(3)

16. Write 750,000 in expanded notation.
(5)

Simplify:

17.
(1)

$$\begin{array}{r} 37{,}428 \\ +\ 59{,}775 \\ \hline \end{array}$$

18.
(1)

$$\begin{array}{r} 31{,}014 \\ -\ 24{,}767 \\ \hline \end{array}$$

19. $45 + 362 + 7 + 4319$
(1)

20. $\$64.59 + \$124 + \$6.30 + 37¢$
(1)

21. $144 \div (12 \div 3)$ (2)

22. $(144 \div 12) \div 3$ (2)

23. $40(500)$ (1)

24. $8505 \div 21$ (1)

25. $10 – ($4.60 – 39¢) (2)

26. 29¢ × 36 (1)

27. (2) Which property can we use to find each missing number?

(a) $365n = 365$ (b) $52 \cdot 7 = 7m$

28. (2) Use words to describe the rule of the following sequence. Then find the next three terms of the sequence.

..., 10, 8, 6, 4, 2, ...

29. (1, 4) Name each set of numbers illustrated:

(a) {1, 2, 3, 4, ...}

(b) {0, 1, 2, 3, ...}

(c) {..., –2, –1, 0, 1, 2, ...}

30. (1, 4) Use braces, an ellipsis, and digits to illustrate the set of negative even numbers.

LESSON

6 Factors • Divisibility

WARM-UP

Facts Practice: 30 Equations (Test B)

Mental Math:

a. \$5.00 + \$2.50 b. \$1.50 × 10 c. \$1.00 − \$0.45

d. 450 + 35 e. 675 − 50 f. 750 ÷ 10

g. 9 × 5, − 1, ÷ 4, + 1, ÷ 4, × 5, + 1, ÷ 4†

Problem Solving:

If there are twelve glubs in a lorn and four lorns in a dort, then how many glubs are in half a dort?

NEW CONCEPTS

Factors Recall that factors are the numbers multiplied to form a product.

$3 \times 5 = 15$ both 3 and 5 are factors of 15

$1 \times 15 = 15$ both 1 and 15 are factors of 15

Therefore, each of the numbers 1, 3, 5, and 15 can serve as a factor of 15.

Notice that 15 can be divided by 1, 3, 5, or 15 without a remainder. This leads us to another definition of factor.

> The **factors** of a number are the whole numbers that divide the number without a remainder.

For example, the numbers 1, 2, 5, and 10 are factors of 10 because each divides 10 without a remainder (that is, with a remainder of zero).

$$\begin{array}{r} 10 \\ 1\overline{)10} \\ \underline{10} \\ 0 \end{array} \qquad \begin{array}{r} 5 \\ 2\overline{)10} \\ \underline{10} \\ 0 \end{array} \qquad \begin{array}{r} 2 \\ 5\overline{)10} \\ \underline{10} \\ 0 \end{array} \qquad \begin{array}{r} 1 \\ 10\overline{)10} \\ \underline{10} \\ 0 \end{array}$$

†As a shorthand, we will use commas to separate operations to be performed sequentially from left to right. This is not standard mathematical notation.

Example 1 List the whole numbers that are factors of 12.

Solution The factors of 12 are the whole numbers that divide 12 with no remainder. They are **1, 2, 3, 4, 6,** and **12.**

Example 2 List the factors of 51.

Solution As we try to think of whole numbers that divide 51 with no remainder, we may think that 51 has only two factors, 1 and 51. However, there are actually four factors of 51. Notice that 3 and 17 are also factors of 51.

$$\text{3 is a factor of 51} \rightarrow 3\overline{)51} \quad 17 \leftarrow \text{17 is a factor of 51}$$

Thus, the four factors of 51 are **1, 3, 17,** and **51.**

From examples 1 and 2 we see that 12 and 51 have two **common factors,** 1 and 3. The **greatest common factor (GCF)** of 12 and 51 is 3, because it is the largest common factor of both numbers.

Example 3 Find the greatest common factor of 18 and 30.

Solution We are asked to find the largest factor (divisor) of both 18 and 30. Here we list the factors of both numbers, circling the common factors.

Factors of 18: (1), (2), (3), (6), 9, 18

Factors of 30: (1), (2), (3), 5, (6), 10, 15, 30

The greatest common factor of 18 and 30 is **6.**

Divisibility As we saw in example 2, the number 51 **can be divided** by 1, 3, 17, and 51 with a remainder of zero. The capability of a whole number to be divided by another whole number with no remainder is called **divisibility.** Thus, 51 is **divisible** by 1, 3, 17, and 51.

There are several methods for testing the divisibility of a number without actually performing the division. Listed below are methods for testing whether a number is divisible by 2, 3, 4, 5, 6, 8, 9, or 10.

TESTS FOR DIVISIBILITY

A number is divisible by ...

2 if the last digit is even.
4 if the last two digits can be divided by 4.
8 if the last three digits can be divided by 8.

5 if the last digit is 0 or 5.
10 if the last digit is 0.

3 if the **sum of the digits** can be divided by 3.
6 if the number can be divided by 2 **and** by 3.
9 if the **sum of the digits** can be divided by 9.

A number ending in ...

one zero is divisible by 2.
two zeros is divisible by 2 and 4.
three zeros is divisible by 2, 4, and 8.

Example 4 Which whole numbers from 1 to 10 are divisors of 9060?

Solution In the sense used in this problem, a **divisor** is a **factor.** The number 1 is a divisor of any whole number. As we apply the tests for divisibility, we find that 9060 passes the tests for 2, 4, 5, and 10, but not for 8. The sum of its digits (9 + 0 + 6 + 0) is 15, which can be divided by 3 but not by 9. Since 9060 is divisible by both 2 and 3, it is also divisible by 6. The only whole number from 1 to 10 we have not tried is 7, for which we have no simple test. We divide 9060 by 7 to find that 7 is not a divisor. We find that the numbers from 1 to 10 that are divisors of 9060 are **1, 2, 3, 4, 5, 6,** and **10.**

LESSON PRACTICE

Practice set* List the whole numbers that are factors of each number:

a. 25 **b.** 24 **c.** 23

List the whole numbers from 1 to 10 that are factors of each number:

d. 1260 **e.** 73,500 **f.** 3600

g. List the single-digit divisors of 1356.

h. The number 7000 is divisible by which single-digit numbers?

i. List all the common factors of 12 and 20.

j. Find the greatest common factor (GCF) of 24 and 40.

MIXED PRACTICE

Problem set

1. If the product of 10 and 20 is divided by the sum of 20 and 30, what is the quotient?
(1)

2. (a) List all the common factors of 30 and 40.
(6)
(b) Find the greatest common factor of 30 and 40.

3. Use braces, an ellipsis, and digits to illustrate the set of negative odd numbers.
(4)

4. Use digits to write four hundred seven million, six thousand, nine hundred sixty-two.
(5)

5. List the whole numbers from 1 to 10 that are divisors of 12,300.
(6)

6. Replace the circle with the proper comparison symbol. Then write the comparison as a complete sentence using words to write the numbers.
(4)

$$-7 \bigcirc -11$$

7. The number 3456 is divisible by which single-digit numbers?
(6)

8. Show this subtraction problem on a number line: $2 - 5$
(4)

9. Write 6400 in expanded notation.
(5)

Find each missing number:

10. $x + \$4.60 = \$10.00$
(3)

11. $p - 3850 = 4500$
(3)

12. $8z = \$50.00$
(3)

13.
(3)

$$\begin{array}{r} 7 \\ 4 \\ 8 \\ 6 \\ 2 \\ 1 \\ 6 \\ 8 \\ 9 \\ + \ n \\ \hline 60 \end{array}$$

14. $1426 - k = 87$
(3)

15. $\dfrac{990}{p} = 45$
(3)

16. $\dfrac{z}{8} = 32$
(3)

Simplify:

17. (1) $\frac{1225}{35}$

18. (1) $\begin{array}{r} 800 \\ \times\ 50 \\ \hline \end{array}$

19. (1) $\begin{array}{r} \$100.00 \\ -\ \$48.37 \\ \hline \end{array}$

20. (1) $\begin{array}{r} 46{,}302 \\ +\ 49{,}998 \\ \hline \end{array}$

21. (1) $45.00 ÷ 20

22. (1) $7 \cdot 11 \cdot 13$

23. (1) $9\overline{)43{,}271}$

24. (1) 3625 + 59 + 570 + 8

25. (1) 48¢ + $8.49 + $14

26. (2) 1000 − (430 − 58)

27. (1) 140(16)

28. (1) $\begin{array}{r} 25¢ \\ \times\ 24 \\ \hline \end{array}$

29. (1) $\frac{\$43.50}{10}$

30. (2) Name the property illustrated by this equation, and describe the meaning of the property.

$$x \cdot 5 = 5x$$

LESSON

7 Lines and Angles

WARM-UP

Facts Practice: 30 Equations (Test B)

Mental Math:

a. 5 − 10 b. \$2.50 × 10 c. \$1.00 − 35¢

d. 340 + 25 e. 565 − 300 f. 480 ÷ 10

g. Start with the number of years in a decade, × 7, + 5, ÷ 3, − 1, ÷ 4.

Problem Solving:

The sum of the counting numbers from 1 through 4 is 10. What is the sum of the counting numbers from 1 through 20?

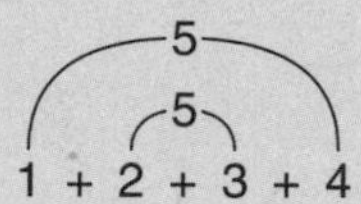

NEW CONCEPT

We live in a world of three dimensions called **space.** We can measure the length, width, and depth of objects that occupy space. We can imagine a two-dimensional world called a **plane,** a flat world having length and width but not depth. Occupants of a two-dimensional world could not pass over or under other objects because, without depth, "over" and "under" would not exist. A one-dimensional world, a **line,** has length but neither width nor depth. Occupants of a one-dimensional world could not pass over, under, or to either side of other objects. They could only move back and forth on their line.

In **geometry** we study figures that have one dimension, two dimensions, and three dimensions, but we begin with a **point,** which has no dimensions. A point is an exact location in space and is unmeasurably small. We represent points with dots and usually name them with uppercase letters. Here we show point *A:*

A
•

A **line** contains an infinite number of points extending in opposite directions without end. A line has one dimension, length. A line has no thickness. We can represent a line by sketching part of a line with two arrowheads. We identify a line by naming two points on the line in either order. Here we show line *AB* (or line *BA*):

Line *AB* or line *BA*

The symbols $\overleftrightarrow{AB}$ and $\overleftrightarrow{BA}$ (read "line *AB*" and "line *BA*") also can be used to refer to the line above.

A **ray** is a part of a line with one endpoint. We identify a ray by naming the endpoint and then one other point on the ray. Here we show ray *AB* ($\overrightarrow{AB}$):

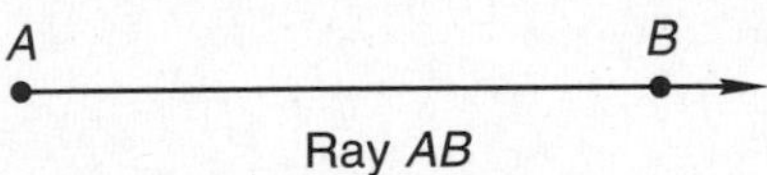

Ray *AB*

A **segment** is a part of a line with two endpoints. We identify a segment by naming the two endpoints in either order. Here we show segment *AB* ($\overline{AB}$):

Segment *AB* or segment *BA*

A segment has a specific length. We may refer to the length of segment *AB* by writing $m\overline{AB}$, which means "the measure of segment *AB*," or by writing the letters *AB* without an overbar. Thus, both *AB* and $m\overline{AB}$ refer to the distance from point *A* to point *B*. We use this notation in the figure below to state that the sum of the lengths of the shorter segments equals the length of the longest segment.

$$AB + BC = AC$$

$$m\overline{AB} + m\overline{BC} = m\overline{AC}$$

Example 1 Use symbols to name a line, two rays, and a segment in the figure at right.

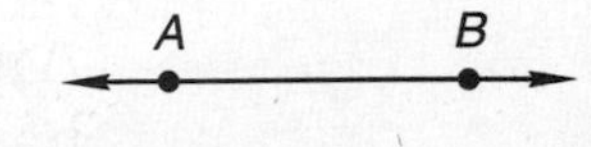

Solution The line is $\boldsymbol{\overleftrightarrow{AB}}$ (or $\boldsymbol{\overleftrightarrow{BA}}$). The rays are $\boldsymbol{\overrightarrow{AB}}$ and $\boldsymbol{\overrightarrow{BA}}$. The segment is $\boldsymbol{\overline{AB}}$ (or $\boldsymbol{\overline{BA}}$).

Example 2 In the figure below, *AB* is 3 cm and *AC* is 7 cm. Find *BC*.

Solution *BC* represents the length of segment *BC*. We are given that *AB* is 3 cm and *AC* is 7 cm. From the figure above, we see that $AB + BC = AC$. Therefore, we find that *BC* is **4 cm**.

A **plane** is a flat surface that extends without end. It has two dimensions, length and width. A desktop occupies a part of a plane.

Two lines in the same plane either cross once or do not cross at all. If two lines cross, we say that they **intersect** at one point. If two lines in a plane do not intersect, they remain the same distance apart and are called **parallel** lines.

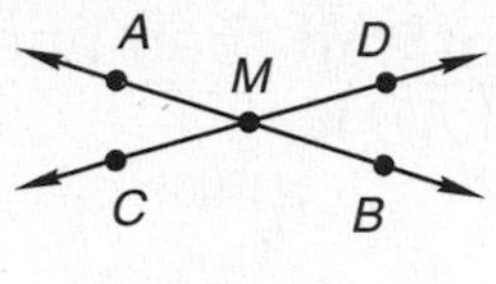

Line *AB* intersects line *CD* at point *M*.

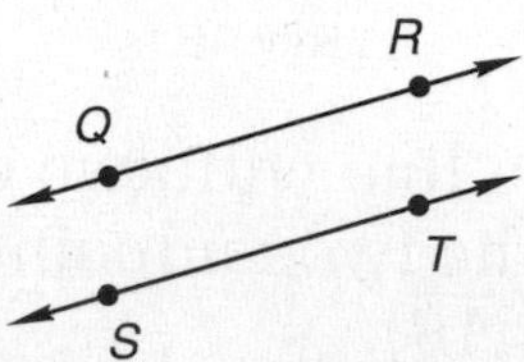

In this figure, line *QR* is parallel to line *ST*. This statement can be written with symbols, as we show here:

$$\overleftrightarrow{QR} \parallel \overleftrightarrow{ST}$$

Lines that intersect and form "square corners" are **perpendicular.** The small square in the figure below indicates a "square corner."

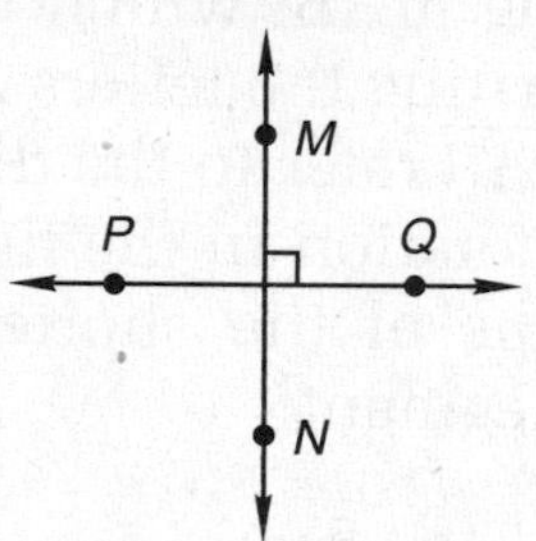

In this figure, line *MN* is perpendicular to line *PQ*. This statement can be written with symbols, as we show here:

$$\overleftrightarrow{MN} \perp \overleftrightarrow{PQ}$$

Lines in a plane that are neither parallel nor perpendicular are **oblique.** In our figure showing intersecting lines, lines *AB* and *CD* are oblique.

An **angle** is formed by two rays that have a common endpoint. The angle at right is formed by the two rays $\overrightarrow{MD}$ and $\overrightarrow{MB}$. The common endpoint is *M*. Point *M* is the **vertex** of the angle. Ray *MD* and ray *MB* are the **sides** of the angle. Angles may be named by listing the following points in order: a point on one ray, the vertex, and then a point on the other ray. So our angle may be named either angle *DMB* or angle *BMD*.

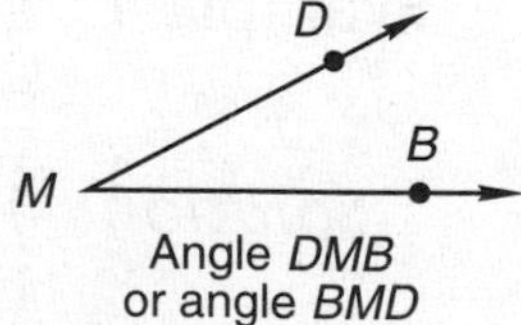

Angle *DMB* or angle *BMD*

When there is no chance of confusion, an angle may be named by only one point, the vertex. At right we have angle A.

Angle A

An angle may also be named by placing a small letter or number near the vertex and between the rays (in the interior of the angle). Here we see angle 1.

Angle 1

The symbol $\angle$ is often used instead of the word *angle.* Thus, the three angles just named could be referred to as:

$\angle DMB$	read as "angle DMB"
$\angle A$	read as "angle A"
$\angle 1$	read as "angle 1"

Angles are classified by their size. An angle formed by perpendicular rays is a **right angle** and is commonly marked with a small square at the vertex. An angle smaller than a right angle is an **acute angle.** An angle that forms a straight line is a **straight angle.** An angle smaller than a straight angle but larger than a right angle is an **obtuse angle.**

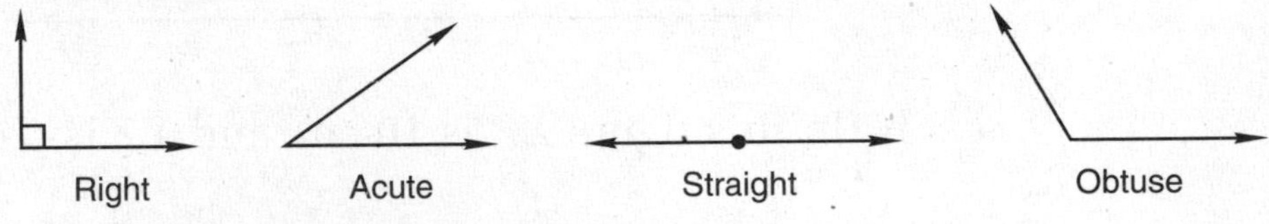

Example 3 (a) Which line is parallel to line AB?

(b) Which line is perpendicular to line AB?

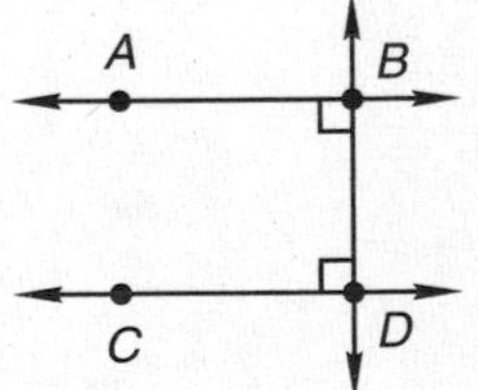

Solution (a) **Line CD** (or $\overleftrightarrow{DC}$) is parallel to line AB.

(b) **Line BD** (or $\overleftrightarrow{DB}$) is perpendicular to line AB.

Example 4 There are several angles in this figure.

(a) Name the straight angle.

(b) Name the obtuse angle.

(c) Name two right angles.

(d) Name two acute angles.

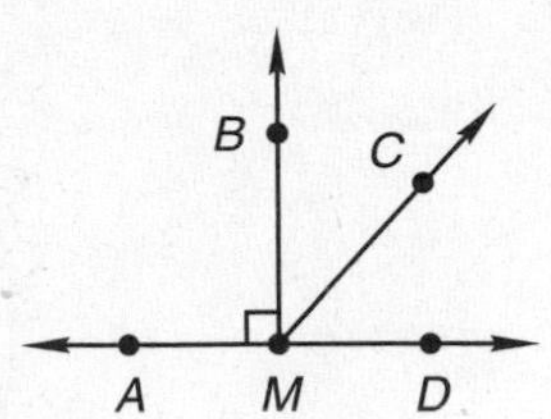

Solution (a) $\angle AMD$ (or $\angle DMA$)

(b) $\angle AMC$ (or $\angle CMA$)

(c) 1. $\angle AMB$ (or $\angle BMA$)
2. $\angle BMD$ (or $\angle DMB$)

(d) 1. $\angle BMC$ (or $\angle CMB$)
2. $\angle CMD$ (or $\angle DMC$)

On earth we refer to objects aligned with the force of gravity as **vertical** and objects aligned with the horizon as **horizontal.**

Example 5 A power pole with two cross pieces can be represented by three segments.

(a) Name a vertical segment.

(b) Name a horizontal segment.

(c) Name a segment perpendicular to $\overline{CD}$.

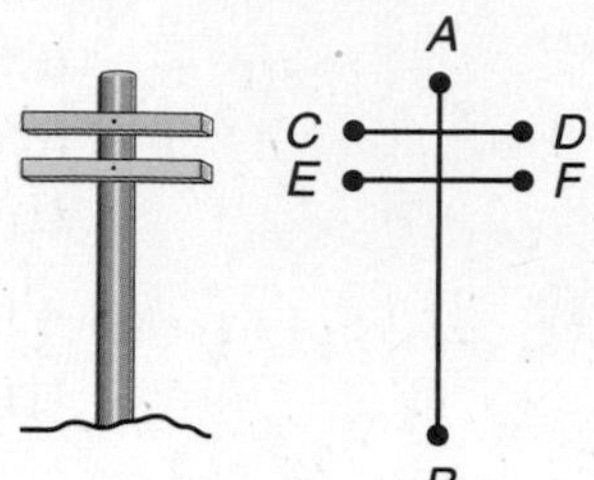

Solution (a) $\overline{AB}$ (or $\overline{BA}$)

(b) $\overline{CD}$ (or $\overline{DC}$) or $\overline{EF}$ (or $\overline{FE}$)

(c) $\overline{AB}$ (or $\overline{BA}$)

LESSON PRACTICE

Practice set **a.** Name a point on this figure that is not on ray *BC:*

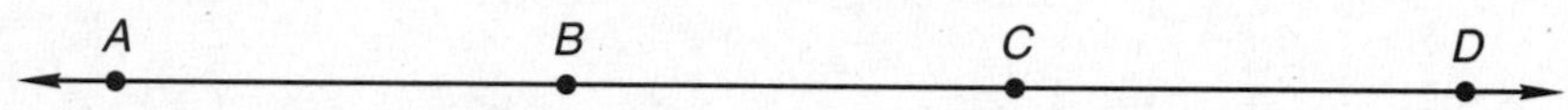

b. In this figure *XZ* is 10 cm, and *YZ* is 6 cm. Find *XY*.

c. Draw two parallel lines.

d. Draw two perpendicular lines.

e. Draw two lines that intersect but are not perpendicular. What word describes the relationship of these lines?

f. Draw a right angle.

g. Draw an acute angle.

h. Draw an obtuse angle.

i. Two intersecting segments are drawn on the board. One segment is vertical and the other is horizontal. Are the segments parallel or perpendicular?

MIXED PRACTICE

Problem set

1. If the product of two one-digit whole numbers is 35, what is the sum of the same two numbers?
(3)

2. Name the property illustrated by this equation:
(2)

$$-5 \cdot 1 = -5$$

3. List the whole number divisors of 50.
(6)

4. Use digits and symbols to write "Two minus five equals negative three."
(4)

5. Use only digits and commas to write 90 million.
(5)

6. List the single-digit factors of 924.
(6)

7. Arrange these numbers in order from least to greatest:
(4)

$$-10, 5, -7, 8, 0, -2$$

8. Use words to describe the following sequence. Then find the next three numbers in the sequence.
(2)

$$\ldots, 49, 64, 81, 100, \ldots$$

9. To build a fence, Megan dug holes in the ground to hold the posts upright. Then she nailed rails to connect the posts. Which fence parts were vertical, the posts or the rails?
(7)

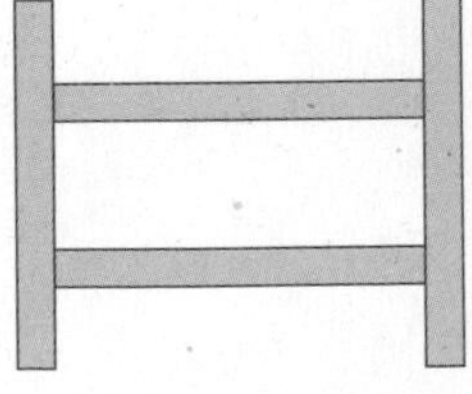

10. (a) List the common factors of 24 and 32.
(6)

(b) Find the greatest common factor of 24 and 32.

11. How many units is it from 3 to −4 on a number line?
(4)

Find each missing number:

12. $6 \cdot 6 \cdot z = 1224$
(3)

13. $\$100.00 - k = \$17.54$
(3)

14. $w - 98 = 432$
(3)

15. $20x = \$36.00$
(3)

16. $\dfrac{w}{20} = 200$ (3)

17. $\dfrac{300}{x} = 30$ (3)

18. Does the quotient of 4554 ÷ 9 have a remainder? How can you tell without dividing? (6)

Simplify:

19. $\begin{array}{r} 36{,}475 \\ +\ 55{,}984 \\ \hline \end{array}$ (1)

20. $\begin{array}{r} 476 \\ \times\ \ 38 \\ \hline \end{array}$ (1)

21. \$80.00 − \$72.45 (1)

22. 49 + 387 + 1579 + 98 (1)

23. \$68.00 ÷ 40 (1)

24. $8 \cdot 7 \cdot 5$ (1)

25. Compare: 4000 ÷ (200 ÷ 10) ◯ (4000 ÷ 200) ÷ 10 (2, 4)

26. Evaluate each expression for $a = 200$ and $b = 400$: (1)

(a) ab (b) $a - b$ (c) $\dfrac{b}{a}$

27. Refer to the figure at right to answer (a) and (b). (7)

(a) Which angle is an acute angle?

(b) Which angle is a straight angle?

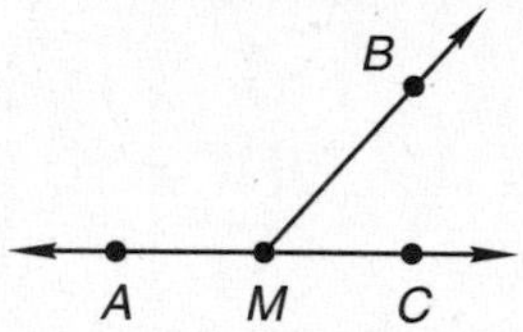

28. What type of angle is formed by perpendicular lines? (7)

Refer to the figure below to answer problems 29 and 30.

29. Name three segments in this figure. (7)

30. If you knew $m\overline{XY}$ and $m\overline{YZ}$, describe how you would find $m\overline{XZ}$. (7)

LESSON

8 Fractions and Percents • Inch Ruler

WARM-UP

Facts Practice: 64 Multiplication Facts (Test A)

Mental Math:

a. 4 – 10 b. $0.25 × 10 c. $1.00 – 65¢
d. 325 + 50 e. 347 – 30 f. 200 × 10
g. Start with a score, + 1, ÷ 3, × 5, + 1, ÷ 4, + 1, ÷ 2, × 6, + 3, ÷ 3.

Problem Solving:

The number 325 contains the three digits 2, 3, and 5. These three digits can be ordered in other ways to make different numbers. Each such ordering is called a **permutation** of the three digits. The smallest permutation of 2, 3, and 5 is 235. Which number is the largest permutation of 2, 3, and 5?

NEW CONCEPTS

Fractions and percents

Fractions and **percents** are commonly used to name parts of a whole or parts of a group.

At right we use a whole circle to represent 1. The circle is divided into four equal parts with one part shaded. One fourth $\left(\frac{1}{4}\right)$ of the circle is shaded, and $\frac{3}{4}$ of the circle is not shaded.

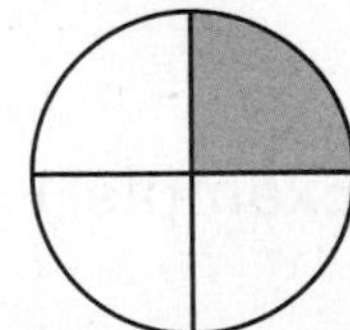

Since the whole circle also represents 100% of the circle, we can divide 100% by 4 to find the percent of the circle that is shaded.

$$100\% \div 4 = 25\%$$

We find that 25% of the circle is shaded, so 75% of the circle is not shaded.

A common fraction is written with two numbers and a division bar. The number below the bar is the **denominator** and shows how many equal parts are in the whole. The number above the bar is the **numerator** and shows how many of the parts have been selected.

numerator → $\dfrac{1}{4}$ ← division bar
denominator →

A percent describes a whole as though there were 100 parts, even though the whole may not actually contain 100 parts. Thus the "denominator" of a percent is always 100.

$$25 \text{ percent means } \frac{25}{100}$$

Instead of writing the denominator, 100, we use the word *percent* or the percent symbol, %.

A whole number plus a fraction is a **mixed number.** To name the number of circles shaded below, we use the mixed number $2\frac{3}{4}$. We see that $2\frac{3}{4}$ means $2 + \frac{3}{4}$. To read a mixed number, we first say the whole number; then we say "and"; then we say the fraction.

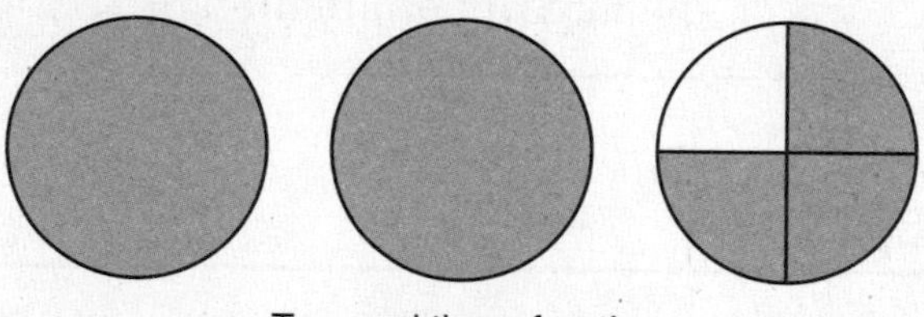

Two and three fourths

It is possible to have percents greater than 100%. If we were to write $2\frac{3}{4}$ as a percent, we would write 275%.

Example 1 Name the shaded part of the circle as a fraction and as a percent.

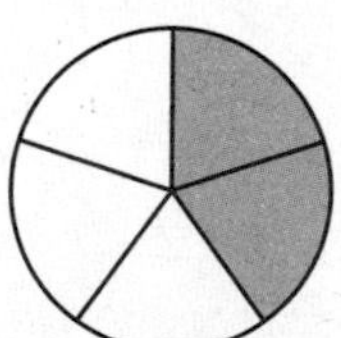

Solution Two of the five equal parts are shaded, so the fraction that is shaded is $\mathbf{\frac{2}{5}}$.

Since the whole circle (100%) is divided into five equal parts, each part is 20%.

$$100\% \div 5 = 20\%$$

Two parts are shaded. So $2 \times 20\%$, or **40%,** is shaded.

Example 2 Which of the following could describe the portion of this rectangle that is shaded?

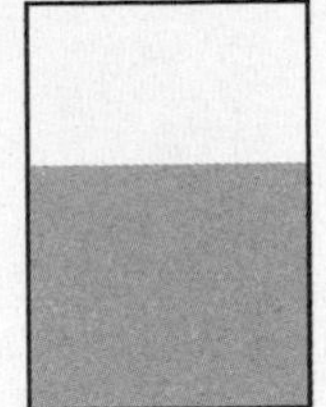

A. $\frac{1}{2}$ B. 40% C. 60%

Solution There is a shaded and an unshaded part of this rectangle, but the parts are not equal. More than $\frac{1}{2}$ of the rectangle is shaded, so the answer is not A. Half of a whole is 50%.

$$100\% \div 2 = 50\%$$

Since more than 50% of the rectangle is shaded, the correct choice must be **C. 60%.**

Between the points on a number line that represent whole numbers are many points that represent fractions and mixed numbers. To identify the fraction or mixed number associated with a point on a number line, it is first necessary to discover the number of segments into which each length has been divided.

Example 3 Point *A* represents what mixed number on this number line?

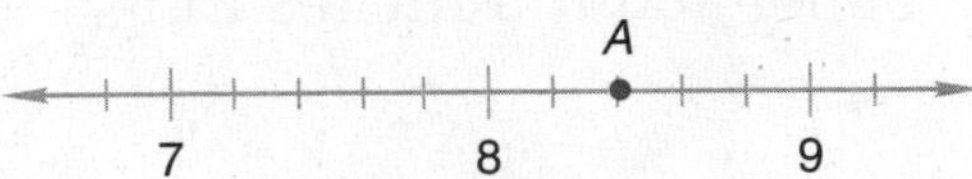

Solution We see that point *A* represents a number greater than 8 but less than 9. It represents 8 plus a fraction. To find the fraction, we first notice that the segment from 8 to 9 has been divided into five smaller segments. The distance from 8 to point *A* crosses two of the five segments. Thus, point *A* represents the mixed number $\mathbf{8\frac{2}{5}}$.

Note: It is important to focus on the *number of segments* and not on the number of vertical tick marks. The four vertical tick marks divide the space between 8 and 9 into five segments, just as four cuts divide a candy bar into five pieces.

Inch ruler A ruler is a practical application of a number line. The units on a ruler are of a standard length and are often divided successively in half. That is, inches are divided in half to show half inches. Then half inches are divided in half to show quarter inches. The divisions may continue in order to show eighths, sixteenths, thirty-seconds, and even sixty-fourths of an inch. In this book we will practice measuring and drawing segments to the nearest sixteenth of an inch.

Here we show a magnified view of an inch ruler with divisions to one sixteenth of an inch. We have labeled each division for reference.

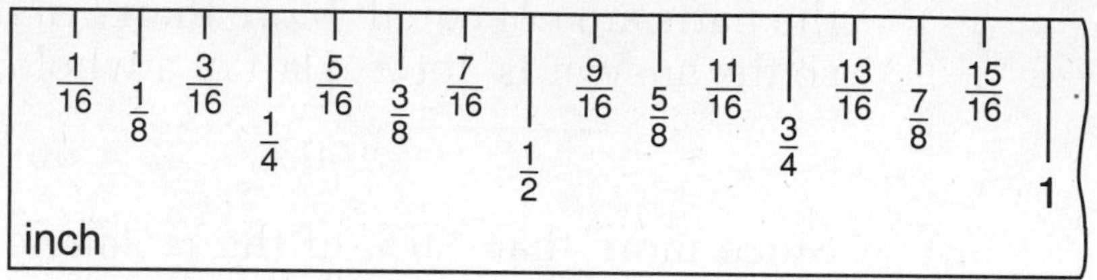

It is important to bear in mind that all measurements are approximate. The quality of a measurement depends upon many conditions, including the care taken in performing the measurement and the precision of the measuring instrument. The finer the gradations are on the instrument, the more precise the measurement can be.

For example, if we measure segments *AB* and *CD* below with an undivided inch ruler, we would describe both segments as being about 3 inches long.

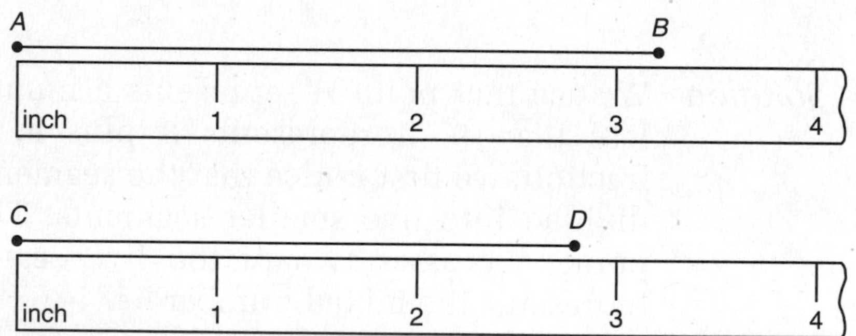

We can say that the measure of each segment is 3 inches $\pm \frac{1}{2}$ inch ("three inches plus or minus one half inch"). This means each segment is within $\frac{1}{2}$ inch of being 3 inches long. In fact, for any measuring instrument, the greatest possible error due to the instrument is one half of the unit that marks the instrument.

We can improve the precision of measurement and reduce the possible error by using an instrument with smaller units. Below we use a ruler divided into quarter inches. We see that *AB* is about $3\frac{1}{4}$ inches and *CD* is about $2\frac{3}{4}$ inches. These measures are precise to the nearest quarter inch. The greatest possible error due to the measuring instrument is one eighth of an inch, which is half of the unit used for the measure.

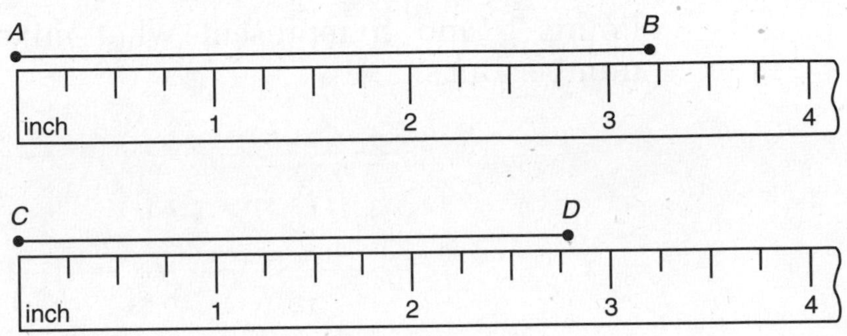

Example 4 Use an inch ruler to find *AB*, *BC*, and *AC* to the nearest sixteenth of an inch.

Solution From point *A* we find *AB* and *AC*. We measure from the center of one dot to the center of the other dot. ***AB* is about $\frac{7}{8}$ inches,** and ***AC* is about $2\frac{1}{2}$ inches.**

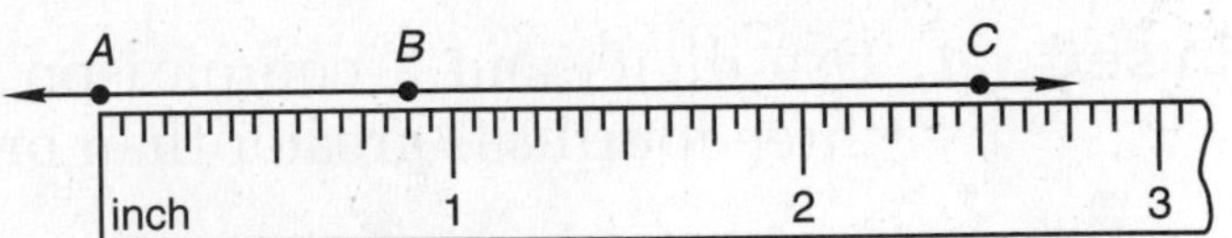

We move the zero mark on the ruler to point *B* to measure *BC*. We find ***BC* is about $1\frac{5}{8}$ inches.**

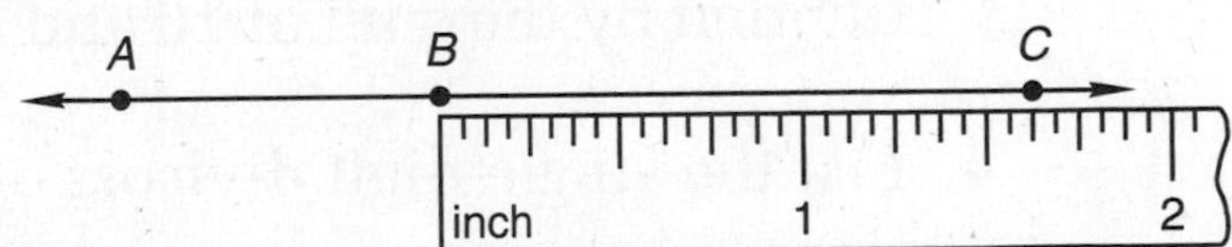

LESSON PRACTICE

Practice set

a. What fraction of this circle is not shaded?

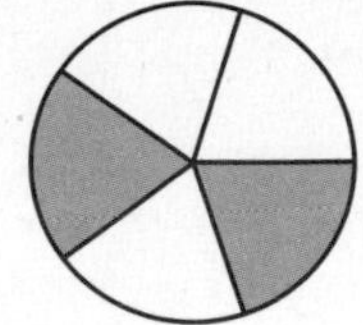

b. What percent of this circle is not shaded?

c. Half of a whole is what percent of the whole?

Draw and shade circles to illustrate each fraction, mixed number, or percent:

d. $\frac{2}{3}$ **e.** 75% **f.** $2\frac{3}{4}$

Points **g** and **h** represent what mixed numbers on these number lines?

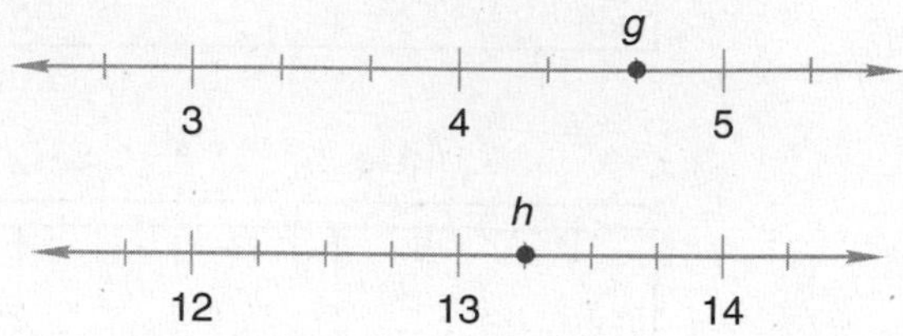

i. Find *XZ* to the nearest sixteenth of an inch.

j. Jack's ruler is divided into eighths of an inch. Assuming the ruler is used correctly, what is the greatest possible measurement error that can be made with Jack's ruler? Express your answer as a fraction of an inch.

MIXED PRACTICE

Problem set

1. (4, 8) Use digits and a comparison symbol to write "One and three fourths is greater than one and three fifths."

2. (8) Refer to practice problem **i** above. Use a ruler to find *XY* and *YZ*.

3. (1) What is the quotient when the product of 20 and 20 is divided by the sum of 10 and 10?

4. (6) List the single-digit divisors of 1680.

5. (8) Point *A* represents what mixed number on this number line?

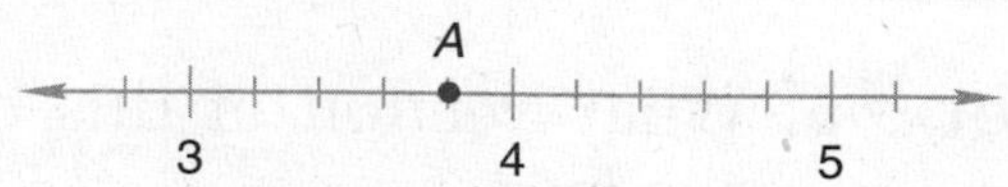

6. (2, 4) (a) Replace the circle with the proper comparison symbol.

$$3 + 2 \bigcirc 2 + 3$$

(b) What property of addition is illustrated by this comparison?

7. Use words to write 32500000000.
(5)

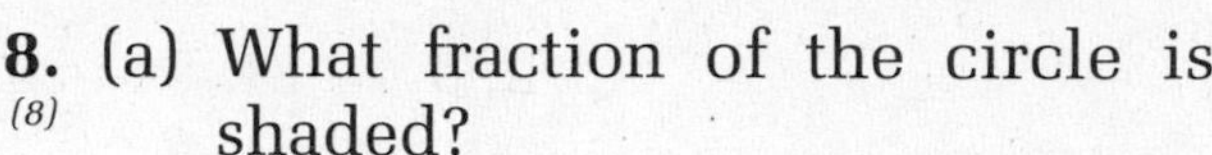

8. (a) What fraction of the circle is shaded?
(8)

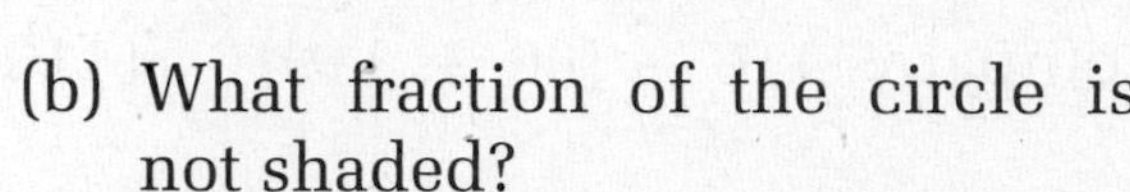

(b) What fraction of the circle is not shaded?

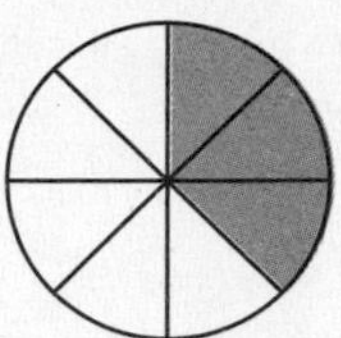

9. (a) What percent of the rectangle is shaded?
(8)

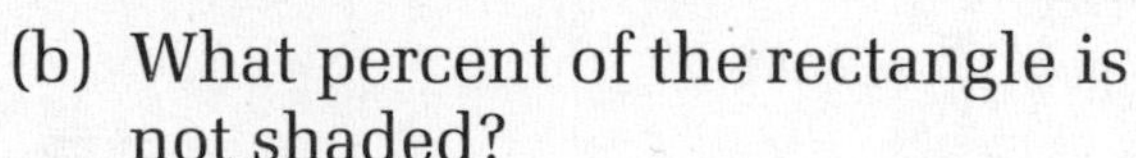

(b) What percent of the rectangle is not shaded?

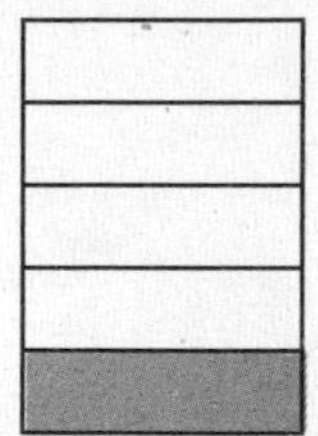

10. What is the name of the part of a fraction that indicates the number of equal parts in the whole?
(8)

Find each missing number:

11. $a - \$4.70 = \$2.35$
(3)

12. $b + \$25.48 = \$60.00$
(3)

13. $8c = \$60.00$
(3)

14. $10{,}000 - d = 5420$
(3)

15. $\frac{e}{15} = 15$
(3)

16. $\frac{196}{f} = 14$
(3)

17. $8 + 9 + 8 + 8 + 9 + 8 + n = 60$
(3)

Simplify:

18. $\begin{array}{r} 400 \\ \times\ 500 \\ \hline \end{array}$
(1)

19. $\begin{array}{r} 79¢ \\ \times\ 30 \\ \hline \end{array}$
(1)

20. $3625 + 431 + 687$
(1)

21. $6000 \div 50$
(1)

22. $20 \cdot 10 \cdot 5$
(1)

23. $\frac{\$27.00}{18}$
(1)

24. $\frac{3456}{6}$
(1)

25. If t is 1000 and v is 11, find
(1)

(a) $t - v$

(b) $v - t$

26. The rule of the following sequence is $k = 3n - 1$. What is the tenth term of the sequence?
(2)

2, 5, 8, 11, ...

27. Compare: 416 − (86 + 119) ◯ (416 − 86) + 119
(2, 4)

Refer to the figure at right to answer problems 28 and 29.

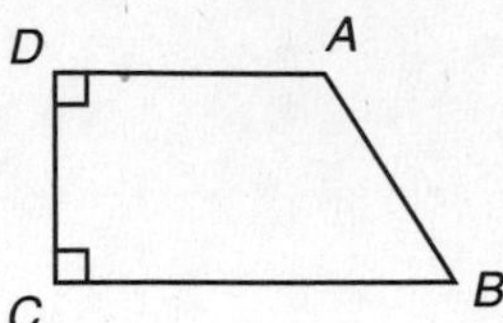

28. Name the acute, obtuse, and right angles.
(7)

29. (a) Name a segment parallel to $\overline{DA}$.
(7)
(b) Name a segment perpendicular to $\overline{DA}$.

30. Referring to the figure below, what is the difference in meaning between the notations $\overline{QR}$ and QR?
(7)

LESSON

9 Adding, Subtracting, and Multiplying Fractions • Reciprocals

WARM-UP

Facts Practice: 30 Equations (Test B)

Mental Math:

a. 3 − 5 **b.** \$0.39 × 10 **c.** \$1.00 − 29¢

d. 342 + 200 **e.** 580 − 40 **f.** 500 ÷ 50

g. Start with half a dozen, + 1, × 6, − 2, ÷ 2, + 4, ÷ 4, − 5, × 15.

Problem Solving:

Find the next four numbers in this sequence:

$$\frac{1}{16}, \frac{1}{8}, \frac{3}{16}, \frac{1}{4}, \ldots$$

NEW CONCEPTS

Adding fractions On the line below, AB is $1\frac{3}{8}$ in. and BC is $1\frac{4}{8}$ in. We can find AC by measuring or by adding $1\frac{3}{8}$ in. and $1\frac{4}{8}$ in.

A B C

$1\frac{3}{8}$ in. $1\frac{4}{8}$ in.

$$1\frac{3}{8}\text{ in.} + 1\frac{4}{8}\text{ in.} = 2\frac{7}{8}\text{ in.}$$

When adding fractions that have the same denominators, we add the numerators and write the sum over the common denominator.

Example 1 Find each sum:

(a) $\frac{1}{7} + \frac{2}{7} + \frac{3}{7}$ (b) $33\frac{1}{3}\% + 33\frac{1}{3}\%$

Solution (a) $\frac{1}{7} + \frac{2}{7} + \frac{3}{7} = \mathbf{\frac{6}{7}}$ (b) $33\frac{1}{3}\% + 33\frac{1}{3}\% = \mathbf{66\frac{2}{3}\%}$

When the numerator and denominator of a fraction are equal (but not zero), the fraction is equal to 1. The illustration shows $\frac{4}{4}$ of a circle, which is one whole circle.

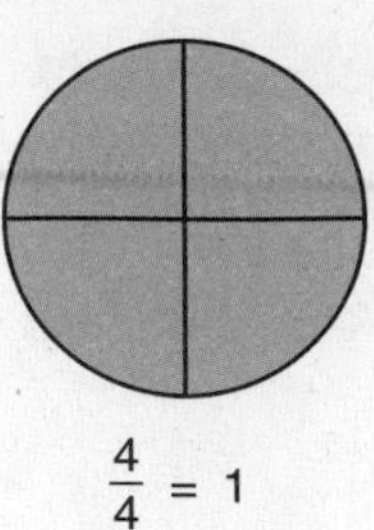

$\frac{4}{4} = 1$

Example 2 Add: $\frac{3}{5} + \frac{2}{5}$

Solution $\frac{3}{5} + \frac{2}{5} = \frac{5}{5} = \mathbf{1}$

Subtracting fractions

To subtract a fraction from a fraction with the same denominator, we write the difference of the numerators over the common denominator.

Example 3 Find each difference:

(a) $3\frac{5}{9} - 1\frac{1}{9}$ (b) $\frac{3}{5} - \frac{3}{5}$

Solution (a) $3\frac{5}{9} - 1\frac{1}{9} = \mathbf{2\frac{4}{9}}$ (b) $\frac{3}{5} - \frac{3}{5} = \frac{0}{5} = \mathbf{0}$

Multiplying fractions

The first illustration shows $\frac{1}{2}$ of a circle. The second illustration shows $\frac{1}{2}$ of $\frac{1}{2}$ of a circle. We see that $\frac{1}{2}$ of $\frac{1}{2}$ is $\frac{1}{4}$. We translate the word *of* into a multiplication symbol and find $\frac{1}{2}$ of $\frac{1}{2}$ by multiplying:

$\frac{1}{2}$

$\frac{1}{2}$ of $\frac{1}{2}$

$$\frac{1}{2} \text{ of } \frac{1}{2} \text{ becomes } \frac{1}{2} \times \frac{1}{2} = \frac{1}{4}$$

To multiply fractions, we multiply the numerators to find the numerator of the product, and we multiply the denominators to find the denominator of the product. Notice that the product of two positive fractions less than 1 is less than either fraction.

Example 4 Find each product:

(a) $\frac{1}{2}$ of $\frac{1}{3}$ (b) $\frac{1}{2} \cdot \frac{3}{4} \cdot \frac{1}{5}$

Solution (a) $\frac{1}{2} \times \frac{1}{3} = \mathbf{\frac{1}{6}}$ (b) $\frac{1}{2} \cdot \frac{3}{4} \cdot \frac{1}{5} = \mathbf{\frac{3}{40}}$

Reciprocals If we **invert** a fraction by switching the numerator and denominator, we form the **reciprocal** of the fraction.

The reciprocal of $\frac{4}{3}$ is $\frac{3}{4}$.

The reciprocal of $\frac{3}{4}$ is $\frac{4}{3}$.

The reciprocal of $\frac{1}{4}$ is $\frac{4}{1}$, which is 4.

The reciprocal of 4 $\left(\text{or } \frac{4}{1}\right)$ is $\frac{1}{4}$.

Note this very important property of reciprocals:

The product of a fraction and its reciprocal is 1.

$$\frac{4}{3} \cdot \frac{3}{4} = \frac{12}{12} = 1$$

$$\frac{1}{4} \cdot \frac{4}{1} = \frac{4}{4} = 1$$

Example 5 Find the reciprocal of each number:

(a) $\frac{3}{5}$ (b) 3

Solution (a) The reciprocal of $\frac{3}{5}$ is $\mathbf{\frac{5}{3}}$.

(b) The reciprocal of 3, which is 3 “wholes” or $\frac{3}{1}$, is $\mathbf{\frac{1}{3}}$.

Example 6 Find the missing number: $\frac{3}{4}n = 1$

Solution The expression $\frac{3}{4}n$ means “$\frac{3}{4}$ times n.” Since the product of $\frac{3}{4}$ and n is 1, the missing number must be the reciprocal of $\frac{3}{4}$, which is $\mathbf{\frac{4}{3}}$.

$$\frac{3}{4} \cdot \frac{4}{3} = \frac{12}{12} = 1 \quad \text{check}$$

Example 7 How many $\frac{3}{4}$'s are in 1?

Solution The answer is the reciprocal of $\frac{3}{4}$, which is $\mathbf{\frac{4}{3}}$.

In Lesson 2 we noted that although multiplication is commutative (6 × 3 = 3 × 6), division is not commutative

($6 \div 3 \neq 3 \div 6$). Now we can say that reversing the order of division results in the reciprocal quotient.

$$6 \div 3 = 2$$

$$3 \div 6 = \frac{1}{2}$$

LESSON PRACTICE

Practice set Simplify:

a. $\frac{5}{6} + \frac{1}{6}$

b. $\frac{4}{5} - \frac{3}{5}$

c. $\frac{3}{5} \times \frac{1}{2} \times \frac{3}{4}$

d. $\frac{3}{3} + \frac{3}{3} + \frac{2}{3}$

e. $\frac{4}{7} \times \frac{2}{3}$

f. $\frac{5}{8} - \frac{5}{8}$

g. $14\frac{2}{7}\% + 14\frac{2}{7}\%$

h. $87\frac{1}{2}\% - 12\frac{1}{2}\%$

Write the reciprocal of each number:

i. $\frac{4}{5}$

j. $\frac{8}{7}$

k. 5

Find each missing number:

l. $\frac{5}{8}a = 1$

m. $6m = 1$

n. Gia's ruler is divided into tenths of an inch. What fraction of an inch represents the greatest possible measurement error due to Gia's ruler? Why?

o. How many $\frac{2}{3}$'s are in 1?

p. If $a \div b$ equals 4, what does $b \div a$ equal?

MIXED PRACTICE

Problem set

1. (1) What is the quotient when the sum of 1, 2, and 3 is divided by the product of 1, 2, and 3?

2. (1) The sign shown is incorrect. Show two ways to correct the sign.

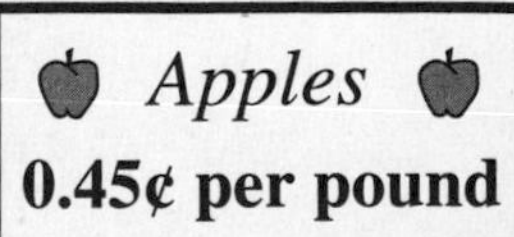

3. (4, 9) Replace each circle with the proper comparison symbol. Then write the comparison as a complete sentence, using words to write the numbers.

(a) $\frac{1}{2} \bigcirc \frac{1}{2} \cdot \frac{1}{2}$ (b) $-2 \bigcirc -4$

4. (5) Write twenty-six thousand in expanded notation.

5. (8) (a) A dime is what fraction of a dollar?

(b) A dime is what percent of a dollar?

6. (8) (a) What fraction of the square is shaded?

(b) What fraction of the square is not shaded?

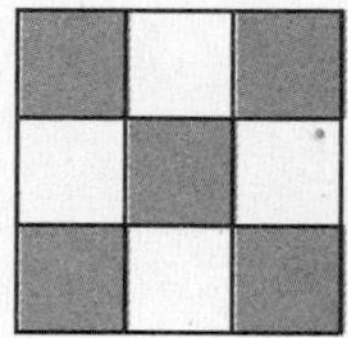

7. (7) Is an imaginary "line" from the Earth to the Moon a line, a ray, or a segment? Why?

8. (8) Use an inch ruler to find *LM, MN,* and *LN* to the nearest sixteenth of an inch.

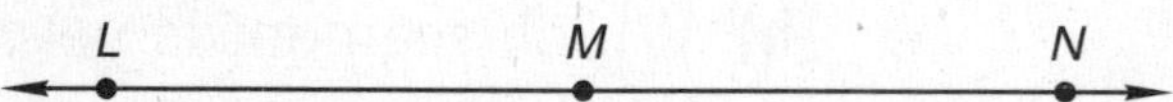

9. (6) (a) List the factors of 18.

(b) List the factors of 24.

(c) Which numbers are factors of both 18 and 24?

(d) Which number is the GCF of 18 and 24?

10. (1, 9) If n is $\frac{2}{5}$, find

(a) $n + n$ (b) $n - n$

Find each missing number:

11. (3) $85{,}000 + b = 200{,}000$

12. (3) $900 \div c = 60$

13. (3) $d + \$5.60 = \$20.00$

14. (3) $e \times 12 = \$30.00$

15. (3) $f - \$98.03 = \$12.47$

16. $5 + 7 + 5 + 7 + 6 + n + 1 + 2 + 3 + 4 = 40$
(3)

Simplify:

17. $3\frac{11}{15} - 1\frac{3}{15}$
(9)

18. $1\frac{3}{8} + 1\frac{4}{8}$
(9)

19. $\frac{3}{4} \times \frac{1}{4}$
(9)

20. $\frac{1802}{17}$
(1)

21. $8.97 + $110 + 53¢
(1)

22. $\begin{array}{r} \$60.00 \\ -\ \$49.49 \\ \hline \end{array}$
(1)

23. $\begin{array}{r} 607 \\ \times\ 78 \\ \hline \end{array}$
(1)

24. $0.09 × 56
(1)

25. $50 \cdot 60 \cdot 70$
(1)

26. $\frac{4}{5} \times \frac{2}{3} \times \frac{1}{3}$
(9)

27. $\frac{1}{9} + \frac{2}{9} + \frac{4}{9}$
(9)

28. Refer to the figure at right to answer (a) and (b).
(7)

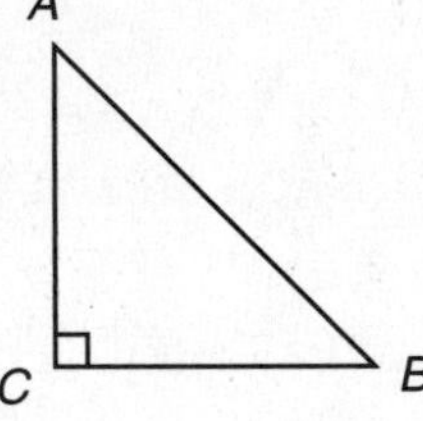

(a) Which angles are acute?

(b) Which segment is perpendicular to $\overline{CB}$?

29. Use words to describe the following sequence. Then find the next number in the sequence.
(2, 8)

$$1, \frac{1}{2}, \frac{1}{4}, \frac{1}{8}, \ldots$$

30. How many $\frac{2}{5}$'s are in 1?
(9)

LESSON

10 Writing Division Answers as Mixed Numbers • Improper Fractions

WARM-UP

Facts Practice: 64 Multiplication Facts (Test A)

Mental Math:

a. $7 - 10$ **b.** $\$1.25 \times 10$ **c.** $\$1.00 - 82¢$

d. $384 + 110$ **e.** $649 - 200$ **f.** $300 \div 30$

g. 3×6, $\div 2$, $\times 5$, $+ 3$, $\div 6$, $- 3$, $\times 4$, $+ 1$, $\div 3$

Problem Solving:

Copy this problem and fill in the missing digits:

```
  _37_
- 2_65
------
  59_7
```

NEW CONCEPTS

Writing division answers as mixed numbers

Alexis cut a 25-inch ribbon into four equal lengths. How long was each piece?

To find the answer to this question, we divide. However, expressing the answer with a remainder does not answer the question.

```
   6 R 1
4)25
  24
  --
   1
```

The answer 6 R 1 means that each of the four pieces of ribbon was 6 inches long and that a piece remained that was 1 inch long. But that would make five pieces of ribbon!

Instead of writing the answer with a remainder, we will write the answer as a mixed number. The remainder becomes the numerator of the fraction, and we use the divisor as the denominator.

$$4\overline{)25} = 6\tfrac{1}{4}$$

$$\begin{array}{r} 6\frac{1}{4} \\ 4\overline{)25} \\ \underline{24} \\ 1 \end{array}$$

This answer means that each piece of ribbon was $6\frac{1}{4}$ inches long, which is the correct answer to the question.

Example 1 What percent of the circle is shaded?

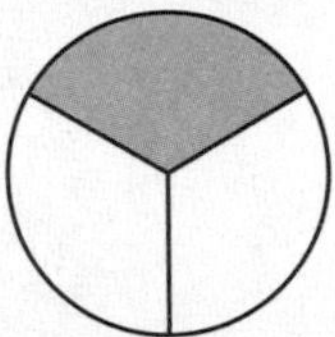

Solution One third of the circle is shaded, so we divide 100% by 3.

$$\begin{array}{r} 33\frac{1}{3}\% \\ 3\overline{)100\%} \\ \underline{9} \\ 10 \\ \underline{9} \\ 1 \end{array}$$

We find that $\mathbf{33\frac{1}{3}\%}$ of the circle is shaded.

Improper fractions A fraction is equal to 1 if the numerator and denominator are equal (and are not zero). Here we show four fractions equal to 1.

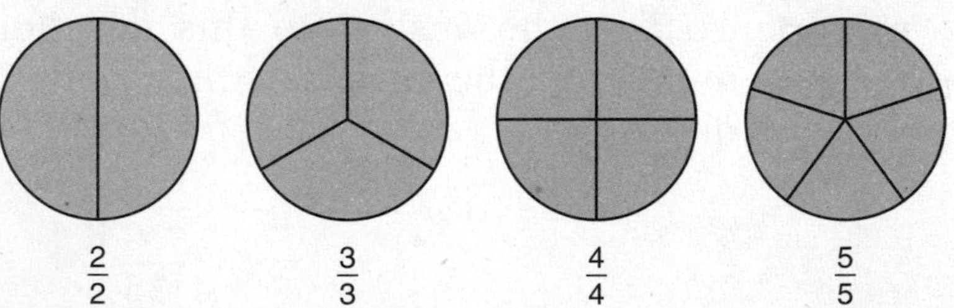

$\frac{2}{2}$ $\frac{3}{3}$ $\frac{4}{4}$ $\frac{5}{5}$

A fraction that is equal to 1 or is greater than 1 is called an **improper fraction.** Improper fractions can be rewritten either as whole numbers or as mixed numbers.

Example 2 Convert each improper fraction to either a whole number or a mixed number:

(a) $\frac{5}{3}$ (b) $\frac{6}{3}$

Solution (a) Since $\frac{3}{3}$ equals 1, the fraction $\frac{5}{3}$ is greater than 1.

$$\frac{5}{3} = \frac{3}{3} + \frac{2}{3}$$
$$= 1 + \frac{2}{3}$$
$$= \mathbf{1\frac{2}{3}}$$

(b) Likewise, $\frac{6}{3}$ is greater than 1.

$$\frac{6}{3} = \frac{3}{3} + \frac{3}{3}$$
$$= 1 + 1$$
$$= \mathbf{2}$$

We can find the whole number within an improper fraction by performing the division indicated by the fraction bar. If there is a remainder, it becomes the numerator of a fraction whose denominator is the same as the denominator in the original improper fraction.

(a) $\frac{5}{3} \longrightarrow 3\overline{)5}$ quotient 1, $\frac{3}{2}$ $\longrightarrow \mathbf{1\frac{2}{3}}$ (b) $\frac{6}{3} \longrightarrow 3\overline{)6}$ quotient $\mathbf{2}$

This picture illustrates that $\frac{5}{3}$ is equivalent to $1\frac{2}{3}$. By shading the remaining section we could illustrate that $\frac{6}{3}$ equals 2.

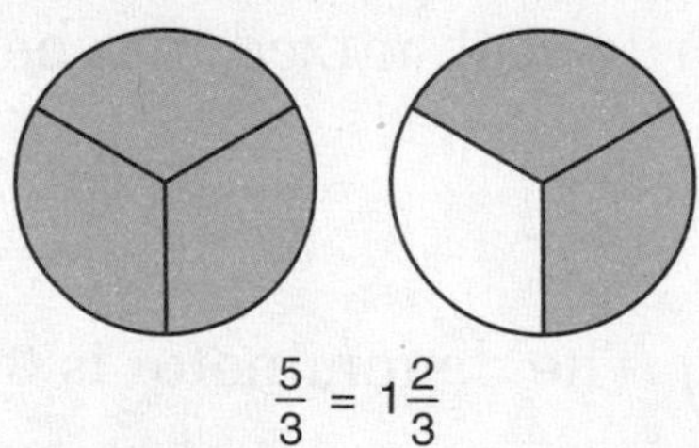

$\frac{5}{3} = 1\frac{2}{3}$

Example 3 Rewrite $3\frac{7}{5}$ with a proper fraction.

Solution The mixed number $3\frac{7}{5}$ means $3 + \frac{7}{5}$. The fraction $\frac{7}{5}$ converts to $1\frac{2}{5}$.

$$\frac{7}{5} = 1\frac{2}{5}$$

Now we combine 3 and $1\frac{2}{5}$.

$$3 + 1\frac{2}{5} = \mathbf{4\frac{2}{5}}$$

When the answer to an arithmetic problem is an improper fraction, we may convert the improper fraction to a mixed number.

Example 4 Simplify:

(a) $\frac{4}{5} + \frac{4}{5}$ (b) $\frac{5}{2} \times \frac{3}{4}$ (c) $1\frac{3}{5} + 1\frac{3}{5}$

Solution (a) $\frac{4}{5} + \frac{4}{5} = \frac{8}{5} = \mathbf{1\frac{3}{5}}$ (b) $\frac{5}{2} \times \frac{3}{4} = \frac{15}{8} = \mathbf{1\frac{7}{8}}$

(c) $1\frac{3}{5} + 1\frac{3}{5} = 2\frac{6}{5} = \mathbf{3\frac{1}{5}}$

Sometimes we need to convert a mixed number to an improper fraction. The illustration below shows $3\frac{1}{4}$ converted to the improper fraction $\frac{13}{4}$.

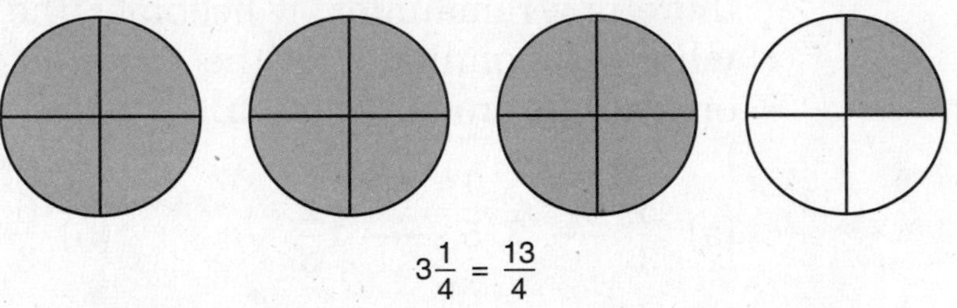

$3\frac{1}{4} = \frac{13}{4}$

We see that every whole circle equals $\frac{4}{4}$. So three whole circles is $\frac{4}{4} + \frac{4}{4} + \frac{4}{4}$, which equals $\frac{12}{4}$. Adding $\frac{1}{4}$ more totals $\frac{13}{4}$.

Example 5 Write each mixed number as an improper fraction:

(a) $3\frac{1}{3}$ (b) $2\frac{3}{4}$ (c) $12\frac{1}{2}$

Solution (a) The denominator is 3, so we use $\frac{3}{3}$ for 1. Thus $3\frac{1}{3}$ is

$$\frac{3}{3} + \frac{3}{3} + \frac{3}{3} + \frac{1}{3} = \mathbf{\frac{10}{3}}$$

(b) The denominator is 4, so we use $\frac{4}{4}$ for 1. Thus $2\frac{3}{4}$ is

$$\frac{4}{4} + \frac{4}{4} + \frac{3}{4} = \mathbf{\frac{11}{4}}$$

(c) The denominator is 2, so we use $\frac{2}{2}$ for 1. If we multiply 12 by $\frac{2}{2}$, we find that 12 equals $\frac{24}{2}$. Thus, $12\frac{1}{2}$ is

$$12\left(\frac{2}{2}\right) + \frac{1}{2} = \frac{24}{2} + \frac{1}{2} = \mathbf{\frac{25}{2}}$$

The solution to example 5(c) suggests a quick way to convert a mixed number to an improper fraction. Multiply the denominator of the fraction by the whole number, add the numerator of the fraction, and put the result over the denominator of the fraction. So, for $12\frac{1}{2}$ we have

$$12\frac{1}{2} = 12\overset{+}{\frown}\underset{\times}{\smile}\frac{1}{2} = \frac{2 \times 12 + 1}{2} = \frac{24 + 1}{2} = \frac{25}{2}$$

LESSON PRACTICE

Practice set

a. Alexis cut a 35-inch ribbon into four equal lengths. How long was each piece?

b. One day is what percent of one week?

Convert each improper fraction to either a whole number or a mixed number:

c. $\frac{12}{5}$ **d.** $\frac{12}{6}$ **e.** $2\frac{12}{7}$

f. Draw and shade circles to illustrate that $2\frac{1}{4} = \frac{9}{4}$.

Simplify:

g. $\frac{2}{3} + \frac{2}{3} + \frac{2}{3}$ **h.** $\frac{7}{3} \times \frac{2}{3}$ **i.** $1\frac{2}{3} + 1\frac{2}{3}$

Convert each mixed number to an improper fraction:

j. $1\frac{2}{3}$ **k.** $3\frac{5}{6}$ **l.** $4\frac{3}{4}$

m. $5\frac{1}{2}$ **n.** $6\frac{3}{4}$ **o.** $10\frac{2}{5}$

MIXED PRACTICE

Problem set

1. (2, 9) Use the fractions $\frac{1}{2}$, $\frac{1}{3}$, and $\frac{1}{6}$ to write an equation that illustrates the associative property of multiplication.

2. (7) Use the words *perpendicular* and *parallel* to complete the following sentence:

In a rectangle, opposite sides are ___(a)___ *and adjacent sides are* ___(b)___.

3. What is the difference when the sum of 2, 3, and 4 is subtracted from the product of 2, 3, and 4?
(1)

4. (a) What percent of the rectangle is shaded?
(8)

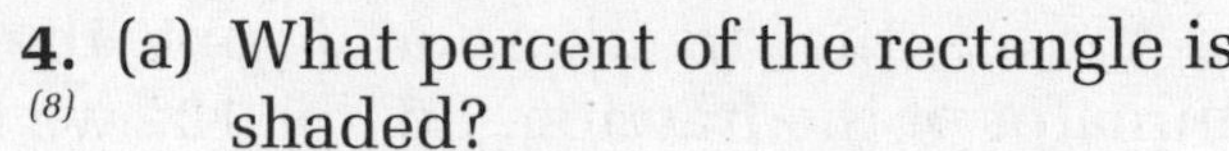

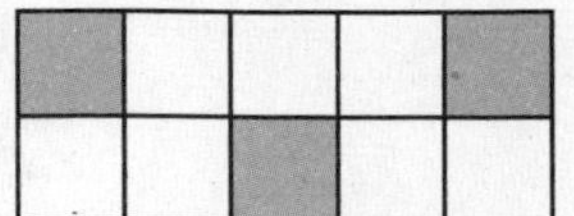

(b) What percent of the rectangle is not shaded?

5. Write $3\frac{2}{3}$ as an improper fraction.
(10)

6. Replace each circle with the proper comparison symbol:
(4, 9)

(a) $2 - 2 \bigcirc 2 \div 2$ (b) $\frac{1}{2} + \frac{1}{2} \bigcirc \frac{1}{2} \times \frac{1}{2}$

7. Point M represents what mixed number on this number line?
(8)

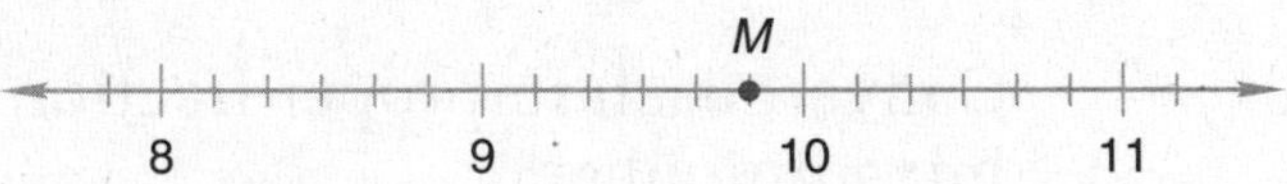

8. Draw and shade circles to show that $1\frac{3}{5} = \frac{8}{5}$.
(10)

9. List the single-digit numbers that are divisors of 420.
(6)

Find each missing number:

10. $12{,}500 + x = 36{,}275$
(3)

11. $18y = 396$
(3)

12. $77{,}000 - z = 39{,}400$
(3)

13. $\frac{a}{8} = \$1.25$
(3)

14. $b - \$16.25 = \$8.75$
(3)

15. $c + \$37.50 = \$75.00$
(3)

16. $8 + 7 + 5 + 6 + 4 + n + 3 + 7 = 50$
(3)

Simplify:

17. $\frac{5}{2} \times \frac{5}{4}$
(10)

18. $\frac{5}{8} - \frac{5}{8}$
(9)

19. $\frac{11}{20} + \frac{18}{20}$
(10)

20. $2000 - (680 - 59)$
(2)

21. $100\% \div 9$
(10)

22. 89¢ + 57¢ + $15.74
(1)

23. 800×300
(1)

24. $2\frac{2}{3} + 2\frac{2}{3}$
(10)

25. $\frac{2}{3} \cdot \frac{2}{3} \cdot \frac{2}{3}$
(9)

26. Describe each figure as a line, ray, or segment. Then use a symbol and letters to name each figure.
(7)

(a) 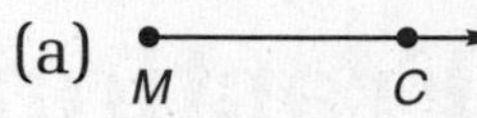

(b) 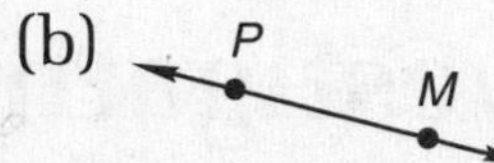

(c) F H

27. How many $\frac{5}{9}$'s are in 1?
(9)

28. What are the next three numbers in this sequence?
(2, 8)

..., 32, 16, 8, 4, 2, ...

29. Which of these numbers is not an integer?
(4)

A. −1 B. 0 C. $\frac{1}{2}$ D. 1

30. (a) If $a - b = 5$, what does $b - a$ equal?
(4, 9)

(b) If $\frac{w}{x} = 3$, what does $\frac{x}{w}$ equal?

INVESTIGATION 1

Focus on

Investigating Fractions and Percents with Manipulatives

In this investigation you will make a set of fraction manipulatives to use in solving problems with fractions.

Materials needed:

- Activity Sheets 1–6 (available in *Saxon Math 8/7—Homeschool Tests and Worksheets*)
- Scissors
- Envelope or zip-top plastic bag
- Colored pencils or markers (optional)

Preparation:

If desired, color each sheet's fraction circles a different color; that is, one color for halves, one color for thirds, and so on. (Color-coding the fraction circles will make sorting the manipulatives easier.) Separate the manipulatives by cutting out the circles and cutting apart the fraction slices along the lines.

Activity: *Using Fraction Manipulatives*

Use your fraction manipulatives to help you with the following exercises:

1. What fraction is half of $\frac{1}{2}$?

2. What fraction is half of $\frac{1}{4}$?

3. What fraction is half of $\frac{1}{3}$?

4. What fraction is half of $\frac{1}{6}$?

5. What fraction is $\frac{1}{3}$ of $\frac{1}{2}$?

6. What fraction is $\frac{1}{3}$ of $\frac{1}{4}$?

7. How many twelfths equal $\frac{1}{2}$?

8. Find a single fraction piece that equals $\frac{3}{12}$.

9. Find a single fraction piece that equals $\frac{4}{8}$.

10. Find a single fraction piece that equals $\frac{4}{12}$.

11. How many sixths equal $\frac{2}{3}$?

12. How many twelfths equal $\frac{3}{4}$?

13. Assemble five $\frac{1}{3}$ pieces to illustrate a mixed number. Draw a picture of your work. Then write an equation that relates the improper fraction to the mixed number.

14. Find a single fraction piece that equals $\frac{3}{6}$.

15. Assemble nine $\frac{1}{6}$ pieces to form $\frac{5}{6}$ of a circle and $\frac{4}{6}$ of a circle. Then demonstrate the addition of $\frac{5}{6}$ and $\frac{4}{6}$ by recombining the pieces to make $1\frac{1}{2}$ circles. Draw a picture to illustrate your work.

16. Two $\frac{1}{4}$ pieces form half of a circle. Which two different manipulative pieces also form half of a circle?

Find a fraction to complete each equation:

17. $\frac{1}{2} + \frac{1}{3} + a = 1$

18. $\frac{1}{6} + b = \frac{1}{4}$

19. $\frac{1}{2} + c = \frac{3}{4}$

20. $\frac{1}{4} + d = \frac{1}{3}$

Find each percent:

21. What percent of a circle is $\frac{2}{3}$ of a circle?

22. What percent of a circle is $\frac{3}{12}$ of a circle?

23. What percent of a circle is $\frac{3}{8}$ of a circle?

24. What percent of a circle is $\frac{3}{6}$ of a circle?

25. What percent of a circle is $\frac{1}{4} + \frac{1}{12}$?

26. Use four $\frac{1}{4}$'s to demonstrate the subtraction $1 - \frac{1}{4}$, and write the answer.

27. What fraction piece, when used twice, will cover $\frac{4}{6}$ of a circle?

28. What fraction piece, when used three times, will cover $\frac{6}{8}$ of a circle?

29. If you subtract $\frac{1}{12}$ of a circle from $\frac{1}{3}$ of a circle, what fraction of the circle is left?

30. Find as many ways as you can to make half of a circle using two or more of the fraction manipulative pieces. Write an equation for each way you find. For example, $\frac{2}{4} = \frac{1}{2}$ and $\frac{1}{3} + \frac{1}{6} = \frac{1}{2}$.

LESSON

11 Problems About Combining • Problems About Separating

WARM-UP

Facts Practice: 30 Improper Fractions and Mixed Numbers (Test C)

Mental Math:

a. \$7.50 + 75¢ **b.** \$40.00 ÷ 10 **c.** \$10.00 − \$5.50
d. (3 × 20) + (3 × 5) **e.** 250 − 1000 **f.** $\frac{1}{2}$ of 28
g. Start with the number of hours in a day, ÷ 2, × 3, ÷ 4, × 5, + 4, ÷ 7.

Problem Solving:

Letha has 7 coins in her hands totaling 50¢. What are the coins?

NEW CONCEPTS

In this lesson we will begin solving one-step story problems by writing and solving appropriate equations for the problems. To write an equation, it is helpful to understand the **plot** of the story. All stories with the same plot can be modeled with the same equation, which is why we say they follow a **pattern.** There are only a small number of story problem plots. In this lesson we deal with two of them.

Problems about combining

One common idea in story problems is that of **combining.** Here is an example of a complete story about combining:

Albert has \$12. Betty has \$15. Together they have \$27.

Stories like this have an **addition pattern.**

some + some more = total

$$S + M = T$$

There are three numbers in this pattern. In a story problem one of the numbers is missing. To write an equation, we use a letter to stand for the missing number. If the total is missing, we add to find the missing number. If an addend is missing, we subtract the known addend from the sum to find the missing addend. **Although we sometimes use subtraction to solve the problem, it is important to recognize that story problems about combining have addition thought patterns.**

We follow four steps when solving story problems:

Step 1: Read the problem and identify its pattern.

Step 2: Write an equation for the given information.

Step 3: Solve and check the equation.

Step 4: Review the question and write the answer.

We will follow these steps as we consider some examples.

Example 1 At the end of the first day of camp, Marissa counted 47 mosquito bites. The next morning she counted 114 mosquito bites. How many new bites did she get during the night?

Solution **Step 1:** We recognize that this problem has an **addition pattern.** Marissa had some mosquito bites, and then she got some more.

Step 2: We write an equation for the given information. Marissa had 47 bites. She got some more bites. Then she had a total of 114 bites.

$$47 + M = 114$$

Step 3: We find *M,* the missing number in the equation. To find a missing addend, we subtract. Then we check our work by substituting the answer into the original equation.

$$\begin{array}{r} 114 \\ -\ \ 47 \\ \hline 67 \end{array} \longrightarrow \begin{array}{r} 47 \text{ bites} \\ +\ 67 \text{ bites} \\ \hline 114 \text{ bites} \end{array} \text{ check}$$

Step 4: Now we review the question and write the answer that completes the story. During the night Marissa got **67 new bites.**

Example 2 The first scout troop encamped in the ravine. A second troop of 137 scouts joined them, making a total of 312 scouts. How many scouts were in the first troop?

Solution **Step 1:** We recognize that this problem is about **combining.** It has an **addition pattern.** There were some scouts. Then some more scouts came.

Step 2: We write an equation using S to stand for the number of scouts in the first troop.

$$S + 137 = 312$$

Step 3: We find S, the missing number in the equation. To find a missing addend, we subtract. Then we check our work by substituting into the original equation.

$$\begin{array}{r} 312 \\ -\ 137 \\ \hline 175 \end{array} \qquad \longrightarrow \qquad \begin{array}{r} 175 \text{ scouts} \\ +\ 137 \text{ scouts} \\ \hline 312 \text{ scouts} \end{array} \quad \text{check}$$

Step 4: Now we review the question and write the answer. There were **175 scouts** in the first troop.

Problems about separating Another common idea in story problems is **separating** an amount into two parts. Often problems about separating involve something "going away." Here is an example:

> *Ted wrote a check to Ned for $37.50. If $824.00 was available in Ted's account before he wrote the check, how much was available after he gave the check to Ned?*

Story problems like this have a **subtraction pattern.**

beginning amount − some went away = what remains

$$B - A = R$$

There are three numbers in this pattern. In a story problem one of the three numbers is missing. To write an equation, we use a letter to represent the missing number. Then we find the missing number and answer the question in the problem. **We follow the same four steps we followed in solving problems with addition patterns.**

Example 3 Tim baked 4 dozen cookies. While they were cooling, he went to answer the phone. When he came back, only 32 cookies remained. His dog was nearby, licking her chops. How many cookies did the dog eat while Tim was answering the phone?

Solution **Step 1:** We recognize that this problem has a **subtraction pattern.** Tim had some cookies. Then some went away.

Step 2: We write an equation using 48 for 4 dozen cookies and A for the number of cookies that went away.

$$48 - A = 32$$

Step 3: We find the missing number. To find the subtrahend in a subtraction pattern, we subtract. Then we check our work by substituting the answer into the original equation.

$$\begin{array}{r} 48 \\ -\ 32 \\ \hline 16 \end{array} \longrightarrow \begin{array}{r} 48 \text{ cookies} \\ -\ 16 \text{ cookies} \\ \hline 32 \text{ cookies} \end{array} \quad \text{check}$$

Step 4: Now we review the question and write the answer. While Tim was answering the phone, his dog ate **16 cookies.**

Example 4 The room was full of boxes when Sharon began. Then she shipped out 56 boxes. Only 88 boxes were left. How many boxes were in the room when Sharon began?

Solution **Step 1:** We recognize that this problem is about **separating.** It has a **subtraction pattern.** There were boxes in a room. Then Sharon shipped some away.

Step 2: We write an equation using B to stand for the number of boxes in the room when Sharon began.

$$B - 56 = 88$$

Step 3: We find the missing number. To find the minuend in a subtraction pattern, we add the subtrahend and the difference. Then we check our work by substituting into the original equation.

$$\begin{array}{r} 88 \\ +\ 56 \\ \hline 144 \end{array} \longrightarrow \begin{array}{r} 144 \text{ boxes} \\ -\ 56 \text{ boxes} \\ \hline 88 \text{ boxes} \end{array} \quad \text{check}$$

Step 4: Now we review the question and write the answer. There were **144 boxes** in the room when Sharon began.

LESSON PRACTICE

Practice set Follow the four-step method shown in this lesson for each problem. Along with each answer, include the equation you used to solve the problem.

a. Billy stood on the scales. Billy weighed 118 pounds. Then both Lola and Billy stood on the scales. Together they weighed 230 pounds. How much did Lola weigh?

b. Lamar cranked for a number of turns. Then Lurdes gave the crank 216 turns. If the total number of turns was 400, how many turns did Lamar give the crank?

c. At dawn 254 horses were in the corral. Later that morning Tex found the gate open and saw that only 126 horses remained. How many horses got away?

d. Cynthia had a lot of paper. After using 36 sheets for a report, only 164 sheets remained. How many sheets of paper did she have at first?

e. Write a story problem about combining that fits this equation:

$$\$15.00 + T = \$16.13$$

f. Write a story problem about separating that fits this equation:

$$32 - S = 25$$

MIXED PRACTICE

Problem set

1. (11) As the day of the festival drew near, there were 200,000 people in the city. If the usual population of the city was 85,000, how many visitors had come to the city?

2. (11) Syd returned from the store with $12.47. He had spent $98.03 on groceries. How much money did he have when he went to the store?

3. (11) Exactly 10,000 runners began the marathon. If only 5420 runners finished the marathon, how many dropped out along the way?

4. (a) What fraction of the group is shaded?
(8, 10)

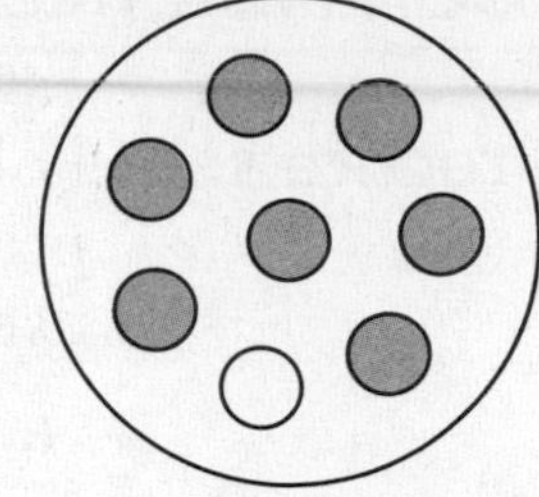

(b) What fraction of the group is not shaded?

(c) What percent of the group is not shaded?

5. (a) Arrange these numbers in order from least to greatest:
(4, 8)

$\frac{1}{2}$, 0, −2, 1

(b) Which of these numbers is not an integer?

6. A 35-inch ribbon was cut into 8 equal lengths. How long was each piece?
(10)

7. Use digits and symbols to write "The product of one and two is less than the sum of one and two."
(4)

8. Subtract 89 million from 100 million. Use words to write the difference.
(5)

9. (a) List the factors of 16.
(6)

(b) List the factors of 24.

(c) Which numbers are factors of both 16 and 24?

(d) What is the GCF of 16 and 24?

Find each missing number:

10. $8000 - k = 5340$
(3)

11. $1320 + m = 1760$
(3)

12. $4 \cdot 9 \cdot n = 720$
(3)

13. $\$126 + r = \$375$
(3)

14. $\frac{169}{s} = 13$
(3)

15. $\frac{t}{40} = \$25.00$
(3)

16. Compare: $100 - (5 \times 20)$ ◯ $(100 - 5) \times 20$
(2, 4)

Simplify:

17. (10) $1\frac{5}{9} + 1\frac{5}{9}$

18. (10) $\frac{5}{3} \times \frac{2}{3}$

19. (1) $\begin{array}{r} 135 \\ \times \ 72 \\ \hline \end{array}$

20. (1) $\frac{1000}{40}$

21. (1) 30($1.49)

22. (1) $140.70 ÷ 35

23. (9) $\frac{5}{9} \cdot \frac{1}{3} \cdot \frac{1}{2}$

24. (9) $\frac{5}{8} + \left(\frac{3}{8} - \frac{1}{8}\right)$

25. (10) Write $3\frac{3}{4}$ as an improper fraction.

26. (8) Which choice below is the best estimate of the portion of the rectangle that is shaded?

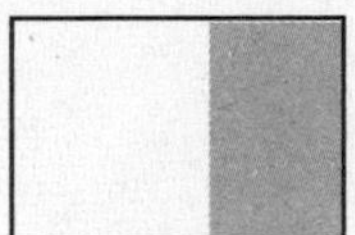

A. $\frac{1}{4}$ B. $\frac{1}{2}$ C. 40% D. 60%

27. (2, 8) What are the next four numbers in this sequence?

$\frac{1}{8}, \frac{1}{4}, \frac{3}{8}, \frac{1}{2}, \ldots$

28. (7) Refer to the figure at right to answer (a) and (b).

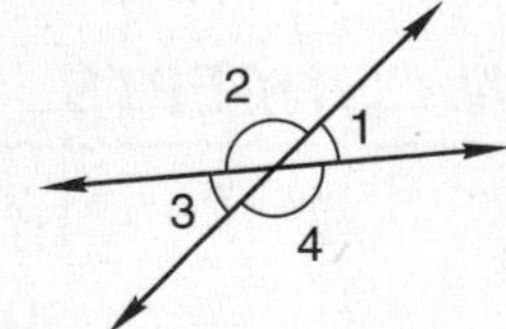

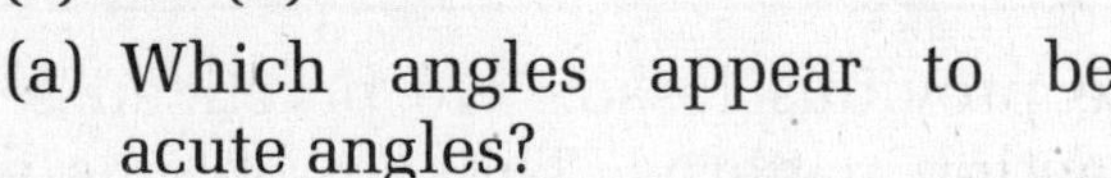

(a) Which angles appear to be acute angles?

(b) Which angles appear to be obtuse angles?

29. (8) Use an inch ruler to draw $\overline{AC}$ $3\frac{1}{2}$ inches long. On $\overline{AC}$ mark point B so that AB is $1\frac{7}{8}$ inches. Now find BC.

30. (9) If $n \div m$ equals $\frac{7}{8}$, what does $m \div n$ equal?

LESSON

12 Problems About Comparing • Elapsed-Time Problems

WARM-UP

Facts Practice: 30 Improper Fractions and Mixed Numbers (Test C)

Mental Math:

a. \$6.50 + 60¢ **b.** \$1.29 × 10

c. \$10.00 − \$2.50 **d.** (4 × 20) + (4 × 3)

e. 500 − 2000 **f.** $\frac{1}{2}$ of 64

g. Start with three score, ÷ 2, + 2, ÷ 2, + 2, ÷ 2, + 2, × 2.

Problem Solving:

The diameter of a circle or a circular object is the distance across the circle through its center. Find the approximate diameter of the penny shown at right.

NEW CONCEPTS

In the previous lesson we practiced solving problems with subtraction patterns. Those problems were about separating an amount into two parts. In this lesson we consider two other types of problems with subtraction patterns.

Problems about comparing Another type of problem that has a subtraction pattern is about **comparing.** In these problems, one amount is larger and one amount is smaller. We not only have to decide which number is greater and which number is less, but also **how much greater** or **how much less.** The number that describes how much greater or how much less is called the *difference.* We write the numbers in the equation in this order:

larger − smaller = difference

$$L - S = D$$

Example 1 During the day 1320 employees work at the toy factory. At night 897 employees work there. How many more employees work at the factory during the day than at night?

Solution **Step 1:** Questions such as "How many more?" or "How many fewer?" indicate that a problem is about **comparison.** Therefore, this problem has a **subtraction pattern.**

Step 2: We write an equation for the given information. We use the letter D in the equation to stand for the difference.

$$1320 - 897 = D$$

Step 3: We find the missing number in the pattern by subtracting.

$$\begin{array}{r} 1320 \text{ employees} \\ -\ 897 \text{ employees} \\ \hline 423 \text{ employees} \end{array}$$

As expected, the difference is less than the larger of the two given numbers.

Step 4: We review the question and write the answer. There are **423 more employees** who work at the factory during the day than who work there at night.

Example 2 The number 620,000 is how much less than 1,000,000?

Solution **Step 1:** The words *how much less* indicate that this is a **comparison** problem. Therefore, it has a **subtraction pattern.**

Step 2: We write an equation using D to stand for the difference between the two numbers.

$$1{,}000{,}000 - 620{,}000 = D$$

Step 3: We subtract to find the missing number.

$$\begin{array}{r} 1{,}000{,}000 \\ -\ 620{,}000 \\ \hline 380{,}000 \end{array}$$

Step 4: We review the question and write the answer. Six hundred twenty thousand is **380,000** less than 1,000,000.

Elapsed-time problems

Elapsed time is the length of time between two points in time. Here we use points on a ray to illustrate elapsed time.

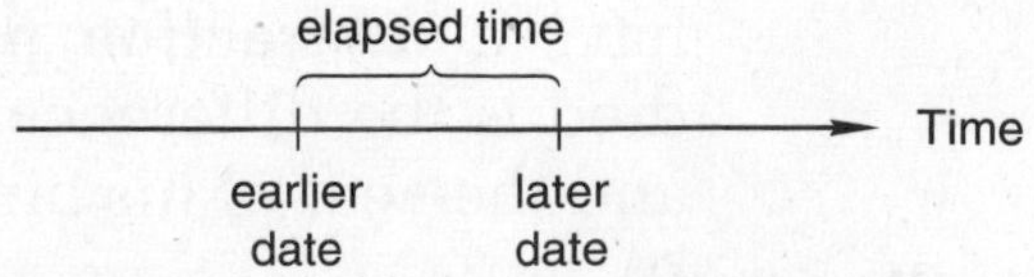

A person's age is an example of elapsed time. Your age is the time that has elapsed since you were born until this present moment. By subtracting the date you were born from today's date you can find your age.

$$\begin{array}{rl} \text{Today's date} & \text{(later)} \\ -\ \underline{\text{Your birth date}} & \text{(earlier)} \\ \text{Your age} & \text{(difference)} \end{array}$$

Elapsed-time problems are like comparison problems. They have a **subtraction pattern.**

$$\text{later} - \text{earlier} = \text{difference}$$

$$L - E = D$$

Notice how similar this is to the larger-smaller-difference equation.

Now it is time to solve some problems.

Example 3 How many years were there from 1492 to 1776? (Unless otherwise specified, years are A.D.)

Solution **Step 1:** We recognize that this is an **elapsed-time problem.** Therefore, it has a **subtraction pattern.**

Step 2: We write an equation for the pattern. The year 1776 is later than 1492. We use Y to represent the number of years between 1492 and 1776. (Y is the difference of these two numbers.)

$$1776 - 1492 = Y$$

Step 3: We subtract to find the difference.

$$\begin{array}{r} 1776 \\ -\ 1492 \\ \hline 284 \end{array}$$

Step 4: Now we review the question and write the answer. There were **284 years** from 1492 to 1776.

Example 4 Martin Luther King Jr. died in 1968 at the age of 39. In what year was he born?

Solution **Step 1:** This is an **elapsed-time problem.** Time problems have a **subtraction pattern.** The age at which King died is the difference of the year of his assassination and the year of his birth.

Step 2: We write an equation using B to stand for the year of King's birth.

$$1968 - B = 39$$

Step 3: To find the subtrahend in a subtraction pattern, we subtract the difference from the minuend. We may check our work by substituting into the original equation.

$$\begin{array}{r} 1968 \\ -\quad 39 \\ \hline 1929 \end{array} \longrightarrow \begin{array}{r} 1968 \\ -\ 1929 \\ \hline 39 \end{array} \text{ check}$$

Step 4: Now we review the question and write the answer. Martin Luther King Jr. was born in **1929.**

LESSON PRACTICE

Practice set Follow the four-step method to solve each problem. Along with each answer, include the equation you used to solve the problem.

a. The number 1,000,000,000 is how much greater than 25,000,000?

b. How many years were there from 1215 to 1791?

c. John F. Kennedy died in 1963 at the age of 46. In what year was he born?

d. Write a story problem about comparing that fits this equation:

$$58 \text{ in.} - 55 \text{ in.} = D$$

e. Write a story problem about elapsed time that fits this equation:

$$2003 - B = 14$$

MIXED PRACTICE

Problem set

1. (11) Seventy-seven thousand fans filled the stadium. As the fourth quarter began, only thirty-nine thousand, four hundred remained. How many fans left before the fourth quarter began?

2. (11) Mary purchased 18 bananas at the store. When she got home, she discovered that she already had some bananas. If she now has 31 bananas, how many did she have before she went to the store?

3. (12) How many years were there from 1066 to 1215?

4. (12) The first week 77,000 fans came to the stadium. Only 49,600 came the second week. How many fewer fans came to the stadium the second week?

5. (11) Write a story problem about separating that fits this equation:

$$\$20.00 - C = \$7.13$$

6. (2) What property is illustrated by this equation?

$$\frac{1}{2} \times 1 = \frac{1}{2}$$

7. (5, 12) Twenty-three thousand is how much less than one million? Use words to write the answer.

8. (4, 8) Replace each circle with the proper comparison symbol:

(a) $2 - 3 \bigcirc -1$ (b) $\frac{1}{2} \bigcirc \frac{1}{3}$

9. (7) Name three segments in the figure below in order of length from shortest to longest.

10. (10) Draw and shade circles to show that $2\frac{1}{4}$ equals $\frac{9}{4}$.

11. (8) (a) What fraction of the triangle is shaded?

(b) What percent of the triangle is not shaded?

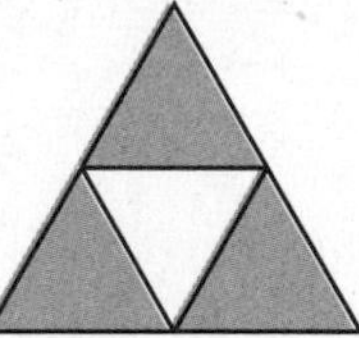

12. (6) The number 100 is divisible by which whole numbers?

Find each missing number:

13. (3) $15x = 630$

14. (3) $y - 2714 = 3601$

15. (3) $2900 - p = 64$

16. (3) $\$1.53 + q = \$5.00$

17. (3) $20r = 1200$

18. (3) $\frac{m}{14} = 16$

Simplify:

19. (1)
$$\begin{array}{r} 72{,}112 \\ -\ 64{,}309 \\ \hline \end{array}$$

20. (1)
$$\begin{array}{r} 453{,}978 \\ +\ 386{,}864 \\ \hline \end{array}$$

21. (9) $\frac{8}{9} - \left(\frac{3}{9} + \frac{5}{9}\right)$

22. (10) $\left(\frac{8}{9} - \frac{3}{9}\right) + \frac{5}{9}$

23. (10) $\frac{9}{2} \times \frac{3}{5}$

24. (1) $37.20 ÷ 15

25. (10) Divide 42,847 by 9 and express the quotient as a mixed number.

26. (1) $4.36 + $15.96 + 76¢ + $35

27. (2, 8) Find the next three numbers in this sequence:

$$\frac{1}{4}, \frac{1}{2}, \frac{3}{4}, \ldots$$

28. (9) How many $\frac{2}{3}$'s are in 1?

29. (10) Write $1\frac{2}{3}$ as an improper fraction, and multiply the improper fraction by $\frac{1}{2}$. What is the product?

30. (7, 8) Using a ruler, draw a triangle that has two perpendicular sides, one that is $\frac{3}{4}$ in. long and one that is 1 in. long. What is the measure of the third side?

LESSON

13 Problems About Equal Groups

WARM-UP

Facts Practice: 30 Improper Fractions and Mixed Numbers (Test C)

Mental Math:

a. \$8.00 − \$0.80 **b.** \$25.00 ÷ 10 **c.** \$10.00 − \$6.75

d. (5 × 30) + (5 × 3) **e.** 250 − 500 **f.** $\frac{1}{2}$ of 86

g. 7 × 8, + 4, ÷ 3, + 1, ÷ 3, + 8, × 2, − 3, ÷ 3

Problem Solving:

Joe, Moe, and Larry stood side by side for a picture. Then they rearranged themselves in the order Moe, Joe, and Larry and took another picture. These arrangements are two of the possible permutations of the three people Joe, Moe, and Larry. Altogether, how many permutations (arrangements) are possible?

NEW CONCEPT

We have used both the addition pattern and the subtraction pattern to solve word problems. In this lesson we will use a **multiplication pattern** to solve word problems. Consider this problem:

Juanita packed 25 marbles in each box. If she filled 32 boxes, how many marbles did she pack in all?

This problem has a pattern that is different from the addition pattern or subtraction pattern. This is a problem about **equal groups,** and it has a **multiplication pattern.**

number of groups × number in each group = total

$$N \times G = T$$

T stands for **total.** To find the total, we multiply. To find an unknown factor, we divide. We will consider three examples.

Example 1 Juanita packed 25 marbles in each box. If she filled 32 boxes, how many marbles did she pack in all?

Solution We use the four-step procedure to solve story problems.

Step 1: Since each box contains the same number of marbles, this problem is about **equal groups.** It has a **multiplication pattern.**

Step 2: We write an equation using T for the total number of marbles. There were 32 groups with 25 marbles in each group.

$$32 \times 25 = T$$

Step 3: To find the missing product, we multiply the factors. The product should be greater than each factor.

```
   32
 × 25
  160
  64
  800
```

Step 4: We review the question and write the answer. Juanita packed **800 marbles** in all.

Example 2 Movie tickets sold for $5 each. The total ticket sales were $820. How many tickets were sold?

Solution **Step 1:** Each ticket sold for the same price. This problem is about **equal groups of money.** Therefore, it has a **multiplication pattern.**

Step 2: We write an equation. In the equation we use N for the number of tickets. Each ticket cost $5 and the total was $820.

$$N \times \$5 = \$820$$

Step 3: To find a missing factor, we divide the product by the known factor. We can check our work by substituting the answer into the original equation.

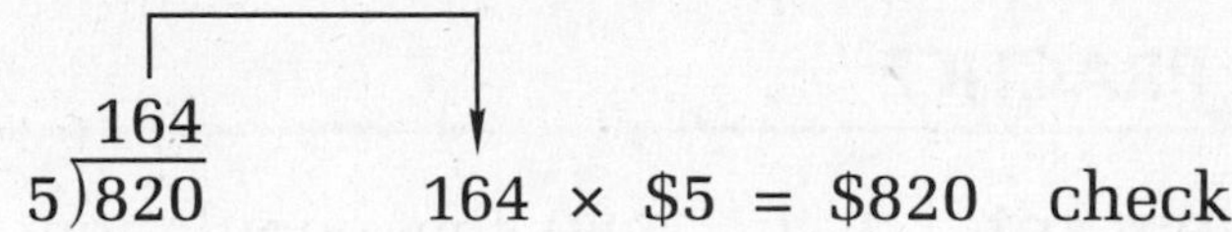

Step 4: We review the question and write the answer: **164 tickets** were sold.

Example 3 Six hundred new cars were delivered to the dealer by 40 trucks. Each truck carried the same number of cars. How many cars were delivered by each truck?

Solution **Step 1:** An equal number of cars were grouped on each truck. The word *each* is a clue that this problem is about **equal groups.** This problem has a **multiplication pattern.**

Step 2: We write an equation using C to stand for the number of cars on each truck.

$$40 \times C = 600$$

Step 3: To find a missing factor, we divide. We check our work by substituting into the original equation.

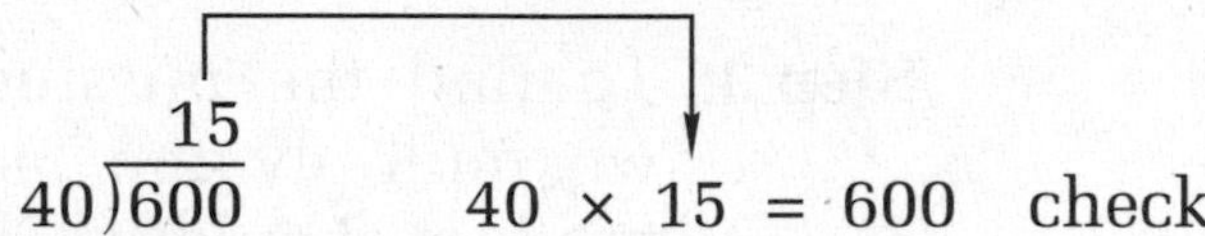

Step 4: We review the question and write the answer: **15 cars** were delivered by each truck.

LESSON PRACTICE

Practice set Follow the four-step method to solve each problem. Along with each answer, include the equation you use to solve the problem.

a. Beverly bought two dozen juice bars for 18¢ each. How much did she pay for all the juice bars?

b. Johnny planted a total of 375 trees. There were 25 trees in each row he planted. How many rows of trees did he plant?

c. Every day Arnold did the same number of push-ups. If he did 1225 push-ups in one week, then how many push-ups did he do each day?

d. Write a story problem about equal groups that fits this equation:

$$12x = \$3.00$$

MIXED PRACTICE

Problem set

1. (12) In 1980 the population of Ashton was 64,309. By the 1990 census, the population had increased to 72,112. The population of Ashton in 1990 was how much greater than in 1980?

2. (11) Huck had five dozen night crawlers in his pockets. He was unhappy when all but 17 escaped through holes in his pockets. How many night crawlers escaped?

3. President Franklin D. Roosevelt died in office in 1945 at the age of 63. In what year was he born?
(12)

4. The beach balls were packed 12 in each case. If 75 cases were delivered, how many beach balls were there in all?
(13)

5. One hundred twenty poles were needed to construct the new pier. If each truckload contained eight poles, how many truckloads were needed?
(13)

6. Write a story problem about equal groups that fits this equation. Then answer the problem.
(13)

$$5t = \$63.75$$

7. The product of 5 and 8 is how much greater than the sum of 5 and 8?
(1, 12)

8. (a) Three quarters make up what fraction of a dollar?
(8)

(b) Three quarters make up what percent of a dollar?

9. How many units is it from –5 to +5 on the number line?
(4)

10. Describe each figure as a line, ray, or segment. Then use a symbol and letters to name each figure.
(7)

(a) (b) (c)

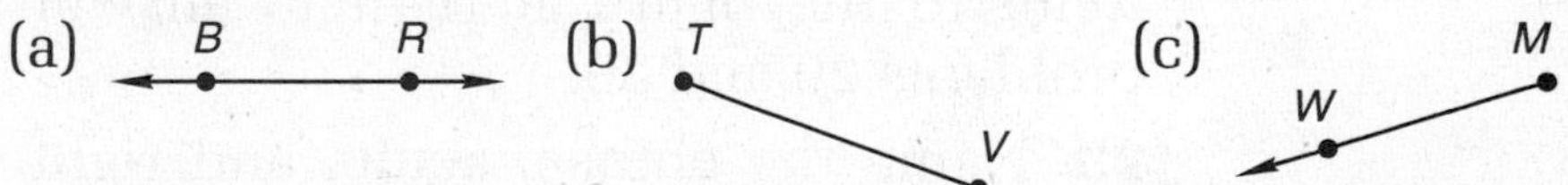

11. (a) What whole numbers are factors of both 24 and 36?
(6)

(b) What is the GCF of 24 and 36?

12. (a) What fractions or mixed numbers are represented by points A and B on this number line?
(8)

(b) Find AB.

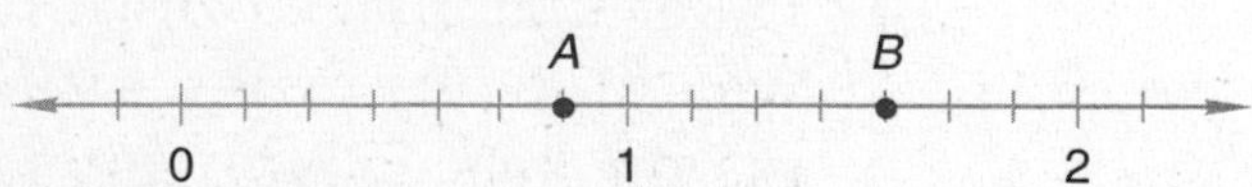

Find each missing number:

13. (3) $36c = 1800$

14. (3) $f - \$1.64 = \$3.77$

15. (3) $\dfrac{d}{7} = 28$

16. (3) $\dfrac{4500}{e} = 30$

17. (3)
$$\begin{array}{r} 4 \\ 7 \\ 6 \\ 8 \\ 4 \\ 5 \\ 5 \\ 7 \\ 9 \\ 6 \\ n \\ +\ 8 \\ \hline 75 \end{array}$$

18. (3) $3674 - a = 2159$

19. (3) $4610 + b = 5179$

Simplify:

20. (1) $363 + 4579 + 86 + 7$

21. (2) $(5 \cdot 4) \div (3 + 2)$

22. (10) $\dfrac{5}{3} \cdot \dfrac{5}{2}$

23. (9) $3\frac{4}{5} - \left(\frac{2}{5} + 1\frac{1}{5}\right)$

24. (1) $\dfrac{600}{25}$

25. (1)
$$\begin{array}{r} 600 \\ \times\ \ 25 \\ \hline \end{array}$$

26. (2, 4) Compare: $1000 \div (100 \div 10) \bigcirc (1000 \div 100) \div 10$

27. (10) Write $2\frac{1}{2}$ as an improper fraction. Then multiply the improper fraction by $\frac{1}{3}$. What is the product?

28. (9) What is the product of $\frac{11}{12}$ and its reciprocal?

Refer to the figure at right to answer problems 29 and 30.

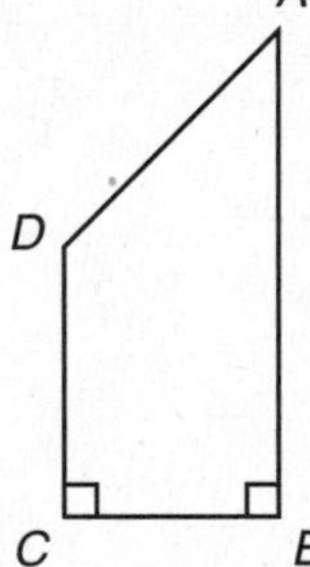

29. (7) Name the obtuse, acute, and right angles.

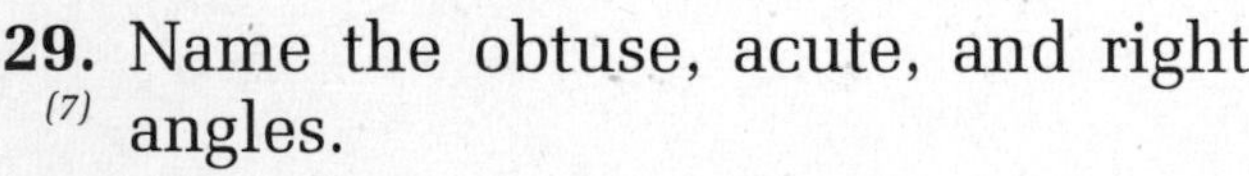

30. (7) (a) $\overline{AB} \parallel$ ___?___

(b) $\overline{AB} \perp$ ___?___

LESSON 14 Problems About Parts of a Whole

WARM-UP

Facts Practice: 30 Equations (Test B)

Mental Math:

a. \$7.50 − 75¢
b. \$0.63 × 10
c. \$10.00 − \$8.25
d. (6 × 20) + (6 × 4)
e. 625 − 500
f. $\frac{1}{2}$ of 36
g. Start with three dozen, ÷ 2, + 2, ÷ 2, + 2, ÷ 2, + 2, ÷ 2, + 2, ÷ 2.

Problem Solving:

Terry folded a square piece of paper in half diagonally to form a triangle. Then he folded the triangle in half as shown, making a smaller triangle. With scissors Terry cut off the upper corner of the triangle. What will the paper look like when it is unfolded?

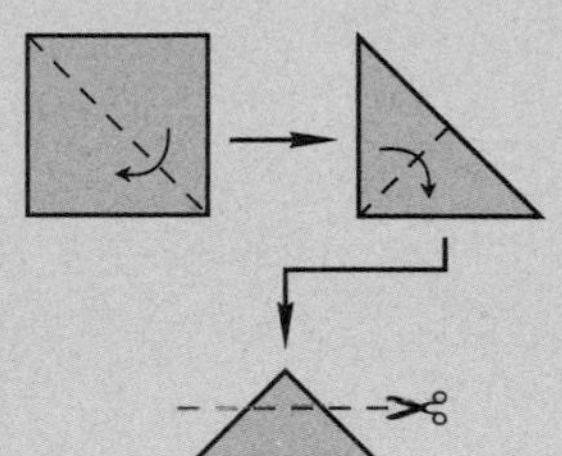

NEW CONCEPT

We remember that problems about combining have an addition pattern. Problems about **parts of a whole** also have an addition pattern.

$$\text{part} + \text{part} = \text{whole}$$

$$P_1 + P_2 = W^\dagger$$

Sometimes the parts are expressed as fractions or percents.

Example 1 One third of the bicycles were on sale. What fraction of the bicycles were not on sale?

Solution We are not given the number of bicycles. We are given only the fraction of bicycles in the store that are on sale. Pictures

†The notations P_1 and P_2 mean "part one" and "part two." Variables with small letters or numbers to the lower right are called *subscripted variables.* A subscripted variable is treated as though it were a single letter.

often help us understand problems about fractions. Here is a picture to help us visualize the problem:

Step 1: We recognize this problem is about **part of a whole.** It has an **addition pattern.**

Step 2: We write an equation for the given information. It may seem as though we are given only one number, $\frac{1}{3}$, but the drawing reminds us that the total group of bicycles in the store is $\frac{3}{3}$. We will use the subscripted variable N_S to stand for the fraction of bicycles not on sale.

$$\frac{1}{3} + N_S = \frac{3}{3}$$

Step 3: We find the missing fraction, N_S, by subtracting. We can check our work by substituting into the original equation.

$$\begin{array}{r} \frac{3}{3} \\ -\ \frac{1}{3} \\ \hline \frac{2}{3} \end{array} \longrightarrow \begin{array}{rl} \frac{1}{3} & \text{bicycles on sale} \\ +\ \frac{2}{3} & \text{bicycles not on sale} \\ \hline \frac{3}{3} & \text{total bicycles} \end{array} \quad \text{check}$$

Step 4: We review the question and write the answer. Of the bicycles in the store, $\frac{2}{3}$ **are not on sale.**

Example 2 Shemp was excited that 61% of his answers were correct. What percent of Shemp's answers were not correct?

Solution **Step 1:** Part of Shemp's answers were correct, and part were not correct. This problem is about **part of a whole.** It has an **addition pattern.** Here we show a picture:

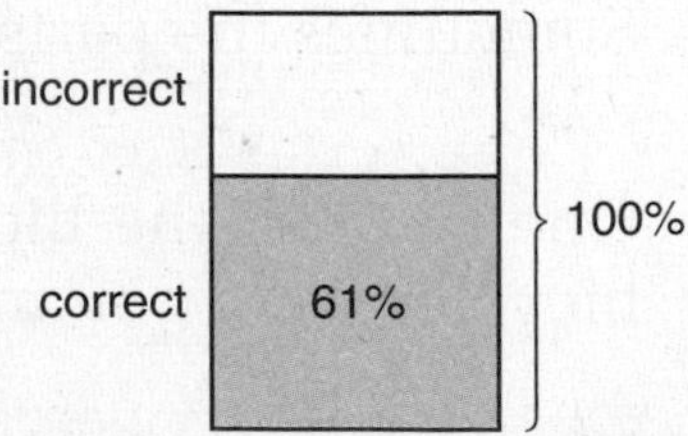

Step 2: We write an equation. The whole is represented by 100%. We use N_C in the equation to stand for the percent not correct.

$$61\% + N_C = 100\%$$

Step 3: We find the missing percent, N_C, by subtracting. We can check our work by substituting into the original equation.

$$100\% - 61\% = 39\%$$

$$61\% + 39\% = 100\% \quad \text{check}$$

Step 4: We review the question and write the answer. Of Shemp's answers, **39% were not correct.**

LESSON PRACTICE

Practice set Follow the four-step method to solve each problem. Along with each answer, include the equation you use to solve the problem.

a. Only 39% of the lights were on. What percent of the lights were off?

b. Two fifths of the pioneers did not survive the journey. What fraction of the pioneers did survive the journey?

c. Write a story problem about parts of a whole that fits this equation:

$$45\% + G = 100\%$$

MIXED PRACTICE

Problem set

1. (11) Beth fed the baby 65 grams of cereal. The baby wanted to eat 142 grams of cereal. How many additional grams of cereal did Beth need to feed the baby?

2. (14) Seven tenths of the new recruits did not like their first haircut. What fraction of the new recruits did like their first haircut?

3. (12) How many years were there from 1776 to 1789?

4. (13) Write a story problem that fits this equation:

$$12p = \$2.40$$

5. (14) If 24% of the players scored a goal, what percent of the players did not score a goal?

6. (10) Draw and shade circles to show that $3\frac{1}{3} = \frac{10}{3}$.

7. (5) Use digits to write four hundred seven million, forty-two thousand, six hundred three.

8. (2) What property is illustrated by this equation?

$$3 \cdot 2 \cdot 1 \cdot 0 = 0$$

9. (6) (a) List the common factors of 40 and 72.

(b) What is the greatest common factor of 40 and 72?

10. (7) Name three segments in the figure below in order of length from shortest to longest.

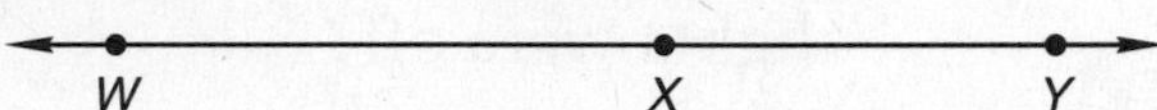

11. (8) Describe how to find the fraction of the group that is shaded.

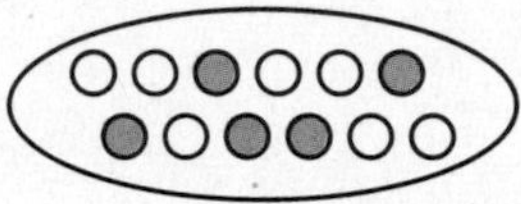

Find each missing number:

12. (3) $b - 407 = 623$

13. (3) $\$20 - e = \$3.47$

14. (3) $7 \cdot 5f = 7070$

15. (3) $\frac{m}{25} = 25$

16. (3)
$$\begin{array}{r} 5 \\ 8 \\ 7 \\ 6 \\ 5 \\ 9 \\ 4 \\ 3 \\ 6 \\ 4 \\ 7 \\ 8 \\ 5 \\ n \\ +\ 6 \\ \hline 89 \end{array}$$

17. (3) $a + 295 = 1000$

Simplify:

18. (10) $3\frac{3}{5} + 2\frac{4}{5}$

19. (10) $\frac{5}{2} \cdot \frac{3}{2}$

20. (1) $3.63 + $0.87 + 96¢

21. (1) $5 \cdot 4 \cdot 3 \cdot 2 \cdot 1$

22. (9) $\frac{2}{3} \cdot \frac{2}{3} \cdot \frac{2}{3}$

23. (1) $\frac{900}{20}$

24. (1) $\begin{array}{r} 145 \\ \times\ 74 \\ \hline \end{array}$

25. (1) 30(65¢)

26. (2) (5)(5 + 5)

27. (4) 9714 − 13,456

28. (2, 4) Compare: (1000 − 100) − 10 ◯ 1000 − (100 − 10)

29. (7) Name each type of angle illustrated:

(a) 

(b) 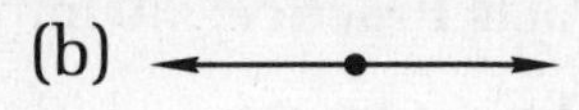

(c) 

30. (9) How many $\frac{4}{5}$'s are in 1?

LESSON

15 Equivalent Fractions • Reducing Fractions, Part 1

WARM-UP

Facts Practice: 30 Improper Fractions and Mixed Numbers (Test C)

Mental Math:

a. \$3.50 + \$1.75
b. \$4.00 ÷ 10
c. \$10.00 − \$4.98
d. (7 × 30) + (7 × 2)
e. 125 − 50
f. $\frac{1}{2}$ of 52
g. 10 − 9, + 8, − 7, + 6, − 5, + 4, − 3, + 2, − 1

Problem Solving:

Copy this problem and fill in the missing digits:

```
   36
 × __
  __0
  _6
  ___
```

NEW CONCEPTS

Equivalent fractions

Different fractions that name the same number are called **equivalent fractions.** Here we show four equivalent fractions:

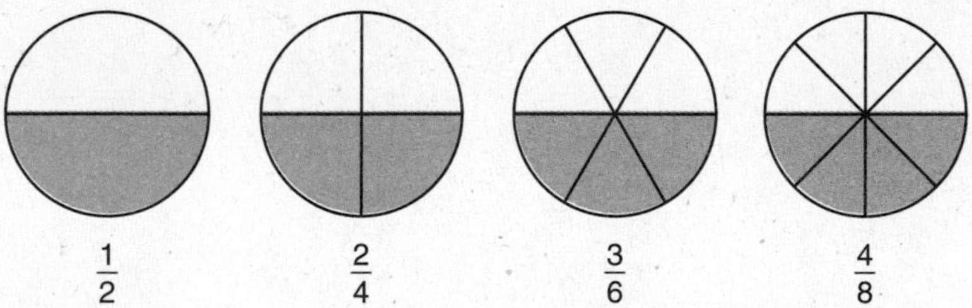

$\frac{1}{2}$ $\frac{2}{4}$ $\frac{3}{6}$ $\frac{4}{8}$

As we can see from the pictures, **equivalent fractions have the same value.**

$$\frac{1}{2} = \frac{2}{4} = \frac{3}{6} = \frac{4}{8}$$

We can form equivalent fractions by multiplying a fraction by fractions equal to 1. Here we multiply $\frac{1}{2}$ by $\frac{2}{2}$, $\frac{3}{3}$, and $\frac{4}{4}$ to form fractions equivalent to $\frac{1}{2}$:

$$\frac{1}{2} \times \frac{2}{2} = \frac{2}{4} \qquad \frac{1}{2} \times \frac{3}{3} = \frac{3}{6} \qquad \frac{1}{2} \times \frac{4}{4} = \frac{4}{8}$$

Example 1 Find an equivalent fraction for $\frac{2}{3}$ that has a denominator of 12.

Solution The denominator of $\frac{2}{3}$ is 3. To make an equivalent fraction with a denominator of 12, we multiply by $\frac{4}{4}$.

$$\frac{2}{3} \times \frac{4}{4} = \mathbf{\frac{8}{12}}$$

Example 2 Find a fraction equivalent to $\frac{1}{3}$ that has a denominator of 6. Next find a fraction equivalent to $\frac{1}{2}$ with a denominator of 6. Then add the two fractions you found.

Solution We multiply $\frac{1}{3}$ by $\frac{2}{2}$ and $\frac{1}{2}$ by $\frac{3}{3}$ to find the fractions equivalent to $\frac{1}{3}$ and $\frac{1}{2}$ that have denominators of 6. Then we add.

$$\frac{1}{3} \times \frac{2}{2} = \mathbf{\frac{2}{6}}$$

$$\frac{1}{2} \times \frac{3}{3} = \mathbf{\frac{3}{6}}$$

$$\mathbf{\frac{5}{6}}$$

Reducing fractions, part 1

An inch ruler provides another example of equivalent fractions. The segment in the figure below is $\frac{1}{2}$ inch long. By counting the tick marks on the ruler, we see that there are several equivalent names for $\frac{1}{2}$ inch.

inch 1

$$\frac{1}{2}\text{ in.} = \frac{2}{4}\text{ in.} = \frac{4}{8}\text{ in.} = \frac{8}{16}\text{ in.}$$

We say that the fractions $\frac{2}{4}$, $\frac{4}{8}$, and $\frac{8}{16}$ each **reduce** to $\frac{1}{2}$. We can reduce some fractions by dividing the fraction to be reduced by a fraction equal to 1.

$$\frac{4}{8} \div \frac{4}{4} = \frac{1}{2} \qquad \begin{matrix}(4 \div 4 = 1)\\(8 \div 4 = 2)\end{matrix}$$

By dividing $\frac{4}{8}$ by $\frac{4}{4}$, we have reduced $\frac{4}{8}$ to $\frac{1}{2}$.

The numbers we use when we write a fraction are called the **terms** of the fraction. To reduce a fraction, we divide both terms of the fraction by a factor of both terms.

$$\frac{4 \div 2}{8 \div 2} = \frac{2}{4} \qquad \frac{4 \div 4}{8 \div 4} = \frac{1}{2}$$

Dividing each term of $\frac{4}{8}$ by 4 instead of by 2 results in a fraction with lower terms, since the terms of $\frac{1}{2}$ are lower than the terms of $\frac{2}{4}$. It is customary to reduce fractions to **lowest terms.** As we see in the next example, fractions can be reduced to lowest terms in one step by dividing the terms of the fraction by the greatest common factor of the terms.

Example 3 Reduce $\frac{18}{24}$ to lowest terms.

Solution Both 18 and 24 are divisible by 2, so we divide both terms by 2.

$$\frac{18}{24} = \frac{18 \div 2}{24 \div 2} = \frac{9}{12}$$

This is not in lowest terms, because 9 and 12 are divisible by 3.

$$\frac{9}{12} = \frac{9 \div 3}{12 \div 3} = \mathbf{\frac{3}{4}}$$

We could have used just one step had we noticed that the greatest common factor of 18 and 24 is 6.

$$\frac{18}{24} = \frac{18 \div 6}{24 \div 6} = \mathbf{\frac{3}{4}}$$

Both methods are correct. One method took two steps, and the other took just one step.

Example 4 Reduce $3\frac{8}{12}$ to lowest terms.

Solution To reduce a mixed number, we reduce the fraction and leave the whole number unchanged.

$$\frac{8}{12} = \frac{8 \div 4}{12 \div 4} = \frac{2}{3}$$

$$3\frac{8}{12} = \mathbf{3\frac{2}{3}}$$

Example 5 Write $\frac{12}{9}$ as a mixed number with the fraction reduced.

Solution There are two steps to reduce and convert to a mixed number. Either step may be taken first.

Reduce First	Convert First
Reduce: $\frac{12}{9} = \frac{4}{3}$	Convert: $\frac{12}{9} = 1\frac{3}{9}$
Convert: $\frac{4}{3} = \mathbf{1\frac{1}{3}}$	Reduce: $1\frac{3}{9} = \mathbf{1\frac{1}{3}}$

Example 6 Simplify: $\frac{7}{9} - \frac{1}{9}$

Solution First we subtract. Then we reduce.

SUBTRACT	REDUCE
$\frac{7}{9} - \frac{1}{9} = \frac{6}{9}$	$\frac{6 \div 3}{9 \div 3} = \mathbf{\frac{2}{3}}$

Example 7 Write 70% as a reduced fraction.

Solution Recall that a percent is a fraction with a denominator of 100.

$$70\% = \frac{70}{100}$$

We can reduce the fraction by dividing each term by 10.

$$\frac{70}{100} \div \frac{10}{10} = \mathbf{\frac{7}{10}}$$

LESSON PRACTICE

Practice set*

a. Form three equivalent fractions for $\frac{3}{4}$ by multiplying by $\frac{5}{5}$, $\frac{7}{7}$, and $\frac{3}{3}$.

b. Find an equivalent fraction for $\frac{3}{4}$ that has a denominator of 16.

Find the number that makes the two fractions equivalent.

c. $\frac{4}{5} = \frac{?}{20}$ **d.** $\frac{3}{8} = \frac{9}{?}$

e. Find a fraction equivalent to $\frac{3}{5}$ that has a denominator of 10. Next find a fraction equivalent to $\frac{1}{2}$ with a denominator of 10. Then subtract the second fraction you found from the first fraction.

Reduce each fraction to lowest terms:

f. $\frac{3}{6}$ **g.** $\frac{8}{10}$ **h.** $\frac{8}{16}$ **i.** $\frac{12}{16}$

j. $4\frac{4}{8}$ **k.** $6\frac{9}{12}$ **l.** $12\frac{8}{15}$ **m.** $8\frac{16}{24}$

Perform each indicated operation and reduce the result:

n. $\frac{5}{12} + \frac{5}{12}$ **o.** $3\frac{7}{10} - 1\frac{1}{10}$ **p.** $\frac{5}{8} \cdot \frac{2}{3}$

Write each percent as a reduced fraction:

q. 90% **r.** 75% **s.** 5%

t. Find a fraction equivalent to $\frac{2}{3}$ that has a denominator of 6. Subtract $\frac{1}{6}$ from the fraction you found and reduce the answer.

MIXED PRACTICE

Problem set

1. (12) Great-Grandpa celebrated his seventy-fifth birthday in 1998. In what year was he born?

2. (11) Austin watched the geese fly south. He counted 27 in the first flock, 38 in the second flock, and 56 in the third flock. How many geese did Austin see in all three flocks?

3. (15) If 40% of the eggs were cracked, what fraction of the eggs were cracked?

4. (13) The farmer harvested 9000 bushels of grain from 60 acres. The crop produced an average of how many bushels of grain for each acre?

5. (8) With a ruler, draw a segment $2\frac{1}{2}$ inches long. Draw a second segment $1\frac{7}{8}$ inches long. The first segment is how much longer than the second segment?

6. (4) Use digits and symbols to write "The product of three and five is greater than the sum of three and five."

7. (6) List the single-digit divisors of 2100.

8. (15) Reduce each fraction or mixed number:

(a) $\frac{6}{8}$ (b) $2\frac{6}{10}$

9. (15) Find three equivalent fractions for $\frac{2}{3}$ by multiplying by $\frac{3}{3}$, $\frac{5}{5}$, and $\frac{6}{6}$.

10. (15) For each fraction, find an equivalent fraction that has a denominator of 20:

(a) $\frac{3}{5}$ (b) $\frac{1}{2}$ (c) $\frac{3}{4}$

11. Refer to this figure to answer (a)–(c):
(7)

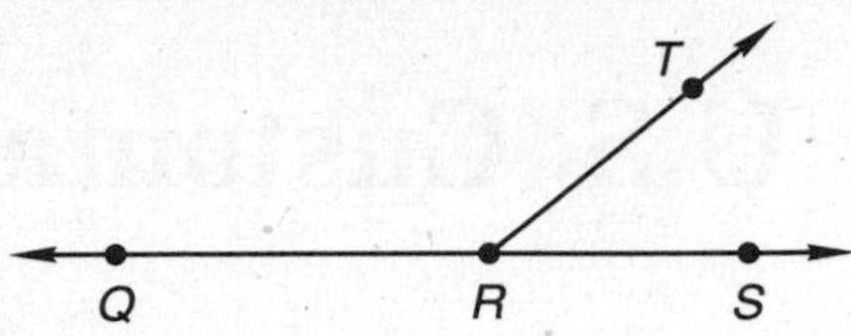

(a) Name the line.

(b) Name three rays originating at point *R*.

(c) Name an acute angle.

12. Convert each fraction to either a whole number or a mixed number:
(10)

(a) $\frac{11}{3}$ (b) $\frac{12}{3}$ (c) $\frac{13}{3}$

13. Compare: (11)(6 + 7) ◯ 66 + 77
(4)

Find each missing number:

14. $39 + b = 50$
(3)

15. $6a = 300$
(3)

16. $c - \$5 = 5¢$
(3)

17. $\frac{w}{35} = 35$
(3)

Write each percent as a reduced fraction:

18. 80%
(15)

19. 35%
(15)

20. How many $\frac{1}{8}$'s are in 1?
(9)

Simplify:

21. $\frac{2}{5} + \frac{3}{5} + \frac{4}{5}$
(10)

22. $3\frac{5}{8} - 1\frac{3}{8}$
(15)

23. $\frac{4}{3} \cdot \frac{3}{4}$
(9)

24. $\frac{3}{4} + \frac{3}{4}$
(15)

25. $\frac{7}{5} + \frac{8}{5}$
(10)

26. $\frac{11}{12} - \frac{1}{12}$
(15)

27. $\frac{5}{6} \cdot \frac{2}{3}$
(15)

28. Evaluate each expression for $a = 4$ and $b = 8$:
(1, 9)

(a) $\frac{a}{b} + \frac{a}{b}$ (b) $\frac{a}{b} - \frac{a}{b}$

29. Find a fraction equal to $\frac{1}{3}$ that has a denominator of 6. Add the fraction to $\frac{1}{6}$ and reduce the answer.
(15)

30. Write $2\frac{2}{3}$ as an improper fraction. Then multiply the improper fraction by $\frac{1}{4}$ and reduce the product.
(10, 15)

LESSON

16 U.S. Customary System

WARM-UP

Facts Practice: 40 Fractions to Reduce (Test D)

Mental Math:

a. 10 − 20 **b.** 15¢ × 10 **c.** $1.00 − 18¢

d. 4 × 23 **e.** 875 − 750 **f.** $\frac{1}{2}$ of $\frac{1}{3}$

g. Start with 2 score and 10, ÷ 2, × 3, − 3, ÷ 9, + 2, ÷ 5.

Problem Solving:

Fiona has 8 coins totaling 50¢. What combinations of coins could Fiona have?

NEW CONCEPT

One of the characteristics of civilization is the use of an agreed-upon system of measurement. The fair exchange of goods and services requires consistent units of weight, volume, and length. In a technological society the necessity for a standard system of measurement is even greater.

There are two systems of measurement currently used in the United States. The traditional system of measurement, with units such as feet, gallons, and pounds, was adopted from England. This system was once known as the English system but is now referred to as the **U.S. Customary System.**

The second system of measurement used in the United States is the system used by almost every country in the world. It is known as the **International System of Units** (or SI, for *Système International d'Unités*) or the **metric system.** The metric system has units such as meters, liters, and kilograms.

We will consider both systems of measurement over many lessons. In this lesson we will consider units of the U.S. Customary System. We can measure many characteristics of objects, including their dimensions, weight, volume, and temperature. Each type of measurement has a set of units. We should remember equivalent measures and have a "feel" for the units so that we can estimate measurements reasonably.

The following table shows the common weight equivalences in the U.S. Customary System:

UNITS OF WEIGHT

16 ounces (oz)	=	1 pound (lb)
2000 pounds	=	1 ton (tn)

Example 1 A $\frac{1}{2}$-ton pickup truck can carry a load of $\frac{1}{2}$ of a ton. What is the load capacity in pounds of a $\frac{1}{2}$-ton pickup truck?

Solution One ton is 2000 pounds, so half of a ton is **1000 pounds.**

The following table shows the common length equivalences in the U.S. Customary System:

UNITS OF LENGTH

12 inches (in.)	=	1 foot (ft)
3 feet	=	1 yard (yd)
1760 yards	=	1 mile (mi)
5280 feet	=	1 mile

Example 2 One yard is how many inches?

Solution One yard equals 3 feet. One foot equals 12 inches. Thus 1 yard is 36 inches.

1 yard = 3 × 12 inches = **36 inches**

Example 3 A mountain bicycle is about how many feet long?

Solution We should develop a feel for various units of measure. Most mountain bicycles are about $5\frac{1}{2}$ feet long, so a good estimate would be **about 5 or 6 feet.**

Just as an inch ruler is divided successively in half, so units of liquid measure are divided successively in half. Half of a gallon is a half gallon. Half of a half gallon is a quart. Half of a quart is a pint. Half of a pint is a cup.

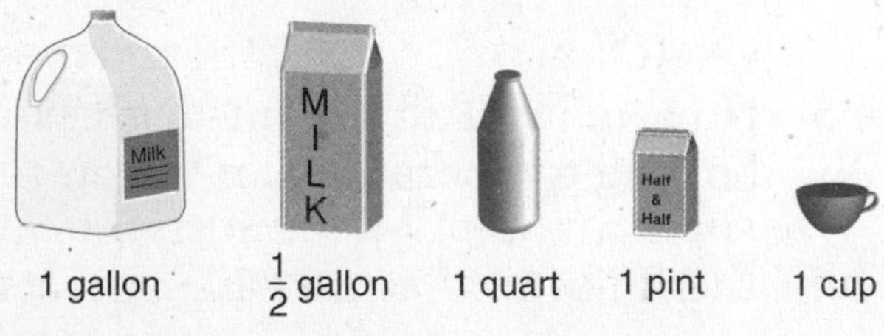

The capacity of each of these containers is half the capacity of the next larger container.

The following table and example demonstrate using units of liquid measure.

UNITS OF LIQUID MEASURE

8 ounces (oz)	=	1 cup (c)
2 cups	=	1 pint (pt)
2 pints	=	1 quart (qt)
4 quarts	=	1 gallon (gal)

Example 4 Steve always drinks at least 8 cups of water every day. How many quarts is that?

Solution Two cups is a pint, so 8 cups is 4 pints. Two pints is a quart, so 4 pints is **2 quarts.**

The following table and example demonstrate using units of temperature.

Fahrenheit Temperature Scale

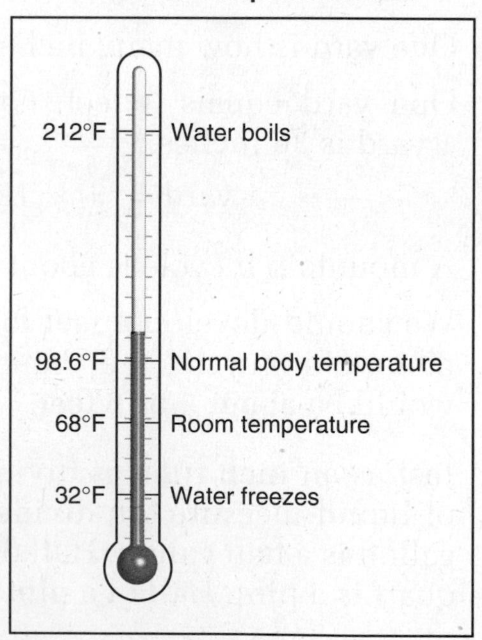

Example 5 How many Fahrenheit degrees are between the freezing and boiling temperatures of water?

Solution 212°F − 32°F = **180°F**

LESSON PRACTICE

Practice set

a. A tall man may be over how many yards tall?

b. How many quarts is half a gallon?

c. When Chad was born, he weighed 8 lb 7 oz. Is that weight closer to 8 lb or to 9 lb?

Simplify:

d. $\frac{3}{8}$ in. + $\frac{5}{8}$ in.

e. 32°F + 180°F

f. 2(3 ft + 4 ft)

g. 1 ton − 1000 pounds

h. A measuring cup is marked at every ounce. Assuming the measuring cup is used correctly, what is the greatest possible measurement error that can be made with the cup? Express your answer as a fraction of an ounce.

MIXED PRACTICE

Problem set

1. (14) Thirty-five of the one hundred eighteen sports cars were convertibles. How many of the sports cars were not convertibles?

2. (13) At Henry's egg ranch 18 eggs are packaged in each carton. How many cartons would be needed to package 4500 eggs?

3. (11) Three hundred twenty-four ducks floated peacefully on the lake. As the first shot rang out, all but 27 of the ducks flew away. How many ducks flew away?

4. (5) Write the integer 250 in expanded notation.

5. (10, 15) Replace each circle with the proper comparison symbol:

(a) $\frac{8}{10} \bigcirc \frac{4}{5}$

(b) $\frac{8}{5} \bigcirc 1\frac{2}{5}$

6. Use an inch ruler to find *AB, CB,* and *CA* to the nearest sixteenth of an inch.
(8)

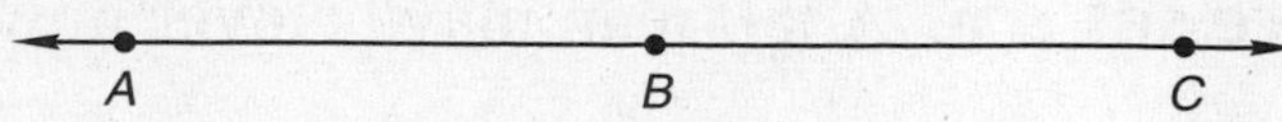

7. Write each number as a reduced fraction or mixed number:
(15)

(a) $\frac{8}{12}$ (b) 40% (c) $6\frac{10}{12}$

8. Draw and shade circles to show that $3\frac{1}{3}$ equals $\frac{10}{3}$.
(10)

9. For each fraction, find an equivalent fraction that has a denominator of 24:
(15)

(a) $\frac{5}{6}$ (b) $\frac{3}{8}$ (c) $\frac{1}{4}$

10. (a) What percent of a yard is a foot?
(16)

(b) What fraction of a gallon is a quart?

11. The number 630 is divisible by which single-digit numbers?
(6)

12. Convert each improper fraction to either a whole number or a mixed number:
(10)

(a) $\frac{16}{7}$ (b) $3\frac{16}{8}$ (c) $2\frac{16}{9}$

13. Which property is illustrated by this equation?
(2)

$$\frac{1}{2} \times 1 = \frac{1}{2}$$

Find each missing number:

14. $m - 1776 = 87$
(3)

15. $\$16.25 - b = \$10.15$
(3)

16. $\frac{1001}{n} = 13$
(3)

17. $42d = 1764$
(3)

Simplify:

18. $3\frac{3}{4} - 1\frac{1}{4}$
(15)

19. $\frac{3}{10}$ in. + $\frac{8}{10}$ in.
(10, 16)

20. $\frac{3}{4} \times \frac{1}{3}$
(15)

21. $\frac{4}{3} \cdot \frac{3}{2}$
(10)

22. $\frac{10{,}000}{16}$ (1)

23. $\frac{100\%}{8}$ (10, 15)

24. $9\overline{)70{,}000}$ (10)

25. $45 \cdot 45$ (1)

26. Describe the rule of this sequence, and find the next three terms: (2, 8)

$$\frac{1}{16}, \frac{1}{8}, \frac{3}{16}, \ldots$$

27. If two intersecting lines are not perpendicular, then they form which two types of angles? (7)

28. Convert $2\frac{1}{2}$ and $1\frac{2}{3}$ to improper fractions, and multiply the improper fractions. Then convert the product to a mixed number. (10)

29. Find a fraction equivalent to $\frac{2}{3}$ that has a denominator of 6. Then add that fraction to $\frac{1}{6}$. What is the sum? (15)

30. How many $\frac{3}{8}$'s are in 1? (9)

LESSON

17 Measuring Angles with a Protractor

WARM-UP

Facts Practice: 40 Fractions to Reduce (Test D)

Mental Math:

a. \$3.50 + \$1.50 **b.** \$3.60 ÷ 10 **c.** \$10.00 − \$6.40

d. 5 × 33 **e.** 250 − 125 **f.** $\frac{1}{2}$ of 32

g. Start with 3 score and 15, ÷ 3, × 2, ÷ 5, × 10, ÷ 2, − 25, ÷ 5.

Problem Solving:

Nelson held a dot cube between two fingers so that he covered opposite faces of the cube. On two of the faces that Nelson could see were 3 dots and 5 dots. How many dots were on each of the two faces his fingers covered?

NEW CONCEPT

In Lesson 7 we discussed angles and classified them as acute, right, obtuse, or straight. In this lesson we will begin measuring angles.

Angles are commonly measured in units called **degrees.** The abbreviation for *degrees* is a small circle written above and to the right of the number. One full rotation, a full circle, measures 360 degrees.

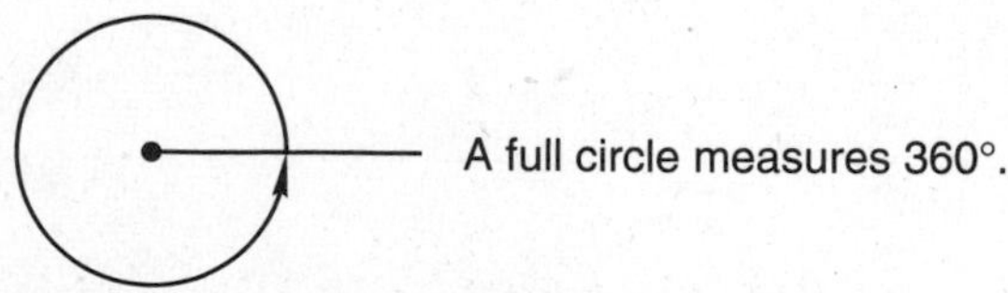

A half circle measures half of 360°, which is 180°.

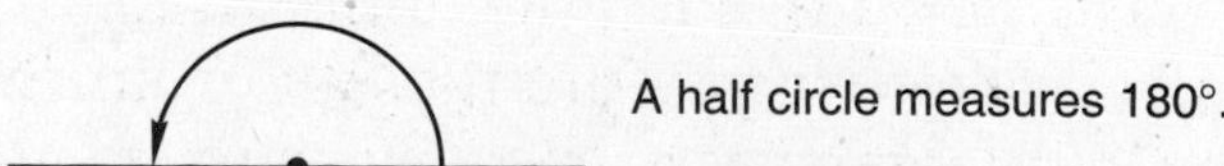

One fourth of a full rotation is a right angle. A right angle measures one fourth of 360°, which is 90°.

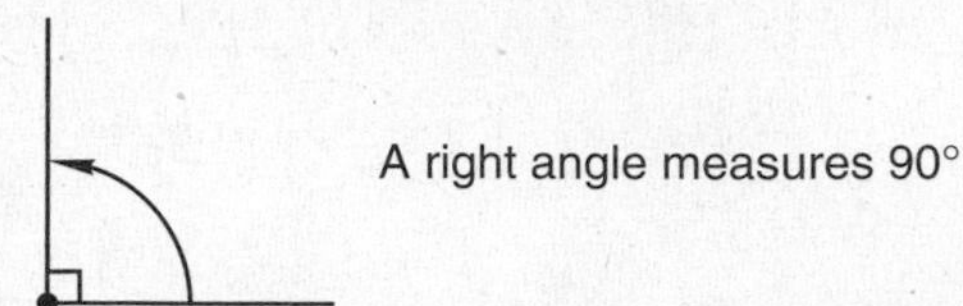

Thus, the measure of an acute angle is less than 90°, and the measure of an obtuse angle is greater than 90° but less than 180°. An angle that measures 180° is a straight angle. The chart below summarizes the types of angles and their measures.

Angle Type	Measure
Acute	Greater than 0° but less than 90°
Right	Exactly 90°
Obtuse	Greater than 90° but less than 180°
Straight	Exactly 180°

A **protractor** can be used to measure angles. The protractor is placed on the angle to be measured so the vertex is under the dot, circle, or crossmark of the protractor, and one side of the angle is under the zero mark at either end of the scale of the protractor.

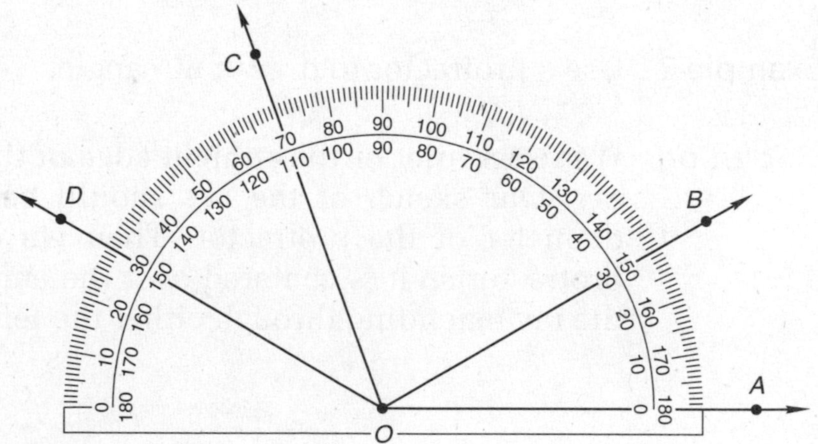

The measures of the three angles shown are as follows:

$\angle AOB = 30°$ $\angle AOC = 110°$ $\angle AOD = 150°$

Notice there are two scales on a protractor, one starting from the left side, the other from the right. One way to check whether you are reading from the correct scale is to consider whether you are measuring an acute angle or an obtuse angle.

Example 1 Find the measure of each angle.

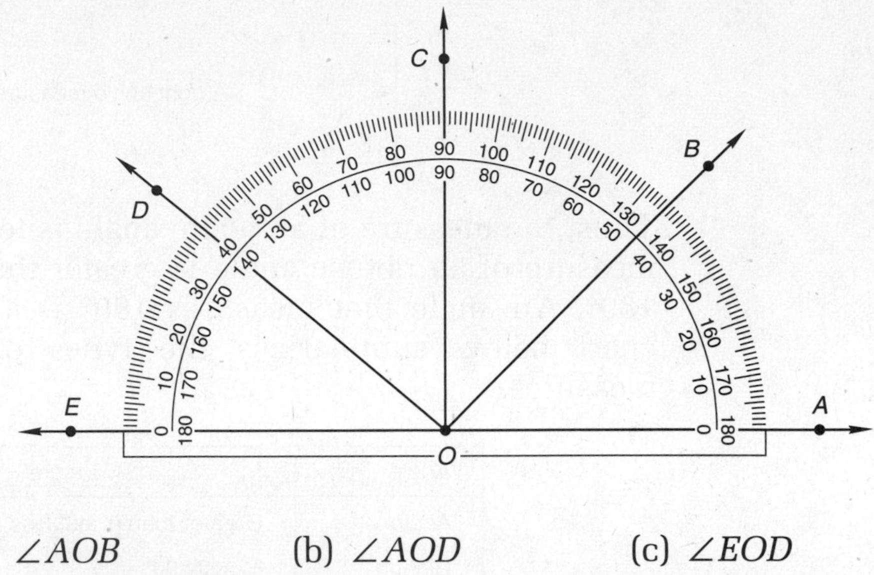

(a) $\angle AOB$ (b) $\angle AOD$ (c) $\angle EOD$

Solution (a) Since $\angle AOB$ is acute, we read the numbers less than 90. Ray *OB* passes through the mark halfway between 40 and 50. Thus, the measure of $\angle AOB$ is **45°**.

(b) Since $\angle AOD$ is obtuse, we read the numbers greater than 90. The measure of $\angle AOD$ is **140°**.

(c) Angle *EOD* is acute. The measure of $\angle EOD$ is **40°**.

Example 2 Use a protractor to draw a 60° angle.

Solution We use a ruler or the straight edge of the protractor to draw a ray. Our sketch of the ray should be longer than half the diameter of the protractor. Then we carefully position the protractor so it is centered over the endpoint of the ray, with the ray extending through either the left or right 0° mark.

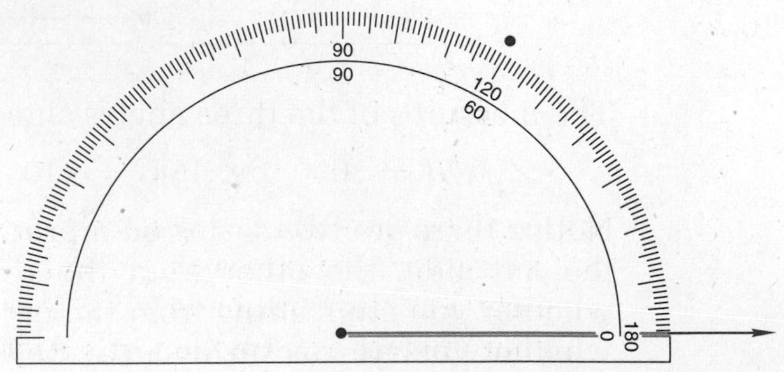

From the 0° mark we follow the curve of the protractor to the 60° mark and make a dot on the paper. Then we remove the protractor and use a straightedge or ruler to draw the second ray of the angle from the endpoint of the first ray through the dot. This completes the 60° angle.

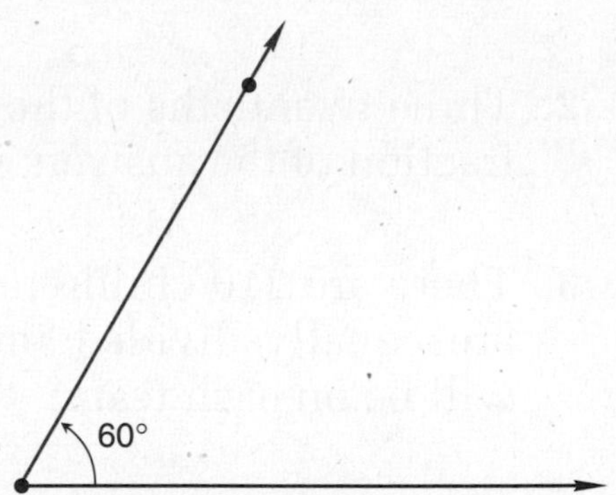

LESSON PRACTICE

Practice set Find the measure of each angle named in problems **a–f.**

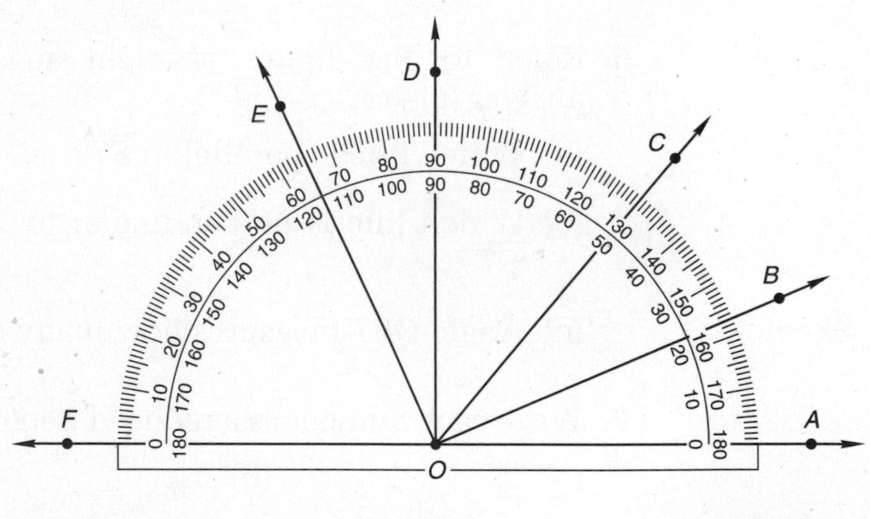

a. $\angle AOD$ **b.** $\angle AOC$ **c.** $\angle AOE$

d. $\angle FOE$ **e.** $\angle FOC$ **f.** $\angle AOB$

Use your protractor to draw each of these angles:

g. 45° **h.** 120° **i.** 100° **j.** 80°

k. Perry's protractor is marked at each degree. Assuming the protractor is used correctly, what is the greatest possible error that can be made with Perry's protractor? Express your answer as a fraction of a degree.

MIXED PRACTICE

Problem set

1. *(11)* Prince Caspian assembled his soldiers on the bank of the river. Two thousand, four hundred twenty had gathered by noon. An additional five thousand, ninety arrived after noon. How many soldiers arrived in all?

2. *(14)* Three twentieths of the test answers were incorrect. What fraction of the answers were correct?

3. *(13)* There are 210 children in the city's soccer league. If they are equally divided into 15 teams, how many children will be on each team?

4. *(12)* How many years were there from 1492 to 1620?

5. *(10, 15)* Which of the following does not equal $1\frac{1}{3}$?

A. $\frac{4}{3}$ B. $1\frac{2}{6}$ C. $\frac{5}{3}$ D. $1\frac{4}{12}$

6. *(7, 17)* Refer to the figure at right to answer (a)–(c):

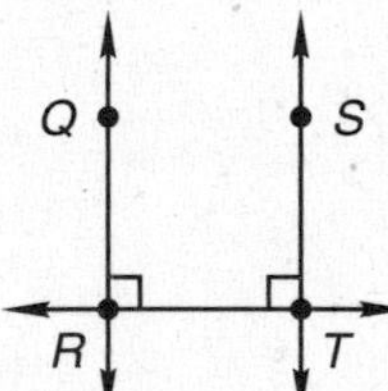

(a) Which line is parallel to $\overleftrightarrow{ST}$?

(b) Which line is perpendicular to $\overleftrightarrow{ST}$?

(c) Angle QRT measures how many degrees?

7. *(15)* Write each number as a reduced fraction or mixed number:

(a) $\frac{12}{16}$ (b) $3\frac{12}{18}$ (c) 25%

8. *(16)* How many ounces is 2 lb 8 oz?

9. *(15)* Complete each equivalent fraction:

(a) $\frac{2}{9} = \frac{?}{18}$ (b) $\frac{1}{3} = \frac{?}{18}$ (c) $\frac{5}{6} = \frac{?}{18}$

10. *(17)* Use a protractor to draw a 30° angle.

11. *(6)* (a) What factors of 20 are also factors of 50?

(b) What is the GCF of 20 and 50?

12. Draw $\overline{RS}$ $1\frac{3}{4}$ in. long. Then draw $\overrightarrow{ST}$ perpendicular to $\overline{RS}$.
(7, 8)

13. If $x = 4$ and $y = 8$, find
(1, 9)

(a) $\frac{y}{x} - \frac{x}{y}$ (b) $x - \frac{x}{y}$

Find each missing number:

14. $x - 231 = 141$
(3)

15. $\$6.30 + y = \$25$
(3)

16. $8w = \$30.00$
(3)

17. $\frac{100\%}{m} = 20\%$
(3)

Simplify:

18. $3\frac{5}{6} - 1\frac{1}{6}$
(15)

19. $\frac{1}{2} \cdot \frac{2}{3}$
(15)

20. $\frac{\$100.00}{40}$
(1)

21. $55 \cdot 55$
(1)

22. 2(8 in. + 6 in.)
(2, 16)

23. $\frac{3}{4}$ in. + $\frac{3}{4}$ in.
(15, 16)

24. $\frac{15}{16}$ in. − $\frac{3}{16}$ in.
(15, 16)

25. $\frac{1}{2} \cdot \frac{4}{3} \cdot \frac{9}{2}$
(10)

26. The cost of the dinner was \$15.17. Loretha gave the cashier a \$20 bill and a quarter. Name the coins and bills she probably received in change.
(1)

27. (a) Compare: $\left(\frac{1}{2} \cdot \frac{3}{4}\right) \cdot \frac{2}{3} \bigcirc \frac{1}{2}\left(\frac{3}{4} \cdot \frac{2}{3}\right)$
(2, 9)

(b) What property is illustrated by the comparison?

28. Write a story problem about parts of a whole that fits this equation:
(14)

$$85\% + W = 100\%$$

29. Write $3\frac{3}{4}$ as an improper fraction. Then write its reciprocal.
(9, 10)

30. Find a fraction equal to $\frac{3}{4}$ with a denominator of 8. Add the fraction to $\frac{5}{8}$. Write the sum as a mixed number.
(15)

LESSON

18 Polygons • Similar and Congruent

WARM-UP

Facts Practice: 40 Fractions to Reduce (Test D)

Mental Math:

a. \$3.75 + \$1.75 b. \$1.65 × 10 c. \$20.00 − \$12.50

d. 6 × 24 e. 375 − 250 f. $\frac{1}{2}$ of $\frac{1}{4}$

g. Start with two score, × 2, + 1, ÷ 9, × 3, + 1, ÷ 4.

Problem Solving:

Sarah remembered that the three numbers to her combination lock were 17, 32, and 8, but she could not remember the order. She knew to turn the dial clockwise, then counterclockwise, then clockwise. List all permutations of the three numbers Sarah could try.

NEW CONCEPTS

Polygons

When three or more line segments are connected to enclose a portion of a plane, a **polygon** is formed. The word *polygon* comes from the ancient Greeks and means "many angles." The name of a polygon tells how many angles and sides the polygon has.

Names of Polygons

Name of Polygon	Number of Sides	Name of Polygon	Number of Sides
Triangle	3	Octagon	8
Quadrilateral	4	Nonagon	9
Pentagon	5	Decagon	10
Hexagon	6	Undecagon	11
Heptagon	7	Dodecagon	12

A polygon with more than 12 sides may be referred to as an *n*-gon, with *n* being the number of sides. Thus, a polygon with 15 sides is a 15-gon.

The point where two sides of a polygon meet is called a **vertex.** (The plural of vertex is **vertices.**) A particular polygon may be identified by naming the letters of its vertices in order. Any letter may be first. The rest of the letters can be named clockwise or counterclockwise. The polygon below has eight names, which are listed to its right.

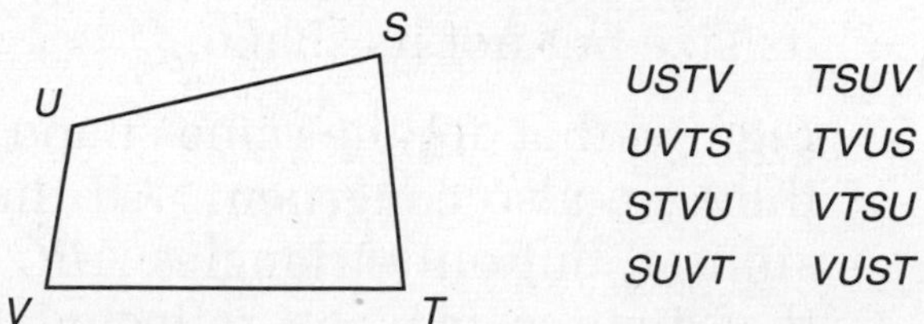

If all the sides of a polygon have the same length and all the angles have the same measure, then the polygon is a **regular polygon**.

Regular and Irregular Polygons

Type	Regular	Irregular
Triangle		
Quadrilateral		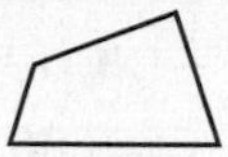
Pentagon	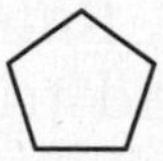	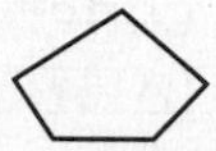
Hexagon	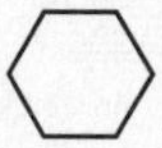	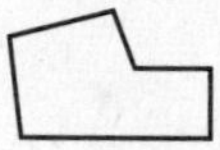

Example 1 (a) Name this polygon.

(b) Is the polygon regular or irregular?

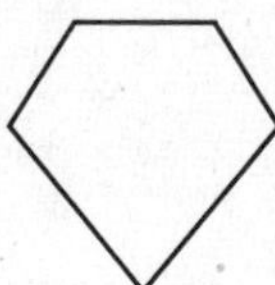

Solution (a) **pentagon** (b) **irregular**

Similar and congruent Two figures are **similar** if they have the same shape even though they may vary in size. In the illustration below, triangles I, II, and III are similar. To see this, we can imagine enlarging (dilating) triangle II as though we were looking through a magnifying glass. By enlarging triangle II, we could make it the same size as triangle I or triangle III. Likewise, we could reduce triangle III to the same size as triangle I or triangle II.

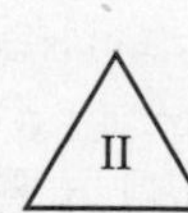

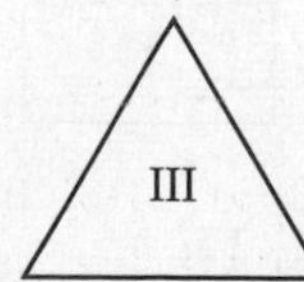

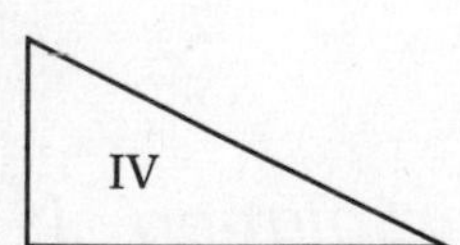

Although triangle IV is a triangle, it is not similar to the other three triangles, because its shape is different. Viewing triangle IV through a reducing or enlarging lens will change its size but not its shape.

Figures that are the same shape and size are not only similar, they are also **congruent.** All three of the triangles below are similar, but only triangles *ABC* and *DEF* are congruent. Note that figures may be reflected (flipped) or rotated (turned) without affecting their similarity or congruence.

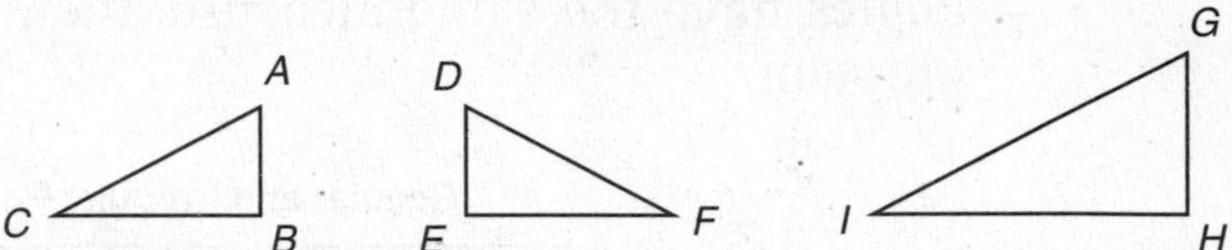

When inspecting polygons to determine whether they are similar or congruent, we compare their **corresponding parts.** A triangle has six parts—three sides and three angles. If the six parts of one triangle have the same measures as the six corresponding parts of another triangle, the triangles are congruent. Referring back to the illustration of triangle *ABC* (ΔABC) and triangle *DEF* (ΔDEF), we identify the following corresponding parts:

$\angle A$ corresponds to $\angle D$

$\angle B$ corresponds to $\angle E$

$\angle C$ corresponds to $\angle F$

$\overline{AB}$ corresponds to $\overline{DE}$

$\overline{BC}$ corresponds to $\overline{EF}$

$\overline{CA}$ corresponds to $\overline{FD}$

Notice that the corresponding angles of similar figures have the same measure even though the corresponding sides may be different lengths.

Example 2 (a) Which of these quadrilaterals appear to be similar?

(b) Which of these quadrilaterals appear to be congruent?

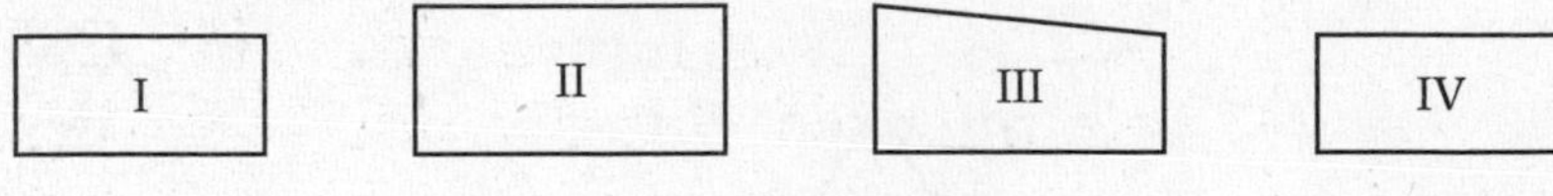

Solution (a) **I, II, IV** (b) **I, IV**

Example 3 (a) Which angle in ΔXYZ corresponds to $\angle A$ in ΔABC?

(b) Which side in ΔXYZ corresponds to $\overline{BC}$ in ΔABC?

Z A C B X Y

Solution (a) $\angle \mathbf{X}$ (b) $\overline{\mathbf{YZ}}$

LESSON PRACTICE

Practice set

a. What is the shape of a stop sign?

b. What do we usually call a regular quadrilateral?

c. What kind of angle is each angle of a regular triangle?

d. Are all squares similar?

e. Are all squares congruent?

f. Referring to example 3, which angle in ΔABC corresponds to $\angle Y$ in ΔXYZ?

g. Referring to example 2, are the angles in figure II larger in measure, smaller in measure, or equal in measure to the corresponding angles in figure I?

h. Draw a triangle that contains a right angle. Label the vertices A, B, and C so that the right angle is at vertex C.

MIXED PRACTICE

Problem set

1. (13) The Collins family completed the 3300-mile, coast-to-coast drive in 6 days. They traveled an average of how many miles each day?

2. (11) On their return trip the Collins family drove four hundred fifty-six miles the first day and five hundred seventeen miles the second day. How far did they travel on the first two days of their return trip?

3. (14) Albert ran 3977 meters of the 5000-meter race but walked the rest of the way. How many meters of the race did Albert walk?

4. (5, 12) One billion is how much greater than ten million? Use words to write the answer.

5. (4, 10) (a) Arrange these numbers in order from least to greatest:

$$\frac{5}{3}, -1, \frac{3}{4}, 0, 1$$

(b) Which of these numbers are not positive?

6. (7) In rectangle *ABCD,* which side is parallel to side *BC?*

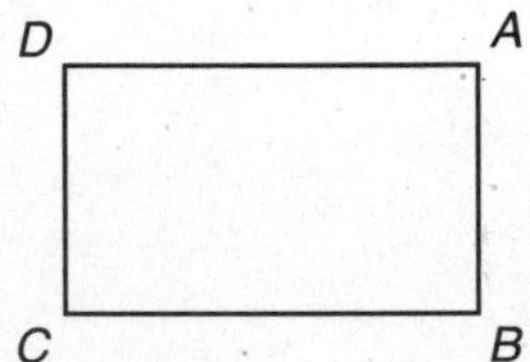

7. (4) Refer to this number line to answer (a) and (b):

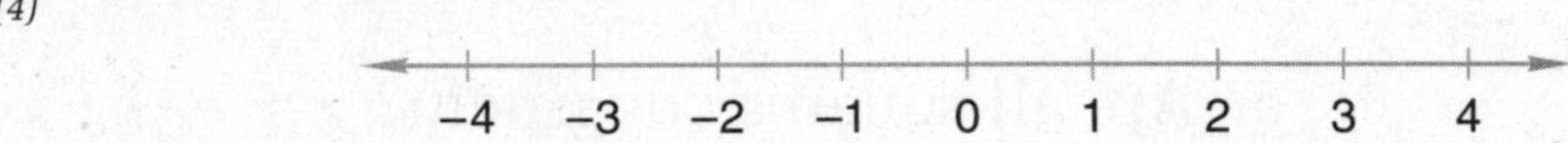

(a) What integer is two units to the left of the origin?

(b) What integer is seven units to the right of −3?

8. (15) Write each number as a reduced fraction or mixed number:

(a) 2% (b) $\frac{12}{20}$ (c) $6\frac{15}{20}$

9. (15) For each fraction, find an equivalent fraction that has a denominator of 30:

(a) $\frac{4}{5}$ (b) $\frac{2}{3}$ (c) $\frac{1}{6}$

10. (18) An octagon has how many more sides than a pentagon?

11. (7, 18) (a) Draw a triangle that has one obtuse angle.

(b) What kind of angles are the other two angles of the triangle?

12. (8, 15) (a) What percent of the circle is shaded?

(b) What fraction of the circle is not shaded?

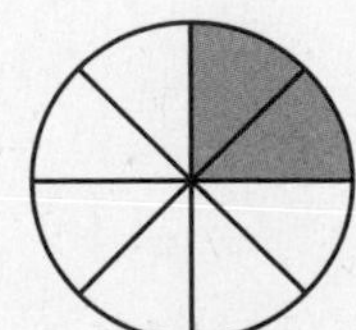

13. Which property is illustrated by this equation?
(2, 15)

$$\frac{1}{2} \times \frac{3}{3} = \frac{3}{6}$$

Find each missing number:

14. $x - \frac{3}{8} = \frac{5}{8}$
(9)

15. $y + \frac{3}{10} = \frac{7}{10}$
(9, 15)

16. $\frac{5}{6} - m = \frac{1}{6}$
(9, 15)

17. $\frac{3}{4}x = 1$
(9)

Simplify:

18. $5\frac{7}{10} - \frac{3}{10}$
(15)

19. $\frac{3}{2} \cdot \frac{2}{4}$
(15)

20. $\frac{2025}{45}$
(1)

21. $\begin{array}{r} 750 \\ \times\ 80 \\ \hline \end{array}$
(1)

22. $21 \cdot 21$
(1)

23. 2(50 in. + 40 in.)
(2, 16)

24. What percent of a pound is 8 ounces?
(16)

25. (a) How many degrees is $\frac{1}{4}$ of a circle or $\frac{1}{4}$ of a full turn?
(17)

(b) How many degrees is $\frac{1}{6}$ of a circle or $\frac{1}{6}$ of a full turn?

26. (a) Use a protractor to draw a 135° angle.
(17)

(b) A 135° angle is how many degrees less than a straight angle?

27. Refer to the triangles below to answer (a)–(c).
(18)

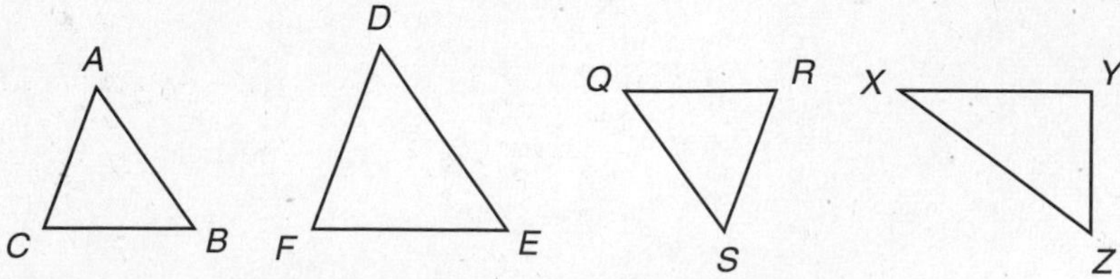

(a) Which triangle appears to be congruent to ΔABC?

(b) Which triangle is not similar to ΔABC?

(c) Which angle in ΔDEF corresponds to $\angle R$ in ΔSQR?

28. Write a fraction equal to $\frac{1}{2}$ with a denominator of 6 and a fraction equal to $\frac{1}{3}$ with a denominator of 6. Then add the fractions.
(15)

29. Write $2\frac{1}{4}$ as an improper fraction, and multiply the improper fraction by the reciprocal of $\frac{3}{4}$.
(9, 10)

30. Use a ruler to draw a triangle with each side 1 inch long. Is the triangle regular or irregular?
(8, 18)

LESSON

19 Perimeter

WARM-UP

Facts Practice: 30 Improper Fractions and Mixed Numbers (Test C)

Mental Math:

a. \$8.25 + \$1.75 **b.** \$12.00 ÷ 10 **c.** \$1.00 − 76¢

d. 7 × 32 **e.** 625 − 250 **f.** $\frac{1}{2}$ of 120

g. Start with 4 dozen, ÷ 6, × 5, + 2, ÷ 6, × 7, + 1, ÷ 2, − 1, ÷ 2.

Problem Solving:

Yin has 25 tickets, Bobby has 12 tickets, and Mary has 8 tickets. How many tickets should Yin give to Bobby and to Mary so that they all have the same number of tickets?

NEW CONCEPT

The distance around a polygon is the **perimeter** of the polygon. To find the perimeter of a polygon, we add the lengths of its sides.

Example 1 What is the perimeter of this rectangle?

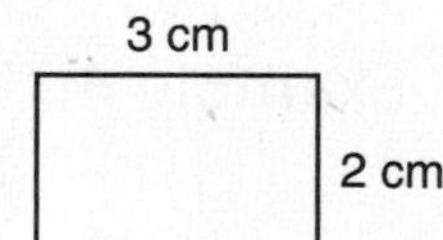

Solution The opposite sides of a rectangle are equal in length. Tracing around the rectangle, our pencil travels 3 cm, then 2 cm, then 3 cm, then 2 cm. Thus, the perimeter is

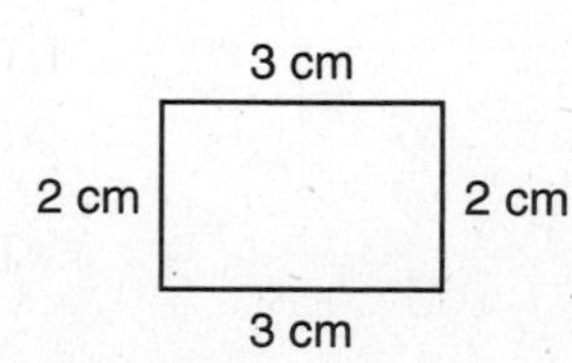

3 cm + 2 cm + 3 cm + 2 cm = **10 cm**

Example 2 What is the perimeter of this regular hexagon?

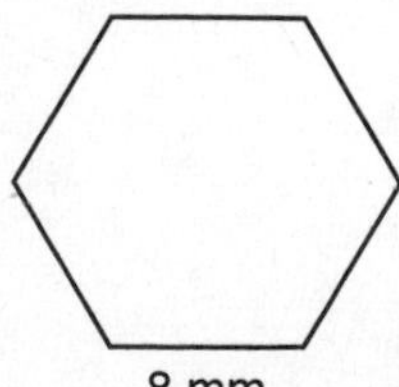

Solution All sides of a regular polygon are equal in length. Thus the perimeter of this hexagon is

$$8 \text{ mm} + 8 \text{ mm} + 8 \text{ mm} + 8 \text{ mm} + 8 \text{ mm} + 8 \text{ mm} = \mathbf{48 \text{ mm}}$$

or

$$6 \times 8 \text{ mm} = \mathbf{48 \text{ mm}}$$

Example 3 Find the perimeter of this polygon. All angles are right angles. Dimensions are in feet.

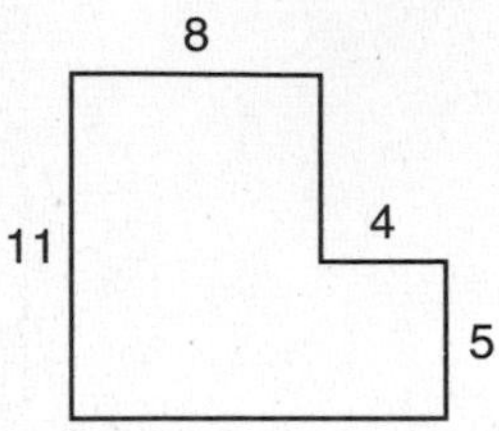

Solution We will use the letters a and b to refer to the unmarked sides. Notice that the lengths of side a and the side marked 5 total 11 feet.

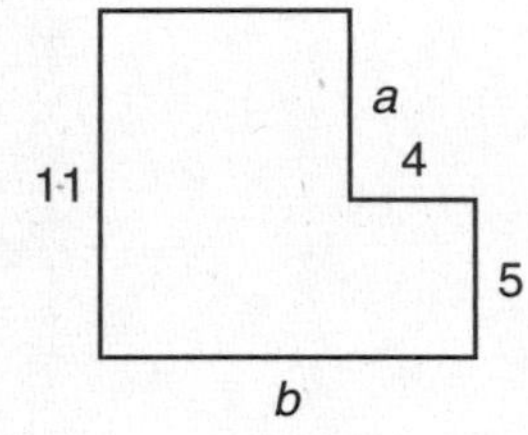

$a + 5 = 11$ So side a is 6 ft.

Also notice that the length of side b equals the total lengths of the sides marked 8 and 4.

$8 + 4 = b$ So side b is 12 ft.

The perimeter of the figure in feet is

$$8 \text{ ft} + 6 \text{ ft} + 4 \text{ ft} + 5 \text{ ft} + 12 \text{ ft} + 11 \text{ ft} = \mathbf{46 \text{ ft}}$$

Example 4 The perimeter of a square is 48 ft. How long is each side of the square?

Solution A square has four sides whose lengths are equal. The sum of the four lengths is 48 ft. Here are two ways to think about this problem:

1. The sum of what four identical addends is 48?

   ____ + ____ + ____ + ____ = 48 ft

2. What number multiplied by 4 equals 48?

   4 × ____ = 48 ft

As we think about the problem the second way, we see that we can divide 48 ft by 4 to find the length of each side.

$$4\overline{)48}\quad 12$$

The length of each side of the square is **12 ft.**

Example 5 Isabel wants to fence some grazing land for her sheep. She made this sketch of her pasture. How many feet of wire fence does she need?

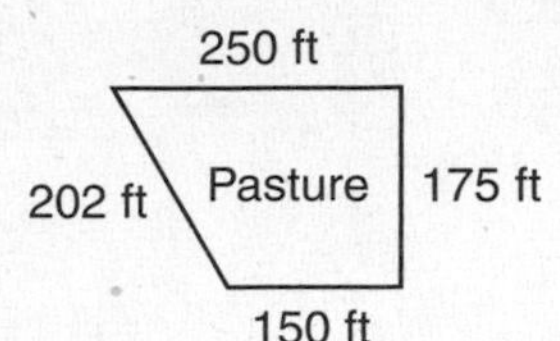

Solution We add the lengths of the sides to find how many feet of fence Isabel needs.

$$250 \text{ ft} + 175 \text{ ft} + 150 \text{ ft} + 202 \text{ ft} = 777 \text{ ft}$$

We see that Isabel needs **777 ft** of wire fence.

LESSON PRACTICE

Practice set* **a.** What is the perimeter of this quadrilateral?

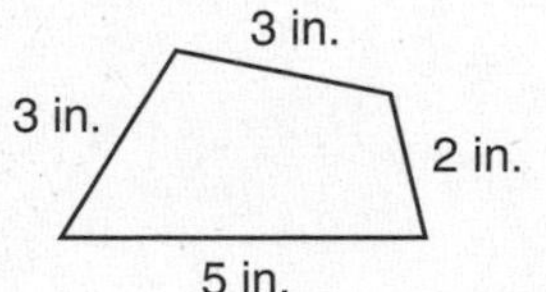

b. What is the perimeter of this regular pentagon?

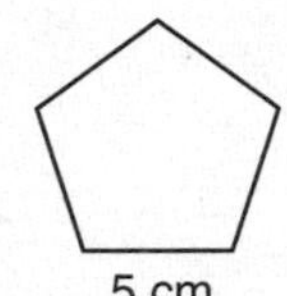

c. If each side of a regular octagon measures 12 inches, what is its perimeter?

d. What is the perimeter of this hexagon?

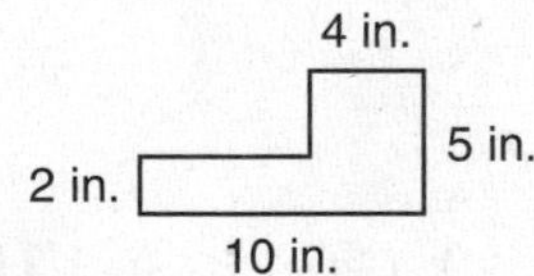

e. MacGregor has 100 feet of wire fence that he plans to use to enclose a square garden. Each side of his garden will be how many feet long?

f. Draw a quadrilateral with each side $\frac{3}{4}$ inch long. What is the perimeter of the quadrilateral?

MIXED PRACTICE

Problem set

1. (14) One eighth of the tomatoes on the vine were ripe. What fraction of the tomatoes were not yet ripe?

2. (11) The theater was full when the horror film began. Seventy-six people left before the movie ended. One hundred twenty-four people remained. How many people were in the theater when it was full?

3. (13) The Pie King restaurant cuts each pie into 6 slices. The restaurant served 84 pies one week. How many slices of pie were served?

4. (1) President Lincoln began his speech, "Four score and seven years ago ..." How many years is four score and seven?

5. (5) (a) Use words to write 18700000.

(b) Write 874 in expanded notation.

6. (4) Use digits and other symbols to write "Three minus seven equals negative four."

7. (16) At what temperatures on the Fahrenheit scale does water freeze and boil?

8. (19) Find the perimeter of this rectangle:

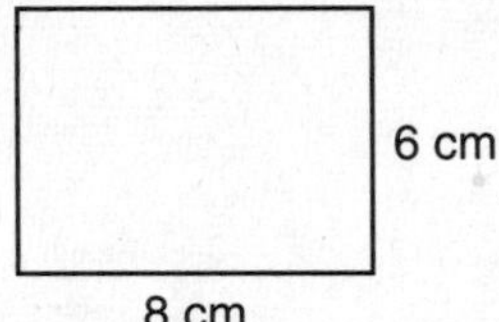

9. (15) Write each number as a reduced fraction or mixed number:

(a) $3\frac{16}{24}$ (b) $\frac{15}{24}$ (c) 4%

10. (15) Find a and b to complete each equivalent fraction:

(a) $\frac{3}{4} = \frac{a}{36}$ (b) $\frac{4}{9} = \frac{b}{36}$

11. (18) Draw a regular pentagon.

12. (18) What is the name of a polygon that has twice as many sides as a quadrilateral?

13. (17) (a) Each angle of a rectangle measures how many degrees?

(b) The four angles of a rectangle total how many degrees?

14. (1, 9) The rule of this sequence is $k = \frac{1}{8}n$. Find the eighth term of the sequence.

$$\frac{1}{8}, \frac{1}{4}, \frac{3}{8}, \frac{1}{2}, \ldots$$

Find the missing number in each equation.

15. (3) $a + 1547 = 8998$

16. (3) $30b = \$41.10$

17. (3) $\$0.32c = \$7.36$

18. (3) $\$26.57 + d = \$30.10$

Simplify:

19. (10) $\frac{2}{3} + \frac{2}{3} + \frac{2}{3}$

20. (15) $3\frac{7}{8} - \frac{5}{8}$

21. (15) $\frac{2}{3} \cdot \frac{3}{7}$

22. (15) $3\frac{7}{8} + \frac{5}{8}$

23. (1) $50 \cdot 50$

24. (1) $\frac{100{,}100}{11}$

25. (9) (a) How many $\frac{1}{2}$'s are in 1?

(b) Use the answer to (a) to find the number of $\frac{1}{2}$'s in 5.

26. (7, 8) Use your ruler to draw $\overline{AB}$ $1\frac{1}{2}$ in. long. Then draw $\overline{BC}$ perpendicular to $\overline{AB}$ 2 in. long. Draw a segment from point A to point C to complete ΔABC. What is the length of $\overline{AC}$?

27. (9, 10) Write $3\frac{1}{3}$ as an improper fraction, and multiply it by the reciprocal of $\frac{2}{3}$.

28. (15) Find a fraction equal to $\frac{1}{2}$ that has a denominator of 10. Subtract this fraction from $\frac{9}{10}$. Write the difference as a reduced fraction.

29. (8, 16) What percent of a yard is a foot?

30. (19) What is the perimeter of this hexagon?

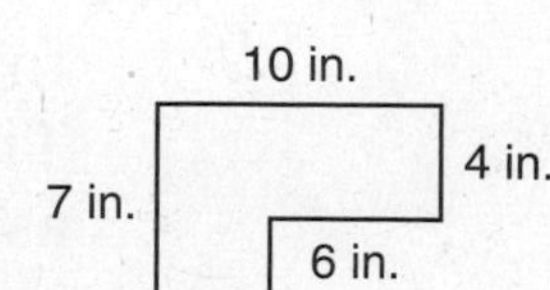

LESSON 20 Exponents • Rectangular Area, Part 1 • Square Root

WARM-UP

Facts Practice: 40 Fractions to Reduce (Test D)

Mental Math:

a. \$4.75 + \$2.50 **b.** 36¢ × 10 **c.** \$5.00 − \$4.32

d. 5 × 43 **e.** 625 − 125 **f.** $\frac{1}{2}$ of $\frac{3}{4}$

g. 10 × 10, − 10, ÷ 10, + 1, − 10, × 10, + 10, ÷ 10

Problem Solving:

Copy this problem and fill in the missing digits:

```
     ___
  8)___
    _
    __
    _
    4_
    _8
     0
```

NEW CONCEPTS

Exponents

We remember that we can show repeated addition by using multiplication.

5 + 5 + 5 + 5 has the same value as 4 × 5

There is also a way to show repeated multiplication. We can show repeated multiplication by using an **exponent.**

$$5 \cdot 5 \cdot 5 \cdot 5 = 5^4$$

In the expression 5^4, the exponent is 4 and the **base** is 5. The exponent shows how many times the base is to be used as a factor.

base → 5^4 ← exponent

The following examples show how we read expressions with exponents, which we call **exponential expressions.**

4^2 "four squared" or "four to the second power"

2^3 "two cubed" or "two to the third power"

5^4 "five to the fourth power"

10^5 "ten to the fifth power"

To find the value of an expression with an exponent, we use the base as a factor the number of times shown by the exponent.

$$5^4 = 5 \cdot 5 \cdot 5 \cdot 5 = 625$$

Example 1 Simplify:

(a) 4^2 (b) 2^3 (c) 10^5 (d) $\left(\frac{2}{3}\right)^2$

Solution (a) $4^2 = 4 \cdot 4 = \mathbf{16}$

(b) $2^3 = 2 \cdot 2 \cdot 2 = \mathbf{8}$

(c) $10^5 = 10 \cdot 10 \cdot 10 \cdot 10 \cdot 10 = \mathbf{100{,}000}$

(d) $\left(\frac{2}{3}\right)^2 = \frac{2}{3} \cdot \frac{2}{3} = \mathbf{\frac{4}{9}}$

Example 2 Simplify: $4^2 - 2^3$

Solution We first find the value of each exponential expression. Then we subtract.

$$4^2 - 2^3$$

$$16 - 8 = \mathbf{8}$$

Example 3 Find the missing number in this equation:

$$2^3 \cdot 2^3 = 2^n$$

Solution We are asked to find a missing exponent. Consider the meaning of each exponent.

$$\underbrace{2 \cdot 2 \cdot 2}_{2^3} \cdot \underbrace{2 \cdot 2 \cdot 2}_{2^3} = 2^n$$

We see that 2 appears as a factor 6 times. So the missing exponent is **6**.

We can use exponents to indicate units that have been multiplied. Recall that when we add or subtract measures with like units, the units do not change.

$$4 \text{ ft} + 8 \text{ ft} = 12 \text{ ft}$$

The units of the addends are the same as the units of the sum.

However, when we multiply or divide measures, the units do change.

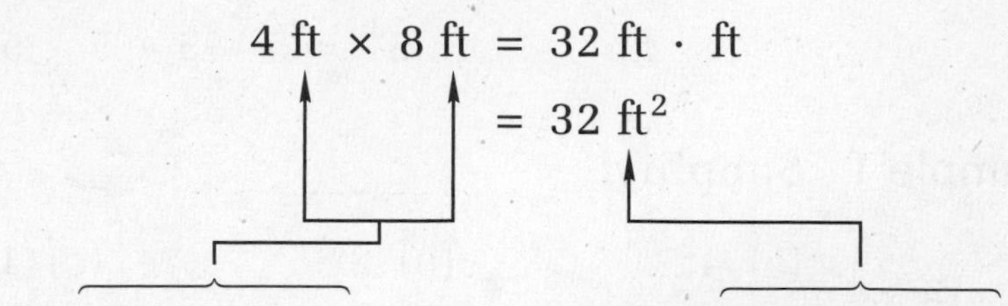

The result of multiplying feet by feet is **square feet,** which we can abbreviate sq. ft or ft^2. Square feet are units used to measure area, as we see in the next section of this lesson.

Rectangular area, part 1

The diagram below represents the floor of a hallway that has been covered with square tiles that are 1 foot on each side. How many 1-ft square tiles does it take to cover the floor of the hallway?

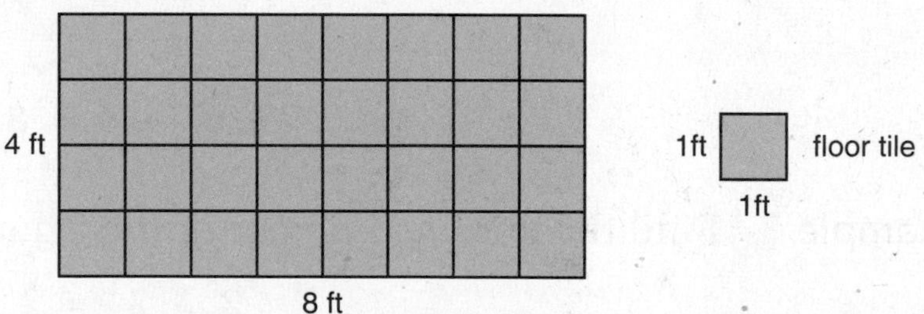

We see that there are 4 rows and 8 floor tiles in each row. So there are 32 1-ft square tiles.

The floor tiles cover the **area** of the hallway. Area is an amount of surface. Floors, ceilings, walls, sheets of paper, and polygons all have areas. If a square is 1 foot on each side, it is a **square foot.** Thus the area of the hallway is 32 square feet. Other standard square units in the U.S. system include square inches, square yards, and square miles.

It is important to distinguish between a unit of length and a unit of area. Units of length, such as inches or feet, are used for measuring distances, not for measuring areas. To measure area, we use units that occupy area. Square inches and square feet occupy area and are used to measure area.

We include the word *square* or the exponent 2 when we designate units of area.

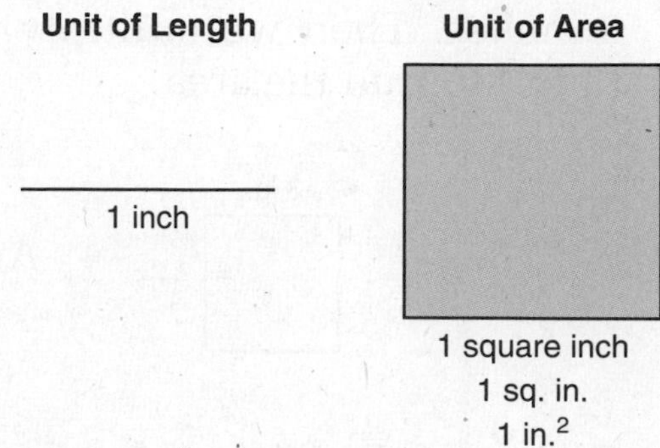

Notice that the area of the rectangular hallway equals the length of the hallway times the width.

Area of a rectangle = length × width

We often abbreviate this formula as

$$A = lw$$

Example 4 What is the area of this rectangle?

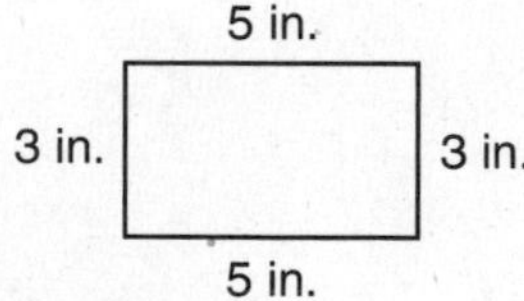

Solution The area of the rectangle is the number of square inches needed to cover the rectangle.

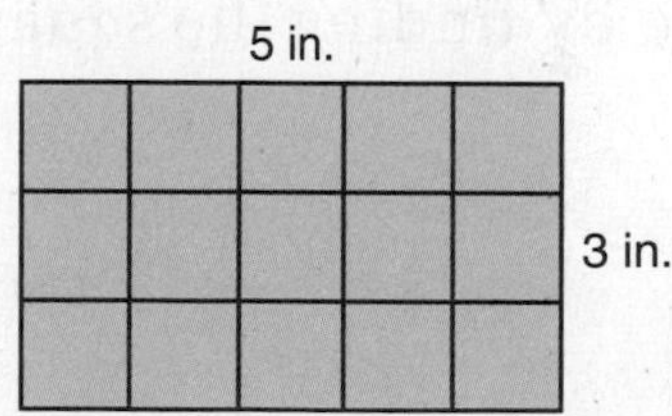

We can find this number by multiplying the length (5 in.) by the width (3 in.).

$$\begin{aligned} \text{Area of rectangle} &= 5 \text{ in.} \cdot 3 \text{ in.} \\ &= \mathbf{15\ in.^2} \end{aligned}$$

Example 5 The perimeter of a certain square is 12 inches. What is the area of the square?

Solution To find the area of the square, we first need to know the length of the sides. The sides of a square are equal in length, so we divide 12 inches by 4 and find that each side is 3 inches. Then we multiply the length (3 in.) by the width (3 in.) to find the area.

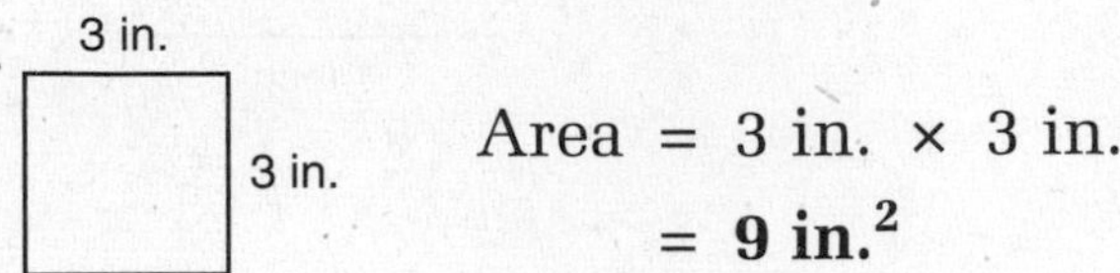

$$\text{Area} = 3 \text{ in.} \times 3 \text{ in.}$$
$$= \mathbf{9 \text{ in.}^2}$$

Example 6 Dickerson Ranch is a level plot of land 4 miles square. The area of Dickerson Ranch is how many square miles?

Solution "Four miles square" does not mean "4 square miles." A plot of land that is 4 miles square is square and has sides 4 miles long. So the area is

$$4 \text{ mi} \times 4 \text{ mi} = \mathbf{16 \text{ mi}^2}$$

Square root The area of a square and the length of its side are related by "squaring." If we know the length of a side of a square, we square the length to find the area.

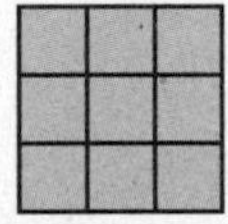

3 units squared is 9 square units.

If we know the area of a square, we can find the length of a side by finding the **square root** of the area.

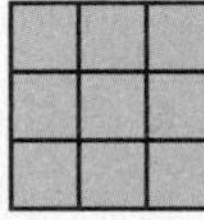

The square root of 9 square units is 3 units.

We often indicate square root with the radical symbol, $\sqrt{\ }$. Here we show "The square root of 9 equals 3."

$$\sqrt{9} = 3$$

Example 7 Simplify:

(a) $\sqrt{121}$ (b) $\sqrt{8^2}$

Solution (a) To find the square root of 121 we may ask, "What number multiplied by itself equals 121?" Since $10 \times 10 = 100$, we try 11×11 and find that $11^2 = 121$. Therefore, $\sqrt{121}$ equals **11.**

(b) Squaring and finding a square root are inverse operations, so one operation "undoes" the other operation.

$$\sqrt{8^2} = \sqrt{64} = \mathbf{8}$$

LESSON PRACTICE

Practice set* Use words to show how each exponential expression is read. Then find the value of each expression.

a. 4^3

b. $\left(\frac{1}{2}\right)^2$

c. 10^6

d. In the expression 10^3, what number is the base and what number is the exponent?

Find each missing exponent:

e. $2^3 \cdot 2^2 = 2^n$ **f.** $\dfrac{2^6}{2^2} = 2^m$

Find each square root:

g. $\sqrt{100}$ **h.** $\sqrt{400}$ **i.** $\sqrt{15^2}$

Find the area of each rectangle:

j. **k.**

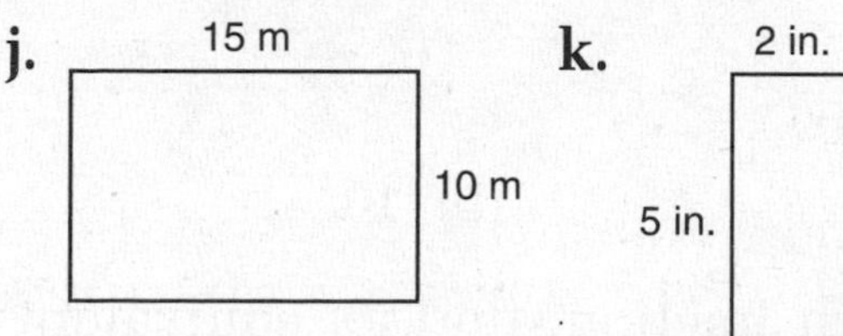

l.

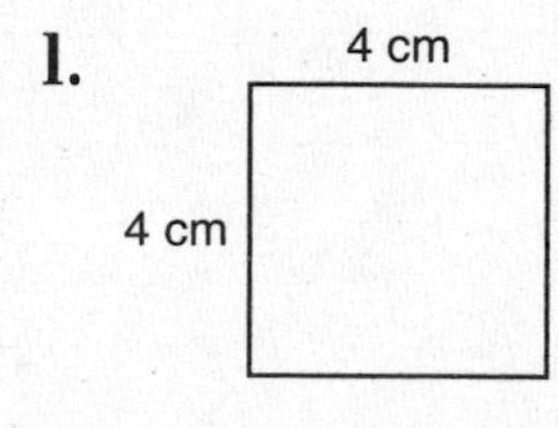

m. If the perimeter of a square is 20 cm, what is its area?

n. What is the area of a park that is 100 yards square?

MIXED PRACTICE

Problem set

1. (13) There were 628 tenants in 4 apartment buildings. Each building housed the same number of tenants. How many tenants were housed in each apartment building?

2. (11) Thirty-six bright green parrots flew away while 46 parrots remained in the tree. How many parrots were in the tree before the 36 parrots flew away?

3. (14) Two hundred twenty-five of the six hundred fish in the lake were trout. How many of the fish were not trout?

4. (11) Twenty-one thousand, fifty swarmed in through the front door. Forty-eight thousand, nine hundred seventy-two swarmed in through the back door. How many swarmed in through both doors?

5. (2, 20) The rule of the following sequence is $k = 2^n$. Find the sixth term of the sequence.

$$2, 4, 8, 16, \ldots$$

6. (4, 8) (a) Arrange these numbers in order from least to greatest:

$$\frac{1}{3}, -2, 1, -\frac{1}{2}, 0$$

(b) Which of these numbers are not integers?

7. (8) Which is the best estimate of how much of this rectangle is shaded?

A. 50% B. $33\frac{1}{3}$%
C. 25% D. 60%

8. (7) Each angle of a rectangle is a right angle. Which two sides are perpendicular to side *BC*?

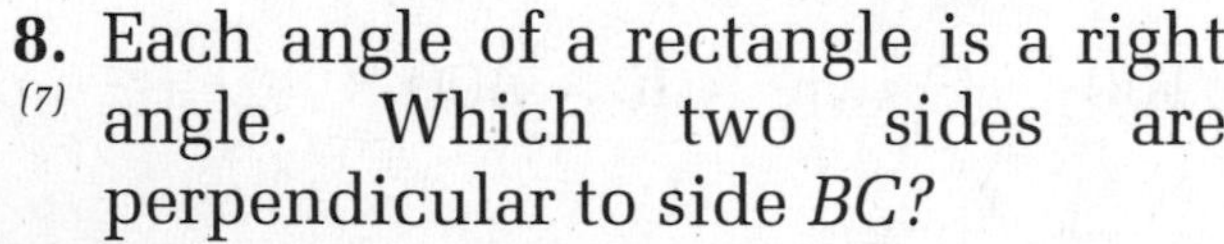

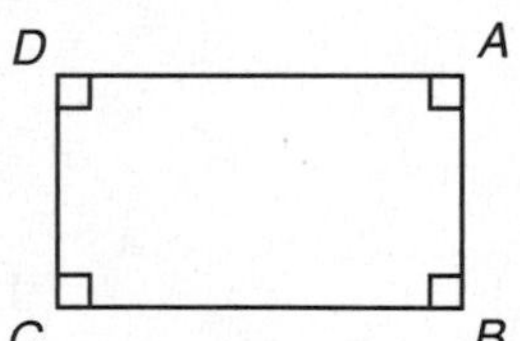

9. (20) Simplify:

(a) $\left(\frac{1}{3}\right)^3$ (b) 10^4 (c) $\sqrt{12^2}$

10. (15) For each fraction, find an equivalent fraction that has a denominator of 36:

(a) $\frac{2}{9}$ (b) $\frac{3}{4}$

11. (6) List the factors of each number:

(a) 10 (b) 7 (c) 1

12. The perimeter of a certain square is 2 feet. How many inches long is each side of the square?
(16, 19)

Solve each equation:

13. $36 + a = 54$
(3)

14. $46 - w = 20$
(3)

15. $5x = 60$
(3)

16. $100 = m + 64$
(3)

17. $5^4 \cdot 5^2 = 5^n$
(20)

18. $\frac{60}{y} = 4$
(3)

Simplify:

19. $1\frac{8}{9} + 1\frac{7}{9}$
(10, 15)

20. $\frac{5}{2} \cdot \frac{5}{6}$
(10)

21. $\frac{6345}{9}$
(1)

22. 360×25
(1)

23. $\frac{3}{4} - \left(\frac{1}{4} + \frac{2}{4}\right)$
(9)

24. $\left(\frac{3}{4} - \frac{1}{4}\right) + \frac{2}{4}$
(10)

25. Evaluate the following expressions for $m = 3$ and $n = 10$:
(1, 9)

(a) $\frac{m}{n} + \frac{m}{n}$

(b) $\frac{m}{n} \cdot \frac{m}{n}$

26. Find a fraction equivalent to $\frac{1}{2}$ that has a denominator of 10. Add $\frac{3}{10}$ to that fraction and reduce the sum.
(15)

27. Write $1\frac{4}{5}$ as an improper fraction. Then multiply the improper fraction by $\frac{1}{3}$ and reduce the product.
(10, 15)

28. Which property is illustrated by this equation?
(2, 15)

$$\frac{1}{3} \cdot \frac{2}{2} = \frac{2}{6}$$

29. A common floor tile is 12 inches square.
(19, 20)

(a) What is the perimeter of a common floor tile?

(b) What is the area of a common floor tile?

30. What is the perimeter of this hexagon?
(19)

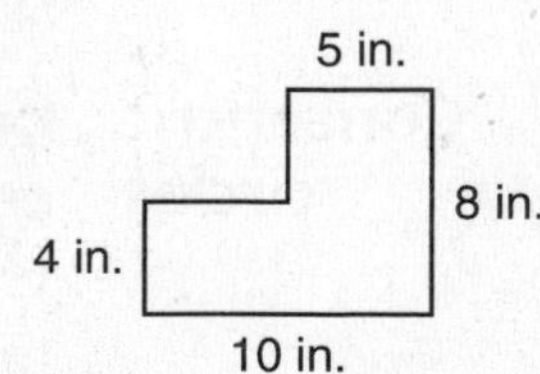

INVESTIGATION 2

Focus on

Using a Compass and Straightedge, Part 1

Materials needed:

- Compass
- Ruler or straightedge
- Protractor

A **compass** is a tool used to draw **circles** and portions of circles called **arcs.** Compasses are manufactured in various forms. Here we show two forms:

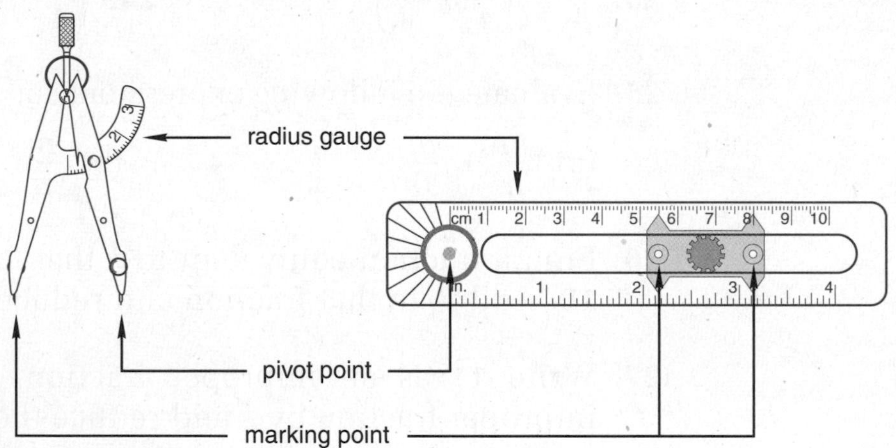

The **marking point** of a compass is the pencil point that draws circles and arcs. The marking point rotates around the **pivot point,** which is placed at the **center** of the desired circle or arc. The **radius** (plural, **radii**) of the circle, which is the distance from every point on the circle to the center of the circle, is set by the **radius gauge.** The radius gauge identifies the distance between the pivot point and the marking point of the compass.

Concentric circles

Concentric circles are two or more circles with a common center. When a pebble is dropped into a quiet pool of water, waves forming concentric circles can be seen. A bull's-eye target is another example of concentric circles.

To draw concentric circles with a compass, we begin by swinging the compass a full turn to make one circle. Then we make additional circles using the same center, changing the radius for each new circle.

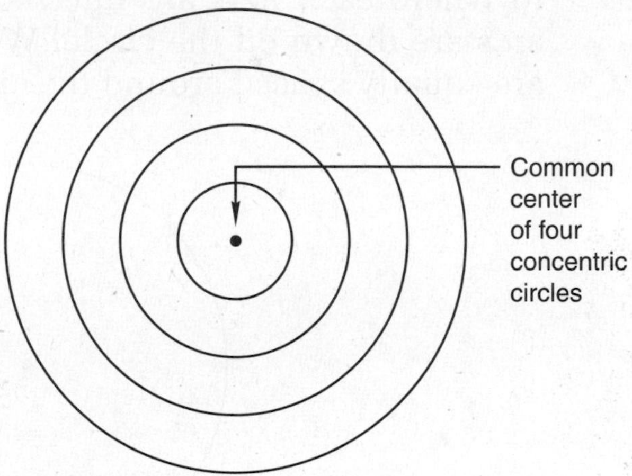

1. Practice drawing several concentric circles.

Regular hexagon and regular triangle

Recall that all the sides of a regular polygon are equal in length and all the angles are equal in measure. Due to their uniform shape, regular polygons can be **inscribed** in circles. A polygon is inscribed in a circle if all of its vertices are on the circle and all of the other points of the polygon are within the circle. We will inscribe a regular hexagon and a regular triangle.

First we fix the compass at a comfortable setting that will not change until the project is finished. We swing the compass a full turn to make a circle. Then we lift the compass without changing the radius and place the pivot point anywhere on the circle. With the pivot point on the circle, we swing a small arc that intersects the circle, as shown below.

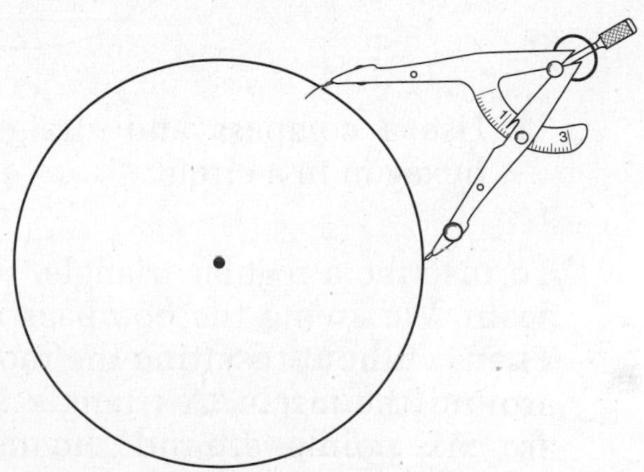

Again we lift the compass without changing the radius and place the pivot point at the point where the arc intersects the circle. From this location we swing another small arc that intersects the circle. We continue by moving the pivot point to where each new arc intersects the circle, until six small arcs are drawn on the circle. We find that the six small arcs are equally spaced around the circle.

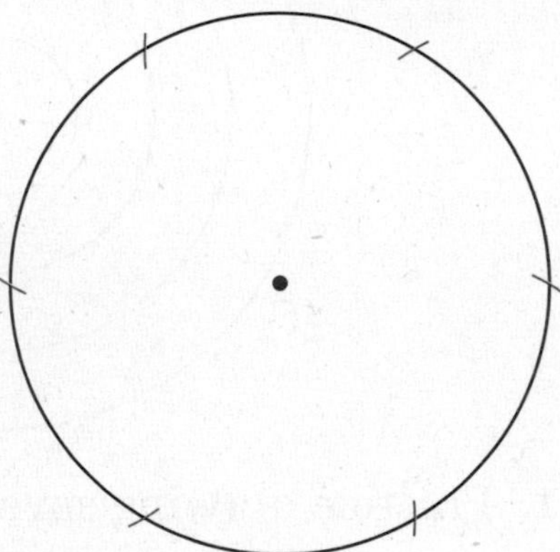

Now, to inscribe a regular hexagon, we draw line segments connecting each point where an arc intersects the circle to the next point where an arc intersects the circle.

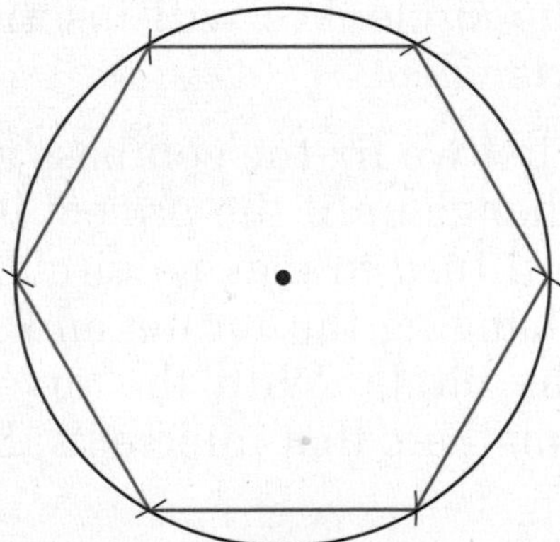

2. Use a compass and straightedge to inscribe a regular hexagon in a circle.

To inscribe a regular triangle, we will start the process over again. We swing the compass a full turn to make a circle. Then, without resetting the radius, we swing six small arcs around the circle. A triangle has three vertices, but there are six points around the circle where the small arcs intersect the circle. Therefore, to inscribe a regular triangle, we draw segments between *every other* point of intersection. In other words, we skip one point of intersection for each side of the triangle.

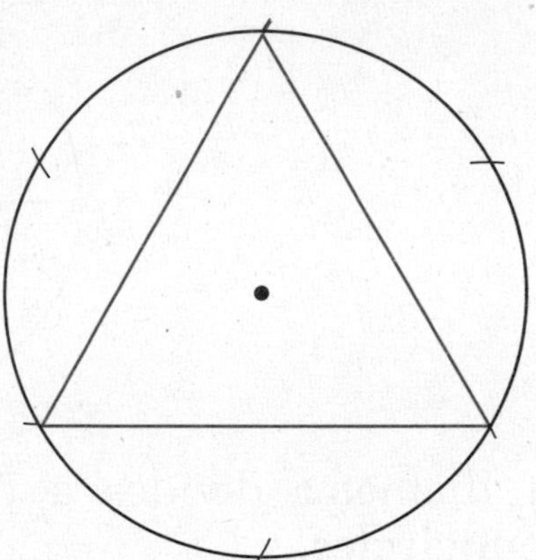

3. Use your tools to inscribe a regular triangle in a circle.

With a protractor we can measure each angle of the triangle. Since the vertex of each angle is on the circle and the angle opens to the interior of the circle, the angle is called an **inscribed angle.**

4. What is the measure of each inscribed angle? (If necessary, extend the rays of each angle to perform the measurements.)

5. What is the sum of the measures of all three angles of the triangle?

6. What shape will we make if we now draw segments between the remaining three points of intersection?

Dividing a circle into sectors We can use a compass and straightedge to divide a circle into equal parts. First we swing the compass a full turn to make a circle. Next we draw a segment across the circle through the center of the circle. A segment with both endpoints on a circle is a **chord.** The longest chord of a circle passes through the center and is called a **diameter** of the circle. Notice that a diameter equals two radii. Thus the length of a diameter of a circle is twice the length of a radius of the circle. The **circumference** is the distance around the circle and is determined by the length of the radius and diameter, as we will see in a later lesson.

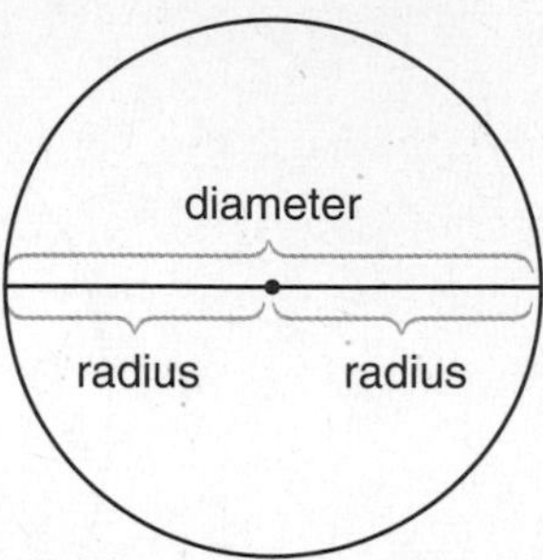

A diameter divides a circle into two half circles called **semicircles.**

To divide a circle into thirds, we begin with the process we used to inscribe a hexagon. We draw a circle and swing six small arcs. Then we draw three segments from the center of the circle to *every other* point where an arc intersects the circle. These segments divide the circle into three congruent **sectors.** A sector of a circle is a region bounded by an arc of the circle and two of its radii. A model of a sector is a slice of pie.

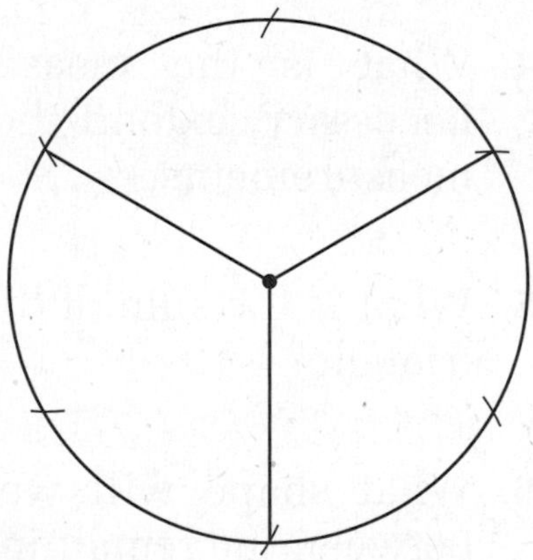

7. Use a compass and straightedge to draw a circle and to divide the circle into thirds.

The segments we drew from the center to the circle formed angles. Each angle that has its vertex at the center of the circle is a **central angle.** We can measure a central angle with a protractor. We may extend the rays of the central angle if necessary in order to use the protractor.

8. What is the measure of each central angle of a circle divided into thirds?

9. Each sector of a circle divided into thirds occupies what percent of the area of the whole circle?

To divide a circle into sixths, we again begin with the process we used to inscribe a hexagon. We divide the circle by drawing a segment from the center of the circle to the point of intersection of each small arc.

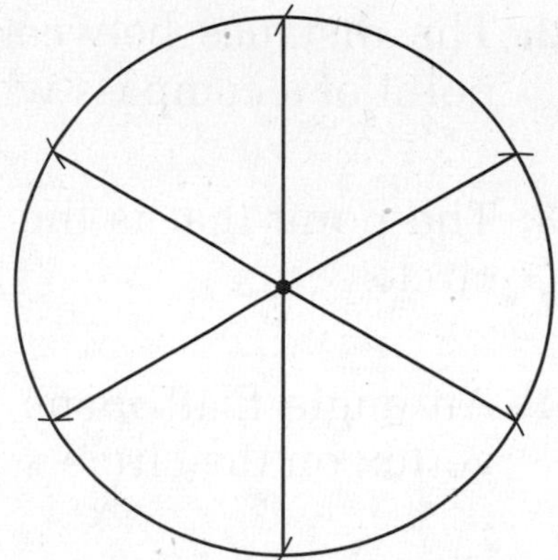

10. What is the measure of each central angle of a circle divided into sixths?

11. Each sector of a circle divided into sixths occupies what percent of the area of the whole circle?

In problems 12–24 we provide definitions of terms presented in this investigation. Find the term for each definition:

12. The distance around a circle

13. The distance across a circle through its center

14. The distance from the center of a circle to every point on the circle

15. Part of the circumference of a circle

16. A region bounded by an arc of a circle and two radii

17. Two or more circles with the same center

18. A segment that passes through the interior of a circle and has both endpoints on the circle

19. A polygon whose vertices are on a circle and whose other points are inside the circle

20. A half circle

21. An angle whose vertex is the center of a circle

22. The distance between the pivot point and the marking point of a compass when drawing a circle

23. The point that is the same distance from any point on a circle

24. An angle that opens to the interior of the circle from a vertex on the circle

The following paragraphs summarize important facts about circles. A copy of this summary is available as Facts Practice Test E in *Saxon Math 8/7—Homeschool Tests and Worksheets.*

The distance around a circle is its circumference. Every point on the circle is the same distance from the center of the circle. The distance from the center to a point on the circle is the radius. The distance across the circle through its center is the diameter, which equals two radii. A diameter divides a circle into two half circles called semicircles. A diameter, as well as any other segment between two points on a circle, is a chord of the circle. Two or more circles with the same center are concentric circles.

An angle formed by two radii of a circle is called a central angle. A central angle opens to a portion of a circle called an arc, which is part of the circumference of a circle. The region enclosed by an arc and its central angle is called a sector.

An angle whose vertex is on the circumference of a circle and whose sides are chords of the circle is an inscribed angle. A polygon is inscribed in a circle if all of its vertices are on the circumference of the circle.

LESSON

21 Prime and Composite Numbers • Prime Factorization

WARM-UP

Facts Practice: Circles (Test E)

Mental Math:

a. \$1.25 + 99¢ **b.** \$6.50 ÷ 10 **c.** \$20.00 − \$15.75

d. 6 × 34 **e.** $1\frac{2}{3} + 2\frac{1}{3}$ **f.** $\frac{1}{3}$ of 36

g. Start with the number of sides of a hexagon, × 5, + 2, ÷ 8, + 1, ÷ 5.

Problem Solving:

If Sam can read 20 pages in 30 minutes, how many hours will it take Sam to read 200 pages?

NEW CONCEPTS

Prime and composite numbers

We remember that the counting numbers (or natural numbers) are the numbers we use to count. They are

1, 2, 3, 4, 5, 6, 7, 8, 9, 10, ...

Counting numbers greater than 1 are either **prime numbers** or **composite numbers.** A prime number has exactly two different factors, and a composite number has three or more factors. In the following table we list the factors of the first ten counting numbers. The numbers 2, 3, 5, and 7 each have exactly two factors, so they are prime numbers.

Factors of Counting Numbers 1–10

Number	Factors
1	1
2	1, 2
3	1, 3
4	1, 2, 4
5	1, 5
6	1, 2, 3, 6
7	1, 7
8	1, 2, 4, 8
9	1, 3, 9
10	1, 2, 5, 10

We see that the factors of each of the prime numbers are 1 and the number itself. So we define a prime number as follows:

> A **prime number** is a counting number greater than 1 whose only factors are 1 and the number itself.

From the table we can also see that 4, 6, 8, 9, and 10 each have three or more factors, so they are composite numbers. Each composite number is divisible by a number other than 1 and itself.

Note: Because the number 1 has only one factor, itself, it is neither a prime number nor a composite number.

Example 1 Make a list of the prime numbers that are less than 16.

Solution First we list the counting numbers from 1 to 15.

1, 2, 3, 4, 5, 6, 7, 8, 9, 10, 11, 12, 13, 14, 15

A prime number must be greater than 1, so we cross out 1. The next number, 2, has only two divisors (factors), so 2 is a prime number. However, all the even numbers greater than 2 are divisible by 2, so they are not prime. We cross these out.

~~1~~, 2, 3, ~~4~~, 5, ~~6~~, 7, ~~8~~, 9, ~~10~~, 11, ~~12~~, 13, ~~14~~, 15

The numbers that are left are

2, 3, 5, 7, 9, 11, 13, 15

The numbers 9 and 15 are divisible by 3, so we cross them out.

2, 3, 5, 7, ~~9~~, 11, 13, ~~15~~

The only divisors of each remaining number are 1 and the number itself. So the prime numbers less than 16 are **2, 3, 5, 7, 11,** and **13.**

Example 2 List the composite numbers between 40 and 50.

Solution First we write the counting numbers between 40 and 50.

41, 42, 43, 44, 45, 46, 47, 48, 49

Any number that is divisible by a number besides 1 and itself is composite. All the even numbers in this list are composite since they are divisible by 2. That leaves the odd numbers to consider. We quickly see that 45 is divisible by 5, and 49 is divisible by 7. So both 45 and 49 are composite.

The remaining numbers, 41, 43, and 47, are prime. So the composite numbers between 40 and 50 are **42, 44, 45, 46, 48,** and **49.**

Prime factorization Every composite number can be *composed* (formed) by multiplying two or more prime numbers. Here we show each of the first nine composite numbers written as a product of prime number factors.

$4 = 2 \cdot 2$	$6 = 2 \cdot 3$	$8 = 2 \cdot 2 \cdot 2$
$9 = 3 \cdot 3$	$10 = 2 \cdot 5$	$12 = 2 \cdot 2 \cdot 3$
$14 = 2 \cdot 7$	$15 = 3 \cdot 5$	$16 = 2 \cdot 2 \cdot 2 \cdot 2$

Notice that we **factor** 8 as $2 \cdot 2 \cdot 2$ and not $2 \cdot 4$, because 4 is not prime.

When we write a composite number as a product of prime numbers, we are writing the **prime factorization** of the number.

Example 3 Write the prime factorization of each number.

(a) 30 (b) 81 (c) 420

Solution We will write each number as the product of two or more prime numbers.

(a) $30 = \mathbf{2 \cdot 3 \cdot 5}$ We do not use $5 \cdot 6$ or $3 \cdot 10$, because neither 6 nor 10 is prime.

(b) $81 = \mathbf{3 \cdot 3 \cdot 3 \cdot 3}$ We do not use $9 \cdot 9$, because 9 is not prime.

(c) $420 = \mathbf{2 \cdot 2 \cdot 3 \cdot 5 \cdot 7}$ Two methods for finding this are shown after example 4.

Example 4 Write the prime factorization of 100 and of $\sqrt{100}$.

Solution The prime factorization of 100 is $\mathbf{2 \cdot 2 \cdot 5 \cdot 5}$. We find that $\sqrt{100}$ is 10, and the prime factorization of 10 is $\mathbf{2 \cdot 5}$. Notice that 100 and $\sqrt{100}$ have the same prime factors, 2 and 5, but that each factor appears half as often in the prime factorization of $\sqrt{100}$.

There are two commonly used methods for factoring composite numbers. One method uses a factor tree. The other method uses division by primes. We will factor 420 using both methods.

To factor a number using a **factor tree,** we first write the number. Below the number we write any two whole numbers greater than 1 that multiply to equal the number. If these numbers are not prime, we continue the process until there is a prime number at the end of each "branch" of the factor tree. These numbers are the prime factors of the original number. We write them in order from least to greatest.

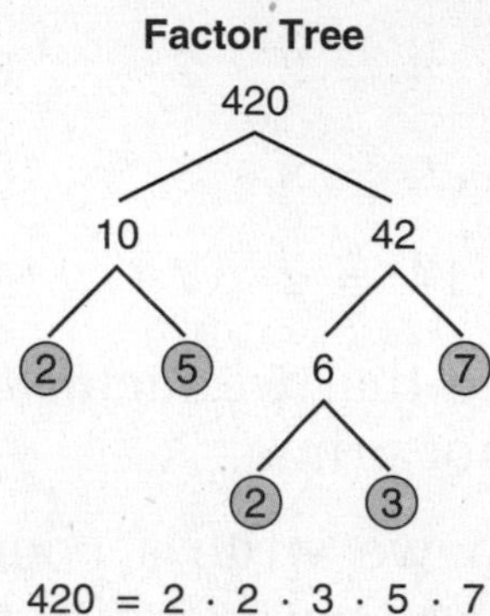

To factor a number using **division by primes,** we write the number in a division box and divide by the smallest prime number that is a factor. Then we divide the resulting quotient by the smallest prime number that is a factor. We repeat this process until the quotient is 1.[†] The divisors are the prime factors of the number.

Division by Primes

```
      1
   7)7
  5)35
 3)105
 2)210
 2)420
```

$420 = 2 \cdot 2 \cdot 3 \cdot 5 \cdot 7$

LESSON PRACTICE

Practice set* **a.** List the first ten prime numbers.

b. If a whole number greater than 1 is not prime, then what kind of number is it?

c. Write the prime factorization of 81 using a factor tree.

[†] Some people prefer to divide until the quotient is a prime number. In this case, the final quotient is included in the list of prime factors.

d. Write the prime factorization of 360 using division by primes.

e. Write the prime factorization of 64 and of $\sqrt{64}$.

MIXED PRACTICE

Problem set

1. (14) Two thirds of the people wore green on St. Patrick's Day. What fraction of the people did not wear green on St. Patrick's Day?

2. (13) Three hundred forty-three quills were carefully placed into 7 compartments. If each compartment held the same number of quills, how many quills were in each compartment?

3. (5, 12) How much less than 2 billion is 21 million? Use words to write the answer.

4. (11) Last year the price was $14,289. This year the price increased $824. What is the price this year?

5. (15) Write each number as a reduced fraction or mixed number:

(a) $3\frac{12}{21}$ (b) $\frac{12}{48}$ (c) 12%

6. (21) List the prime numbers between 50 and 60.

7. (21) Write the prime factorization of each number:

(a) 50 (b) 60 (c) 300

8. (4) Which point could represent 1610 on this number line? How did you decide?

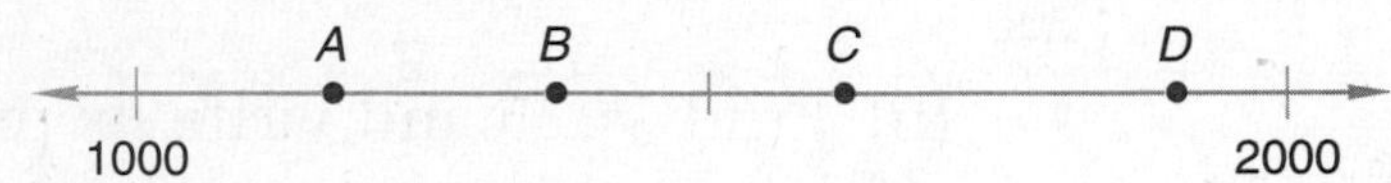

9. (15) Complete each equivalent fraction:

(a) $\frac{2}{3} = \frac{?}{15}$ (b) $\frac{3}{5} = \frac{?}{15}$ (c) $\frac{?}{3} = \frac{8}{12}$

10. (9) (a) How many $\frac{1}{3}$'s are in 1?

(b) How many $\frac{1}{3}$'s are in 3?

11. (20) The perimeter of a regular quadrilateral is 12 inches. What is its area?

12. (8, 19) Use a ruler to draw a rectangle that is $\frac{3}{4}$ in. wide and twice as long as it is wide.

(a) How long is the rectangle?

(b) What is the perimeter of the rectangle?

13. (19) Find the perimeter of this hexagon:

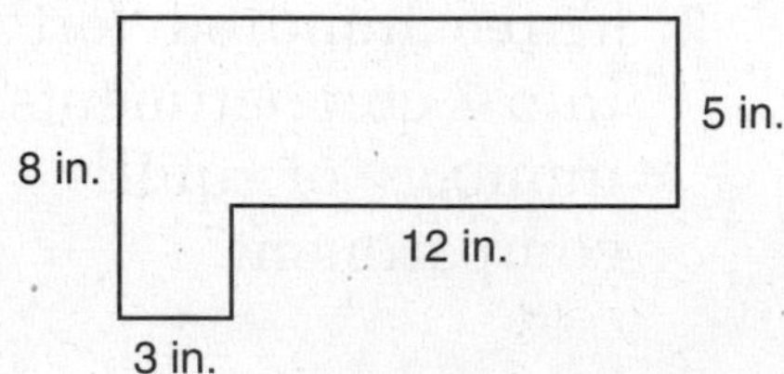

14. (18) Draw a pentagon.

Solve:

15. (9) $p + \frac{3}{5} = 1$

16. (9) $\frac{3}{5}q = 1$

17. (3) $\frac{w}{25} = 50$

18. (9, 15) $\frac{1}{6} + f = \frac{5}{6}$

19. (10) $m - 3\frac{2}{3} = 1\frac{2}{3}$

20. (3) $51 = 3c$

Simplify:

21. (9) $\frac{2}{3} + \frac{2}{3} + \frac{2}{3}$

22. (20) $\left(\frac{2}{3}\right)^3$

23. (21) (a) Write the prime factorization of 225.

(b) Find $\sqrt{225}$ and write its prime factorization.

24. (15) Describe how finding the greatest common factor of the numerator and denominator of a fraction can help reduce the fraction.

25. (17) Draw $\overline{AB}$ $2\frac{1}{2}$ inches long. Then draw $\overline{BC}$ $2\frac{1}{2}$ inches long perpendicular to $\overline{AB}$. Complete the triangle by drawing $\overline{AC}$. Use a protractor to find the measure of $\angle A$.

26. Write $1\frac{3}{4}$ as an improper fraction. Multiply the improper fraction by the reciprocal of $\frac{2}{3}$. Then write the product as a mixed number.
(9, 10)

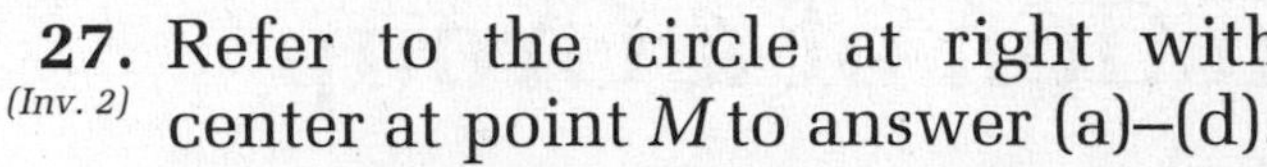

27. Refer to the circle at right with center at point M to answer (a)–(d).
(Inv. 2)

(a) Which segment is a diameter?

(b) Which segment is a chord but not a diameter?

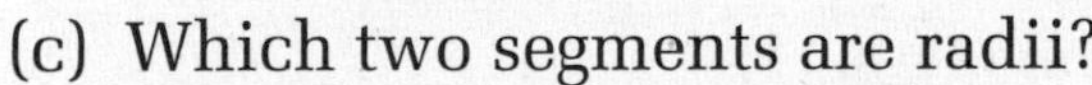

(c) Which two segments are radii?

(d) Which angle is an inscribed angle?

28. A quart is what percent of a gallon?
(16)

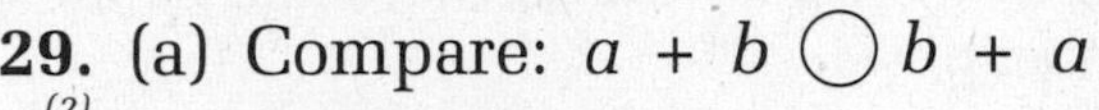

29. (a) Compare: $a + b \bigcirc b + a$
(2)

(b) What property of operations applies to part (a) of this problem?

30. Refer to the triangles below to answer (a)–(c).
(18)

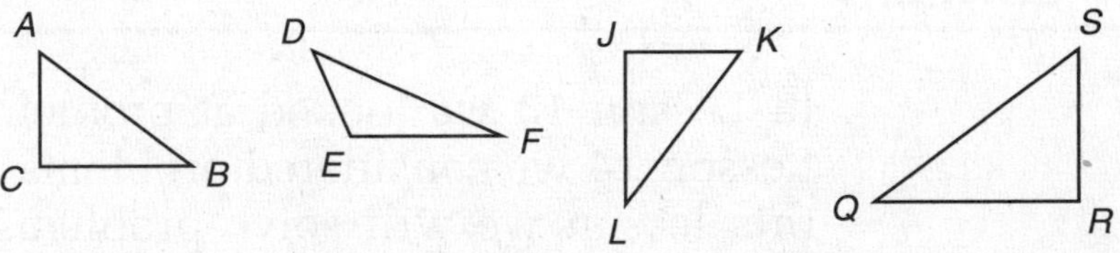

(a) Which triangle appears to be congruent to $\triangle ABC$?

(b) Which triangle is not similar to $\triangle ABC$?

(c) Which angle in $\triangle QRS$ corresponds to $\angle A$ in $\triangle ABC$?

LESSON

22 Problems About a Fraction of a Group

WARM-UP

Facts Practice: Circles (Test E)

Mental Math:

a. \$1.54 + 99¢ b. 8¢ × 100 c. \$10.00 − \$7.89

d. 7 × 53 e. $3\frac{3}{4} + 1\frac{1}{4}$ f. $\frac{1}{4}$ of 24

g. Start with the number of years in half a century. Add the number of inches in half a foot; then divide by the number of days in a week. What is the name of the polygon with this number of sides?

Problem Solving:

How many inches longer than $2\frac{2}{3}$ feet is $2\frac{2}{3}$ yards?

NEW CONCEPT

In Lesson 13 we looked at problems about equal groups. In Lesson 14 we considered problems about parts of a whole. In this lesson we will solve problems that involve both equal groups and parts of a whole. Many of the problems will require two or more steps to solve.

Consider the following statement:

Two thirds of the fans wore green to the game.

We can diagram this statement. We use a rectangle to represent all the fans at the game. Next we divide the rectangle into three equal parts. Then we describe the parts.

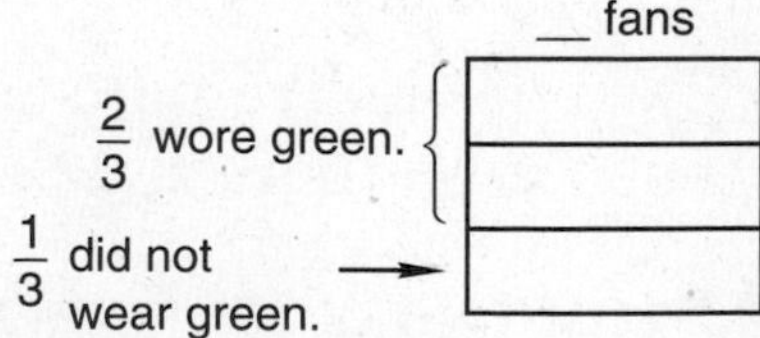

If we know how many fans are at the game, we can figure out how many fans are in each part.

Two thirds of the 270 fans wore green to the game.

There are 270 fans in all. If we divide 270 fans into three equal parts, there will be 90 fans in each part. We write these numbers on our diagram.

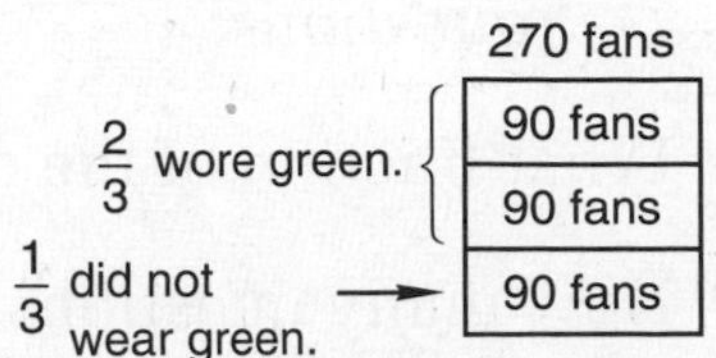

Since $\frac{2}{3}$ of the fans wore green, we add two of the parts and find that 180 fans wore green. Since $\frac{1}{3}$ of the fans did not wear green, we find that 90 fans did not wear green.

Example 1 Diagram this statement. Then answer the questions that follow.

Two fifths of the 30 singers in the choir are boys.

(a) How many boys are in the choir?

(b) How many girls are in the choir?

Solution We draw a rectangle to represent all 30 singers. Since the statement uses fifths to describe a part of the choir, we divide the 30 singers into five equal parts. Since 30 ÷ 5 is 6, there are 6 singers in each part.

30 singers

$\frac{2}{5}$ are boys. { 6 singers, 6 singers

$\frac{3}{5}$ are girls. { 6 singers, 6 singers, 6 singers

Now we can answer the questions.

(a) Two of the five parts are boys. Since there are 6 singers in each part, there are **12 boys.**

(b) Since two of the five parts are boys, three of the five parts must be girls. Thus there are **18 girls.**

Another way to find the answer to (b) after finding the answer to (a) is to subtract. Since 12 of the 30 singers are boys, the rest of the singers (30 − 12 = 18) are girls.

Example 2 In the following statement, change the percent to a fraction. Then diagram the statement and answer the questions.

Britt correctly answered 80% of the 40 questions.

(a) What fraction of the questions did Britt answer correctly?

(b) How many questions did Britt answer correctly?

Solution This problem is about a fraction of a group, but the fraction is disguised as a percent. We write 80% as 80 over 100 and reduce.

$$\frac{80}{100} \div \frac{20}{20} = \frac{4}{5}$$

So 80% is equivalent to the fraction $\frac{4}{5}$.

Now we draw a rectangle to represent all 40 questions, dividing the rectangle into five equal parts. Since 40 ÷ 5 is 8, there are 8 questions in each part.

	40 questions
$\frac{1}{5}$ are incorrect.	8 questions
$\frac{4}{5}$ are correct.	8 questions
	8 questions
	8 questions
	8 questions

Now we can answer the questions.

(a) Britt answered $\frac{4}{5}$ **of the questions** correctly.

(b) Britt correctly answered 4 × 8 questions, which is **32 questions.**

LESSON PRACTICE

Practice set Diagram each statement. Then answer the questions that follow.

First statement:

Three fourths of the 60 pumpkins were ripe.

a. How many pumpkins were ripe?

b. How many pumpkins were not ripe?

Second statement:

Sixty percent of the 20 tomatoes were green.

c. What fraction of the tomatoes were not green?

d. How many tomatoes were green?

e. For the following statement, write and answer two questions:

Three fifths of the thirty bicycles were blue.

MIXED PRACTICE

Problem set

1. (11) There are 28 books on the top shelf. There are 30 books on the middle shelf. There are 23 books on the bottom shelf. How many books are on all three shelves?

2. (13) If all the books in problem 1 were equally divided among the three shelves, how many books would be on each shelf?

3. (11) One hundred twenty-six thousand scurried through the colony before the edentate attacked. Afterward only seventy-nine thousand remained. How many were lost when the edentate attacked?

4. (5, 12) Two thousand, seven hundred is how much less than ten thousand, three hundred thirteen? Use words to write the answer.

5. (22) Diagram this statement. Then answer the questions that follow.

Five ninths of the 36 spectators were happy with the outcome.

(a) How many spectators were happy with the outcome?

(b) How many spectators were not happy with the outcome?

6. (22) In the following statement, change the percent to a reduced fraction. Then diagram the statement and answer the questions.

Twenty-five percent of the three dozen eggs were cracked.

(a) What fraction of the eggs were not cracked?

(b) How many eggs were not cracked?

7. (15) (a) What fraction of the rectangle is shaded?

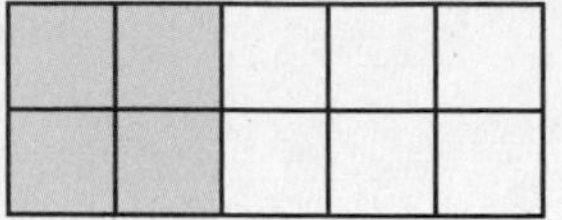

(b) What percent of the rectangle is not shaded?

8. (9) (a) How many $\frac{1}{4}$'s are in 1?

(b) Use the answer to part (a) to find the number of $\frac{1}{4}$'s in 3.

9. (2) (a) Multiply: $6 \cdot 5 \cdot 4 \cdot 3 \cdot 2 \cdot 1 \cdot 0$

(b) What property is illustrated by the multiplication in part (a)?

10. (9) Simplify and compare: $\frac{3}{3} - \left(\frac{1}{3} \cdot \frac{3}{1}\right) \bigcirc \left(\frac{3}{3} - \frac{1}{3}\right) \cdot \frac{3}{1}$

11. (19, 20) Draw a rectangle *ABCD* so that *AB* is 2 in. and *BC* is 1 in.

(a) What is the perimeter of rectangle *ABCD?*

(b) What is the area of the rectangle?

(c) What is the sum of the measures of all four angles of the rectangle?

12. (21) Write the prime factorization of each number:

(a) 32 (b) 900 (c) $\sqrt{900}$

13. (15) For each fraction, write an equivalent fraction that has a denominator of 60.

(a) $\frac{5}{6}$ (b) $\frac{3}{5}$ (c) $\frac{7}{12}$

14. (10) Add the three fractions with denominators of 60 from problem 13, and write their sum as a mixed number.

15. (4, 10) (a) Arrange these numbers in order from least to greatest:

$$0, -\frac{2}{3}, 1, \frac{3}{2}, -2$$

(b) Which of these numbers are positive?

Find the missing number in each equation:

16. (9, 15) $\frac{5}{12} + a = \frac{11}{12}$ **17.** (3) $\frac{900}{c} = 90$ **18.** (3) $121 = 11x$

19. (10) $2\frac{2}{3} = y - 1\frac{1}{3}$ **20.** (20) $10^2 \cdot 10^5 = 10^n$

Simplify:

21. $\frac{5}{6} + \frac{5}{6} + \frac{5}{6}$
(15)

22. $\frac{15}{2} \cdot \frac{10}{3}$
(10)

23. $\left(\frac{5}{6}\right)^2$
(20)

24. $\sqrt{30^2}$
(20)

25. How many $\frac{5}{9}$'s are in 1?
(9)

26. Write $1\frac{1}{2}$ and $1\frac{2}{3}$ as improper fractions. Then multiply the improper fractions, and write the product as a mixed number.
(10, 15)

27. A package that weighs 1 lb 5 oz weighs how many ounces?
(16)

28. Use a protractor to draw a 45° angle.
(17)

29. Find the next number in this sequence:
(2, 9)

$$\ldots, 100, 10, 1, \frac{1}{10}, \ldots$$

30. Write an odd negative integer greater than –3.
(4)

LESSON

23 Subtracting Mixed Numbers with Regrouping

WARM-UP

Facts Practice: Circles (Test E)

Mental Math:

a. \$3.65 + 98¢ b. \$25.00 ÷ 100 c. 449 − 500

d. 8 × 62 e. $1\frac{1}{2} + 2\frac{1}{2}$ f. $\frac{1}{2}$ of 76

g. 8 × 8, − 1, ÷ 9, × 4, − 1, ÷ 3, × 2, + 2, ÷ 4

Problem Solving:

There are two routes Imani can take to the park. There are three routes Samantha can take to the park. If Imani is going from her house to the park and then on to Samantha's house, how many different routes is it possible for Imani to take? Draw a diagram that illustrates the problem.

NEW CONCEPT

In this lesson we will practice subtracting mixed numbers that require regrouping. Regrouping that involves fractions differs from regrouping with whole numbers. When regrouping with whole numbers, we know that each unit equals ten of the next-smaller unit. However, when regrouping from a whole number to a fraction, we need to focus on the denominator of the fraction to determine how to regroup. We will use illustrations to help explain the process.

Example 1 There are $3\frac{1}{5}$ pies on the shelf. If the baker takes away $1\frac{2}{5}$ pies, how many pies will be on the shelf?

Solution To answer this question, we subtract $1\frac{2}{5}$ from $3\frac{1}{5}$. Before we subtract, however, we will draw a picture to see how the baker solves the problem.

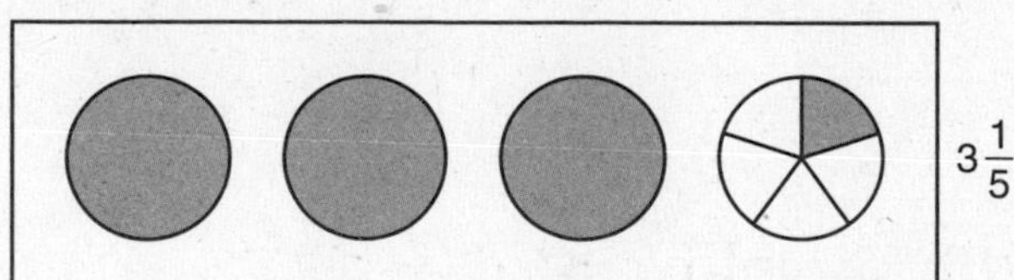

In order for the baker to remove $1\frac{2}{5}$ pies, it will be necessary to slice one of the whole pies into fifths. After cutting one pie into fifths, there are 2 whole pies plus $\frac{5}{5}$ plus $\frac{1}{5}$, which is $2\frac{6}{5}$ pies. Then the baker can remove $1\frac{2}{5}$ pies, as we illustrate.

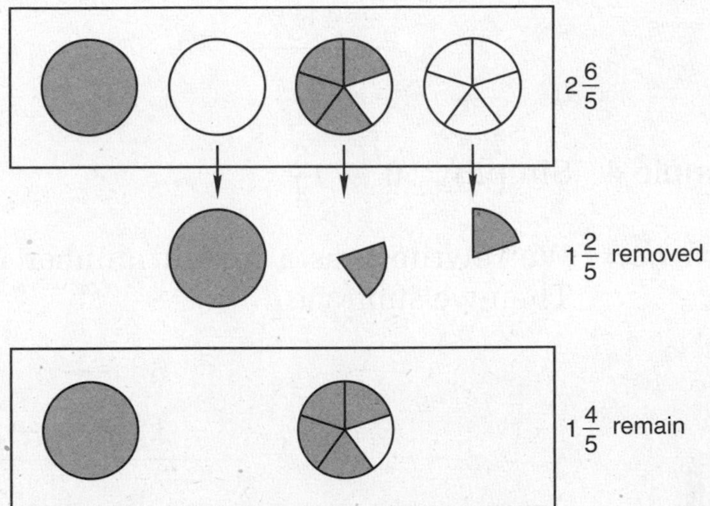

As we can see from the picture, **$1\frac{4}{5}$ pies** will be left on the shelf.

To perform the subtraction on paper, we first rename $3\frac{1}{5}$ as $2\frac{6}{5}$, as shown below. Then we can subtract.

$$\begin{array}{r} 3\frac{1}{5} \\ -\ 1\frac{2}{5} \\ \hline \end{array} \xrightarrow{2 + \frac{5}{5} + \frac{1}{5}} \begin{array}{r} 2\frac{6}{5} \\ -\ 1\frac{2}{5} \\ \hline 1\frac{4}{5} \end{array}$$

Example 2 Simplify: $3\frac{5}{8} - 1\frac{7}{8}$

Solution We need to regroup in order to subtract. The mixed number $3\frac{5}{8}$ equals $2 + 1 + \frac{5}{8}$, which equals $2 + \frac{8}{8} + \frac{5}{8}$. Combining $\frac{8}{8}$ and $\frac{5}{8}$ gives us $\frac{13}{8}$, so we use $2\frac{13}{8}$. Now we can subtract and reduce.

$$\begin{array}{r} 3\frac{5}{8} \\ -\ 1\frac{7}{8} \\ \hline \end{array} \xrightarrow{2 + \frac{8}{8} + \frac{5}{8}} \begin{array}{r} 2\frac{13}{8} \\ -\ 1\frac{7}{8} \\ \hline 1\frac{6}{8} \end{array} = \mathbf{1\frac{3}{4}}$$

Example 3 Simplify: $83\frac{1}{3}\% - 41\frac{2}{3}\%$

Solution The fraction in the subtrahend is greater than the fraction in the minuend, so we rename $83\frac{1}{3}\%$.

$$\begin{array}{r} 83\frac{1}{3}\% \\ -\ 41\frac{2}{3}\% \\ \hline \end{array} \xrightarrow{\left(82\ +\ \frac{3}{3}\ +\ \frac{1}{3}\right)\%} \begin{array}{r} 82\frac{4}{3}\% \\ -\ 41\frac{2}{3}\% \\ \hline \mathbf{41\frac{2}{3}\%} \end{array}$$

Example 4 Simplify: $6 - 1\frac{3}{4}$

Solution We rewrite 6 as a mixed number with a denominator of 4. Then we subtract.

$$\begin{array}{r} 6 \\ -\ 1\frac{3}{4} \\ \hline \end{array} \longrightarrow \begin{array}{r} 5\frac{4}{4} \\ -\ 1\frac{3}{4} \\ \hline \mathbf{4\frac{1}{4}} \end{array}$$

Example 5 Simplify: $100\% - 16\frac{2}{3}\%$

Solution We rename 100% as $99\frac{3}{3}\%$ and subtract.

$$\begin{array}{r} 100\% \\ -\ 16\frac{2}{3}\% \\ \hline \end{array} \longrightarrow \begin{array}{r} 99\frac{3}{3}\% \\ -\ 16\frac{2}{3}\% \\ \hline \mathbf{83\frac{1}{3}\%} \end{array}$$

LESSON PRACTICE

Practice set* Simplify:

a. $7 - 2\frac{1}{3}$

b. $6\frac{2}{5} - 1\frac{4}{5}$

c. $5\frac{1}{6} - 1\frac{5}{6}$

d. $100\% - 12\frac{1}{2}\%$

e. $83\frac{1}{3}\% - 16\frac{2}{3}\%$

MIXED PRACTICE

Problem set

1. (13) Willie shot eighteen rolls of film for the local newspaper. If there were thirty-six exposures in each roll, how many exposures were there in all?

2. (5, 12) Fifty million is how much greater than two hundred fifty thousand? Use words to write the answer.

3. (11) Two hundred fifty-nine people attended on opening night. On the second night 269 attended, and 307 attended on the third night. How many people attended on the first three nights?

4. (13) The 16-pound turkey cost $14.24. What was the price per pound?

5. (22) Diagram this statement. Then answer the questions that follow.

Three eighths of the 56 restaurants in town were closed on Monday.

(a) How many of the restaurants in town were closed on Monday?

(b) How many of the restaurants in town were open on Monday?

6. (22) In the following statement, write the percent as a reduced fraction. Then diagram the statement and answer the questions.

Forty percent of the 30 children at the picnic were boys.

(a) How many boys were at the picnic?

(b) How many girls were at the picnic?

7. (16) After contact was made, the spheroid sailed four thousand, one hundred forty inches. How many yards did the spheroid sail after contact was made?

8. (9) (a) How many $\frac{1}{5}$'s are in 1?

(b) How many $\frac{1}{5}$'s are in 3?

9. (9, 10) Describe how to find the reciprocal of a mixed number.

10. (15) Replace each circle with the proper comparison symbol:

(a) $\frac{2}{3} \cdot \frac{3}{2} \bigcirc \frac{5}{5}$

(b) $\frac{12}{36} \bigcirc \frac{12}{24}$

11. Write $2\frac{1}{4}$ and $3\frac{1}{3}$ as improper fractions. Then multiply the improper fractions, and write the product as a reduced mixed number.
(10, 15)

12. Complete each equivalent fraction:
(15)

(a) $\frac{3}{4} = \frac{?}{40}$ (b) $\frac{2}{5} = \frac{?}{40}$ (c) $\frac{?}{8} = \frac{15}{40}$

13. The prime factorization of 100 is $2 \cdot 2 \cdot 5 \cdot 5$. We can write the prime factorization of 100 using exponents this way:
(21)

$$2^2 \cdot 5^2$$

(a) Write the prime factorization of 400 using exponents.

(b) Write the prime factorization of $\sqrt{400}$ using exponents.

14. Refer to this figure to answer (a)–(d):
(7)

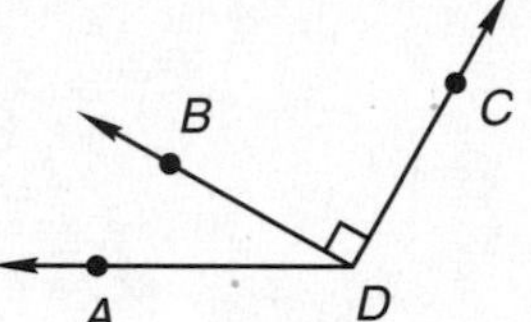

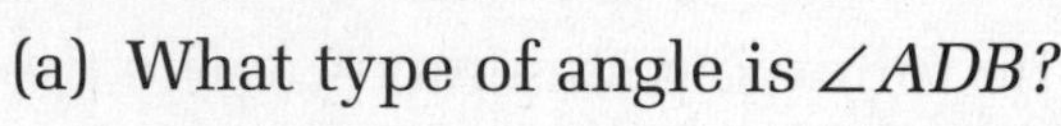

(a) What type of angle is $\angle ADB$?

(b) What type of angle is $\angle BDC$?

(c) What type of angle is $\angle ADC$?

(d) Which ray is perpendicular to $\overrightarrow{DB}$?

15. Draw an octagon. (A stop sign is a physical example of an octagon.)
(18)

Solve:

16. $\frac{105}{w} = 7$
(3)

17. $2x = 10^2$
(3, 20)

18. $x + 1\frac{1}{4} = 6\frac{3}{4}$
(9, 15)

19. $m - 4\frac{1}{8} = 1\frac{5}{8}$
(9, 15)

Simplify:

20. $5 - 3\frac{1}{3}$
(23)

21. $83\frac{1}{3}\% - 66\frac{2}{3}\%$
(23)

22. $\frac{7}{12} + \left(\frac{1}{4} \cdot \frac{1}{3}\right)$
(9, 15)

23. $\frac{7}{8} - \left(\frac{3}{4} \cdot \frac{1}{2}\right)$
(9, 15)

24. (19) Draw $\overline{AB}$ $1\frac{3}{4}$ inches long. Then draw $\overline{BC}$ 1 inch long perpendicular to $\overline{AB}$. Complete the triangle by drawing $\overline{AC}$. Use a ruler to find the approximate length of $\overline{AC}$. Use that length to find the perimeter of $\triangle ABC$.

25. (17) Use a protractor to find the measure of $\angle A$ in problem 24. If necessary, extend the sides to measure the angle.

26. (19) Mary wants to apply a strip of wallpaper along the walls of the dining room just below the ceiling. If the room is a 14-by-12-ft rectangle, then the strip of wallpaper needs to be at least how long?

27. (9, 15) Multiply $\frac{3}{4}$ by the reciprocal of 3 and reduce the product.

28. (15) Find fractions equivalent to $\frac{3}{4}$ and $\frac{2}{3}$ with denominators of 12. Then subtract the smaller fraction from the larger fraction.

29. (20) A sequence of perfect cubes ($k = n^3$) may be written as in (a) or as in (b). Find the next two terms of both sequences.

(a) $1^3, 2^3, 3^3, \ldots$

(b) 1, 8, 27, ...

30. (Inv. 2) The figure shows a circle with the center at point *M*.

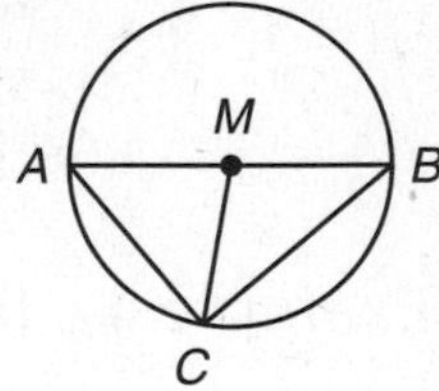

(a) Which chord is a diameter?

(b) Which central angle appears to be obtuse?

(c) Name an inscribed angle that appears to be a right angle.

LESSON

24 Reducing Fractions, Part 2

WARM-UP

Facts Practice: 40 Fractions to Reduce (Test D)

Mental Math:

a. \$5.74 + 98¢ **b.** \$1.50 × 10 **c.** \$1.00 − 36¢

d. 4 × 65 **e.** $3\frac{1}{3} + 1\frac{2}{3}$ **f.** $\frac{1}{3}$ of 24

g. What number is 3 more than half the product of 4 and 6?

Problem Solving:

A card with a triangle on it in the position shown is rotated 90° clockwise three times. Sketch the pattern shown and draw the triangle in the correct position.

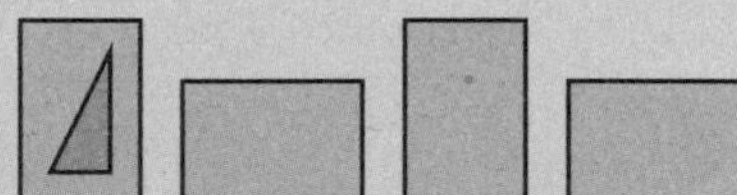

NEW CONCEPT

Using prime factorization to reduce We have been practicing reducing fractions by dividing the numerator and the denominator by a common factor. In this lesson we will practice a method of reducing that uses prime factorization to find the common factors of the terms. If we write the prime factorization of the numerator and of the denominator, we can see how to reduce a fraction easily.

Example 1 (a) Use prime factorization to reduce $\frac{420}{1050}$.

(b) Find the greatest common factor of 420 and 1050.

Solution (a) We rewrite the numerator and the denominator as products of prime numbers.

$$\frac{420}{1050} = \frac{2 \cdot 2 \cdot 3 \cdot 5 \cdot 7}{2 \cdot 3 \cdot 5 \cdot 5 \cdot 7}$$

Next we look for pairs of factors that equal 1. A fraction equals 1 if the numerator and denominator are equal. In this fraction there are four pairs of factors that equal 1. They are $\frac{2}{2}$, $\frac{3}{3}$, $\frac{5}{5}$, and $\frac{7}{7}$. Below we have indicated each of these pairs.

$$\frac{2 \cdot 2 \cdot 3 \cdot 5 \cdot 7}{2 \cdot 3 \cdot 5 \cdot 5 \cdot 7}$$

Each pair of factors reduces to $\frac{1}{1}$.

$$\frac{\overset{1}{\cancel{2}} \cdot 2 \cdot \overset{1}{\cancel{3}} \cdot \overset{1}{\cancel{5}} \cdot \overset{1}{\cancel{7}}}{\underset{1}{\cancel{2}} \cdot \underset{1}{\cancel{3}} \cdot 5 \cdot \underset{1}{\cancel{5}} \cdot \underset{1}{\cancel{7}}}$$

The reduced fraction equals $1 \cdot 1 \cdot 1 \cdot 1 \cdot \frac{2}{5}$, which is $\mathbf{\frac{2}{5}}$.

(b) In (a) we found the common prime factors of 420 and 1050. The common prime factors are 2, 3, 5, and 7. The product of these prime factors is the greatest common factor of 420 and 1050.

$$2 \cdot 3 \cdot 5 \cdot 7 = \mathbf{210}$$

Reducing before multiplying When multiplying fractions, we often get a product that can be reduced even though the individual factors could not be reduced. Consider this multiplication:

$$\frac{3}{8} \cdot \frac{2}{3} = \frac{6}{24} \qquad \frac{6}{24} \text{ reduces to } \frac{1}{4}$$

We see that neither $\frac{3}{8}$ nor $\frac{2}{3}$ can be reduced. The product, $\frac{6}{24}$, can be reduced. We can avoid reducing after we multiply by reducing before we multiply. Reducing before multiplying is also known as **canceling.** To reduce, any numerator may be paired with any denominator. Below we have paired the 3 with 3 and the 2 with 8.

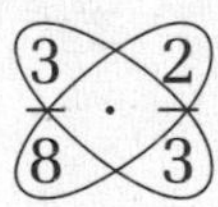

Then we reduce these pairs: $\frac{3}{3}$ reduces to $\frac{1}{1}$, and $\frac{2}{8}$ reduces to $\frac{1}{4}$, as we show below. Then we multiply the reduced terms.

$$\frac{\overset{1}{\cancel{3}}}{\underset{4}{\cancel{8}}} \cdot \frac{\overset{1}{\cancel{2}}}{\underset{1}{\cancel{3}}} = \frac{1}{4}$$

Example 2 Simplify: $\frac{9}{16} \cdot \frac{2}{3}$

Solution Before multiplying, we pair 9 with 3 and 2 with 16 and reduce these pairs. Then we multiply the reduced terms.

$$\frac{\overset{3}{\cancel{9}}}{\underset{8}{\cancel{16}}} \cdot \frac{\overset{1}{\cancel{2}}}{\underset{1}{\cancel{3}}} = \mathbf{\frac{3}{8}}$$

Example 3 Simplify: $\frac{8}{9} \cdot \frac{3}{10} \cdot \frac{5}{4}$

Solution We mentally pair 8 with 4, 3 with 9, and 5 with 10 and reduce.

$$\frac{\overset{2}{\cancel{8}}}{\underset{3}{\cancel{9}}} \cdot \frac{\overset{1}{\cancel{3}}}{\underset{2}{\cancel{10}}} \cdot \frac{\overset{1}{\cancel{5}}}{\underset{1}{\cancel{4}}}$$

We can still reduce by pairing 2 with 2. Then we multiply.

$$\frac{\overset{\overset{1}{\cancel{2}}}{\cancel{8}}}{\underset{3}{\cancel{9}}} \cdot \frac{\overset{1}{\cancel{3}}}{\underset{\underset{1}{\cancel{2}}}{\cancel{10}}} \cdot \frac{\overset{1}{\cancel{5}}}{\underset{1}{\cancel{4}}} = \mathbf{\frac{1}{3}}$$

Example 4 Simplify: $\frac{27}{32} \cdot \frac{20}{63}$

Solution To give us easier numbers to work with, we factor the terms of the fractions before we reduce and multiply.

$$\frac{3 \cdot \overset{1}{\cancel{3}} \cdot \overset{1}{\cancel{3}}}{2 \cdot 2 \cdot 2 \cdot \underset{1}{\cancel{2}} \cdot \underset{1}{\cancel{2}}} \cdot \frac{\overset{1}{\cancel{2}} \cdot \overset{1}{\cancel{2}} \cdot 5}{\underset{1}{\cancel{3}} \cdot \underset{1}{\cancel{3}} \cdot 7} = \mathbf{\frac{15}{56}}$$

LESSON PRACTICE

Practice set Use prime factorization to reduce each fraction:

a. $\frac{48}{144}$ **b.** $\frac{90}{324}$

c. Find the greatest common factor of 90 and 324.

Reduce before multiplying:

d. $\frac{5}{8} \cdot \frac{3}{10}$ **e.** $\frac{8}{15} \cdot \frac{5}{12} \cdot \frac{9}{10}$ **f.** $\frac{8}{3} \cdot \frac{6}{7} \cdot \frac{5}{16}$

g. Factor and reduce before multiplying: $\frac{36}{45} \cdot \frac{25}{24}$

MIXED PRACTICE

Problem set

1. (12) From Hartford to Los Angeles is two thousand, eight hundred ninety-five miles. From Hartford to Portland is three thousand, twenty-six miles. The distance from Hartford to Portland is how much greater than the distance from Hartford to Los Angeles?

2. (13) Hal ordered 15 boxes of microprocessors. If each box contained two dozen microprocessors, how many microprocessors did Hal order?

3. (22) In the following statement, write the percent as a fraction. Then diagram the statement and answer the questions.

Ashanti went to the store with $30.00 and spent 75% of the money.

(a) What fraction of the money did she spend?

(b) How much money did she spend?

4. (16, Inv. 2) If the diameter of a wheel is one yard, then its radius is how many inches?

5. (22) Nancy descended the 30 steps that led to the floor of the cellar. One third of the way down she paused. How many more steps were there to the cellar floor?

6. (9) (a) How many $\frac{1}{8}$'s are in 1?

(b) How many $\frac{1}{8}$'s are in 3?

7. (9) (a) Write the reciprocal of 3.

(b) What fraction of 3 is 1?

8. (24) (a) Use prime factorization to reduce $\frac{540}{600}$.

(b) What is the greatest common factor of 540 and 600?

9. *(17)* What type of angle is formed by the hands of a clock at
(a) 2 o'clock? (b) 3 o'clock? (c) 4 o'clock?

10. *(15)* Describe how to complete this equivalent fraction:

$$\frac{3}{5} = \frac{?}{30}$$

11. *(21)* The prime factorization of 1000 using exponents is $2^3 \cdot 5^3$.

(a) Write the prime factorization of 10,000 using exponents.

(b) Write the prime factorization of $\sqrt{10{,}000}$ using exponents.

12. *(7)* (a) Draw two parallel lines that are intersected by a third line perpendicular to the parallel lines.

(b) What type of angles are formed?

13. *(20)* The perimeter of a square is one yard.
(a) How many inches long is each side of the square?

(b) What is the area of the square in square inches?

14. *(2)* This equation illustrates that which property does not apply to division?

$$10 \div 5 \neq 5 \div 10$$

Solve:

15. *(9, 15)* $4\frac{7}{12} = x + 1\frac{1}{12}$

16. *(9, 15)* $w - 3\frac{3}{4} = 2\frac{3}{4}$

17. *(3, 10)* $8m = 100$

18. *(3)* $\frac{n}{12} = 28¢$

Simplify:

19. *(20)* $10^5 \div 10^2$

20. *(20)* $\sqrt{9} - \sqrt{4^2}$

21. *(23)* $100\% - 66\frac{2}{3}\%$

22. *(23)* $5\frac{1}{8} - 1\frac{7}{8}$

23. *(20)* $\left(\frac{5}{6}\right)^2$

24. *(24)* $\frac{3}{4} \cdot \frac{1}{2} \cdot \frac{8}{9}$

25. (18) Kevin clipped three corners from a square sheet of paper. What is the name of the polygon that was formed?

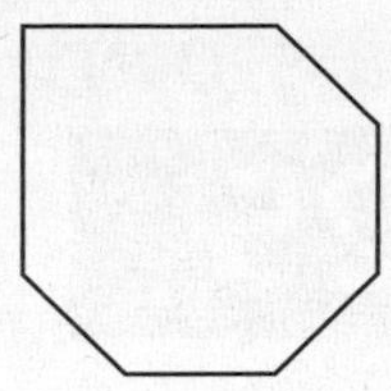

26. (1, 4) Evaluate the following expressions for $a = 10$ and $b = 100$:

(a) ab (b) $a - b$ (c) $\frac{a}{b}$

27. (19) Find the perimeter of the figure at right. Dimensions are in yards. All angles are right angles.

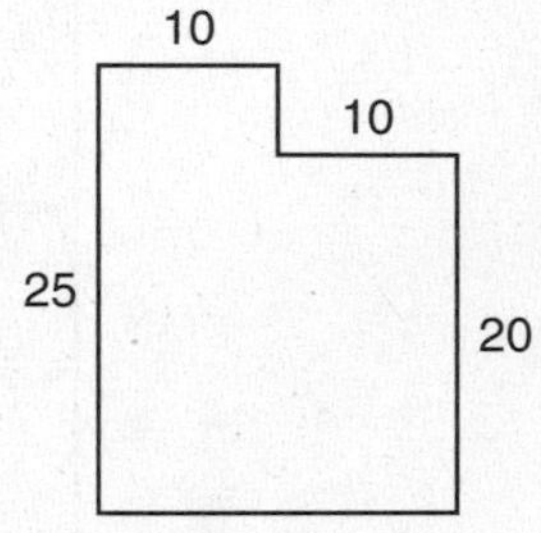

28. (15) Find equivalent fractions for $\frac{1}{4}$ and $\frac{1}{6}$ that have denominators of 12. Then add them.

29. (18) Segment AC divides rectangle $ABCD$ into two congruent triangles. Angle ADC corresponds to $\angle CBA$. Name two more pairs of corresponding angles.

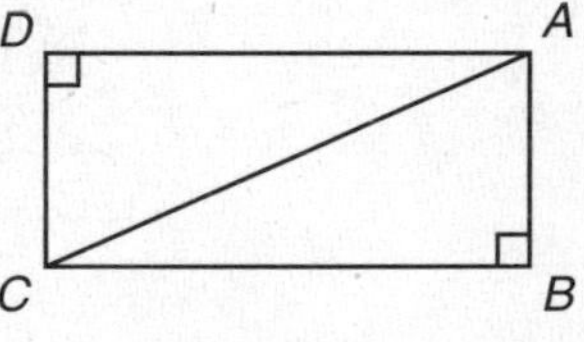

30. (4, 9) (a) Arrange these numbers in order from least to greatest:

$$0, 1, -1, \tfrac{1}{2}, -\tfrac{1}{2}$$

(b) The ordered numbers in the answer to part (a) form a sequence. What are the next three positive numbers in the sequence?

LESSON

25 Dividing Fractions

WARM-UP

Facts Practice: Lines, Angles, Polygons (Test F)

Mental Math:

a. \$2.65 + \$1.99 **b.** \$60.00 ÷ 10 **c.** \$2.00 − \$1.24

d. 7 × 36 **e.** $1\frac{3}{4} + 4\frac{1}{4}$ **f.** $\frac{1}{4}$ of 36

g. What number is 3 less than half the sum of 8 and 12?

Problem Solving:

Copy this problem and fill in the missing digits:

$$\begin{array}{r} \_6 \\ \times\ \_\_ \\ \hline \_\_ \\ \_\_\ \\ \hline 2\_6 \end{array}$$

NEW CONCEPT

When we ask the question "How many $\frac{1}{4}$'s are in 1?" we are asking a division question that can be expressed by writing:

$$1 \div \frac{1}{4}$$

The question can also be modeled with fraction manipulatives.

How many $\frac{1}{4}$ are in 1?

With manipulatives we see that the answer is 4. Recall that 4 $\left(\text{or } \frac{4}{1}\right)$ is the reciprocal of $\frac{1}{4}$.

Likewise, when we ask the question, "How many $\frac{1}{4}$'s are in 3?" we are again asking a division question.

$$3 \div \frac{1}{4}$$

How many $\frac{1}{4}$ are in 1 1 1?

We can use the answer to the first question to help us answer the second question. There are four $\frac{1}{4}$'s in 1, so there must be *three times as many* $\frac{1}{4}$'s in 3. Thus, there are twelve $\frac{1}{4}$'s in 3.

We found the answer to the second question by multiplying 3 by 4, the answer to the first question. We will follow this same line of thinking in the next few examples.

Example 1 (a) How many $\frac{2}{3}$'s are in 1? $\left(1 \div \frac{2}{3}\right)$

(b) How many $\frac{2}{3}$'s are in 3? $\left(3 \div \frac{2}{3}\right)$

Solution (a) We may model the question with manipulatives.

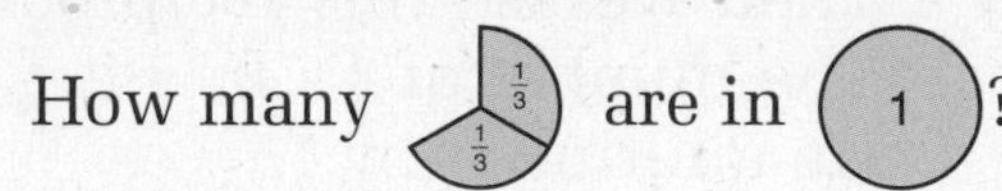

We see from the manipulatives that the answer is more than 1 but less than 2. If we think of the two $\frac{1}{3}$ pieces as one piece, we see that another *half* of the $\frac{2}{3}$ piece would make a whole. Thus there are $\mathbf{\frac{3}{2}}$ (or $\mathbf{1\frac{1}{2}}$) $\frac{2}{3}$'s in 1.

(b) We use the answer to (a) to help us answer (b). There are $\frac{3}{2}$ $\left(\text{or } 1\frac{1}{2}\right)$ $\frac{2}{3}$'s in 1, so there are three times as many $\frac{2}{3}$'s in 3. Thus, we answer the question by multiplying 3 by $\frac{3}{2}$ $\left(\text{or } 3 \text{ by } 1\frac{1}{2}\right)$.

$$3 \times \frac{3}{2} = \frac{9}{2} = 4\frac{1}{2}$$

$$3 \times 1\frac{1}{2} = 1\frac{1}{2} + 1\frac{1}{2} + 1\frac{1}{2} = 4\frac{1}{2}$$

The number of $\frac{2}{3}$'s in 3 is $\mathbf{4\frac{1}{2}}$. We found the answer by multiplying 3 by the reciprocal of $\frac{2}{3}$.

Example 2 (a) $1 \div \frac{2}{5}$ (b) $\frac{3}{4} \div \frac{2}{5}$

Solution (a) The problem $1 \div \frac{2}{5}$ means, "How many $\frac{2}{5}$'s are in 1?" The answer is the reciprocal of $\frac{2}{5}$, which is $\mathbf{\frac{5}{2}}$.

$$1 \div \frac{2}{5} = \frac{5}{2}$$

(b) We use the answer to (a) to help us answer (b). There are $\frac{5}{2}$ $\left(\text{or } 2\frac{1}{2}\right)$ $\frac{2}{5}$'s in 1, so there are $\frac{3}{4}$ times as many $\frac{2}{5}$'s in $\frac{3}{4}$. Thus we multiply $\frac{3}{4}$ by $\frac{5}{2}$.

$$\frac{3}{4} \times \frac{5}{2} = \frac{15}{8} = 1\frac{7}{8}$$

The number of $\frac{2}{5}$'s in $\frac{3}{4}$ is $\mathbf{1\frac{7}{8}}$. We found the answer by multiplying $\frac{3}{4}$ by the reciprocal of $\frac{2}{5}$.

Example 3 $\frac{2}{3} \div \frac{3}{4}$

Solution To find how many $\frac{3}{4}$'s are in $\frac{2}{3}$, we take two steps. First we find how many $\frac{3}{4}$'s are in 1. The answer is the reciprocal of $\frac{3}{4}$.

$$1 \div \frac{3}{4} = \frac{4}{3}$$

Then we use this reciprocal to find the number of $\frac{3}{4}$'s in $\frac{2}{3}$. The number of $\frac{3}{4}$'s in $\frac{2}{3}$ is $\frac{2}{3}$ times as many $\frac{3}{4}$'s as are in 1. So we multiply $\frac{2}{3}$ by $\frac{4}{3}$.

$$\frac{2}{3} \times \frac{4}{3} = \mathbf{\frac{8}{9}}$$

This means there is slightly less than one $\frac{3}{4}$ in $\frac{2}{3}$. We found the answer by multiplying $\frac{2}{3}$ by the reciprocal of $\frac{3}{4}$.

Example 4 $\frac{3}{4} \div \frac{9}{10}$

Solution **Step 1:** Find the number of $\frac{9}{10}$'s in 1. $1 \div \frac{9}{10} = \frac{10}{9}$

Step 2: Use the number of $\frac{9}{10}$'s in 1 to find the number of $\frac{9}{10}$'s in $\frac{3}{4}$. $\frac{\overset{1}{\cancel{3}}}{\underset{2}{\cancel{4}}} \times \frac{\overset{5}{\cancel{10}}}{\underset{3}{\cancel{9}}} = \mathbf{\frac{5}{6}}$

Working on paper, we often move from the original problem directly to step 2 by multiplying the dividend (first number) by the reciprocal of the divisor (second number).

$$\frac{3}{4} \div \frac{9}{10}$$

$$\downarrow \quad \downarrow$$

$$\frac{\overset{1}{\cancel{3}}}{\underset{2}{\cancel{4}}} \times \frac{\overset{5}{\cancel{10}}}{\underset{3}{\cancel{9}}} = \mathbf{\frac{5}{6}}$$

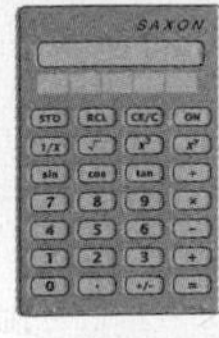

The reciprocal function on a calculator is the 1/x key. Pressing this key changes the previously entered number to its reciprocal (in decimal form). If we press 2 then 1/x, the calculator display changes from 2 to 0.5, which is the decimal form of $\frac{1}{2}$, the reciprocal of 2. The 1/x key can be helpful when dividing. Consider this division problem.

$$144\overline{)\$10{,}461.60}$$

The divisor is 144. You could choose to divide \$10,461.60 by 144 or to multiply \$10,461.60 by the reciprocal of 144. Since multiplication is commutative, using the reciprocal allows you to enter the numbers in either order. The following multiplication yields the answer even though the entry begins with the divisor. (Notice that we drop the terminal zero from \$10,461.60, since it does not affect the value.)

1 4 4 1/x × 1 0 4 6 1 . 6 =

Whether we choose to divide \$10,461.60 by 144 or to multiply by the reciprocal of 144, the answer is \$72.65.

LESSON PRACTICE

Practice set **a.** How many $\frac{2}{3}$'s are in 1? How many $\frac{2}{3}$'s are in $\frac{3}{4}$?

b. How many $\frac{3}{4}$'s are in 3?

c. Describe how to use the reciprocal of the divisor to find the answer to a division problem.

d. Describe the function of the 1/x key on a calculator.

Use the two-step method described in this lesson to find each quotient:

e. $\frac{3}{5} \div \frac{2}{3}$ **f.** $\frac{7}{8} \div \frac{1}{4}$ **g.** $\frac{5}{6} \div \frac{2}{3}$

MIXED PRACTICE

Problem set

1. (13) Three hundred twenty-four ice cream bars were needed for the town picnic. If half a dozen ice cream bars are in a box, then how many boxes of ice cream bars were needed?

2. (17, 19) Use a ruler to draw square *ABCD* with sides $2\frac{1}{2}$ in. long. Then divide the square into two congruent triangles by drawing $\overline{AC}$.

(a) What is the perimeter of square *ABCD*?

(b) What is the measure of each angle of the square?

(c) What is the measure of each acute angle in $\triangle ABC$?

(d) What is the sum of the measures of the three angles in $\triangle ABC$?

Use this information to answer problems 3–5:

The family picnic was a success, as 56 relatives attended. Half of those who attended played in the big game. However, the number of players on the two teams was not equal since one team had only 7 players.

3. How many relatives played in the game?
(22)

4. If one team had 7 players, how many players did the other team have?
(11)

5. If the teams were rearranged so that the number of players on each team was equal, how many players would be on each team?
(13)

6. In the following statement, write the percent as a reduced fraction. Then diagram the statement and answer the questions.
(22)

Jason has read 70% of the 310 pages in the book.

(a) How many pages has Jason read?

(b) How many pages has Jason not read?

7. (a) How many $\frac{3}{4}$'s are in 1?
(25)

(b) How many $\frac{3}{4}$'s are in $\frac{7}{8}$?

8. Which is the best estimate of how much of this rectangle is shaded? Why?
(8)

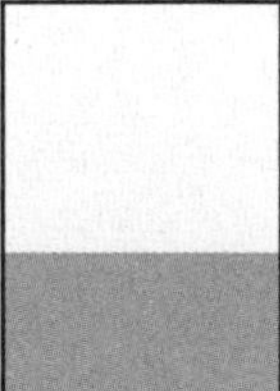

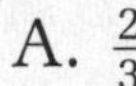

A. $\frac{2}{3}$ B. $\frac{2}{4}$ C. $\frac{2}{5}$

9. Write 84 and 210 as products of prime numbers. Then reduce $\frac{84}{210}$.
(24)

10. Write the reciprocal of each number:
(9, 10)

(a) $\frac{9}{10}$ (b) 8 (c) $2\frac{3}{8}$

11. (15) Find fractions equivalent to $\frac{3}{4}$ and $\frac{4}{5}$ with denominators of 20. Then add the two fractions you found, and write the sum as a mixed number.

12. (21) The prime factorization of 40 is $2^3 \cdot 5$. Write the prime factorization of 640 using exponents.

13. (10, 24) Write $2\frac{2}{3}$ and $2\frac{1}{4}$ as improper fractions. Then find the product of the improper fractions.

14. (8, 15) (a) Points *A* and *B* represent what mixed numbers on this number line?

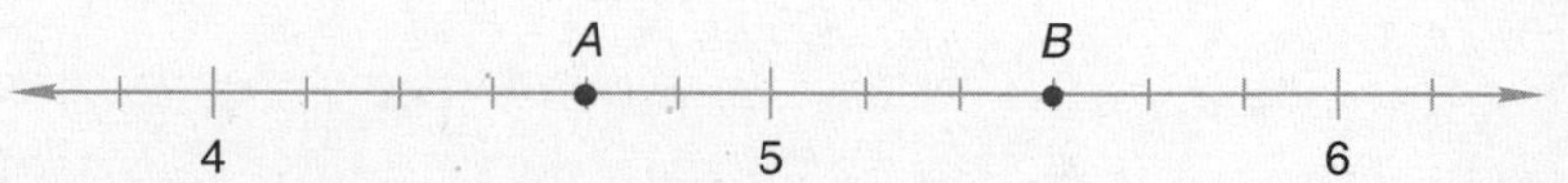

(b) Find the difference between the numbers represented by points *A* and *B*.

15. (17) (a) Draw line *AB*. Then draw ray *BC* so that angle *ABC* measures 30°. Use a protractor.

(b) What type of angle is angle *ABC*?

Solve:

16. (23) $1\frac{7}{12} + y = 3$

17. (9, 15) $5\frac{7}{8} = x - 4\frac{5}{8}$

18. (3) $8n = 360°$

19. (9, 20) $\frac{4}{3}m = 1^3$

Simplify:

20. (10) $6\frac{1}{6} + 1\frac{5}{6}$

21. (24) $\frac{3}{4} \cdot \frac{5}{9} \cdot \frac{8}{15}$

22. (25) $\frac{4}{5} \div \frac{2}{1}$

23. (25) $\frac{8}{5} \div \frac{6}{5}$

24. (25) $\frac{3}{7} \div \frac{5}{6}$

25. (10) $\frac{100\%}{8}$

26. (25) In the division $5 \div \frac{3}{5}$, instead of dividing 5 by $\frac{3}{5}$, we can find the answer by multiplying 5 by what number?

27. (20) (a) Simplify and compare: $2^2 \cdot 2^3 \bigcirc 2^3 \cdot 2^2$

(b) Simplify: $\sqrt{2^2}$

28. (19) A regular hexagon is inscribed in a circle. If one side of the hexagon is 6 inches long, then the perimeter of the hexagon is how many feet?

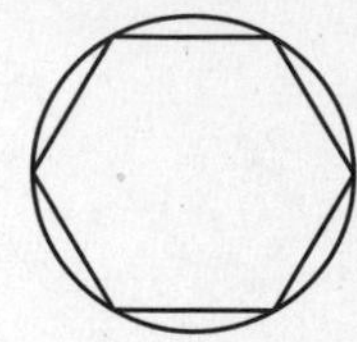

29. (19) A 2-in. square was cut from a 4-in. square as shown in the figure. What is the perimeter of the resulting polygon?

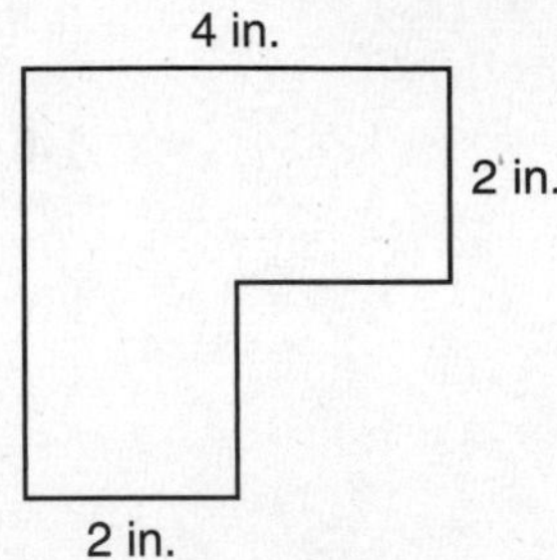

30. (4, 21) Which negative integer is the opposite of the third prime number?

LESSON

26 Multiplying and Dividing Mixed Numbers

WARM-UP

Facts Practice: Lines, Angles, Polygons (Test F)

Mental Math:

a. \$8.56 + 98¢ **b.** 30¢ × 100 **c.** \$1.00 − 7¢

d. 3 × 74 **e.** $\frac{2}{3} + \frac{2}{3}$ **f.** $\frac{2}{3}$ of 24

g. 7 × 7, + 1, × 2, ÷ 5, + 5, ÷ 5, − 5, × 5

Problem Solving:

The restaurant bill was \$20.00. Nelda left a tip equal to $\frac{1}{5}$ of the bill. What was the total cost of the bill and the tip?

NEW CONCEPT

One way to multiply or divide mixed numbers is to first rewrite the mixed numbers as improper fractions. Then we multiply or divide the improper fractions as indicated.

Example 1 Simplify: $3 \times 2\frac{1}{2}$

Solution We will show two ways to find the answer. One way is to recognize that $3 \times 2\frac{1}{2}$ equals three $2\frac{1}{2}$'s, which we add.

$$3 \times 2\frac{1}{2} = 2\frac{1}{2} + 2\frac{1}{2} + 2\frac{1}{2}$$

$$= \mathbf{7\frac{1}{2}}$$

Another way to find the product is to write 3 and $2\frac{1}{2}$ as improper fractions and multiply. We can write 3 as $\frac{3}{1}$, since 3 divided by 1 is 3.

$$3 \times 2\frac{1}{2}$$

$$\downarrow \quad \downarrow$$

$$\frac{3}{1} \times \frac{5}{2} = \frac{15}{2} = \mathbf{7\frac{1}{2}}$$

Example 2 Simplify:

(a) $3\frac{2}{3} \times 1\frac{1}{2}$ (b) $\left(1\frac{1}{2}\right)^2$

Solution (a) We first rewrite $3\frac{2}{3}$ as $\frac{11}{3}$ and $1\frac{1}{2}$ as $\frac{3}{2}$. Then we multiply and simplify.

$$\frac{11}{\cancel{3}_1} \times \frac{\overset{1}{\cancel{3}}}{2} = \frac{11}{2} = \mathbf{5\frac{1}{2}}$$

(b) The expression $\left(1\frac{1}{2}\right)^2$ means $1\frac{1}{2} \times 1\frac{1}{2}$. We write each factor as an improper fraction and multiply.

$$1\frac{1}{2} \times 1\frac{1}{2}$$
$$\downarrow \quad \downarrow$$
$$\frac{3}{2} \times \frac{3}{2} = \frac{9}{4} = \mathbf{2\frac{1}{4}}$$

Example 3 Find the area of a square with sides $2\frac{1}{2}$ inches long.

Solution If we draw the square on a grid, we see a physical representation of the area of the square. We see four whole square inches, four half square inches, and one quarter square inch within the shaded figure. We can calculate the area by adding.

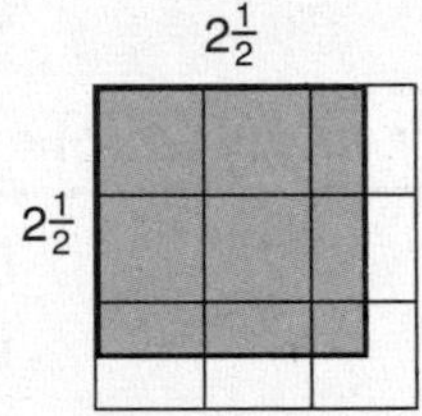

$$4 \text{ in.}^2 + \frac{4}{2} \text{ in.}^2 + \frac{1}{4} \text{ in.}^2 = \mathbf{6\frac{1}{4} \text{ in.}^2}$$

If we multiply $2\frac{1}{2}$ inches by $2\frac{1}{2}$ inches, we obtain the same result.

$$2\frac{1}{2} \text{ in.} \times 2\frac{1}{2} \text{ in.}$$
$$= \frac{5}{2} \text{ in.} \times \frac{5}{2} \text{ in.}$$
$$= \frac{25}{4} \text{ in.}^2 = \mathbf{6\frac{1}{4} \text{ in.}^2}$$

Example 4 Simplify: $3\frac{2}{3} \div 2$

Solution As we think about the problem, we see that by dividing $3\frac{2}{3}$ by 2, we will be finding *half of* $3\frac{2}{3}$. We can find half of a number either by dividing by 2 or by multiplying by $\frac{1}{2}$. In other words, the following are equivalent expressions:

$$3\frac{2}{3} \div 2 \qquad 3\frac{2}{3} \times \frac{1}{2}$$

Notice that multiplying by $\frac{1}{2}$ can be thought of as multiplying by the *reciprocal* of 2. We will write $3\frac{2}{3}$ as an improper fraction and multiply by $\frac{1}{2}$.

$$3\frac{2}{3} \times \frac{1}{2}$$

$$\downarrow$$

$$\frac{11}{3} \times \frac{1}{2} = \frac{11}{6} = \mathbf{1\frac{5}{6}}$$

Example 5 Simplify: $3\frac{1}{3} \div 2\frac{1}{2}$

Solution First we write $3\frac{1}{3}$ and $2\frac{1}{2}$ as improper fractions. Then we multiply by the reciprocal of the divisor and simplify.

$$3\frac{1}{3} \div 2\frac{1}{2} \quad \text{original problem}$$

$$\frac{10}{3} \div \frac{5}{2} \quad \text{changed mixed numbers to improper fractions}$$

$$\frac{\overset{2}{\cancel{10}}}{3} \times \frac{2}{\underset{1}{\cancel{5}}} = \frac{4}{3} \quad \text{multiplied by reciprocal of the divisor}$$

$$= \mathbf{1\frac{1}{3}} \quad \text{simplified}$$

LESSON PRACTICE

Practice set* **a.** Find the area of a rectangle that is $1\frac{1}{2}$ in. wide and $2\frac{1}{2}$ in. long. Draw the rectangle on a grid.

Simplify:

b. $6\frac{2}{3} \times \frac{3}{5}$ **c.** $2\frac{1}{3} \times 3\frac{1}{2}$ **d.** $3 \times 3\frac{3}{4}$

e. $1\frac{2}{3} \div 3$ **f.** $2\frac{1}{2} \div 3\frac{1}{3}$ **g.** $5 \div \frac{2}{3}$

h. $2\frac{2}{3} \div 1\frac{1}{3}$ **i.** $1\frac{1}{3} \div 2\frac{2}{3}$ **j.** $4\frac{1}{2} \times 1\frac{2}{3}$

MIXED PRACTICE

Problem set

1. (11) After the first hour of the monsoon, 23 millimeters of precipitation had fallen. After the second hour a total of 61 millimeters of precipitation had fallen. How many millimeters of precipitation fell during the second hour?

2. (13) Each enlargement cost 85¢ and Willie needed 26 enlargements. What was the total cost of the enlargements Willie needed?

3. (12) The Byzantine Empire lasted from 330 to 1453. How many years did the Byzantine Empire last?

4. (11) Dolores went to the theater with $20 and came home with $11.25. How much money did Dolores spend at the theater?

5. (13) A gross is a dozen dozens. A gross of pencils is how many pencils?

6. (22) Diagram this statement and answer the questions that follow. Begin by changing the percent to a reduced fraction.

Forty percent of the 60 marbles in the bag were blue.

(a) How many of the marbles in the bag were blue?

(b) How many of the marbles in the bag were not blue?

7. (16) Roan estimated that the weight of the water in a full bathtub is a quarter ton. How many pounds is a quarter of a ton?

8. (8) (a) What fraction of this square is shaded?

(b) What percent of this square is not shaded?

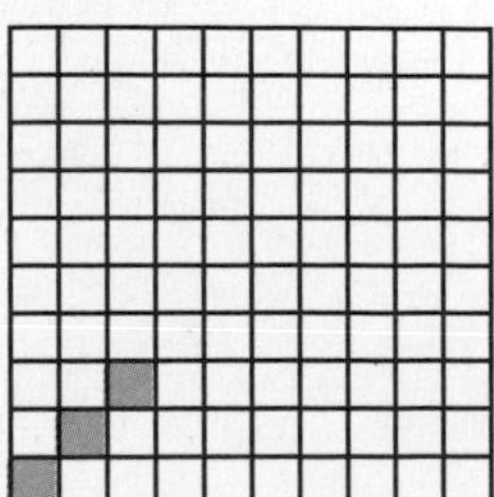

9. (a) Write 210 and 252 as products of prime numbers. Then reduce $\frac{210}{252}$.
(24)

(b) Find the GCF of 210 and 252.

10. Write the reciprocal of each number:
(9, 10)

(a) $\frac{5}{9}$ (b) $5\frac{3}{4}$ (c) 7

11. Find the number that makes the two fractions equivalent.
(15)

(a) $\frac{5}{8} = \frac{?}{24}$ (b) $\frac{5}{12} = \frac{?}{24}$

(c) Add the fractions you found in (a) and (b).

12. Draw a heptagon.
(18)

13. Draw $\overline{AB}$ 2 in. long. Then draw $\overline{BC}$ $1\frac{1}{2}$ in. long perpendicular to $\overline{AB}$. Complete ΔABC by drawing $\overline{AC}$. How long is $\overline{AC}$?
(8)

14. (a) Arrange these numbers in order from least to greatest:
(4, 10)

$$1, -3, \frac{5}{6}, 0, \frac{4}{3}$$

(b) Which of these numbers are whole numbers?

Solve:

15. $x - 8\frac{11}{12} = 6\frac{5}{12}$
(10, 15)

16. $180 - y = 75$
(3)

17. $12w = 360°$
(3)

18. $w + 58\frac{1}{3} = 100$
(23)

19. (a) Find the area of the square.
(20)

(b) Find the area of the shaded part of the square.

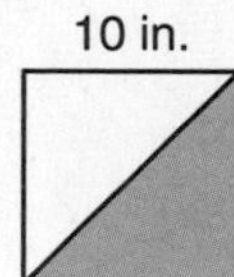

Simplify:

20. $9\frac{1}{9} - 4\frac{4}{9}$
(23)

21. $\frac{5}{8} \cdot \frac{3}{10} \cdot \frac{1}{6}$
(24)

22. $\left(2\frac{1}{2}\right)^2$
(20, 26)

23. $1\frac{3}{5} \div 2\frac{2}{3}$
(26)

24. (26) $3\frac{1}{3} \div 4$

25. (26) $5 \cdot 1\frac{3}{4}$

26. (20) $\sqrt{10^2 \cdot 10^4}$

27. (1) $\frac{16,524}{36}$

28. (1, 9) Evaluate the following expressions for $x = 3$ and $y = 6$:

(a) $x - \frac{y}{x}$ (b) $\frac{xy}{y}$ (c) $\frac{x}{y} \cdot \frac{y}{x}$

29. (2) The rule of the following sequence is $k = 3n - 2$. Find the ninth term.

1, 4, 7, 10, ...

30. (Inv. 2) The central angle of a half circle is 180°. The central angle of a quarter circle is 90°. How many degrees is the central angle of an eighth of a circle?

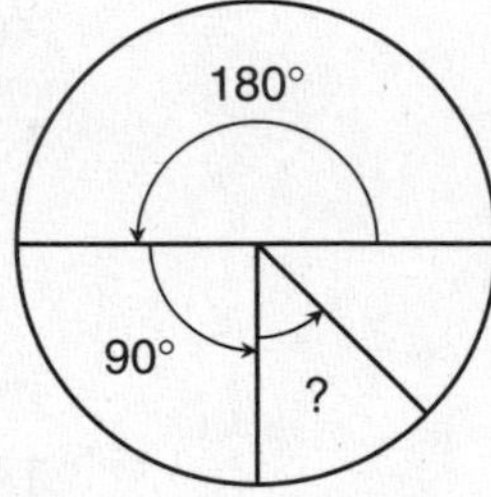

LESSON

27 Multiples • Least Common Multiple • Equivalent Division Problems

WARM-UP

Facts Practice: Circles (Test E)

Mental Math:

a. \$3.75 + \$1.98 **b.** \$125.00 ÷ 10 **c.** 10 × 42

d. 5 × 42 **e.** $\frac{3}{4} + \frac{3}{4}$ **f.** $\frac{3}{4}$ of 24

g. Start with a score. Add a dozen; then add the number of feet in a yard. Divide by half the number of years in a decade; then subtract the number of days in a week. What is the answer?

Problem Solving:

Simon held a dot cube so that he could see the dots on three adjoining faces. Simon said he could see a total of 8 dots. Could Simon have been telling the truth? Why or why not?

NEW CONCEPTS

Multiples

The **multiples** of a number are produced by multiplying the number by 1, by 2, by 3, by 4, and so on. Thus the multiples of 4 are

4, 8, 12, 16, 20, 24, 28, 32, 36, ...

The multiples of 6 are

6, 12, 18, 24, 30, 36, 42, 48, 54, ...

If we inspect these two lists, we see that some of the numbers in both lists are the same. A number appearing in both of these lists is a **common multiple** of 4 and 6. Below we have circled some of the common multiples of 4 and 6.

Multiples of 4: 4, 8, (12), 16, 20, (24), 28, 32, (36), ...

Multiples of 6: 6, (12), 18, (24), 30, (36), 42, 48, 54, ...

We see that 12, 24, and 36 are common multiples of 4 and 6. If we continued both lists, we would find many more common multiples.

Least common multiple Of particular interest is the least (smallest) of the common multiples. The **least common multiple** of 4 and 6 is 12. Twelve is the smallest number that is a multiple of both 4 and 6. The term *least common multiple* is often abbreviated **LCM.**

Example 1 Find the least common multiple of 6 and 8.

Solution We will list some multiples of 6 and of 8 and circle common multiples.

Multiples of 6: 6, 12, 18, (24), 30, 36, 42, (48), ...

Multiples of 8: 8, 16, (24), 32, 40, (48), 56, 64, ...

We find that the least common multiple of 6 and 8 is **24.**

It is unnecessary to list multiples each time. Often the search for the least common multiple can be conducted mentally.

Example 2 Find the LCM of 3, 4, and 6.

Solution To find the least common multiple of 3, 4, and 6, we can mentally search for the smallest number divisible by 3, 4, and 6. We can conduct the search by first thinking of multiples of the largest number, 6.

6, 12, 18, 24, ...

Then we mentally test these multiples for divisibility by 3 and by 4. We find that 6 is divisible by 3 but not by 4, while 12 is divisible by both 3 and 4. Thus the LCM of 3, 4, and 6 is **12.**

We can use prime factorization to help us find the least common multiple of a set of numbers. The LCM of a set of numbers is the product of *all the prime factors necessary to form any number in the set.*

Example 3 Use prime factorization to help you find the LCM of 18 and 24.

Solution We write the prime factorization of 18 and of 24.

$$18 = 2 \cdot 3 \cdot 3 \qquad 24 = 2 \cdot 2 \cdot 2 \cdot 3$$

The prime factors of 18 and 24 are 2's and 3's. From a pool of three 2's and two 3's, we can form either 18 or 24. So the LCM of 18 and 24 is the product of three 2's and two 3's.

$$\text{LCM of 18 and 24} = 2 \cdot 2 \cdot 2 \cdot 3 \cdot 3$$
$$= \mathbf{72}$$

Equivalent division problems Tricia's teacher asked this question:

> *If sixteen flavored icicles cost \$4.00, what was the price for each flavored icicle?*

Tricia quickly gave the correct answer, 25¢, and then explained how she found the answer.

> *I knew I had to divide \$4.00 by 16, but I did not know the answer. So I mentally found half of each number, which made the problem \$2.00 ÷ 8. I still couldn't think of the answer, so I found half of each of those numbers. That made the problem \$1.00 ÷ 4, and I knew the answer was 25¢.*

How did Tricia's mental technique work? Recall from Lesson 15 that we can form equivalent fractions by multiplying or dividing a fraction by a fraction equal to 1.

$$\frac{3}{4} \times \frac{10}{10} = \frac{30}{40} \qquad \frac{6}{9} \div \frac{3}{3} = \frac{2}{3}$$

We can form equivalent division problems in a similar way. We multiply (or divide) the dividend and divisor by the same number to form a new division problem that is easier to calculate mentally. The new division problem will produce the same quotient, as we show below.

$$\frac{\$4.00 \div 2}{16 \div 2} = \frac{\$2.00}{8} = \frac{\$2.00 \div 2}{8 \div 2} = \frac{\$1.00}{4} = \$0.25$$

Example 4 Instead of dividing 220 by 5, double both numbers and mentally calculate the quotient.

Solution We double the two numbers in 220 ÷ 5 and get 440 ÷ 10. We mentally calculate the new quotient to be **44,** which is also the quotient of the original problem.

Example 5 Instead of dividing 6000 by 200, divide both numbers by 100, and then mentally calculate the quotient.

Solution We mentally divide by 100 by removing two places (two zeros) from each number. This forms the equivalent division problem 60 ÷ 2. We mentally calculate the quotient as **30.**

LESSON PRACTICE

Practice set Find the least common multiple (LCM) of each pair or group of numbers:

a. 8 and 10 **b.** 4, 6, and 10

Use prime factorization to help you find the LCM of these pairs of numbers:

c. 24 and 40 **d.** 30 and 75

e. Instead of dividing $7\frac{1}{2}$ by $1\frac{1}{2}$, double each number and mentally calculate the quotient.

Mentally calculate each quotient by finding an equivalent division problem. Discuss your strategy with the class.

f. $24{,}000 \div 400$ **g.** $\$6.00 \div 12$ **h.** $140 \div 5$

MIXED PRACTICE

Problem set

1. *(11)* There were three towns in the valley. The population of Brenton was 11,460. The population of Elton was 9420. The population of Jennings was 8916. What was the total population of the three towns in the valley?

2. *(13, 16)* Norman is 6 feet tall. How many inches tall is Norman?

3. *(27)* If the cost of one dozen eggs was $1.80, what was the cost per egg? Write an equivalent division problem that is easier to calculate mentally, and find the quotient.

4. *(5, 20)* Which of the following equals one billion?

A. 10^3 B. 10^6 C. 10^9 D. 10^{12}

5. *(22)* Diagram this statement. Then answer the questions that follow.

Three eighths of the 712 fruit flies had red eyes.

(a) How many fruit flies had red eyes?

(b) How many fruit flies did not have red eyes?

6. *(19, 20)* The perimeter of this rectangle is 30 inches.

6 in.

(a) What is the length of the rectangle?

(b) What is the area of the rectangle?

7. (27) Use prime factorization to find the least common multiple of 25 and 45.

8. (4) What number is halfway between 3000 and 4000?

9. (15, 24) (a) Write 24% as a reduced fraction.

(b) Use prime factorization to reduce $\frac{36}{180}$.

10. (16) It was a "scorcher." The temperature was 102°F in the shade.

(a) The temperature was how many degrees above the freezing point of water?

(b) The temperature was how many degrees below the boiling point of water?

11. (2, 15) For each fraction, write an equivalent fraction that has a denominator of 36.

(a) $\frac{5}{12}$ (b) $\frac{1}{6}$ (c) $\frac{7}{9}$

(d) What property do we use when we find equivalent fractions?

12. (21) (a) Write the prime factorization of 576 using exponents.

(b) Find $\sqrt{576}$.

13. (26) Write $5\frac{5}{6}$ and $6\frac{6}{7}$ as improper fractions and find their product.

In the figure below, quadrilaterals *ABCF* and *FCDE* are squares. Refer to the figure to answer problems 14–16.

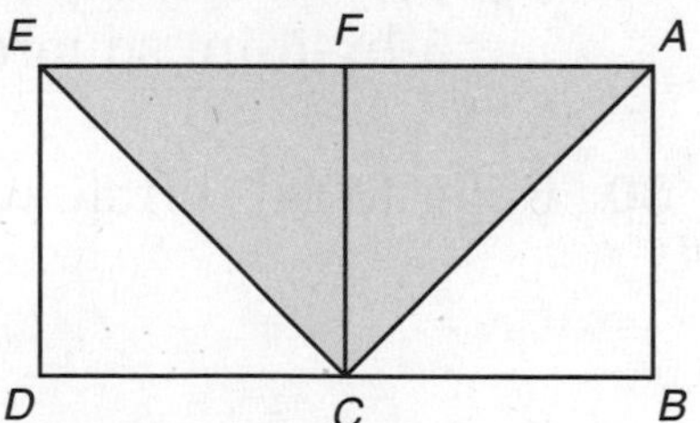

14. (7) (a) What kind of angle is $\angle ACD$?

(b) Name two segments parallel to $\overline{FC}$.

15. (8) (a) What fraction of square *CDEF* is shaded?

(b) What fraction of square *ABCF* is shaded?

(c) What fraction of rectangle *ABDE* is shaded?

16. (19, 20) If *AB* is 3 ft,

(a) what is the perimeter of rectangle *ABDE*?

(b) what is the area of rectangle *ABDE*?

Solve:

17. (3) $10y = 360°$

18. (3, 20) $p + 2^4 = 12^2$

19. (23) $5\frac{1}{8} - n = 1\frac{3}{8}$

20. (10) $m - 6\frac{2}{3} = 4\frac{1}{3}$

Simplify:

21. (23) $10 - 1\frac{3}{5}$

22. (26) $5\frac{1}{3} \cdot 1\frac{1}{2}$

23. (26) $3\frac{1}{3} \div \frac{5}{6}$

24. (26) $5\frac{1}{4} \div 3$

25. (24) $\frac{5}{6} \cdot \frac{9}{8} \cdot \frac{4}{15}$

26. (9, 15) $\frac{8}{9} - \left(\frac{7}{9} - \frac{5}{9}\right)$

27. (Inv. 2) If the diameter of a circle is half of a yard, then its radius is how many inches?

28. (27) Divide \$12.00 by 16 or find the quotient of an equivalent division problem.

29. (19, 20) A 3-by-3-in. paper square is cut from a 5-by-5-in. paper square as shown.

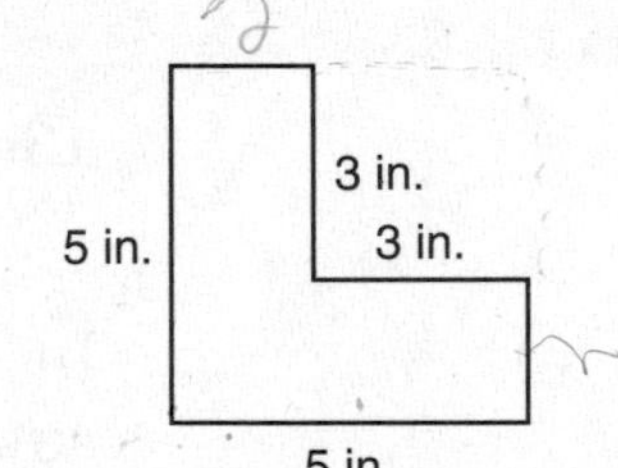

(a) What is the perimeter of the resulting polygon?

(b) How many square inches of the 5-by-5-in. square remain?

30. (Inv. 2) Refer to this circle with center at point *M* to answer (a)–(e):

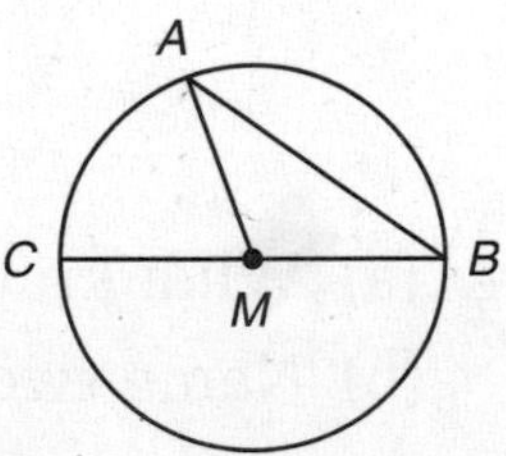

(a) Which chord is a diameter?

(b) Which chord is not a diameter?

(c) What angle is an acute central angle?

(d) Which angles are inscribed angles?

(e) Which two sides of triangle *AMB* are equal in length?

LESSON

28 Two-Step Word Problems • Average, Part 1

WARM-UP

Facts Practice: Lines, Angles, Polygons (Test F)

Mental Math:

a. \$6.23 + \$2.99 **b.** \$1.75 × 100 **c.** \$5.00 − \$1.29

d. 8 × 53 **e.** $\frac{5}{8} + \frac{5}{8}$ **f.** $\frac{2}{5}$ of 25

g. Think of an easier equivalent division for \$56.00 ÷ 14. Then find the quotient.

Problem Solving:

When Bill, Phil, Jill, and Gil entered the room, they found four chairs waiting for them. They each took a seat, and one minute later they traded seats. Then they traded seats again and again. If they don't move the chairs but only move themselves, how many seating arrangements (permutations) are possible?

NEW CONCEPTS

Two-step word problems

Thus far we have considered six one-step word-problem themes:

1. Combining
2. Separating
3. Comparing
4. Elapsed Time
5. Equal Groups
6. Part of a Whole

Word problems often require more than one step to solve. In this lesson we will continue practicing problems that require multiple steps to solve. These problems involve two or more of the themes mentioned above.

Example 1 Julie went to the store with \$20. If she bought 8 cans of dog food for 67¢ per can, how much money did she have left?

Solution This is a two-step problem. First we find out how much Julie spent. This first step is an "equal groups" problem.

Number in group →	\$0.67 each can
Number of groups →	× 8 cans
Total →	\$5.36

Now we can find out how much money Julie had left. This second step is about separating.

$$\begin{array}{r} \$20.00 \\ -\ \$5.36 \\ \hline \$14.64 \end{array}$$

After spending \$5.36 of her \$20 on dog food, Julie had **\$14.64** left.

Average, part 1 Calculating an **average** is often a two-step process. As an example, consider these five stacks of coins:

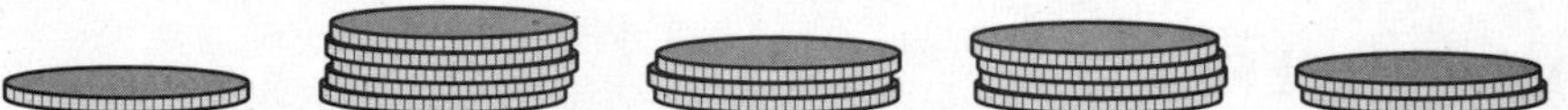

There are 15 coins in all. If we made all the stacks the same size, there would be 3 coins in each stack.

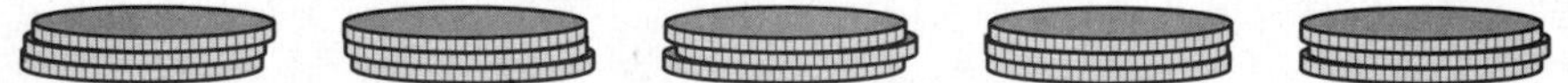

We say the average number of coins in each stack is 3. Now look at the following problem:

> *There are 4 teams participating in the scavenger hunt. Team A has 7 players, team B has 9 players, team C has 6 players, and team D has 10 players. What is the average number of players per team?*

The average number of players per team is the number of players that would be on each team if all of the teams had the same number of players. To find the average of a group of numbers, we begin by finding the sum of the numbers.

$$\begin{array}{r} 7 \text{ players} \\ 9 \text{ players} \\ 6 \text{ players} \\ +\ 10 \text{ players} \\ \hline 32 \text{ players} \end{array}$$

Then we divide the sum of the numbers by the number of numbers. There are 4 teams, so we divide by 4.

$$\frac{\text{sum of numbers}}{\text{number of numbers}} = \frac{32 \text{ players}}{4 \text{ teams}}$$

$$= 8 \text{ players per team}$$

Finding the average took two steps. First we added the numbers to find the total. Then we divided the total to make equal groups.

Example 2 When people were seated, there were 3 in the first row, 7 in the second row, and 20 in the third row. What was the average number of people in each of the first three rows?

Solution The average number of people in the first three rows is the number of people that would be in each row if the numbers were equal. First we add to find the total number of people.

$$\begin{array}{r} 3 \text{ people} \\ 7 \text{ people} \\ +\ 20 \text{ people} \\ \hline 30 \text{ people} \end{array}$$

Then we divide by 3 to separate the total into 3 equal groups.

$$\frac{30 \text{ people}}{3 \text{ rows}} = 10 \text{ people per row}$$

The average was **10 people** in each of the first 3 rows. Notice that the average of a set of numbers is *greater than the smallest number* in the set but *less than the largest number* in the set.

Another name for the average is the **mean.** We find the mean of a set of numbers by adding the numbers and then dividing the sum by the number of numbers.

Example 3 Robert took 20 math tests. On five tests he scored 100, on four tests he scored 95, on six tests he scored 90, and on five tests he scored 80. What was the mean of his scores?

Solution First we find the total of Robert's scores.

$$\begin{aligned} 5 \times 100 &= 500 \\ 4 \times 95 &= 380 \\ 6 \times 90 &= 540 \\ 5 \times 80 &= \underline{400} \\ & \quad 1820 \end{aligned}$$

Next we divide the total by 20 because there were 20 scores in all.

$$\frac{\text{sum of numbers}}{\text{number of numbers}} = \frac{1820}{20} = 91$$

We find that the mean of Robert's scores was **91.**

LESSON PRACTICE

Practice set Work each problem as a two-step problem:

a. Jody went to the store with $20 and returned home with $5.36. If all she bought was 3 bags of dog food, how much did she pay for each bag?

b. Three eighths of the 32 children in the club were girls. How many boys were in the club?

c. There were 28 passengers on the first bus, 29 passengers on the second bus, 30 passengers on the third bus, and 25 passengers on the fourth bus. What was the average number of passengers per bus?

d. What is the mean of 46, 37, 34, 31, 29, and 24?

e. What is the average of 40 and 70? What number is halfway between 40 and 70?

f. Willis has taken eight tests. His lowest score was 80, and his highest score was 95. Which of the following *could be* his average test score? Why?

A. 80 B. 84 C. 95 D. 96

MIXED PRACTICE

Problem set

1. (28) The 5 players on the front line weighed 242 pounds, 236 pounds, 248 pounds, 268 pounds, and 226 pounds. What was the average weight of the players on the front line?

2. (28) Yuko ran a mile in 5 minutes 14 seconds. How many seconds did it take Yuko to run a mile?

3. (28) Luisa bought a pair of pants for $24.95 and 3 blouses for $15.99 each. Altogether, how much did she spend?

4. (12) The Italian navigator Christopher Columbus was 41 years old when he reached the Americas in 1492. In what year was he born?

5. (22) In the following statement, change the percent to a reduced fraction. Then diagram the statement and answer the questions.

Salma led for 75% of the 5000-meter race.

(a) Salma led the race for how many meters?

(b) Salma did not lead the race for how many meters?

6. (19, 20) This rectangle is twice as long as it is wide.

8 in.

(a) What is the perimeter of the rectangle?

(b) What is the area of the rectangle?

7. (27) (a) List the first six multiples of 3.

(b) List the first six multiples of 4.

(c) What is the LCM of 3 and 4?

(d) Use prime factorization to find the least common multiple of 27 and 36.

8. (27) On the number line below, 283 is closest to

(a) which multiple of 10?

(b) which multiple of 100?

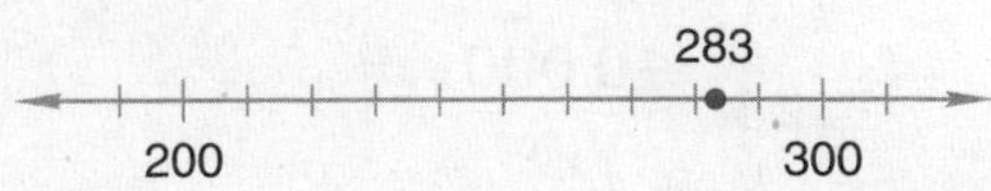

9. (24) Write 56 and 240 as products of prime numbers. Then reduce $\frac{56}{240}$.

10. (16) A mile is five thousand, two hundred eighty feet. Three feet equals a yard. So a mile is how many yards?

11. (15) For (a) and (b), find an equivalent fraction that has a denominator of 24.

(a) $\frac{7}{8}$ (b) $\frac{11}{12}$

(c) Add the two fractions you found.

12. (21) (a) Write the prime factorization of 3600 using exponents.

(b) Find $\sqrt{3600}$.

13. (28) Describe how to find the mean of 45, 36, 42, 29, 16, and 24.

14. (8, 20) (a) Draw square *ABCD* so that each side is 1 inch long. What is the area of the square?

(b) Draw segments *AC* and *BD*. Label the point at which they intersect point *E*.

(c) Shade triangle *CDE*.

(d) What percent of the area of the square did you shade?

15. (4, 10) (a) Arrange these numbers in order from least to greatest:

$$-1, \frac{1}{10}, 1, \frac{11}{10}, 0$$

(b) Which of these numbers are odd integers?

Solve:

16. (3) $12y = 360°$ **17.** (3, 20) $10^2 = m + 8^2$ **18.** (3) $\frac{180}{w} = 60$

Simplify:

19. (9, 15) $4\frac{5}{12} - 1\frac{1}{12}$ **20.** (10, 15) $8\frac{7}{8} + 3\frac{3}{8}$

21. (23) $12 - 8\frac{1}{8}$ **22.** (26) $6\frac{2}{3} \cdot 1\frac{1}{5}$

23. (20, 26) $\left(1\frac{1}{2}\right)^2 \div 7\frac{1}{2}$ **24.** (26) $8 \div 2\frac{2}{3}$

25. (1) $\frac{10{,}000}{80}$ **26.** (25) $\frac{3}{4} - \left(\frac{1}{2} \div \frac{2}{3}\right)$

27. Evaluate the following expressions for $x = 3$ and $y = 4$:
(1, 20)
(a) x^y (b) $x^2 + y^2$

28. Draw a decagon.
(18)

29. In the figure below the two triangles are congruent.
(18)

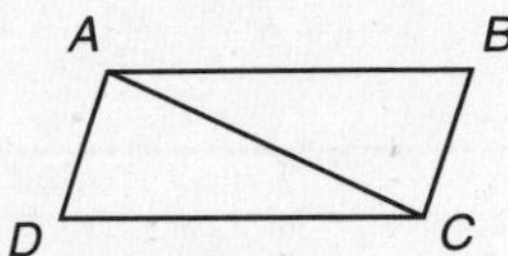

(a) Which angle in ΔACD corresponds to $\angle CAB$ in ΔABC?

(b) Which segment in ΔABC corresponds to $\overline{AD}$ in ΔACD?

(c) If the area of ΔABC is $7\frac{1}{2}$ in.^{2}, what is the area of figure $ABCD$?

30. With a ruler draw $\overline{PQ}$ $2\frac{3}{4}$ in. long. Then with a protractor draw $\overrightarrow{QR}$ so that $\angle PQR$ measures 30°. Then, from point P, draw a ray perpendicular to $\overline{PQ}$ that intersects $\overrightarrow{QR}$. (You may need to extend $\overrightarrow{QR}$ to show the intersection.) Label the point where the rays intersect point M. Use a protractor to measure $\angle PMQ$.
(17)

LESSON

29 Rounding Whole Numbers • Rounding Mixed Numbers • Estimating Answers

WARM-UP

Facts Practice: Circles (Test E)

Mental Math:

a. \$4.32 + \$2.98 **b.** \$12.50 ÷ 10 **c.** \$10.00 − \$8.98
d. 9 × 22 **e.** $\frac{5}{6} + \frac{5}{6}$ **f.** $\frac{3}{5}$ of 20
g. 6 × 6, ÷ 4, × 3, + 1, ÷ 4, × 8, − 1, ÷ 5, × 2, − 2, ÷ 2

Problem Solving:

Huck followed the directions on the treasure map. Starting at the big tree, he walked six paces north, turned left, and walked seven more paces. He turned left and walked five paces, turned left again, and walked four more paces. He then turned right and took one pace. In which direction was Huck facing, and how many paces was he from the big tree?

NEW CONCEPTS

Rounding whole numbers

The first sentence below uses an exact number to state the size of a crowd. The second sentence uses a round number.

There were 3947 fans at the game.

There were about 4000 fans at the game.

Round numbers are often used instead of exact numbers. One way to round numbers is to consider where the number is located on the number line.

Example 1 Use a number line to

(a) round 283 to the nearest hundred.

(b) round 283 to the nearest ten.

Solution (a) We draw a number line showing multiples of 100 and mark the estimated location of 283.

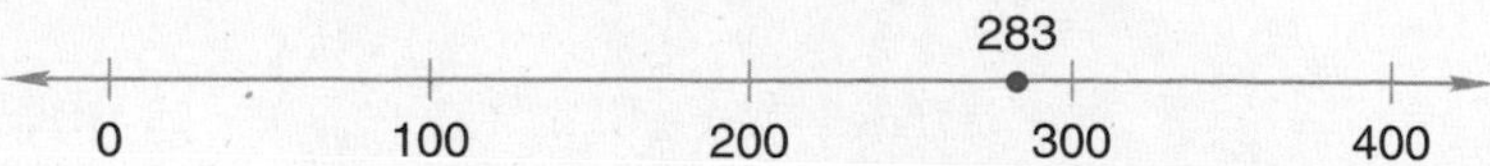

We see that 283 is between 200 and 300 and is closer to 300. To the nearest hundred, 283 rounds to **300.**

(b) We draw a number line showing the tens from 200 to 300 and mark the estimated location of 283.

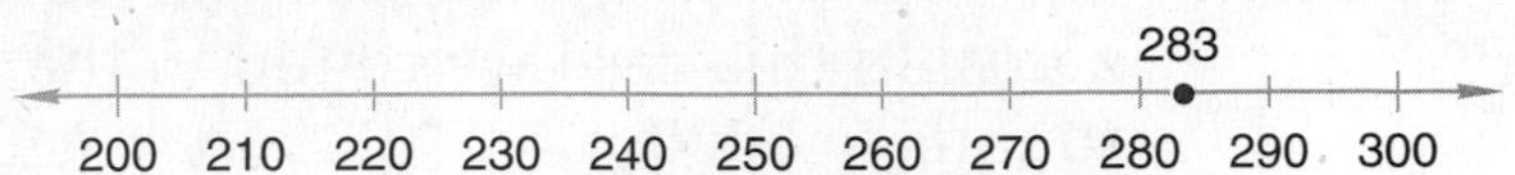

We see that 283 is between 280 and 290 and is closer to 280. To the nearest ten, 283 rounds to **280.**

Sometimes we are asked to round a number to a certain place value. We can use an underline and a circle to help us do this. We will underline the digit in the place to which we are rounding, and we will circle the next place to the right. Then we will follow these rules:

1. If the circled digit is 5 or more, we add 1 to the underlined digit. If the circled digit is less than 5, we leave the underlined digit unchanged.
2. We replace the circled digit and all digits to the right of the circled digit with zeros.

This rounding strategy is sometimes called the "4-5 split," because if the circled digit is 4 or less we round down, and if it is 5 or more we round up.

Example 2 (a) Round 283 to the nearest hundred.

(b) Round 283 to the nearest ten.

Solution (a) We underline the 2 since it is in the hundreds place. Then we circle the digit to its right.

2⑧3

Since the circled digit is 5 or more, we add 1 to the underlined digit, changing it from 2 to 3. Then we replace the circled digit and all digits to its right with zeros and get

300

(b) Since we are rounding to the nearest ten, we underline the tens digit and circle the digit to its right.

28③

Since the circled digit is less than 5, we leave the 8 unchanged. Then we replace the 3 with a zero and get

280

Example 3 Round 5280 so that there is one nonzero digit.

Solution We round the number so that all but one of the digits are zeros. In this case we round to the nearest thousand, so 5280 rounds to **5000.**

Example 4 Round 93,167,000 to the nearest million.

Solution To the nearest million, 93,①67,000 rounds to **93,000,000.**

Rounding mixed numbers When rounding a mixed number to a whole number, we need to determine whether the fraction part of the mixed number is greater than, equal to, or less than $\frac{1}{2}$. If the fraction is greater than or equal to $\frac{1}{2}$, the mixed number rounds up to the next whole number. If the fraction is less than $\frac{1}{2}$, the mixed number rounds down.

A fraction is greater than $\frac{1}{2}$ if the numerator of the fraction is more than half of the denominator. A fraction is less than $\frac{1}{2}$ if the numerator is less than half of the denominator.

Example 5 Round $14\frac{7}{12}$ to the nearest whole number.

Solution The mixed number $14\frac{7}{12}$ is between the consecutive whole numbers 14 and 15. We study the fraction to decide which is nearer. The fraction $\frac{7}{12}$ is greater than $\frac{1}{2}$ because 7 is more than half of 12. So $14\frac{7}{12}$ rounds to **15.**

Estimating answers Rounding can help us **estimate** the answers to arithmetic problems. Estimating is a quick and easy way to get close to an exact answer. Sometimes a close answer is "good enough," but even when an exact answer is necessary, estimating can help us determine whether our exact answer is reasonable. One way to estimate is to round the numbers before calculating.

Example 6 Mentally estimate:

(a) $5\frac{7}{10} \times 3\frac{1}{3}$ (b) 396×312 (c) $4160 \div 19$

Solution (a) We round each mixed number to the nearest whole number before we multiply.

$$5\frac{7}{10} \times 3\frac{1}{3}$$
$$\downarrow \quad \downarrow$$
$$6 \times 3 = \mathbf{18}$$

(b) We can round each number to the nearest ten or to the nearest hundred.

Problem	Rounded to tens	Rounded to hundreds
$\begin{array}{r} 396 \\ \times\ 312 \\ \hline \end{array}$	$\begin{array}{r} 400 \\ \times\ 310 \\ \hline \end{array}$	$\begin{array}{r} 400 \\ \times\ 300 \\ \hline \end{array}$

When mentally estimating we often round the numbers to one nonzero digit so that the calculation is easier to perform. In this case we round to the nearest hundred.

$$\begin{array}{r} 400 \\ \times\ 300 \\ \hline \mathbf{120{,}000} \end{array}$$

(c) We round each number so there is one nonzero digit before we divide.

$$\frac{4160}{19} \rightarrow \frac{4000}{20} = \mathbf{200}$$

Performing a quick mental estimate helps us determine whether the result of a more complicated calculation is reasonable.

LESSON PRACTICE

Practice set

a. Round 1760 to the nearest hundred.

b. Round 5489 to the nearest thousand.

c. Round 186,282 to the nearest thousand.

Estimate each answer:

d. 7986 − 3074

e. 297 × 31

f. 5860 ÷ 19

g. $12\frac{1}{4} \div 3\frac{7}{8}$

h. Calculate the area of this rectangle. After calculating, check the reasonableness of your answer by using round numbers to estimate the area.

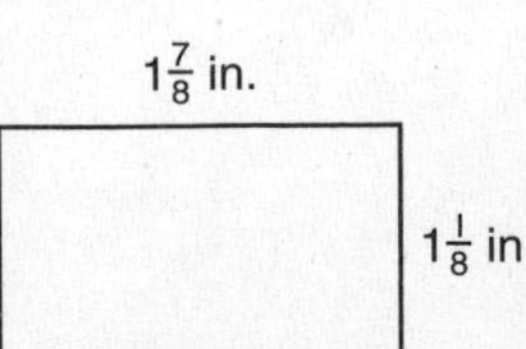

MIXED PRACTICE

Problem set

1. (16, 28) Lorenz jumped 16 feet 8 inches on his first try. How many inches did he jump on his first try?

2. (13) If 8 pounds of bananas cost $3.68, how can we find the cost per pound?

3. (28) On her first six tests Sandra's scores were 75, 70, 80, 80, 85, and 90. Find the mean of these six scores.

4. (5, 12) Two hundred nineteen billion, eight hundred million is how much less than one trillion? Use words to write your answer.

5. (22) In the following statement, change the percent to a reduced fraction. Then diagram the statement and answer the questions.

Forty percent of the 80 chips were blue.

(a) How many of the chips were blue?

(b) How many of the chips were not blue?

6. (27) (a) What is the least common multiple (LCM) of 4, 6, and 8?

(b) Use prime factorization to find the LCM of 16 and 36.

7. (19, 20) (a) What is the perimeter of this square?

(b) What is the area of this square?

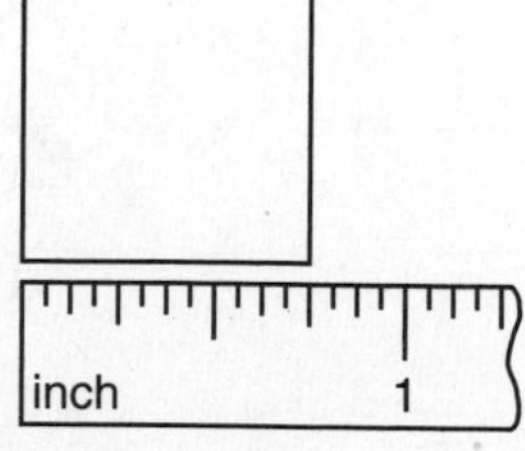

8. (29) (a) Round 366 to the nearest hundred.

(b) Round 366 to the nearest ten.

9. (29) Mentally estimate the sum of 6143 and 4952 by rounding each number to the nearest thousand before adding.

10. (26, 29) (a) Mentally estimate the following product by rounding each number to the nearest whole number before multiplying:

$$\frac{3}{4} \cdot 5\frac{1}{3} \cdot 1\frac{1}{8}$$

(b) Now find the exact product of these fractions and mixed numbers.

11. (15) Complete each equivalent fraction:

(a) $\frac{2}{3} = \frac{?}{30}$ (b) $\frac{?}{6} = \frac{25}{30}$

12. (20, 21) The prime factorization of 1000 is $2^3 \cdot 5^3$. Write the prime factorization of one billion using exponents.

In the figure below, quadrilaterals *ACDF, ABEF,* and *BCDE* are rectangles. Refer to the figure to answer problems 13–15.

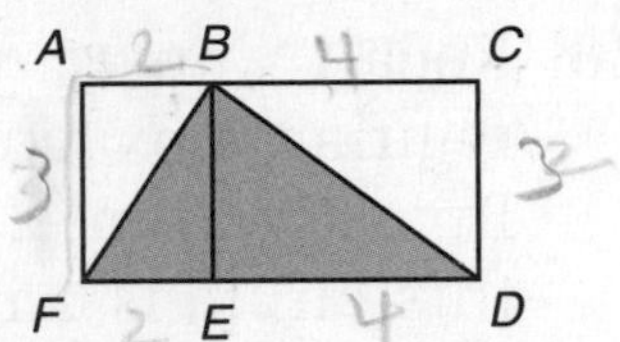

13. (8) (a) What percent of rectangle *ABEF* is shaded?

(b) What percent of rectangle *BCDE* is shaded?

(c) What percent of rectangle *ACDF* is shaded?

14. (19, 20) The relationships between the lengths of the sides of the rectangles are as follows:

$$AB + FE = BC$$

$$AF + CD = AC$$

$$AB = 2 \text{ in.}$$

(a) Find the perimeter of rectangle *ABEF*.

(b) Find the area of rectangle *BCDE*.

15. (18) Triangle *ABF* is congruent to ΔEFB.

(a) Which angle in ΔABF corresponds to $\angle EBF$ in ΔEFB?

(b) What is the measure of $\angle A$?

Solve:

16. $8^2 = 4m$ (3, 20)

17. $x + 4\frac{4}{9} = 15$ (23)

18. $3\frac{5}{9} = n - 4\frac{7}{9}$ (10, 15)

Simplify:

19. $6\frac{1}{3} - 5\frac{2}{3}$ (23)

20. $6\frac{2}{3} \div 5$ (26)

21. $1\frac{2}{3} \div 3\frac{1}{2}$ (26)

22. $\$7.49 \times 24$ (1)

23. Describe how to estimate the product of $5\frac{1}{3}$ and $4\frac{7}{8}$. (29)

24. Find the missing exponents. (20)

(a) $10^3 \cdot 10^3 = 10^m$ (b) $\frac{10^6}{10^3} = 10^n$

25. The rule of the following sequence is $k = 2^n + 1$. Find the fifth term of the sequence. (2)

3, 5, 9, 17, ...

26. Recall how you inscribed a regular hexagon in a circle in Investigation 2. If the radius of this circle is 1 inch, (19, Inv. 2)

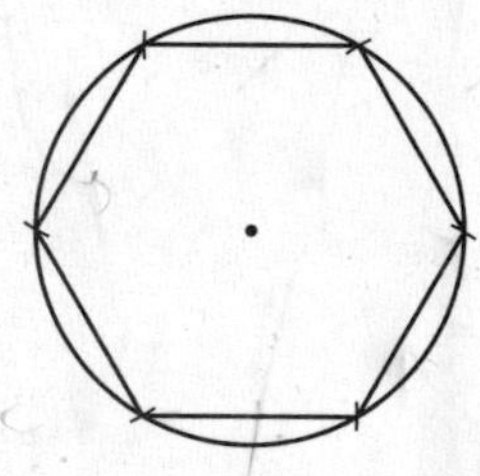

(a) what is the diameter of the circle?

(b) what is the perimeter of the hexagon?

27. Find fractions equivalent to $\frac{2}{3}$ and $\frac{1}{2}$ with denominators of 6. Subtract the smaller fraction you found from the larger fraction. (15)

28. What type of angle is (7)

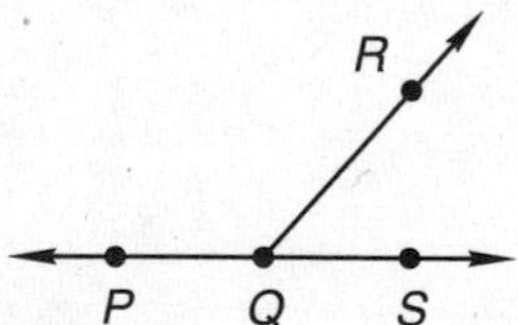

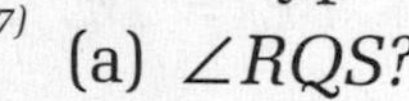
(a) $\angle RQS$?

(b) $\angle PQR$?

(c) $\angle PQS$?

29. If two full cups of water are poured from a full quart container, how many ounces of water would be left in the quart container? (16)

30. Find the perimeter of the hexagon at right. (19)

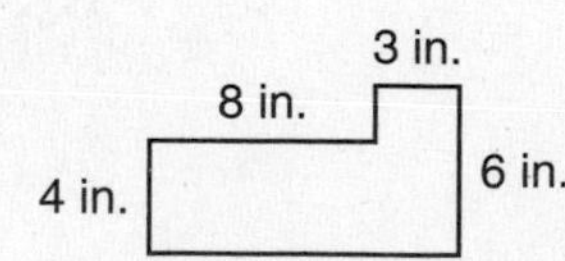

LESSON

30 Common Denominators • Adding and Subtracting Fractions with Different Denominators

WARM-UP

Facts Practice: Lines, Angles, Polygons (Test F)

Mental Math:

a. \$1.99 + \$1.99 b. \$0.15 × 1000 c. $\frac{3}{4} = \frac{?}{12}$

d. 5 × 84 e. $1\frac{2}{3} + 2\frac{2}{3}$ f. $\frac{3}{4}$ of 20

g. Find $\frac{1}{2}$ of 88, + 4, ÷ 8, × 5, − 5, double that number, − 2, ÷ 2, ÷ 2, ÷ 2.

Problem Solving:

Copy this problem and fill in the missing digits:

```
     3_
_3)_6_6
   1__
    _0_
    ___
      0
```

NEW CONCEPTS

Common denominators

When two fractions have the same denominator, we say they have **common denominators.**

$\frac{3}{8}$ $\frac{6}{8}$

These two fractions have common denominators.

$\frac{3}{8}$ $\frac{3}{4}$

These two fractions do not have common denominators.

If two fractions do not have common denominators, then one or both fractions can be renamed so both fractions do have common denominators. We remember that we can rename a fraction by multiplying it by a fraction equal to 1. Thus by multiplying by $\frac{2}{2}$, we can rename $\frac{3}{4}$ so that it has a denominator of 8.

$$\frac{3}{4} \cdot \frac{2}{2} = \frac{6}{8}$$

Example 1 Rename $\frac{2}{3}$ and $\frac{1}{4}$ so that they have common denominators.

Solution The denominators are 3 and 4. A common denominator for these two fractions would be any common multiple of 3 and 4. The **least common denominator** would be the least common multiple of 3 and 4, which is 12. We want to rename each fraction so that the denominator is 12.

$$\frac{2}{3} = \frac{}{12} \qquad \frac{1}{4} = \frac{}{12}$$

We multiply $\frac{2}{3}$ by $\frac{4}{4}$ and multiply $\frac{1}{4}$ by $\frac{3}{3}$.

$$\frac{2}{3} \cdot \frac{4}{4} = \frac{8}{12} \qquad \frac{1}{4} \cdot \frac{3}{3} = \frac{3}{12}$$

Thus $\frac{2}{3}$ and $\frac{1}{4}$ can be written with common denominators as

$$\mathbf{\frac{8}{12}} \quad \text{and} \quad \mathbf{\frac{3}{12}}$$

Fractions written with common denominators can be compared by simply comparing the numerators.

Example 2 Write these fractions with common denominators and then compare them.

$$\frac{5}{6} \bigcirc \frac{7}{9}$$

Solution The least common denominator for these fractions is the LCM of 6 and 9, which is 18.

$$\frac{5}{6} \cdot \frac{3}{3} = \frac{15}{18} \qquad \frac{7}{9} \cdot \frac{2}{2} = \frac{14}{18}$$

In place of $\frac{5}{6}$ we may write $\frac{15}{18}$, and in place of $\frac{7}{9}$ we may write $\frac{14}{18}$. Then we compare the renamed fractions.

$$\frac{15}{18} \bigcirc \frac{14}{18} \quad \text{renamed}$$

$$\mathbf{\frac{15}{18} > \frac{14}{18}} \quad \text{compared}$$

Adding and subtracting fractions with different denominators

To add or subtract two fractions that do not have common denominators, we first rename one or both fractions so they do have common denominators. Then we can add or subtract.

Example 3 Add: $\frac{3}{4} + \frac{3}{8}$

Solution First we write the fractions so they have common denominators. The denominators are 4 and 8. The least common multiple of 4 and 8 is 8. We rename $\frac{3}{4}$ so the denominator is 8 by multiplying by $\frac{2}{2}$. We do not need to rename $\frac{3}{8}$. Then we add the fractions and simplify.

$$\frac{3}{4} \cdot \frac{2}{2} = \frac{6}{8} \quad \text{renamed } \tfrac{3}{4}$$
$$+\ \frac{3}{8} \qquad = \frac{3}{8}$$
$$\frac{9}{8} \quad \text{added}$$

We finish by simplifying $\frac{9}{8}$.

$$\frac{9}{8} = \mathbf{1\frac{1}{8}}$$

Example 4 Subtract: $\frac{5}{6} - \frac{3}{4}$

Solution First we write the fractions so they have common denominators. The LCM of 6 and 4 is 12. We multiply $\frac{5}{6}$ by $\frac{2}{2}$ and multiply $\frac{3}{4}$ by $\frac{3}{3}$ so that both denominators are 12. Then we subtract the renamed fractions.

$$\frac{5}{6} \cdot \frac{2}{2} = \frac{10}{12} \quad \text{renamed } \tfrac{5}{6}$$
$$-\ \frac{3}{4} \cdot \frac{3}{3} = \frac{9}{12} \quad \text{renamed } \tfrac{3}{4}$$
$$\mathbf{\frac{1}{12}} \quad \text{subtracted}$$

Example 5 Subtract: $8\frac{2}{3} - 5\frac{1}{6}$

Solution We first write the fractions so that they have common denominators. The LCM of 3 and 6 is 6. We multiply $\frac{2}{3}$ by $\frac{2}{2}$ so that the denominator is 6. Then we subtract and simplify.

$$8\frac{2}{3} = 8\frac{4}{6} \quad \text{renamed } 8\tfrac{2}{3}$$
$$-\ 5\frac{1}{6} = 5\frac{1}{6}$$
$$3\frac{3}{6} = \mathbf{3\frac{1}{2}} \quad \text{subtracted and simplified}$$

Example 6 Add: $\frac{1}{2} + \frac{2}{3} + \frac{3}{4}$

Solution The denominators are 2, 3, and 4. The LCM of 2, 3, and 4 is 12. We rename each fraction so that the denominator is 12. Then we add and simplify.

$$\frac{1}{2} \cdot \frac{6}{6} = \frac{6}{12} \quad \text{renamed } \tfrac{1}{2}$$

$$\frac{2}{3} \cdot \frac{4}{4} = \frac{8}{12} \quad \text{renamed } \tfrac{2}{3}$$

$$+\ \frac{3}{4} \cdot \frac{3}{3} = \frac{9}{12} \quad \text{renamed } \tfrac{3}{4}$$

$$\frac{23}{12} = \mathbf{1\frac{11}{12}} \quad \text{added and simplified}$$

Example 7 Use prime factorization to help you add these fractions:

$$\frac{5}{32} + \frac{7}{24}$$

Solution We write the prime factorization of the denominators for both fractions.

$$\frac{5}{32} = \frac{5}{2 \cdot 2 \cdot 2 \cdot 2 \cdot 2} \qquad \frac{7}{24} = \frac{7}{2 \cdot 2 \cdot 2 \cdot 3}$$

The least common denominator of the two fractions is the least common multiple of the denominators. So the least common denominator is

$$2 \cdot 2 \cdot 2 \cdot 2 \cdot 2 \cdot 3 = 96$$

To rename the fractions with common denominators, we multiply $\frac{5}{32}$ by $\frac{3}{3}$, and we multiply $\frac{7}{24}$ by $\frac{2 \cdot 2}{2 \cdot 2}$.

$$\frac{5}{32} \cdot \frac{3}{3} = \frac{15}{96}$$

$$+\ \frac{7}{24} \cdot \frac{2 \cdot 2}{2 \cdot 2} = \frac{28}{96}$$

$$\mathbf{\frac{43}{96}}$$

LESSON PRACTICE

Practice set* Write the fractions so that they have common denominators. Then compare the fractions.

a. $\frac{3}{5} \bigcirc \frac{7}{10}$ **b.** $\frac{5}{12} \bigcirc \frac{7}{15}$

Add or subtract:

c. $\frac{3}{4} + \frac{5}{6} + \frac{3}{8}$

d. $7\frac{5}{6} - 2\frac{1}{2}$

e. $4\frac{3}{4} + 5\frac{5}{8}$

f. $4\frac{1}{6} - 2\frac{5}{9}$

g. Use prime factorization to help you subtract these fractions:

$$\frac{3}{25} - \frac{2}{45}$$

MIXED PRACTICE

Problem set

1. The 5 starters on the basketball team were tall. Their heights were 76 inches, 77 inches, 77 inches, 78 inches, and 82 inches. What was the average height of the 5 starters?
(28)

2. Marie bought 6 pounds of apples for $0.87 per pound and paid for them with a $10 bill. How much did she get back in change?
(28)

3. Fred lifted 317 rocks averaging 38 pounds each. He calculated that he had lifted over 120,000 pounds in all. Barney thought Fred's calculation was unreasonable. Do you agree or disagree with Barney? Why?
(29)

4. One hundred forty of the two hundred sixty flowers in the garden were purple. What fraction of the flowers in the garden were not purple?
(14)

5. In the following statement, change the percent to a reduced fraction. Then diagram the statement and answer the questions.
(22)

   The Daltons completed 30% of their 2140-mile trip the first day.

   (a) How many miles did they travel the first day?

   (b) How many miles of their trip do they still have to travel?

6. If the perimeter of a square is 5 feet, how many inches long is each side of the square?
(16, 19)

7. Use prime factorization to subtract these fractions:
(30)

$$\frac{1}{18} - \frac{1}{30}$$

8. (a) Round 36,467 to the nearest thousand.
(29)

(b) Round 36,467 to the nearest hundred.

9. Mentally estimate the quotient when 29,376 is divided by 49.
(29)

10. (a) Write 32% as a reduced fraction.
(15, 24)

(b) Use prime factorization to reduce $\frac{48}{72}$.

11. Write these fractions so that they have common denominators. Then compare the fractions.
(30)

$$\frac{5}{6} \bigcirc \frac{7}{8}$$

In the figure below, a 3-by-3-in. square is joined to a 4-by-4-in. square. Refer to the figure to answer problems 12 and 13.

12. (a) What is the area of the smaller square?
(20)

(b) What is the area of the larger square?

(c) What is the total area of the figure?

13. (a) What is the perimeter of the hexagon that is formed by joining the two squares?
(19)

(b) The perimeter of the hexagon is how many inches less than the combined perimeter of the two squares? Why?

14. (a) Write the prime factorization of 5184 using exponents.
(21)

(b) Use the answer to (a) to find $\sqrt{5184}$.

15. What is the mean of 5, 7, 9, 11, 12, 13, 24, 25, 26, and 28?
(28)

16. List the single-digit divisors of 5670.
(6)

Solve:

17. $6w = 6^3$
(3, 20)

18. $90° + 30° + a = 180°$
(3)

19. $\$45.00 = 36p$
(3)

20. $\frac{t}{32} = \$3.75$
(3)

Simplify:

21. $\frac{1}{2} + \frac{1}{3}$
(30)

22. $\frac{3}{4} - \frac{1}{3}$
(30)

23. $2\frac{5}{6} - 1\frac{1}{2}$
(30)

24. $\frac{4}{5} \cdot 1\frac{2}{3} \cdot 1\frac{1}{8}$
(26)

25. $1\frac{3}{4} \div 2\frac{2}{3}$
(26)

26. $3 \div 1\frac{7}{8}$
(26)

For problems 27 and 28, record an estimated answer and an exact answer.

27. $3\frac{2}{3} + 1\frac{5}{6}$
(30)

28. $5\frac{1}{8} - 1\frac{3}{4}$
(23, 30)

29. Draw a circle with a compass, and label the center point *O*. Draw chord *AB* through point *O*. Draw chord *CB* not through point *O*. Draw segment *CO*.
(Inv. 2)

30. Refer to the figure drawn in problem 29 to answer (a)–(c).
(Inv. 2)

(a) Which chord is a diameter?

(b) Which segments are radii?

(c) Which central angle is an angle of ΔOBC?

INVESTIGATION 3

Focus on

Coordinate Plane

By drawing two perpendicular number lines and extending the tick marks, we can create a grid over an entire plane called the **coordinate plane.** We can identify any point on the coordinate plane with two numbers.

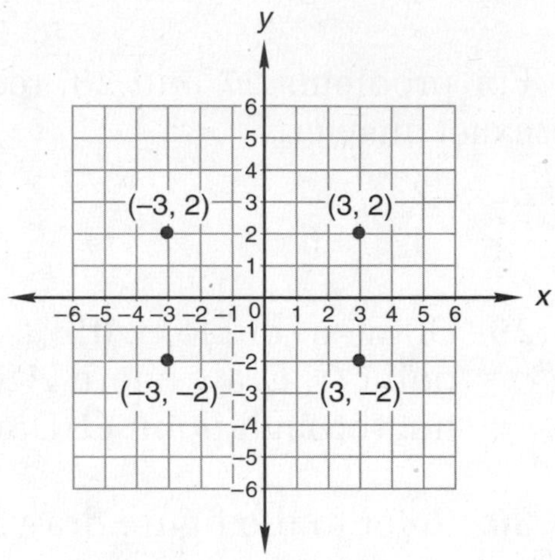

The horizontal number line is called the ***x*-axis.** The vertical number line is called the ***y*-axis.** The point at which the *x*-axis and the *y*-axis intersect is called the **origin.** The two numbers that indicate the location of a point are the **coordinates** of the point. The coordinates are written as a pair of numbers in parentheses, such as (3, 2). The first number shows the horizontal (↔) direction and distance from the origin. The second number shows the vertical (↕) direction and distance from the origin. The sign of the number indicates the direction. Positive coordinates are to the right or up. Negative coordinates are to the left or down. The origin is at point (0, 0).

The two axes divide the plane into four regions called **quadrants,** which are numbered counterclockwise, beginning with the upper right, as first, second, third, and fourth. The signs of the coordinates of each quadrant are shown below. Every point on a plane is either in a quadrant or on an axis.

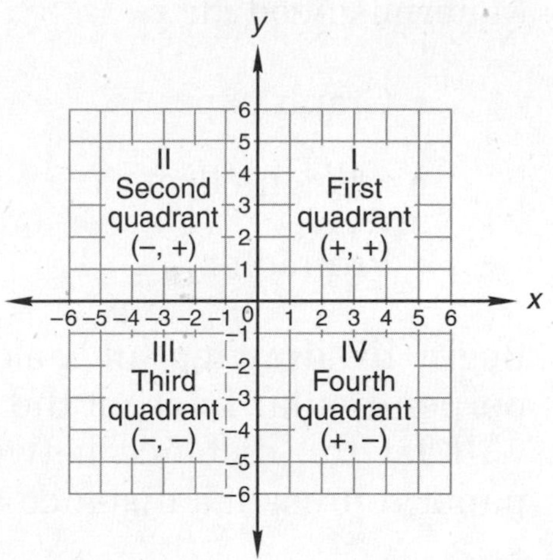

Example 1 Find the coordinates for points *A*, *B*, and *C* on this coordinate plane.

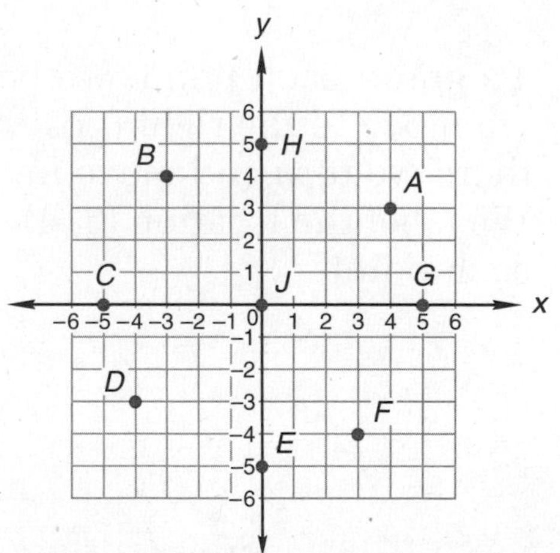

Solution We first find the point on the *x*-axis that is directly above, below, or on the designated point. That number is the first coordinate. Then we determine how many units above or below the *x*-axis the point is. That number is the second coordinate.

Point *A* **(4, 3)**

Point *B* **(–3, 4)**

Point *C* **(–5, 0)**

Activity: *Coordinate Plane*

Materials needed:

- Graph paper
- Straightedge
- Protractor

Begin by drawing an x-axis and y-axis by darkening two perpendicular lines on the graph paper. For this activity we will let the distance between adjacent lines on the graph paper represent a distance of one unit.

Example 2 Graph the following points on a coordinate plane:

(a) (3, 4) (b) (2, −3) (c) (−1, 2) (d) (0, −4)

Solution To graph each point, we begin at the origin. To graph (3, 4), we move to the right (positive) 3 units along the x-axis. From there we turn and move up (positive) 4 units and make a dot. We label the location (3, 4). We follow a similar procedure for each point.

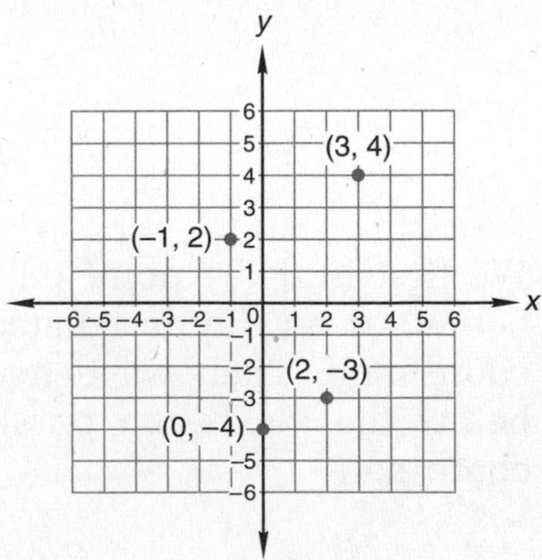

Example 3 The vertices of a square are located at (2, 2), (2, −1), (−1, −1), and (−1, 2). Draw the square and find its perimeter and area.

Solution We graph the vertices and draw the square.

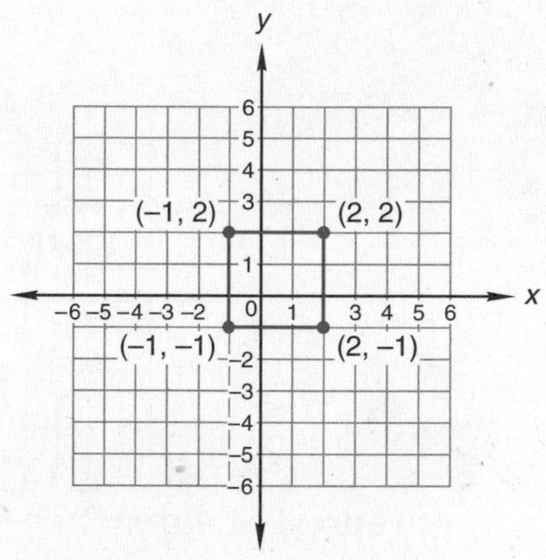

We find that each side of the square is 3 units long. So its perimeter is **12 units,** and its area is **9 square units.**

Example 4 Three vertices of a rectangle are located at (2, 1), (2, −1), and (−2, −1). Find the coordinates of the fourth vertex and the perimeter and area of the rectangle.

Solution We graph the given coordinates.

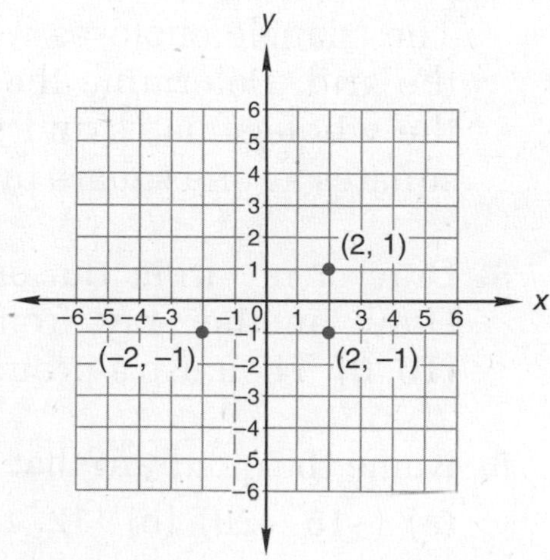

We see that the location of the fourth vertex is **(−2, 1),** which we graph. Then we draw the rectangle and find that it is 4 units long and 2 units wide. So its perimeter is **12 units,** and its area is **8 square units.**

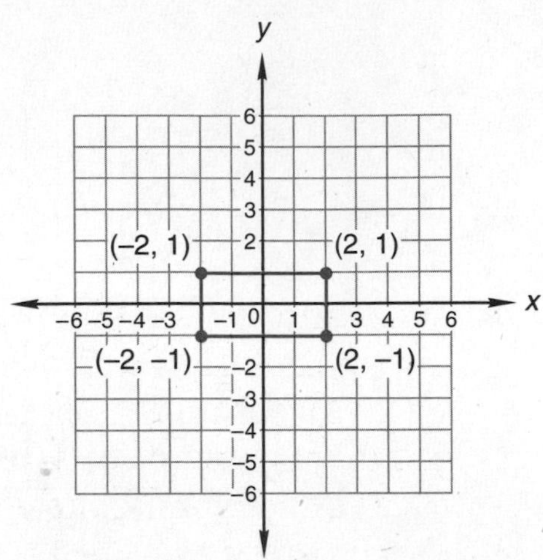

1. Graph these three points: (2, 4), (0, 2), and (–3, –1). Then draw a line that passes through these points. Name a point in the second quadrant that is on the line.

2. One vertex of a square is the origin. Two other vertices are located at (–2, 0) and (0, –2). What are the coordinates of the fourth vertex?

3. Find the perimeter and area of a rectangle whose vertices are located at (3, –1), (–2, –1), (–2, –4), and (3, –4).

4. Points (4, 4), (4, 0), and (0, 0) are the vertices of a triangle. The triangle encloses whole squares and half squares on the grid. Determine the area of the triangle by counting the whole squares and the half squares. (Count two half squares as one square unit.)

5. Draw a ray from the origin through the point (10, 10). Draw another ray from the origin through the point (10, 0). Then use a protractor to measure the angle.

6. Name the quadrant that contains each of these points:
(a) (–15, –20) (b) (12, 1) (c) (20, –20) (d) (–3, 5)

7. Draw ΔABC with vertices at A (0, 0), B (8, –8), and C (–8, –8). Use a protractor to find the measure of each angle of the triangle.

8. Shae wrote these directions for a dot-to-dot drawing. To complete the drawing, draw segments from point to point in the order given.

1. (0, 4)
2. (–3, –4)
3. (5, 1)
4. (–5, 1)
5. (3, –4)
6. (0, 4)

9. Plan and create a straight-segment drawing on graph paper. Determine the coordinates of the vertices. Then write directions for completing the dot-to-dot drawing for another person to follow. Include the directions "lift pencil" between consecutive coordinates of points not to be connected. If desired, have another person use your directions to create the drawing. Did the drawing turn out as planned?

LESSON

31 Reading and Writing Decimal Numbers

WARM-UP

Facts Practice: + – × ÷ Fractions (Test G)

Mental Math:

a. \$4.00 – 99¢ b. 7 × 35¢ c. $\frac{2}{3} = \frac{?}{12}$

d. Reduce $\frac{18}{24}$. e. $\sqrt{100} + 3^2$ f. $\frac{3}{4}$ of 60

g. Start with the number of degrees in a right angle, ÷ 2, + 5, ÷ 5, – 1, find the square root.

Problem Solving:

Find the next three numbers in this sequence:

..., 100, 121, 144, ...

NEW CONCEPT

We have used fractions and percents to name parts of a whole. We remember that a fraction has a numerator and a denominator. The denominator indicates the number of equal parts in the whole. The numerator indicates the number of parts that are selected.

$$\frac{\text{Number of parts selected}}{\text{Number of equal parts in the whole}} = \frac{3}{10}$$

Parts of a whole can also be named by using **decimal fractions.** In a decimal fraction we can see the numerator, but we cannot see the denominator. **The denominator of a decimal fraction is indicated by place value.** Below is the decimal fraction three tenths. We know the denominator is 10 because only one place is shown to the right of the decimal point.

0.3

The decimal fraction 0.3 and the common fraction $\frac{3}{10}$ are equivalent. Both are read "three tenths."

$$0.3 = \frac{3}{10} \qquad \text{three tenths}$$

A decimal fraction written with two digits after the decimal point (two decimal places) is understood to have a denominator of 100, as we show here:

$$0.03 = \frac{3}{100} \qquad \text{three hundredths}$$

$$0.21 = \frac{21}{100} \qquad \text{twenty-one hundredths}$$

A number that contains a decimal fraction is called a **decimal number** or just a **decimal.**

decimal point — decimal fraction

12.345

decimal number
or
decimal

Example 1 Write seven tenths (a) as a fraction and (b) as a decimal.

Solution (a) $\mathbf{\frac{7}{10}}$ (b) **0.7**

Example 2 Name the shaded part of this square

(a) as a fraction.

(b) as a decimal.

Solution (a) $\mathbf{\frac{23}{100}}$ (b) **0.23**

In our number system the place a digit occupies has a value called **place value.** We remember that places to the left of the decimal point have values of 1, 10, 100, 1000, and so on, becoming greater and greater. Places to the right of the decimal point have values of $\frac{1}{10}$, $\frac{1}{100}$, $\frac{1}{1000}$, and so on, becoming less and less. This chart shows decimal place values from the millions place through the millionths place:

Decimal Place Values

millions		hundred thousands	ten thousands	thousands		hundreds	tens	ones	decimal point	tenths	hundredths	thousandths	ten-thousandths	hundred-thousandths	millionths
__	,	__	__	__	,	__	__	__	.	__	__	__	__	__	__
1,000,000		100,000	10,000	1000		100	10	1		$\frac{1}{10}$	$\frac{1}{100}$	$\frac{1}{1000}$	$\frac{1}{10,000}$	$\frac{1}{100,000}$	$\frac{1}{1,000,000}$

Example 3 In the number 12.34579, which digit is in the thousandths place?

Solution The thousandths place is the third place to the right of the decimal point and is occupied by the **5.**

Example 4 Name the place occupied by the 7 in 4.63471.

Solution The 7 is in the fourth place to the right of the decimal point. This is the **ten-thousandths place.**

To read a decimal number, we first read the whole-number part, and then we read the fraction part. To read the fraction part of a decimal number, we read the digits to the right of the decimal point as though we were reading a whole number. This number is the numerator of the decimal fraction. Then we say the name of the last decimal place. This number is the denominator of the decimal fraction.

Example 5 Read this decimal number: 123.123

Solution First we read the whole-number part. *When we come to the decimal point, we say "and."* Then we read the fraction part, ending with the name of the last decimal place.

We say "and" for the decimal point.

123.12<u>3</u>

We say "thousandths" to conclude naming the number.

One hundred twenty-three and one hundred twenty-three thousandths

Example 6 Use digits to write these decimal numbers:

(a) Seventy-five thousandths

(b) One hundred and eleven hundredths

Solution (a) The last word tells us the last place in the decimal number. "Thousandths" means there are three places to the right of the decimal point.

. ___ ___ ___

We fit the digits of 75 into the places so the 5 is in the last place. We write zero in the remaining place.

. $\underline{0}$ $\underline{7}$ $\underline{5}$

Decimal numbers without a whole-number part are usually written with a zero in the ones place. Therefore, we will write the decimal number "seventy-five thousandths" as follows:

0.075

(b) To write "one hundred and eleven hundredths," we remember that the word *and* separates the whole-number part from the fraction part. First we write the whole-number part followed by a decimal point for "and":

100.

Then we write the fraction part. We shift our attention to the last word to find out how many decimal places there are. "Hundredths" means there are two decimal places.

100. ___ ___

Now we fit "eleven" into the two decimal places, as follows:

100.11

LESSON PRACTICE

Practice set* **a.** Write three hundredths as a fraction. Then write three hundredths as a decimal.

b. Name the shaded part of the circle both as a fraction and as a decimal.

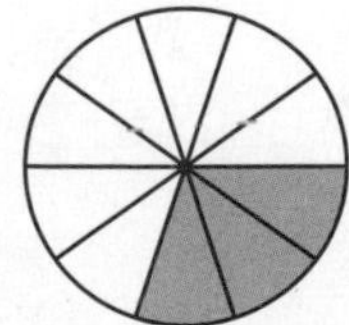

c. In the number 16.57349, which digit is in the thousandths place?

d. The number 36.4375 has how many decimal places?

Use words to write each decimal number:

e. 25.134

f. 100.01

Use digits to write each decimal number:

g. one hundred two and three tenths

h. one hundred twenty-five ten-thousandths

i. three hundred and seventy-five thousandths

MIXED PRACTICE

Problem set

1. (28) James and his brother are putting their money together to buy a radio that costs \$89.89. James has \$26.47. His brother has \$32.54. How much more money do they need to buy the radio?

2. (28) Norton read 4 books during his vacation. The first book had 326 pages, the second had 288 pages, the third had 349 pages, and the fourth had 401 pages. The 4 books he read had an average of how many pages per book?

3. (13) A one-year subscription to the monthly magazine costs \$15.96. At this price, what is the cost for each issue?

4. (12) The settlement at Jamestown began in 1607. This was how many years after Columbus reached the Americas in 1492?

5. (19) A square and a regular hexagon share a common side. The perimeter of the square is 24 in. Describe how to find the perimeter of the hexagon.

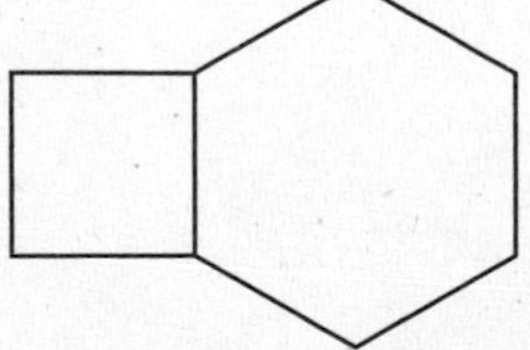

6. (22) In the following statement, change the percent to a reduced fraction. Then diagram the statement and answer the questions.

Kelly correctly answered 80% of the 20 questions on the test.

(a) How many questions did Kelly answer correctly?

(b) How many questions did Kelly answer incorrectly?

7. (29) Round 481,462

(a) to the nearest hundred thousand.

(b) to the nearest thousand.

8. (29) Mentally estimate the difference between 49,623 and 20,162.

9. (31) Name the shaded part of this square

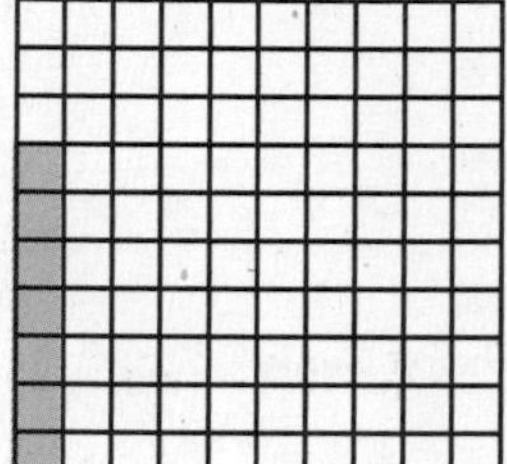

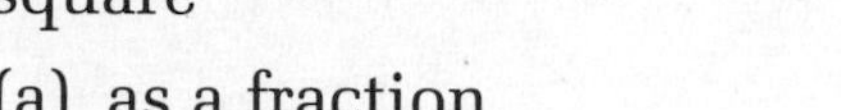

(a) as a fraction.

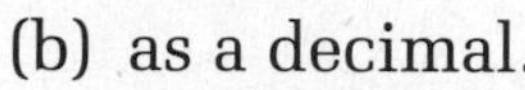

(b) as a decimal.

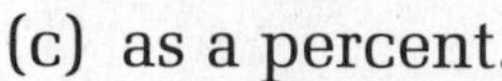

(c) as a percent.

10. (31) In the number 9.87654, which digit is in the hundredths place?

11. (31) Replace each circle with the proper comparison symbol:

(a) $\frac{3}{10} \bigcirc 0.3$ (b) $\frac{3}{100} \bigcirc 0.3$

12. (Inv. 3) The vertices of a square are located at (3, 3), (3, –3), (–3, –3), and (–3, 3).

(a) What is the perimeter of the square?

(b) What is the area of the square?

13. (15) Complete each equivalent fraction:

(a) $\frac{5}{?} = \frac{15}{24}$ (b) $\frac{7}{12} = \frac{?}{24}$ (c) $\frac{?}{6} = \frac{4}{24}$

14. (21) (a) Write the prime factorization of 2025 using exponents.

(b) Find $\sqrt{2025}$.

15. (18) Draw two parallel lines. Then draw two more parallel lines that are perpendicular to the first pair of lines. Label the points of intersection *A*, *B*, *C*, and *D* consecutively in a counterclockwise direction. Draw segment *AC*. Refer to the figure to answer (a) and (b):

(a) What kind of quadrilateral is figure *ABCD*?

(b) Triangles *ABC* and *CDA* are congruent. Which angle in $\triangle ABC$ corresponds to $\angle DAC$ in $\triangle CDA$?

Solve:

16. (3) $9n = 6 \cdot 12$

17. (3) $90° + 45° + b = 180°$

18. (3) $\$98.75 + w = \$220.15$

19. (3) $\frac{m}{48} = \$4.65$

Simplify:

20. (30) $\frac{1}{2} + \frac{2}{3}$

21. (9, 24) $\frac{1}{2} - \left(\frac{3}{4} \cdot \frac{2}{3}\right)$

22. (30) $3\frac{5}{6} - \frac{1}{3}$

23. (26) $\frac{5}{8} \cdot 2\frac{2}{5} \cdot \frac{4}{9}$

24. (26) $2\frac{2}{3} \div 1\frac{3}{4}$

25. (26) $1\frac{7}{8} \div 3$

26. (30) $3\frac{1}{2} + 1\frac{5}{6}$

27. (23, 30) $5\frac{1}{4} - 1\frac{5}{8}$

28. (1, 30) Evaluate this expression for $a = 3$ and $b = 4$:

$$\frac{b}{a} + \frac{a}{b}$$

29. (2, 20) The rule of the following sequence is $k = 10^n$. Use words to name the sixth term.

10, 100, 1,000, ...

30. (17) (a) A half circle or half turn measures how many degrees?

(b) A quarter of a circle measures how many degrees?

(c) An eighth of a circle measures how many degrees?

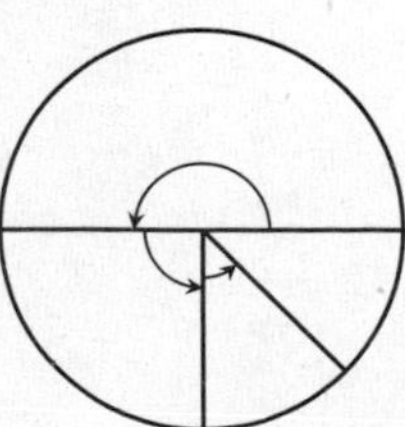

LESSON

32 Metric System

WARM-UP

Facts Practice: + − × ÷ Fractions (Test G)

Mental Math:

a. \$3.76 − 99¢ **b.** 8 × 25¢ **c.** $\frac{5}{6} = \frac{?}{24}$

d. Reduce $\frac{12}{20}$. **e.** $3^2 + 4^2$ **f.** $\frac{2}{5}$ of 30

g. Start with the number of sides of an octagon, × 5, + 2, ÷ 6, × 5, + 1, $\sqrt{\ }$, ÷ 3.

Problem Solving:

The diameter of a penny is $\frac{3}{4}$ inch. How many pennies placed side by side would it take to make a row of pennies 1 foot long?

NEW CONCEPT

The U.S. Customary System of measurement, with units like inches, feet, miles, pounds, and gallons, was once used throughout the former British Empire and was known as the English system. Over the years countries have converted to the **metric system** of measurement, which is now formally called the International System of Units (or SI, for *Système International d'Unités*). The metric system has two primary advantages over the U.S. Customary System: it is a decimal system, and the units of one category of measurement are linked to units of other categories of measurement.

The metric system is a decimal system in that units within a category of measurement differ by a factor, or power, of 10. The U.S. Customary System is not a decimal system, so converting between units is more difficult. Here we show some equivalent measures of length in the metric system:

Units of Length

10 millimeters (mm)	=	1 centimeter (cm)
1000 millimeters (mm)	=	1 meter (m)
100 centimeters (cm)	=	1 meter (m)
1000 meters (m)	=	1 kilometer (km)

Notice that the basic unit of length in the metric system is the meter. Units larger than a meter or smaller than a meter are indicated by prefixes that are consistent across the categories of measurement. These prefixes are shown in the table below with their numerical relationship to the basic unit:

Examples of Metric Prefixes

Prefix	Unit	Relationship
kilo-	kilometer (km)	1000 meters
hecto-	hectometer (hm)	100 meters
deka-	dekameter (dam)	10 meters
	meter (m)	
deci-	decimeter (dm)	$\frac{1}{10}$ meter
centi-	centimeter (cm)	$\frac{1}{100}$ meter
milli-	millimeter (mm)	$\frac{1}{1000}$ meter

As we move up the table, the units become larger and the number of units needed to describe a length decreases:

$$1000 \text{ mm} = 100 \text{ cm} = 10 \text{ dm} = 1 \text{ m}$$

As we move down the table, the units become smaller and the number of units required to describe a length increases:

$$1 \text{ km} = 10 \text{ hm} = 100 \text{ dam} = 1000 \text{ m}$$

Note: One meter is about 3 inches longer than a yard, and 1 inch is exactly 2.54 centimeters.

Example 1 (a) Five kilometers is how many meters?

(b) Three hundred centimeters is how many meters?

Solution (a) One kilometer is 1000 meters, so 5 kilometers is **5000 meters.**

(b) A centimeter is $\frac{1}{100}$ of a meter (just as a cent is $\frac{1}{100}$ of a dollar). One hundred centimeters equals 1 meter, so 300 centimeters equals **3 meters.**

The liter is the basic unit of capacity in the metric system. We are familiar with 2-liter bottles of soft drink. Each 2-liter bottle can hold a little more than a half gallon, so a liter is a little more than a quart. A milliliter is $\frac{1}{1000}$ of a liter.

Units of Capacity

1000 milliliters (mL) = 1 liter (L)

Example 2 A 2-liter bottle can hold how many milliliters of beverage?

Solution One liter is 1000 mL, so 2 L is **2000 mL.**

The basic unit of mass in the metric system is the kilogram. For scientific purposes, we distinguish between mass and weight. The weight of an object varies with the gravitational force, while its mass does not vary. The weight of an object on the Moon is about $\frac{1}{6}$ of its weight on Earth, yet its mass is the same. The mass of this book is about one kilogram. A gram is $\frac{1}{1000}$ of a kilogram—about the mass of a paperclip. A milligram, $\frac{1}{1000}$ of a gram and $\frac{1}{1,000,000}$ of a kilogram, is a unit used for measuring the mass of smaller quantities of matter, such as the amount of vitamins in the food we eat.

Units of Mass/Weight

1000 grams (g) = 1 kilogram (kg)
1000 milligrams (mg) = 1 gram

Although a kilogram is a unit for measuring mass anywhere in the universe, here on Earth shopkeepers use kilograms to measure the weights of the goods they buy and sell. A kilogram mass weighs about 2.2 pounds.

Example 3 How many 250-mg tablets of vitamin C equal one gram of vitamin C?

Solution A gram is 1000 mg, so **four** 250-mg tablets of vitamin C total one gram of vitamin C.

The **Celsius** and **Kelvin** scales are used by scientists to measure temperature. Both are **centigrade** scales because there are 100 gradations, or degrees, between the freezing and boiling temperatures of water. The Celsius scale places 0°C at the freezing point of water. The Kelvin scale places 0 K at **absolute zero,** which is 273 centigrade degrees below the freezing temperature of water (–273°C). The Celsius scale is more commonly used by the general population. Below we show frequently referenced temperatures on the Celsius scale.

Celsius Temperature Scale

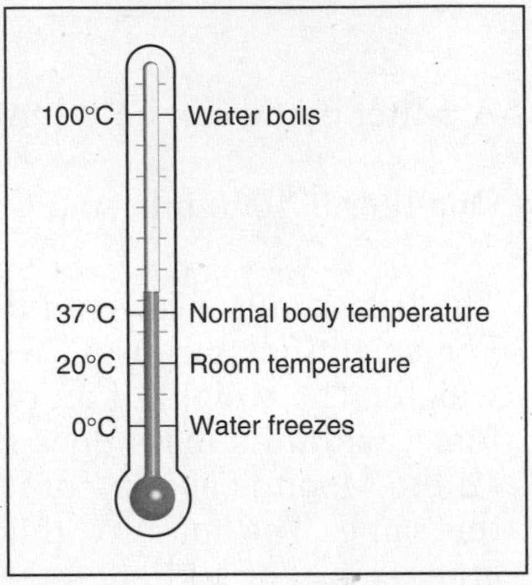

Example 4 A temperature increase of 100° on the Celsius scale is an increase of how many degrees on the Fahrenheit scale?

Solution The Celsius and Fahrenheit scales are different scales. An increase of 1°C is not equivalent to an increase of 1°F. On the Celsius scale there are 100° between the freezing point of water and the boiling point of water. On the Fahrenheit scale water freezes at 32° and boils at 212°, a difference of 180°. So an increase of 100°C is an increase of **180°F.**

LESSON PRACTICE

Practice set

a. The closet door is 2 meters tall. How many centimeters is that?

b. A 1-gallon plastic jug can hold about how many liters of milk?

c. A metric ton is 1000 kilograms, so a metric ton is about how many pounds?

d. A temperature increase of 10° on the Celsius scale is equivalent to an increase of how many degrees on the Fahrenheit scale? (See example 4.)

e. After running 800 meters of a 3-kilometer race, Michelle still had how many meters to run?

f. A 30-cm ruler broke into two pieces. One piece was 120 mm long. How long was the other piece? Express your answer in millimeters.

MIXED PRACTICE

Problem set

1. (28) There were 3 towns on the mountain. The population of Hazelhurst was 4248. The population of Baxley was 3584. The population of Jesup was 9418. What was the average population of the 3 towns on the mountain?

2. (28) The film was a long one, lasting 206 minutes. How many hours and minutes long was the film?

3. (14) A mile is 1760 yards. Claudia ran 440 yards. What fraction of a mile did she run?

4. (18, 19) A square and a regular pentagon share a common side. The perimeter of the square is 20 cm. What is the perimeter of the pentagon?

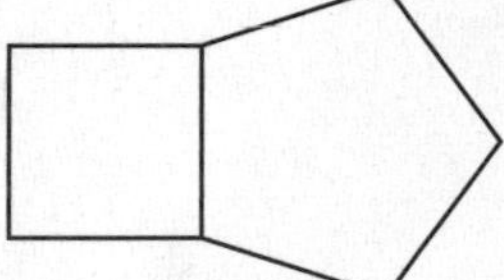

5. (29) Round 3,197,270

(a) to the nearest million.

(b) to the nearest hundred thousand.

6. (29) Mentally estimate the product of 313 and 489.

7. (22) Diagram this statement. Then answer the questions that follow.

Five eighths of the troubadour's 200 songs were about love and chivalry.

(a) How many of the songs were about love and chivalry?

(b) How many of the songs were not about love and chivalry?

8. (31) (a) What fraction of the rectangle is not shaded?

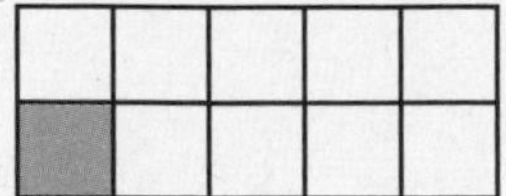

(b) What decimal part of the rectangle is not shaded?

(c) What percent of the rectangle is not shaded?

9. (31) Use words to write 3.025.

10. (31) Use digits to write the decimal number seventy-six and five hundredths.

11. (27) Instead of dividing \$15.00 by $2\frac{1}{2}$, double both numbers and then find the quotient.

12. (5, 21) (a) Write 2500 in expanded notation.

(b) Write the prime factorization of 2500 using exponents.

(c) Find $\sqrt{2500}$.

13. (13) If 35 liters of petrol cost \$21.00, what is the price per liter?

14. (17) Use a protractor to draw a triangle that has a 90° angle and a 45° angle.

In the figure below, a 6-by-6-cm square is joined to an 8-by-8-cm square. Refer to the figure to answer problems 15 and 16.

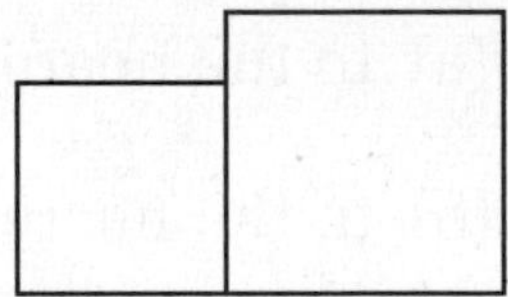

15. (20) (a) What is the area of the smaller square?

(b) What is the area of the larger square?

(c) What is the total area of the figure?

16. (19) What is the perimeter of the hexagon that is formed by joining the squares?

Solve:

17. (3) $10 \cdot 6 = 4w$

18. (3) $180° - s = 65°$

Simplify:

19. (30) $\frac{1}{4} + \frac{3}{8} + \frac{1}{2}$

20. (30) $\frac{5}{6} - \frac{3}{4}$

21. (30) $\frac{5}{16} - \frac{3}{20}$

22. (26) $\frac{8}{9} \cdot 1\frac{1}{5} \cdot 10$

23. (23, 30) $6\frac{1}{6} - 2\frac{1}{2}$

24. (30) $4\frac{5}{8} + 1\frac{1}{2}$

25. (25) $\frac{2}{3} + \left(\frac{2}{3} \div \frac{1}{2}\right)$

26. (24) $\frac{25}{36} \cdot \frac{9}{10} \cdot \frac{8}{15}$

For problems 27 and 28, record an estimated answer and an exact answer:

27. (26) $5\frac{2}{5} \div \frac{9}{10}$

28. (30) $7\frac{3}{4} + 1\frac{7}{8}$

29. (Inv. 3) The coordinates of three vertices of a rectangle are (–5, 3), (–5, –2), and (2, –2).

(a) What are the coordinates of the fourth vertex?

(b) What is the area of the rectangle?

30. (Inv. 2) Refer to the figure below to answer (a)–(c).

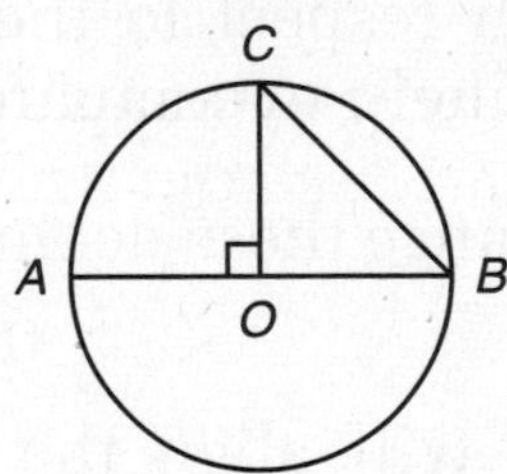

(a) Which chord is not a diameter?

(b) Name a central angle that is a right angle.

(c) Name an inscribed angle.

LESSON

33 Comparing Decimals • Rounding Decimals

WARM-UP

Facts Practice: Lines, Angles, Polygons (Test F)

Mental Math:

a. \$2.84 − 99¢ **b.** 6 × 55¢ **c.** $\frac{3}{8} = \frac{?}{24}$

d. Reduce $\frac{24}{30}$. **e.** $5^2 - \sqrt{25}$ **f.** $\frac{5}{6}$ of 30

g. Think of an equivalent division problem for 600 ÷ 50. Then find the quotient.

Problem Solving:

When Bill, Phil, Jill, and Gil entered the room, they all shook hands with each other. How many handshakes were there in all? Diagram the situation. (If three partners are available, you may wish to act out the story.)

NEW CONCEPTS

Comparing decimals When comparing decimal numbers, it is necessary to consider place value. The value of a place is determined by its position with respect to the decimal point. Aligning decimal points can help to compare decimal numbers digit by digit.

Example 1 Arrange these decimal numbers in order from least to greatest:

0.13 0.128 0.0475

Solution We will align the decimal points and consider the digits column by column. First we look at the tenths place.

↓
0.13
0.128
0.0475

Two of the decimal numbers have a 1 in the tenths place, while the third number has a 0. So we can determine that 0.0475 is the least of the three numbers. Now we look at the hundredths place to compare the remaining two numbers.

↓
0.13
0.128

Since 0.128 has a 2 in the hundredths place, it is less than 0.13, which has a 3 in the hundredths place. So from least to greatest the order is

0.0475, 0.128, 0.13

Note that terminal zeros on a decimal number add no value to the decimal number.

$$1.3 = 1.30 = 1.300 = 1.3000$$

When we compare two decimal numbers, it may be helpful to insert terminal zeros so that both numbers will have the same number of digits to the right of the decimal point. We will practice this technique in the next few examples.

Example 2 Compare: 0.12 ◯ 0.012

Solution So that each number has the same number of decimal places, we insert a terminal zero in the number on the left and get

0.120 ◯ 0.012

One hundred twenty thousandths is greater than twelve thousandths, so we write our answer this way:

0.12 > 0.012

Example 3 Compare: 0.4 ◯ 0.400

Solution We can delete two terminal zeros from the number on the right and get

$$0.4 = 0.4$$

We could have added terminal zeros to the number on the left to get

$$0.400 = 0.400$$

We write our answer this way:

0.4 = 0.400

Example 4 Compare: 1.232 ◯ 1.23185

Solution We insert two terminal zeros in the number on the left and get

1.23200 ◯ 1.23185

Since 1.23200 is greater than 1.23185, we write

1.232 > 1.23185

Rounding decimals To round decimal numbers, we can use the same procedure that we use to round whole numbers.

Example 5 Round 3.1<u>4</u>159 to the nearest hundredth.

Solution The hundredths place is two places to the right of the decimal point. We underline the digit in that place and circle the digit to its right.

$$3.1\underline{4}\enclose{circle}{1}59$$

Since the circled digit is less than 5, we leave the underlined digit unchanged. Then we replace the circled digit and all digits to the right of it with zeros.

$$3.14000$$

Terminal zeros to the right of the decimal point do not serve as placeholders as they do in whole numbers. After rounding decimal numbers, we should remove terminal zeros to the right of the decimal point.

3.14~~000~~ ⟶ **3.14**

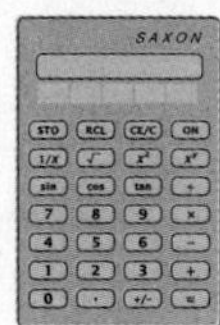

Note that a calculator simplifies decimal numbers by omitting from the display extraneous (unnecessary) zeros. For example, enter the following sequence of keystrokes:

[1] [0] [.] [2] [0] [3] [0] [0]

Notice that all entered digits are displayed. Now press the [=] key, and observe that the unnecessary zeros disappear from the display.

Example 6 Round 4396.4315 to the nearest hundred.

Solution We are rounding to the nearest hundred, not to the nearest hundredth.

$$4\underline{3}\enclose{circle}{9}6.4315$$

Since the circled digit is 5 or more, we increase the underlined digit by 1. All the following digits become zeros.

$$4400.0000$$

Zeros at the end of the whole-number part are needed as placeholders. Terminal zeros to the right of the decimal point are not needed as placeholders. We remove these zeros.

4400.~~0000~~ ⟶ **4400**

Example 7 Round 38.62 to the nearest whole number.

Solution To round a number to the nearest whole number, we round to the ones place.

$$3\underline{8}.\textcircled{6}2 \longrightarrow 39.\cancel{00} \longrightarrow \mathbf{39}$$

Example 8 Estimate the product of 12.21 and 4.9 by rounding each number to the nearest whole number before multiplying.

Solution We round 12.21 to 12 and 4.9 to 5. Then we multiply 12 and 5 and find that the estimated product is **60.** (The actual product is 59.829.)

LESSON PRACTICE

Practice set* Compare:

a. 10.30 ◯ 10.3

b. 5.06 ◯ 5.60

c. 1.1 ◯ 1.099

d. Round 3.14159 to the nearest ten-thousandth.

e. Round 365.2418 to the nearest hundred.

f. Round 57.432 to the nearest whole number.

g. Simplify 10.2000 by removing extraneous zeros.

h. Estimate the sum of 8.65, 21.7, and 11.038 by rounding each decimal number to the nearest whole number before adding.

MIXED PRACTICE

Problem set

1. (28) The high jumper set a new personal record when she cleared 5 feet 8 inches. How can we find how many inches high 5 feet 8 inches is?

2. (28) During the first week of November the daily high temperatures in degrees Fahrenheit were 42°F, 43°F, 38°F, 47°F, 51°F, 52°F, and 49°F. What was the average daily high temperature during the first week of November?

3. (11) In 10 years the population increased from 87,196 to 120,310. By how many people did the population increase in 10 years?

4. (2) Find the next two numbers in this sequence:

..., 120, 105, 90, 75, ...

5. (19) A regular hexagon and a regular octagon share a common side. If the perimeter of the hexagon is 24 cm, what is the perimeter of the octagon?

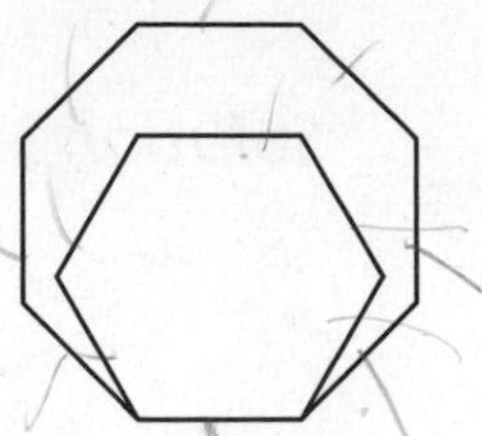

6. (22) Diagram this statement. Then answer the questions that follow.

One third of the 60 questions on the test were true-false.

(a) How many of the questions on the test were true-false?

(b) How many of the questions on the test were not true-false?

(c) What percent of the questions were true-false?

7. (Inv. 3) Find the area of a square whose vertices have the coordinates (3, 6), (3, 1), (−2, 1), and (−2, 6).

8. (33) (a) Round 15.73591 to the nearest hundredth.

(b) Estimate the product of 15.73591 and 3.14 by rounding each decimal number to the nearest whole number before multiplying.

9. (31) Use words to write each of these decimal numbers:

(a) 150.035

(b) 0.0015

10. (31) Use digits to write each of these decimal numbers:

(a) one hundred twenty-five thousandths

(b) one hundred and twenty-five thousandths

11. (33) Replace each circle with the proper comparison symbol:

(a) 0.128 ◯ 0.14 (b) 0.03 ◯ 0.0015

12. (32) Find the length of this segment

cm 1 2 3 4 5

(a) in centimeters.

(b) in millimeters.

13. (17) Draw the straight angle *AOC*. Then use a protractor to draw ray *OD* so that angle *COD* measures 60°.

14. (2) If we multiply one integer by another integer that is a whole number but not a counting number, what is the product?

15. (27) Use prime factorization to find the least common multiple of 63 and 49.

Solve:

16. (3) $8m = 4 \cdot 18$

17. (3) $135° + a = 180°$

Simplify:

18. (30) $\frac{3}{4} + \frac{5}{8} + \frac{1}{2}$

19. (30) $\frac{3}{4} - \frac{1}{6}$

20. (30) $4\frac{1}{2} - \frac{3}{8}$

21. (26) $\frac{3}{8} \cdot 2\frac{2}{5} \cdot 3\frac{1}{3}$

22. (26) $2\frac{7}{10} \div 5\frac{2}{5}$

23. (26) $5 \div 4\frac{1}{6}$

24. (23, 30) $6\frac{1}{2} - 2\frac{5}{6}$

25. (25) $\frac{3}{4} + \left(\frac{1}{2} \div \frac{2}{3}\right)$

26. (1) $40.00 ÷ 16

27. (a) Solve: $54 = 54 + y$
(2)
(b) What property is illustrated by the equation in (a)?

28. Consider the following division problem. Without
(29) dividing, decide whether the quotient will be greater than 1 or less than 1. How did you decide?

$$5 \div 4\frac{1}{6}$$

29. When Kelly saw the following addition problem, she
(29) knew that the sum would be greater than 13 and less than 15. How did she know?

$$8\frac{7}{8} + 5\frac{2}{3}$$

30. Use a protractor to draw a triangle that has a 30° angle
(17) and a 60° angle.

LESSON

34 Decimal Numbers on the Number Line

WARM-UP

Facts Practice: + − × ÷ Fractions (Test G)

Mental Math:

a. \$6.48 − 98¢ b. 5 × 48¢ c. $\frac{3}{5} = \frac{?}{30}$

d. Reduce $\frac{16}{24}$. e. $\sqrt{36} \cdot \sqrt{49}$ f. $\frac{2}{3}$ of 36

g. Square the number of sides on a pentagon, double that number, − 1, $\sqrt{\ }$, × 4, − 1, ÷ 3, $\sqrt{\ }$.

Problem Solving:

Jamaal glued 27 small blocks together to make this cube. Then he painted the six faces of the cube. Later the cube broke apart into 27 blocks. How many of the small blocks had 3 painted faces? ... 2 painted faces? ... 1 painted face? ... no painted faces?

NEW CONCEPT

If the distance between consecutive whole numbers on a number line is divided by tick marks into 10 equal units, then numbers corresponding to these marks can be named using decimal numbers with one decimal place. An example of this kind of number line is a centimeter ruler.

If each centimeter segment on a centimeter scale is divided into 10 equal segments, then each segment is 1 millimeter long. Each segment is also one tenth of a centimeter long.

Example 1 Find the length of this segment

(a) in millimeters.

(b) in centimeters.

cm 1 2 3

Solution (a) Each centimeter is 10 mm. Thus, each small segment on the scale is 1 mm. The length of the segment is **23 mm.**

(b) Each centimeter on the scale has been divided into 10 equal parts. The length of the segment is 2 centimeters plus three tenths of a centimeter. In the metric system we use decimals rather than common fractions to indicate parts of a unit. So the length of the segment is **2.3 cm.**

If the distance between consecutive whole numbers on a number line is divided into 100 equal units, then numbers corresponding to the marks on the number line can be named using two decimal places. For instance, a meter is 100 cm. So each centimeter segment on a meterstick is 0.01 $\left(\text{or } \frac{1}{100}\right)$ of the length of the meterstick. This means that an object 25 cm long is also 0.25 m long.

Example 2 Find the perimeter of this rectangle in meters.

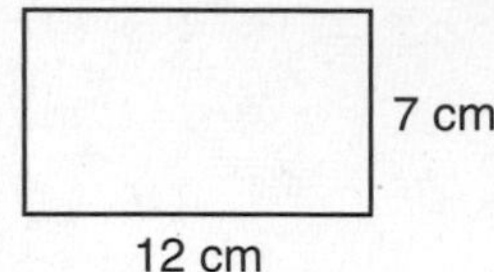

Solution The perimeter of the rectangle is 38 cm. Each centimeter is $\frac{1}{100}$ of a meter. So 38 cm is $\frac{38}{100}$ of a meter, which we write as **0.38 m.**

Example 3 Find the number on the number line indicated by each arrow:

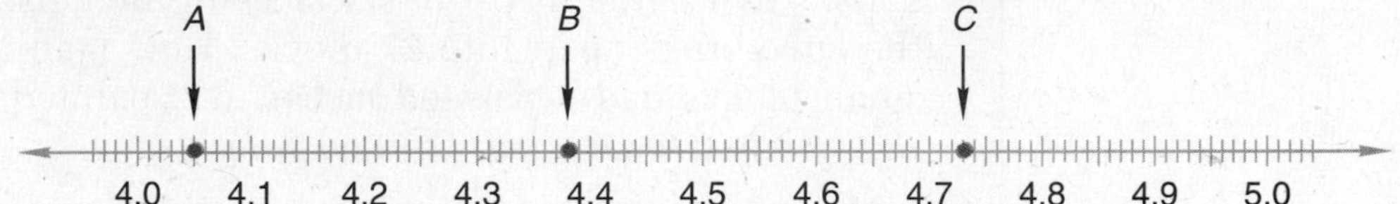

Solution We are considering a portion of the number line from 4 to 5. The distance from 4 to 5 has been divided into 100 equal segments. Tenths have been identified. The point 4.1 is one tenth of the distance from 4 to 5. However, it is also ten hundredths of the distance from 4 to 5, so 4.1 equals 4.10.

Arrow *A* indicates **4.05.**

Arrow *B* indicates **4.38.**

Arrow *C* indicates **4.73.**

Example 4 Find the following sum

(a) in millimeters.

(b) in centimeters.

4.2 cm + 24 mm

Solution (a) We express 4.2 cm as 42 mm and add.

42 mm + 24 mm = **66 mm**

(b) We express 24 mm as 2.4 cm and add.

4.2 cm + 2.4 cm = **6.6 cm**

LESSON PRACTICE

Practice set Refer to the figure below to answer problems **a–c.**

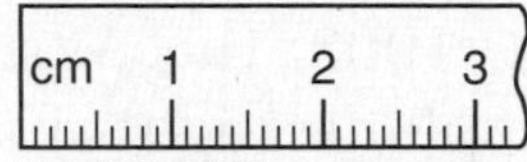

a. Find the length of the segment in centimeters.

b. Find the length of the segment to the nearest millimeter.

c. What is the greatest possible error of the measurement in problem **b?** Express your answer as a fraction of a millimeter.

d. Seventy-five centimeters is how many meters?

e. Alfredo is 1.57 meters tall. How many centimeters tall is Alfredo?

f. What point on a number line is halfway between 2.6 and 2.7?

g. What decimal number names the point marked *A* on this number line?

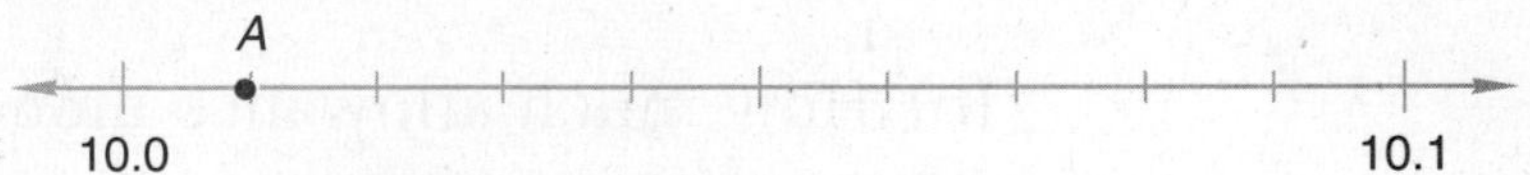

h. Estimate the length of this segment in centimeters. Then use a centimeter ruler to measure its length.

i. 3.5 cm + 12 mm = _____ cm

j. 4 cm − 12 mm = _____ mm

MIXED PRACTICE

Problem set

1. (28) In 3 boxes of cereal, Jeff counted 188 raisins, 212 raisins, and 203 raisins. What was the average number of raisins in each box of cereal?

2. (11) The pollen count had increased from 497 parts per million to 1032 parts per million. By how much had the pollen count increased?

3. Sylvia spent \$3.95 for lunch but still had \$12.55. How much money did she have before she bought lunch?
(11)

4. In 1903 the Wright brothers made the first powered airplane flight. Just 66 years later astronauts first landed on the Moon. In what year did astronauts first land on the Moon?
(12)

5. The perimeter of the square equals the perimeter of the regular hexagon. If each side of the hexagon is 6 inches long, how long is each side of the square?
(19)

6. In the following statement, write the percent as a reduced fraction. Then diagram the statement and answer the questions.
(22)

Each week Jessica saves 40% of her \$4.00 allowance.

(a) How much allowance money does she save each week?

(b) How much allowance money does she not save each week?

7. Describe how to estimate the product of 396 and 71.
(29)

8. Round 7.49362 to the nearest thousandth.
(33)

9. Use words to write each of these decimal numbers:
(31)

(a) 200.02

(b) 0.001625

10. Use digits to write each of these decimal numbers:
(31)

(a) one hundred seventy-five millionths

(b) three thousand, thirty and three hundredths

11. Replace each circle with the proper comparison symbol:
(33)
(a) 6.174 ◯ 6.17401 (b) 14.276 ◯ 1.4276

12. Find the length of this segment
(34)
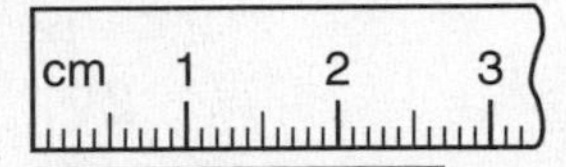

(a) in centimeters.

(b) in millimeters.

13. What decimal number names the point marked X on this number line?
(34)

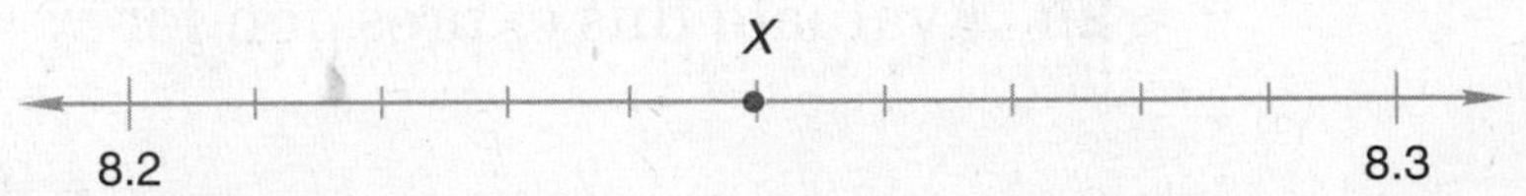

14. The coordinates of three vertices of a square are (0, 0), (0, 3), and (3, 3).
(Inv. 3)

(a) What are the coordinates of the fourth vertex?

(b) What is the area of the square?

15. (a) What decimal number is halfway between 7 and 8?
(34)

(b) What number is halfway between 0.7 and 0.8?

Solve:

16. $15 \cdot 20 = 12y$
(3)

17. $180° = 74° + c$
(3)

Simplify:

18. $\frac{5}{6} + \frac{2}{3} + \frac{1}{2}$
(30)

19. $\frac{5}{36} - \frac{1}{24}$
(30)

20. $5\frac{1}{6} - 1\frac{2}{3}$
(23, 30)

21. $\frac{1}{10} \cdot 2\frac{2}{3} \cdot 3\frac{3}{4}$
(26)

22. $5\frac{1}{4} \div 1\frac{2}{3}$
(26)

23. $3\frac{1}{5} \div 4$
(26)

24. $6\frac{7}{8} + 4\frac{1}{4}$
(30)

25. $\frac{1}{8} + \left(\frac{5}{6} \cdot \frac{3}{4}\right)$
(9, 24)

26. Express the following difference (a) in centimeters and (b) in millimeters.
(34)

$$3.6 \text{ cm} - 24 \text{ mm}$$

27. Which is equivalent to $2^2 \cdot 2^3$?
(20)

A. 2^5 B. 2^6 C. 12 D. 24

28. Arrange these numbers in order from least to greatest:
(33)

0.365, 0.3575, 0.36

29. Evaluate this expression for $x = 5$ and $y = 10$:
(1, 4)

$$\frac{y}{x} - x$$

30. Use a compass to draw three concentric circles. Make the radii 1 inch, $1\frac{1}{2}$ inches, and 2 inches.
(Inv. 2)

LESSON

35 Adding, Subtracting, Multiplying, and Dividing Decimal Numbers

WARM-UP

Facts Practice: Measurement Facts (Test H)

Mental Math:

a. \$7.50 − \$1.99 **b.** 5 × 64¢ **c.** $\frac{9}{10} = \frac{?}{30}$

d. Reduce $\frac{15}{24}$. **e.** $4^2 - \sqrt{4}$ **f.** $\frac{5}{12}$ of 24

g. Start with the number of inches in two feet, + 1, × 4, $\sqrt{\ }$. What do we call this many years?

Problem Solving:

Copy this problem and fill in the missing digits:

```
       ____ R 5
   _)____
     8
     __
     16
     __
      24
      __
      __
       _
```

NEW CONCEPTS

Adding and subtracting decimal numbers Adding and subtracting decimal numbers is similar to adding and subtracting money. **We align the decimal points to ensure that we are adding or subtracting digits that have the same place value.**

Example 1 Add: 3.6 + 0.36 + 36

Solution We align the decimal points vertically. A number written without a decimal point is a whole number, so the decimal point is to the right of 36.

```
   3.6
   0.36
+ 36.
 -----
  39.96
```

Example 2 Add: 0.1 + 0.2 + 0.3 + 0.4

Solution We align the decimal points vertically and add. The sum is 1.0, not 0.10. Since 1.0 equals 1, we can simplify the answer to 1.

$$\begin{array}{r} 0.1 \\ 0.2 \\ 0.3 \\ +\ 0.4 \\ \hline 1.0 \end{array} = \mathbf{1}$$

Example 3 Subtract: 12.3 − 4.567

Solution We write the first number above the second number, aligning the decimal points. We write zeros in the empty places and subtract.

$$\begin{array}{r} \overset{0}{\cancel{1}}\overset{1}{\cancel{2}}.\overset{2}{\cancel{3}}\overset{9}{\cancel{0}}\overset{1}{0} \\ -\quad 4.567 \\ \hline \mathbf{7.733} \end{array}$$

Example 4 Subtract: 5 − 4.32

Solution We write the whole number 5 with a decimal point and write zeros in the two empty decimal places. Then we subtract.

$$\begin{array}{r} \overset{4}{\cancel{5}}.\overset{9}{\cancel{0}}\overset{1}{0} \\ -\ 4.32 \\ \hline \mathbf{0.68} \end{array}$$

Multiplying decimal numbers

If we multiply the fractions three tenths and seven tenths, the product is twenty-one hundredths.

$$\frac{3}{10} \times \frac{7}{10} = \frac{21}{100}$$

Likewise, if we multiply the decimal numbers three tenths and seven tenths, the product is twenty-one hundredths.

$$0.3 \times 0.7 = 0.21$$

Here we use an area model to illustrate this multiplication:

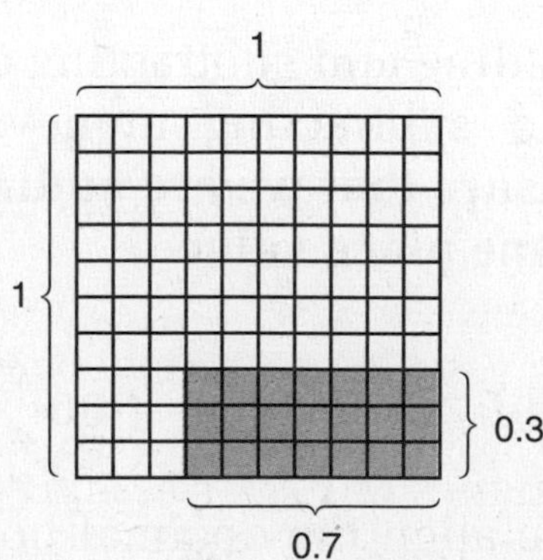

Each side of the square is one unit in length. We multiply three tenths of one side by seven tenths of a perpendicular side. The product is an area that contains twenty-one hundredths of the square.

$$0.3 \times 0.7 = 0.21$$

Notice that the factors each have one decimal place and the product has two decimal places. **When we multiply decimal numbers, the product has as many decimal places as there are in all the factors combined.**

Example 5 Multiply: (0.23)(0.4)

Solution We need not align decimal points to multiply. We set up the problem as though we were multiplying whole numbers. After multiplying, we count the number of decimal places in both factors. There are a total of three decimal places, so we write the product with three decimal places. We count from right to left, writing one or more zeros in front as necessary. The product of 0.23 and 0.4 is **0.092.**

$$\begin{array}{r} 0.23 \\ \times\ 0.4 \\ \hline 92 \end{array}$$

$$\begin{array}{rl} 0.23 & \text{2 places} \\ \times\ 0.4 & \text{1 place} \\ \hline 0.092 & \text{3 places} \end{array}$$

Example 6 Multiply: 35 × 0.4

Solution We set up the problem as though we were multiplying whole numbers. After multiplying, we count the total number of decimal places in the factors. Then we place a decimal point in the product so that the product has the same number of decimal places as there are in the factors combined. After placing the decimal point, we simplify the result.

$$\begin{array}{rl} 35 & \text{0 places} \\ \times\ 0.4 & \text{1 place} \\ \hline 14.0 & \text{1 place} \end{array}$$

$$14.0 = \mathbf{14}$$

Example 7 Multiply: (0.2)(0.3)(0.04)

Solution Sometimes we can perform the multiplication mentally. First we multiply as though we were multiplying whole numbers: 2 · 3 · 4 = 24. Then we count decimal places. There are four decimal places in the three factors. Starting from the right side of 24, we count to the left four places. We write zeros in the empty places.

$$.\_\_24 \longrightarrow \mathbf{0.0024}$$

Dividing decimal numbers Dividing a decimal number by a whole number is similar to dividing money. The decimal point in the answer is straight up from the decimal point in the division box.

Example 8 Divide: 3.425 ÷ 5

Solution We rewrite the problem with a division box. We place a decimal point in the answer directly above the decimal point in the division box. Then we divide as though we were dividing whole numbers. The answer is **0.685.**

$$\begin{array}{r} 0.685 \\ 5\overline{)3.425} \\ \underline{3\,0} \\ 42 \\ \underline{40} \\ 25 \\ \underline{25} \\ 0 \end{array}$$

Example 9 Divide: 0.0144 ÷ 8

Solution We place the decimal point in the answer directly above the decimal point inside the division box. We write a digit in every place following the decimal point until the division is completed. If we cannot perform a division, we write a zero in that place. The answer is **0.0018.**

$$\begin{array}{r} 0.0018 \\ 8\overline{)0.0144} \\ \underline{8} \\ 64 \\ \underline{64} \\ 0 \end{array}$$

Example 10 Divide: 1.2 ÷ 5

Solution We do not write a decimal division answer with a remainder. Since a decimal point fixes place values, we may write a zero in the next decimal place. This zero does not change the value of the number, but it does let us continue dividing. The answer is **0.24.**

$$\begin{array}{r} 0.24 \\ 5\overline{)1.2\underline{0}} \\ \underline{1\,0} \\ 20 \\ \underline{20} \\ 0 \end{array}$$

LESSON PRACTICE

Practice set* Simplify:

a. 1.2 + 3.45 + 23.6 **b.** 4.5 + 0.51 + 6 + 12.4

c. 0.2 + 0.4 + 0.6 + 0.8 **d.** 36.274 – 5.39

e. 16.7 – 1.936 **f.** 12 – 0.875

g. 4.2 × 0.24

h. (0.12)(0.06)

i. 5.4 × 7

j. 0.3 × 0.2 × 0.1

k. (0.04)(10)

l. 0.045 × 0.6

m. 14.4 ÷ 6

n. 0.048 ÷ 8

o. 3.4 ÷ 5

p. 0.3 ÷ 6

MIXED PRACTICE

Problem set

1. (28) During the first six months of the year, the Montgomerys' monthly electric bills were $128.45, $131.50, $112.30, $96.25, $81.70, and $71.70. How can the Montgomerys find their average monthly electric bill for the first six months of the year?

2. (23, 30) There were $2\frac{1}{2}$ gallons of milk in the refrigerator before breakfast. There were $1\frac{3}{4}$ gallons after dinner. How many gallons of milk were consumed during the day?

3. (28) A one-year subscription to a monthly magazine costs $15.60. The regular newsstand price is $1.75 per issue. How much is saved per issue by paying the subscription price?

4. (28) Carlos ran one lap in 1 minute 3 seconds. Orlando ran one lap 5 seconds faster than Carlos. How many seconds did it take Orlando to run one lap?

5. (19) The perimeter of the square equals the perimeter of the regular pentagon. Each side of the pentagon is 16 cm long. How long is each side of the square?

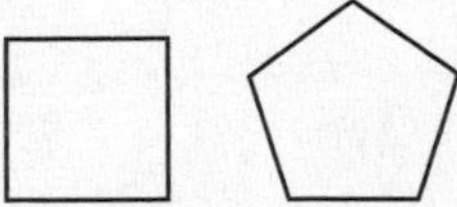

6. (22) Diagram this statement. Then answer the questions that follow.

   Two ninths of the 54 fish in the tank were guppies.

   (a) How many of the fish were guppies?

   (b) How many of the fish were not guppies?

7. (20) A 6-by-6-cm square is cut from a 10-by-10-cm square sheet of paper as shown below. Refer to this figure to answer (a)–(c):

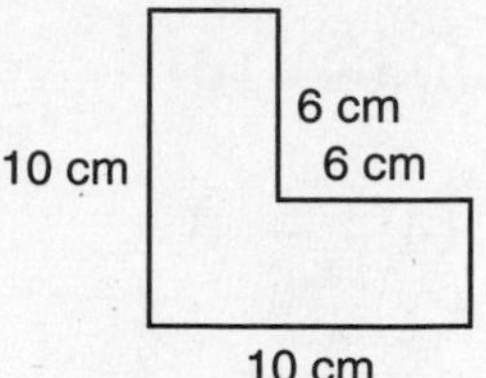

(a) What was the area of the original square?

(b) What was the area of the square that was cut out?

(c) What is the area of the remaining figure?

8. (31) (a) In the square at right, what fraction is not shaded?

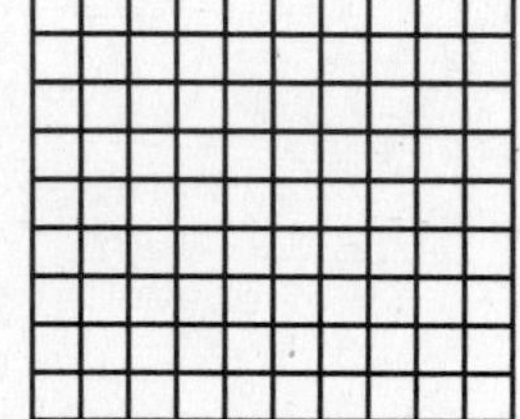

(b) What decimal part of the square is not shaded?

(c) What percent of the square is not shaded?

9. (Inv. 3) The coordinates of three vertices of a rectangle are (–3, 2), (3, –2), and (–3, –2).

(a) What are the coordinates of the fourth vertex?

(b) What is the area of the rectangle?

10. (31) (a) Use words to write 100.075.

(b) Use digits to write the decimal number twenty-five hundred-thousandths.

11. (34) Find the length of this segment

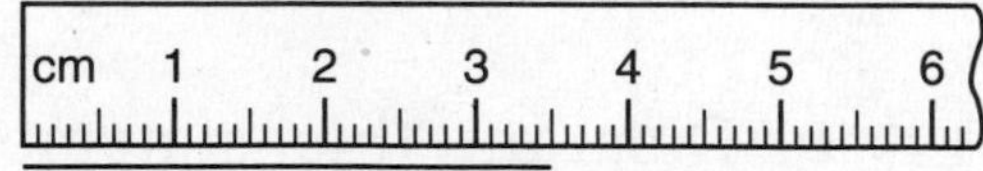

(a) in centimeters.

(b) in millimeters.

12. (33) Miss Edwards bought 11.92 gallons of gasoline at $\$1.49\frac{9}{10}$ per gallon. Estimate how much she paid for the gasoline.

13. (34) What decimal number names the point marked with an arrow on this number line?

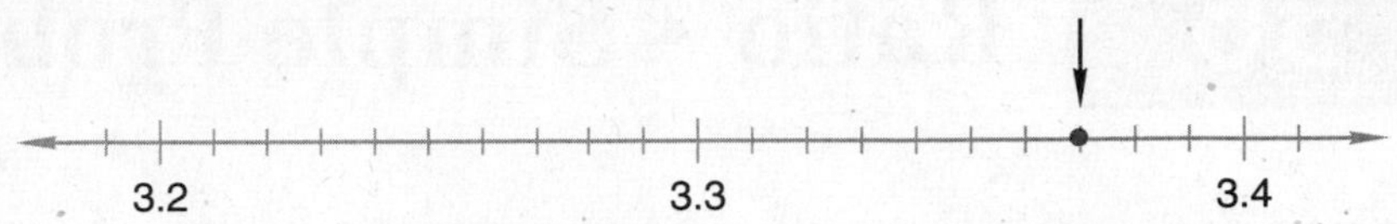

14. (35) This figure illustrates the multiplication of which two decimal numbers? What is their product?

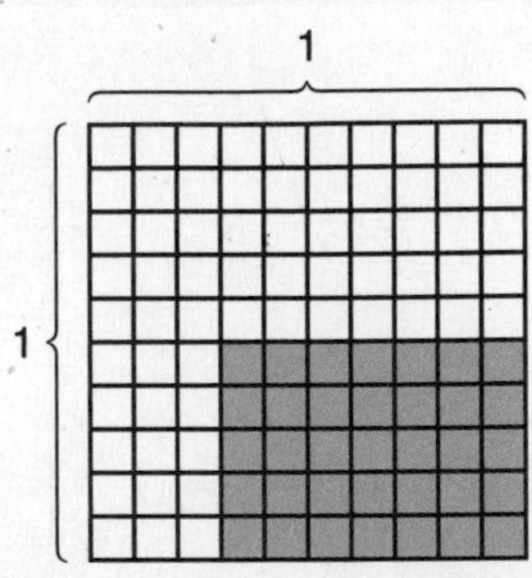

15. (34) What decimal number is halfway between 1.2 and 1.3?

Solve:

16. (3) $15x = 9 \cdot 10$

17. (3, 35) $f + 4.6 = 5.83$

18. (3, 35) $8y = 46.4$

19. (3, 35) $w - 3.4 = 12$

Simplify:

20. (35) $3.65 + 0.9 + 8 + 15.23$

21. (30) $1\frac{1}{2} + 2\frac{2}{3} + 3\frac{3}{4}$

22. (26) $1\frac{1}{2} \cdot 2\frac{2}{3} \cdot 3\frac{3}{4}$

23. (23, 30) $1\frac{1}{6} - \left(\frac{1}{2} + \frac{1}{3}\right)$

24. (23, 30) $3\frac{1}{12} - 1\frac{3}{4}$

25. (35) $1.2 \div 10$

26. (35) $(0.3)(0.4)(0.5)$

27. (26, 30) $\left(3\frac{1}{2} + 1\frac{3}{4}\right) \div \left(4 - 3\frac{1}{8}\right)$

For problems 28 and 29, record an estimated answer and an exact answer.

28. (33, 35) $36.45 - 4.912$

29. (33, 35) 4.2×0.9

30. (17) Use a protractor to draw a triangle that has two 45° angles.

LESSON

36 Ratio • Simple Probability

WARM-UP

Facts Practice: + – × ÷ Fractions (Test G)

Mental Math:

a. \$1.45 × 10
b. 4 × \$1.50
c. $\frac{4}{5} = \frac{?}{20}$
d. Reduce $\frac{24}{30}$.
e. $\sqrt{144} - 3^2$
f. $\frac{9}{10}$ of 40
g. Find the square root of three dozen, × 5, ÷ 3, square that number, – 20, + 1, $\sqrt{\ }$.

Problem Solving:

Before lunch, Matt studied math, English, science, and history, though not necessarily in that order. How many different permutations of the four subjects are possible?

NEW CONCEPTS

Ratio A **ratio** is a way to describe a relationship between two numbers. For instance, if there are 12 boys and 16 girls in the choir, the ratio of boys to girls is 12 to 16, which reduces to 3 to 4. The ratio 3 to 4 can be written several ways.

with the word *to*	3 to 4
as a fraction	$\frac{3}{4}$
as a decimal number	0.75
with a colon	3:4

The numbers used to express a ratio are stated in the same order as the items are named. If the boy-girl ratio is 3 to 4, then the girl-boy ratio is 4 to 3. Note that we reduce ratios, but we do not express them as mixed numbers.

Most ratios involve three numbers, even though only two numbers may be stated. When the ratio of boys to girls is 3 to 4, the unstated number is 7, which is the total.

$$\begin{array}{r} 3 \text{ boys} \\ + \ 4 \text{ girls} \\ \hline 7 \text{ total} \end{array}$$

Sometimes the total is given and one of the parts is unstated, as we show in the following two examples.

Example 1 In a group of 28 children, there are 12 boys.

(a) What is the boy-girl ratio?

(b) What is the girl-boy ratio?

Solution We will begin by writing all three ratio numbers.

$$\begin{array}{r} 12 \text{ boys} \\ + \ ?\ \text{girls} \\ \hline 28 \text{ total} \end{array} \longrightarrow \begin{array}{r} 12 \text{ boys} \\ + \ 16 \text{ girls} \\ \hline 28 \text{ total} \end{array}$$

Now we can write the answers.

(a) The boy-girl ratio is $\frac{12}{16}$, which reduces to $\frac{3}{4}$, a ratio of 3 to 4.

(b) The girl-boy ratio is $\frac{16}{12}$, which reduces to $\frac{4}{3}$, a ratio of 4 to 3. Remember, we do not change the ratio to a mixed number.

Example 2 The team won $\frac{4}{7}$ of its games and lost the rest. What was the team's win-loss ratio?

Solution We are not told how many games the team played. However, we are told that the team won $\frac{4}{7}$ of its games. Therefore, the team lost $\frac{3}{7}$ of its games. Thus, on average, the team won 4 out of every 7 games. So the three ratio numbers are

$$\begin{array}{r} 4 \text{ won} \\ + \ 3 \text{ lost} \\ \hline 7 \text{ total} \end{array}$$

The team's win-loss ratio was 4 to 3, which we write as $\frac{4}{3}$.

Example 3 In the bag are red marbles and green marbles. If the ratio of red marbles to green marbles is 4 to 5, what fraction of the marbles are red?

Solution First we write all three ratio numbers.

$$\begin{array}{r} 4 \text{ red} \\ + \ 5 \text{ green} \\ \hline ?\ \text{total} \end{array} \longrightarrow \begin{array}{r} 4 \text{ red} \\ + \ 5 \text{ green} \\ \hline 9 \text{ total} \end{array}$$

We are asked to determine what fraction of the marbles are red. There are 4 red marbles and a total of 9 marbles, so the fraction that are red is $\frac{4}{9}$.

Simple probability

Probability is the likelihood that a particular event will occur. Probability is often written as a ratio.

$$\text{Probability} = \frac{\text{number of favorable outcomes}}{\text{number of possible outcomes}}$$

If we know a bag contains 4 red marbles and 5 green marbles, we can calculate the probability—the likelihood—of drawing a marble of a specified color from the bag. If one marble is to be drawn from the bag without looking, then the probability that the marble will be

$$\text{red is } \frac{\text{number of red}}{\text{number of marbles}} = \frac{4}{9} \text{ (4 in 9)}$$

$$\text{green is } \frac{\text{number of green}}{\text{number of marbles}} = \frac{5}{9} \text{ (5 in 9)}$$

$$\text{blue is } \frac{\text{number of blue}}{\text{number of marbles}} = \frac{0}{9} = 0$$

A probability of 0 means that the event cannot occur. A probability of 1 means that the event is certain to occur. A probability of $\frac{1}{2}$ means that the event is as likely to occur as it is not to occur.

Example 4 A single dot cube is rolled. What is the probability of rolling

(a) a 4?

(b) a number greater than 4?

(c) a number greater than 6?

(d) a number less than 7?

Solution There are six different faces on a dot cube, so there are six equally likely outcomes. Thus 6 will be the bottom number of each probability ratio.

(a) There is only one way to roll a 4 with one dot cube, so the probability of rolling a 4 is one in six, $\frac{1}{6}$.

(b) The numbers greater than 4 on a dot cube are 5 and 6, so there are two ways to roll a number greater than 4. Thus the probability is two in six, $\frac{2}{6}$, which we reduce to $\frac{1}{3}$.

(c) On a dot cube, there are no numbers greater than 6, so there is no way to roll a number greater than 6. Thus the probability is $\frac{0}{6}$, which is **0.** An event that cannot happen has a probability of 0.

(d) There are six numbers less than 7 on a dot cube, so there are six ways to roll a number less than 7. Thus the probability is six out of six, $\frac{6}{6}$, which is **1.** An event that is certain to happen has a probability of 1.

Example 5 The face of a spinner is divided into five equal sectors and is numbered as shown. If this spinner is spun, what is the probability of the spinner

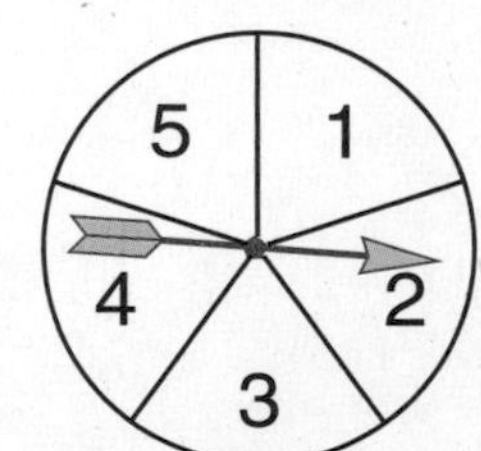

(a) stopping on 3?

(b) not stopping on 3?

Solution There are five equally likely outcomes, so 5 is the bottom number of each ratio.

(a) There is one way for the spinner to stop on 3, so the probability is one in five, $\mathbf{\frac{1}{5}}$.

(b) There are four ways for the spinner not to stop on 3, so the probability is four out of five, $\mathbf{\frac{4}{5}}$.

Notice that the probability of an event happening plus the probability of the event not happening is 1.

Example 6 The face of a spinner is divided into one half and two quarters as shown. What is the probability of this spinner stopping on 3?

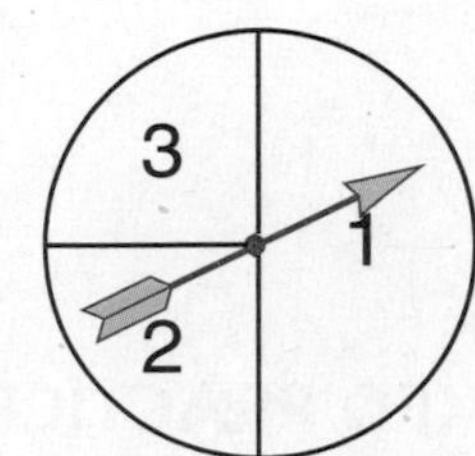

Solution There are three possible outcomes, but the outcomes are not equally likely. Since region 1 occupies half the area, the probability of the spinner stopping in region 1 is $\frac{1}{2}$. Regions 2 and 3 each occupy $\frac{1}{4}$ of the area. Thus, the probability of the spinner stopping on 2 is $\frac{1}{4}$, and the probability of it stopping on 3 is also $\mathbf{\frac{1}{4}}$.

LESSON PRACTICE

Practice set

a. In the pond were 240 little fish and 90 big fish. What was the ratio of big fish to little fish?

b. Fourteen of the 30 children at the party were girls. What was the boy-girl ratio at the party?

c. The team won $\frac{3}{8}$ of its games and lost the rest. What was the team's win-loss ratio?

d. The bag contained red marbles and blue marbles. If the ratio of red marbles to blue marbles was 5 to 3, what fraction of the marbles were blue?

e. What is the probability of rolling a number less than 4 with one toss of a dot cube?

The face of this spinner is divided into four equal parts.

f. What is the probability of this spinner stopping on 3?

g. What is the probability of this spinner stopping on 5?

h. What is the probability of this spinner stopping on a number less than 6?

The face of this spinner is divided into one half and two fourths.

i. What is the probability of this spinner stopping on A?

j. What is the probability of this spinner not stopping on B?

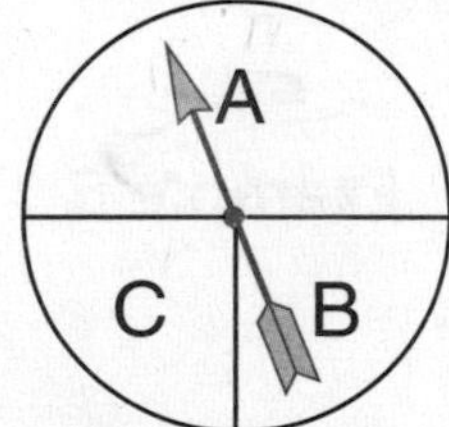

MIXED PRACTICE

Problem set

1. (36) Fourteen of the 32 marbles in the bag were blue. The rest were red. What was the ratio of red marbles to blue marbles in the bag?

2. (28) During the last 3 years, the annual rainfall has been 23 inches, 21 inches, and 16 inches. What has been the average (mean) annual rainfall during the last 3 years?

3. (13) Pilar reads 35 pages each night. At this pace, how many pages can Pilar read in a week?

4. (35) Shannon swam 100 meters in 56.24 seconds. Donna swam 100 meters in 59.48 seconds. Donna took how many seconds longer to swim 100 meters than Shannon?

5. (22, 36) In the following statement, change the percent to a reduced fraction. Then diagram the statement and answer the questions.

Forty percent of the 30 players in the game had never played rugby.

(a) How many of the players had never played rugby?

(b) What was the ratio of those who had played rugby to those who had not played rugby?

6. (34) *AB* is 4 cm. *AC* is 9.5 cm. Describe how to find *BC* in millimeters.

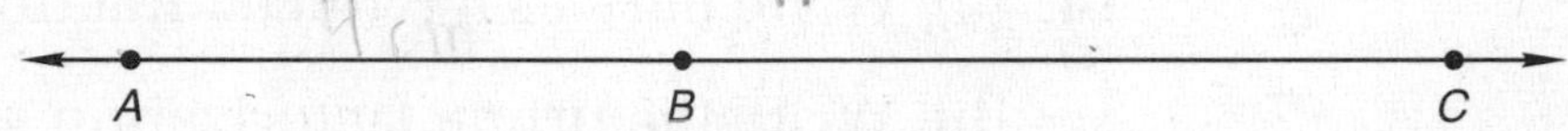

7. (19, 20) The length of the rectangle is 5 cm greater than its width.

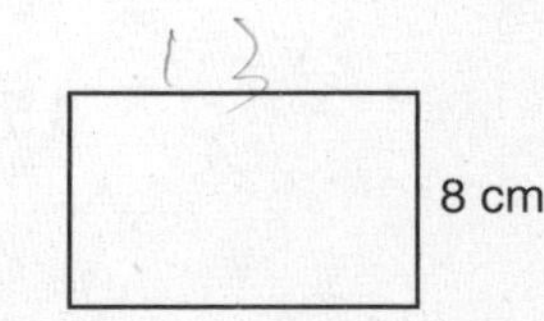

(a) What is the area of the rectangle?

(b) What is the perimeter of the rectangle?

8. (29) Estimate the sum of 3624, 2889, and 896 by rounding each number to the nearest hundred before adding.

9. (33) (a) Round 6.857142 to three decimal places.

(b) Estimate the product of 6.8571420 and 1.9870 by rounding each number to the nearest whole number before multiplying.

10. (31) Use digits to write each number:

(a) twelve million

(b) twelve millionths

11. (36) A bag contains 3 red marbles, 4 white marbles, and 5 blue marbles. If one marble is drawn from the bag, what is the probability that the marble will be

(a) red? (b) white?

(c) blue? (d) green?

12. (34) Find the length of this segment

mm 10 20 30 40 50 60

(a) in centimeters.

(b) in millimeters.

13. (34) What decimal number names the point marked *M* on this number line?

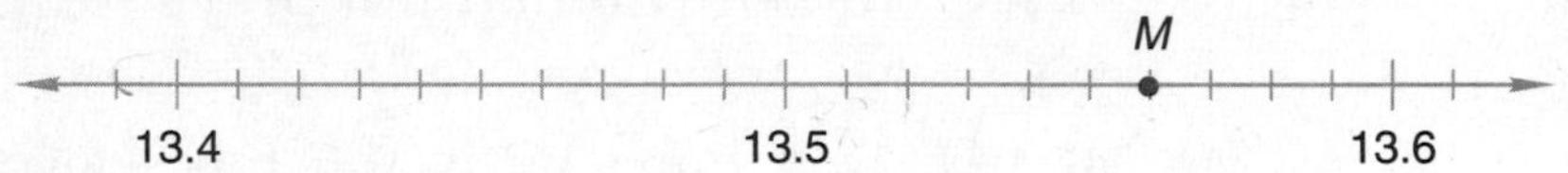

14. (24) (a) Write 85% as a reduced fraction.

(b) Write the prime factorization and reduce:

15. (29) Alba worked for 6 hr 45 min at \$7.90 per hour. What numbers could she use to estimate how much money she earned? Estimate the amount she earned, and state whether you think the exact amount is a little more or a little less than your estimate.

16. (7) In this figure, which angle is

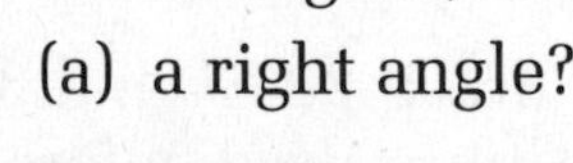

(a) a right angle?

(b) an acute angle?

(c) an obtuse angle?

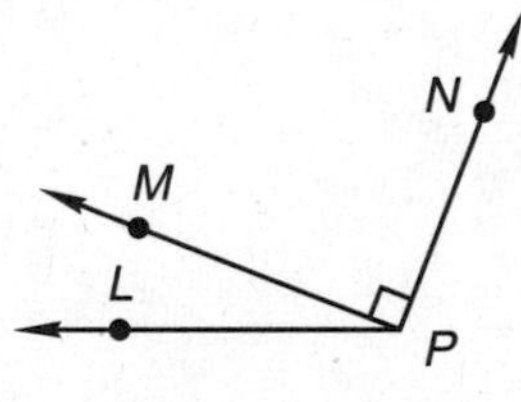

Solve:

17. (20) $8y = 12^2$

18. (35) $\frac{w}{4} = 1.2$

Estimate each answer to the nearest whole number. Then perform the calculation.

19. (35) $4.27 + 16.3 + 10$

20. (35) $4.2 - 0.42$

Simplify:

21. $3\frac{1}{2} + 1\frac{1}{3} + 2\frac{1}{4}$
(30)

22. $3\frac{1}{2} \cdot 1\frac{1}{3} \cdot 2\frac{1}{4}$
(26)

23. $3\frac{5}{6} - \left(\frac{2}{3} - \frac{1}{2}\right)$
(30)

24. $8\frac{5}{12} - 3\frac{2}{3}$
(23, 30)

25. $2\frac{3}{4} \div 4\frac{1}{2}$
(26)

26. $5 - \left(\frac{2}{3} \div \frac{1}{2}\right)$
(23, 25)

27. $1.4 \div 8$
(35)

28. $(0.2)(0.3)(0.4)$
(35)

29. (a) 12.25×10
(35)
(b) $12.25 \div 10$

30. On a coordinate plane draw a square that has an area of 25 units^2. Then write the coordinates of the vertices of the square on your paper.
(Inv. 3)

LESSON

37 Area of a Triangle • Rectangular Area, Part 2

WARM-UP

Facts Practice: Measurement Facts (Test H)

Mental Math:

a. \$3.67 + \$0.98
b. 5 × \$1.25
c. $\frac{7}{8} = \frac{?}{24}$
d. Reduce $\frac{18}{30}$.
e. $\frac{\sqrt{144}}{\sqrt{36}}$
f. $\frac{3}{10}$ of 60
g. What number is 5 less than the sum of 5^2 and $\sqrt{100}$?

Problem Solving:

If the last page of a section of a large newspaper is page 36, what is the fewest number of sheets of paper that could be in that section?

NEW CONCEPTS

Area of a triangle

A triangle has a **base** and a **height** (or **altitude**).

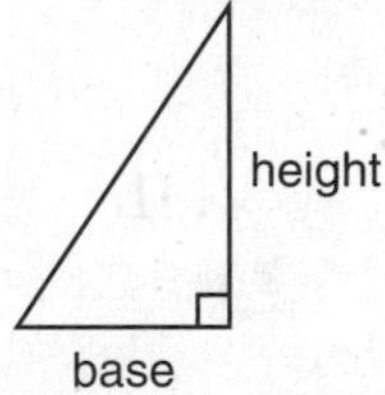

The base is one of the sides of the triangle. The height (or altitude) is the perpendicular distance between the base (or baseline) and the opposite vertex of the triangle. Since a triangle has three sides and any side can be the base, a triangle can have three base-height orientations, as we show by rotating this triangle.

One Right Triangle Rotated to Three Positions

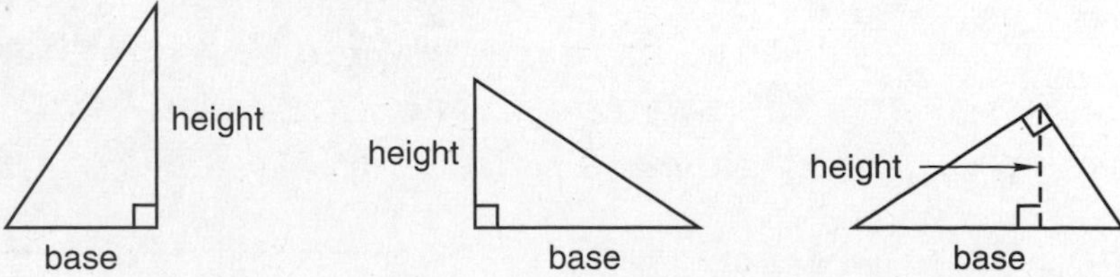

If one angle of a triangle is a right angle, the height may be a side of the triangle, as we see above. If none of the angles of a triangle are right angles, then the height will not be a side of the triangle. When the height is not a side of the triangle, a dashed line segment will represent it, as in the right-hand figure above. If one angle of a triangle is an obtuse angle, then the height is shown outside the triangle in two of the three orientations, as shown below.

One Obtuse Triangle Rotated to Three Positions

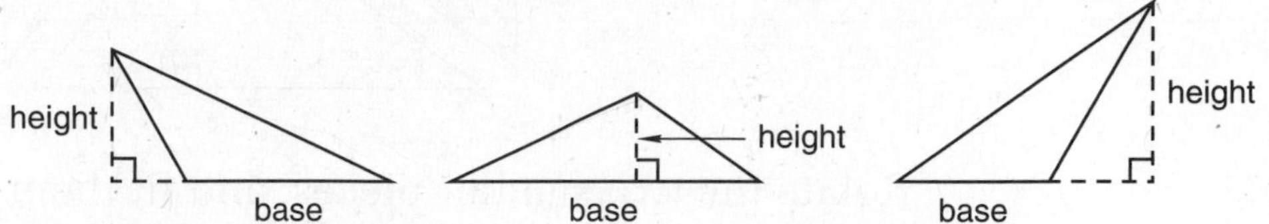

The area of a triangle is half the area of a rectangle with the same base and height, as the following activity illustrates.

Activity: *Area of a Triangle*

Materials needed:

- Paper
- Ruler or straightedge
- Protractor
- Scissors

Use a ruler or straightedge to draw a triangle. Determine which side of the triangle is the longest side. The longest side will be the base of the triangle for this activity. To represent the height (altitude) of the triangle, draw a series of dashes from the topmost vertex of the triangle to the base. Use a protractor to make sure the dashes are perpendicular to the base, as in the figure below.

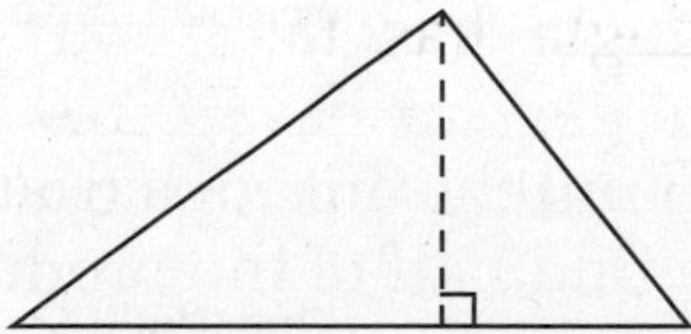

Now we draw a rectangle that contains the triangle. The base of the triangle is one side of the rectangle. The height of the triangle equals the height (width) of the rectangle.

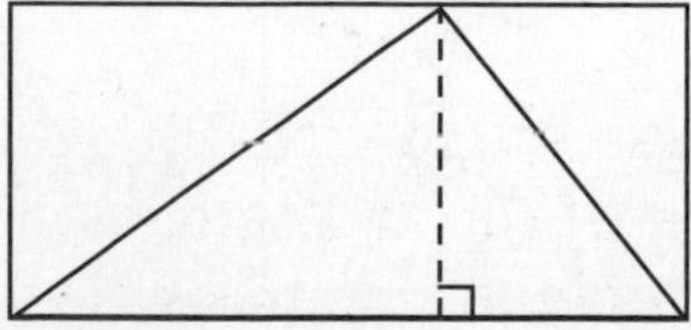

Cut out the rectangle and set the scraps aside. Next, carefully cut out the triangle you drew from the rectangle. Save all three pieces.

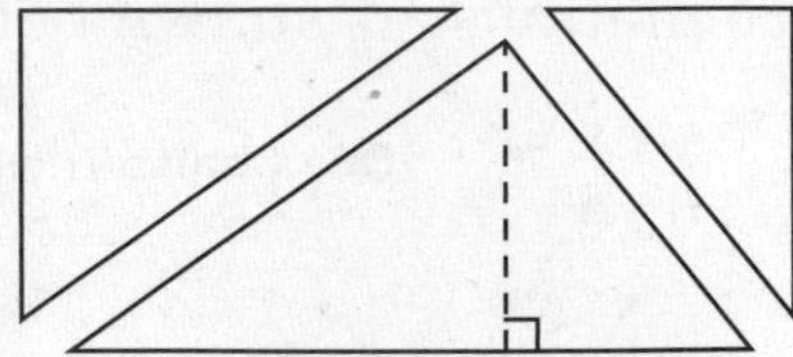

Rotate the two smaller pieces, and fit them together to make a triangle congruent to the triangle you drew.

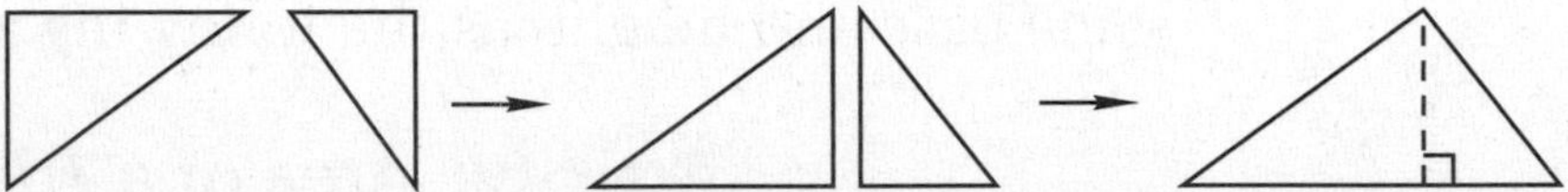

Notice that each of the congruent triangles is half the area of the rectangle with the same base and height (length and width).

When we multiply two perpendicular dimensions, the product is the area of a rectangle with those dimensions.

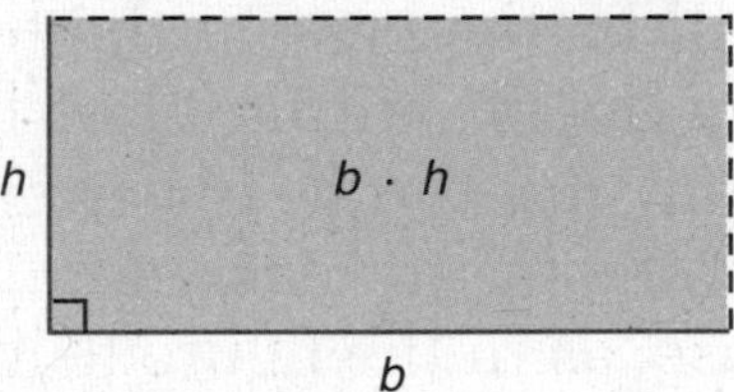

The product $b \cdot h$ equals the area of a b-by-h rectangle.

To find the area of a triangle with a base of b and a height of h, we find half of the product of b and h. We show two formulas for finding the area of a triangle. How are the formulas different? Why do both formulas yield the same result?

$$\textbf{Area of a triangle} = \tfrac{1}{2}bh$$

$$\textbf{Area of a triangle} = \frac{bh}{2}$$

Example 1 Find the area of this triangle. (Use $A = \frac{bh}{2}$.)

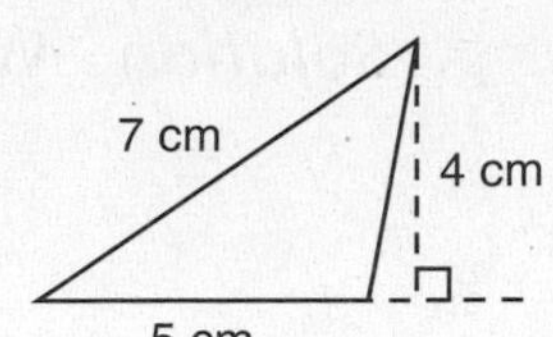

Solution We find the area of the triangle by multiplying the base by the height then dividing the product by 2. The base and height are perpendicular dimensions. In this figure the base is 5 cm, and the height is 4 cm.

$$\text{Area} = \frac{5 \text{ cm} \times 4 \text{ cm}}{2}$$
$$= \frac{20 \text{ cm}^2}{2}$$
$$= \mathbf{10\ cm^2}$$

Example 2 Find the area of this triangle. (Use $A = \frac{1}{2}bh$.)

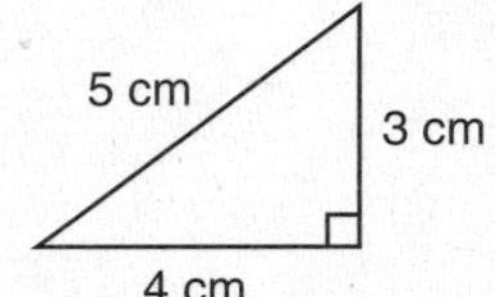

Solution The base and height are perpendicular dimensions. Since one angle of this triangle is a right angle, the base and height are the perpendicular sides, which are 4 cm and 3 cm long.

$$\text{Area} = \frac{1}{2} \cdot 4 \text{ cm} \cdot 3 \text{ cm}$$
$$= \mathbf{6\ cm^2}$$

Rectangular area, part 2 We have practiced finding the areas of rectangles. Sometimes we can find the area of a more complex shape by dividing the shape into rectangular parts. We find the area of each part and then add the areas of the parts to find the total area.

Example 3 Find the area of the figure below. Dimensions are in centimeters. All angles are right angles.

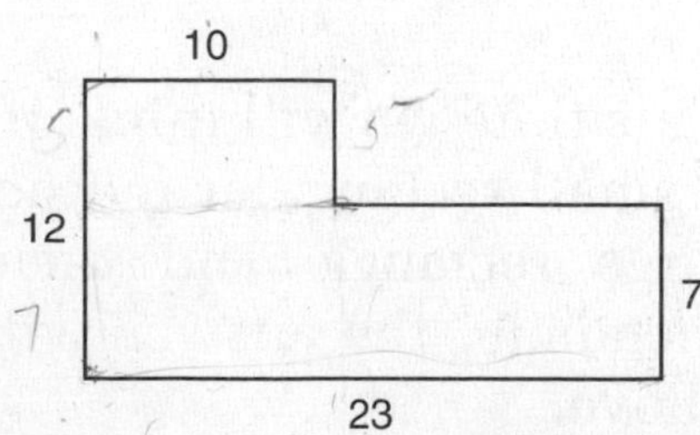

Solution We show two ways to solve this problem.

SOLUTION 1

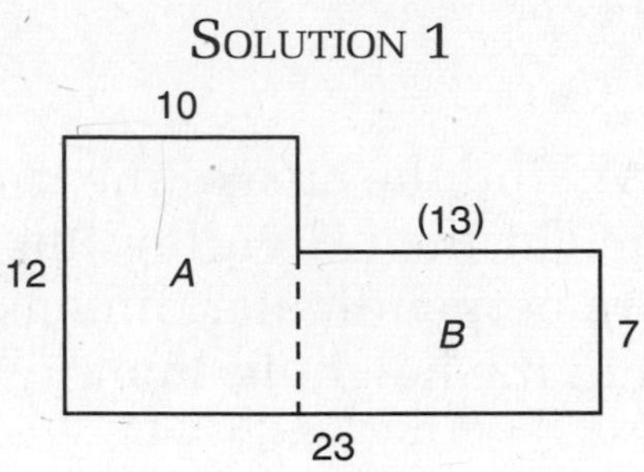

$$\text{Total area} = \text{area } A + \text{area } B$$

$$\begin{array}{rl} \text{Area } A = 10 \text{ cm} \cdot 12 \text{ cm} = & 120 \text{ cm}^2 \\ + \text{ Area } B = 13 \text{ cm} \cdot 7 \text{ cm} = & 91 \text{ cm}^2 \\ \hline \text{Total area} \qquad\qquad\qquad = & \mathbf{211 \text{ cm}^2} \end{array}$$

SOLUTION 2

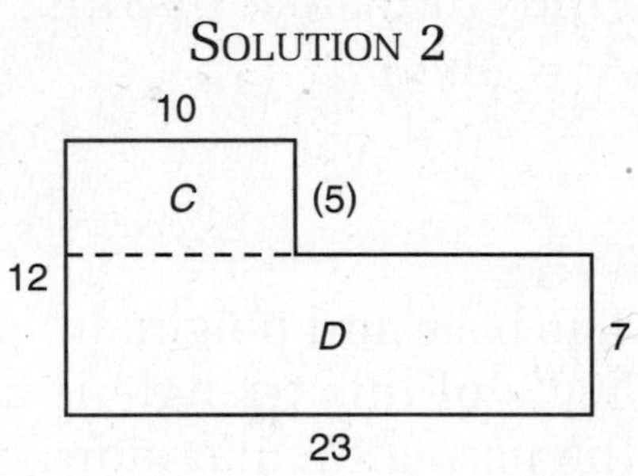

$$\text{Total area} = \text{area } C + \text{area } D$$

$$\begin{array}{rl} \text{Area } C = 10 \text{ cm} \cdot 5 \text{ cm} = & 50 \text{ cm}^2 \\ + \text{ Area } D = 23 \text{ cm} \cdot 7 \text{ cm} = & 161 \text{ cm}^2 \\ \hline \text{Total area} \qquad\qquad\qquad = & \mathbf{211 \text{ cm}^2} \end{array}$$

Example 4 Find the area of the figure at right. Dimensions are in meters. All angles are right angles.

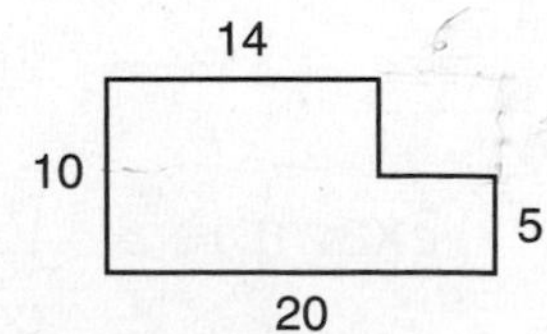

Solution This time we will think of our figure as a large rectangle with a small rectangular piece removed. If we find the area of the large rectangle and then *subtract* the area of the small rectangle, the answer will be the area of the figure shown above.

Here we show the figure redrawn and the calculations:

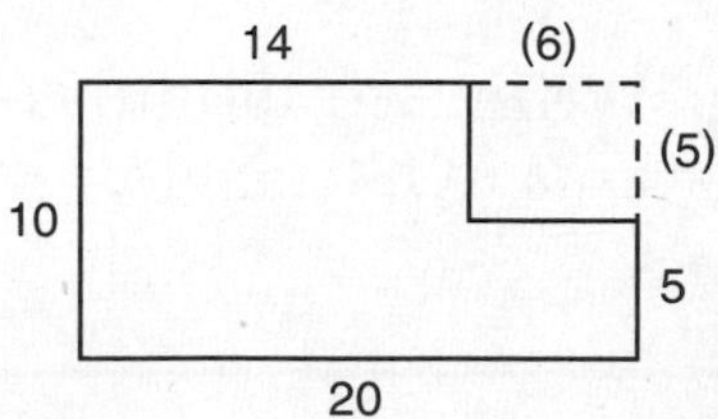

Area of figure = area of large rectangle − area of small rectangle

$$\begin{array}{lcl} \text{Area of large rectangle} & = 20 \text{ m} \cdot 10 \text{ m} = & 200 \text{ m}^2 \\ -\ \text{Area of small rectangle} & = 6 \text{ m} \cdot 5 \text{ m} = & 30 \text{ m}^2 \\ \hline \text{Area of figure} & = & \mathbf{170 \text{ m}^2} \end{array}$$

We did not need to subtract to find the area. We could have added the areas of two smaller rectangles as we did in example 3. Often addition is the easier method; however, the subtraction method is quicker in some situations.

LESSON PRACTICE

Practice set* Find the area of each triangle. Dimensions are in centimeters.

a.

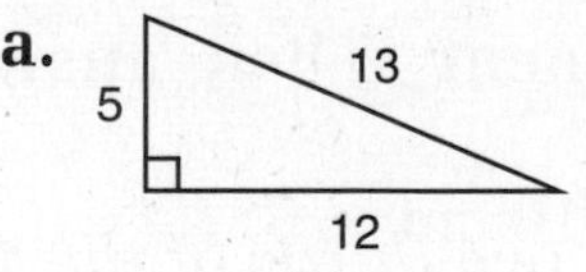

b.

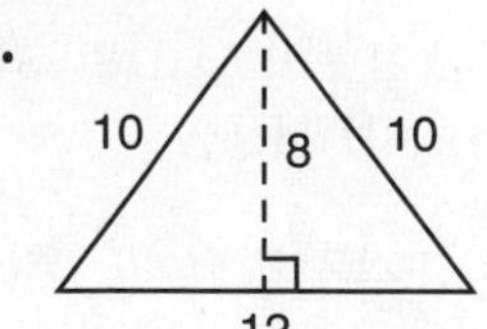

c.

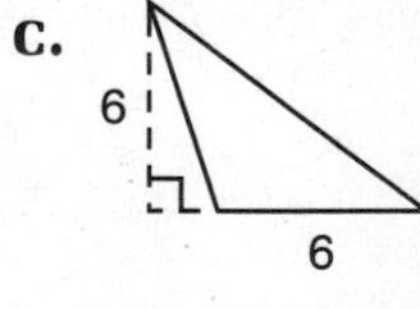

d. Copy the figure in example 4, and find its area by dividing the shape into two rectangles and adding the areas.

e. Copy the figure in example 3. Imagine that a small rectangle was cut out of a large rectangle to make this shape. Use dashes to complete the outline of these rectangles. Then find the area of both the smaller rectangle and the larger rectangle. Subtract the area of the smaller rectangle from the area of the larger rectangle to find the area of the remaining figure.

f. Find the area of the figure at right. Dimensions are in inches. All angles are right angles.

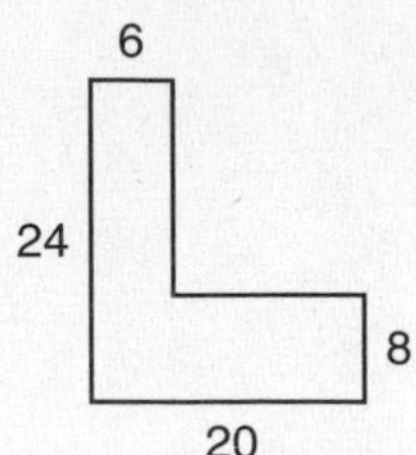

g. Write two formulas for finding the area of a triangle.

MIXED PRACTICE

Problem set

1. (36) The team won $\frac{2}{3}$ of its games and lost the rest. What was the team's win-loss ratio?

2. (28) During the first six months of the year, the car dealership sold 47 cars, 53 cars, 62 cars, 56 cars, 46 cars, and 48 cars. What was the average number of cars sold per month during the first six months of the year?

3. (31, 35) The relay team carried the baton around the track. Hope ran her leg of the relay in eleven and six tenths seconds. Robert ran his leg in eleven and three tenths seconds. Maria ran her leg in eleven and two tenths seconds. Takeshi ran his leg in ten and nine tenths seconds. What was the team's total time?

4. (28) Jenny went to the store with $10 and returned home with 3 gallons of milk and $1.30 in change. How can she find the cost of each gallon of milk?

5. (22) Diagram this statement. Then answer the questions that follow.

Aziz shot par on two thirds of the 18 holes.

(a) On how many holes did Aziz shoot par?

(b) On how many holes did Aziz not shoot par?

Copy this hexagon on your paper, and find the length of each unmarked side. Dimensions are in inches. All angles are right angles. Refer to the figure to answer problems 6 and 7.

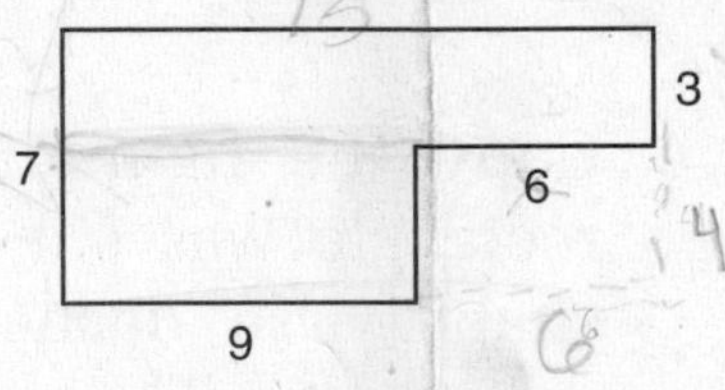

6. (19) What is the perimeter of the hexagon?

7. (37) What is the area of the hexagon?

8. Complete each equivalent fraction:
(15)

(a) $\frac{5}{6} = \frac{?}{18}$ (b) $\frac{?}{8} = \frac{9}{24}$ (c) $\frac{3}{4} = \frac{15}{?}$

9. (a) What decimal part of this square is shaded?
(31)

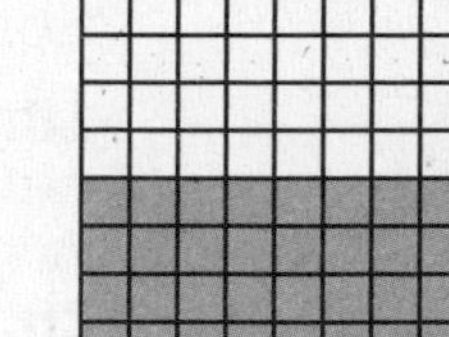

(b) What decimal part of this square is not shaded?

(c) What percent of this square is not shaded?

10. Round 3184.5641
(33)

(a) to two decimal places.

(b) to the nearest hundred.

11. (a) Name 0.00025.
(31)

(b) Use digits to write sixty and seven hundredths.

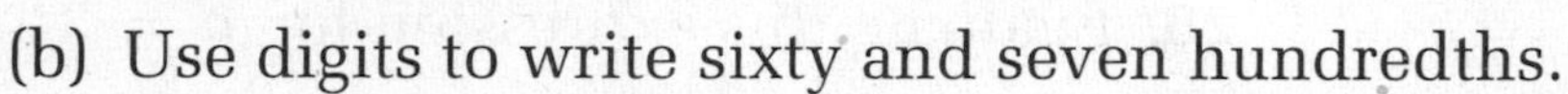

12. (a) Write 2% as a reduced fraction.
(24)

(b) Reduce $\frac{720}{1080}$.

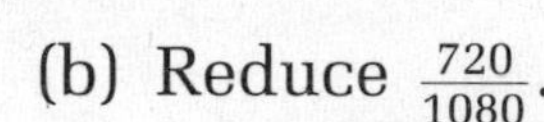

13. Find the length of segment *BC*.
(8)

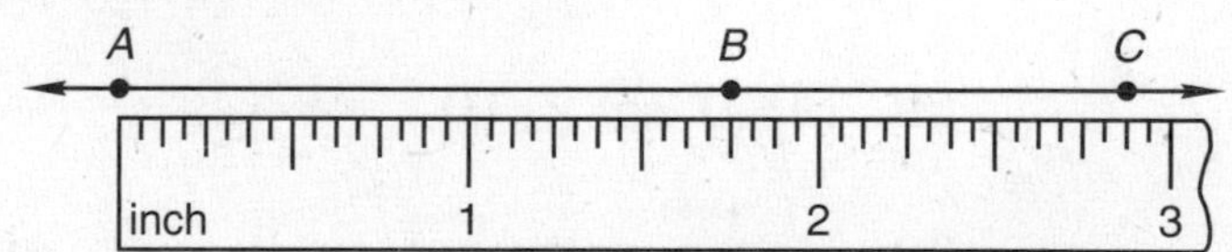

14. Draw a pair of parallel lines. Next draw another pair of parallel lines that intersect the first pair of lines but are not perpendicular to them. Then shade the region enclosed by the intersecting pairs of lines.
(7)

15. Refer to the triangle below to answer (a) and (b).
(37)

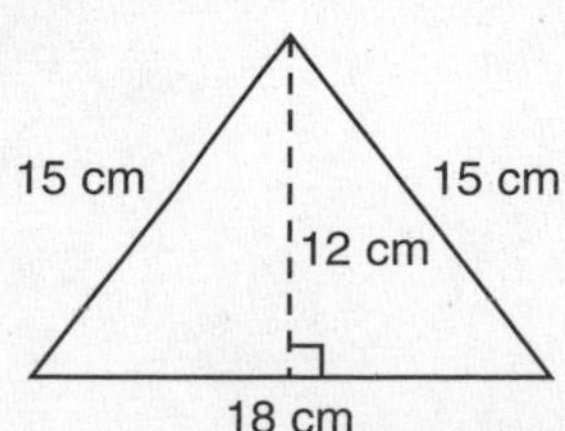

(a) What is the perimeter of the triangle?

(b) What is the area of the triangle?

16. (33, 35) Simplify and compare: $0.2 + 0.3 \bigcirc 0.2 \times 0.3$

17. (36) What is the probability of tossing heads with one toss of a fair coin?

Solve:

18. (3) $7 \cdot 8 = 4x$

19. (35) $4.2 = 1.7 + y$

20. (35) $m - 3.6 = 0.45$

21. (35) $\frac{4.5}{w} = 3$

Simplify:

22. (26) $\frac{3}{5} \cdot 12 \cdot 4\frac{1}{6}$

23. (30) $\frac{5}{6} + 1\frac{3}{4} + 2\frac{1}{2}$

24. (30) $\frac{5}{8} + \left(\frac{1}{2} + \frac{3}{8}\right)$

25. (30) $3\frac{9}{20} - 1\frac{5}{12}$

26. (1, 26) Evaluate this expression for $a = 3\frac{1}{3}$ and $b = 5$:

$$\frac{a}{b}$$

27. (20) Find the missing exponent.

$$2^2 \cdot 2^2 \cdot 2^2 = 2^n$$

28. (35) (a) 0.25×10 (b) $0.25 \div 10$

29. (Inv. 2) Use a compass to draw a circle. Then use the compass and a straightedge to inscribe a regular hexagon.

30. (Inv. 3) A point with the coordinates (3, –3) lies in which quadrant of the coordinate plane?

LESSON

38 Interpreting Graphs

WARM-UP

Facts Practice: + − × ÷ Fractions (Test G)

Mental Math:

a. \$7.43 − \$0.99 **b.** 3 × \$2.50 **c.** $\frac{5}{6} = \frac{?}{30}$

d. Reduce $\frac{18}{36}$. **e.** $\sqrt{121} + 7^2$ **f.** $\frac{7}{10}$ of 50

g. 8 × 4, − 2, ÷ 3, × 4, ÷ 5, + 1, $\sqrt{}$, × 6, + 2, × 2, + 2, ÷ 6, × 5, + 1, $\sqrt{}$

Problem Solving:

Javier used a two-yard length of string to make a rectangle that was twice as long as it was wide. What was the area of the rectangle in square feet?

NEW CONCEPT

We use **graphs** to help us understand quantitative information. A graph can use pictures, bars, lines, or parts of circles to help the reader visualize comparisons or changes. In this lesson we will practice interpreting graphs.

Example 1 Refer to the pictograph below to answer the questions that follow.

Doughnut Sales

Jan.

Feb.

Mar.

Represents 10,000 doughnuts

(a) About how many doughnuts were sold in March?

(b) About how many doughnuts were sold in the first three months of the year?

Solution The key at the bottom of the graph shows us that each picture of a doughnut represents 10,000 doughnuts.

(a) For March we see 5 whole doughnuts, which represent 50,000 doughnuts, and half a doughnut, which represents 5000 doughnuts. Thus the $5\frac{1}{2}$ doughnuts pictured mean that **about 55,000 doughnuts** were sold in March.

(b) We see a total of $15\frac{1}{2}$ doughnuts pictured for the first three months of the year. Fifteen times 10,000 is 150,000. Half of 10,000 is 5000. Thus **about 155,000 doughnuts** were sold in the first three months of the year.

Example 2 Refer to the bar graph below to answer the questions that follow.

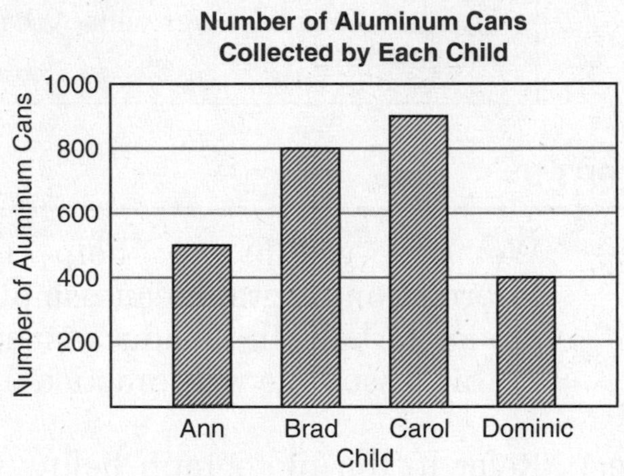

(a) About how many cans did Brad collect?

(b) Carol collected about as many cans as what other two children combined?

Solution We look at the scale on the left side of the graph. We see that the distance between two horizontal lines on the scale represents 200 cans. Thus, halfway from one line to the next represents 100 cans.

(a) Brad collected **about 800 cans.**

(b) Carol collected about 900 cans. This was about as many cans as **Ann** and **Dominic** combined.

Example 3 This line graph shows Paul's test scores during the year.

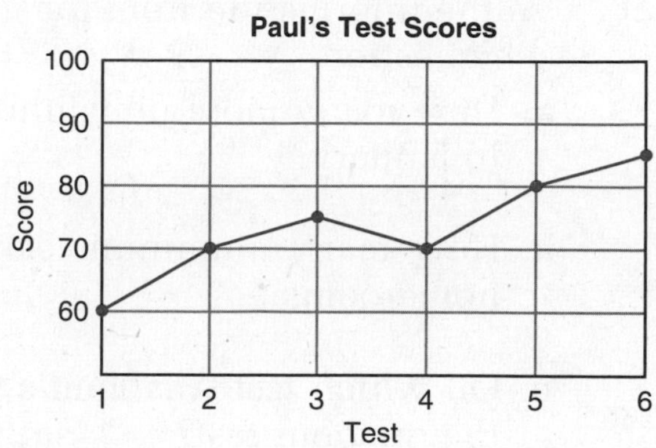

(a) What was Paul's score on Test 3?

(b) In general, were Paul's scores improving or getting worse?

Solution (a) We find Test 3 on the scale across the bottom of the graph and go up to the point that represents Paul's score. We see that the point is halfway between the lines that represent 70 and 80. Thus Paul's score on Test 3 was about **75.**

(b) With only one exception, Paul scored higher on each succeeding test. So, in general, Paul's scores were **improving.**

Example 4 Use the information in this circle graph to answer the following questions:

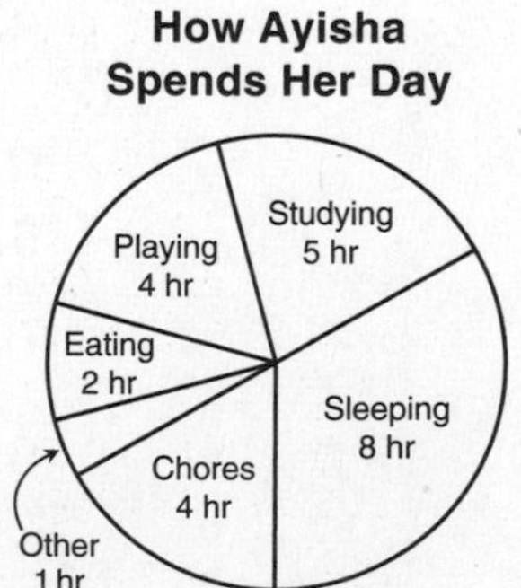

(a) Altogether, how many hours are included in this graph?

(b) What fraction of Ayisha's day is spent sleeping?

Solution A circle graph (sometimes called a pie graph) shows the relationship between parts of a whole. This graph shows parts of a whole day.

(a) This graph includes **24 hours,** one whole day.

(b) Ayisha spends 8 of the 24 hours sleeping. We reduce $\frac{8}{24}$ to $\mathbf{\frac{1}{3}}$.

LESSON PRACTICE

Practice set Use the information from the graphs in this lesson to answer each question.

a. How many more doughnuts were sold in February than in January?

b. How many aluminum cans were collected by all four homerooms?

c. On which test was Paul's score lower than his score on the previous test?

d. What fraction of Ayisha's day is spent performing chores?

MIXED PRACTICE

Problem set

1. (36) The ratio of soldiers to civilians at the outpost was 3 to 7. What fraction of the people at the outpost were soldiers?

2. (28) Denise read a 345-page book in 3 days. What was the average number of pages she read each day?

3. (28) Christine ran a mile in 5 minutes 52 seconds. How many seconds did it take Christine to run a mile?

Refer to the graphs in this lesson to answer problems 4–6.

4. (12, 38) How many fewer cans were collected by Dominic than by Carol?

5. (28, 38) If Paul scores 85 on Test 7, what will be his test score average for all 7 tests?

6. (12, 38) Use the information in the graph in example 3 to write a problem about comparing.

7. (22) Diagram this statement. Then answer the questions that follow.

Mira read three eighths of the 384-page book before she could put it down.

(a) How many pages did she read?

(b) How many more pages does she need to read to be halfway through the book?

8. Refer to the figure at right to answer (a) and (b). Dimensions are in inches. All angles are right angles.
(19, 37)

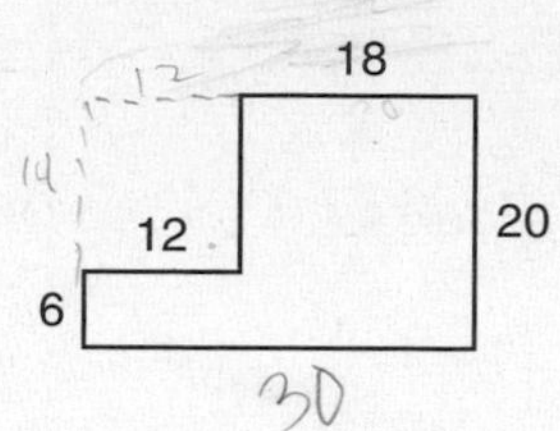

(a) What is the area of the hexagon?

(b) What is the perimeter of the hexagon?

9. Complete each equivalent fraction:
(15)

(a) $\frac{7}{9} = \frac{?}{18}$ (b) $\frac{?}{9} = \frac{20}{36}$ (c) $\frac{4}{5} = \frac{24}{?}$

10. Round 2986.34157
(33)

(a) to the nearest thousand.

(b) to three decimal places.

11. The face of this spinner is divided into eight congruent sectors. If the spinner is spun once, on which number is it
(36)

(a) most likely to stop?

(b) least likely to stop?

12. Find the length of this segment
(34)

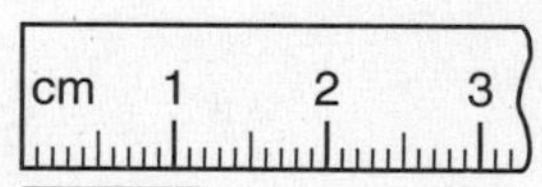

(a) in centimeters.

(b) in millimeters.

13. Find the perimeter of this rectangle in centimeters.
(32, 34)

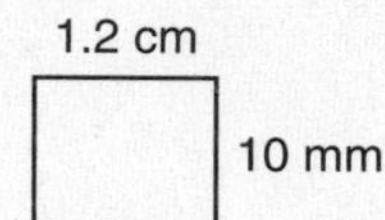

14. Which point marked on this number line could represent 3.4? Why?
(34)

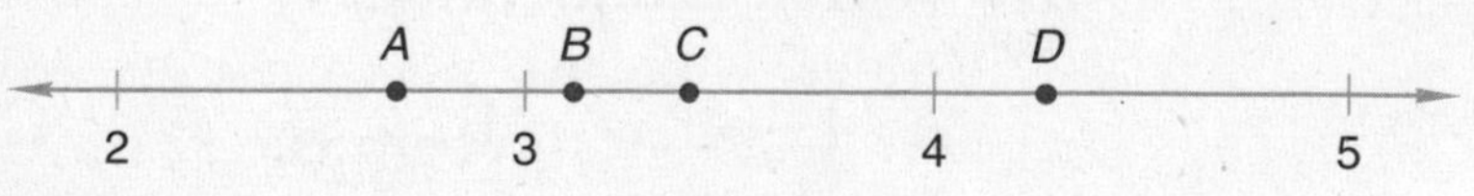

In the figure below, diagonal AC divides quadrilateral $ABCD$ into two congruent triangles. Refer to the figure to answer problems 15 and 16.

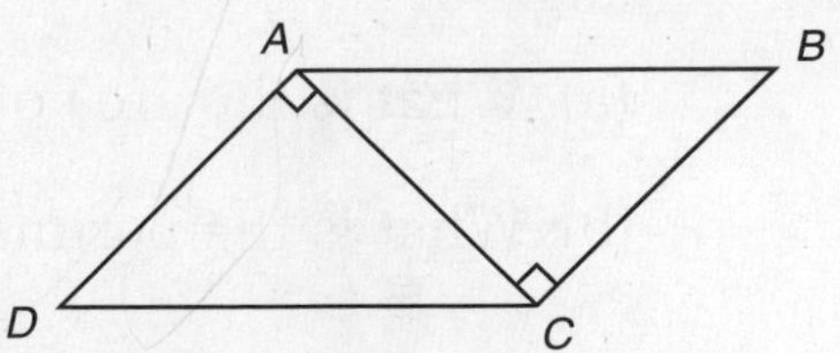

15. (7) (a) Which segment is perpendicular to $\overline{AD}$?

(b) Which segment appears to be parallel to $\overline{AD}$?

16. (37) The perpendicular sides of ΔACD measure 6 cm each.

(a) What is the area of ΔACD?

(b) What is the area of ΔCAB?

(c) What is the area of the quadrilateral?

Solve:

17. (35) $4.3 + a = 6.7$

18. (35) $m - 3.6 = 4.720$

19. (35) $10w = 4.5$

20. (35) $\frac{x}{2.5} = 2.5$

Simplify:

21. (35) $5.37 + 27.7 + 4$

22. (35) $1.25 \div 5$

23. (26) $\frac{5}{9} \cdot 6 \cdot 2\frac{1}{10}$

24. (30) $\frac{5}{8} + \frac{3}{4} + \frac{1}{2}$

25. (26) $5 \div 3\frac{1}{3}$

26. (30) $\frac{3}{10} - \left(\frac{1}{2} - \frac{1}{5}\right)$

27. (20) Which is equivalent to $2^2 \cdot 2^4$?

A. $4 \cdot 4^2$ B. 2^8 C. 4^8 D. 4^6

28. (32) (a) About how many milliliters of liquid are in this container?

(b) The amount of liquid in this container is how much less than a liter?

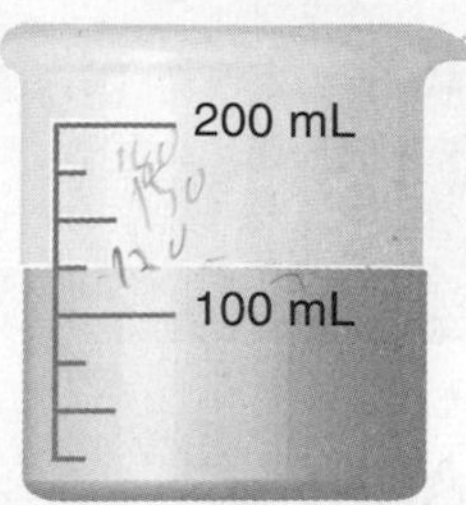

29. (33) Five books on the library shelf were in the order shown below. Which two books should be switched so that they are arranged in the correct order?

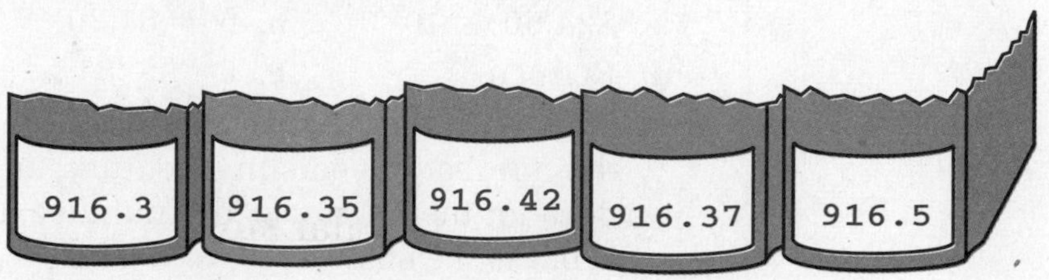

30. (Inv. 3) On graph paper draw a ray from the origin through the point (−10, −10). Then draw a ray from the origin through the point (−10, 10). Use a protractor to measure the angle formed by the two rays, and write the measure on the graph paper.

LESSON

39 Proportions

WARM-UP

Facts Practice: Measurement Facts (Test H)

Mental Math:

a. $24.50 ÷ 10 b. 6 × $1.20 c. $\frac{7}{12} = \frac{?}{60}$

d. Reduce $\frac{24}{32}$. e. $5^2 - \sqrt{81}$ f. $\frac{4}{5}$ of 40

g. Start with the number of degrees in a straight angle. Subtract the number of years in a century; add the number of years in a decade; then subtract the number of degrees in a right angle. What is the answer?

Problem Solving:

Is it possible to hold a die in a position such that the three adjoining faces that can be seen each have an even number of dots? An odd number of dots?

NEW CONCEPT

A **proportion** is a statement that two ratios are equal.

$$\frac{16}{20} = \frac{4}{5}$$

One way to test whether two ratios are equal is to compare their cross products. If we multiply the upper term of one ratio by the lower term of the other ratio, we form a **cross product.** The cross products of equal ratios are equal. We illustrate by finding the cross products of this proportion:

$$5 \cdot 16 = 80 \qquad 20 \cdot 4 = 80$$

$$\frac{16}{20} = \frac{4}{5}$$

We find that both cross products equal 80. Whenever the cross products are equal, the ratios are equal as well.

We can use cross products to help us find missing terms in proportions. We will follow a two-step process.

Step 1: Find the cross products.

Step 2: Divide the known product by the known factor.

Example 1 Solve the following proportion: $\frac{12}{20} = \frac{n}{30}$

Solution We solve a proportion by finding the missing term.

Step 1: First we find the cross products. Since we are completing a proportion, the cross products are equal.

$$\frac{12}{20} = \frac{n}{30}$$

$$20 \cdot n = 30 \cdot 12 \quad \text{equal cross products}$$

$$20n = 360 \quad \text{simplified}$$

Step 2: Divide the known product (360) by the known factor (20). The result is the missing term.

$$n = \frac{360}{20} \quad \text{divided by 20}$$

$$n = \mathbf{18} \quad \text{simplified}$$

The complete proportion is

$$\frac{12}{20} = \frac{18}{30}$$

We can check our work by noting that both ratios reduce to $\frac{3}{5}$.

Example 2 Solve: $\frac{15}{x} = \frac{20}{32}$

Solution **Step 1:** $20x = 480$ equal cross products

Step 2: $x = \mathbf{24}$ divided by 20

We check our work and see that both ratios reduce to $\frac{5}{8}$.

LESSON PRACTICE

Practice set Solve each proportion:

a. $\frac{a}{12} = \frac{6}{8}$

b. $\frac{30}{b} = \frac{20}{16}$

c. $\frac{14}{21} = \frac{c}{15}$

d. $\frac{30}{25} = \frac{24}{d}$

e. $\frac{30}{100} = \frac{n}{40}$

f. $\frac{m}{100} = \frac{9}{12}$

MIXED PRACTICE

Problem set Eduardo made a line graph to show his height on each birthday. Refer to the graph to answer problems 1 and 2.

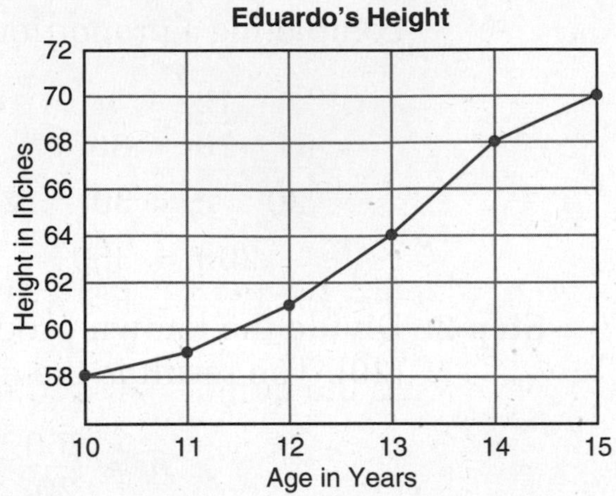

1. (38) How many inches did Eduardo grow between his twelfth and thirteenth birthdays?

2. (38) Between which two birthdays did Eduardo grow the most?

3. (36) There were 12 princes and 16 princesses in the palace. What was the ratio of princes to princesses in the palace?

4. (28) On the first 4 days of their trip, the Curtis family drove 497 miles, 513 miles, 436 miles, and 410 miles. What was the average number of miles they drove per day on the first 4 days of their trip?

5. (13) Don receives a weekly allowance of \$4.50. How much allowance does he receive in a year (52 weeks)?

6. (22) Diagram this statement. Then answer the questions that follow.

Three sevenths of the 105 adults in the Khoikhoi clan were less than 5 feet tall.

(a) How many of the adults were less than 5 feet tall?

(b) How many of the adults were 5 feet tall or taller?

Refer to this figure to answer problems 7 and 8. Dimensions are in millimeters. All angles are right angles.

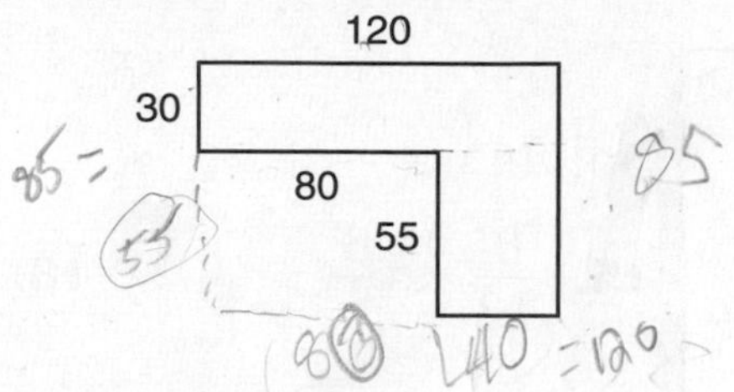

7. (37) What is the area of the figure?

8. (19) What is the perimeter of the figure?

9. (31) Name the number of shaded circles

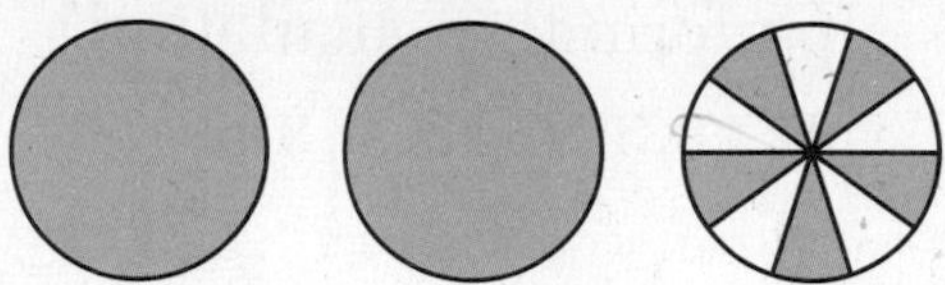

(a) as a decimal number.

(b) as a mixed number.

10. (33) Round 0.9166666

(a) to the nearest hundredth.

(b) to the nearest hundred-thousandth.

11. (33) Sharon pulled into the service station and bought 9.16 gallons of gasoline priced at $\$1.39\frac{9}{10}$ per gallon. Estimate the total cost.

12. (31) Use digits to write each number:

(a) one hundred and seventy-five thousandths

(b) one hundred seventy-five thousandths

13. (7) Refer to the figure at right to name

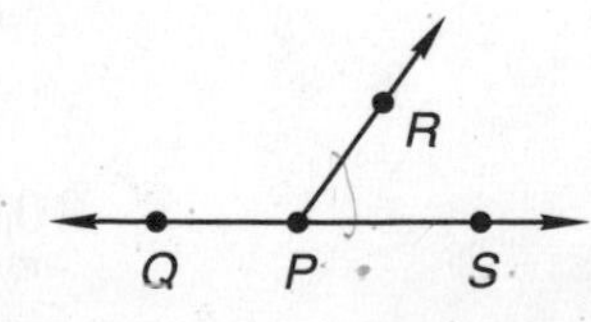

(a) an acute angle.

(b) an obtuse angle.

(c) a straight angle.

14. (2, 35) Find the next three numbers in this sequence, and state the rule of the sequence:

$$\ldots, 100, 10, 1, 0.1, \ldots$$

Solve:

15. (39) $\frac{8}{12} = \frac{6}{x}$

16. (39) $\frac{16}{y} = \frac{2}{3}$

17. (39) $\frac{21}{14} = \frac{n}{4}$

18. (35) $m + 0.36 = 0.75$

19. (35) $1.4 - w = 0.8$

20. (35) $8x = 7.2$

21. (35) $\frac{y}{0.4} = 1.2$

Estimate each answer to the nearest whole number. Then perform the calculation.

22. (35) $9.6 + 12 + 8.59$

23. (35) $3.15 - (2.1 - 0.06)$

Simplify:

24. (30) $4\frac{5}{12} + 6\frac{5}{8}$

25. (23, 30) $4\frac{1}{4} - 1\frac{3}{5}$

26. (26) $8\frac{1}{3} \cdot 1\frac{4}{5}$

27. (26) $5\frac{5}{6} \div 7$

28. (37) Refer to this triangle to answer (a) and (b):

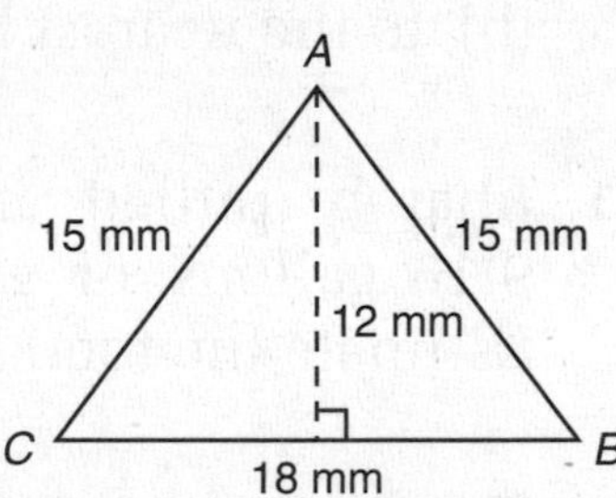

(a) What is the perimeter of ΔABC?

(b) What is the area of ΔABC?

29. (36) What is the probability of rolling an odd prime number with one toss of a dot cube?

30. (30) Use common denominators to arrange these numbers in order from least to greatest:

$$\frac{2}{3}, \frac{1}{2}, \frac{5}{6}, \frac{7}{12}$$

LESSON

40 Sum of the Angle Measures of a Triangle • Angle Pairs

WARM-UP

Facts Practice: + − × ÷ Fractions (Test G)

Mental Math:

a. \$0.18 × 100 **b.** 4 × \$1.25 **c.** $\frac{3}{4} = \frac{?}{24}$

d. Reduce $\frac{12}{32}$. **e.** $\sqrt{144} + \sqrt{121}$ **f.** $\frac{2}{3}$ of 60

g. Start with \$10.00. Divide by 4; add two quarters; multiply by 3; find half of that amount; then subtract two dimes.

Problem Solving:

Copy this problem and fill in the missing digits:

```
      2_
 _7)4__
    __
     _8
     __
      0
```

NEW CONCEPTS

Sum of the angle measures of a triangle

A square has four angles that measure 90° each. If we draw a segment from one vertex of a square to the opposite vertex, we have drawn a diagonal that divides the square into two congruent triangles. We show this in the figure below.

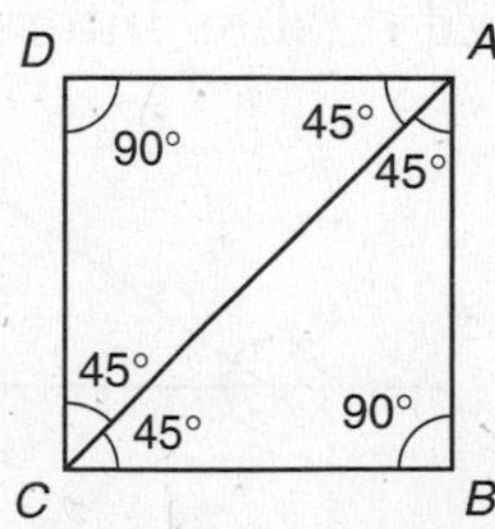

Segment AC divides the right angle at A into two angles, each measuring 45° (90° ÷ 2 = 45°). Segment AC also divides the right angle at C into two 45° angles. Each triangle has angles that measure 90°, 45°, and 45°. The sum of these angles is 180°.

$$90° + 45° + 45° = 180°$$

The three angles of every triangle have measures that total 180°. We illustrate this fact with the following activity.

Activity: *Sum of the Angle Measures of a Triangle*

Materials needed:

- Paper
- Ruler or straightedge
- Protractor
- Scissors

With a ruler or straightedge, draw two or three triangles of various shapes large enough to easily fold. Let the longest side of each triangle be the base of that triangle, and indicate the height (altitude) of the triangle by drawing a series of dashes perpendicular to the base. The dashes should extend from the base to the opposite vertex as in the figure below. Use the corner of a paper or a protractor if necessary to ensure the indicated height is perpendicular to the base.

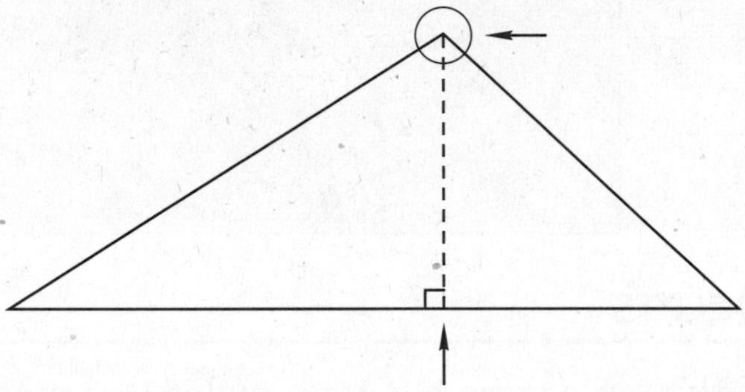

After the drawings are completed, carefully cut out each triangle. Then select a triangle for folding. First fold the vertex with the dash down to the point on the base where the row of dashes intersects the base.

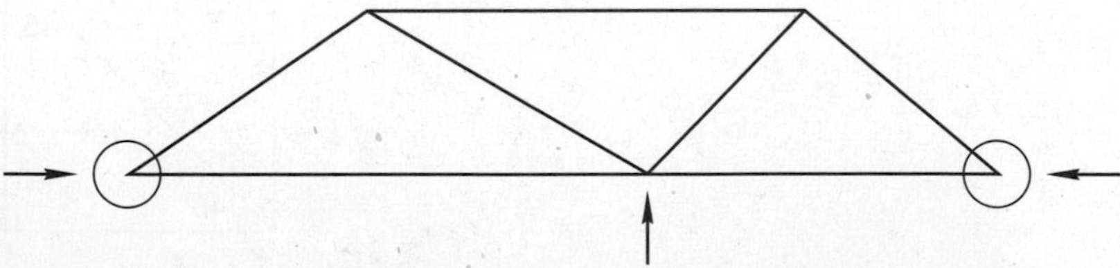

Then fold each of the other two vertices to the same point. When finished, your folded triangle should look like this:

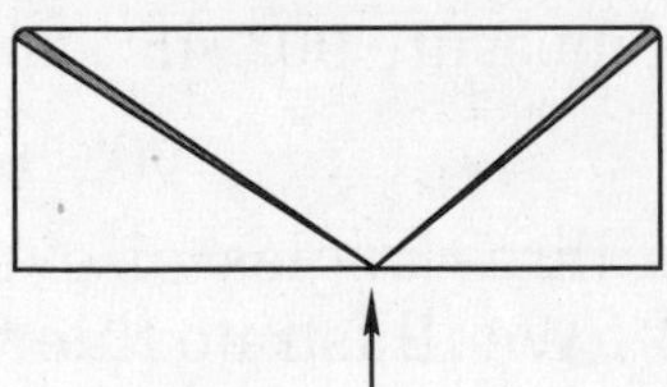

If you sketch a semicircle about the meeting point, you will see that the three angles of the triangle together form a half circle. That is, the sum of their measures is 180°.

Repeat the folding activity with the other triangle(s) you drew.

Example 1 Find the measure of $\angle A$ in ΔABC.

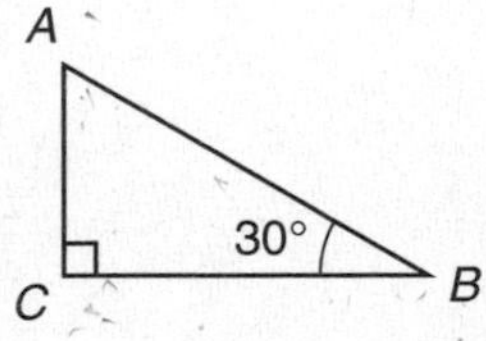

Solution The sum of the measures of the angles is 180°. Angle B measures 30° and angle C is a right angle that measures 90°. Using this information, we can write the following equation:

$$m\angle A + 90° + 30° = 180°$$

$$m\angle A = \mathbf{60°}$$

Note: The abbreviation $m\angle A$ is read "the measure of angle A."

Angle pairs Two intersecting lines form four angles. We have labeled these four angles $\angle 1$, $\angle 2$, $\angle 3$, and $\angle 4$ for easy reference.

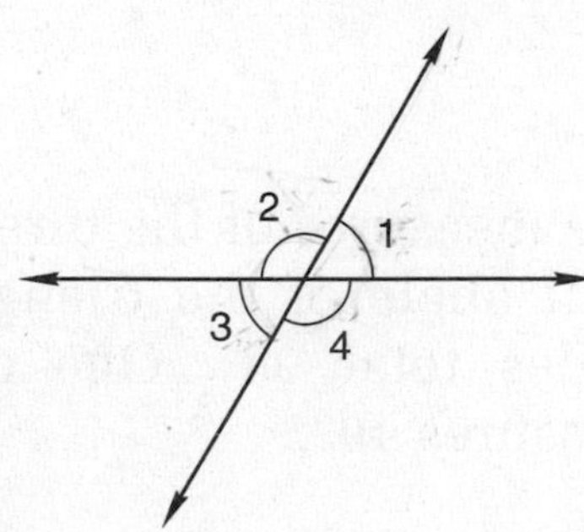

Angle 1 and $\angle 2$ are **adjacent angles,** sharing a common side. Notice that together they form a straight angle. A straight angle measures 180°, so the sum of the measures of $\angle 1$ and $\angle 2$ is 180°. Two angles whose sum is 180° are called **supplementary angles.** We say that $\angle 1$ is a supplement of $\angle 2$ and that $\angle 2$ is a supplement of $\angle 1$.

Notice that $\angle 1$ and $\angle 4$ are also supplementary angles. If we know that $\angle 1$ measures 60°, then we can calculate that $\angle 2$ measures 120° (60° + 120° = 180°) and that $\angle 4$ measures 120°. So $\angle 2$ and $\angle 4$ have the same measure.

Another pair of supplementary angles is $\angle 2$ and $\angle 3$, and the fourth pair of supplementary angles is $\angle 3$ and $\angle 4$. Knowing that $\angle 2$ or $\angle 4$ measures 120°, we can calculate that $\angle 3$ measures 60°. So $\angle 1$ and $\angle 3$ have the same measure.

Angles 1 and 3 are not adjacent angles; they are **vertical angles.** Likewise, $\angle 2$ and $\angle 4$ are vertical angles. Vertical angles are a pair of nonadjacent angles formed by a pair of intersecting lines. Vertical angles have the same measure.

Two angles whose measures total 90° are called **complementary angles.** In the triangle below, $\angle A$ and $\angle B$ are complementary because the sum of their measures is 90°. So $\angle A$ is a complement of $\angle B$, while $\angle B$ is a complement of $\angle A$.

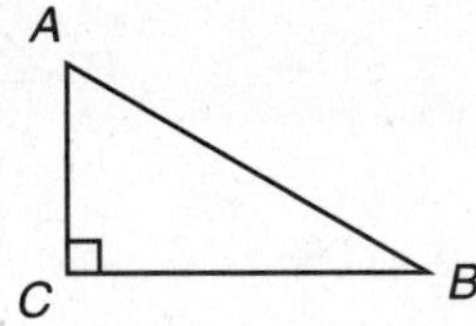

Example 2 Find the measures of $\angle x$, $\angle y$, and $\angle z$.

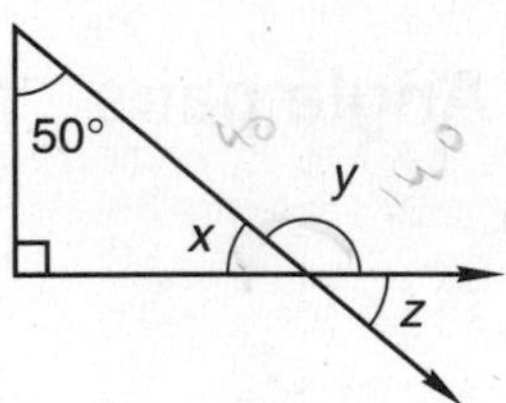

Solution The measures of the three angles of a triangle total 180°. The right angle of the triangle measures 90°. So the two acute angles total 90°. One of the acute angles is 50°, so $\angle x$ measures **40°**.

Together $\angle x$ and $\angle y$ form a straight angle measuring 180°, so they are supplementary. Since $\angle x$ measures 40°, $\angle y$ measures the rest of 180°, which is **140°**.

Angle z and $\angle y$ are supplementary. Also, $\angle z$ and $\angle x$ are vertical angles. Since vertical angles have the same measure, $\angle z$ measures **40°**.

Example 3 Figure *ABCD* is a parallelogram. Segment *BD* divides the parallelogram into two congruent triangles. Angle *CBD* measures 40°. Which other angle measures 40°?

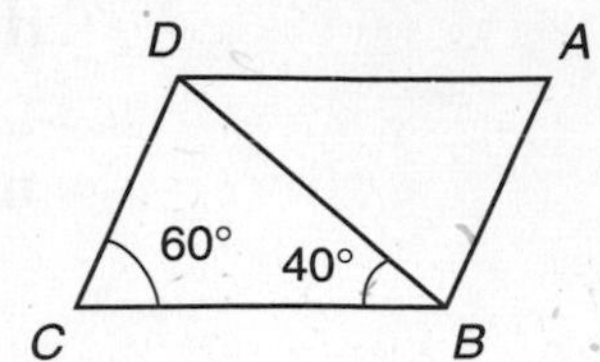

Solution A segment that passes through a polygon to connect two vertices is a diagonal. A diagonal divides any parallelogram into two congruent triangles. Segment *BD* is a diagonal. Rotating (turning) one of the triangles 180° (a half turn) positions the triangle in the same orientation as the other triangle. We illustrate rotating ΔABD 180°.

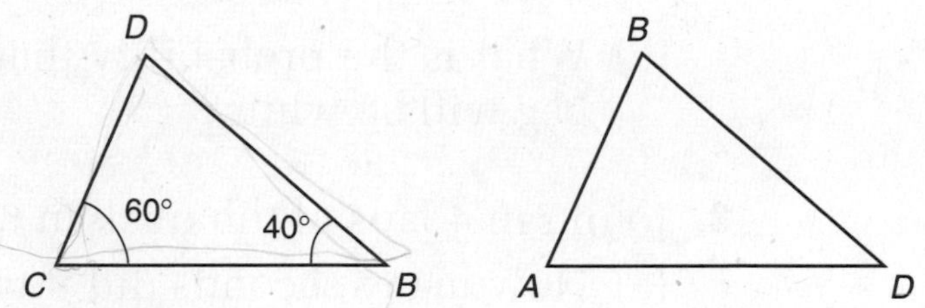

Each side and angle of ΔABD now has the same orientation as its **corresponding** side or angle in ΔCDB. (Notice that segments *BD* and *DB* are the same segment in both triangles. They have the same orientation in both triangles, but their vertices are reversed due to the rotation.) Angle *CBD* measures 40°. The angle in ΔABD that corresponds to this angle is $\angle ADB$. So $\angle ADB$ **measures 40°.**

LESSON PRACTICE

Practice set **a.** The sides of a regular triangle are equal in length, and the angles are equal in measure. What is the measure of each angle of a regular triangle and why?

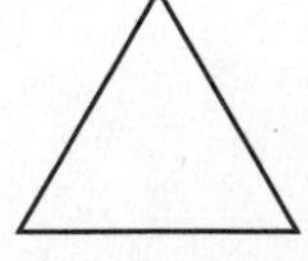

Refer to rectangle *ABCD* to answer problems **b–d.**

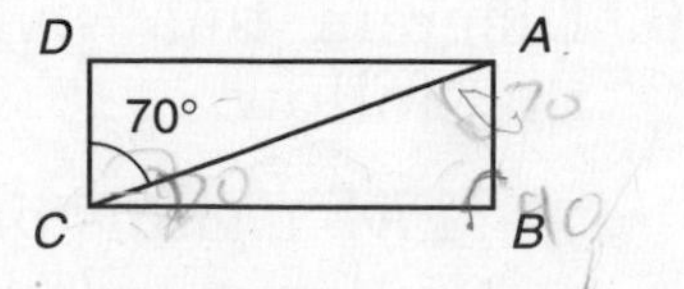

b. What is the measure of $\angle ACB$ and why?

c. What is the measure of $\angle CAB$ and why?

d. Are angles *ACD* and *CAB* vertical angles? Why or why not?

e. Find the measures of $\angle x$, $\angle y$, and $\angle z$ in this figure.

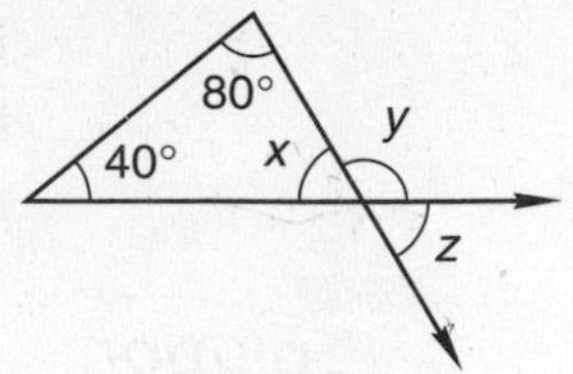

MIXED PRACTICE

Problem set

1. (36) The bag contained only red marbles and white marbles in the ratio of 3 red to 2 white.

(a) What fraction of the marbles were white?

(b) What is the probability that a marble drawn from the bag will be white?

2. (28) John ran 4 laps of the track in 6 minutes 20 seconds.

(a) How many seconds did it take John to run 4 laps?

(b) John's average time for running each lap was how many seconds?

3. (13) The Curtises' car traveled an average of 24 miles per gallon of gas. At that rate how far could the car travel on a full tank of 18 gallons?

4. (12, 35) Normal body temperature is 98.6°F. Allan's temperature was 103.4°F. His temperature was how many degrees above normal?

5. (19, 20) The length of the rectangle at right is twice its width.

70 mm

(a) What is the perimeter of the rectangle?

(b) What is the area of the rectangle?

6. (22) Diagram this statement. Then answer the questions that follow.

Five eighths of the 200 sheep in the flock grazed in the meadow. The rest drank from the brook.

(a) How many of the sheep grazed in the meadow?

(b) How many of the sheep drank from the brook?

7. (34) AB is 30 mm. CD is 45 mm. AD is 100 mm. Find BC in centimeters.

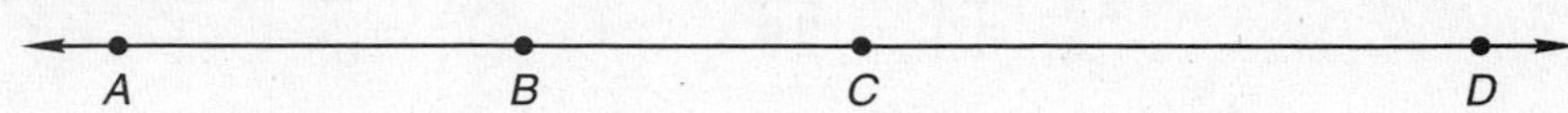

8. (33) Round 0.083333

(a) to the nearest thousandth.

(b) to the nearest tenth.

9. (31) Use words to write each number:

(a) 12.054 (b) $10\frac{11}{100}$

10. (Inv. 3, 37) The coordinates of the three vertices of a triangle are (–2, 5), (4, 0), and (–2, 0). What is the area of the triangle?

11. (34) What decimal number names the point marked B on this number line?

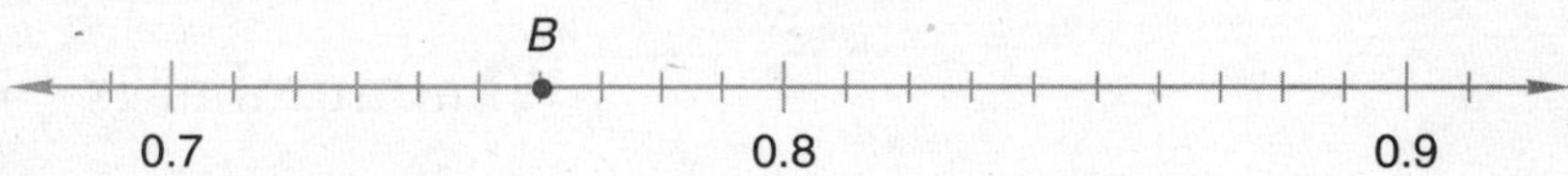

Refer to the figure below to answer problems 12 and 13.

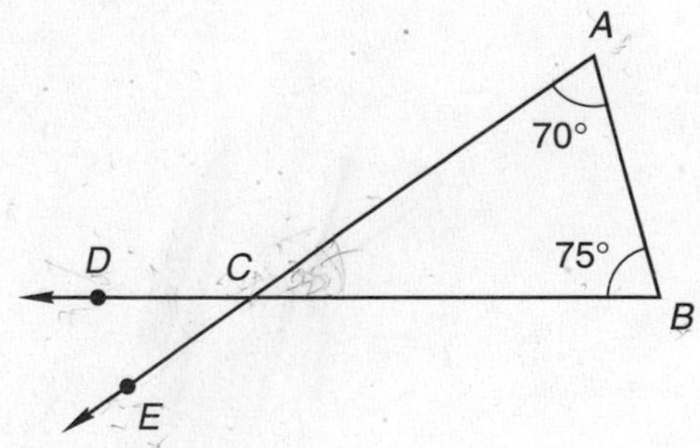

12. (40) Find the measure of each angle:

(a) $\angle ACB$ (b) $\angle ACD$ (c) $\angle DCE$

13. (40) Angle ACD is supplementary to $\angle ACB$. Name another angle supplementary to $\angle ACB$.

14. (1, 2) (a) Which property is illustrated by this equation?

$$\frac{12}{0.5} \cdot \frac{10}{10} = \frac{120}{5}$$

(b) Find the quotient of $\frac{120}{5}$.

Solve:

15. $\frac{8}{10} = \frac{w}{25}$ *(39)*

16. $\frac{n}{1.5} = \frac{6}{9}$ *(39)*

17. $\frac{9}{12} = \frac{15}{m}$ *(39)*

18. $4 = a + 1.8$ *(35)*

19. $3.9 = t - 0.39$ *(35)*

Simplify:

20. 1.2 m − 12 cm = ____ m *(34)*

21. (0.15)(0.05) *(35)*

22. 15 × 1.5 *(35)*

23. 14.4 ÷ 12 *(35)*

24. 5.6 − (4 − 1.25) *(35)*

25. 5 − (3.14 + 1.2) *(35)*

26. $6\frac{1}{4} \cdot 1\frac{3}{5}$ *(26)*

27. $7 \div 5\frac{5}{6}$ *(26)*

28. $\frac{8}{15} + \frac{12}{25}$ *(30)*

29. $4\frac{2}{5} - 1\frac{3}{4}$ *(23, 30)*

30. Estimate the answer to the following question: *(33)*

What number is $\frac{1}{4}$ of 31.975?

INVESTIGATION 4

Focus on

Stem-and-Leaf Plots, Box-and-Whisker Plots

Stem-and-leaf plots

A college counselor evaluated the results of a math placement test administered to incoming students to help determine the appropriate math course for each student. The scores of one group of students are listed below.

40, 30, 43, 48, 26, 50, 55, 40, 34, 42, 47, 47, 52, 25, 32, 38, 41, 36, 32, 21, 35, 43, 51, 58, 26, 30, 41, 45, 23, 36, 41, 51, 53, 39, 28

To organize the scores, the counselor created a **stem-and-leaf plot.** Noticing that the scores ranged from a low of 21 to a high of 58, the counselor chose the initial digits of 20, 30, 40, and 50 to serve as the stem digits.

Stem
2
3
4
5

Then the counselor used the ones place digits of the scores as the leaves of the stem-and-leaf plot. (The "stem" of a stem-and-leaf plot may have more than one digit. Each "leaf" is only one digit.)

3 | 2 represents a score of 32

Stem	Leaf
2	1 3 5 6 6 8
3	0 0 2 2 4 5 6 6 8 9
4	0 0 1 1 1 2 3 3 5 7 7 8
5	0 1 1 2 3 5 8

The counselor included a key to the left to help a reader interpret the plot. The top row of leaves indicates the six scores 21, 23, 25, 26, 26, and 28.

1. Looking at this stem-and-leaf plot, we see that there is one score of 21, one 23, one 25, two 26's, and so on. Scanning through all of the scores, which score occurs more than twice?

The number that occurs most frequently in a set of numbers is the **mode.** The mode of these 35 scores is 41.

2. Looking at the plot, we immediately see that the lowest score is 21 and the highest score is 58. What is the difference of these scores?

The difference between the least and greatest numbers in a set is the **range** of the numbers. We find the range of this set of scores by subtracting 21 from 58.

3. The **median** of a set of numbers is the middle number of the set when the numbers are arranged in order. The counselor drew a vertical segment through the median on the stem-and-leaf plot. Which score was the median score?

Half of the scores are at or below the median score, and half of the scores are at or above the median score. There are 35 scores and half of 35 is $17\frac{1}{2}$. This means there are 17 whole scores below the median and 17 whole scores above the median. The $\frac{1}{2}$ means that the median is one of the scores on the list. (The median of an even number of scores is the mean—the average—of the two middle scores.) We can count 17 scores up from the lowest score or 17 scores down from the highest score. The next score is the median. We find that the median score is 40.

Stem	Leaf
2	1 3 5 6 6 8
3	0 0 2 2 4 5 6 6 8 9
4	0 0 1 1 1 2 3 3 5 7 7 8
5	0 1 1 2 3 5 8

3 | 2 represents a score of 32

median

Next the counselor found the middle number of the lower 17 scores and the middle number of the upper 17 scores. The middle number of the lower half of scores is the **first quartile** or **lower quartile.** The middle number of the upper half of scores is the **third quartile** or **upper quartile.** The second quartile is the median.

4. What are the first and third quartiles of these 35 scores?

There are 17 scores below the median. Half of 17 is $8\frac{1}{2}$. We count up 8 whole scores from the lowest score. The next score is the lower quartile score. Likewise, since there are also 17 scores above the median, we count down 8 whole scores from the highest score. The next score is the upper quartile score. Note that if there is an even number of numbers above and below the median, the quartiles are the mean of the two central numbers in each half.

Stem	Leaf
2	1 3 5 6 6 8
3	0 0 2 2 4 5 6 6 8 9
4	0 0 1 1 1 2 3 3 5 7 7 8
5	0 1 1 2 3 5 8

3 | 2 represents a score of 32

lower quartile

median

upper quartile

We count 8 scores below the first quartile, 8 scores between the first quartile and the median, 8 scores between the median and the third quartile, and 8 scores above the third quartile. The median and quartiles have "quartered" the scores.

Box-and-whisker plots

After locating the median and quartiles, the counselor created a **box-and-whisker plot** of the scores, which shows the location of certain scores compared to a number line. The five dots on a box-and-whisker plot show the **extremes** of the scores—the lowest score and highest score—as well as the lower quartile, median, and upper quartile. A *box* that is split at the median shows the location of the middle half of the scores. The *whiskers* show the scores below the first quartile and above the third quartile.

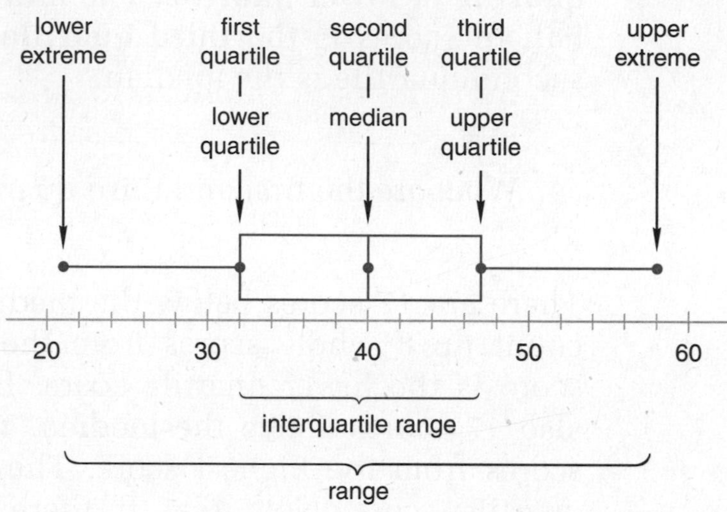

5. Create a stem-and-leaf plot for the following set of scores. Then draw vertical segments on the plot to indicate the median and the first and third quartiles.

   15, 26, 26, 27, 28, 29, 29, 30, 32, 33, 35, 36, 38, 38, 38, 38, 40, 41, 42, 43, 45, 45, 46, 47, 47, 48, 50, 52, 54, 55, 57, 58

6. What is the lower quartile, median, and upper quartile of this set of scores?

7. What is the mode of this set of scores?

8. What are the upper and lower extremes of these scores?

9. What is the range of the scores?

10. The **interquartile range** is the difference between the upper and lower quartiles. What is the interquartile range of these scores?

11. Create a box-and-whisker plot for this set of scores by using the calculations you have made for the median and the quartiles.

12. An **outlier** is a number in a set of numbers that is distant from the other numbers in the set. In this set of scores there is an outlier. Which score is the outlier?

LESSON

41 Using Formulas • Distributive Property

WARM-UP

Facts Practice: Proportions (Test I)

Mental Math:

a. 5×140

b. 1.54×10

c. $\frac{3}{5} = \frac{15}{x}$

d. $5^2 - 4^2$

e. Estimate: 39×29

f. $\frac{3}{10}$ of 70

g. Find the sum, difference, product, and quotient of $\frac{2}{3}$ and $\frac{1}{2}$.

Problem Solving:

Six blocks were used to build this three-step shape. How many blocks would be needed to build a nine-step shape?

NEW CONCEPTS

Using formulas In Lesson 20 it was stated that the area (A) of a rectangle is related to the length (l) and width (w) of the rectangle by this formula:

$$A = lw$$

This formula means "the area of a rectangle equals the product of its length and width." If we are given measures for l and w, we can replace the letters in the formula with numbers and calculate the area.

Example 1 Find A in $A = lw$ when l is 8 ft and w is 4 ft.

Solution We replace l and w in the formula with 8 ft and 4 ft respectively. Then we simplify.

$$A = lw$$

$$A = (8 \text{ ft})(4 \text{ ft})$$

$$A = \mathbf{32\ ft^2}$$

Notice the effect on the units when the calculation is performed. Multiplying two units of length results in a unit of area.

Example 2 Evaluate $2(l + w)$ when l is 8 cm and w is 4 cm.

Solution In place of l and w we substitute 8 cm and 4 cm. Then we simplify.

$$2(l + w)$$
$$2(8\text{ cm} + 4\text{ cm})$$
$$2(12\text{ cm})$$
$$\mathbf{24\text{ cm}}$$

Distributive property There are two formulas commonly used to relate the perimeter (p) of a rectangle to its length and width.

$$p = 2(l + w)$$
$$p = 2l + 2w$$

l

w

Both formulas describe how to find the perimeter of a rectangle if we are given its length and width. The first formula means "add the length and width and then double this sum." The second formula means "double the length and double the width and then add."

Example 3 Use the two perimeter formulas to find the perimeter of this rectangle.

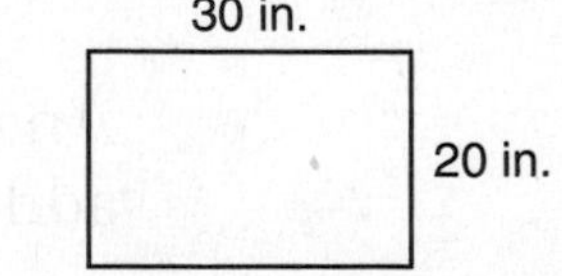

Solution In both formulas we replace l with 30 in. and w with 20 in. Then we simplify.

$p = 2(l + w)$	$p = 2l + 2w$
$p = 2(30\text{ in.} + 20\text{ in.})$	$p = 2(30\text{ in.}) + 2(20\text{ in.})$
$p = 2(50\text{ in.})$	$p = 60\text{ in.} + 40\text{ in.}$
$p =$ **100 in.**	$p =$ **100 in.**

Both formulas in example 3 yield the same result because the two formulas are equivalent.

$$2(l + w) = 2l + 2w$$

These equivalent expressions illustrate the **distributive property of multiplication over addition,** often called simply the **distributive property.** Applying the distributive property, we distribute, or "spread," the multiplication over the terms that are being added (or subtracted) within the parentheses. In this case we multiply l by 2, giving us $2l$, and we multiply w by 2, giving us $2w$.

$$2(l + w) = 2l + 2w$$

The distributive property is often expressed in equation form using variables:

$$a(b + c) = ab + ac$$

The distributive property also applies over subtraction.

$$a(b - c) = ab - ac$$

Example 4 Show two ways to simplify this expression:

$$6(20 + 5)$$

Solution One way is to add 20 and 5 and then multiply the sum by 6.

$$\mathbf{6(20 + 5)}$$

$$\mathbf{6(25)}$$

$$\mathbf{150}$$

Another way is to multiply 20 by 6 and multiply 5 by 6. Then add the products.

$$\mathbf{6(20 + 5)}$$

$$\mathbf{(6 \cdot 20) + (6 \cdot 5)}$$

$$\mathbf{120 + 30}$$

$$\mathbf{150}$$

LESSON PRACTICE

Practice set **a.** Find A in $A = bh$ when b is 15 in. and h is 8 in.

b. Evaluate $\frac{ab}{2}$ when a is 6 ft and b is 8 ft.

c. Write an equation using the letters x, y, and z that illustrates the distributive property of multiplication over addition.

d. Show two ways to simplify this expression:

$$6(20 - 5)$$

e. Write two formulas for finding the perimeter of a rectangle.

f. Describe two ways to simplify this expression:

$$2(6 + 4)$$

MIXED PRACTICE

Problem set

1. (36) Two hundred wildebeests and 150 gazelles grazed on the savannah. What was the ratio of gazelles to wildebeests grazing on the savannah?

2. (28) In their first 5 games the Celtics scored 105 points, 112 points, 98 points, 113 points, and 107 points. What was the average number of points the Celtics scored in their first 5 games?

3. (28) The crowd watched with anticipation as the pole vault bar was set to 19 feet 6 inches. How many inches is 19 feet 6 inches?

4. (2, 41) Which property is illustrated by each of these equations?

(a) $(a + b) + c = a + (b + c)$

(b) $a(bc) = (ab)c$

(c) $a(b + c) = ab + ac$

5. (35) Draw a sketch to help with this problem. From Tracey's house to John's house is 2.3 kilometers. From John's house to the library is 0.8 kilometer. Tracey rode from her house to John's house and then to the library. Later she rode from the library to John's house to her house. Altogether, how far did Tracey ride?

6. (36) About 70% of the earth's surface is water.

(a) About what fraction of the earth's surface is land?

(b) On the earth's surface, what is the ratio of water area to land area?

The stem-and-leaf plot below shows the distribution of test scores for a class of 20 students. Refer to the stem-and-leaf plot to answer problems 7 and 8.

2 | 4 represents 24 correct answers

Stem	Leaf
1	1
2	2 4 5 6 6 7 8 9
3	0 0 0 1 3 3 5 6 7 9
4	0

7. *(Inv. 4)* For the test scores find

(a) the median.

(b) the first quartile.

(c) the third quartile.

(d) any outliers.

8. *(Inv. 4)* Make a box-and-whisker plot of the test scores in the stem-and-leaf plot.

9. *(19, 37)* Refer to the figure at right to answer (a) and (b). All angles are right angles. Dimensions are in feet.

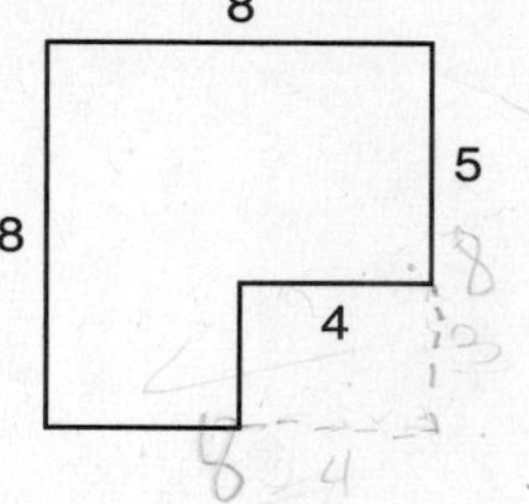

(a) What is the area of the figure?

(b) What is the perimeter of the figure?

10. *(34)* Name the point marked M on this number line:

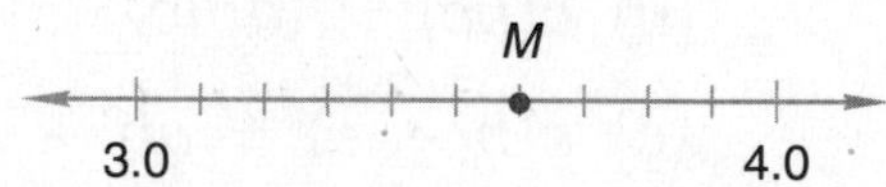

(a) as a decimal number.

(b) as a mixed number.

11. *(21)* What is the sum of the first four prime numbers?

12. *(37)* Dimensions of the triangle at right are in millimeters.

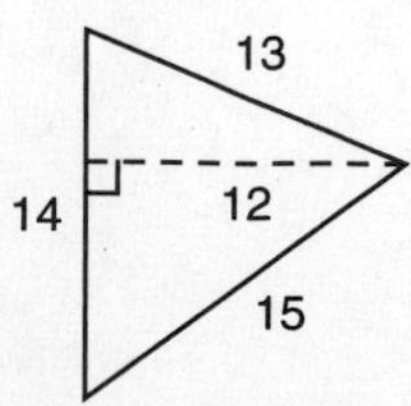

(a) What is the perimeter of the triangle?

(b) What is the area of the triangle?

13. (31) Use digits to write each number:

(a) sixty-seven hundred-thousandths

(b) one hundred and twenty-three thousandths

14. (41) Evaluate $2\pi r$ when π is 3.14 and r is 10.

15. (30) Write $\frac{3}{5}$, $\frac{1}{2}$, and $\frac{5}{7}$ with a common denominator, and arrange the renamed fractions in order from least to greatest.

16. (20) Find the area of the rectangle at right.

5.6 cm

3.4 cm

Solve:

17. (39) $\frac{x}{2.4} = \frac{10}{16}$

18. (39) $\frac{18}{8} = \frac{m}{20}$

19. (35) $3.45 + a = 7.6$

20. (35) $3y = 0.144$

Simplify:

21. (35) $7.4 \div 8$

22. (35) $(0.4)(0.6)(0.02)$

23. (35) $4.315 \div 5$

24. (35) $\frac{6.5}{100}$

25. (30) $3\frac{1}{3} + 1\frac{5}{6} + \frac{7}{12}$

26. (23, 30) $4\frac{1}{6} - \left(4 - 1\frac{1}{4}\right)$

27. (26) $3\frac{1}{5} \cdot 2\frac{5}{8} \cdot 1\frac{3}{7}$

28. (26) $4\frac{1}{2} \div 6$

29. (41) (a) Compare: $(12 \cdot 7) + (12 \cdot 13) \bigcirc 12(7 + 13)$

(b) Which property of operations is illustrated by this comparison?

30. (40) Find the measures of $\angle x$, $\angle y$, and $\angle z$ in this figure:

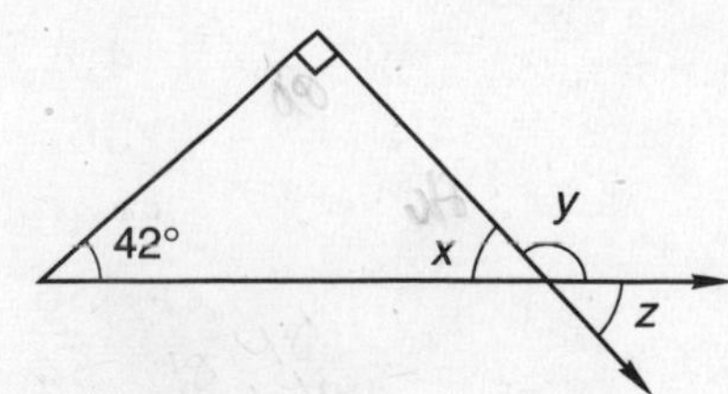

LESSON

42 Repeating Decimals

WARM-UP

Facts Practice: Measurement Facts (Test H)

Mental Math:

a. $3 \times 78¢$

b. 0.4×100

c. $\frac{4}{y} = \frac{20}{25}$

d. $\sqrt{121} - 3^2$

e. Estimate: $1\frac{7}{8} \times 3\frac{1}{8}$

f. $\frac{4}{5}$ of 35

g. Start with three score and 10, + 2, ÷ 8, $\sqrt{\ }$, × 5, + 1, $\sqrt{\ }$, + 1, square that number.

Problem Solving:

Half of a gallon is a half gallon. Half of a half gallon is a quart. Half of a quart is a pint. Half of a pint is a cup. If milk from a full gallon container is used to fill empty half-gallon, quart, pint, and cup containers, how much milk will be left in the gallon container?

1 gallon

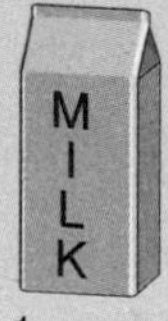

$\frac{1}{2}$ gallon

1 quart

1 pint

1 cup

NEW CONCEPT

When a decimal number is divided, the answer is sometimes a decimal number that will not end with a remainder of zero. Instead the answer will have one or more digits in a pattern that repeats indefinitely. Here we show two examples:

```
    7.1666...            0.31818...
 6)43.0000...        11)3.50000...
   42                    3 3
    1 0                    20
      6                    11
     40                     90
     36                     88
      40                     20
      36                     11
       40                     90
       36                     88
        4                      2
```

The repeating digits of a decimal number are called the **repetend.** In 7.1666..., the repetend is 6. In 0.31818..., the repetend is 18 (not 81). One way to indicate that a decimal number has repeating digits is to write the number with a bar over the repetend where it first appears to the right of the decimal point. For example,

$$7.1666... = 7.1\overline{6} \qquad 0.31818... = 0.3\overline{18}$$

Example 1 Rewrite each of these repeating decimals with a bar over the repetend:

(a) 0.0833333...

(b) 5.14285714285714...

(c) 454.5454545...

Solution (a) The repeating digit is 3.

$$\mathbf{0.08\overline{3}}$$

(b) This is a six-digit repeating pattern.

$$\mathbf{5.\overline{142857}}$$

(c) The repetend is always to the right of the decimal point. We do not write a bar over a whole number.

$$\mathbf{454.\overline{54}}$$

Example 2 Round each number to five decimal places:

(a) $5.31\overline{6}$ (b) $25.\overline{405}$

Solution (a) We remove the bar and write the repeating digits to the right of the desired decimal place.

$$5.31\overline{6} = 5.316666...$$

Then we round to five places.

$$5.3166\underline{6}⑥... \longrightarrow \mathbf{5.31667}$$

(b) We remove the bar and continue the repeating pattern beyond the fifth decimal place.

$$25.\overline{405} \longrightarrow 25.405405...$$

Then we round to five places.

$$25.4054\underline{0}⑤... \longrightarrow \mathbf{25.40541}$$

Example 3 Divide 1.5 by 11 and write the quotient

(a) with a bar over the repetend.

(b) rounded to the nearest hundredth.

Solution (a) Since place value is fixed by the decimal point, we can write zeros in the "empty" places to the right of the decimal point. We continue dividing until the repeating pattern is apparent. The repetend is 36 (not 63). We write the quotient with a bar over 36 where it first appears.

$$0.13636\ldots = \mathbf{0.1\overline{36}}$$

```
     0.13636...
 11)1.50000...
    1 1
      40
      33
       70
       66
        40
        33
         70
         66
          4
```

(b) The hundredths place is the second place to the right of the decimal point.

$$0.1\underline{3}\textcircled{6}36\ldots \longrightarrow \mathbf{0.14}$$

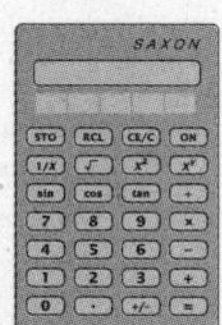

When a division problem is entered into a calculator and the display is filled with a decimal number, it is likely that the quotient is a repeating decimal. However, since a calculator either truncates (cuts off) or rounds the number displayed, the repetend may not be obvious. For example, to convert the fraction $\frac{1}{7}$ to a decimal, we enter

[1] [÷] [7] [=]

An eight-digit display shows

[0.1428571]

We might wonder whether the final digit, 1, is the beginning of another 142857 pattern. We can get a peek at the next digit by shifting the digits that are displayed one place to the left. We can do this by multiplying the numerator by 10 and dividing again. This time we divide 10 by 7.

[1] [0] [÷] [7] [=]

Each digit shifts one place to the left, replacing the zero in the original display with a 1. The display shows

1.4285714

Seeing the final digit, 4, following the 1 increases our certainty that the 142857 pattern is repeating.

LESSON PRACTICE

Practice set Write each repeating decimal with a bar over the repetend:

a. 2.72727... **b.** 0.816666...

Round each number to the nearest thousandth:

c. $0.\overline{6}$ **d.** $5.\overline{381}$

Divide 1.7 by 12 and write the quotient

e. with a bar over the repetend.

f. rounded to four decimal places.

MIXED PRACTICE

Problem set

1. (36) Two fifths of the children in the nursery were boys. What was the ratio of boys to girls in the nursery?

2. (13, 28) Four hundred thirty-two magazines were stored on 16 shelves. What was the average number of magazines per shelf?

3. (13) The migrating birds flew for 7 hours at an average rate of 23 miles per hour. How far did the birds travel in 7 hours?

4. (22) Diagram this statement. Then answer the questions that follow.

Seven ninths of the 450 people in the audience were enthralled by the speaker.

(a) How many people were enthralled?

(b) How many people were not enthralled?

5. (42) Round each number to four decimal places:

(a) $5.1\overline{6}$ (b) $5.\overline{27}$

6. Refer to the pie graph below to answer (a) and (b).
(38)

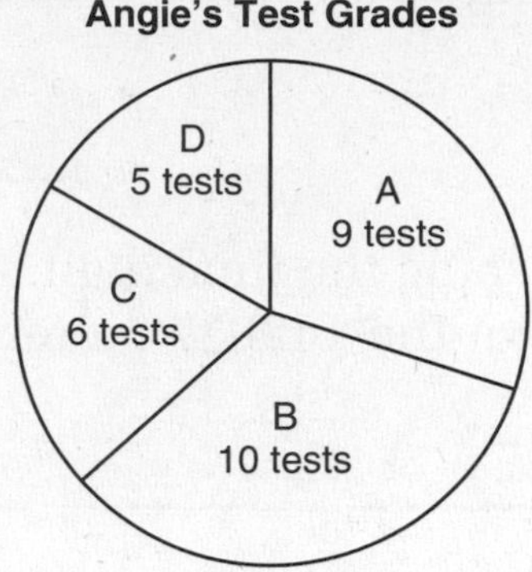

(a) On how many more tests did Angie earn an A or B than a C or D?

(b) On what fraction of the tests did Angie earn an A?

7. The coordinates of the vertices of a triangle are (−6, 0), (0, −6), and (0, 0). What is the area of the triangle?
(Inv. 3, 37)

8. All angles in the figure are right angles. Dimensions are in inches.
(19, 37)

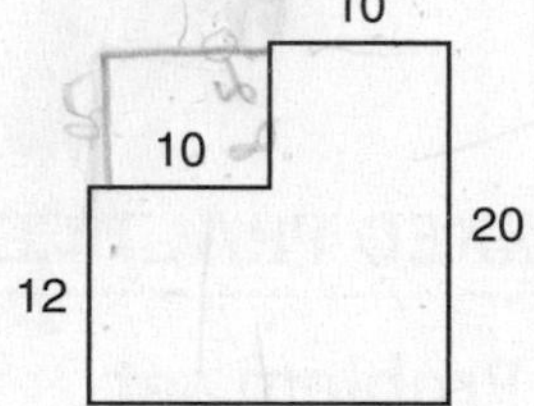

(a) Find the perimeter of the figure.

(b) Find the area of the figure.

9. Divide 1.7 by 11 and write the quotient
(42)

(a) with a bar over the repetend.

(b) rounded to three decimal places.

10. Use digits to write the sum of the decimal numbers twenty-seven thousandths and fifty-eight hundredths.
(31, 35)

11. What is the probability of rolling a composite number with one roll of a die (dot cube)?
(36)

12. (a) Make a factor tree showing the prime factorization of 7200. (Start with the factors 72 and 100.)
(20, 21)

(b) Write the prime factorization of 7200 using exponents.

13. Use a protractor and a ruler to draw a triangle with three 60° angles and sides 5 cm long.
(17)

14. What is the least common multiple of 12 and 15?
(27)

Solve:

15. (39) $\frac{21}{24} = \frac{w}{40}$

16. (39) $\frac{1.2}{x} = \frac{9}{6}$

17. (35) $m + 9.6 = 14$

18. (35) $n - 4.2 = 1.63$

19. (41) Evaluate $\frac{1}{2}bh$ when $b = 12$ and $h = 10$.

20. (41) Show two ways to simplify this expression:

$$4(5 + 6)$$

21. (27) Instead of dividing 686 by 14, we can find the quotient by dividing what number by 7?

22. (33, 35) Multiply \$4.56 by 0.08 and round the product to the nearest cent.

23. (33) Estimate the quotient of $23.8 \div 5.975$ by rounding each number to the nearest whole number before dividing.

24. (Inv. 2) What are the missing words in the following sentence?

The longest chord of a circle is the ___(a)___, *which is twice the length of the* ___(b)___.

Simplify:

25. (35) $7.1 \div 4$

26. (30) $6\frac{1}{4} + 5\frac{5}{12} + \frac{2}{3}$

27. (23, 30) $4 - \left(4\frac{1}{6} - 1\frac{1}{4}\right)$

28. (26) $6\frac{2}{5} \cdot 2\frac{5}{8} \cdot 2\frac{6}{7}$

29. (26) Before dividing, determine whether the quotient is greater than or less than 1 and state why. Then perform the calculation.

$$6 \div 4\frac{1}{2}$$

30. (40) Find the measures of $\angle a$, $\angle b$, and $\angle c$ in this figure:

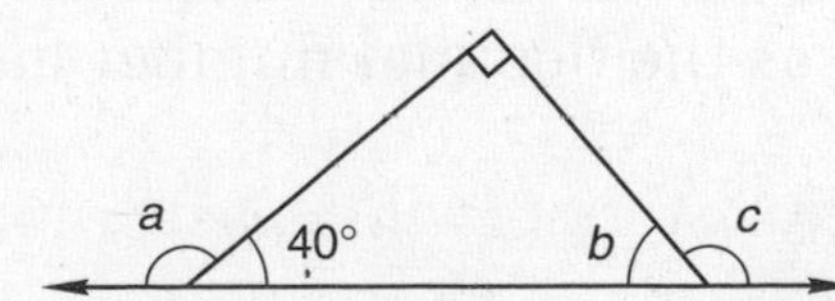

LESSON

43 Converting Decimals to Fractions • Converting Fractions to Decimals • Converting Percents to Decimals

WARM-UP

Facts Practice: Proportions (Test I)

Mental Math:

a. $6 \times 48¢$

b. $3.5 \div 100$

c. $\frac{n}{4} = \frac{21}{12}$

d. $7^2 - \sqrt{100}$

e. Estimate: $\$9.95 \times 6$

f. $\frac{1}{5}$ of 300

g. Find the sum, difference, product, and quotient of $\frac{2}{3}$ and $\frac{1}{4}$.

Problem Solving:

You can roll six different numbers (1, 2, ..., 6) with one number cube. You can roll eleven different numbers (2, 3, ..., 12) with two number cubes. If four number cubes are rolled at once, how many different totals are possible?

NEW CONCEPTS

Converting decimals to fractions

To write a decimal number as a fraction, we write the digits after the decimal point as the numerator of the fraction. For the denominator of the fraction, we write the place value of the last digit. Then we reduce.

Example 1 Write 0.125 as a fraction.

Solution The digits 125 form the numerator of the fraction. The denominator of the fraction is 1000 because 5, the last digit, is in the thousandths place.

$$0.125 = \frac{125}{1000}$$

Notice that the denominator of the fraction has as many zeros as the decimal number has decimal places. Now we reduce.

$$\frac{125}{1000} = \mathbf{\frac{1}{8}}$$

Example 2 Write 11.42 as a mixed number.

Solution The number 11 is the whole-number part. The numerator of the fraction is 42, and the denominator is 100 because 2 is in the hundredths place.

$$11.42 = 11\frac{42}{100}$$

Now we reduce the fraction.

$$11\frac{42}{100} = \mathbf{11\frac{21}{50}}$$

Converting fractions to decimals

To change a fraction to a decimal number, we perform the division indicated by the fraction. The fraction $\frac{1}{4}$ indicates that 1 is divided by 4.

$$4\overline{)1}$$

It may appear that we cannot perform this division. However, if we fix place values with a decimal point and write zeros in the decimal places to the right of the decimal point, we can perform the division. The result is a decimal number that equals the fraction $\frac{1}{4}$.

$$\begin{array}{r} 0.25 \\ 4\overline{)1.00} \\ \underline{8} \\ 20 \\ \underline{20} \\ 0 \end{array}$$

Thus $\frac{1}{4} = 0.25$.

Some fractions convert to repeating decimals. We convert $\frac{1}{3}$ to a decimal by dividing 1 by 3.

$$\begin{array}{r} 0.33\ldots \\ 3\overline{)1.00\ldots} \\ \underline{9} \\ 10 \\ \underline{9} \\ 1 \end{array}$$

Thus $\frac{1}{3} = 0.\overline{3}$.

Every fraction of whole numbers converts to either a terminating decimal (like 0.25) or a repeating decimal (like $0.\overline{3}$).

Example 3 Write each of these numbers as a decimal number:

(a) $\frac{23}{100}$ (b) $\frac{7}{4}$ (c) $3\frac{4}{5}$ (d) $\frac{2}{3}$

Solution (a) Fractions with denominators of 10, 100, 1000, and so on can be written directly as decimal numbers, without dividing. The decimal part will have the same number of places as the number of zeros in the denominator.

$$\frac{23}{100} = \mathbf{0.23}$$

(b) An improper fraction is equal to or greater than 1. When we change an improper fraction to a decimal number, the decimal number will be greater than or equal to 1.

$$\frac{7}{4} \longrightarrow \begin{array}{r} 1.75 \\ 4\overline{)7.00} \\ \underline{4} \\ 3\,0 \\ \underline{2\,8} \\ 20 \\ \underline{20} \\ 0 \end{array} \qquad \frac{7}{4} = \mathbf{1.75}$$

(c) To change a mixed number to a decimal number, we can change the mixed number to an improper fraction and then divide. Another way is to separate the fraction from the whole number and change the fraction to a decimal number. Then we write the whole number and the decimal number as one number. Here we show both ways.

$$3\frac{4}{5} = \frac{19}{5} \quad \text{or} \quad 3\frac{4}{5} = 3 + \frac{4}{5}$$

$$\begin{array}{r} 3.8 \\ 5\overline{)19.0} \\ \underline{15} \\ 4\,0 \\ \underline{4\,0} \\ 0 \end{array} \qquad\qquad \begin{array}{r} 0.8 \\ 5\overline{)4.0} \\ \underline{4\,0} \\ 0 \end{array}$$

$$3\frac{4}{5} = \mathbf{3.8} \qquad\qquad 3\frac{4}{5} = \mathbf{3.8}$$

(d) To change $\frac{2}{3}$ to a decimal number, we divide.

$$\frac{2}{3} \longrightarrow 3\overline{)2.000\ldots} \quad \text{quotient } 0.666\ldots \qquad \frac{2}{3} = \mathbf{0.\overline{6}}$$

$$\begin{array}{r} 18 \\ \hline 20 \\ 18 \\ \hline 20 \\ 18 \\ \hline 2 \end{array}$$

Note: We will write repeating decimal numbers with a bar over the repetend unless directed otherwise.

Converting percents to decimals Recall that *percent* means "per hundred" or "hundredths." So 75% means 75 hundredths, which can be written as a fraction or as a decimal.

$$75\% = \frac{75}{100} = 0.75$$

Likewise, 5% means 5 hundredths.

$$5\% = \frac{5}{100} = 0.05$$

We see that a percent may be written as a decimal using the same digits but with the decimal point shifted two places to the left.

Example 4 Write each percent as a decimal number:

(a) 25% (b) 125% (c) 2.5%

(d) 50% (e) $7\frac{1}{2}\%$

Solution (a) **0.25** (b) **1.25** (c) **0.025**

(d) 0.50 = **0.5** (e) 7.5% = **0.075**

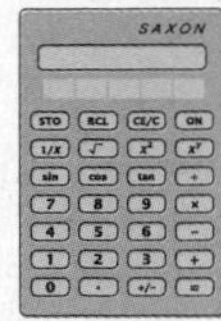

Many scientific calculators do not have a percent key. Designers of these calculators assume the user will mentally convert percents to decimals before entering the calculation.

If your calculator does have a percent key, you may find the decimal equivalent of a percent by entering [1] [×] the percent. For example, enter

The calculator displays the decimal equivalent 0.25.

LESSON PRACTICE

Practice set* Change each decimal number to a reduced fraction or to a mixed number:

a. 0.24 **b.** 45.6 **c.** 2.375

Change each fraction or mixed number to a decimal number:

d. $\frac{23}{4}$ **e.** $4\frac{3}{5}$ **f.** $\frac{5}{8}$ **g.** $\frac{5}{6}$

Convert each percent to a decimal number:

h. 8% **i.** 12.5% **j.** 150% **k.** $6\frac{1}{2}$%

MIXED PRACTICE

Problem set

1. (36) The ratio of Celtic soldiers to Roman soldiers was 2 to 5. What fraction of the soldiers were Celtic?

2. (28) Eric ran 8 laps in 11 minutes 44 seconds.

(a) How many seconds did it take Eric to run 8 laps?

(b) What is the average number of seconds it took Eric to run each lap?

3. (11, 35) Some gas was still in the tank. Paloma added 13.3 gallons of gas, which filled the tank. If the tank held a total of 21.0 gallons of gas, how much gas was in the tank before Paloma added the gas?

4. (5, 12) From 1750 to 1850, the estimated population of the world increased from seven hundred twenty-five million to one billion, two hundred thousand. How many more people were living in the world in 1850 than in 1750?

5. (22, 36) Diagram this statement. Then answer the questions that follow.

The Jets won two thirds of their 15 games and lost the rest.

(a) How many games did the Jets win?

(b) What was the Jets' win-loss ratio?

The stem-and-leaf plot below shows the distribution of finish times in a 100-meter sprint. Refer to the stem-and-leaf plot to answer problems 6 and 7.

Stem	Leaf
11	2
12	3 4 8
13	0 3 4 5 6
14	1 4 7 8
15	2 5

12 | 3 represents 12.3 seconds

6. *(Inv. 4)* For the 100-meter finish times, find the

(a) median.

(b) lower quartile.

(c) upper quartile.

7. *(Inv. 4)* Make a box-and-whisker plot of the 100-meter finish times in the stem-and-leaf plot.

8. *(19)* A square and a regular hexagon share a common side, as shown. The perimeter of the hexagon is 120 mm. What is the perimeter of the square?

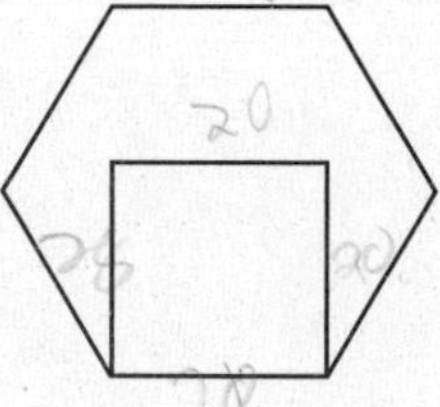

9. *(43)* Write each of these numbers as a reduced fraction or mixed number:

(a) 0.375 (b) 5.55

10. *(43)* Write each of these numbers as a decimal number:

(a) $2\frac{2}{5}$ (b) $\frac{1}{8}$

11. *(42)* Round each number to the nearest thousandth:

(a) $0.\overline{45}$

(b) $3.\overline{142857}$

12. *(42)* Divide 1.9 by 12 and write the quotient

(a) with a bar over the repetend.

(b) rounded to three decimal places.

13. *(31, 35)* Four and five hundredths is how much greater than one hundred sixty-seven thousandths?

14. Draw $\overline{AB}$ 1 inch long. Then draw $\overline{AC}$ $\frac{3}{4}$ inch long perpendicular to $\overline{AB}$. Complete ΔABC by drawing $\overline{BC}$. How long is $\overline{BC}$?
(8)

15. A normal deck of cards is composed of four suits (red heart, red diamond, black spade, and black club) of 13 cards each (2 through 10, jack, queen, king, and ace) for a total of 52 cards. If one card is drawn from a normal deck of cards, what is the probability that the card will be a red card?
(36)

16. (a) Make a factor tree showing the prime factorization of 900. (Start with the factors 30 and 30.)
(20, 21)

(b) Write the prime factorization of 900 using exponents.

(c) Write the prime factorization of $\sqrt{900}$.

17. The eyedropper held 2 milliliters of liquid. How many eyedroppers of liquid would it take to fill a 1-liter container?
(32)

18. (a) Write 8% as a decimal number.
(43)

(b) Find 8% of \$8.90 by multiplying \$8.90 by the answer to (a). Round the answer to the nearest cent.

19. (a) What is the perimeter of this triangle?
(19, 37)

(b) What is the area of this triangle?

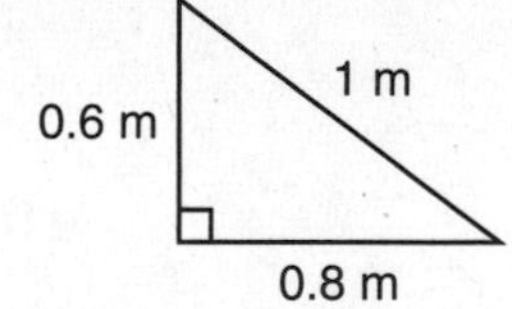

20. Compare and explain the reason for your answer:
(27)

$$\frac{32}{2} \bigcirc \frac{320}{20}$$

21. Evaluate $a(b + c)$ if $a = 2$, $b = 3$, and $c = 4$.
(41)

Solve:

22. $\frac{10}{18} = \frac{c}{4.5}$
(39)

23. $1.9 = w + 0.42$
(35)

Simplify:

24. $6.5 \div 4$
(35)

25. $3\frac{3}{10} - 1\frac{11}{15}$
(23, 30)

26. $5\frac{1}{2} + 6\frac{3}{10} + \frac{4}{5}$
(30)

27. $7\frac{1}{2} \cdot 3\frac{1}{3} \cdot \frac{4}{5} \div 5$
(26)

28. Find the next coordinate pair in this sequence:
(Inv. 3)

(1, 2), (2, 4), (3, 6), (4, 8), ...

29. Find the measures of $\angle a$, $\angle b$, and $\angle c$ in the figure at right.
(40)

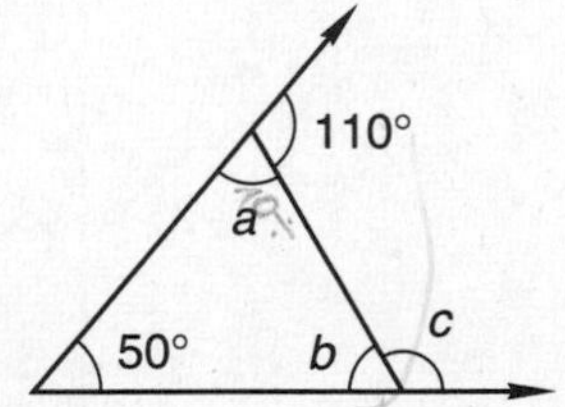

30. Refer to the figure at right to answer (a)–(c):
(Inv. 2)

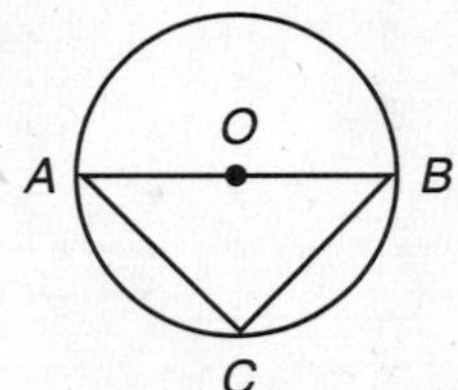

(a) What is the measure of central angle *AOB*?

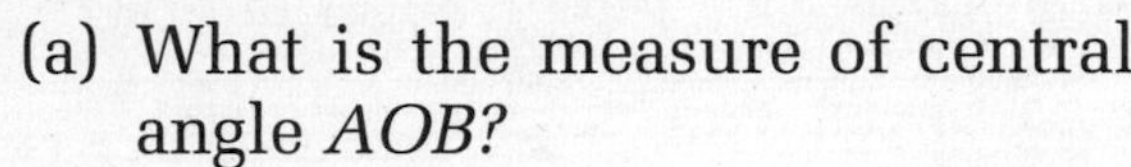

(b) What appears to be the measure of inscribed angle *ACB*?

(c) Chord *AC* is congruent to chord *BC*. What appears to be the measure of inscribed angle *ABC*?

LESSON

44 Division Answers

WARM-UP

Facts Practice: Measurement Facts (Test H)

Mental Math:

a. $5 \times 64¢$

b. $0.5 \div 10$

c. $\frac{3}{m} = \frac{12}{24}$

d. Estimate: $596 \div 11$

e. $\frac{\sqrt{144}}{12}$

f. $\frac{3}{4}$ of 200

g. Start with the number of meters in a kilometer, $\div$ 10, $\sqrt{\ }$, $\times$ 5, $-$ 1, $\sqrt{\ }$, $\times$ 5, $+$ 1, $\sqrt{\ }$.

Problem Solving:

The prime number 7 is the average of which two different prime numbers?

NEW CONCEPT

We can write answers to division problems with remainders in different ways. We can write them with a remainder or as a mixed number.

$$\begin{array}{r} 6\text{ R }3 \\ 4\overline{)27} \\ \underline{24} \\ 3 \end{array} \qquad \begin{array}{r} 6\frac{3}{4} \\ 4\overline{)27} \\ \underline{24} \\ 3 \end{array}$$

We can also write the answer as a decimal number. We fix place values with a decimal point, affix zeros to the right of the decimal point, and continue dividing.

$$\begin{array}{r} 6.75 \\ 4\overline{)27.00} \\ \underline{24} \\ 3\,0 \\ \underline{2\,8} \\ 20 \\ \underline{20} \\ 0 \end{array}$$

Example 1 Divide 54 by 4 and write the answer

(a) with a remainder.

(b) as a mixed number.

(c) as a decimal.

Solution (a) We divide and find the result is **13 R 2.**

(b) The remainder is the numerator of a fraction, and the divisor is the denominator. Thus this answer can be written as $13\frac{2}{4}$, which reduces to $\mathbf{13\frac{1}{2}}$.

```
   13 R 2
 4)54
   4
   14
   12
    2
```

(c) We fix place values by placing the decimal point to the right of 54. Then we can write zeros in thefollowing places and continue dividing until the remainder is zero. The result is **13.5.**

```
   13.5
 4)54.0
   4
   14
   12
    2 0
    2 0
      0
```

Sometimes a division answer written as a decimal number will be a repeating decimal number or will have more decimal places than the problem requires. In this book we show the complete division of the number unless the problem states that the answer is to be rounded.

Example 2 Divide 37.4 by 9 and round the quotient to the nearest thousandth.

Solution We continue dividing until the answer has four decimal places. Then we round to the nearest thousandth.

$4.15\underline{5}⑤\ldots \longrightarrow$ **4.156**

```
   4.1555...
 9)37.4000...
   36
    1 4
      9
     50
     45
      50
      45
       50
       45
        5
```

Problems involving division often require us to interpret the results of the division and express our answer in other ways. Consider the following example.

Example 3 Vans will be used to transport 27 scouts on a camping trip. Each van can carry 6 scouts.

(a) How many vans can be filled?

(b) How many vans will be needed?

(c) If all but one van will be full, then how many scouts will be in the van that will not be full?

Solution The quotient when 27 is divided by 6 can be expressed in three forms.

$$6\overline{)27}^{\;4\text{ R }3} \qquad 6\overline{)27}^{\;4\frac{1}{2}} \qquad 6\overline{)27.0}^{\;4.5}$$

The questions require us to interpret the results of the division.

(a) The whole number 4 in the quotient means that **4 vans** can be filled to capacity.

(b) Four vans will hold 24 scouts. Since 27 scouts are going on the camping trip, another van is needed. So **5 vans** will be needed.

(c) The fifth van will carry the remaining **3 scouts.**

LESSON PRACTICE

Practice set Divide 55 by 4 and write the answer

a. with a remainder.

b. as a mixed number.

c. as a decimal number.

d. Divide 5.5 by 3 and round the answer to three decimal places.

e. Ninety-three horses are kept in four stables as equally as possible. How many horses are in each of the four stables?

MIXED PRACTICE

Problem set

1. The rectangle was 24 inches long and 18 inches wide. What was the ratio of its length to its width?
(36)

2. Lakeisha's test scores were 90, 95, 90, 85, 80, 85, 90, 80, 95, and 100. What was her mean (average) test score?
(28)

3. The report stated that two out of every five young people were unable to find a job. What fraction of the young people were able to find a job?
(14)

4. Rachel bought a sheet of fifty 34-cent stamps from the post office. She paid for the stamps with a $20 bill. How much money should she get back?
(28)

5. Ninety-seven thousandths is how much less than two and ninety-eight hundredths? Write the answer in words.
(31, 35)

6. Diagram this statement. Then answer the questions that follow.
(22, 36)

Five sixths of the 300 runners finished the marathon.

(a) How many runners did not finish the marathon?

(b) What was the ratio of runners who finished the marathon to runners who did not finish the marathon?

7. Copy this figure on your paper. Find the length of each unmarked side, and find the perimeter of the polygon. Dimensions are in meters. All angles are right angles.
(19)

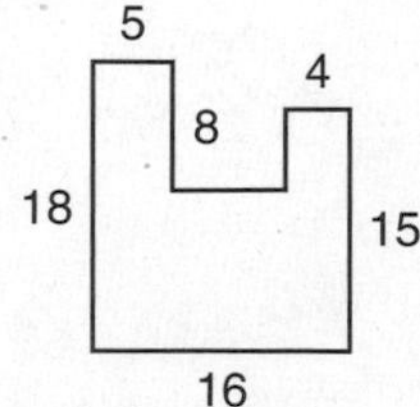

8. (a) Write 0.75 as a reduced fraction.
(43)

(b) Write $\frac{5}{8}$ as a decimal number.

(c) Write 125% as a decimal number.

9. If a card is drawn from a normal deck of cards, what is the probability that the card will be a heart?
(36)

10. (41) The expression $2(3 + 4)$ equals which of the following?

A. $(2 \cdot 3) + 4$ B. $(2 \cdot 3) + (2 \cdot 4)$

C. $2 + 7$ D. $23 + 24$

11. (2) Find the next three numbers in this sequence:

1, 3, 6, 10, ...

12. (42, 44) Divide 5.4 by 11 and write the answer

(a) with a bar over the repetend.

(b) rounded to the nearest thousandth.

13. (21) What composite number is equal to the product of the first four prime numbers?

14. (4, 33) (a) Arrange these numbers in order from least to greatest:

$1.2, -12, 0.12, 0, \frac{1}{2}$

(b) Which numbers in (a) are integers?

15. (26) Each math book is $1\frac{1}{2}$ inches thick.

(a) A stack of 12 math books would stand how many inches tall?

(b) How many math books would make a stack 1 yard tall?

16. (34, 35) What is the sum of the numbers represented by points *M* and *N* on this number line?

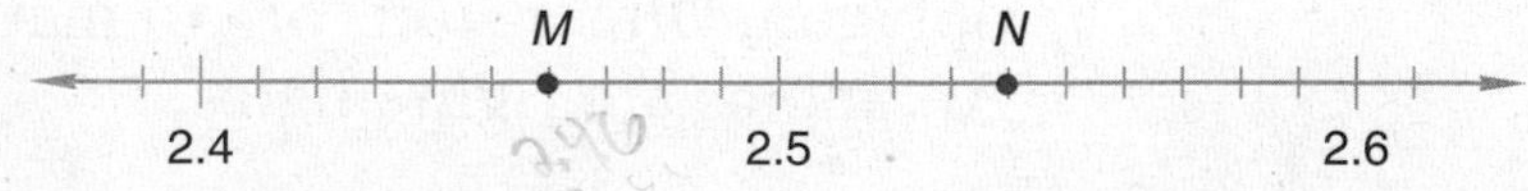

17. (33, 41) Estimate the value of πd when π is 3.14159 and d is 9.847 meters.

18. (34, 35) Draw a square with sides 2.5 cm long.

(a) What is the area of the square?

(b) What is the perimeter of the square?

19. (Inv. 3, 37) The coordinates of the vertices of a triangle are (–2, 0), (4, 0), and (3, 3). What is the area of the triangle?

Solve:

20. (39) $\frac{25}{15} = \frac{n}{1.2}$

21. (39) $\frac{p}{90} = \frac{4}{18}$

22. (35) $4 = 3.14 + x$

23. (35) $0.1 = 1 - z$

Simplify:

24. (35) $16.42 \div 8$

25. (35) $0.153 \div 9$

26. (30) $5\frac{3}{4} + \frac{5}{6} + 2\frac{1}{2}$

27. (23, 30) $3\frac{1}{3} - \left(5 - 1\frac{5}{6}\right)$

28. (26) $3\frac{3}{4} \cdot 3\frac{1}{3} \cdot 8$

29. (26) $7 \div 10\frac{1}{2}$

30. (40) Figure *ABCD* is a rectangle. The measure of ∠*ADB* is 35°. Find the measure of

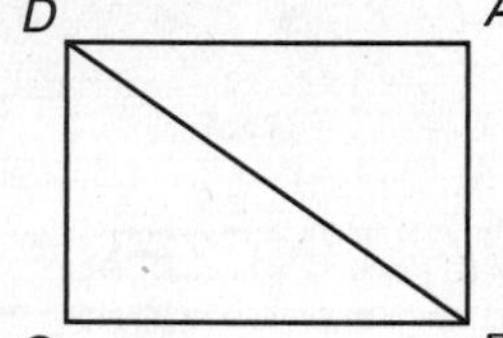

(a) ∠*ABD*.

(b) ∠*CBD*.

(c) ∠*BDC*.

State the reason for each answer.

LESSON

45 Dividing by a Decimal Number

WARM-UP

Facts Practice: Proportions (Test I)

Mental Math:

a. $7 \times \$1.50$

b. 1.25×10^2

c. $\frac{4}{6} = \frac{10}{w}$

d. $5^2 \cdot \sqrt{16}$

e. Estimate: $4\frac{1}{8} \times 2\frac{6}{7}$

f. $\frac{2}{3}$ of 75

g. Find the sum, difference, product, and quotient of $\frac{3}{4}$ and $\frac{2}{3}$.

Problem Solving:

Copy this problem and fill in the missing digits:

```
  _8_
×   _
 8__8
```

NEW CONCEPT

Dan has \$2.00 and wants to buy flavored icicles for his friends. If flavored icicles cost \$0.25 each, how many can Dan buy?

This is a problem we may be able to solve quickly by mental calculation. Dan can buy 4 flavored icicles priced at \$0.25 for \$1.00, so he can buy 8 flavored icicles for \$2.00. But how do we get an answer of "8 flavored icicles" from what seems to be a division problem?

$$\$0.25\overline{)\$2.00}$$

In this lesson we will consider how to get the "8." In a later lesson we will consider how to get the "flavored icicles." Notice that dividing \$2.00 by \$0.25 is dividing by a decimal number (\$0.25).

$$\frac{\$2.00}{\$0.25}$$

If we wish to divide by a whole number instead of by a decimal number, we can convert to an equivalent division problem using cents rather than dollars.

$$\frac{200¢}{25¢}$$

Changing from dollars to cents shifts the decimal point two places to the right. The units (cents over cents) cancel, and 200 divided by 25 is 8.

$$\frac{200¢}{25¢} = 8$$

Recall that we can form equivalent division problems by multiplying (or dividing) the dividend and divisor by the same number. We use this method to change "division by a decimal" problems to "division by a whole number" problems.

If we want to divide 1.36 by 0.4, we have

$$\frac{1.36}{0.4}$$

We can change the divisor to the whole number 4 by multiplying both the dividend and divisor by 10.

$$\frac{1.36}{0.4} \times \frac{10}{10} = \frac{13.6}{4}$$

The quotient of 13.6 divided by 4 is the same as the quotient of 1.36 divided by 0.4. This means that both of these division problems have the same answer.

$0.4\overline{)1.36}$ is equivalent to $4\overline{)13.6}$

To divide by a decimal number, we move the decimal point in the divisor to the right to make the divisor a whole number. Then we move the decimal point in the dividend the same number of places to the right.

Example 1 Divide: 3.36 ÷ 0.06

Solution We use a division box and write

$0.06\overline{)3.36}$

First we move the decimal point in 0.06 two places to the right to make it 6.

$006.\overline{)3.36}$

Then we move the decimal point in 3.36 the same number of places to the right. This forms an equivalent division problem.

$006.\overline{)336.}$

The decimal point in the answer is just above the new location of the decimal point.

$$6\overline{)336.}$$

Now we divide.

```
     56.
  6)336.
    30
     36
     36
      0
```

Thus 3.36 ÷ 0.06 = **56.**

Example 2 Divide: 0.144 ÷ 0.8

Solution We want the divisor, 0.8, to be a whole number. Moving the decimal point one place to the right changes the divisor to the whole number 8. To do this, we must also move the decimal point in the dividend one place to the right.

```
      0.18
  08.)1.44
       8
       64
       64
        0
```

Example 3 Divide: 15.4 ÷ 0.07

Solution We move both decimal points two places. This makes an empty place in the division box, which we fill with a zero. We keep dividing until we reach the decimal point. We find 220 as the answer.

```
        220.
  007.)1540.
       14
        14
        14
         00
          0
          0
```

Example 4 Divide: 21 ÷ 0.5

Solution We move the decimal point in 0.5 one place to the right. The decimal point on 21 is to the right of the 1. We shift this decimal point one place to the right to form the equivalent division problem 210 ÷ 5.

```
       42.
  05.)210.
      20
       10
       10
        0
```

Example 5 Divide: 1.54 ÷ 0.8

Solution We do not write a remainder. We write zeros in the places to the right of the 4. We continue dividing until the remainder is zero, until the digits begin repeating, or until we have divided to the desired number of decimal places.

```
      1.925
08.)15.400
    8
    7 4
    7 2
      20
      16
       40
       40
        0
```

Example 6 How many $0.35 erasers can be purchased with $7.00?

Solution We record the problem as 7.00 divided by 0.35. We shift both decimal points two places and divide. The quotient is 20 and the answer to the question is **20 erasers.**

```
        20.
035.)700.
     70
      00
       0
       0
```

LESSON PRACTICE

Practice set* Divide:

a. 5.16 ÷ 0.6

b. 0.144 ÷ 0.09

c. 23.8 ÷ 0.07

d. 24 ÷ 0.08

e. How many $0.75 pens can be purchased with $12.00?

f. Explain why these division problems are equivalent:

$$\frac{0.25}{0.5} = \frac{2.5}{5}$$

MIXED PRACTICE

Problem set

1. (36) Raisins and nuts were mixed in a bowl. If nuts made up five eighths of the mixture, what was the ratio of raisins to nuts?

2. (28) The taxi ride cost $1 plus 40¢ more for each quarter mile traveled. What was the total cost for a 2-mile trip?

3. (31, 35) Fifty-four and five hundredths is how much greater than fifty and forty thousandths? Use words to write the answer.

4. Refer to the election tally sheet below to answer (a) and (b).
(38)

Vote Totals

Judy	𝍸 𝍸 𝍸 |
Carlos	𝍸 𝍸 ||||
Yolanda	𝍸 𝍸 𝍸 𝍸 ||
Khanh	𝍸 𝍸 𝍸 |||

(a) The winner of the election received how many more votes than the runner-up?

(b) What fraction of the votes did Carlos receive?

5. Diagram this statement. Then answer the questions that follow.
(22, 36)

Four sevenths of those who rode the Giant Gyro at the fair were euphoric. All the rest were vertiginous.

(a) What fraction of the riders were vertiginous?

(b) What was the ratio of euphoric to vertiginous riders?

6. What is the least common multiple of 10 and 16?
(27)

7. Find the product of 5^2 and 10^2.
(20)

8. The perimeter of this rectangle is 56 cm:
(19, 20)

10 cm

(a) What is the length of the rectangle?

(b) What is the area of the rectangle?

9. (a) Write 62.5 as a mixed number.
(43)

(b) Write $\frac{9}{100}$ as a decimal number.

(c) Write 7.5% as a decimal number.

10. Round each number to five decimal places:
(42)

(a) $23.\overline{54}$

(b) $0.91\overline{6}$

11. A 2-liter bottle of water has a mass of 2 kilograms. How many grams is that?
(32)

12. Find 6.5% of $5.00 by multiplying $5.00 by 0.065. Round the answer to the nearest cent.
(35)

13. (42, 44) Divide 5.1 by 9 and write the quotient

(a) rounded to the nearest thousandth.

(b) with a bar over the repetend.

14. (36) If a card is drawn from a normal deck of cards, what is the probability that the card will be an ace?

15. (34) Draw $\overline{XY}$ 2 cm long. Then draw $\overline{YZ}$ 1.5 cm long perpendicular to $\overline{XY}$. Complete ΔXYZ by drawing $\overline{XZ}$. How long is $\overline{XZ}$?

16. (19, 37) Find the (a) perimeter and (b) area of the triangle drawn in problem 15.

Solve:

17. (39) $\frac{3}{w} = \frac{25}{100}$

18. (39) $\frac{1.2}{4.4} = \frac{3}{a}$

19. (35) $m + 0.23 = 1.2$

20. (35) $r - 1.97 = 0.65$

Simplify:

21. (35) $(0.15)(0.15)$

22. (35) $1.2 \times 2.5 \times 4$

23. (35) $14.14 \div 5$

24. (45) $0.096 \div 0.12$

25. (30) $\frac{5}{8} + \frac{5}{6} + \frac{5}{12}$

26. (30) $4\frac{1}{2} - \left(2\frac{1}{3} - 1\frac{1}{4}\right)$

27. (26) $\frac{7}{15} \cdot 10 \cdot 2\frac{1}{7}$

28. (26) $6\frac{3}{5} \div 1\frac{1}{10}$

29. (45) How many \$0.21 pencils can be purchased with \$7.00?

30. (40) Amanda cut out a triangle and labeled the corners *a*, *b*, and *c* as shown. Then she tore off the three corners of the triangle and fit the pieces together to form the semicircular shape shown at right.

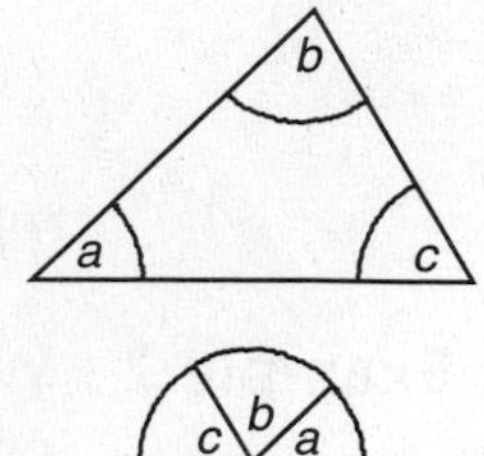

(a) Try the activity described in this problem, and tape or glue the fitted corners to your paper.

(b) Describe the characteristic of triangles demonstrated by this activity.

LESSON

46 Unit Price • Rates • Sales Tax

WARM-UP

Facts Practice: + − × ÷ Decimals (Test J)

Mental Math:

a. $9 \times \$0.82$

b. $3.6 \div 10^2$

c. $\frac{4}{8} = \frac{a}{20}$

d. Estimate: 4.97×1.9

e. $\sqrt{16} + 2^3$

f. $\frac{9}{10}$ of 80

g. Start with the number of vertices on a quadrilateral. Add the number of years in a decade; subtract a half dozen; then multiply by the number of feet in a yard. What is the answer?

Problem Solving:

Think of a three-digit number composed of three different digits and write it down. Create another three-digit number by reversing the order of the digits. Then subtract the smaller number from the larger number. Is the difference divisible by 9?

NEW CONCEPTS

Unit price As an aid to grocery shoppers, the unit prices for various products are often posted. The **unit price** is the cost for a single-unit measurement of the product. The unit price can be found by dividing the price by the number of units in the measurement.

Example 1 What is the unit price for a 24-ounce box of cereal that costs \$3.60?

Solution The cereal is measured in ounces. The unit price is the cost for 1 ounce. We divide the price by 24 ounces.

$$\frac{\$3.60}{24 \text{ ounces}} = \frac{\$0.15}{1 \text{ ounce}}$$

The unit price is \$0.15 for one ounce, which is **15¢ per ounce.**

Example 2 What is the unit price for a 36-ounce box of cereal that costs \$4.50?

Solution The unit price for the cereal is the price per ounce. We divide the price by 36 ounces.

$$\frac{\$4.50}{36 \text{ ounces}} = \frac{\$0.125}{1 \text{ ounce}}$$

The unit price is **12.5¢ per ounce.**

Unit pricing helps customers determine which brand or which size package provides the better buy. From the two previous examples, we see that the larger box of cereal is the better buy because it costs less per ounce.

Rates Unit price is an example of a **rate.** A rate is a ratio of two measurements. The two measurements in unit-price ratios are price and quantity. Other examples of rates include speed (distance divided by time) and mileage (distance divided by quantity of fuel used).

Example 3 Hans pedaled 84 kilometers in 4 hours. What was his average speed?

Solution Speed is the ratio of distance to time.

$$\frac{84 \text{ km}}{4 \text{ hr}} = 21 \frac{\text{km}}{\text{hr}}$$

By dividing we find that Hans's average speed was **21 km/hr.**

Example 4 Rodric rode his motorcycle on an interstate trip. He traveled 243 miles on 4.5 gallons of gas. Rodric's motorcycle averaged how many miles per gallon (mpg) on the trip?

Solution Miles per gallon means "miles divided by gallons."

$$\frac{243 \text{ mi}}{4.5 \text{ gal}} = \mathbf{54 \frac{mi}{gal}}$$

We may choose to reverse the terms of a ratio. For instance, if Laura can walk 6 miles in 2 hours, we can write these two rates:

$$\frac{6 \text{ mi}}{2 \text{ hr}} = 3 \frac{\text{mi}}{\text{hr}} \quad \text{(3 miles per hour)}$$

$$\frac{2 \text{ hr}}{6 \text{ mi}} = \frac{1}{3} \frac{\text{hr}}{\text{mi}} \quad \left(\frac{1}{3} \text{ hour per mile}\right)$$

Example 5 For a 5% service fee, a merchant agreed to exchange 20 dollars for 2400 yen or 2400 yen for 20 dollars. Disregarding the fee, write two reduced rates for the stated exchange rate.

Solution We write and reduce the two rates.

$$\frac{2400 \text{ yen}}{20 \text{ dollars}} = \mathbf{120 \frac{yen}{dollar}}$$

$$\frac{20 \text{ dollars}}{2400 \text{ yen}} = \mathbf{\frac{1}{120} \frac{dollar}{yen}}$$

Notice that the two rates are reciprocals.

Sales tax To find the amount of **sales tax** on a purchase, we multiply the full price of the purchase by the tax rate.

Example 6 A bicycle is on sale for $119.95. The tax rate is 6 percent.

(a) What is the tax on the bicycle?

(b) What is the total price including tax?

Solution (a) To find the tax, we change 6 percent to the decimal 0.06 and multiply $119.95 by 0.06. We round the result to the nearest cent.

$$\begin{array}{r} \$119.95 \\ \times \quad 0.06 \\ \hline \$7.1970 \end{array} \longrightarrow \mathbf{\$7.20}$$

(b) To find the total price, including tax, we add the tax to the initial price.

$119.95	price
+ $7.20	tax
$127.15	total

Example 7 Find the total price, including tax, of an $18.95 book, a $1.89 pen, and a $2.29 pad of paper when the tax rate is 5 percent.

Solution We begin by finding the combined price of the items.

$18.95	book
$1.89	pen
+ $2.29	paper
$23.13	

Next we multiply the combined price by 0.05 (5 percent) and round the product to the nearest cent.

$$\begin{array}{r} \$23.13 \\ \times \quad 0.05 \\ \hline \$1.1565 \end{array} \longrightarrow \$1.16 \text{ tax}$$

Then we add the tax to the combined price to find the total.

$23.13	price
+ $1.16	tax
$24.29	total

Example 8 The restaurant bill was about \$20. Inez wants to leave a 15% tip for the server. How much money should Inez leave for the tip?

Solution A tip, or gratuity, is an amount of money paid to acknowledge the service of others. For restaurant service the tip is customarily 15%–20% of the price of the food that is served. To find 15% of \$20, we multiply \$20 by 0.15.

$$\begin{array}{r} \$20 \\ \times\ 0.15 \\ \hline \$3.00 \end{array}$$

Inez should leave a tip of **\$3.**

One way to mentally estimate a 15% tip is to estimate 10% (one tenth) of the bill plus half that amount. So for a \$20 food bill we can estimate the tip this way:

One tenth (10%) of \$20 is \$2.

Half of \$2 is \$1.

So a 15% tip is \$3.

Example 9 Mentally estimate a 15% tip on a \$39.45 restaurant bill.

Solution We round \$39.45 to \$40. Ten percent of \$40 is \$4, and half of \$4 is \$2. We add \$4 and \$2, and estimate a tip of **\$6.**

LESSON PRACTICE

Practice set

a. What is the unit price for a 28-ounce box of cereal that costs \$1.12?

b. What is the unit price for an 11-ounce can of soup that costs 55¢?

c. Which is the better buy: an 18-ounce jar of jelly that costs \$1.98, or a 24-ounce jar of jelly that costs \$2.28?

d. The Smiths drove 416 miles in 8 hours. What was their average speed?

e. Their car traveled the first 322 miles of the trip on 14 gallons of gas, which is an average of how many miles per gallon?

f. When Monica landed in Belgium, she exchanged 40 dollars for 44 euros.

(a) What was the rate of exchange in euros per dollar?

(b) What was the rate of exchange in dollars per euro?

g. Find the sales tax on a $36.89 radio when the tax rate is 7 percent.

h. Find the total price of the radio in problem **g,** including tax.

i. Find the total price, including 6 percent tax, for a $6.95 dinner, a 95¢ beverage, and a $2.45 dessert.

j. After paying a restaurant bill of about $15, Hector returned to the table and left a tip of 15%. About how much money did Hector leave for a tip?

k. Mentally estimate a 15% tip on an $11.95 bill.

l. How could you estimate a 20% tip?

MIXED PRACTICE

Problem set

1. *(46)* Brand X costs $2.40 for 16 ounces. Brand Y costs $1.92 for 12 ounces. Find the unit price of each brand. Which brand is the better buy?

2. *(46)* The new coupe traveled 702 kilometers down the autobahn in 6 hours. The new coupe averaged how many kilometers per hour?

3. *(36)* Forty-eight sheep were on the farm. Thirty-six cows were also on the farm. What was the ratio of sheep to cows?

4. *(28)* At four different stores the price of 1 gallon of milk was $2.86, $2.83, $2.98, and $3.09. Find the average price per gallon rounded to the nearest cent.

5. *(31, 35)* Two and three hundredths is how much less than three and two tenths? Write the answer in words.

6. *(26)* A math book is $1\frac{1}{2}$ inches thick. How many math books will fit on a shelf that is 2 feet long?

7. (22) Diagram this statement. Then answer the questions that follow.

Three eighths of the 48 roses were red.

(a) How many roses were red?

(b) How many roses were not red?

(c) What fraction of the roses were not red?

8. (33) Replace each circle with the proper comparison symbol:

(a) 3.0303 ◯ 3.303 (b) 0.6 ◯ 0.600

9. (16) From goal line to goal line, a football field is 100 yards long. How many feet long is a football field?

10. (43) (a) Write 0.080 as a fraction.

(b) Write $37\frac{1}{2}\%$ as a decimal.

(c) Write $\frac{1}{11}$ as a decimal with a bar over the repetend.

11. (46) The price of a CD is $14.95. The sales-tax rate is 7%.

(a) What is the tax on the CD?

(b) What is the price of the CD including tax?

12. (Inv. 3, 37) The coordinates of three vertices of a triangle are (4, 0), (5, 3), and (0, 0). What is the area of the triangle?

13. (36) If a card is drawn from a normal deck of cards, what is the probability that the card will be a face card (jack, queen, or king)?

14. (21, 28) What is the average of the first five prime numbers?

15. (41) Show two ways to evaluate $x(y + z)$ for $x = 0.3$, $y = 0.4$, and $z = 0.5$.

16. (19, 37) In this figure all angles are right angles. Dimensions are in inches.

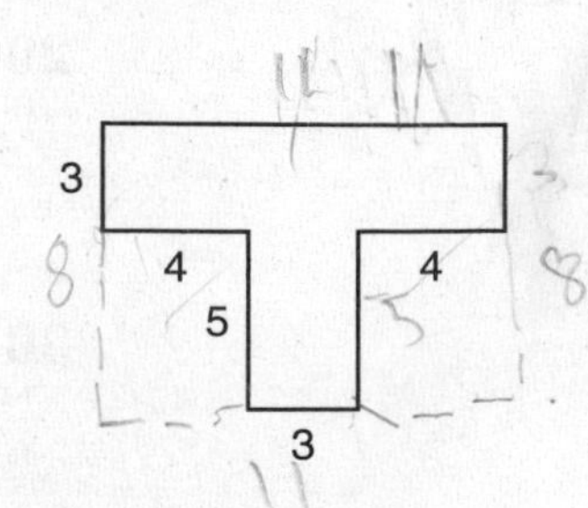

(a) What is the perimeter of the figure?

(b) What is the area of the figure?

17. (Inv. 2) The circle with center at point O has been divided into three sectors as shown. Find the measure of each of these central angles.

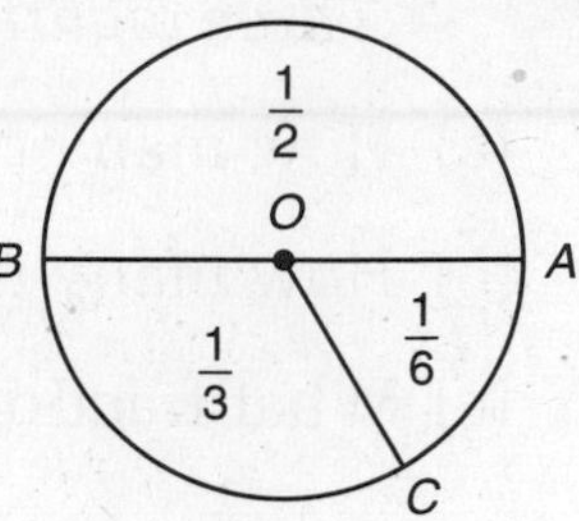

(a) $\angle AOB$ (b) $\angle BOC$ (c) $\angle AOC$

Solve:

18. (39) $\frac{10}{12} = \frac{2.5}{a}$

19. (39) $\frac{6}{8} = \frac{b}{100}$

20. (35) $4.7 - w = 1.2$

21. (3, 20) $10x = 10^2$

Estimate each answer to the nearest whole number. Then perform the calculation.

22. (30) $1\frac{11}{18} + 2\frac{11}{24}$

23. (30) $5\frac{5}{6} - \left(3 - 1\frac{1}{3}\right)$

Simplify:

24. (26) $\frac{2}{3} \times 4 \times 1\frac{1}{8}$

25. (26) $6\frac{2}{3} \div 4$

26. (35) $3.45 + 6 + (5.2 - 0.57)$

27. (45) $2.4 \div 0.016$

28. (29) Describe how to estimate the product of $6\frac{7}{8}$ and $5\frac{1}{16}$.

In the figure below, ΔABC is congruent to ΔCDA. Refer to the figure for problems 29 and 30.

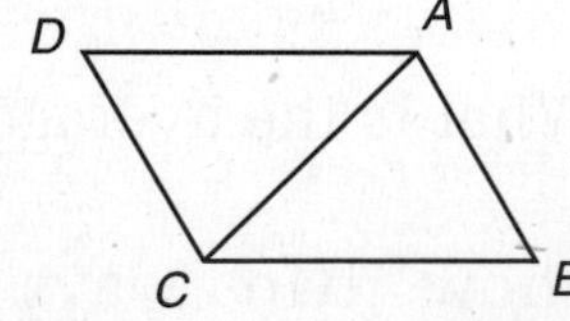

29. (18) Name the angle or side in ΔABC that corresponds to the following angle and side from ΔCDA:

(a) $\angle ACD$ (b) $\overline{DC}$

30. (40) The measure of $\angle ACB$ is 45°, and the measure of $\angle ADC$ is 60°. Find the measure of

(a) $\angle B$. (b) $\angle CAB$. (c) $\angle CAD$.

LESSON

47 Powers of 10

WARM-UP

Facts Practice: Proportions (Test I)

Mental Math:

a. 5 × \$8.20

b. 0.015×10^3

c. $\frac{c}{10} = \frac{9}{15}$

d. Estimate: \$4.95 × 19

e. $2^2 \cdot 2^3$

f. $\frac{5}{6}$ of 60

g. Find the sum, difference, product, and quotient of $\frac{1}{2}$ and $\frac{2}{5}$.

Problem Solving:

If 2 chickens can lay a total of 2 eggs in 2 days, then how many eggs can 4 chickens lay in 4 days?

NEW CONCEPTS

Place value as powers of 10 The positive powers of 10 are easy to write. The exponent matches the number of zeros in the product.

$$10^2 = 10 \cdot 10 = 100 \quad \text{(two zeros)}$$

$$10^3 = 10 \cdot 10 \cdot 10 = 1000 \quad \text{(three zeros)}$$

$$10^4 = 10 \cdot 10 \cdot 10 \cdot 10 = 10{,}000 \quad \text{(four zeros)}$$

Notice that when we multiply powers of 10, the exponent of the product equals the sum of the exponents of the factors.

$$10^3 \times 10^3 = 10^6$$

$$1000 \times 1000 = 1{,}000{,}000$$

Also, when we divide powers of 10, the exponent of the quotient equals the difference of the exponents of the dividend and divisor.

$$10^6 \div 10^3 = 10^3$$

$$1{,}000{,}000 \div 1000 = 1000$$

We can use powers of 10 to show place value, as we see in the chart below. Notice that 10^0 equals 1.

Trillions			Billions			Millions			Thousands			Units (Ones)			
hundreds	tens	ones	hundreds	tens	ones	hundreds	tens	ones	hundreds	tens	ones	hundreds	tens	ones	Decimal point
10^{14}	10^{13}	10^{12}	10^{11}	10^{10}	10^9	10^8	10^7	10^6	10^5	10^4	10^3	10^2	10^1	10^0	.

Powers of 10 are sometimes used to write numbers in expanded notation.

Example 1 Write 5206 in expanded notation using powers of 10.

Solution The number 5206 means 5000 + 200 + 6. We will write each number as a digit times its place value.

$$5000 \quad + \quad 200 \quad + \quad 6$$

$$\mathbf{(5 \times 10^3) + (2 \times 10^2) + (6 \times 10^0)}$$

Multiplying by powers of 10 When we multiply a decimal number by a power of 10, the answer has the same digits in the same order. Only their place values are changed.

Example 2 Multiply: 46.235×10^2

Solution This time we will write 10^2 as 100 and multiply.

$$\begin{array}{r} 46.235 \\ \times \quad 100 \\ \hline 4623.500 \end{array} = \mathbf{4623.5}$$

We see that the same digits occur in the same order. Only the place values have changed as the decimal point has been shifted two places to the right. **To multiply a decimal number by a positive power of 10, we shift the decimal point to the right the number of places indicated by the exponent.**

Example 3 Multiply: 3.14×10^4

Solution The power of 10 shows us the number of places to move the decimal point to the right. We move the decimal point four places to the right.

$$3.14 \times 10^4 = \mathbf{31{,}400}$$

Sometimes powers of 10 are written with words instead of with digits. For example, we might read that 1.5 million spectators lined the parade route. The expression 1.5 million means $1.5 \times 1{,}000{,}000$, which is 1,500,000.

Example 4 Write $2\frac{1}{2}$ billion in standard form.

Solution First we write $2\frac{1}{2}$ as the decimal number 2.5. Then we multiply by one billion (10^9), which shifts the decimal point 9 places to the right.

$$2.5 \text{ billion} = 2.5 \times 10^9 = \mathbf{2{,}500{,}000{,}000}$$

Dividing by powers of 10 When dividing by positive powers of 10, the quotient has the same digits as the dividend, only with smaller place values.

$$4.75 \div 10^3 \longrightarrow 1000\overline{)4.75000} \quad (0.00475)$$

To divide a number by a positive power of 10, we shift the decimal point to the left the number of places indicated by the exponent.

Example 5 Divide: $3.5 \div 10^4$

Solution The decimal point of the quotient is 4 places to the left of the decimal point in 3.5.

$$3.5 \div 10^4 = \mathbf{0.00035}$$

LESSON PRACTICE

Practice set Write each number in expanded notation using powers of 10:

a. 456

b. 1760

c. 186,000

Simplify:

d. 24.25×10^3

e. 25×10^6

f. $12.5 \div 10^3$

g. $4.8 \div 10^4$

Find each missing exponent:

h. $10^3 \cdot 10^4 = 10^{\square}$

i. $10^8 \div 10^2 = 10^{\square}$

Write each of the following numbers in standard form:

j. $2\frac{1}{2}$ million

k. 15 billion

l. 1.6 trillion

MIXED PRACTICE

Problem set Refer to the graph to answer problems 1–3.

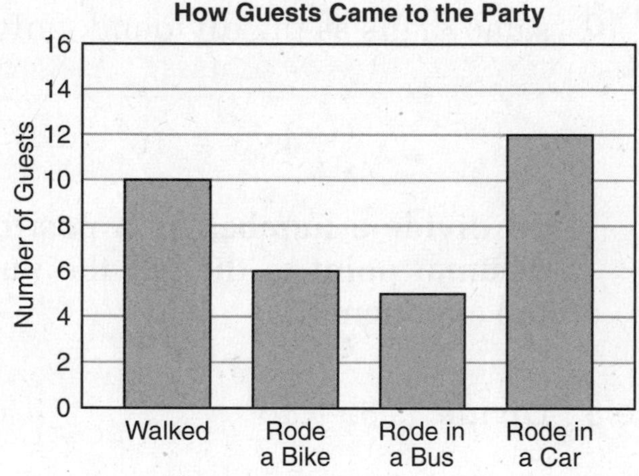

1. Answer true or false:
(38)

(a) Twice as many guests walked to the party as rode to the party in a bus.

(b) The majority of the guests rode to the party in either a bus or car.

2. What is the ratio of those who walked to the party to those who rode in a car?
(36, 38)

3. What fraction of the guests rode in a car?
(38)

4. What is the mean (average) of these numbers?
(28, 35)

1.2, 1.4, 1.5, 1.7, 2

5. (a) The newspaper reported that 134.8 million viewers watched the Super Bowl. Write the number of viewers in standard form.
(47)

(b) Write 5280 in expanded notation using powers of 10.

6. Diagram this statement. Then answer the questions that follow.
(22)

Only one eighth of the 40 contestants correctly answered the question.

(a) How many contestants correctly answered the question?

(b) How many contestants did not correctly answer the question?

7. A gallon of punch (128 ounces) is poured into 12-ounce glasses.
(44)

(a) How many glasses can be filled to the top?

(b) How many glasses are needed to hold all of the punch?

8. A cubit is an ancient unit of measure equal to the distance from the elbow to the fingertips.
(8)

(a) Estimate the number of inches from your elbow to your fingertips.

(b) Measure the distance from your elbow to your fingertips to the nearest inch.

9. (a) Write 0.375 as a fraction.
(43)

(b) Write $62\frac{1}{2}\%$ as a decimal.

10. Find the tax on a \$56.40 purchase if the sales-tax rate is 8%.
(46)

11. Round $53{,}714.\overline{54}$ to the nearest
(42)

(a) thousandth.

(b) thousand.

12. (47) Find each missing exponent:

(a) $10^5 \cdot 10^2 = 10^{\square}$ (b) $10^8 \div 10^4 = 10^{\square}$

13. (34) The point marked by the arrow represents what decimal number?

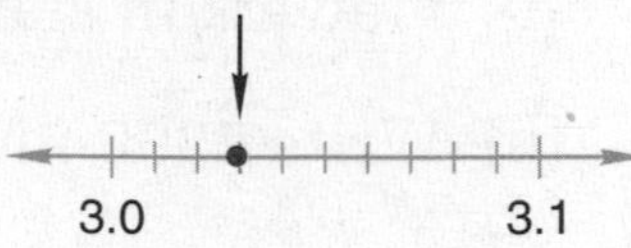

14. (19) In figure *ABCDEF* all angles are right angles and $AF = AB = BC$. Segment BC is twice the length of $\overline{CD}$. If CD is 3 cm, what is the perimeter of the figure?

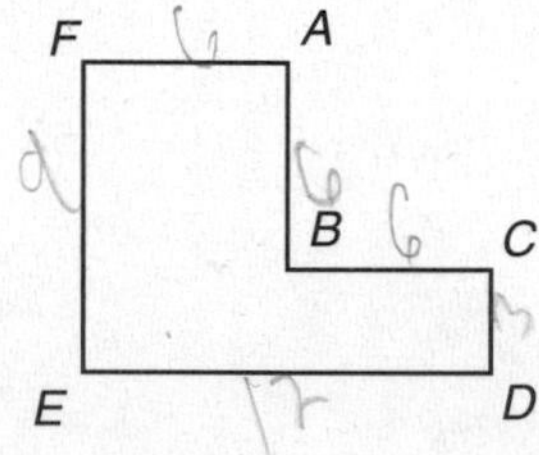

15. (Inv. 2) Use a compass to draw a circle with a radius of 1 inch. Then inscribe a regular hexagon in the circle.

(a) What is the diameter of the circle?

(b) What is the perimeter of the regular hexagon?

Solve:

16. (39) $\dfrac{6}{10} = \dfrac{w}{100}$

17. (39) $\dfrac{3.6}{x} = \dfrac{16}{24}$

18. (35) $\dfrac{a}{1.5} = 1.5$

19. (35) $9.8 = x + 8.9$

Estimate each answer to the nearest whole number. Then perform the calculation.

20. (30) $4\frac{1}{5} + 5\frac{1}{3} + \frac{1}{2}$

21. (23, 30) $6\frac{1}{8} - \left(5 - 1\frac{2}{3}\right)$

Simplify:

22. (20) $\sqrt{16 \cdot 25}$

23. (47) 3.6×10^3

24. (26) $8\frac{1}{3} \times 3\frac{3}{5} \times \frac{1}{3}$

25. (26) $3\frac{1}{8} \div 6\frac{1}{4}$

26. (35) $26.7 + 3.45 + 0.036 + 12 + 8.7$

27. The figures below illustrate one triangle rotated into three different positions. Dimensions are in inches.
(19, 37)

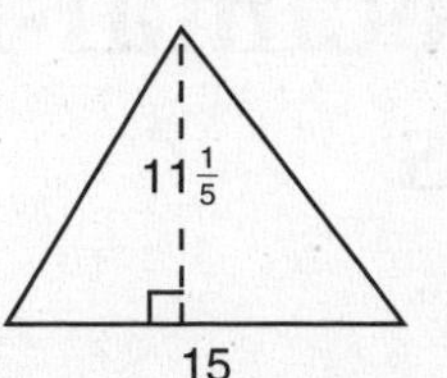

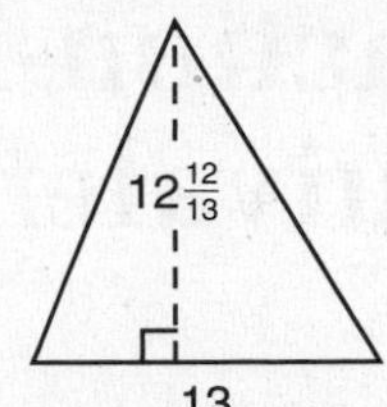

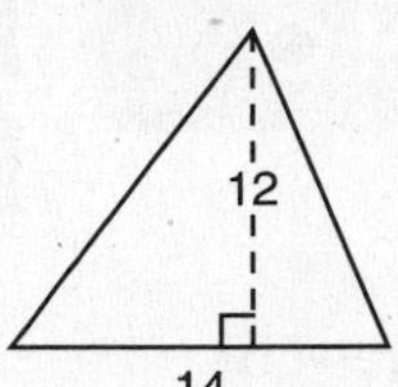

(a) What is the perimeter of the triangle?

(b) What is the area of the triangle?

28. Simplify and compare: $125 \div 10^2 \bigcirc 0.125 \times 10^2$
(47)

29. Arrange these numbers in order from least to greatest:
(30)

$$\frac{2}{3}, \frac{1}{2}, \frac{7}{12}, \frac{5}{6}$$

30. In this figure find the measure of
(40)

(a) $\angle a$.

(b) $\angle b$.

(c) Describe how to find the measure of $\angle c$.

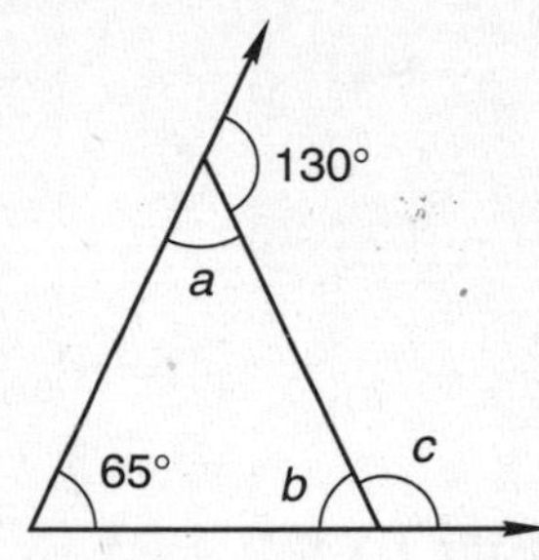

LESSON

48 Fraction-Decimal-Percent Equivalents

WARM-UP

Facts Practice: + – × ÷ Decimals (Test J)

Mental Math:

a. 7 × \$35.00

b. 12.75 ÷ 10

c. $\frac{6}{4} = \frac{9}{n}$

d. $\frac{10^2}{5^2}$

e. Estimate: $6\frac{1}{6} \times 3\frac{4}{5}$

f. $\frac{3}{8}$ of 80

g. 10 × 8, + 1, $\sqrt{}$, + 2, × 4, – 2, ÷ 6, × 9, + 1, $\sqrt{}$, ÷ 2, ÷ 2, ÷ 2

Problem Solving:

The counting numbers 1 through 9 are arranged in three columns. Each column contains three numbers, and the sum of the numbers in each column is the same. Describe how to find the sum of the numbers in each column. Then find that sum.

NEW CONCEPT

We may describe part of a whole using a fraction, a decimal, or a percent.

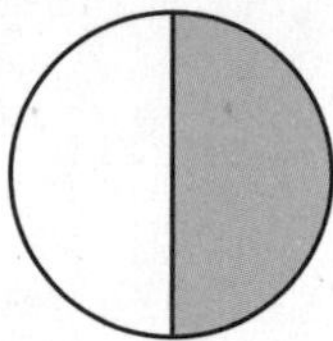

$\frac{1}{2}$ of the circle is shaded.
0.5 of the circle is shaded.
50% of the circle is shaded.

Recall that when we rename a fraction, we multiply by a form of 1 such as $\frac{2}{2}$, $\frac{5}{5}$, or $\frac{100}{100}$. Another form of 1 is 100%, so to convert a fraction or a decimal to a percent, we multiply the number by 100%.

Example 1 Write $\frac{7}{10}$ as a percent.

Solution To change a number to its percent equivalent, we multiply the number by 100%.

$$\frac{7}{10} \times 100\% = \frac{700\%}{10} = \mathbf{70\%}$$

Example 2 Write $\frac{2}{3}$ as a percent.

Solution We multiply by 100 percent.

$$\frac{2}{3} \times 100\% = \frac{200\%}{3} = \mathbf{66\frac{2}{3}\%}$$

Notice the mixed-number form of the percent.

Example 3 Write 0.8 as a percent.

Solution We multiply 0.8 by 100%.

$$0.8 \times 100\% = \mathbf{80\%}$$

Example 4 Complete the table.

FRACTION	DECIMAL	PERCENT
$\frac{1}{3}$	(a)	(b)
(c)	1.5	(d)
(e)	(f)	60%

Solution For (a) and (b) we find decimal and percent equivalents of $\frac{1}{3}$.

(a) $3\overline{)1.00}$ with quotient $\mathbf{0.\overline{3}}$

(b) $\frac{1}{3} \times 100\% = \frac{100\%}{3} = \mathbf{33\frac{1}{3}\%}$

For (c) and (d) we find a fraction (or a mixed number) and a percent equivalent to 1.5.

(c) $1.5 = 1\frac{5}{10} = \mathbf{1\frac{1}{2}}$

(d) $1.5 \times 100\% = \mathbf{150\%}$

For (e) and (f) we find fraction and decimal equivalents of 60%.

(e) $60\% = \frac{60}{100} = \mathbf{\frac{3}{5}}$

(f) $60\% = \frac{60}{100} = \mathbf{0.6}$

LESSON PRACTICE

Practice set* Complete the table.

FRACTION	DECIMAL	PERCENT
$\frac{2}{3}$	**a.**	**b.**
c.	1.1	**d.**
e.	**f.**	4%

MIXED PRACTICE

Problem set

1. Ling pedaled hard. She traveled 80 kilometers in 2.5 hours. What was her average speed in kilometers per hour?
(46)

2. Write the prime factorization of 1008 and 1323. Then reduce $\frac{1008}{1323}$.
(24)

3. In 1803 the United States purchased the Louisiana Territory from France for $15 million. In 1867 the United States purchased Alaska from Russia for $7.2 million. The purchase of Alaska occurred how many years after the purchase of the Louisiana Territory?
(12)

4. Red and blue marbles were in the bag. Five twelfths of the marbles were red.
(36)

   (a) What fraction of the marbles were blue?

   (b) What was the ratio of red marbles to blue marbles?

5. A 6-ounce can of peaches sells for 90¢. A 9-ounce can of peaches sells for $1.26. Find the unit price for each size. Which size is the better buy?
(46)

6. The average of two numbers is the number halfway between the two numbers. What number is halfway between two thousand, five hundred fifty and two thousand, nine hundred?
(28)

7. Diagram this statement. Then answer the questions that follow.
(22)

   Van has read five eighths of the 336-page novel.

   (a) How many pages has Van read?

   (b) How many more pages does he have to read?

8. Complete the table.
(48)

FRACTION	DECIMAL	PERCENT
$\frac{1}{2}$	(a)	(b)
(c)	0.1	(d)
(e)	(f)	25%

9. The graph shows how one family spends their annual
(38) income. Use this graph to answer (a)–(c).

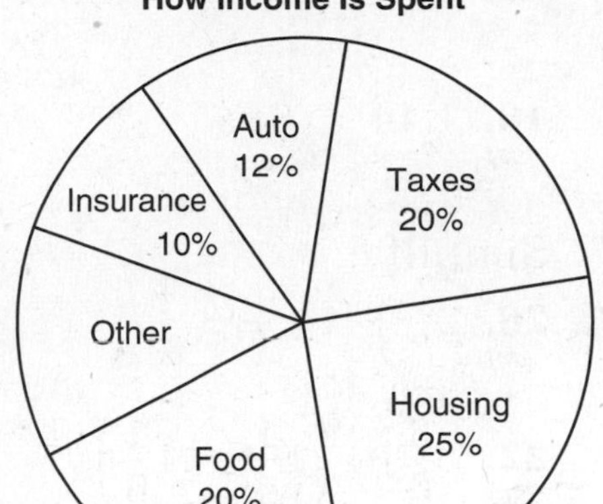

(a) What percent of the family's income is spent on "other"?

(b) What fraction of the family's income is spent on food?

(c) If $3200 is spent on insurance, how much is spent on taxes?

10. Write $0.\overline{54}$ as a decimal rounded to three decimal places.
(42)

11. (a) Estimate the length of $\overline{AB}$ in centimeters.
(32)

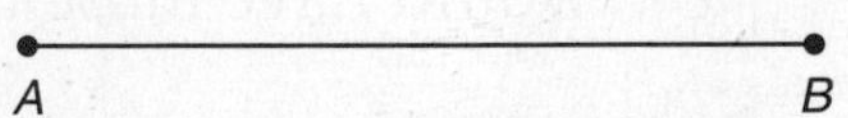

(b) Use a centimeter scale to find the length of $\overline{AB}$ to the nearest centimeter.

12. (a) Identify the exponent and the base in the expression 5^3.
(20, 47)

(b) Find the missing exponent: $10^4 \cdot 10^4 = 10^{\square}$

13. If the perimeter of a regular hexagon is 1 foot, each side is
(18, 19) how many inches long?

14. Copy this figure on your paper.
(19) Find the length of the unmarked sides, and find the perimeter of the polygon. Dimensions are in centimeters. All angles are right angles.

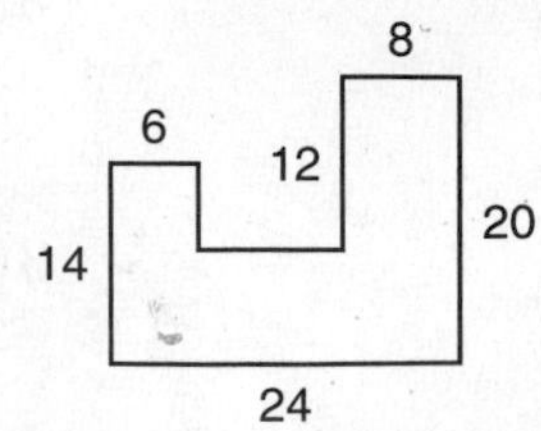

15. The moped traveled 78 miles on 1.2 gallons of gas. The
(46) moped averaged how many miles per gallon?

Solve:

16. $\dfrac{6}{100} = \dfrac{15}{w}$ *(39)*

17. $\dfrac{20}{x} = \dfrac{15}{12}$ *(39)*

18. $1.44 = 6m$ *(35)*

19. $\dfrac{1}{2} = \dfrac{1}{3} + f$ *(30)*

Simplify:

20. $2^5 + 1^4 + 3^3$ *(20)*

21. $\sqrt{10^2 \cdot 6^2}$ *(20)*

22. $3\dfrac{5}{6} - \left(1\dfrac{1}{4} + 1\dfrac{1}{6}\right)$ *(30)*

23. $8\dfrac{3}{4} + \left(4 - \dfrac{2}{3}\right)$ *(23, 30)*

24. $\dfrac{15}{16} \cdot \dfrac{24}{25} \cdot 1\dfrac{1}{9}$ *(26)*

25. $1\dfrac{1}{3} \div \left(2\dfrac{2}{3} \div 4\right)$ *(26)*

26. *(41, 45)* Find the value of $\frac{a}{b}$ when a = \$13.93 and b = 0.07.

27. *(Inv. 3, 37)* The coordinates of three vertices of a triangle are (–1, –1), (5, –1), and (5, –4). What is the area of the triangle?

28. *(36, 38)* Jane asked her friends how many siblings they had, and she tallied their answers. If Jane selects one of her friends at random, what is the probability that the selected friend would have more than one sibling?

Number of Siblings	Number of Jane's Friends
0	|||
1	𝍸 ||||
2	𝍸
3	||
4 or more	|

29. *(46)* What is the total price of a \$50.00 item including 7.5% sales tax?

30. *(40)* Find the measures of $\angle a$, $\angle b$, and $\angle c$ in the figure at right.

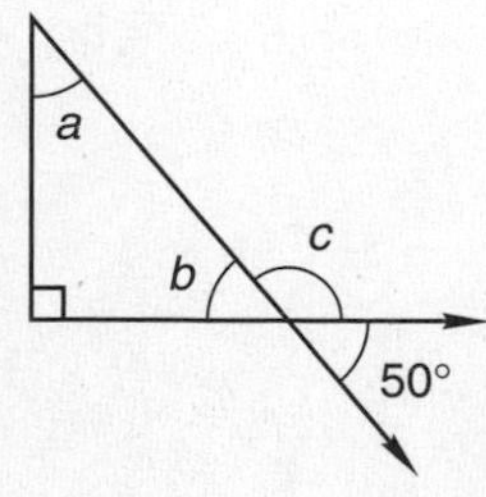

LESSON 49 Adding Mixed Measures

WARM-UP

Facts Practice: + – × ÷ Decimals (Test J)

Mental Math:

a. 8 × \$6.50

b. 25.75 × 10

c. $\frac{4}{x} = \frac{40}{100}$

d. Estimate: 12.11 ÷ 1.9

e. $\sqrt{400}$

f. $\frac{3}{10}$ of 200

g. Find the sum, difference, product, and quotient of $\frac{3}{5}$ and $\frac{1}{3}$.

Problem Solving:

The coach asked for two volunteers, and Adam, Blanca, and Chad raised their hands. From these three children, list the possible combinations of two children the coach could select.

NEW CONCEPT

A mixed measure is a measurement that includes different units from the same category (length, volume, time, etc.).

Ivan is 5 feet 8 inches tall.

The movie was 1 hour 48 minutes long.

To add mixed measures, we align the numbers in order to add units that are the same. Then we simplify when possible.

Example 1 Add and simplify: 1 yd 2 ft 7 in. + 2 yd 2 ft 8 in.

Solution We add like units, and then we simplify from right to left.

$$\begin{array}{r} 1\text{ yd } 2\text{ ft } 7\text{ in.} \\ +\ 2\text{ yd } 2\text{ ft } 8\text{ in.} \\ \hline 3\text{ yd } 4\text{ ft } 15\text{ in.} \end{array}$$

We change 15 in. to 1 ft 3 in. and add to 4 ft. Now we have

3 yd 5 ft 3 in.

Then we change 5 ft to 1 yd 2 ft and add to 3 yd. Now we have

4 yd 2 ft 3 in.

Example 2 Add and simplify:

$$\begin{array}{r} 2\text{ hr } 40\text{ min } 35\text{ s} \\ +\ 1\text{ hr } 45\text{ min } 50\text{ s} \\ \hline \end{array}$$

Solution We add. Then we simplify from right to left.

$$\begin{array}{r} 2\text{ hr } 40\text{ min } 35\text{ s} \\ +\ 1\text{ hr } 45\text{ min } 50\text{ s} \\ \hline 3\text{ hr } 85\text{ min } 85\text{ s} \end{array}$$

We change 85 s to 1 min 25 s and add to 85 min. Now we have

3 hr 86 min 25 s

Then we simplify 86 min to 1 hr 26 min and combine hours.

4 hr 26 min 25 s

LESSON PRACTICE

Practice set*

a. Change 70 inches to feet and inches.

b. Change 6 feet 3 inches to inches.

c. Simplify: 5 ft 20 in.

d. Add: 2 yd 1 ft 8 in. + 1 yd 2 ft 9 in.

e. Add: 5 hr 42 min 53 s + 6 hr 17 min 27 s

MIXED PRACTICE

Problem set

1. (35, 45) What is the quotient when the sum of 0.2 and 0.05 is divided by the product of 0.2 and 0.05?

2. (44) Darren carried the football 20 times and gained a total of 184 yards. What was the average number of yards he gained on each carry? Write the answer as a decimal number.

3. (46) Artemis bought two dozen arrows for six dollars. What was the cost of each arrow?

4. (18) Jeffrey counted the sides on three octagons, two hexagons, a pentagon, and two quadrilaterals. Altogether, how many sides did he count?

5. (28, 35) What is the mean of these numbers?

6.21, 4.38, 7.5, 6.3, 5.91, 8.04

6. (22, 36) Diagram this statement. Then answer the questions that follow.

Only two ninths of the 72 billy goats were gruff. The rest were cordial.

(a) How many of the billy goats were cordial?

(b) What was the ratio of gruff billy goats to cordial billy goats?

7. (42) Arrange these numbers in order from least to greatest:

$$0.\overline{5},\ 0.5,\ 0.\overline{54}$$

8. (8) (a) Estimate the length of segment AB in inches.

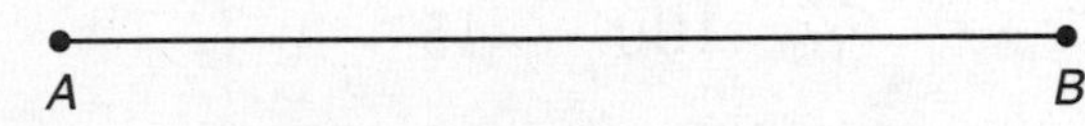

(b) Measure the length of segment AB to the nearest eighth of an inch.

9. (48) Write each of these numbers as a percent:

(a) 0.9 (b) $1\frac{3}{5}$ (c) $\frac{5}{6}$

10. (48) Complete the table.

FRACTION	DECIMAL	PERCENT
(a)	(b)	75%
(c)	(d)	5%

11. (13) Mathea's resting heart rate is 62 beats per minute. While she is resting, about how many times will her heart beat in an hour?

12. (36) What is the probability of rolling an even prime number with one roll of a die (dot cube)?

13. (37) A $\frac{1}{2}$-by-$\frac{1}{2}$-inch square was cut from a 1-by-1-inch square.

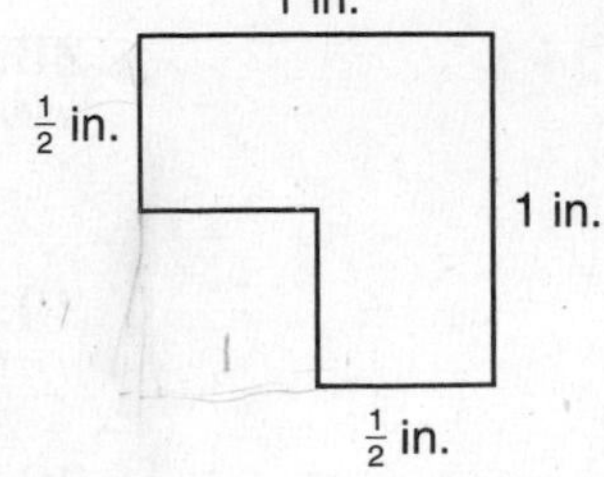

(a) What was the area of the original square?

(b) What is the area of the square that was removed?

(c) What is the area of the remaining figure?

14. (19) What is the perimeter of the figure in problem 13?

15. (37) The figures below show a triangle with sides 6 cm, 8 cm, and 10 cm long in three orientations. What is the height of the triangle when the base is

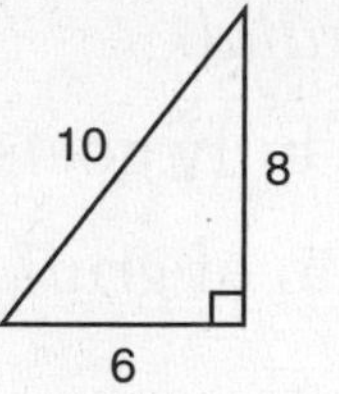

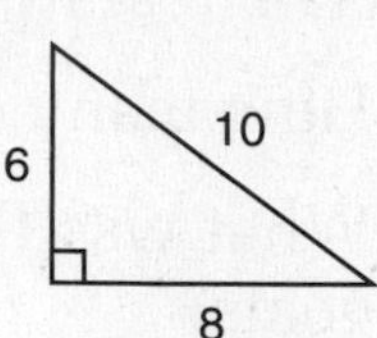

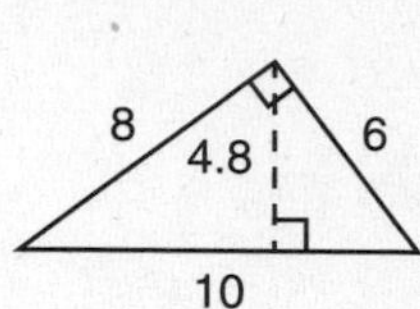

(a) 6 cm? (b) 8 cm? (c) 10 cm?

Solve:

16. (39) $\dfrac{y}{100} = \dfrac{18}{45}$

17. (39) $\dfrac{35}{40} = \dfrac{1.4}{m}$

18. (30) $\dfrac{1}{2} - n = \dfrac{1}{6}$

19. (35) $9d = 2.61$

Simplify:

20. (20) $\sqrt{100} + 4^3$

21. (47) 3.14×10^4

22. (23, 30) $3\frac{3}{4} + \left(4\frac{1}{6} - 2\frac{1}{2}\right)$

23. (26) $6\frac{2}{3} \cdot \left(3\frac{3}{4} \div 1\frac{1}{2}\right)$

24. (49)

	3 days	8 hr	15 min
+	2 days	15 hr	45 min

25. (49)

	1 yd	2 ft	6 in.
+	2 yd	1 ft	9 in.

26. (45) $18.00 ÷ 0.06

27. (29, 33) Describe how to estimate the quotient when 35.675 is divided by $2\frac{7}{8}$.

28. (46) The bat cost $18.50. The ball cost $3.50. What was the total price of the bat and ball including 6% sales tax?

29. (41) Evaluate: LWH if $L = 0.5$, $W = 0.2$, and $H = 0.1$

30. (40) This quadrilateral is a rectangle. Find the measures of $\angle a$, $\angle b$, and $\angle c$.

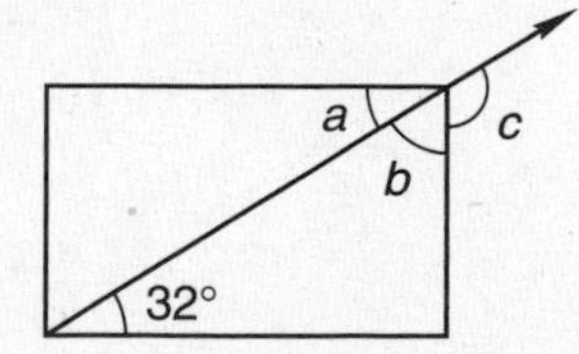

LESSON

50 Unit Multipliers and Unit Conversion

WARM-UP

Facts Practice: Proportions (Test I)

Mental Math:

a. 5 × \$48.00 **b.** 0.0125×10^2

c. $\frac{y}{20} = \frac{40}{100}$ **d.** $\sqrt{225}$

e. Estimate: $4\frac{3}{4} \times 1\frac{7}{8}$ **f.** $\frac{2}{5}$ of 40

g. Start with a half dozen, + 4, square that number, ÷ 2, + 6, ÷ 8, × 7, + 1, ÷ 10, − 10.

Problem Solving:

Copy this problem and fill in the missing digits:

$$\begin{array}{r} \_\,\_\,\_ \\ \times \quad \_ \\ \hline 1001 \end{array}$$

NEW CONCEPT

Let's take a moment to review the procedure for reducing a fraction. When we reduce a fraction, we can replace factors that appear in both the numerator and denominator with 1's, since each pair reduces to 1.

$$\frac{24}{36} = \frac{\overset{1}{\cancel{2}} \cdot \overset{1}{\cancel{2}} \cdot 2 \cdot \overset{1}{\cancel{3}}}{\underset{1}{\cancel{2}} \cdot \underset{1}{\cancel{2}} \cdot 3 \cdot \underset{1}{\cancel{3}}} = \frac{2}{3}$$

Also, recall that we can reduce before we multiply. This is sometimes called **canceling.**

$$\frac{2}{\underset{1}{\cancel{3}}} \cdot \frac{\overset{1}{\cancel{3}}}{5} = \frac{2}{5}$$

We can apply this procedure to units as well. We may cancel units before we multiply.

$$\frac{5\ \cancel{\text{ft}}}{1} \cdot \frac{12\text{ in.}}{1\ \cancel{\text{ft}}} = 60\text{ in.}$$

In this instance we performed the division $5 \text{ ft} \div 1 \text{ ft}$, which means, "How many feet are in 5 feet?" The answer is simply 5. Then we multiplied 5 by 12 in.

We remember that we change the name of a number by multiplying by a fraction whose value equals 1. Here we change the name of 3 to $\frac{12}{4}$ by multiplying by $\frac{4}{4}$:

$$3 \cdot \frac{4}{4} = \frac{12}{4}$$

The fraction $\frac{12}{4}$ is another name for 3 because $12 \div 4 = 3$.

When the numerator and denominator of a fraction are equal (and are not zero), the fraction equals 1. There is an unlimited number of fractions that equal 1. A fraction equal to 1 may have units, such as

$$\frac{12 \text{ inches}}{12 \text{ inches}}$$

Since 12 inches equals 1 foot, we can write two more fractions that equal 1.

$$\frac{12 \text{ inches}}{1 \text{ foot}} \qquad \frac{1 \text{ foot}}{12 \text{ inches}}$$

Because these fractions have units and are equal to 1, we call them **unit multipliers.** Unit multipliers are very useful for converting from one unit of measure to another. For instance, if we want to convert 5 feet to inches, we can multiply 5 feet by a multiplier that has inches on top and feet on bottom. The feet units cancel and the product is 60 inches.

$$5 \cancel{\text{ft}} \cdot \frac{12 \text{ in.}}{1 \cancel{\text{ft}}} = 60 \text{ in.}$$

Note that 5 ft and $\frac{5 \text{ ft}}{1}$ are equivalent. You may use either form.

$$5 \text{ ft} = \frac{5 \text{ ft}}{1}$$

If we want to convert 96 inches to feet, we can multiply 96 inches by a unit multiplier that has a numerator of feet and a denominator of inches. The inch units cancel and the product is 8 feet.

$$96 \cancel{\text{in.}} \cdot \frac{1 \text{ ft}}{12 \cancel{\text{in.}}} = 8 \text{ ft}$$

Notice that we selected a unit multiplier that canceled the unit we wanted to remove and kept the unit we wanted in the answer.

When we set up unit conversion problems, we will write the numbers in this order:

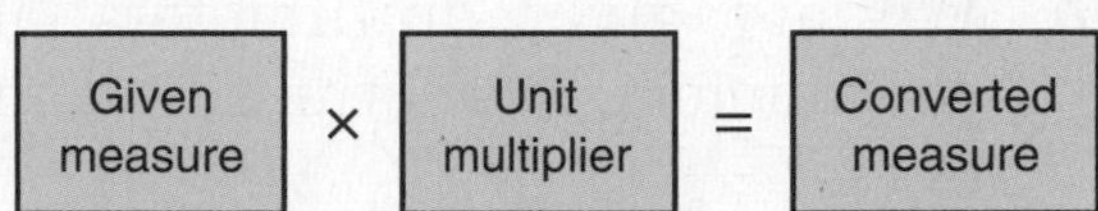

Example 1 Write two unit multipliers for these equivalent measures:

$$3 \text{ ft} = 1 \text{ yd}$$

Solution We write one measure as the numerator and its equivalent as the denominator.

$$\frac{\mathbf{3\ ft}}{\mathbf{1\ yd}} \quad \text{and} \quad \frac{\mathbf{1\ yd}}{\mathbf{3\ ft}}$$

Example 2 Use one of the unit multipliers from example 1 to convert

(a) 240 yards to feet.

(b) 240 feet to yards.

Solution (a) We are given a measure in yards. We want the answer in feet. So we write the following:

$$240 \text{ yd} \cdot \boxed{\text{Unit multiplier}} = \quad \text{ft}$$

We want to cancel the unit "yd" and keep the unit "ft," so we select the unit multiplier that has a numerator of ft and a denominator of yd. Then we multiply and cancel units.

$$240 \cancel{\text{yd}} \cdot \frac{3 \text{ ft}}{1 \cancel{\text{yd}}} = \mathbf{720\ ft}$$

We know our answer is reasonable because feet are shorter units than yards, and therefore it takes more feet than yards to measure the same distance.

(b) We are given the measure in feet, and we want the answer in yards. We choose the unit multiplier that has a numerator of yd.

$$240 \cancel{\text{ft}} \cdot \frac{1 \text{ yd}}{3 \cancel{\text{ft}}} = \mathbf{80\ yd}$$

We know our answer is reasonable because yards are longer units than feet, and therefore it takes fewer yards than feet to measure the same distance.

Example 3 Convert 350 millimeters to centimeters (1 cm = 10 mm).

Solution We are given millimeters and are asked to convert to centimeters. We form a unit multiplier that has a numerator of cm.

$$350 \cancel{\text{mm}} \cdot \frac{1 \text{ cm}}{10 \cancel{\text{mm}}} = \mathbf{35\ cm}$$

LESSON PRACTICE

Practice set* Write two unit multipliers for each pair of equivalent measures:

a. 1 yd = 36 in.

b. 100 cm = 1 m

c. 16 oz = 1 lb

Use unit multipliers to answer problems **d–f.**

d. Convert 10 yards to inches.

e. Twenty-four feet is how many yards (1 yd = 3 ft)?

f. In old England 12 pence equaled 1 shilling. Merlin had 24 shillings. This was the same as how many pence?

MIXED PRACTICE

Problem set

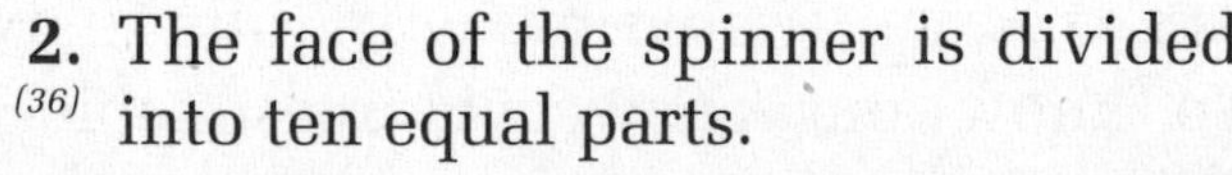

1. (35) When the product of 3.5 and 0.4 is subtracted from the sum of 3.5 and 0.4, what is the difference?

2. (36) The face of the spinner is divided into ten equal parts.

(a) What fraction of this circle is marked with a 1?

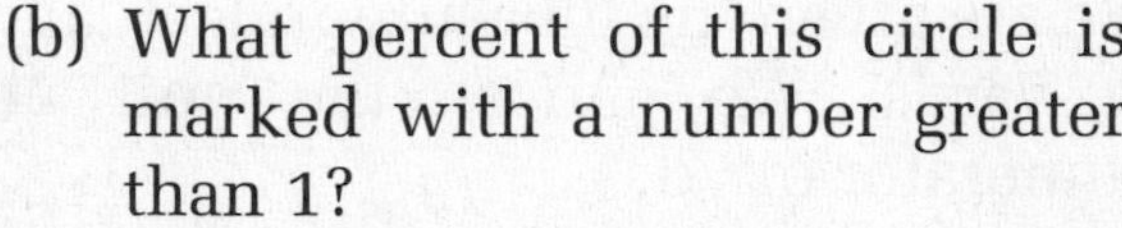

(b) What percent of this circle is marked with a number greater than 1?

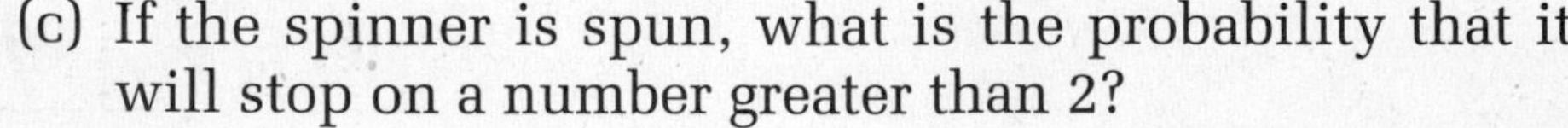

(c) If the spinner is spun, what is the probability that it will stop on a number greater than 2?

3. (46) The 13-ounce box of cooked cereal costs \$1.17, while the 18-ounce box costs \$1.44. Find the unit price for each size. Which size is the better buy?

4. (46) Nelson covered the first 20 miles in $2\frac{1}{2}$ hours. What was his average speed in miles per hour?

5. (28) The parking lot charges \$2 for the first hour plus 50¢ for each additional half hour or part thereof. What is the total charge for parking in the lot for 3 hours 20 minutes?

6. (13) The train traveled at an average speed of 60 miles per hour. How long did it take the train to travel 420 miles?

7. (22) Diagram this statement. Then answer the questions that follow.

Forty percent of the 30 football players were endomorphic.

(a) How many of the football players were endomorphic?

(b) What percent of the football players were not endomorphic?

8. (8) Which percent best identifies the shaded part of this circle?

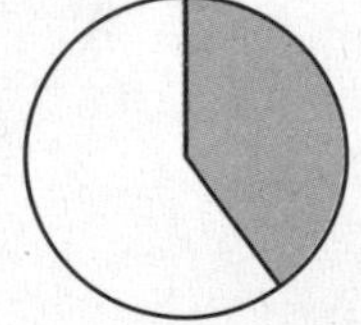

A. 25% B. 40%

C. 50% D. 60%

9. (43) Write $3\frac{5}{6}$ as a decimal number rounded to four decimal places.

10. (47) Use exponents to write 7.5 million in expanded notation.

11. (48) Write each number as a percent:

(a) 0.6 (b) $\frac{1}{6}$ (c) $1\frac{1}{2}$

12. (48) Complete the table.

FRACTION	DECIMAL	PERCENT
(a)	(b)	30%
(c)	(d)	250%

13. List the prime numbers between 90 and 100.
(21)

14. The dashes divide this figure into a rectangle and a triangle.
(37)

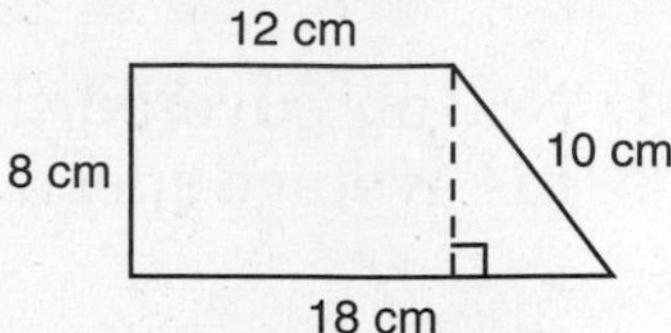

(a) What is the area of the rectangle?

(b) What is the area of the triangle?

(c) What is the combined area of the rectangle and triangle?

15. Use a compass to draw a circle with a radius of $1\frac{1}{2}$ in. Then use a protractor to draw a central angle that measures 60°. Shade the sector that is formed by the 60° central angle.
(Inv. 2)

Solve:

16. $\frac{10}{x} = \frac{7}{42}$
(39)

17. $\frac{1.5}{1} = \frac{w}{4}$
(39)

18. $3.56 = 5.6 - y$
(35)

19. $\frac{3}{20} = w + \frac{1}{15}$
(30)

20. Which property is illustrated by each of the following equations?
(2, 41)

(a) $x(y + z) = xy + xz$

(b) $x + y = y + x$

(c) $1x = x$

21. Which is equivalent to $\frac{10^6}{10^2}$?
(47)

A. 10^3 B. 10^4 C. 1000 D. 30

22. The coordinates of three vertices of a square are (2, 0), (0, –2), and (–2, 0).
(Inv. 3)

(a) What are the coordinates of the fourth vertex?

(b) Counting whole square units and half square units, find the area of the square.

23. If 10 cookies are shared equally by 4 children, how many cookies will each child receive?
(44)

24. Below is a box-and-whisker plot of test scores. Refer to the plot to answer (a)–(c).
(Inv. 4)

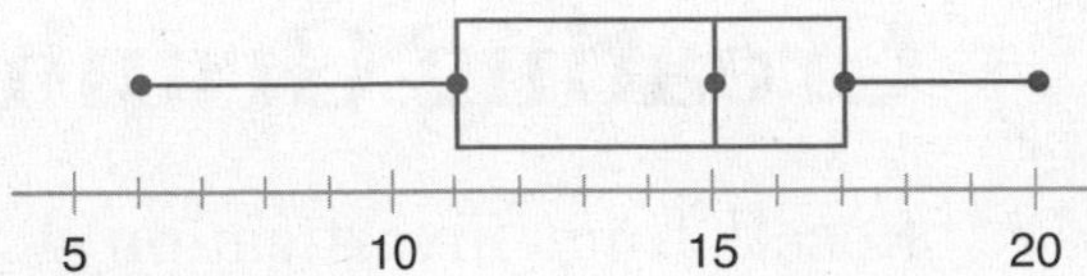

(a) What is the range of scores?

(b) What is the median score?

(c) Write another question that can be answered by referring to the plot. Then answer the question.

25. Write two unit multipliers for the conversion 10 mm = 1 cm. Then use one of the unit multipliers to convert 160 mm to centimeters.
(50)

26. 4 yd 2 ft 7 in. + 3 yd 5 in.
(49)

27. $5\frac{1}{6} - \left(1\frac{3}{4} \div 2\frac{1}{3}\right)$
(26, 30)

28. $3\frac{5}{7} + \left(3\frac{1}{8} \cdot 2\frac{2}{5}\right)$
(26, 30)

29. In the figure at right, ΔABC is congruent to ΔDCB. Find the measure of
(40)

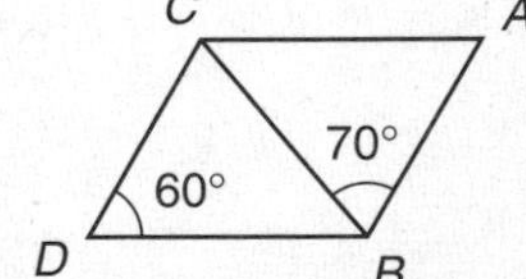

(a) $\angle BAC$.

(b) $\angle BCA$.

(c) $\angle CBD$.

30. Show two ways to evaluate $a(b - c)$ for $a = 4$, $b = 5$, and $c = 3$.
(41)

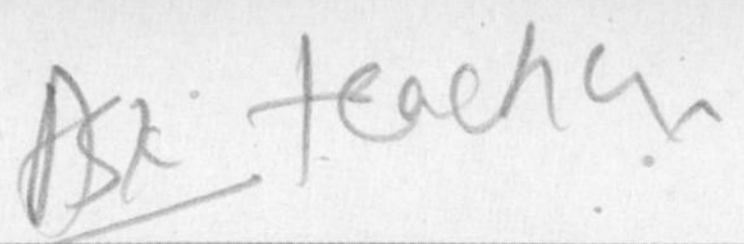

INVESTIGATION 5

Focus on

Creating Graphs

Recall from Investigation 4 that we considered a stem-and-leaf plot that a counselor created to display student test scores. If we rotate that plot 90°, the display resembles a vertical bar graph, or **histogram.**

```
2 | 1 3 5 6 6 8
3 | 0 0 2 2 4 5 6 6 8 9
4 | 0 0 1 1 1 2 3 3 5 7 7 8
5 | 0 1 1 2 3 5 8
```

A histogram is a special type of bar graph that displays data in equal-sized intervals. There are no spaces between the bars. The height of the bars in this histogram show the number of test scores in each interval.

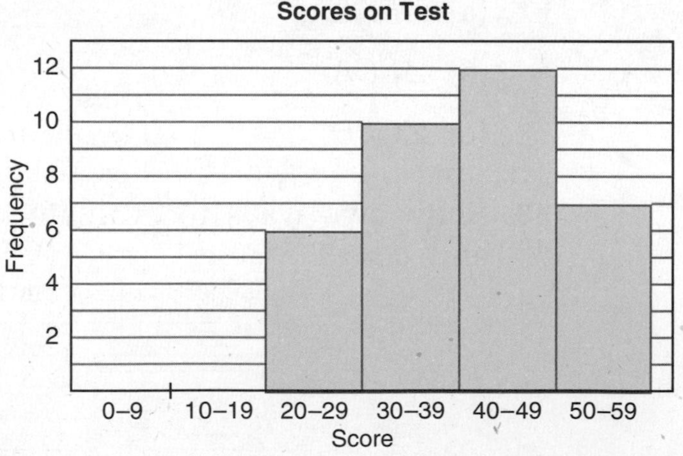

1. Changing the intervals can change the appearance of a histogram. Create a new histogram for the test scores itemized in the stem-and-leaf plot using the following intervals: 21–28, 29–36, 37–44, 45–52, and 53–60. Draw a break in the horizontal scale (⁄\/⁄) between 0 and 21.

Histograms and other bar graphs are useful for showing comparisons, but sometimes the visual effect can be misleading. When viewing a graph, it is important to carefully note the scale. Compare these two bar graphs that display the same information.

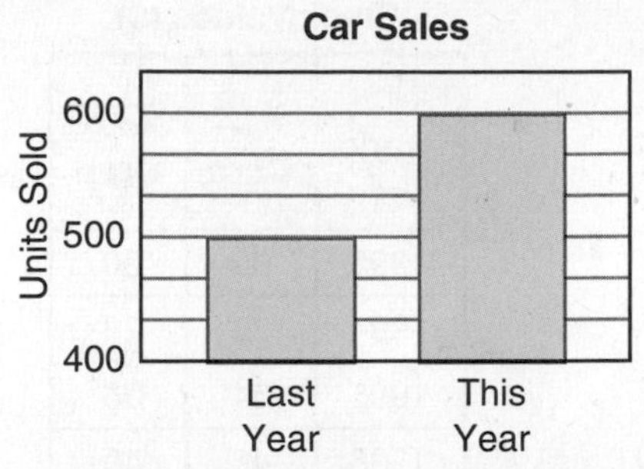

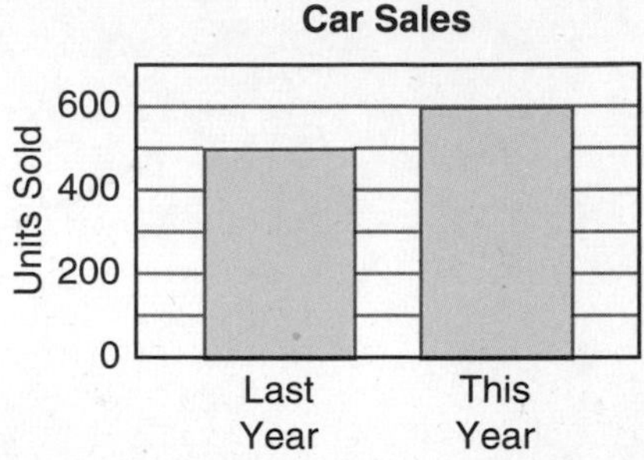

2. Which of the two graphs visually exaggerates the growth in sales from one year to the next? How was the exaggerated visual effect created?
3. Larry made the bar graph below that compares his test score to Moe's test score. Create another bar graph that shows the same information in a less misleading way.

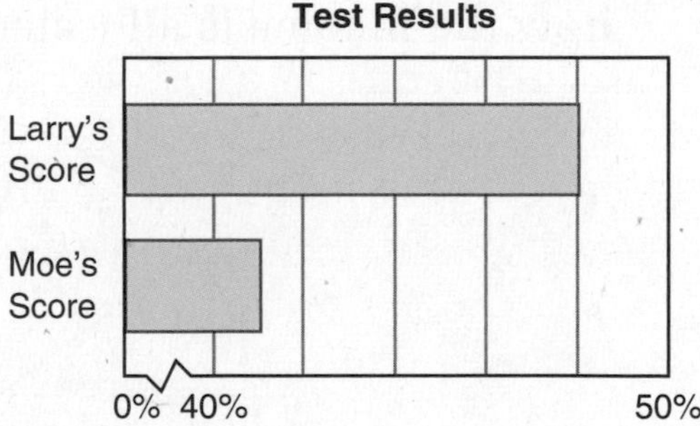

Changes over time are often displayed by line graphs. A **double-line graph** may compare two performances over time. The graph below illustrates the differences in the growing value of a $1000 investment compounded at 7% and at 10% annual interest rates.

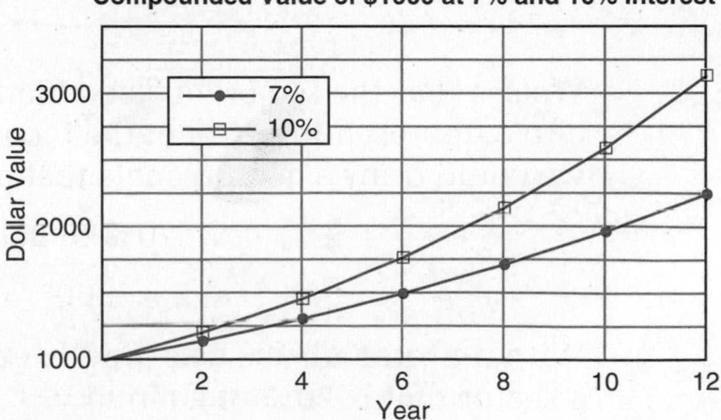

4. Create a double-line graph using the information in the table below. Label the axes; then select and number the scales. Make a legend (or key) so that the reader can distinguish between the two graphed lines.

Stock Values ($)

First Trade Of	XYZ Corp	ZYX Corp
1993	30	30
1994	36	28
1995	34	36
1996	46	40
1997	50	46
1998	50	42

A **circle graph** (or pie graph) is commonly used to show components of a budget. The entire circle, 100%, may represent monthly income. The sectors of the circle show how the income is allocated.

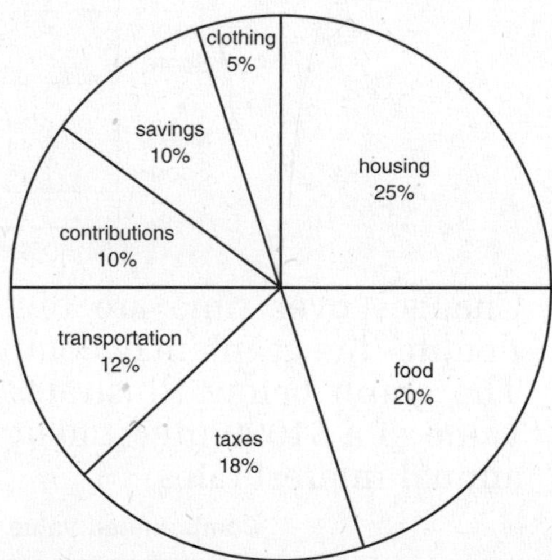

We see that the sector labeled "food" is 20% of the area of the circle, representing 20% of the income. To make a 20% sector, we could draw a central angle that measures 20% of 360°.

$$20\% \text{ of } 360°$$

$$0.2 \times 360° = 72°$$

With a protractor we can draw a central angle of 72° to make a sector that is 20% of a circle.

5. Create a pie graph for the table below to show how Kerry spends a weekday. First calculate the number of degrees in the central angle for each sector of the pie graph. Next use a compass to draw a circle with a radius of about $2\frac{1}{2}$ inches. Then, with a protractor and straightedge, divide the circle into sectors of the correct size and label each sector.

How Kerry Spends a Day

Activity	% of Day
Studies	25%
Recreation	10%
Music lessons	10%
Eating	10%
Sleeping	40%
Other	5%

Extensions

a. Create a circle graph showing the percentages of your friends and family with various eye colors.

b. Explore the graph-creating capabilities of database computer programs.

LESSON

51 Scientific Notation for Large Numbers

WARM-UP

Facts Practice: + – × ÷ Decimals (Test J)

Mental Math:

a. $4 \times \$3.50$

b. 4.5×10^2

c. $\frac{5}{20} = \frac{3}{x}$

d. Convert 5 km to m.

e. $15^2 - 5^2$

f. $\frac{5}{9}$ of 45

g. Find the sum, difference, product, and quotient of $\frac{7}{8}$ and $\frac{1}{2}$.

Problem Solving:

Beginning with 1, the first five perfect squares are 1, 4, 9, 16, and 25. What number is the 1000th perfect square?

NEW CONCEPT

The numbers used in scientific measurement are often very large or very small and occupy many places when written in standard form. For example, a light-year is about

$$9{,}461{,}000{,}000{,}000 \text{ km}$$

Scientific notation is a way of expressing numbers as a product of a decimal number and a power of 10. In scientific notation a light-year is

$$9.461 \times 10^{12} \text{ km}$$

In the table below we use scientific notation to approximate some common distances. Measurements are in millimeters.

Scientific notation	Standard form	Length
2.0×10^0 mm	2 mm	width of pencil lead
2.4×10^1 mm	24 mm	diameter of a quarter
1.6×10^2 mm	160 mm	length of a dollar bill
4.5×10^3 mm	4500 mm	length of average car
2.9×10^4 mm	29,000 mm	length of basketball court
1.1×10^5 mm	110,000 mm	length of football field
1.6×10^6 mm	1,600,000 mm	one mile
4.2×10^7 mm	42,000,000 mm	distance of runner's marathon

In scientific notation the power of 10 indicates where the decimal point is located when the number is written in standard form. Consider this number expressed in scientific notation:

$$4.62 \times 10^6$$

Multiplying 4.62 by 10^6 (one million) has the effect of shifting the decimal point six places (note the exponent in 10^6) to the right. We use zeros as placeholders.

$$4620000. \longrightarrow 4{,}620{,}000$$

Example 1 Write 2.46×10^8 in standard form.

Solution We shift the decimal point in 2.46 eight places to the right, using zeros as placeholders.

$$246000000. \longrightarrow \mathbf{246{,}000{,}000}$$

To write a number in scientific notation, it is customary to place the decimal point to the right of the first nonzero digit. Then we use a power of 10 to indicate the actual location of the decimal point. To write

$$405{,}700{,}000$$

in scientific notation, we begin by placing the decimal point to the right of 4 and then counting the places from the original decimal point.

$$4.\underbrace{05700000}_{\text{8 places}}$$

We see that the original decimal point was eight places to the right of where we put it. We omit the terminal zeros and write

$$4.057 \times 10^8$$

Example 2 Write 40,720,000 in scientific notation.

Solution We begin by placing the decimal point after the 4.

$$4.\underbrace{0720000}_{\text{7 places}}$$

Now we discard the terminal zeros and write 10^7 to show that the original decimal point is really seven places to the right.

$$\mathbf{4.072 \times 10^7}$$

Example 3 Compare: $1.2 \times 10^4 \bigcirc 2.1 \times 10^3$

Solution Since 1.2×10^4 equals 12,000 and 2.1×10^3 equals 2100, we see that

$$\mathbf{1.2 \times 10^4 > 2.1 \times 10^3}$$

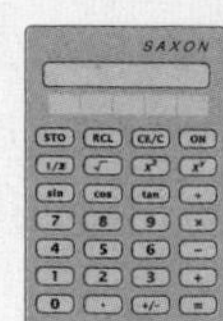

Scientific calculators will display the results of an operation in scientific notation if the number would otherwise exceed the display capabilities of the calculator. For example, to multiply one million by one million, we would enter

The answer, one trillion, contains more digits than can be displayed by many calculators. Instead of displaying one trillion in standard form, the calculator displays one trillion in some modified form of scientific notation such as

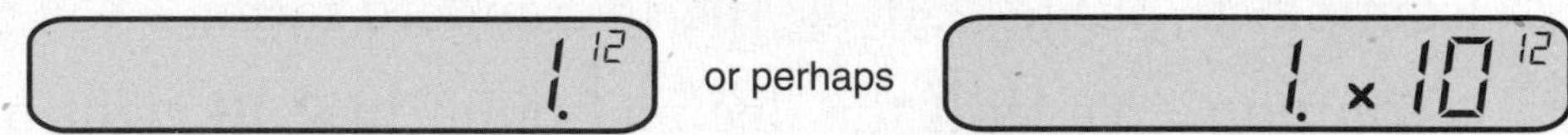

LESSON PRACTICE

Practice set Write each number in scientific notation:

a. 15,000,000

b. 400,000,000,000

c. 5,090,000

d. two hundred fifty billion

Write each number in standard form:

e. 3.4×10^6

f. 5×10^8

g. 1×10^5

Compare:

h. $1.5 \times 10^5 \bigcirc 1.5 \times 10^6$

i. one million $\bigcirc$ 1×10^6

MIXED PRACTICE

Problem set Refer to the double-line graph below to answer problems 1 and 2.

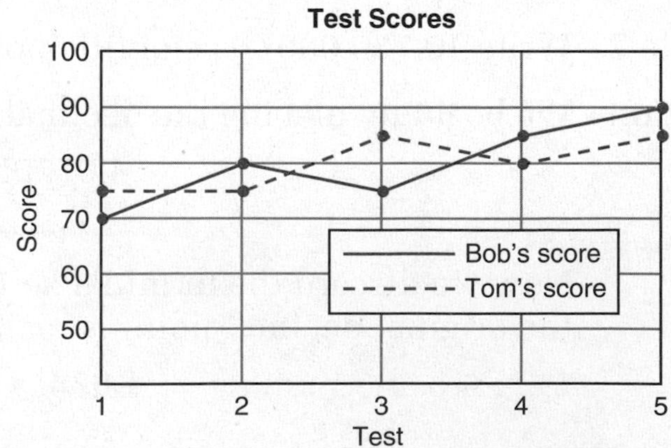

1. On how many tests was Bob's score better than Tom's score?
(38)

2. What was Bob's average score on these five tests?
(28, 38)

3. (19) In the pattern on a soccer ball, a regular hexagon and a regular pentagon share a common side. If the perimeter of the hexagon is 9 in., what is the perimeter of the pentagon?

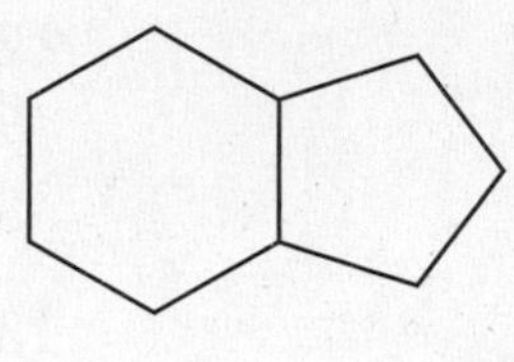

4. (46) The store sold juice for 40¢ per can or 6 cans for $1.98. How much can be saved per can by buying 6 cans at the 6-can price?

5. (14, 36) Five sevenths of the people who saw the phenomenon were convinced. The rest were unconvinced.

(a) What fraction of those who saw the phenomenon were unconvinced?

(b) What was the ratio of the convinced to the unconvinced?

6. (51) (a) Write twelve million in scientific notation.

(b) Write 17,600 in scientific notation.

7. (51) (a) Write 1.2×10^4 in standard form.

(b) Write 5×10^6 in standard form.

8. (43) Write each number as a decimal:

(a) $\frac{1}{8}$ (b) $87\frac{1}{2}\%$

9. (33) Round each number to the nearest thousand:

(a) 29,647 (b) 5280.08

10. (48) Complete the table.

FRACTION	DECIMAL	PERCENT
(a)	(b)	40%
(c)	(d)	4%

11. (Inv. 5) Find the number of degrees in the central angle of each sector of the circle shown.

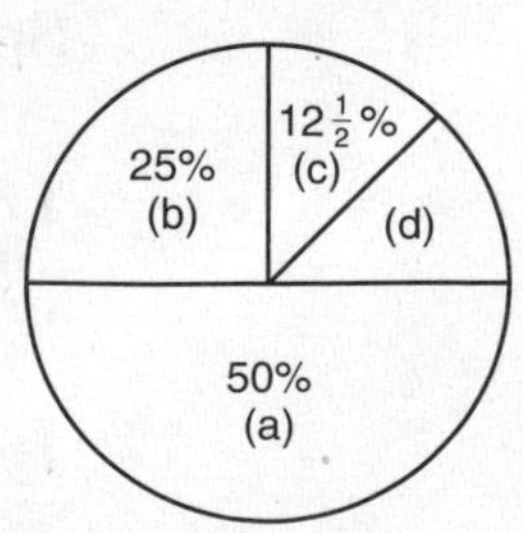

12. (46) What is the total price, including 5% sales tax, of a \$15.80 item?

13. (36) Layla is thinking of a positive, single-digit, even number. Lou guesses it is 7. What is the probability that Lou's guess is correct?

14. (18) These two quadrilaterals are congruent. Refer to these figures to answer (a) and (b).

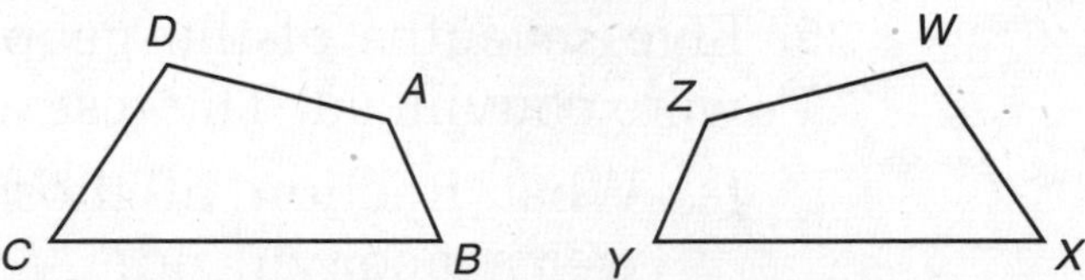

(a) Which angle in *WXYZ* is congruent to $\angle A$ in *ABCD?*

(b) Which segment in *ABCD* is congruent to $\overline{WX}$ in *WXYZ?*

Refer to the figure below to answer problems 15 and 16. Dimensions are in meters. All angles are right angles.

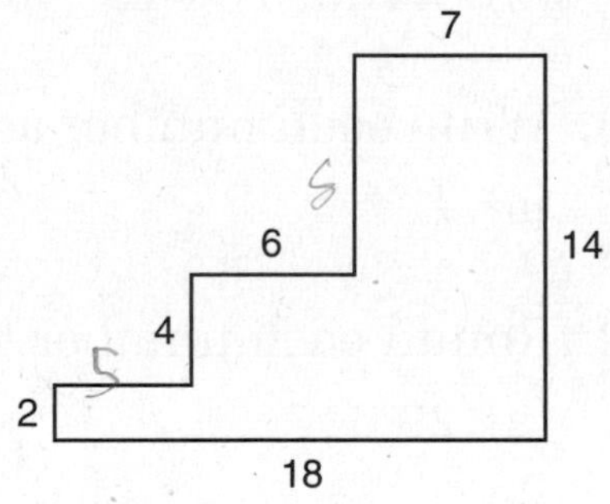

15. (19) What is the perimeter of the figure?

16. (37) What is the area of the figure?

Solve:

17. (39) $\dfrac{24}{x} = \dfrac{60}{40}$

18. (39) $\dfrac{6}{4.2} = \dfrac{n}{7}$

19. (35) $5m = 8.4$

20. (35) $6.5 - y = 5.06$

Simplify:

21. $5^2 + 3^3 + \sqrt{64}$
(20)

22. $16 \text{ cm} \cdot \dfrac{10 \text{ mm}}{1 \text{ cm}}$
(50)

23.
(49)
$$\begin{array}{r} 5 \text{ days } 18 \text{ hr } 50 \text{ min} \\ + \ 2 \text{ days } \ 8 \text{ hr } 25 \text{ min} \\ \hline \end{array}$$

24.
(49)
$$\begin{array}{r} 3 \text{ yd } 2 \text{ ft } 5 \text{ in.} \\ + \ 1 \text{ yd } \qquad 9 \text{ in.} \\ \hline \end{array}$$

25. $6\frac{2}{3} + \left(5\frac{1}{4} - 3\frac{7}{8}\right)$
(23, 30)

26. $3\frac{1}{3} \cdot \left(2\frac{2}{3} \div 1\frac{1}{2}\right)$
(26)

27. Show two ways to evaluate $x(x + y)$ for $x = 0.5$ and $y = 0.6$.
(41)

The coordinates of three vertices of a triangle are A (−4, 0), B (0, −4), and C (−8, −4). Graph the triangle and refer to it to answer problems 28 and 29.

28. Use a protractor to find the measures of $\angle A$, $\angle B$, and $\angle C$.
(17)

29. What is the area of ΔABC?
(37)

30. When the temperature increases from the freezing temperature of water to the boiling temperature of water, it is an increase of 100 degrees on the Celsius scale. The same increase in temperature is how many degrees on the Fahrenheit scale?
(32)

LESSON

52 Order of Operations

WARM-UP

Facts Practice: Powers and Roots (Test K)

Mental Math:

a. $6 \times 75¢$

b. $4.5 \div 10^2$

c. $\frac{15}{5} = \frac{m}{6}$

d. Convert 250 cm to m.

e. $10^3 - 20^2$

f. $\frac{9}{10}$ of 200

g. At 80 km per hour, how far will a car travel in $2\frac{1}{2}$ hours?

Problem Solving:

Find x if $3x + 5 = 80$. Explain your thinking.

NEW CONCEPT

Recall that the four fundamental operations of arithmetic are addition, subtraction, multiplication, and division. We can also raise numbers to powers or find their roots. When more than one operation occurs in the same expression, we perform the operations in the order listed below.

ORDER OF OPERATIONS

1. Simplify powers and roots.
2. Multiply and divide in order from left to right.
3. Add and subtract in order from left to right.

Note: If there are parentheses (or other enclosures), we simplify within the parentheses, from innermost to outermost, before simplifying outside the parentheses.

The initial letter of each word in the sentence "Please excuse my dear Aunt Sally" reminds us of the order of operations: parentheses (or other **symbols of inclusion**), exponents (and roots), multiplication and division, addition and subtraction.

Example 1 Simplify: $2 + 4 \times 3 - 4 \div 2$

Solution We multiply and divide in order from left to right before we add or subtract.

$$2 + 4 \times 3 - 4 \div 2 \quad \text{problem}$$

$$2 + 12 - 2 \quad \text{multiplied and divided}$$

$$\mathbf{12} \quad \text{added and subtracted}$$

Example 2 Simplify: $\dfrac{3^2 + 3 \cdot 5}{2}$

Solution A division bar can serve as a symbol of inclusion, like parentheses. We simplify above and below the bar before dividing.

$$\frac{3^2 + 3 \cdot 5}{2} \quad \text{problem}$$

$$\frac{9 + 3 \cdot 5}{2} \quad \text{simplified power}$$

$$\frac{9 + 15}{2} \quad \text{multiplied above}$$

$$\frac{24}{2} \quad \text{added above}$$

$$\mathbf{12} \quad \text{divided}$$

Example 3 Evaluate: $a + ab$ if $a = 3$ and $b = 4$

Solution We will begin by writing parentheses in place of each variable. This step may seem unnecessary, but many errors can be avoided if this is always our first step.

$$a + ab$$

$$(\ ) + (\ )(\ ) \quad \text{parentheses}$$

Then we replace a with 3 and b with 4.

$$a + ab$$

$$(3) + (3)(4) \quad \text{substituted}$$

We follow the order of operations, multiplying before adding.

$$(3) + (3)(4) \quad \text{problem}$$

$$3 + 12 \quad \text{multiplied}$$

$$\mathbf{15} \quad \text{added}$$

Example 4 Evaluate: $xy - \frac{x}{2}$ if $x = 9$ and $y = \frac{2}{3}$

Solution First we replace each variable with parentheses.

$$xy - \frac{x}{2}$$

$$(\ )(\ ) - \frac{(\ )}{2} \qquad \text{parentheses}$$

Then we write 9 in place of x and $\frac{2}{3}$ in place of y.

$$xy - \frac{x}{2}$$

$$(9)\left(\frac{2}{3}\right) - \frac{(9)}{2} \qquad \text{substituted}$$

We follow the order of operations, multiplying and dividing before we subtract.

$$(9)\left(\frac{2}{3}\right) - \frac{(9)}{2} \qquad \text{problem}$$

$$6 - 4\frac{1}{2} \qquad \text{multiplied and divided}$$

$$\mathbf{1\frac{1}{2}} \qquad \text{subtracted}$$

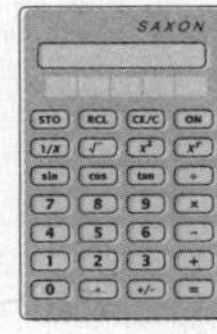

Calculators with *algebraic-logic* circuitry are designed to perform calculations according to the order of operations. Calculators without algebraic-logic circuitry perform calculations in sequence. You can test a calculator's design by selecting a problem such as that in example 1 and entering the numbers and operations from left to right, concluding with an equal sign. If the problem in example 1 is used, a displayed answer of 12 indicates an algebraic-logic design.

LESSON PRACTICE

Practice set* Simplify:

a. $5 + 5 \cdot 5 - 5 \div 5$

b. $50 - 8 \cdot 5 + 6 \div 3$

c. $24 - 8 - 6 \cdot 2 \div 4$

d. $\frac{2^3 + 3^2 + 2 \cdot 5}{3}$

Evaluate:

e. $ab - bc$ if $a = 5$, $b = 3$, and $c = 4$

f. $ab + \frac{a}{c}$ if $a = 6$, $b = 4$, and $c = 2$

g. $x - xy$ if $x = \frac{2}{3}$ and $y = \frac{3}{4}$

MIXED PRACTICE

Problem set

1. (21) If the product of the first three prime numbers is divided by the sum of the first three prime numbers, what is the quotient?

2. (18) Sean counted a total of 100 sides on the heptagons and nonagons. If there were 4 heptagons, how many nonagons were there?

3. (31, 35) Twenty-five and two hundred seventeen thousandths is how much less than two hundred two and two hundredths?

4. (46) Jermaine bought a pack of 3 blank tapes for \$5.95.

(a) What was the price per tape to the nearest cent?

(b) The sales-tax rate was 7%. What was the total cost of the three tapes including tax?

5. (28) Ginger is starting a 330-page book. Suppose she reads for 4 hours and averages 35 pages per hour.

(a) How many pages will she read in 4 hours?

(b) After 4 hours, how many pages will she still have to read to finish the book?

6. (22) In the following statement, convert the percent to a reduced fraction. Then diagram the statement and answer the questions.

Seventy-five percent of the 60 passengers disembarked at the terminal.

(a) How many passengers disembarked at the terminal?

(b) What percent of the passengers did not disembark at the terminal?

7. (51) (a) Write 3,750,000 in scientific notation.

(b) Write eighty million in scientific notation.

8. (51) (a) Write 2.05×10^6 in standard form.

(b) Write 4×10^1 in standard form.

9. (43) Write each number as a decimal:

(a) $\frac{3}{8}$ (b) 6.5%

10. (42) Write $3.\overline{27}$ as a decimal number rounded to the nearest thousandth.

11. (48) Complete the table.

FRACTION	DECIMAL	PERCENT
(a)	(b)	250%
(c)	(d)	25%

12. (44) Divide 70 by 9 and write the answer

(a) as a mixed number.

(b) as a decimal number with a bar over the repetend.

13. (34) What decimal number names the point marked by the arrow?

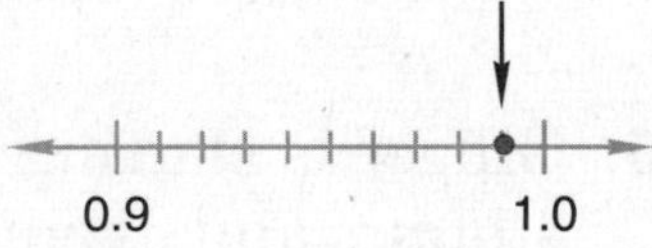

14. (32) Draw a rectangle that is 3 cm long and 2 cm wide. Then answer (a) and (b).

(a) What is the perimeter of the rectangle in millimeters?

(b) What is the area of the rectangle in square centimeters?

15. (37) In quadrilateral *ABCD*, $\overline{AD}$ is parallel to $\overline{BC}$. Dimensions are in centimeters.

(a) Find the area of $\triangle ABC$.

(b) Find the area of $\triangle ACD$.

(c) What is the combined area of the two triangles?

D 8 A
6 6
C 12 B

Solve:

16. (39) $\frac{8}{f} = \frac{56}{105}$

17. (39) $\frac{12}{15} = \frac{w}{2.5}$

18. (35) $p + 6.8 = 20$

19. (35) $q - 3.6 = 6.4$

Simplify:

20. (20) $5^3 - 10^2 - \sqrt{25}$

21. (52) $4 + 4 \cdot 4 - (4 \div 4)$

22. (35, 45) $\frac{4.8 - 0.24}{(0.2)(0.6)}$

23. (49)

$$\begin{array}{r} 5 \text{ hr } 45 \text{ min } 30 \text{ s} \\ + \ 2 \text{ hr } 53 \text{ min } 55 \text{ s} \\ \hline \end{array}$$

24. (26, 30) $6\frac{3}{4} + \left(5\frac{1}{3} \cdot 2\frac{1}{2}\right)$

25. (26, 30) $5\frac{1}{2} - \left(3\frac{3}{4} \div 2\right)$

26. (35) Estimate the sum to the nearest whole number. Then perform the calculation.

$$8.575 + 12.625 + 8.4 + 70.4$$

27. (47) $0.8 \times 1.25 \times 10^6$

28. (52) Evaluate: $ab + \frac{a}{b}$ if $a = 4$ and $b = 0.5$

29. (50) Convert 1.4 meters to centimeters (1 m = 100 cm).

30. (36) Thirty children were asked to name their favorite sport. Twelve said football, 10 said basketball, and 8 said baseball. If a child is selected at random, what is the probability that the child's favorite sport is basketball?

LESSON

53 Multiplying Rates

WARM-UP

Facts Practice: Powers and Roots (Test K)

Mental Math:

a. $8 \times \$1.25$

b. 12.75×10

c. $2x + 5 = 75$

d. Convert 35 cm to mm.

e. $\left(\frac{1}{2}\right)^2$

f. $\frac{3}{5}$ of 45

g. 10×6, $+ 4$, $\sqrt{\ }$, $\times 3$, double that number, $+ 1$, $\sqrt{\ }$, $\times 8$, $- 1$, $\div 5$, square that number

Problem Solving:

Allen wanted to form a triangle out of straws that were 5 in., 7 in., and 12 in. long. He threaded a piece of string through the three straws, pulled the string tight, and tied it. What was the area of the figure?

NEW CONCEPT

If we are traveling in a car at 50 miles per hour (50 mph), we can calculate how far we would travel in 4 hours by multiplying.

$$4\,\cancel{\text{hr}} \times \frac{50 \text{ mi}}{1\,\cancel{\text{hr}}} = 200 \text{ mi}$$

We can find out how long it will take us to travel 300 miles by dividing.

$$300 \text{ mi} \div \frac{50 \text{ mi}}{1 \text{ hr}}$$

$$300\,\cancel{\text{mi}} \times \frac{1 \text{ hr}}{50\,\cancel{\text{mi}}} = \frac{300 \text{ hr}}{50} = 6 \text{ hr}$$

Notice that dividing by a rate is similar to dividing by a fraction. We actually multiply by the reciprocal of the original rate. There are two forms of every rate and they are reciprocals. Thus we may solve rate problems by multiplying by the correct form of the rate. The following statement expresses a rate:

There were 5 chairs in each row.

We can express this rate in two forms.

(a) $\dfrac{5 \text{ chairs}}{1 \text{ row}}$ (b) $\dfrac{1 \text{ row}}{5 \text{ chairs}}$

Using rate (a) we can find the number of chairs in 6 rows.

$$6 \cancel{\text{rows}} \times \frac{5 \text{ chairs}}{1 \cancel{\text{row}}} = 30 \text{ chairs}$$

Using rate (b) we can find the number of rows needed for 20 chairs.

$$20 \cancel{\text{chairs}} \times \frac{1 \text{ row}}{5 \cancel{\text{chairs}}} = 4 \text{ rows}$$

Example 1 Eight ounces of the solution costs 40 cents.

(a) Write two forms of the rate given by this statement.

(b) Find the cost of 32 ounces of the solution.

(c) How many ounces can be purchased for $1.20?

Solution (a) The two forms of the rate are

$$\mathbf{\frac{8 \text{ oz}}{40 \text{ cents}}} \qquad \mathbf{\frac{40 \text{ cents}}{8 \text{ oz}}}$$

(b) To find the cost, we use the form that has money on top.

$32 \cancel{\text{oz}} \times \dfrac{40 \text{ cents}}{8 \cancel{\text{oz}}}$ canceled ounces

$\dfrac{1280}{8}$ cents multiplied

160 cents simplified

We usually write answers equal to a dollar or more by using a dollar sign. Thus the cost is **$1.60.**

(c) To find the number of ounces, we use the form that has ounces on top.

$120 \cancel{\text{cents}} \times \dfrac{8 \text{ oz}}{40 \cancel{\text{cents}}}$ canceled cents

$\dfrac{960}{40}$ oz multiplied

24 oz simplified

Note that rates, like fractions, can be reduced. Both forms of the rate in part (a) of example 1 can be reduced.

$$\frac{\overset{1}{\cancel{8}}\text{ oz}}{\underset{5}{\cancel{40}}\text{ cents}} = \frac{1}{5}\,\frac{\text{oz}}{\text{cents}} \qquad \frac{\overset{5}{\cancel{40}}\text{ cents}}{\underset{1}{\cancel{8}}\text{ oz}} = 5\,\frac{\text{cents}}{\text{oz}}$$

However, rates can be used to solve problems without first reducing as we saw in parts (b) and (c).

Example 2 Jennifer's speed was 60 miles per hour.

(a) Write the two forms of the rate given by this statement.

(b) How far did she drive in 5 hours?

(c) How long would it take her to drive 300 miles?

Solution (a) The two forms of the rate are

$$\mathbf{\frac{60\text{ miles}}{1\text{ hour}} \qquad \frac{1\text{ hour}}{60\text{ miles}}}$$

(b) To find how far, we use the form with miles on top.

$$5\ \cancel{\text{hours}} \times \frac{60\text{ miles}}{1\ \cancel{\text{hour}}} = \mathbf{300\text{ miles}}$$

(c) To find how much time, we use the form with time on top.

$$300\ \cancel{\text{miles}} \times \frac{1\text{ hour}}{60\ \cancel{\text{miles}}} = \mathbf{5\text{ hours}}$$

Example 3 If pencils cost \$0.25 each, how many pencils can Carol Ann buy for \$2.00?

Solution We can use rates to solve this problem. The rate of \$0.25 per pencil has two forms.

$$\frac{\$0.25}{1\text{ pencil}} \text{ and } \frac{1\text{ pencil}}{\$0.25}$$

To find how many, we use the form with pencils on top.

$$\cancel{\$}2.00 \times \frac{1\text{ pencil}}{\cancel{\$}0.25} = \mathbf{8\text{ pencils}}$$

LESSON PRACTICE

Practice set In the lecture hall there were 18 rows. Fifteen chairs were in each row.

a. Write the two forms of the rate given by this statement.

b. Find the total number of chairs in the lecture hall.

A car could travel 24 miles on one gallon of gas.

c. Write the two forms of the rate given by this statement.

d. How many gallons would it take to travel 160 miles?

MIXED PRACTICE

Problem set Refer to this double-bar graph to answer problems 1–3:

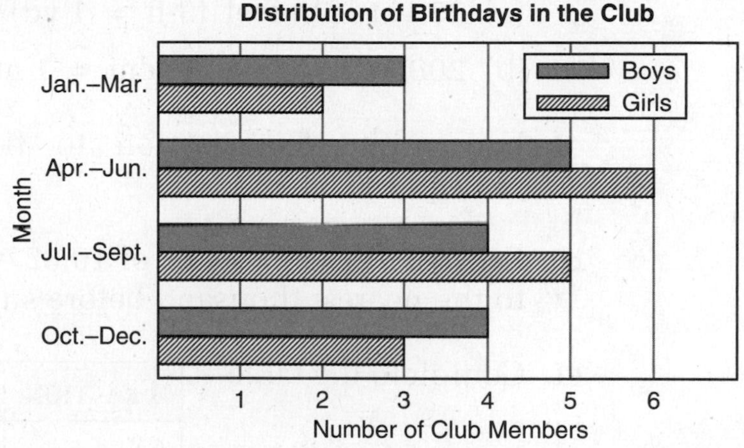

1. (38) (a) How many boys are in the club?

(b) How many girls are in the club?

2. (38) What percent of the club members have a birthday in one of the months from January through June?

3. (38) What fraction of the boys have a birthday in one of the months from April through June?

4. (28, 46) At the book fair Muhammad bought 4 books. One book cost \$3.95. Another book cost \$4.47. The other 2 books cost \$4.95 each.

(a) What was the average price per book?

(b) What was the total price of the books including 8% sales tax?

5. (22) Diagram this statement. Then answer the questions that follow.

Seven twelfths of the 840 gerbils were hiding in their burrows.

(a) What fraction of the gerbils were not hiding in their burrows?

(b) How many gerbils were not hiding in their burrows?

6. (51) (a) Write one trillion in scientific notation.

(b) Write 475,000 in scientific notation.

7. (51) (a) Write 7×10^2 in standard form.

(b) Compare: $2.5 \times 10^6 \bigcirc 2.5 \times 10^5$

8. (50) Use unit multipliers to perform the following conversions:

(a) 35 yards to feet (3 ft = 1 yd)

(b) 2000 cm to m (100 cm = 1 m)

9. (27) Use prime factorization to find the least common multiple of 54 and 36.

10. (29) Estimate the difference of 19,827 and 12,092 by rounding to the nearest thousand before subtracting.

11. (48) Complete the table.

FRACTION	DECIMAL	PERCENT
(a)	(b)	150%
(c)	(d)	15%

12. (48) Write each number as a percent:

(a) $\frac{4}{5}$ (b) 0.06

13. (32) Big Bill is 2 m tall. Stephanie is 165 cm tall. Big Bill is how many centimeters taller than Stephanie?

14. (19, 37) Refer to this figure to answer (a) and (b). Dimensions are in feet. All angles are right angles.

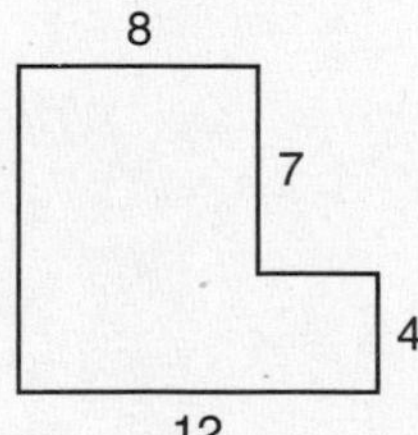

(a) What is the area of the figure?

(b) What is the perimeter of the figure?

15. (53) The bank would exchange 1.6 Canadian dollars (C$) for 1 U.S. dollar (US$).

(a) Write two forms of the exchange rate.

(b) How many Canadian dollars would the bank exchange for 160 U.S. dollars?

Solve:

16. (39) $\dfrac{18}{100} = \dfrac{90}{p}$

17. (39) $\dfrac{6}{9} = \dfrac{t}{1.5}$

18. (35) $8 = 7.25 + m$

19. (35) $1.5 = 10n$

Simplify:

20. (20) $\sqrt{81} + 9^2 - 2^5$

21. (52) $16 \div 4 \div 2 + 3 \times 4$

22. (49)

$$\begin{array}{r} 3 \text{ yd } 1 \text{ ft } 7\frac{1}{2} \text{ in.} \\ + \quad 2 \text{ ft } 6\frac{1}{2} \text{ in.} \\ \hline \end{array}$$

23. (26, 30) $12\frac{2}{3} + \left(5\frac{5}{6} \div 2\frac{1}{3}\right)$

24. (26, 30) $8\frac{3}{5} - \left(1\frac{1}{2} \cdot 3\frac{1}{5}\right)$

25. (35, 47) $10.6 + 4.2 + 16.4 + (3.875 \times 10^1)$

26. (29, 33) Estimate: $6.85 \times 4\frac{1}{16}$

27. (52) Evaluate: $\dfrac{ab}{bc}$ if $a = 6$, $b = 0.9$, and $c = 5$

28. (44) Petersen needed to pack 1000 eggs into flats that held $2\frac{1}{2}$ dozen eggs. How many flats could he fill?

29. (36) If there is one chance in five of guessing the correct answer, then what is the probability of not guessing the correct answer?

30. (40) Find the measures of angles a, b, and c in this figure:

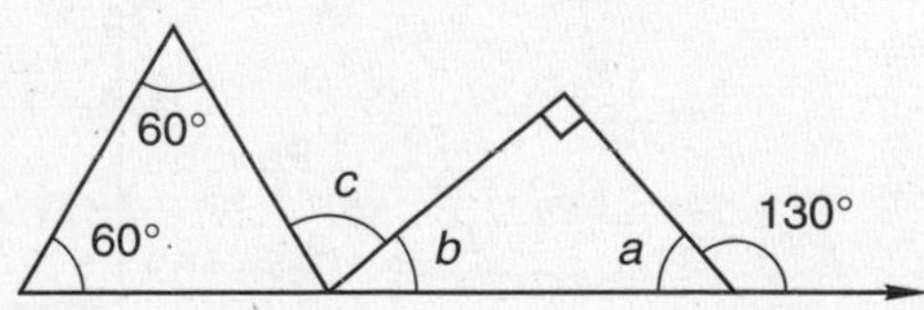

LESSON

54 Ratio Word Problems

WARM-UP

Facts Practice: Fraction-Decimal-Percent Equivalents (Test L)

Mental Math:

a. $4 \times \$4.50$ **b.** $12.75 \div 10$

c. $\frac{12}{w} = \frac{9}{6}$ **d.** Convert 1.5 m to cm.

e. $\sqrt{900} - 3^3$ **f.** $\frac{3}{10}$ of 90

g. Mentally perform each calculation:

$\frac{3}{4} + \frac{2}{5}$ $\frac{3}{4} - \frac{2}{5}$ $\frac{3}{4} \cdot \frac{2}{5}$ $\frac{3}{4} \div \frac{2}{5}$

Problem Solving:

At first $\frac{1}{3}$ of the children in the room were girls. When 3 boys left, $\frac{1}{2}$ of the remaining children were girls. How many girls were in the room?

NEW CONCEPT

In this lesson we will use proportions to solve ratio word problems. Consider the following ratio word problem:

The ratio of parrots to macaws was 5 to 7. If there were 750 parrots, how many macaws were there?

In this problem there are two kinds of numbers, ratio numbers and actual count numbers. The ratio numbers are 5 and 7. The number 750 is an actual count of parrots. We will arrange these numbers into two columns and two rows to form a ratio box. Practicing the use of ratio boxes now will pay dividends in later lessons when we extend their application to more complex problems.

	Ratio	Actual Count
Parrots	5	750
Macaws	7	M

We were not given the actual count of macaws, so we have used M to stand for the number of macaws. The numbers in this ratio box can be used to write a proportion. By solving the proportion, we find the actual count of macaws.

	Ratio	Actual Count
Parrots	5	750
Macaws	7	M

$$\frac{5}{7} = \frac{750}{M}$$
$$5M = 5250$$
$$M = 1050$$

We find that the actual count of macaws was 1050.

Example At the amusement park the ratio of boys to girls was 5 to 4. If there were 200 girls at the amusement park, how many boys were there?

Solution We begin by making a ratio box.

	Ratio	Actual Count
Boys	5	B
Girls	4	200

We use the numbers in the ratio box to write a proportion. Then we solve the proportion and answer the question.

	Ratio	Actual Count
Boys	5	B
Girls	4	200

$$\frac{5}{4} = \frac{B}{200}$$
$$4B = 1000$$
$$B = 250$$

There were **250 boys** at the amusement park.

LESSON PRACTICE

Practice set Solve each of these ratio word problems. Begin by making a ratio box.

a. The girl-boy ratio was 9 to 7. If 63 girls attended, how many boys attended?

b. The ratio of sparrows to blue jays in the yard was 5 to 3. If there were 15 blue jays in the yard, how many sparrows were in the yard?

c. The ratio of tagged fish to untagged fish was 2 to 9. Ninety fish were tagged. How many fish were untagged?

d. Calculate the ratio of males to females in your family. Then calculate the ratio of females to males.

MIXED PRACTICE

Problem set

1. (12, 28) Thomas Jefferson died on the fiftieth anniversary of the signing of the Declaration of Independence. He was born in 1743. The Declaration of Independence was signed in 1776. How many years did Thomas Jefferson live?

2. (28) The heights of the five basketball players are 190 cm, 195 cm, 197 cm, 201 cm, and 203 cm. What is the average height of the players to the nearest centimeter?

3. (54) Use a ratio box to solve this problem. The ratio of winners to losers was 5 to 4. If there were 1200 winners, how many losers were there?

4. (53) What is the cost of 2.6 pounds of cheese at $1.75 per pound?

5. (6, 27) What is the quotient when the least common multiple of 4 and 6 is divided by the greatest common factor of 4 and 6?

6. (22) Diagram this statement. Then answer the questions that follow.

Eighty percent of the 80 trees were infested.

(a) How many trees were infested?

(b) How many trees were not infested?

7. (51) (a) Write 405,000 in scientific notation.

(b) Write 0.04×10^5 in standard form.

8. (47) Find each missing exponent:

(a) $10^6 \cdot 10^2 = 10^{\square}$ (b) $10^6 \div 10^2 = 10^{\square}$

9. Use unit multipliers to perform the following conversions:
(50)
(a) 5280 feet to yards (3 ft = 1 yd)

(b) 300 cm to mm (1 cm = 10 mm)

10. Write 3.1415926 as a decimal number rounded to four decimal places.
(33)

11. Find the number of degrees in the central angle of each sector of the circle shown.
(Inv. 5)

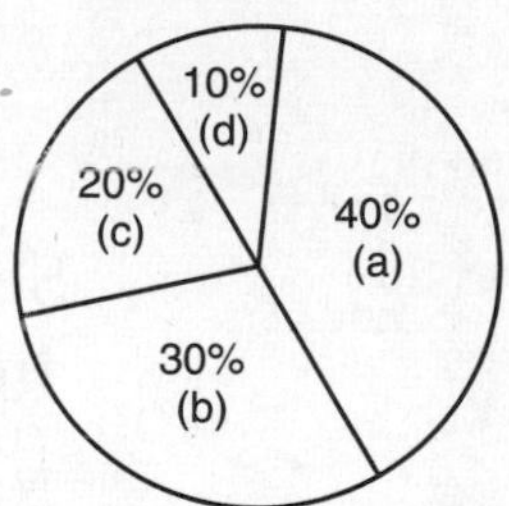

12. A train is traveling at a steady speed of 60 miles per hour.
(53)
(a) How far will the train travel in four hours?

(b) How long will it take the train to travel 300 miles?

13. Which is equivalent to $\frac{2^6}{2^2}$?
(20)

A. 2^3 B. 2^4 C. 1^3 D. 3

Refer to the figure below to answer problems 14 and 15. Dimensions are in centimeters. All angles are right angles.

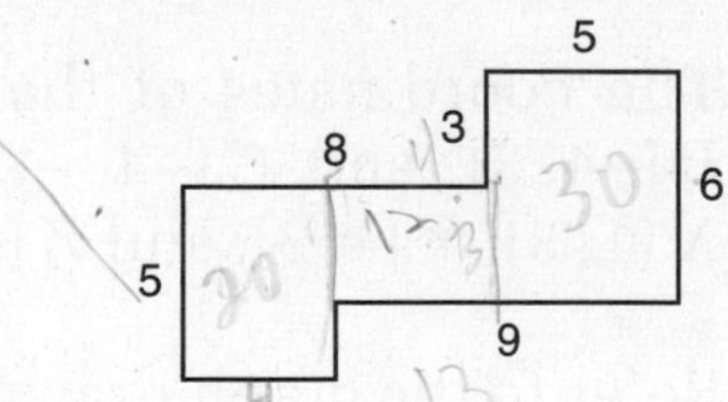

14. What is the perimeter of the figure?
(19)

15. What is the area of the figure?
(37)

16. Name each property illustrated:
(2, 41)
(a) $\frac{1}{2} + 0 = \frac{1}{2}$

(b) $5(6 + 7) = 30 + 35$

(c) $(5 + 6) + 4 = 5 + (6 + 4)$

17. Draw a square with sides 0.5 inch long.
(34, 35)
(a) What is the perimeter of the square?

(b) What is the area of the square?

18. (Inv. 4) The box-and-whisker plot below was created from Terrence's math test scores (number of correct out of 20). Do you think that Terrence's mean (average) test score is likely to be above, at, or below his median score? Explain your answer.

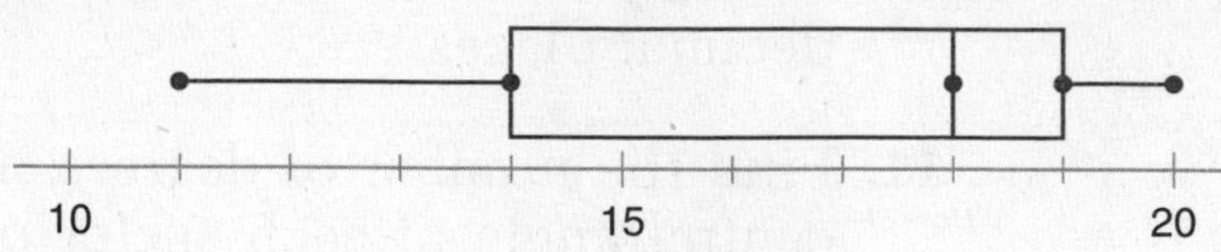

Solve:

19. (35) $6.2 = x + 4.1$

20. (35) $1.2 = y - 0.21$

21. (39) $\frac{24}{r} = \frac{36}{27}$

22. (35) $\frac{w}{0.16} = 6.25$

Simplify:

23. (20) $11^2 + 1^3 - \sqrt{121}$

24. (52) $24 - 4 \times 5 \div 2 + 5$

25. (35) $\frac{(2.5)^2}{2(2.5)}$

26. (49)

	1 week	5 days	14 hr
+	2 weeks	6 days	10 hr

27. (23, 30) $3\frac{5}{10} + \left(9\frac{1}{2} - 6\frac{2}{3}\right)$

28. (26) $7\frac{1}{3} \cdot \left(6 \div 3\frac{2}{3}\right)$

29. (Inv. 3) The coordinates of the vertices of ΔABC are $A\,(-1, 3)$, $B\,(-4, 3)$, and $C\,(-4, -1)$. The coordinates of ΔXYZ are $X\,(1, 3)$, $Y\,(4, 3)$, and $Z\,(4, -1)$. Graph ΔABC and ΔXYZ.

30. (18) Refer to the graph drawn in problem 29 to answer (a)–(c).

(a) Are ΔABC and ΔXYZ similar?

(b) Are ΔABC and ΔXYZ congruent?

(c) Which angle in ΔABC corresponds to $\angle Z$ in ΔXYZ?

LESSON 55 Average, Part 2

WARM-UP

Facts Practice: + − × ÷ Decimals (Test J)

Mental Math:

a. 20 × \$0.25

b. 0.375×10^2

c. $2x - 5 = 75$

d. Convert 3000 m to km.

e. $\left(\frac{2}{3}\right)^2$

f. $\frac{3}{4}$ of 100

g. At 30 pages an hour, how many pages can Mike read in $2\frac{1}{2}$ hours?

Problem Solving:

Copy this problem and fill in the missing digits:

```
    _3_
 ×   __
 -------
   3___
   _3_
 -------
   9__9
```

NEW CONCEPT

If we know the average of a group of numbers and how many numbers are in the group, we can determine the sum of the numbers.

Example 1 The average of three numbers is 17. What is their sum?

Solution We are not told what the numbers are, only their average. Each of these sets of three numbers has an average of 17:

$$\frac{16 + 17 + 18}{3} = \frac{51}{3} = 17$$

$$\frac{10 + 11 + 30}{3} = \frac{51}{3} = 17$$

$$\frac{1 + 1 + 49}{3} = \frac{51}{3} = 17$$

Notice that for each set the sum of the three numbers is 51. Since average tells us what the numbers would be if they were "equalized," the sum of the three numbers is the same as if each of the numbers were 17.

$$17 + 17 + 17 = \mathbf{51}$$

Thus, the number of numbers times their average equals the sum of the numbers.

Example 2 The average of four numbers is 25. If three of the numbers are 16, 26, and 30, what is the fourth number?

Solution If the average of four numbers is 25, their sum is 100.

$$4 \times 25 = 100$$

We are given three of the numbers. The sum of these three numbers plus the fourth number, n, must equal 100.

$$16 + 26 + 30 + n = 100$$

The sum of the first three numbers is 72. Since the sum of the four numbers must be 100, the fourth number is **28.**

$$16 + 26 + 30 + (28) = 100$$

Example 3 After four tests Annette's average score was 89. What score does Annette need on her fifth test to bring her average up to 90?

Solution Although we do not know the specific scores on the first four tests, the total is the same as if each of those scores were 89. Thus the total after four tests is

$$4 \times 89 = 356$$

The total of her first four scores is 356. However, to have an average of 90 after five tests, she needs a five-test total of 450.

$$5 \times 90 = 450$$

Therefore she needs to raise her total from 356 to 450 on the fifth test. To do this, she needs to score **94.**

$$\begin{array}{rl} 356 & \text{four-test total} \\ +\ \ 94 & \text{fifth test} \\ \hline 450 & \text{five-test total} \end{array}$$

LESSON PRACTICE

Practice set

a. Ralph scored an average of 18 points in each of his first five games. Altogether, how many points did Ralph score in his first five games?

b. The average of four numbers is 45. If three of the numbers are 24, 36, and 52, what is the fourth number?

c. After five tests, Tisha's average score was 91. After six tests, her average score was 89. What was her score on the sixth test?

MIXED PRACTICE

Problem set

1. (54) Use a ratio box to solve this problem. The ratio of sailboats to rowboats in the bay was 7 to 4. If there were 56 sailboats in the bay, how many rowboats were there?

2. (55) The average of four numbers is 85. If three of the numbers are 76, 78, and 81, what is the fourth number?

3. (46) A one-quart container of oil costs 89¢. A case of 12 one-quart containers costs $8.64. How much is saved per container by buying the oil by the case?

4. (8) Segment *BC* is how much longer than segment *AB*?

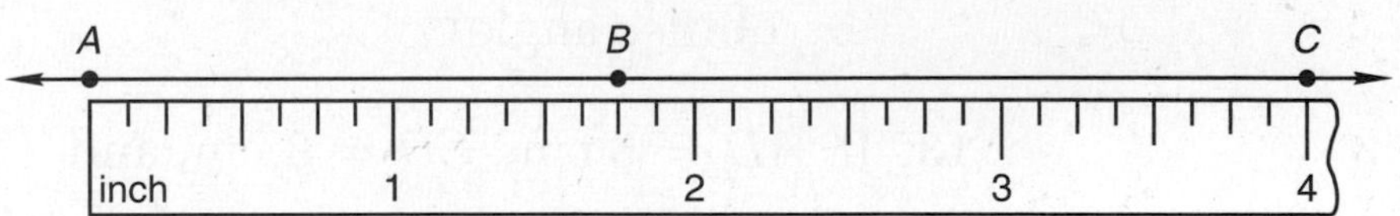

5. (22) Diagram this statement. Then answer the questions that follow.

Three tenths of the 30 wolves howled at the moon.

(a) How many wolves howled at the moon?

(b) What percent of the wolves howled at the moon?

6. (51) (a) Write 675 million in scientific notation.

(b) Write 1.86×10^5 in standard form.

7. (47) Find each missing exponent:

(a) $10^8 \cdot 10^2 = 10^{\square}$ (b) $10^8 \div 10^2 = 10^{\square}$

8. Use unit multipliers to perform the following conversions:
(50) (a) 24 feet to inches
(b) 500 millimeters to centimeters

9. Use digits and other symbols to write "The product of
(31) two hundredths and twenty-five thousandths is five ten-thousandths."

10. What is the total price of a \$3.25 sandwich and a \$1.10
(46) drink including 7% sales tax?

11. Complete the table.
(48)

FRACTION	DECIMAL	PERCENT
$\frac{1}{5}$	(a)	(b)
(c)	0.1	(d)
(e)	(f)	75%

Refer to the figure at right to answer problems 12 and 13.

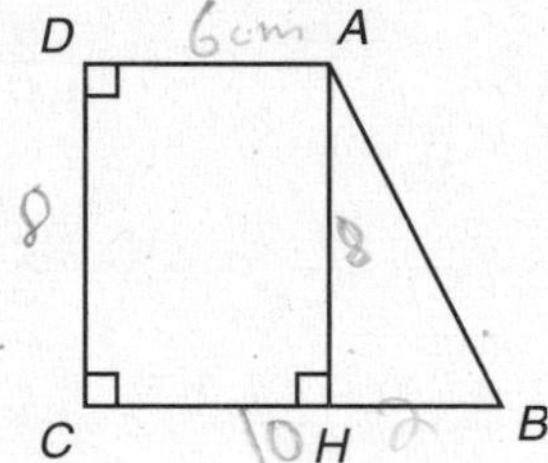

12. (a) Which segment is parallel to
(7) $\overline{BC}$?

(b) Which two segments are perpendicular to $\overline{BC}$?

(c) Angle *ABC* is an acute angle. Which angle is an obtuse angle?

13. If *AD* = 6 cm, *CD* = 8 cm, and *CB* = 10 cm, then
(37) (a) what is the area of rectangle *AHCD?*
(b) what is the area of triangle *ABH?*
(c) what is the area of figure *ABCD?*

14. Donato is 6 feet 2 inches tall. Bob is 68 inches tall. Donato
(28) is how many inches taller than Bob?

15. Monte swam 5 laps in 4 minutes.
(53) (a) Write two forms of this rate.
(b) How many laps could Monte swim in 20 minutes at this rate?
(c) How long would it take for Monte to swim 20 laps at this rate?

16. Show two ways to evaluate $b(a + b)$ for $a = \frac{1}{4}$ and $b = \frac{1}{2}$.
(41)

Solve:

17. $\frac{30}{70} = \frac{21}{x}$
(39)

18. $\frac{1000}{w} = 2.5$
(45)

Estimate each answer to the nearest whole number. Then perform the calculation.

19. $2\frac{5}{12} + 6\frac{5}{6} + 4\frac{7}{8}$
(30)

20. $6 - \left(7\frac{1}{3} - 4\frac{4}{5}\right)$
(23, 30)

Simplify:

21. $10 \text{ yd} \cdot \frac{36 \text{ in.}}{1 \text{ yd}}$
(50)

22.
(49)

$$\begin{array}{r} 8 \text{ yd } 2 \text{ ft } 7 \text{ in.} \\ + \quad 5 \text{ in.} \\ \hline \end{array}$$

23. $12^2 - 4^3 - 2^4 - \sqrt{144}$
(20)

24. $50 + 30 \div 5 \cdot 2 - 6$
(52)

25. $6\frac{2}{3} \cdot 5\frac{1}{4} \cdot 2\frac{1}{10}$
(26)

26. $3\frac{1}{3} \div 3 \div 2\frac{1}{2}$
(26)

27. $3.47 + (6 - 1.359)$
(35)

28. $(0.6)(0.28)(0.01)$
(35)

29. $\$1.50 \div 0.075$
(45)

30. This quadrilateral is a rectangle. Find the measures of angles *a*, *b*, and *c*.
(40)

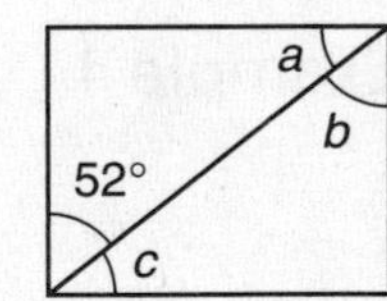

LESSON

56 Subtracting Mixed Measures

WARM-UP

Facts Practice: Fraction-Decimal-Percent Equivalents (Test L)

Mental Math:

a. 30×2.5 **b.** $0.25 \div 10$

c. $3x + 4 = 40$ **d.** Convert 0.5 m to cm.

e. $25^2 - 15^2$ **f.** $\frac{7}{10}$ of \$50.00

g. Square 9, − 1, ÷ 2, − 4, $\sqrt{\ }$, × 3, + 2, ÷ 5, $\sqrt{\ }$, − 5.

Problem Solving:

A palindrome is a word or number that reads the same forward and backward, such as "mom" and "121." How many three-digit numbers are palindromes?

NEW CONCEPT

We have practiced adding mixed measures. In this lesson we will learn to subtract them. When subtracting mixed measures, it may be necessary to convert units.

Example 1 Subtract:

5 days	10 hr	15 min
− 1 day	15 hr	40 min

Solution Before we can subtract minutes, we must convert 1 hour to 60 minutes. We combine 60 minutes and 15 minutes, making 75 minutes. Then we can subtract.

	9 ⟶ (60 min)	
5 days	~~10~~ hr	15 min
− 1 day	15 hr	40 min

⟶

	9	75
5 days	~~10~~ hr	~~15~~ min
− 1 day	15 hr	40 min
		35 min

Next we convert 1 day to 24 hours and complete the subtraction.

⟶

4 ⟶ (24 hr)	9	75
~~5~~ days	~~10~~ hr	~~15~~ min
− 1 day	15 hr	40 min
		35 min

⟶

4	33 ~~9~~	75
~~5~~ days	~~10~~ hr	~~15~~ min
− 1 day	15 hr	40 min
3 days	**18 hr**	**35 min**

Example 2 Subtract: 4 yd 3 in. − 2 yd 1 ft 8 in.

Solution We carefully align the numbers with like units. We convert 1 yard to 3 feet.

```
 3 ⌒→ (3 ft)
 4̸ yd       3 in.
− 2 yd 1 ft 8 in.
```

Next, we convert 1 foot to 12 inches. We combine 12 inches and 3 inches, making 15 inches. Then we can subtract.

```
  3     2    15
  4̸ yd  3̸ ft  3̸ in.
− 2 yd  1 ft  8 in.
  1 yd  1 ft  7 in.
```

LESSON PRACTICE

Practice set* Subtract:

a.
```
   3 hr            3 s
 − 1 hr 15 min   55 s
```

b.
```
   8 yd 1 ft 5 in.
 − 3 yd 2 ft 7 in.
```

c. 2 days 3 hr 30 min − 1 day 8 hr 45 min

MIXED PRACTICE

Problem set

1. (31, 35) Three hundred twenty-nine ten-thousandths is how much greater than thirty-two thousandths? Use words to write the answer.

2. (54) Use a ratio box to solve this problem. The ratio of the length to the width of the rectangle is 4 to 3. If the length of the rectangle is 12 feet,

(a) what is its width?

(b) what is its perimeter?

3. (28) The parking lot charges $2 for the first hour plus 50¢ for each additional half hour or part thereof. What is the total charge for parking a car in the lot from 11:30 a.m. until 2:15 p.m.?

4. (55) After four tests Trudy's average score was 85. If her score is 90 on the fifth test, what will be her average for all five tests?

5. (46) Twelve ounces of Brand X costs $1.50. Sixteen ounces of Brand Y costs $1.92. Find the unit price for each brand. Which brand is the better buy?

6. (36, 48) Five eighths of the rocks in the box were metamorphic. The rest were igneous.

(a) What fraction of the rocks were igneous?

(b) What was the ratio of igneous to metamorphic rocks?

(c) What percent of the rocks were metamorphic?

7. (40) Refer to the figure at right to answer (a) and (b).

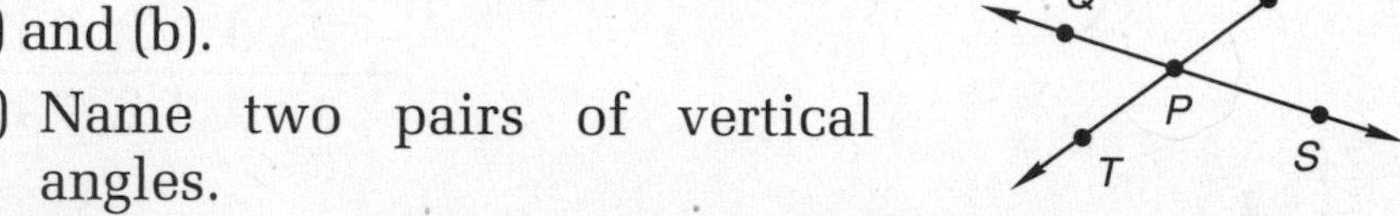

(a) Name two pairs of vertical angles.

(b) Name two angles that are supplemental to $\angle RPS$.

8. (51) (a) Write six hundred ten thousand in scientific notation.

(b) Write 1.5×10^4 in standard form.

9. (50) Use unit multipliers to perform the following conversions:

(a) 216 hours to days

(b) 5 minutes to seconds

10. (43, 48) (a) Write $\frac{1}{6}$ as a decimal number rounded to the nearest hundredth.

(b) Write $\frac{1}{6}$ as a percent.

11. (51) How many pennies equal one million dollars? Write the answer in scientific notation.

12. (51) Compare: 11 million ◯ 1.1×10^6

13. (27) Which even two-digit number is a common multiple of 5 and 7?

14. (32, 54) There are 100° on the Celsius scale from the freezing temperature to the boiling temperature of water. There are 180° on the Fahrenheit scale between these temperatures. So a change in temperature of 10° on the Celsius scale is equivalent to a change of how many degrees on the Fahrenheit scale?

Refer to the figure below to answer problems 15 and 16. Dimensions are in millimeters. All angles are right angles.

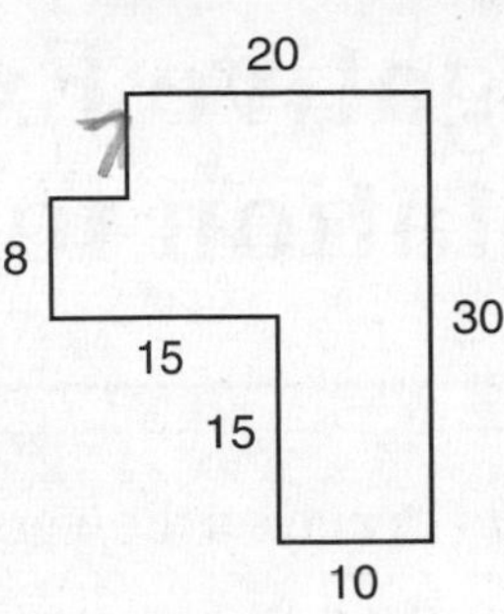

15. What is the perimeter of the figure?
(19)

16. What is the area of the figure?
(37)

Solve:

17. $\frac{3}{2.5} = \frac{48}{c}$
(39)

18. $k - 0.75 = 0.75$
(35)

Simplify:

19. $15^2 - 5^3 - \sqrt{100}$
(20)

20. $6 + 12 \div 3 \cdot 2 - 3 \cdot 4$
(52)

21.
(49)
$$\begin{array}{r} 5 \text{ yd } 2 \text{ ft } 3 \text{ in.} \\ +\ 2 \text{ yd } 2 \text{ ft } 9 \text{ in.} \\ \hline \end{array}$$

22.
(56)
$$\begin{array}{r} 5 \text{ yd } 2 \text{ ft } 3 \text{ in.} \\ -\ 2 \text{ yd } 2 \text{ ft } 9 \text{ in.} \\ \hline \end{array}$$

23. $\frac{88 \text{ km}}{1 \text{ hr}} \cdot 4 \text{ hr}$
(53)

24. $2\frac{3}{4} + \left(5\frac{1}{6} - 1\frac{1}{4}\right)$
(23, 30)

25. $3\frac{3}{4} \cdot 2\frac{1}{2} \div 3\frac{1}{8}$
(26)

26. $3\frac{3}{4} \div 2\frac{1}{2} \cdot 3\frac{1}{8}$
(26)

27. Describe how to find the 99th number in this sequence:
(2)

1, 4, 9, 16, 25, ...

28. Use a protractor and a straightedge to draw a triangle that has a right angle and a 30° angle. Then measure the shortest and longest sides of the triangle to the nearest millimeter. What is the relationship of the two measurements?
(17)

29. If the diameter of a wheel is 0.5 meter, then the radius of the wheel is how many centimeters?
(Inv. 2, 32)

30. Graph points A (0, 0) and B (4, 2). By inspection, find the point halfway between points A and B. What are the coordinates of this "halfway" point?
(Inv. 3)

LESSON 57

Negative Exponents • Scientific Notation for Small Numbers

WARM-UP

Facts Practice: Powers and Roots (Test K)

Mental Math:

a. 40×3.2

b. 4.2×10^3

c. $\frac{n}{20} = \frac{7}{5}$

d. Convert 500 mL to L.

e. $15^2 - 5^3$

f. $\frac{2}{5}$ of \$25.00

g. Start with the number of pounds in a ton, ÷ 2, − 1, ÷ 9, − 11, $\sqrt{\ }$, ÷ 2, ÷ 2.

Problem Solving:

Along the road on which Jesse lives are telephone poles spaced 100 feet apart. If Jesse runs from the first pole to the seventh pole, how many feet does he run? Draw a picture that illustrates the problem.

NEW CONCEPTS

Negative exponents Cantara multiplied 0.000001 by 0.000001 on her scientific calculator. After she pressed [=] the display read

1. × 10⁻¹²

The calculator displayed the product in scientific notation. Notice that the exponent is a negative number. So

$$1 \times 10^{-12} = 0.000000000001$$

Studying the pattern below may help us understand the meaning of a negative exponent.

$$\frac{10^6}{10^3} = \frac{\overset{1}{\cancel{10}} \cdot \overset{1}{\cancel{10}} \cdot \overset{1}{\cancel{10}} \cdot 10 \cdot 10 \cdot 10}{\underset{1}{\cancel{10}} \cdot \underset{1}{\cancel{10}} \cdot \underset{1}{\cancel{10}}} = 10^3 = 1000$$

$$\frac{10^5}{10^3} = \frac{\overset{1}{\cancel{10}} \cdot \overset{1}{\cancel{10}} \cdot \overset{1}{\cancel{10}} \cdot 10 \cdot 10}{\underset{1}{\cancel{10}} \cdot \underset{1}{\cancel{10}} \cdot \underset{1}{\cancel{10}}} = 10^2 = 100$$

$$\frac{10^4}{10^3} = \frac{\overset{1}{\cancel{10}} \cdot \overset{1}{\cancel{10}} \cdot \overset{1}{\cancel{10}} \cdot 10}{\underset{1}{\cancel{10}} \cdot \underset{1}{\cancel{10}} \cdot \underset{1}{\cancel{10}}} = 10^1 = 10$$

Notice that the exponents in the third column can be found by subtracting the exponents in the first column (the exponent of the dividend minus the exponent of the divisor). Now we will continue the pattern.

$$\frac{10^3}{10^3} = \frac{\overset{1}{\cancel{10}} \cdot \overset{1}{\cancel{10}} \cdot \overset{1}{\cancel{10}}}{\underset{1}{\cancel{10}} \cdot \underset{1}{\cancel{10}} \cdot \underset{1}{\cancel{10}}} = 10^0 = 1$$

$$\frac{10^2}{10^3} = \frac{\overset{1}{\cancel{10}} \cdot \overset{1}{\cancel{10}}}{\underset{1}{\cancel{10}} \cdot \underset{1}{\cancel{10}} \cdot 10} = 10^{-1} = \frac{1}{10} = \frac{1}{10^1}$$

$$\frac{10^1}{10^3} = \frac{\overset{1}{\cancel{10}}}{\underset{1}{\cancel{10}} \cdot 10 \cdot 10} = 10^{-2} = \frac{1}{100} = \frac{1}{10^2}$$

Notice especially these results:

$$10^0 = 1$$

$$10^{-1} = \frac{1}{10^1}$$

$$10^{-2} = \frac{1}{10^2}$$

The pattern suggests two facts about exponents, which we express algebraically below.

> If a number a is not zero, then
>
> $$a^0 = 1$$
>
> $$a^{-n} = \frac{1}{a^n}$$

Example 1 Simplify:

(a) 2^0 (b) 3^{-2} (c) 10^{-3}

Solution (a) The exponent is zero and the base is not zero, so 2^0 equals **1.**

(b) We rewrite the expression using the reciprocal of the base with a positive exponent. Then we simplify.

$$3^{-2} = \frac{1}{3^2} = \mathbf{\frac{1}{9}}$$

(c) Again we rewrite the expression with the reciprocal of the base and a positive exponent.

$$10^{-3} = \frac{1}{10^3} = \mathbf{\frac{1}{1000}} \text{ (or } \mathbf{0.001}\text{)}$$

Scientific notation for small numbers

As we saw at the beginning of this lesson, negative exponents can be used to express small numbers in scientific notation. For instance, an inch is 2.54×10^{-2} meters. If we multiply 2.54 by 10^{-2}, the product is 0.0254.

$$2.54 \times 10^{-2} = 2.54 \times \frac{1}{10^2} = 0.0254$$

Notice the product, 0.0254, has the same digits as 2.54 but with the decimal point shifted two places to the left and with zeros used for placeholders. The two-place decimal shift to the left is indicated by the exponent –2. This is similar to the method we have used to change scientific notation to standard form. Note the sign of the exponent. If the exponent is a *positive number,* we shift the decimal point *to the right* to express the number in standard form. In the number

$$6.32 \times 10^7$$

the exponent is *positive* seven, so we shift the decimal point seven places *to the right.*

$$\underbrace{63200000.}_{\text{7 places}} \longrightarrow 63{,}200{,}000$$

If the exponent is a *negative number,* we shift the decimal point *to the left* to write the number in standard form. In the number

$$6.32 \times 10^{-7}$$

the exponent is *negative* seven, so we shift the decimal point seven places *to the left.*

$$\underbrace{.000000632}_{\text{7 places}} \longrightarrow 0.000000632$$

In either case, we use zeros as placeholders.

Example 2 Write 4.63×10^{-8} in standard notation.

Solution The negative exponent indicates that the decimal point is eight places to the left when the number is written in standard form. We shift the decimal point and insert zeros as placeholders.

$$\underbrace{.0000000463}_{\text{8 places}} \longrightarrow \mathbf{0.0000000463}$$

Example 3 Write 0.0000033 in scientific notation.

Solution We place the decimal point to the right of the first digit that is not a zero.

$$\underbrace{000000}3.3$$

6 places

In standard form the decimal point is six places to the left of where we have placed it. So we write

$$\mathbf{3.3 \times 10^{-6}}$$

Example 4 Compare: zero ◯ 1×10^{-3}

Solution The expression 1×10^{-3} equals 0.001. Although this number is less than 1, it is still positive, so it is greater than zero.

$$\mathbf{zero < 1 \times 10^{-3}}$$

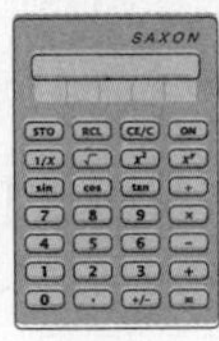

Very small numbers may exceed the display capabilities of a calculator. One millionth of one millionth is more than zero, but it is a very small number. On a calculator we enter

The product, one trillionth, contains more digits than can be displayed by many calculators. Instead of displaying one trillionth in standard form, the calculator displays the number in a modified form of scientific notation such as

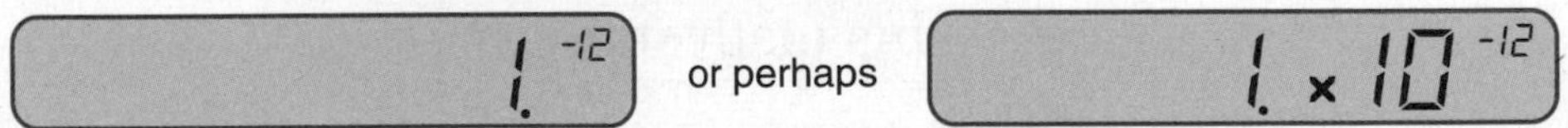

LESSON PRACTICE

Practice set* Simplify:

a. 5^{-2} b. 3^0 c. 10^{-4}

Write each number in scientific notation:

d. 0.00000025 e. 0.000000001 f. 0.000105

Write each number in standard form:

g. 4.5×10^{-7} h. 1×10^{-3} i. 1.25×10^{-5}

Compare:

j. 1×10^{-3} ◯ 1×10^2 k. 2.5×10^{-2} ◯ 2.5×10^{-3}

MIXED PRACTICE

Problem set

1. (54) Use a ratio box to solve this problem. The ratio of walkers to riders was 5 to 3. If 315 were walkers, how many were riders?

2. (55) After five tests Allison's average score was 88. After six tests her average score had increased to 90. What was her score on the sixth test?

3. (28) When Richard rented a car, he paid \$34.95 per day plus 18¢ per mile. If he rented the car for 2 days and drove 300 miles, how much did he pay?

4. (16) If lemonade costs \$0.52 per quart, what is the cost per pint?

5. (22) Diagram this statement. Then answer the questions that follow.

Tyrone finished his math studies in four fifths of an hour.

(a) How many minutes did it take Tyrone to finish his math studies?

(b) What percent of an hour did it take Tyrone to finish his math studies?

6. (51, 57) Write each number in scientific notation:

(a) 186,000 (b) 0.00004

7. (51, 57) Write each number in standard form:

(a) 3.25×10^1 (b) 1.5×10^{-6}

8. (57) Simplify:

(a) 2^{-3} (b) 5^0 (c) 10^{-2}

9. (50) Use a unit multiplier to convert 2000 milliliters to liters.

10. (21, 36) What is the probability of rolling a composite number on one toss of a dot cube?

11. (46) The tickets for two dozen children to enter the amusement park cost \$330. What was the price per ticket?

12. (Inv. 5) The frequency table below shows how Lanora scored on her last 25 quizzes. Create a histogram that illustrates the data in the frequency table.

Lanora's Quiz Scores

% Correct	Tally	Frequency
91–100	𝍸 II	7
81–90	𝍸 IIII	9
71–80	𝍸 I	6
61–70	III	3

13. (57) Compare:

(a) $2.5 \times 10^{-2} \bigcirc 2.5 \div 10^{2}$

(b) one millionth $\bigcirc$ 1×10^{-6}

(c) $3^{0} \bigcirc 2^{0}$

Refer to the figure below to answer problems 14 and 15. Dimensions are in yards. All angles are right angles.

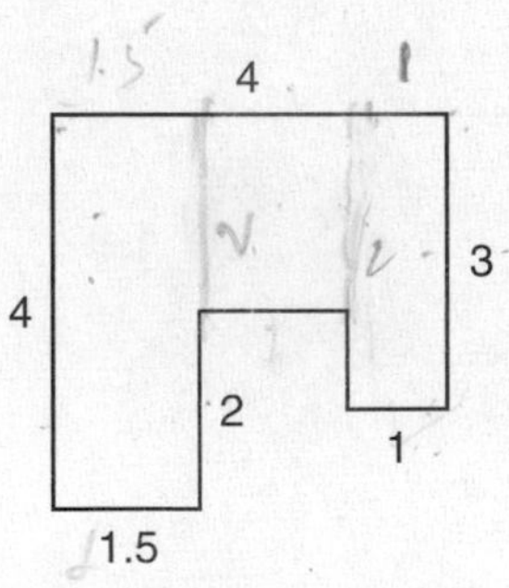

14. (19) What is the perimeter of the figure?

15. (37) What is the area of the figure?

16. (41) Evaluate: $4ac$ if $a = 5$ and $c = 0.5$

17. (33) Estimate the quotient: \$19.89 ÷ 3.987

18. (41) In the following equation, y is 5 more than the product of 3 and x. Find y when x is 12.

$$y = 3x + 5$$

Simplify:

19. $20^2 + 10^3 - \sqrt{36}$ (20)

20. $48 \div 12 \div 2 + 2(3)$ (52)

21. (56)
$$\begin{array}{r} 3 \text{ yd } 2 \text{ ft } 1 \text{ in.} \\ - \ 1 \text{ yd } 2 \text{ ft } 3 \text{ in.} \\ \hline \end{array}$$

22. (49)
$$\begin{array}{r} 4 \text{ gal } 3 \text{ qt } 1 \text{ pt } 6 \text{ oz} \\ + \ 1 \text{ gal } 2 \text{ qt } 1 \text{ pt } 5 \text{ oz} \\ \hline \end{array}$$

23. $48 \text{ oz} \cdot \dfrac{1 \text{ pt}}{16 \text{ oz}}$ (50)

24. $5\frac{1}{3} \cdot \left(7 \div 1\frac{3}{4}\right)$ (26)

25. $5\frac{1}{6} + 3\frac{5}{8} + 2\frac{7}{12}$ (30)

26. $\dfrac{1}{20} - \dfrac{1}{36}$ (30)

27. $(4.6 \times 10^{-2}) + 0.46$ (57)

28. $10 - (2.3 - 0.575)$ (35)

29. $0.24 \times 0.15 \times 0.05$ (35)

30. $10 \div (0.14 \div 70)$ (45)

LESSON

58 Line Symmetry • Functions, Part 1

WARM-UP

Facts Practice: Powers and Roots (Test K)

Mental Math:

a. 50×4.3

b. $4.2 \div 10^3$

c. $3x - 5 = 40$

d. Convert 1.5 kg to g.

e. $10^3 \div 10^2$

f. $\frac{2}{3}$ of \$33.00

g. Find the sum, difference, product, and quotient of 1.2 and 0.6.

Problem Solving:

Four friends met at a party and shook one another's hands. How many handshakes were there in all? Draw a diagram to illustrate the problem. (If three partners are available, you may wish to act out the story.)

NEW CONCEPTS

Line symmetry

A two-dimensional figure has **line symmetry** if it can be divided in half so that the halves are mirror images of each other. Line *r* divides this triangle into two mirror images; so the triangle is symmetrical, and line *r* is a **line of symmetry**.

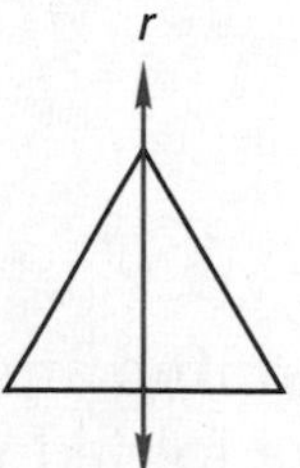

Actually, the regular triangle has three lines of symmetry.

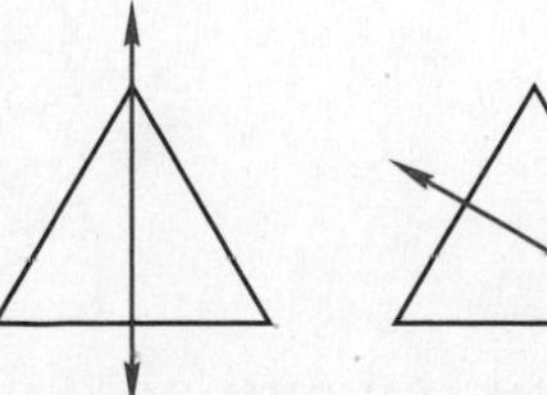

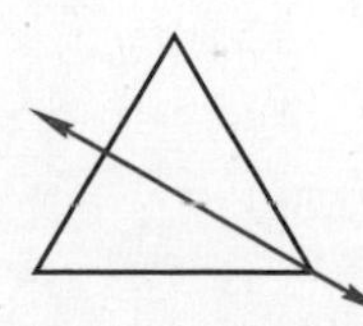

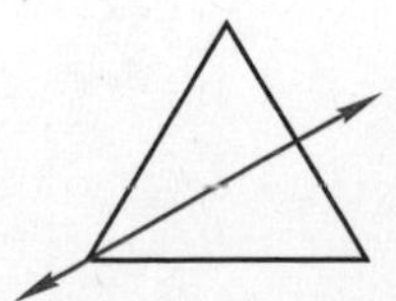

Example 1 Draw a regular quadrilateral and show all lines of symmetry.

Solution A regular quadrilateral is a square. A square has four lines of symmetry.

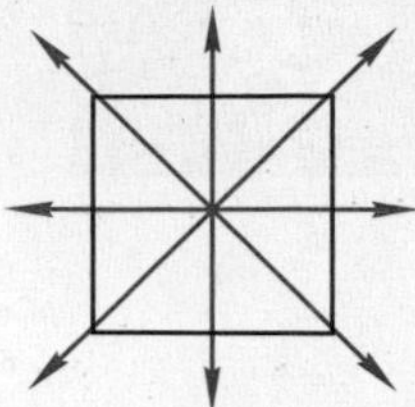

The *y*-axis is a line of symmetry for the figure below. Notice that corresponding points on the two sides of the figure are the same distance from the line of symmetry.

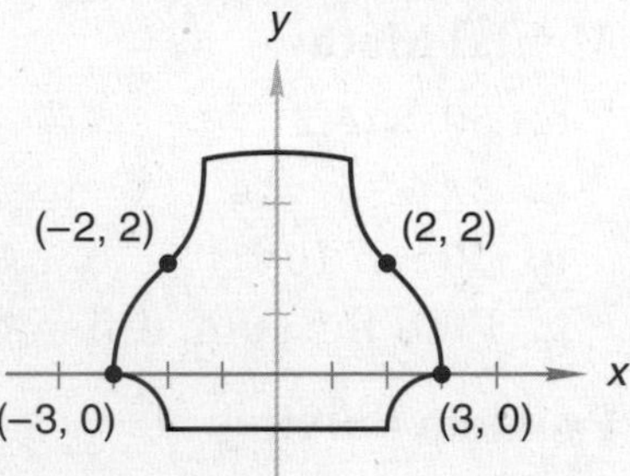

If this figure were folded along the *y*-axis, each point of the figure on one side of the *y*-axis would be folded against its corresponding point on the other side of the *y*-axis.

Activity: *Line Symmetry*

Materials needed:

- Paper and scissors

1. Fold a piece of paper in half.

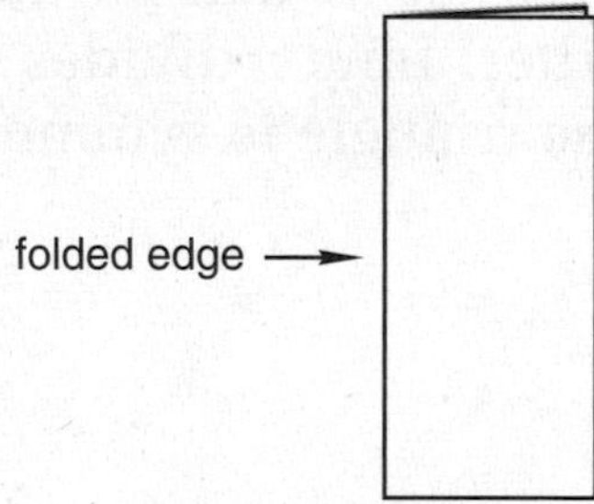

Beginning and ending at the folded edge, cut a pattern out of the folded paper.

Open the cut-out and note its symmetry.

2. Fold a piece of paper twice as shown.

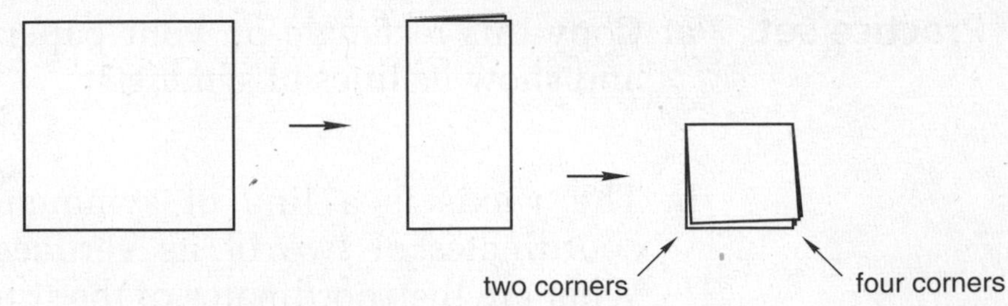

Hold the paper on the corner opposite the "four corners," and cut out a pattern that removes the four corners.

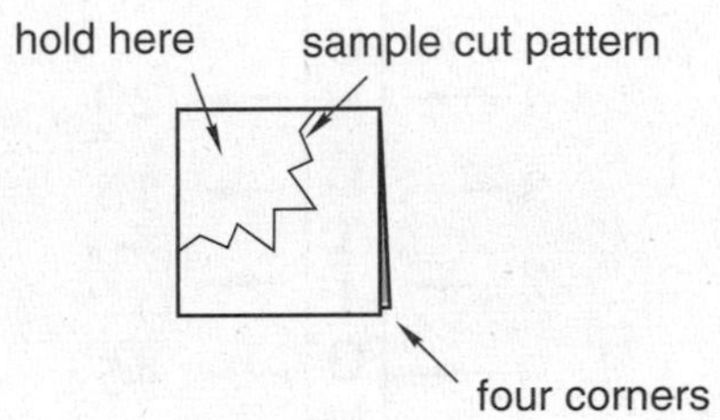

Unfold the cut-out. How many lines of symmetry do you see?

Functions, part 1

A **function** is a set of number pairs that are related by a certain rule. We will study pairs of numbers to determine a rule for the function, and then we will use the rule to find the missing number. Note that for a function there is exactly one "out" number for every "in" number.

Example 2 Find the missing number.

In	FUNCTION	Out
3	→	9
5	→	15
7	→	□
10	→	30

Solution We study each "in-out" number pair to determine the rule for the function. We see that for each complete pair, if the "in" number is multiplied by 3, it equals the "out" number. Thus the rule of the function is "multiply by 3." We use this rule to find the missing number. We multiply 7 by 3 and find that the missing number is **21**.

LESSON PRACTICE

Practice set

a. Copy this rectangle on your paper, and show its lines of symmetry.

b. The *y*-axis is a line of symmetry for a triangle. The coordinates of two of its vertices are (0, 1) and (3, 4). What are the coordinates of the third vertex?

Find the missing number in each diagram:

c.

IN		FUNCTION		OUT
4	→		→	20
3	→		→	15
7	→		→	☐
9	→		→	45

d.

IN		FUNCTION		OUT
0	→		→	4
1	→		→	☐
3	→		→	7
5	→		→	9

MIXED PRACTICE

Problem set

1. (28, 35) It is 1.4 kilometers from Jim's house to his grandmother's house. How far does Jim walk going to and from his grandmother's house once every day for 5 days?

2. (28) The parking lot charges 75¢ for each half hour or part of a half hour. If Edie parks her car in the lot from 10:45 a.m. until 1:05 p.m., how much money will she pay?

3. (41) If the product of the number *n* and 17 is 340, what is the sum of the number *n* and 17?

4. (36) The football team won 3 of its 12 games but lost the rest.

(a) What was the team's win-loss ratio?

(b) What fraction of the games did the team lose?

(c) What percent of the games did the team win?

5. (55) Willis's bowling average after 5 games was 120. In his next 3 games, Willis scored 118, 124, and 142. What was Willis's bowling average after 8 games?

6. (22) Diagram this statement. Then answer the questions that follow.

Three fifths of the 60 questions were multiple-choice.

(a) How many of the 60 questions were multiple-choice?

(b) What percent of the 60 questions were not multiple-choice?

7. (Inv. 2) In the figure at right, the center of the circle is point *O* and *OB* = *CB*. Refer to the figure to answer (a)–(d).

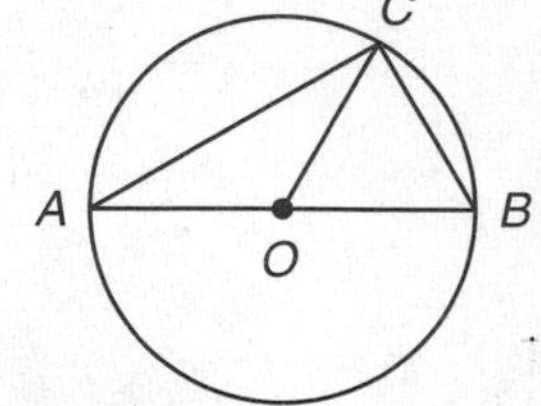

(a) Name three radii.

(b) Name two chords that are not diameters.

(c) Estimate the measure of central angle *BOC*.

(d) Estimate the measure of inscribed angle *BAC*.

8. (51, 57) Write each number in standard form:

(a) 1.5×10^7 (b) 2.5×10^{-4}

(c) 10^{-1}

9. (16) Compare: 20 qt ◯ 5 gal

10. (33, 45) Divide 3.45 by 0.18 and write the answer rounded to the nearest whole number.

11. (2) Find the next three numbers in this sequence:

20, 15, 10, ...

12. (48) Complete the table.

FRACTION	DECIMAL	PERCENT
$\frac{1}{6}$	(a)	(b)
(c)	(d)	16%

13. (58) Find the missing number.

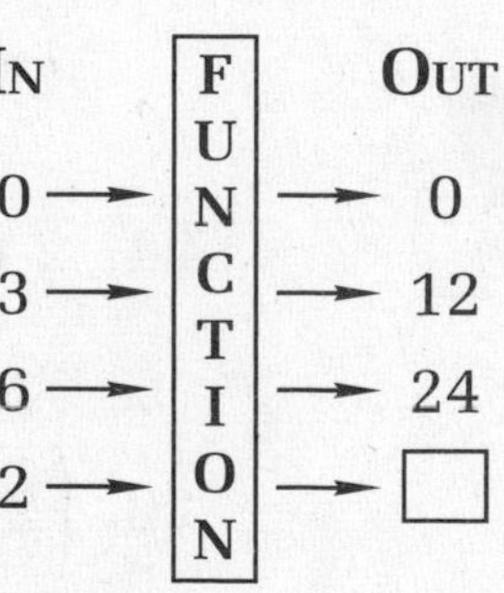

14. (40) In the figure at right, the measure of $\angle D$ is 35° and the measure of $\angle CAB$ is 35°. Find the measure of

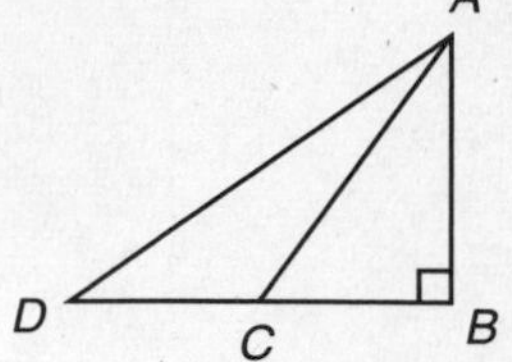

(a) $\angle ACB$.

(b) $\angle ACD$.

(c) $\angle CAD$.

15. (Inv. 3, 58) The y-axis is a line of symmetry for a triangle. The coordinates of two of its vertices are (–3, 2) and (0, 5).

(a) What are the coordinates of the third vertex?

(b) What is the area of the triangle?

16. (58) A regular pentagon has how many lines of symmetry?

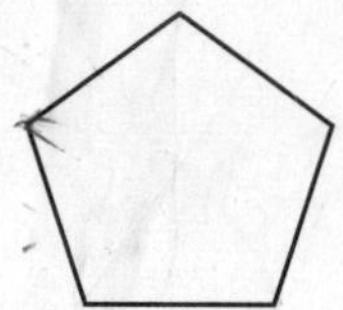

17. (53) (a) Traveling at 60 miles per hour, how long would it take to travel 210 miles?

(b) How long would the same trip take at 70 miles per hour?

Solve:

18. (39) $\frac{1.5}{2} = \frac{7.5}{w}$

19. (35) $1.7 - y = 0.17$

Simplify:

20. (20, 57) $10^3 - 10^2 + 10^1 - 10^0$

21. (52) $6 + 3(2) - 4 - (5 + 3)$

22. (49)

$$\begin{array}{r} 1 \text{ gal } 2 \text{ qt } 1 \text{ pt} \\ + \ 1 \text{ gal } 2 \text{ qt } 1 \text{ pt} \\ \hline \end{array}$$

23. (56)

$$\begin{array}{r} 1 \text{ day } 3 \text{ hr } 15 \text{ min} \\ - \quad 8 \text{ hr } 30 \text{ min} \\ \hline \end{array}$$

24. (50) $2 \text{ mi} \cdot \frac{5280 \text{ ft}}{1 \text{ mi}}$

25. (23, 30) $10 - \left(5\frac{3}{4} - 1\frac{5}{6}\right)$

26. (26, 30) $\left(2\frac{1}{5} + 5\frac{1}{2}\right) \div 2\frac{1}{5}$

27. (26) $3\frac{3}{4} \cdot \left(6 \div 4\frac{1}{2}\right)$

28. Evaluate: $b^2 - 4ac$ if $a = 3.6$, $b = 6$, and $c = 2.5$
(52)

29. (a) Arrange these numbers in order from **greatest** to **least:**
(4, 10)

$$-1, \frac{3}{2}, 2.5, 0, -\frac{1}{2}, 2$$

(b) Which of the numbers in (a) are integers?

30. Lindsey had the following division to perform:
(27)

$$35 \div 2\frac{1}{2}$$

Describe how Lindsey could form an equivalent division problem that would be easier to perform mentally.

LESSON

59 Adding Integers on the Number Line

WARM-UP

Facts Practice: Fraction-Decimal-Percent Equivalents (Test L)

Mental Math:

a. 60×5.4

b. 0.005×10^2

c. $\frac{30}{20} = \frac{3}{t}$

d. Convert 185 cm to m.

e. $2 \cdot 2^3$

f. $\frac{7}{8}$ of \$40.00

g. At \$7.50 an hour, how much money can Shelly earn in 8 hours?

Problem Solving:

A square and a regular pentagon share a common side. If the perimeter of the pentagon is one meter, what is the area of the square in square centimeters?

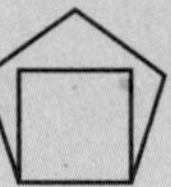

NEW CONCEPT

Recall that **integers** include all the whole numbers and also the opposites of the positive integers (their negatives). All the numbers in this sequence are integers:

$$\ldots, -3, -2, -1, 0, 1, 2, 3, \ldots$$

The dots on this number line mark the integers from –5 through +5:

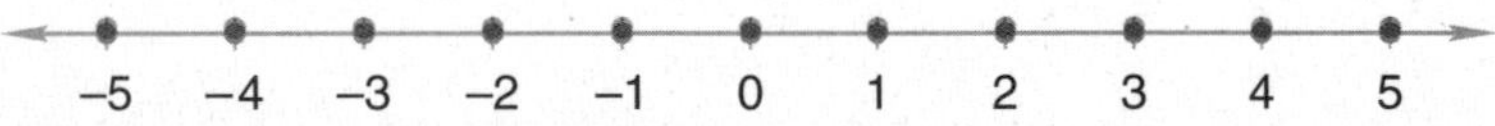

Remember that the numbers between the whole numbers, such as $3\frac{1}{2}$ and 1.3, are not integers.

All numbers on the number line except zero are **signed numbers,** either positive or negative. Zero is neither positive nor negative. Positive and negative numbers have a sign and a value, which is called **absolute value.** The absolute value of a number is its distance from zero.

Numeral	Number	Sign	Absolute Value
+3	Positive three	+	3
–3	Negative three	–	3

The absolute value of both +3 and –3 is 3. Notice on the number line that +3 and –3 are both 3 units from zero. We may use two vertical segments to indicate absolute value.

$|3| = 3$	$|-3| = 3$
The absolute value of 3 equals 3.	The absolute value of –3 equals 3.

Example 1 Simplify: $|3 - 5|$

Solution To find the absolute value of $3 - 5$, we first subtract 5 from 3 and get –2. Then we find the absolute value of –2, which is **2.**

Absolute value can be represented by distance, whereas the sign can be represented by direction. Thus signed numbers are sometimes called **directed numbers** because the sign of the number (+ or –) can be thought of as a direction indicator.

When we add, subtract, multiply, or divide directed numbers, we need to pay attention to the signs as well as the absolute values of the numbers. In this lesson we will practice adding positive and negative numbers.

A number line can be used to illustrate the addition of signed numbers. A positive 3 is indicated by a 3-unit arrow that points to the right. A negative 3 is indicated by a 3-unit arrow that points to the left.

To show the addition of +3 and –3, we begin at zero on the number line and draw the +3 arrow. From its arrowhead we draw the –3 arrow. The sum of +3 and –3 is found at the point on the number line that corresponds to the second arrowhead.

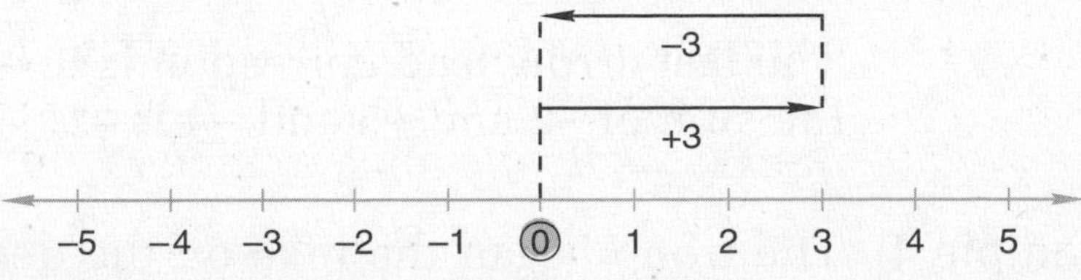

We see that the sum of +3 and –3 is 0. We find the sum of two opposites is always zero.

Example 2 Show each addition problem on a number line:

(a) (−3) + (+5) (b) (−4) + (−2)

Solution (a) We begin at zero and draw an arrow 3 units long that points to the left. From this arrowhead we draw an arrow 5 units long that points to the right. We see that the sum of −3 and +5 is **2.**

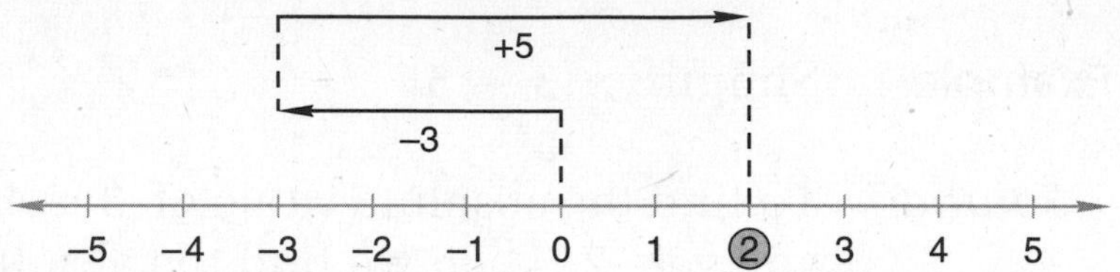

(b) We use arrows to show that the sum of −4 and −2 is **−6.**

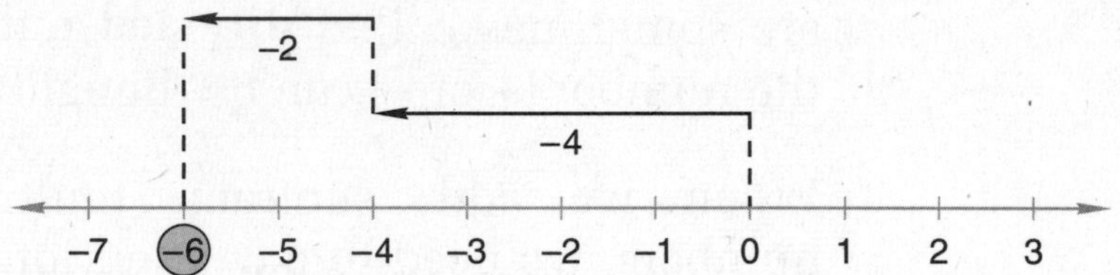

Example 3 Show this addition problem on a number line:

$$(-2) + (+5) + (-4)$$

Solution This time we draw three arrows. We always begin the first arrow at zero. We begin each remaining arrow at the arrowhead of the previous arrow.

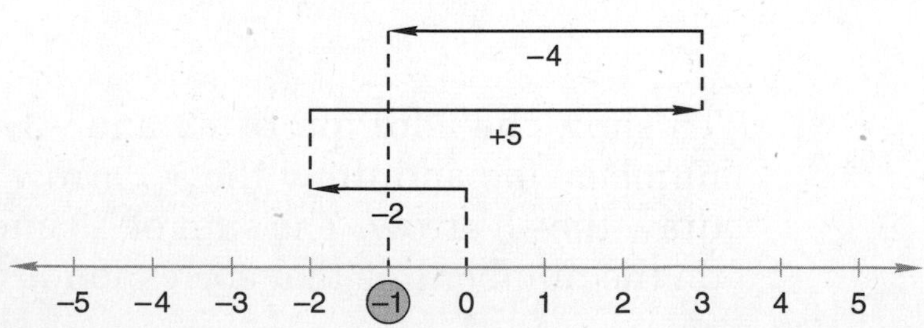

The last arrowhead corresponds to −1 on the number line, so the sum of −2 and +5 and −4 is **−1.**

Example 4 The troop began the hike on the desert floor, 126 feet below sea level. The troop camped for the night on a ridge 2350 feet above sea level. What was the elevation gain from the start of the hike to the campsite?

Solution A number line that is oriented vertically rather than horizontally is more helpful for this problem. The troop climbed 126 feet to reach sea level (zero elevation) and then climbed 2350 feet more to the campsite. We calculate the total elevation gain as shown below.

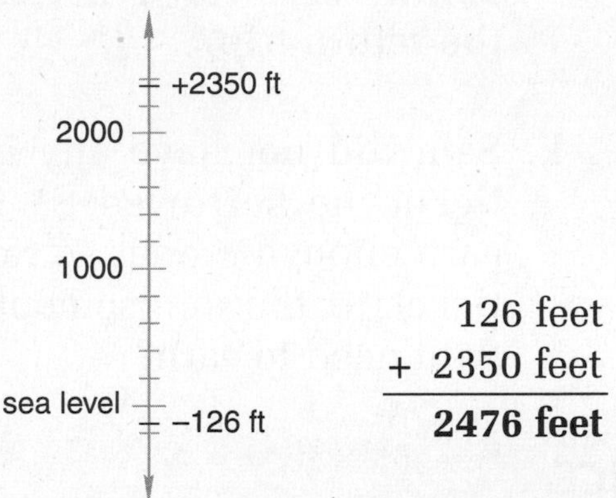

Example 5 Krissie did not have any money. In order to buy a friend's birthday present, Krissie borrowed $5 from her sister. Later Krissie received a check for $25 from her grandmother. After she repays her sister, how much money will Krissie have?

Solution We may use negative numbers to represent debt (borrowed money). After borrowing $5, Krissie had negative five dollars. Then she received $25. We show the addition of these dollar amounts on the number line below.

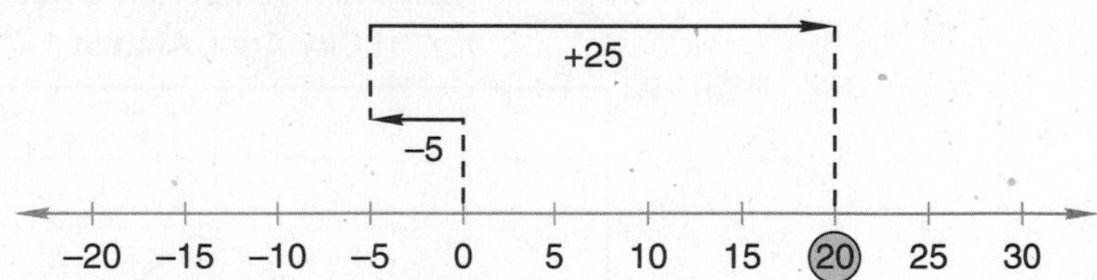

After she repays her sister, Krissie will have **$20.**

LESSON PRACTICE

Practice set Use arrows to show each addition problem on a number line:

a. (−2) + (−3)

b. (+4) + (+2)

c. (−5) + (+2)

d. (+5) + (−2)

e. (−4) + (+4)

f. (−3) + (+6) + (−1)

Simplify:

g. $|-3| + |3|$ **h.** $|3 - 3|$ **i.** $|5 - 3|$

j. On the return trip the troop hiked down the mountain from 4362 ft above sea level to the valley floor 126 ft below sea level. What was the drop in elevation during the return trip?

k. Sam did not have any money. In order to buy a movie ticket, he borrowed $5 from his brother. Sam wants to earn enough money to repay his brother and to buy a $25 ticket to the amusement park. How much money does Sam need to earn?

MIXED PRACTICE

Problem set

1. (28) At the portrait studio, pictures cost $4.25 for an 8-by-10 print. They cost $2.35 for a 5-by-7 print and 60¢ for each wallet-size print. What is the total cost of two 5-by-7 prints and six wallet-size prints?

The double-line graph below compares the daily maximum temperatures for the first seven days of August to the average maximum temperature for the entire month of August. Refer to the graph to answer problems 2 and 3:

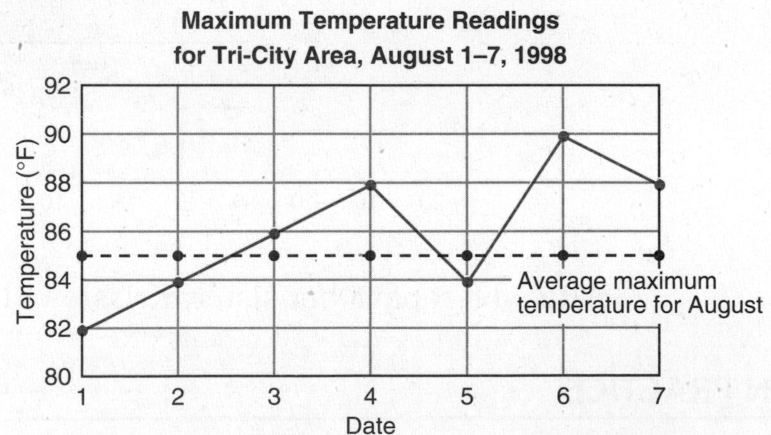

2. (38) The maximum temperature reading on August 6, 1998, was how much greater than the average maximum temperature for the month of August?

3. (28, 38) What was the average maximum temperature for the first seven days of August 1998?

4. (59) On January 1 the temperature at noon was 7°F. By 10 p.m. the temperature had fallen to –9°F. The temperature dropped how many degrees from noon to 10 p.m.?

5. (54) Use a ratio box to solve this problem. The ratio of sonorous to discordant voices in the crowd was 7 to 4. If 56 voices were discordant, how many voices were sonorous?

6. (22) Diagram this statement. Then answer the questions that follow.

The Celts won three fourths of their first 20 games.

(a) How many of their first 20 games did the Celts win?

(b) What percent of their first 20 games did the Celts fail to win?

7. (59) Compare: $|-3| \bigcirc |3|$

8. (51) (a) Write 4,000,000,000,000 in scientific notation.

(b) Pluto's average distance from the Sun is 3.67×10^9 miles. Write that distance in standard form.

9. (57) (a) A micron is 1×10^{-6} meter. Write that number and unit in standard form.

(b) Compare: 1 millimeter $\bigcirc$ 1×10^{-3} meter

10. (50) Use a unit multiplier to convert 300 mm to m.

11. (48) Complete the table.

FRACTION	DECIMAL	PERCENT
(a)	(b)	12%
$\frac{1}{3}$	(c)	(d)

12. (59) Use arrows to show each addition problem on a number line:

(a) (+2) + (–5)

(b) (–2) + (+5)

13. Find the missing number.
(58)

IN	FUNCTION	OUT
2	→ →	14
12	→ →	24
8	→ →	□
0	→ →	12

Refer to the figure below to answer problems 14 and 15. Dimensions are in millimeters. All angles are right angles.

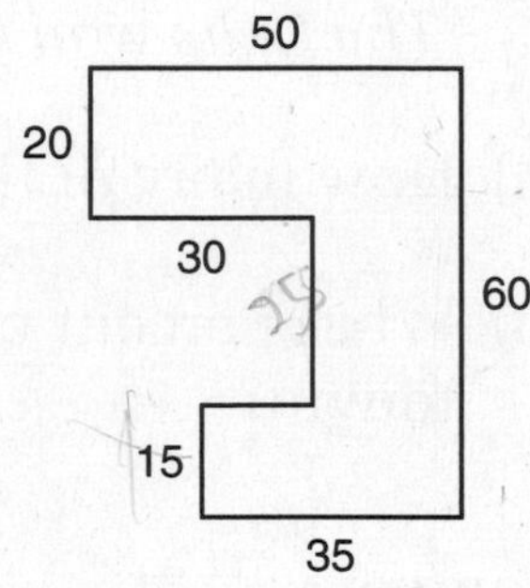

14. What is the perimeter of the figure?
(19)

15. What is the area of the figure?
(37)

Solve:

16. $4.4 = 8w$
(35)

17. $\frac{0.8}{1} = \frac{x}{1.5}$
(39)

18. $n + \frac{11}{20} = \frac{17}{30}$
(30)

19. $\frac{0.364}{m} = 7$
(35)

Simplify:

20. $2^{-1} + 2^{-1}$
(57)

21. $\sqrt{64} - 2^3 + 4^0$
(20, 57)

22.
(49)

$$\begin{array}{lrr} & 3 \text{ yd } 2 \text{ ft} & 7\frac{1}{2} \text{ in.} \\ + & 1 \text{ yd} & 5\frac{1}{2} \text{ in.} \\ \hline \end{array}$$

23.
(56)

$$\begin{array}{lrr} & 1 \text{ qt } 1 \text{ pt} & 6 \text{ oz} \\ - & 1 \text{ pt} & 12 \text{ oz} \\ \hline \end{array}$$

24. $2\frac{1}{2} \text{ hr} \cdot \frac{50 \text{ mi}}{1 \text{ hr}}$
(53)

25. $\left(\frac{5}{9} \cdot 12\right) \div 6\frac{2}{3}$
(26)

Estimate each answer to the nearest whole number. Then perform the calculation.

26. (23, 30) $3\frac{5}{6} - \left(4 - 1\frac{1}{9}\right)$

27. (26, 30) $\left(5\frac{5}{8} + 6\frac{1}{4}\right) \div 6\frac{1}{4}$

28. (52) Evaluate: $a - bc$ if $a = 0.1$, $b = 0.2$, and $c = 0.3$

29. (46) Find the tax on an $18.00 purchase when the sales-tax rate is 6.5%.

30. (36) This table shows the results of a club election. If one club member who voted is selected at random, what is the probability that the club member voted for the candidate who received the most votes?

Vote Tally

Candidate	Votes
Vasquez	𝍸 II
Lam	𝍸 𝍸
Enzinwa	𝍸 III

LESSON

60 Fractional Part of a Number, Part 1 • Percent of a Number, Part 1

WARM-UP

Facts Practice: Metric Conversions (Test M)

Mental Math:

a. 70×2.3

b. $435 \div 10^2$

c. $5x - 1 = 49$

d. Convert 75 mm to cm.

e. $\sqrt{144} - \sqrt{25}$

f. $\frac{4}{5}$ of \$1.00

g. Start with 25¢, double that amount, double that amount, double that amount, × 5, add \$20, ÷ 10, ÷ 10.

Problem Solving:

Copy this problem and fill in the missing digits:

$$\begin{array}{r} \_\_\_ \\ \times \quad \_ \\ \hline 1101 \end{array}$$

NEW CONCEPTS

Fractional part of a number, part 1

We can solve fractional-part-of-a-number problems by translating the question into an equation and then solving the equation. To translate,

we replace the word *is* with =

we replace the word *of* with ×

Example 1 What number is 0.6 of 31?

Solution This problem uses a decimal number to ask the question. We represent *what number* with W_N. We replace *is* with an equal sign. We replace *of* with a multiplication symbol.

What number is 0.6 of 31?	question
$W_N = 0.6 \times 31$	equation

To find the answer, we multiply.

$W_N =$ **18.6** multiplied

Example 2 Three fifths of 120 is what number?

Solution This time the question is phrased by using a common fraction. The procedure is the same: we translate directly.

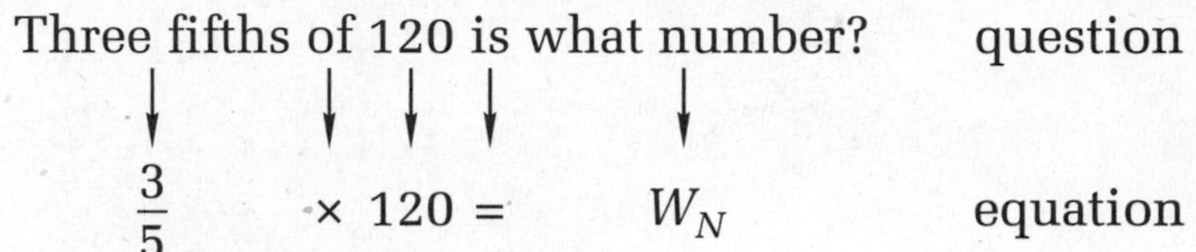

$$\frac{3}{5} \times 120 = W_N \qquad \text{equation}$$

To find the answer, we multiply.

$$W_N = \mathbf{72}$$

Percent of a number, part 1 We can translate percent problems into equations the same way we translate fractional-part-of-a-number problems: we convert the percent to either a fraction or a decimal.

Example 3 The jacket sold for \$75. Forty percent of the selling price was profit. How much money is 40% of \$75?

Solution We translate the question into an equation. We may convert the percent to a fraction or to a decimal. We show both ways.

Percent to Fraction	Percent to Decimal
$W_N = \frac{40}{100} \times \$75$	$W_N = 0.40 \times \$75$
$W_N = \frac{2}{5} \times \$75$	$W_N = 0.4 \times \$75$
$W_N = \mathbf{\$30}$	$W_N = \mathbf{\$30}$

Example 4 A certain used-car salesperson receives a commission of 8% of the selling price of a car. If the salesperson sells a car for \$3600, how much is the salesperson's commission?

Solution We want to find 8% of \$3600. This time we convert the percent to a decimal.

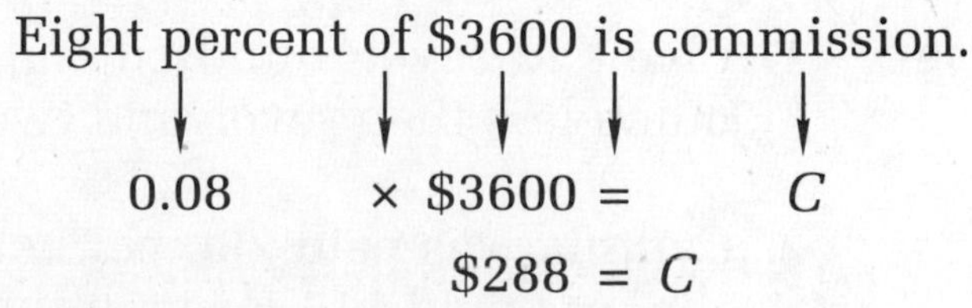

$$0.08 \times \$3600 = C$$

$$\$288 = C$$

The salesperson's commission is **\$288.**

Example 5 What number is 25% of 88?

Solution This time we convert the percent to a fraction.

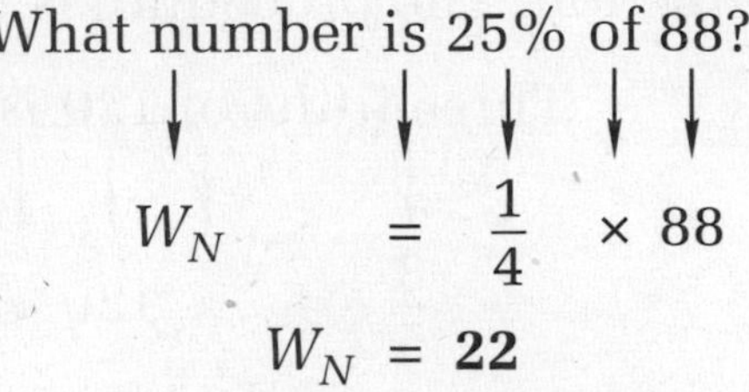

$$\text{What number is } 25\% \text{ of } 88?$$

$$W_N = \frac{1}{4} \times 88$$

$$W_N = \mathbf{22}$$

Whether a percent should be changed to a fraction or to a decimal is up to the person solving the problem. Often one form makes the problem easier to solve than the other form. With practice the choice of which form to use becomes more apparent.

LESSON PRACTICE

Practice set* Write equations to solve each problem:

a. What number is $\frac{4}{5}$ of 71?

b. Three eighths of $3\frac{3}{7}$ is what number?

c. What number is 0.6 of 145?

d. Seventy-five hundredths of 14.4 is what number?

e. What number is 50% of 150?

f. Three percent of $39 is how much money?

g. What number is 25% of 64?

h. If a salesperson receives a commission of 12% of sales, what is the salesperson's commission on $250,000 of sales?

MIXED PRACTICE

Problem set

1. (31, 35) Five and seven hundred eighty-four thousandths is how much less than seven and twenty-one ten-thousandths?

2. (28) Cynthia was paid 20¢ per board for painting the fence. If she was paid $10 for painting half the boards, how many boards were there in all?

3. When 72 is divided by n, the quotient is 12. What is the product when 72 is multiplied by n?
(41)

4. Four fifths of the trumpeters hit the note.
(36)

(a) What percent of the trumpeters did not hit the note?

(b) What was the ratio of trumpeters who hit the note to trumpeters who did not hit the note?

5. The average height of the five players on the basketball team was 77 inches. One of the players was 71 inches tall. Another was 74 inches tall, and two were each 78 inches tall. How tall was the tallest player on the team?
(55)

6. Write each number in scientific notation:
(51, 57)

(a) 0.00000008 (b) 67.5 billion

7. Diagram this statement. Then answer the questions that follow.
(22, 48)

Two thirds of the 96 members approved of the plan.

(a) How many of the 96 members approved of the plan?

(b) What percent of the members did not approve of the plan?

8. The first stage of the rocket fell from a height of 23,000 feet and settled on the ocean floor 9000 feet below sea level. In all, how many feet did the rocket's first stage descend?
(59)

Write equations to solve problems 9 and 10.

9. What number is $\frac{3}{4}$ of 17?
(60)

10. If 40% of the selling price of a $65 sweater is profit, then how many dollars profit does the store make when the sweater is sold?
(60)

11. Compare:
(43, 59)

(a) $\frac{1}{3} \bigcirc 0.33$ (b) $|5 - 3| \bigcirc |3 - 5|$

12. Complete the table.
(48)

FRACTION	DECIMAL	PERCENT
$\frac{1}{8}$	(a)	(b)
(c)	(d)	125%

13. Use arrows to show each addition problem on a number line:
(59)

(a) (−3) + (−1) (b) (−3) + (+1)

14. (a) Write the prime factorization of 3600 using exponents.
(21)

(b) Write the prime factorization of $\sqrt{3600}$.

15. Find the number of degrees in the central angle of each sector of the circle at right.
(Inv. 5)

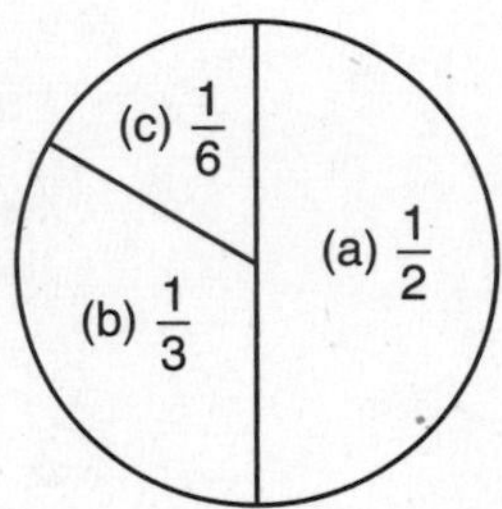

Refer to the figure below to answer problems 16–18. Dimensions between labeled points are in feet. The measure of ∠*EDF* equals the measure of ∠*ECA*.

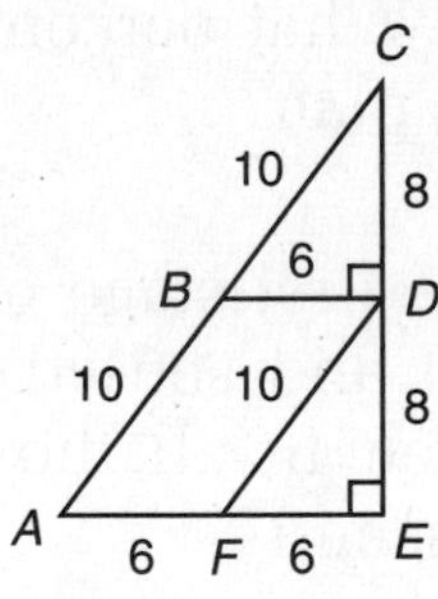

16. (a) Name a triangle congruent to triangle *DEF*.
(18)

(b) Name a triangle similar to Δ*DEF* but not congruent to Δ*DEF*.

17. (a) Find the area of Δ*BCD*.
(37)

(b) Find the area of Δ*ACE*.

18. By subtracting the areas of the two smaller triangles from the area of the large triangle, find the area of the quadrilateral *ABDF*.
(37)

Solve:

19. (30) $p - \frac{1}{30} = \frac{1}{20}$

20. (35) $9m = 0.117$

Simplify:

21. (20, 52) $3^2 + 4(3 + 2) - 2^3 \cdot 2^{-2} + \sqrt{36}$

22. (49)

$$\begin{array}{r} 3 \text{ days } 16 \text{ hr } 48 \text{ min} \\ + \ 1 \text{ day } \ 15 \text{ hr } 54 \text{ min} \\ \hline \end{array}$$

23. (30) $19\frac{3}{4} + 27\frac{7}{8} + 24\frac{5}{6}$

24. (30) $3\frac{3}{5} - \left(\frac{5}{6} \cdot 4\right)$

25. (26) $\left(1\frac{1}{4} \div \frac{5}{12}\right) \div 24$

26. (35) $6.5 - (0.65 - 0.065)$

27. (45) $0.3 \div (3 \div 0.03)$

28. (50) Use a unit multiplier to convert 3.5 centimeters to meters. (1 m = 100 cm)

29. (27, 45) Explain why these two division problems are equivalent. Then give a money example of the two divisions.

$$\frac{1.5}{0.25} = \frac{150}{25}$$

30. (Inv. 3, 58) The x-axis is a line of symmetry for ΔABC. The coordinates of point A are (3, 0), and the coordinates of point B are (0, –2). Find the coordinates of point C.

INVESTIGATION 6

Focus on

Classifying Quadrilaterals

Recall from Lesson 18 that a four-sided polygon is a quadrilateral. Refer to the quadrilaterals shown below to answer the problems that follow.

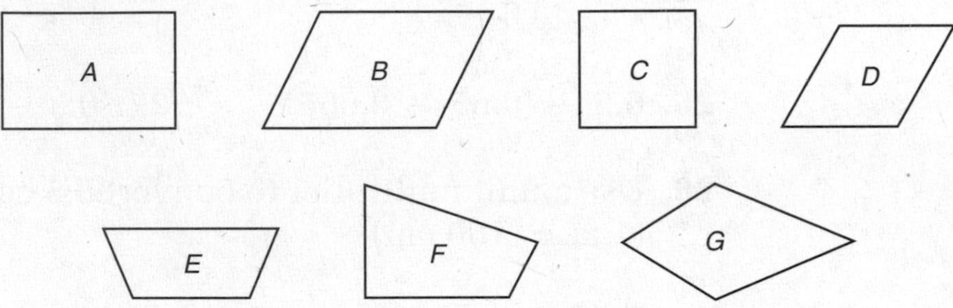

1. Which figures have four right angles?

2. Which figures have four sides of equal length?

3. Which figures have two pairs of parallel sides?

4. Which figure has just one pair of parallel sides?

5. Which figures have no pairs of parallel sides?

6. Which figures have two pairs of equal-length sides?

We can sort quadrilaterals by their characteristics. One way to sort is by the number of pairs of parallel sides. A quadrilateral with two pairs of parallel sides is a **parallelogram.** Here we show four parallelograms.

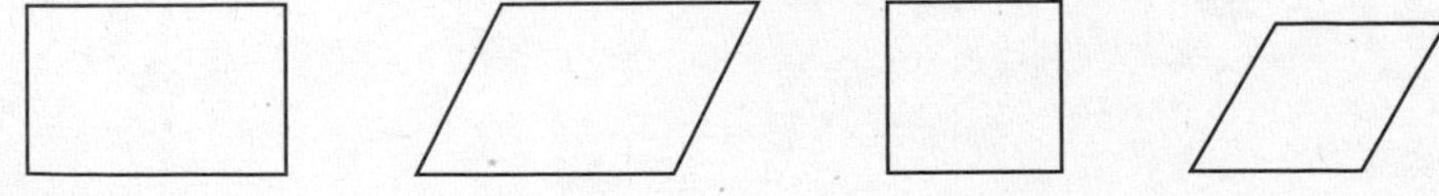

7. Which of the figures *A*–*G* are parallelograms?

A quadrilateral with just one pair of parallel sides is a **trapezoid.** The figures shown below are trapezoids. Can you find the parallel sides? (Notice that the parallel sides are not the same length.)

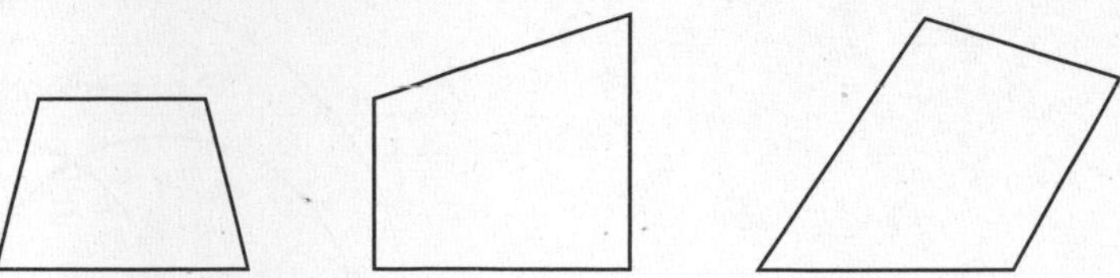

8. Which of the figures *A–G* is a trapezoid?

A quadrilateral with no pairs of parallel sides is a **trapezium.** Here we show two examples:

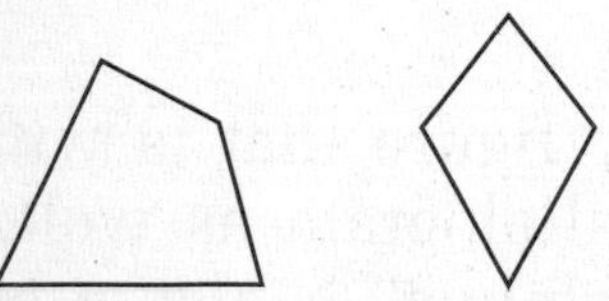

9. Which of the figures *A–G* are trapeziums?

We can sort quadrilaterals by the lengths of their sides. If the four sides are the same length, the quadrilaterals are **equilateral.** An equilateral quadrilateral is a **rhombus.** A rhombus is a type of parallelogram. Here we show two examples.

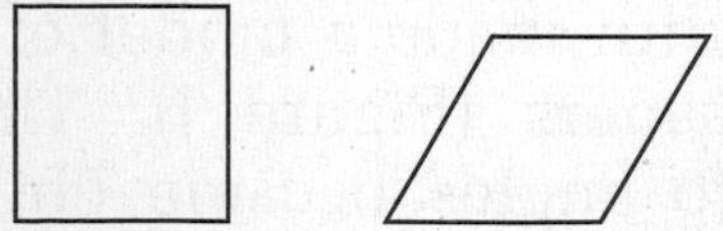

10. Which of the figures *A–G* are rhombuses?

We can sort quadrilaterals by the measures of their angles. If the four angles are of equal measure, then each angle is a right angle, and the quadrilateral is a **rectangle.** A rectangle is a type of parallelogram.

11. Which of the figures *A–G* are rectangles?

Notice that a square is both a rectangle and a rhombus. A square is also a parallelogram. We can use a **Venn diagram** to illustrate the relationships.

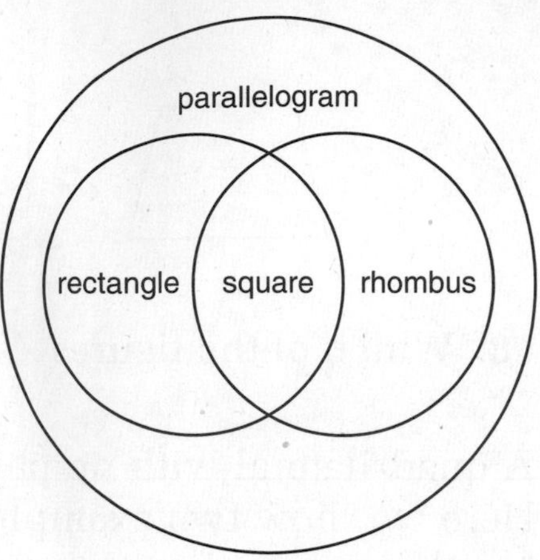

Any figure that is within the circle labeled "rectangle" is a parallelogram as well. Any figure within the circle labeled "rhombus" is also a parallelogram. A figure within both the rectangle and rhombus circles is a square.

12. Copy the Venn diagram above on your paper. Then refer to quadrilaterals *A, B, C, D,* and *E* at the beginning of this investigation. Draw each of the quadrilaterals in the Venn diagram in the proper location. (One of the figures will be outside the parallelogram category.)

Judith made a model of a rectangle out of straws and pipe cleaners (Figure J). Then she shifted the sides so that two angles became obtuse and two angles became acute (Figure K).

Figure J   Figure K

Refer to Figures J and K to answer problems 13–16.

13. Is Figure K a rectangle?

14. Is Figure K a parallelogram?

15. Does the perimeter of Figure K equal the perimeter of Figure J?

16. Does the area of Figure K equal the area of Figure J?

Chris made a model of a rectangle out of straws and pipe cleaners (Figure L). Then he reversed the positions of two of the straws so that the straws that were the same length were adjacent to each other instead of opposite each other (Figure M).

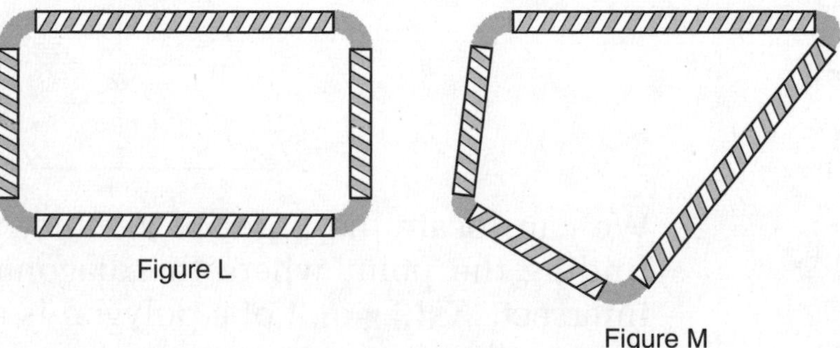

Figure L

Figure M

Figure M does not have a pair of parallel sides, so it is a trapezium. However, it is a special type of trapezium called a **kite**.

17. Which of the figures *A–G* is a kite?

18. If two sides of a kite are 2 ft and 3 ft, what is the perimeter of the kite?

Notice that a kite has a line of symmetry.

19. Draw a kite and show its line of symmetry.

20. Draw a rhombus that is not a square, and show its lines of symmetry.

21. Draw a rectangle that is not a square, and show its lines of symmetry.

22. Draw a rhombus that is a rectangle, and show its lines of symmetry.

23. An **isosceles trapezoid** has a line of symmetry. The nonparallel sides of an isosceles trapezoid are the same length. Draw an isosceles trapezoid and show its line of symmetry.

Not every trapezoid has a line of symmetry. Any parallelogram that is not a rhombus or rectangle does not have line symmetry. However, every parallelogram does have **point symmetry.** A figure is symmetrical about a point if every line drawn through the point intersects the figure at points that are equal distances from the point of symmetry.

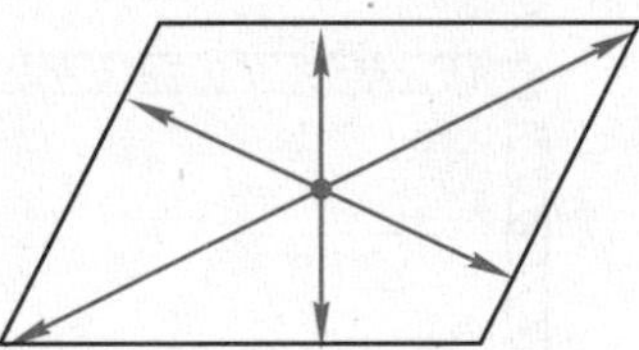

We can locate the point of symmetry of a parallelogram by finding the point where the diagonals of the parallelogram intersect. A **diagonal** of a polygon is a segment between non-consecutive vertices.

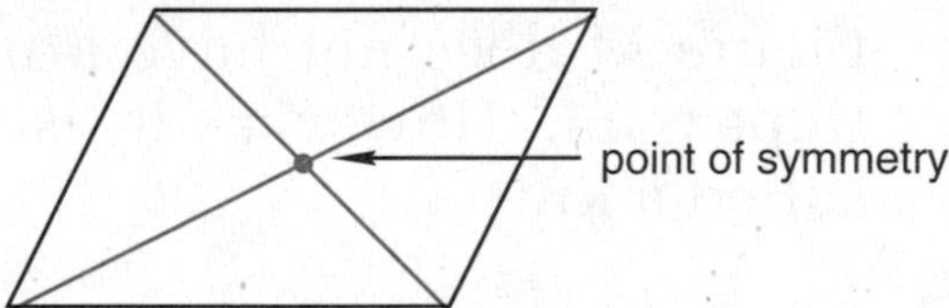

In the following problem we learn a way to test for point symmetry.

24. Draw two or three parallelograms on grid paper. Be sure that one of the parallelograms is a rectangle and one is not a rectangle. Locate and mark the point in each parallelogram where the diagonals intersect. Then carefully cut out the parallelograms. If we rotate a figure with point symmetry a half turn (180°) about its point of symmetry, the figure will appear to be in the same position it was in before it was rotated. On one of the cut-out parallelograms, place the tip of a pencil on the point where the diagonals intersect. Then rotate the parallelogram 180°. Is the point of intersection a point of symmetry?

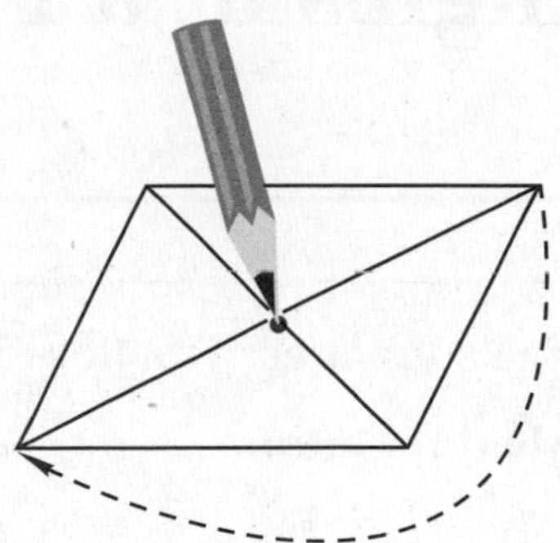

Repeat the rotation with the other parallelogram(s) you cut out.

25. Which of the figures *A–G* have point symmetry?

Below we classify the figures illustrated at the beginning of this investigation. You may refer to them as you answer the remaining problems.

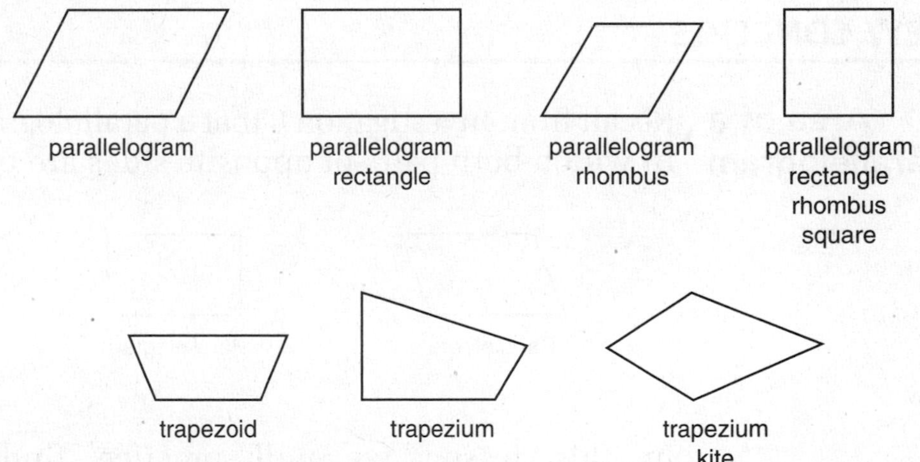

Answer true or false, and state the reason(s) for your answer.

26. A square is a rectangle.

27. All rectangles are parallelograms.

28. Some squares are trapezoids.

29. Some parallelograms are rectangles.

30. Draw a Venn diagram illustrating the relationship of quadrilaterals, parallelograms, and trapezoids.

LESSON

61 Area of a Parallelogram • Angles of a Parallelogram

WARM-UP

Facts Practice: Fraction-Decimal-Percent Equivalents (Test L)

Mental Math:

a. 50×4.6 **b.** 2.4×10^{-1}

c. $\frac{a}{20} = \frac{12}{8}$ **d.** Convert 1.5 km to m.

e. $3^2 - 2^3$ **f.** $\frac{7}{10}$ of \$3.00

g. What is the total cost of a \$20 item plus 6% sales tax?

Problem Solving:

The squares of the first nine counting numbers are each less than 100. Altogether, how many counting numbers have squares that are less than 1000?

NEW CONCEPTS

Area of a parallelogram Recall from Investigation 6 that a parallelogram is a quadrilateral in which both pairs of opposite sides are parallel.

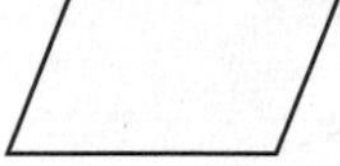
Parallelogram

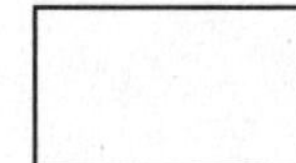
Parallelogram

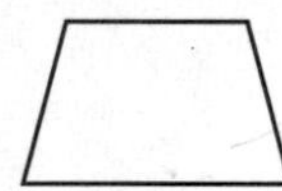
Not a parallelogram

In this lesson we will practice finding the areas of parallelograms. We can use a paper parallelogram and scissors to help us understand the concept.

Activity 1: *Area of a Parallelogram*

Materials needed:

- Paper
- Scissors
- Straightedge

Cut a piece of paper to form a parallelogram as shown. You may use graph paper if available.

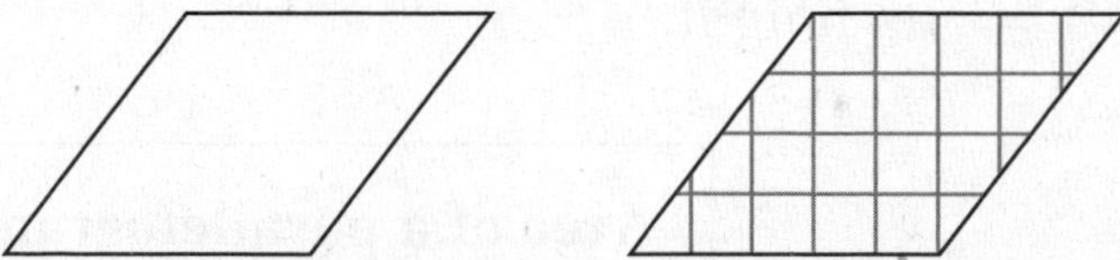

Next, sketch a segment perpendicular to two of the parallel sides of the parallelogram. Cut the parallelogram into two pieces along the segment you drew.

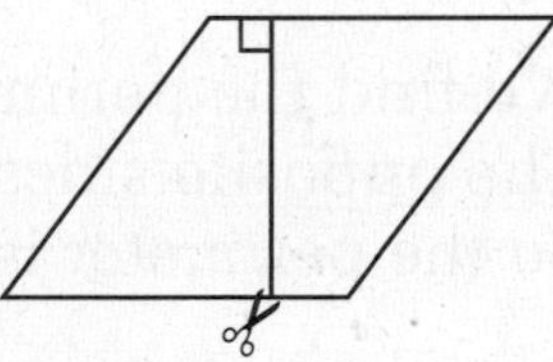

Finally, reverse the positions of the two pieces and fit them together to form a rectangle. The area of the original parallelogram equals the area of this rectangle.

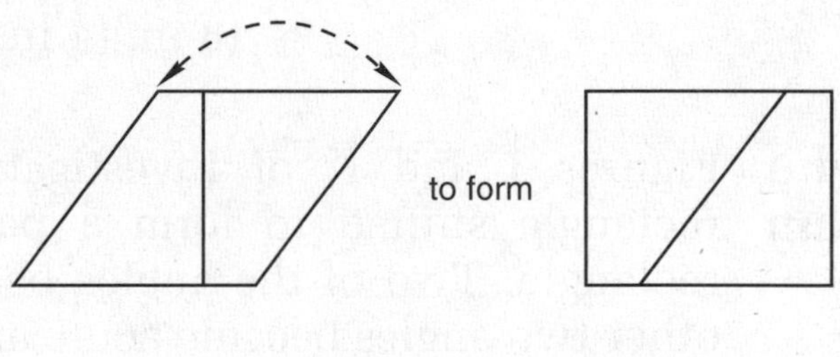

The dimensions of a rectangle are often called the length and the width. When describing a parallelogram, we do not use these terms. Instead we use the terms **base** and **height.**

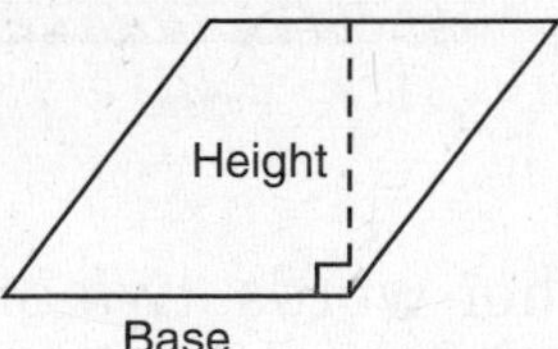

Notice that the height is not one of the sides of the parallelogram (unless the parallelogram is a rectangle). Instead, **the height is perpendicular to the base.** Multiplying the base by the height gives us the area of a rectangle.

However, as we saw in Activity 1, the area of the rectangle equals the area of the parallelogram we are considering. Thus we find the area of a parallelogram by multiplying its base by its height.

Area of a parallelogram = base · height

Example 1 Find the (a) perimeter and (b) area of this parallelogram. Dimensions are in inches.

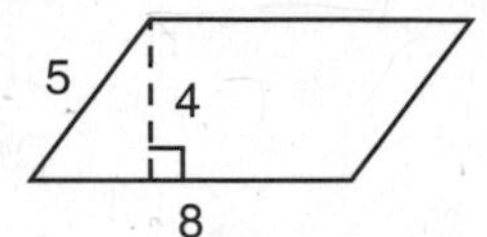

Solution (a) We find the perimeter by adding the lengths of the sides. The opposite sides of a parallelogram are equal in length. So the perimeter is

$$5 \text{ in.} + 8 \text{ in.} + 5 \text{ in.} + 8 \text{ in.} = \mathbf{26 \text{ in.}}$$

(b) We find the area of a parallelogram by multiplying the base by the height. The base is 8 in., and the height is 4 in. So the area is

$$(8 \text{ in.})(4 \text{ in.}) = \mathbf{32 \text{ in.}^2}$$

Angles of a parallelogram

Figures J and K of Investigation 6 illustrated a "straw" rectangle shifted to form a parallelogram that was not a rectangle. Two of the angles became obtuse angles, and the other two angles became acute angles.

Figure J

Figure K

In other words, two of the angles became more than 90°, and two of the angles became less than 90°. Each angle became greater than or less than 90° *by the same amount.* If, by shifting the sides of the straw rectangle, the obtuse angles became 10° *greater than* 90° (100° angles), then the acute angles became 10° *less than* 90° (80° angles). The following activity illustrates this relationship.

Activity 2: *Angles of a Parallelogram*

Materials needed:

- Protractor
- Paper
- Two pairs of plastic straws (The straws within a pair must be the same length. The two pairs may be different lengths.)
- Thread or lightweight string
- Paper clip for threading the straws (optional)

Make a "straw" parallelogram by running a string or thread through two pairs of plastic straws. If the pairs of straws are of different lengths, alternate the lengths as you thread them (long-short-long-short).

Bring the two ends of the string together, pull until the string is snug but not bowing the straws, and tie a knot.

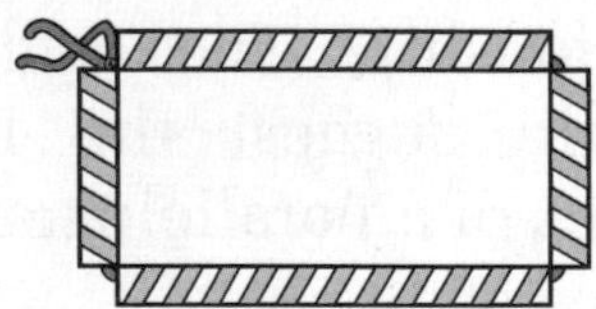

You should be able to shift the sides of the parallelogram to various positions.

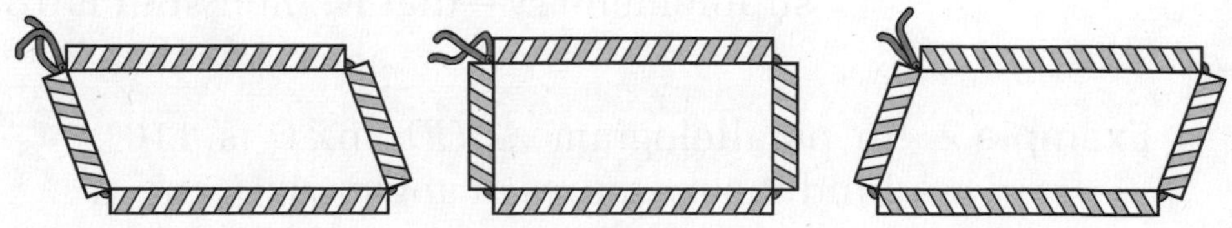

Lay the straw parallelogram on a desktop with a piece of paper under it. On the paper you will trace the parallelogram. Shift the parallelogram into a position you want to measure, hold the straws and paper still (this may require more than two hands), and carefully trace with a pencil around the *inside* of the parallelogram.

Set the straw parallelogram aside, and use a protractor to measure each angle of the traced parallelogram. Write the measure inside each angle. Trace and measure the angles of a second parallelogram with a different shape before answering the questions below.

1. What were the measures of the two obtuse angles of one parallelogram?

2. What were the measures of the two acute angles of the same parallelogram?

3. What was the sum of the measures of one obtuse angle and one acute angle of the same parallelogram?

Answer the three questions again for the second parallelogram. Can you form any general conclusions about parallelograms based on your findings?

The quality of all types of measurement is affected by the quality of the measuring instrument, the material being measured, and the person performing the measurement. However, even rough measurements can suggest underlying relationships. The rough measurements performed in Activity 2 should suggest the following relationships between the angles of a parallelogram:

1. Nonadjacent angles (angles in opposite corners) have equal measures.

2. Adjacent angles (angles that share a common side) are supplementary—that is, their sum is 180°.

Example 2 In parallelogram *ABCD,* m∠*D* is 110°. Find the measures of angles *A, B,* and *C* in the parallelogram.

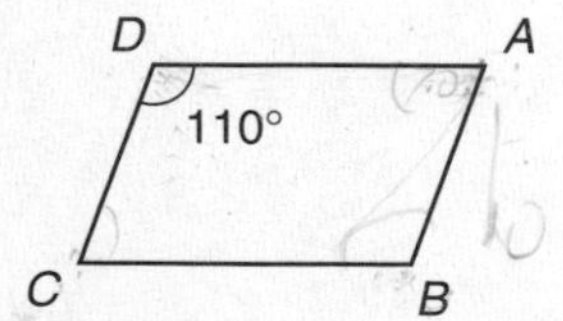

Solution Nonadjacent angles, like ∠*B* and ∠*D,* have equal measures, so **m∠*B* = 110°**. Adjacent angles are supplementary, and both ∠*A* and ∠*C* are adjacent to ∠*D,* so **m∠*A* = 70°** and **m∠*C* = 70°**.

LESSON PRACTICE

Practice set Find the perimeter and area of each parallelogram. Dimensions are in centimeters.

a.

10

8

12

b.

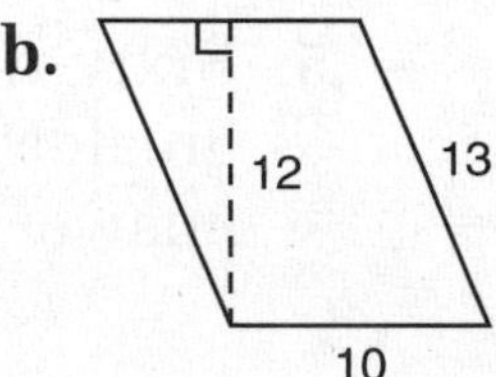

c.

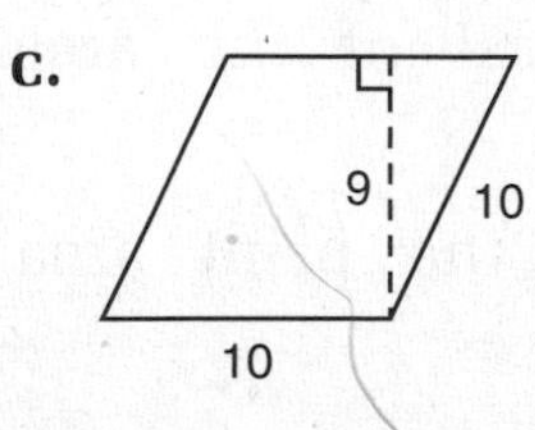

For problems **d–g,** find the measures of the angles marked *d*, *e*, *f*, and *g* in this parallelogram.

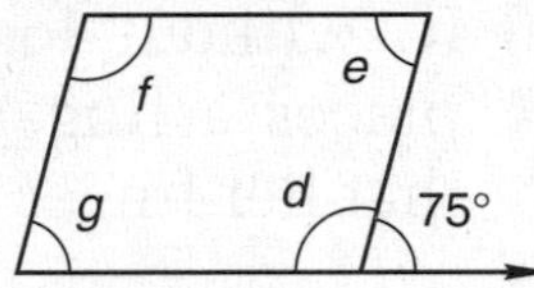

Figure *ABCD* is a parallelogram. Refer to this figure to find the measures of the angles in problems **h–j.**

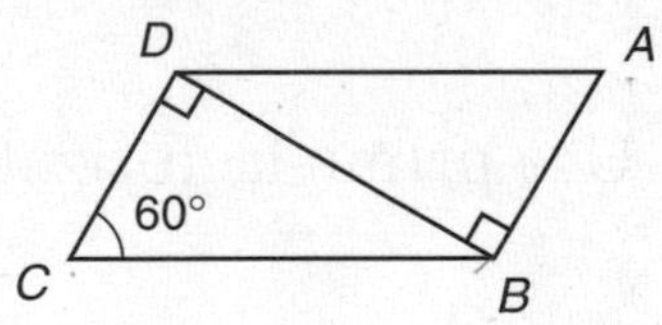

h. $\angle A$ **i.** $\angle ADB$ **j.** $\angle ABC$

MIXED PRACTICE

Problem set

1. If $\frac{1}{2}$ gallon of milk costs \$1.12, what is the cost per pint?
(16, 46)

2. Use a ratio box to solve this problem. The cookie recipe called for oatmeal and brown sugar in the ratio of 2 to 1. If 3 cups of oatmeal were called for, how many cups of brown sugar were needed?
(54)

3. Ricardo ran the 400-meter race 3 times. His fastest time was 54.3 seconds. His slowest time was 56.1 seconds. If his average time was 55.0 seconds, what was his time for the third race?
(55)

4. (46) It is $4\frac{1}{2}$ miles to the end of the trail. If Paula runs to the end of the trail and back in 60 minutes, what is her average speed in miles per hour?

5. (51) Sixty-three million, one hundred thousand is how much greater than seven million, sixty thousand? Write the answer in scientific notation.

6. (36, 48) Only three tenths of the print area of the newspaper carried news. The rest of the area was filled with advertisements.

(a) What percent of the print area was filled with advertisements?

(b) What was the ratio of news area to advertisement area?

(c) If, without looking, Kali opens the newspaper and places a finger on the page, what is the probability that her finger will be on an advertisement?

7. (51, 57) (a) Write 0.00105 in scientific notation.

(b) Write 3.02×10^5 in standard form.

8. (24) Use prime factorization to reduce $\frac{128}{192}$.

9. (50) Use a unit multiplier to convert 1760 yards to feet.

In the figure below, quadrilateral *ABDE* is a rectangle and $\overline{EC} \parallel \overline{FB}$. Refer to the figure to answer problems 10–12.

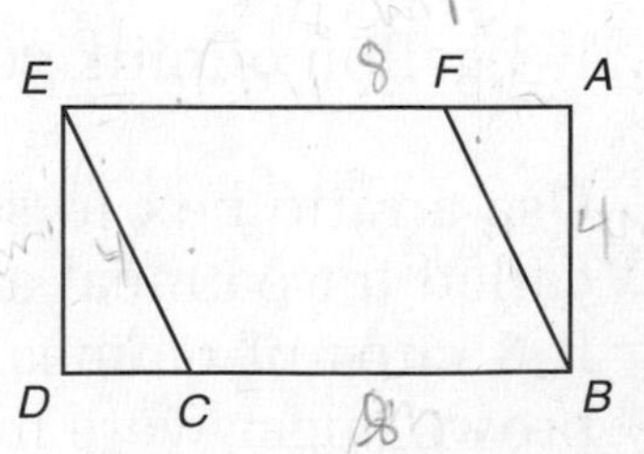

10. (Inv. 6) Classify each of the following quadrilaterals:

(a) *ECBF*

(b) *ECBA*

11. If $AB = ED = 4$ m, $BC = EF = 6$ m, and $BD = AE = 8$ m, then
(37, 61)

(a) what is the area of quadrilateral *BCEF?*

(b) what is the area of triangle *ABF?*

(c) what is the area of quadrilateral *ECBA?*

12. Classify each of the following angles as acute, right, or obtuse:
(7)

(a) $\angle ECB$ (b) $\angle EDC$ (c) $\angle FBA$

13. Following is an ordered list of the per-game point averages of 19 basketball players. Find the (a) median, (b) first quartile, and (c) third quartile of these averages; (d) identify any outliers.
(Inv. 4)

2, 5, 5, 6, 6, 6, 7, 7, 7, 8, 8, 8, 8, 9, 9, 10, 10, 10, 10

14. Refer to the parallelogram below to answer (a)–(c).
(Inv. 6, 61)

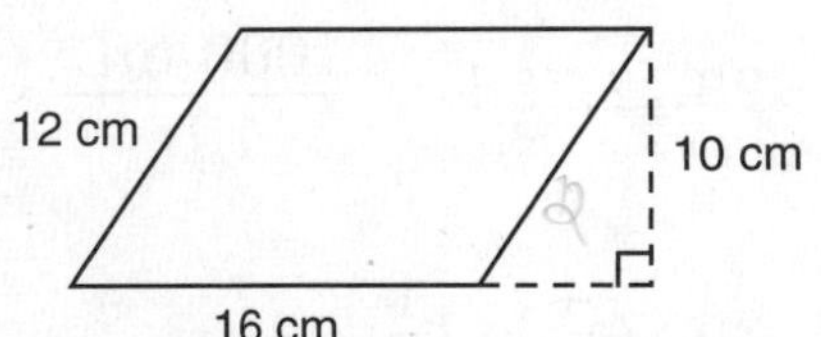

(a) What is the perimeter of this parallelogram?

(b) What is the area of this parallelogram?

(c) Trace the parallelogram on your paper, and locate its point of symmetry.

15. The parallelogram at right is divided by a diagonal into two congruent triangles. Find the measure of
(40, 61)

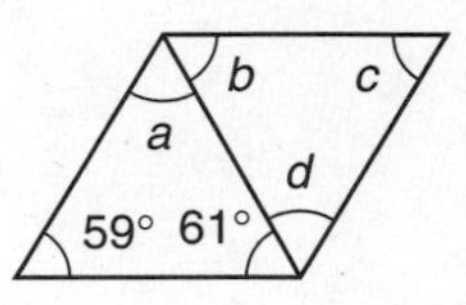

(a) $\angle a$. (b) $\angle b$. (c) $\angle c$. (d) $\angle d$.

16. Tara noticed that the tape she was using to wrap packages was 2 centimeters wide. How many meters wide was the tape?
(50)

17. (Inv. 3) A circle is drawn on a coordinate plane with its center at the origin. The circle intersects the x-axis at (5, 0) and (–5, 0).

(a) At what coordinates does the circle intersect the y-axis?

(b) What is the diameter of the circle?

18. (3) On one tray of a balanced scale was a 50-g mass. On the other tray were a small cube, a 10-g mass, and a 5-g mass. What was the mass of the small cube? Describe how you found your answer.

19. (59) The trail Paula runs begins at 27 feet below sea level and ends at 164 feet above sea level. What is the gain in elevation from the beginning to the end of the trail?

Simplify:

20. (52) $10 + (10 \times 10) - (10 \div 10)$

21. (57) $2^0 - 2^{-3}$

22. (34) 4.5 m + 70 cm = ____ m

23. (50) $2.75 \text{ L} \cdot \dfrac{1000 \text{ mL}}{1 \text{ L}}$

24. (23, 30) $5\frac{7}{8} + \left(3\frac{1}{3} - 1\frac{1}{2}\right)$

25. (26) $4\frac{4}{5} \cdot 1\frac{1}{9} \cdot 1\frac{7}{8}$

26. (26) $6\frac{2}{3} \div \left(3\frac{1}{5} \div 8\right)$

27. (35) $12 - (0.8 + 0.97)$

28. (35) $(2.4)(0.05)(0.005)$

29. (47) $0.2 \div (4 \times 10^2)$

30. (45) $0.36 \div (4 \div 0.25)$

LESSON

62 Classifying Triangles

WARM-UP

Facts Practice: Metric Conversions (Test M)

Mental Math:

a. 5×8.6 **b.** 2.5×10^{-2}

c. $10x + 2 = 32$ **d.** Convert 2500 g to kg.

e. $10^3 \div 10^3$ **f.** $\frac{2}{3}$ of $24.00

g. 8^2, $- 4$, $\div 2$, $\times 3$, $+ 10$, $\sqrt{\ }$, $\times 2$, $+ 5$, $\sqrt{\ }$, $- 4$, square that number, $- 1$

Problem Solving:

Marsha started the 1024-m race, ran half the distance to the finish line, and handed the baton to Greg. Greg ran half the remaining distance and handed off to Alice, who ran half the remaining distance. How far from the finish line did Alice stop? If the team continues this pattern, how many more runners will they need in order to cross the finish line?

NEW CONCEPT

Recall from Lesson 7 that we classify angles as acute angles, right angles, and obtuse angles.

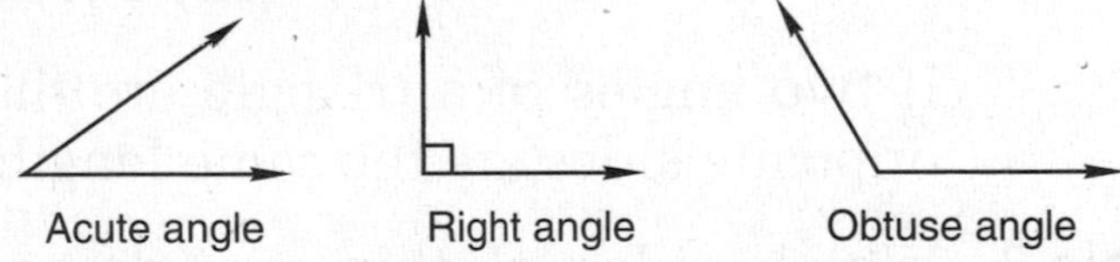

We use the same words to describe triangles that contain these angles. If every angle of a triangle measures less than 90°, then the triangle is an **acute triangle.** If the triangle contains a 90° angle, then the triangle is a **right triangle.** An **obtuse triangle** contains one angle that measures more than 90°.

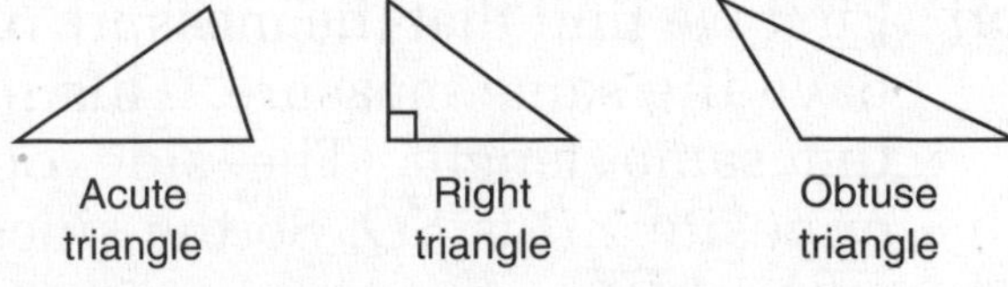

When describing triangles, we can refer to the sides and angles as "opposite" each other. For example, we might say, "The side opposite the right angle is the longest side of a right triangle." The side opposite an angle is the side the angle opens toward. In this right triangle, $\overline{AB}$ is the side opposite $\angle C$, and $\angle C$ is the angle opposite side AB.

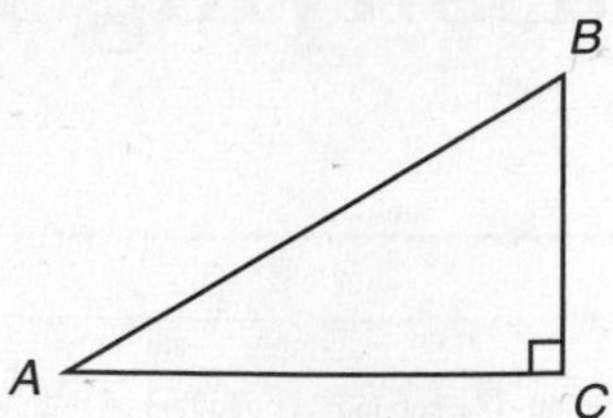

Each angle of a triangle has a side that is opposite that angle. The lengths of the sides of a triangle are in the same order as the measures of their opposite angles. This means that the longest side of a triangle is opposite the largest angle, and the shortest side is opposite the smallest angle.

Example 1 Name the sides of this triangle in order from shortest to longest.

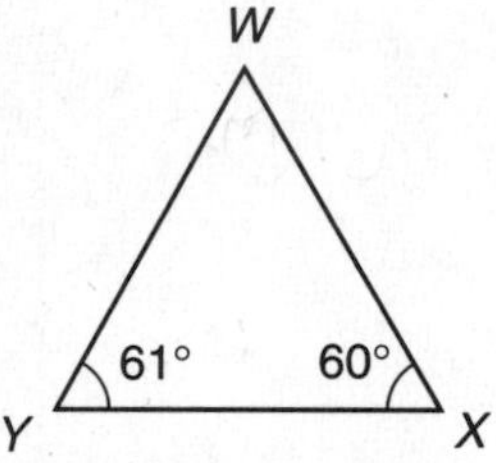

Solution First we note the measures of all three angles. The sum of their measures is 180°, so the measure of $\angle W$ is 59°. Since $\angle W$ is the smallest of the three angles, the side opposite $\angle W$, which is $\overline{XY}$, is the shortest side. The next angle in order of size is $\angle X$, so $\overline{YW}$ is the second longest side. The largest angle is $\angle Y$, so $\overline{WX}$ is the longest side. So the sides in order are

$$\overline{XY}, \overline{YW}, \overline{WX}$$

If two angles of a triangle are the same measure, then their opposite sides are the same length.

Example 2 Which sides of this triangle are the same length?

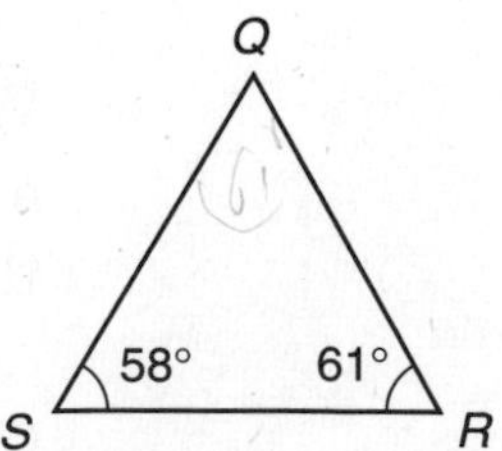

Solution First we find that the measure of $\angle Q$ is 61°. So angles Q and R have the same measure. This means their opposite sides are the same length. The side opposite $\angle Q$ is $\overline{SR}$. The side opposite $\angle R$ is $\overline{SQ}$. So the sides that are the same length are $\overline{SR}$ **and** $\overline{SQ}$.

If all three angles of a triangle are the same measure, then all three sides are the same length.

Example 3 In triangle JKL, $JK = KL = LJ$. Find the measure of $\angle J$.

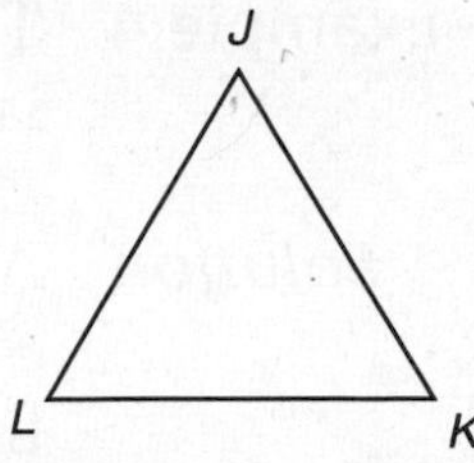

Solution If two or more sides of a triangle are the same length, then the angles opposite those sides are equal in measure. In ΔJKL all three sides are the same length, so all three angles have the same measure. The angles equally share 180°. We find the measure of each angle by dividing 180° by 3.

$$180° \div 3 = 60°$$

We find that the measure of $\angle J$ is **60°**.

The triangle in example 3 is a regular triangle. We usually call a regular triangle an **equilateral triangle.** The three angles of an equilateral triangle each measure 60°, and the three sides are the same length.

If a triangle has at least two sides of the same length (and thus two angles of the same measure), then the triangle is called an **isosceles triangle.** The triangle in example 2 is an isosceles triangle, as are each of these triangles:

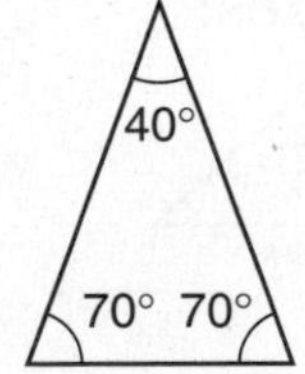

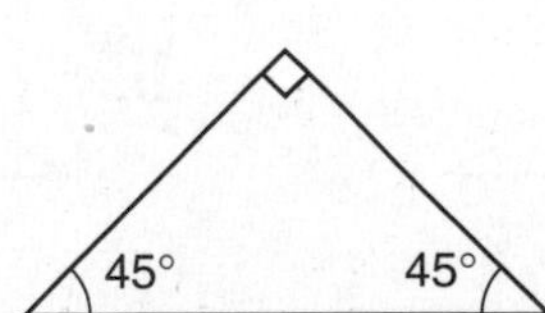

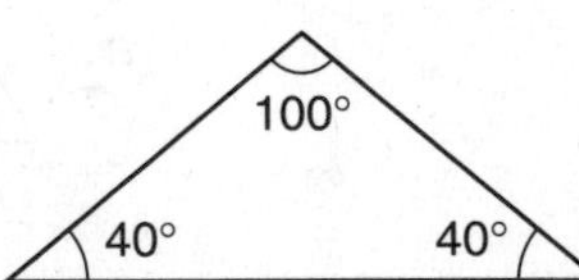

If the three sides of a triangle are all different lengths and the angles are all different measures, then the triangle is called a **scalene triangle.** Here we show a scalene triangle, an isosceles triangle, and an equilateral triangle. The tick marks on the sides indicate sides of equal length, while tick marks on the arcs indicate angles of equal measure.

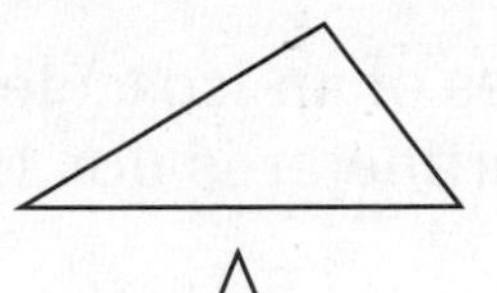

Scalene triangles have three sides that are all different lengths.

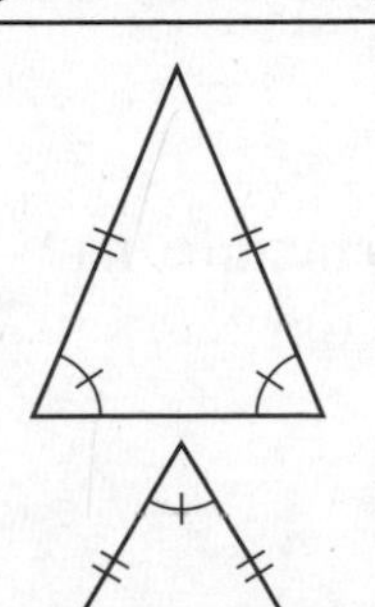

Isosceles triangles have at least two sides that are the same length.

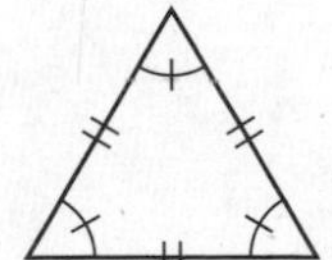

Equilateral triangles have three sides that are the same length. Equilateral triangles are **regular triangles.**

Example 4 The perimeter of an equilateral triangle is 2 feet. How many inches long is each side?

Solution All three sides of an equilateral triangle are equal in length. Since 2 feet equals 24 inches, we divide 24 inches by 3 and find that the length of each side is **8 inches.**

Example 5 Draw an isosceles right triangle.

Solution *Isosceles* means the triangle has at least two sides that are the same length. *Right* means the triangle contains a right angle. We sketch a right angle, making both segments equal in length. Then we complete the triangle.

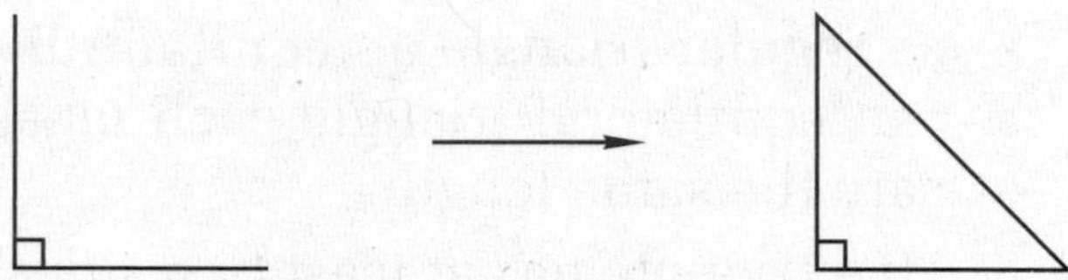

LESSON PRACTICE

Practice set Classify each triangle by its angles:

a. 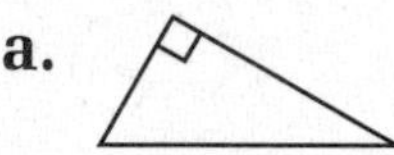

b. 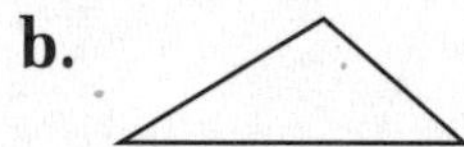

c. 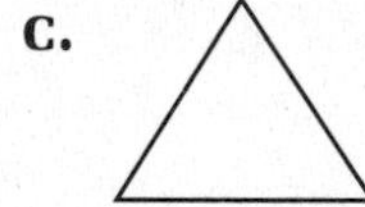

Classify each triangle by its sides:

d. 3, 4, 5

e. 4, 4, 4

f. 5, 5, 8

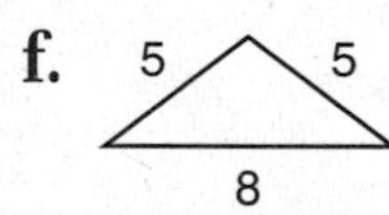

g. If we know that two sides of an isosceles triangle are 3 cm and 4 cm and that its perimeter is not 10 cm, then what is its perimeter?

h. Name the angles of this triangle in order from smallest to largest.

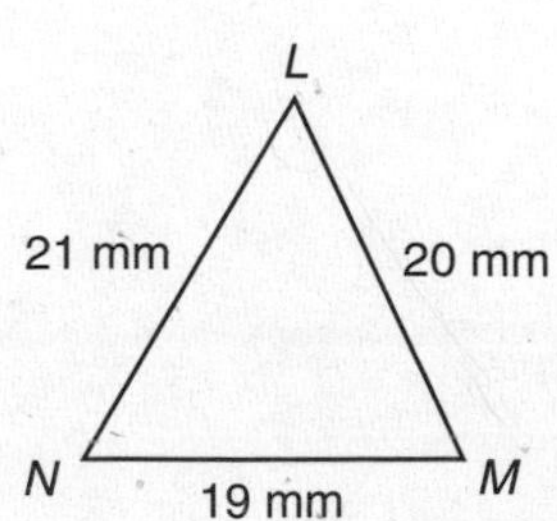

MIXED PRACTICE

Problem set

1. (28) At 1:30 p.m. Dante found a parking meter that still had 10 minutes until it expired. He put 2 dimes into the meter and went to his meeting. If 5 cents buys 15 minutes of parking time, at what time will the meter expire?

Use the information in the paragraph below to answer problems 2 and 3.

The Barkers started their trip with a full tank of gas and a total of 39,872 miles on their car. They stopped and filled the gas tank 4 hours later with 8.0 gallons of gas. At that time the car's total mileage was 40,060.

2. (12) How far did they travel in 4 hours?

3. (46) The Barkers' car traveled an average of how many miles per gallon during the first 4 hours of the trip?

4. (41) When 24 is multiplied by w, the product is 288. What is the quotient when 24 is divided by w?

5. (54) Use a ratio box to solve this problem. There were 144 Bolsheviks in the crowd. If the ratio of Bolsheviks to czarists was 9 to 8, how many czarists were in the crowd?

6. (22, 48) Diagram this statement. Then answer the questions that follow.

Exit polls showed that 7 out of 10 voters cast their ballot for the incumbent.

(a) According to the exit polls, what percent of the voters cast their ballot for the incumbent?

(b) According to the exit polls, what fraction of the voters did not cast their ballot for the incumbent?

7. (60) Write an equation to solve this problem:

What number is $\frac{5}{6}$ of $3\frac{1}{3}$?

8. (46) What is the total price of a $10,000 car plus 8.5% sales tax?

9. (51) Write 1.86×10^5 in standard form. Then use words to write the number.

10. (32) Compare: 1 quart ◯ 1 liter

11. (59) Show this addition problem on a number line:

$$(-3) + (+4) + (-2)$$

12. (48) Complete the table.

FRACTION	DECIMAL	PERCENT
$\frac{5}{8}$	(a)	(b)
(c)	(d)	275%

13. (52) Evaluate: $x + \frac{x}{y} - y$ if $x = 12$ and $y = 3$

14. (47, 57) Find each missing exponent:

(a) $2^5 \cdot 2^3 = 2^{\square}$ (b) $2^5 \div 2^3 = 2^{\square}$

(c) $2^3 \div 2^3 = 2^{\square}$ (d) $2^3 \div 2^5 = 2^{\square}$

15. (62) In the figure below, angle *ZWX* measures 90°.

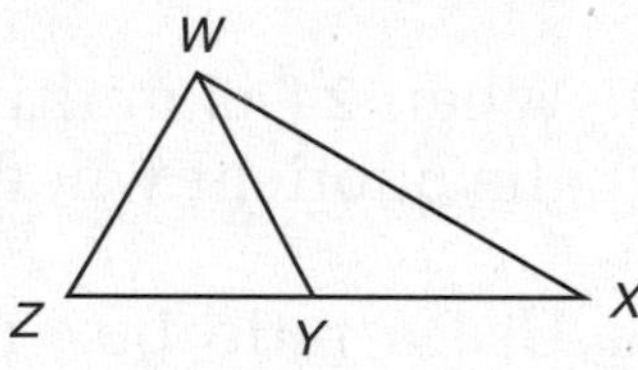

(a) Which triangle is an acute triangle?

(b) Which triangle is an obtuse triangle?

(c) Which triangle is a right triangle?

16. (19, 37) In the figure at right, dimensions are in inches and all angles are right angles.

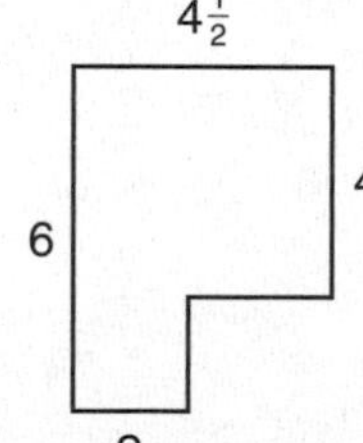

(a) What is the perimeter of the figure?

(b) What is the area of the figure?

17. (37, 62) (a) Classify this triangle by its sides.

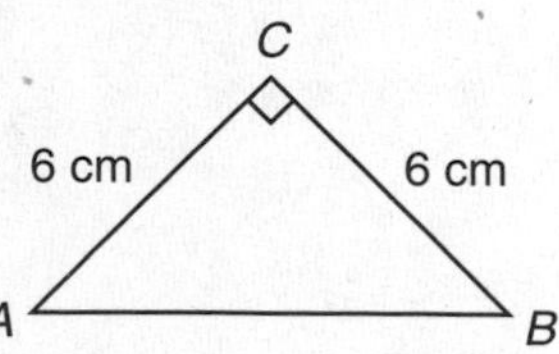

(b) What is the measure of each acute angle of the triangle?

(c) What is the area of the triangle?

(d) The longest side of this triangle is opposite which angle?

Solve:

18. $7q = 1.428$
(35)

19. $\frac{30}{70} = \frac{w}{\$2.10}$
(39)

Simplify:

20. $5^2 + 2^5 - \sqrt{49}$
(20, 52)

21. $3(8) - (5)(2) + 10 \div 2$
(52)

22.
(49)

$$\begin{array}{r} 1 \text{ yd } 2 \text{ ft } 3\frac{3}{4} \text{ in.} \\ + \quad 2 \text{ ft } 6\frac{1}{2} \text{ in.} \\ \hline \end{array}$$

23. 1 L − 50 mL = ____ mL
(32)

24. $\frac{60 \text{ mi}}{1 \text{ hr}} \cdot \frac{1 \text{ hr}}{60 \text{ min}}$
(50)

25. $2\frac{7}{24} + 3\frac{9}{32}$
(30)

26. $2\frac{2}{5} \div \left(4\frac{1}{5} \div 1\frac{3}{4}\right)$
(26)

27. $20 - \left(7\frac{1}{2} \div \frac{2}{3}\right)$
(23, 26)

28. Draw an equilateral triangle and show its lines of symmetry.
(58, 62)

29. Evaluate: $|x - y|$ if $x = 3$ and $y = 4$
(52, 59)

30. On one tray of a balanced scale was a 1-kg mass. On the other tray were a box and a 250-g mass. What was the mass of the box?
(3, 32)

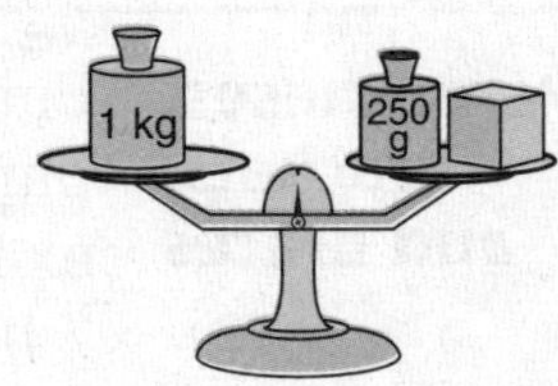

LESSON

63 Symbols of Inclusion

WARM-UP

Facts Practice: Fraction-Decimal-Percent Equivalents (Test L)

Mental Math:

a. 5×246

b. 4×10^{-3}

c. $\frac{15}{20} = \frac{x}{8}$

d. Convert 0.5 L to mL.

e. $\sqrt{196}$

f. $\frac{3}{8}$ of \$24.00

g. Instead of multiplying 50 and 28, double 50, find half of 28, and multiply those numbers.

Problem Solving:

The Smiths traveled the 60-mile road to town at 60 mph. The traffic was heavy on the return trip, and they averaged just 30 mph. What was their average speed for the round trip?

NEW CONCEPTS

Parentheses, brackets, and braces

Parentheses are called **symbols of inclusion.** We have used parentheses to show which operation to perform first. For example, to simplify the following expression, we add 5 and 7 before subtracting their sum from 15.

$$15 - (5 + 7)$$

Brackets, [ ], and **braces,** { }, are also symbols of inclusion. When an expression contains multiple symbols of inclusion, we simplify within the innermost symbols first.

To simplify the expression

$$20 - [15 - (5 + 7)]$$

we simplify within the parentheses first.

$20 - [15 - (12)]$ simplified within parentheses

Next we simplify within the brackets.

$20 - [3]$ simplified within brackets

17 subtracted

Example 1 Simplify: $50 - [20 + (10 - 5)]$

Solution First we simplify within the parentheses.

$50 - [20 + (5)]$	simplified within parentheses
$50 - [25]$	simplified within brackets
25	subtracted

Example 2 Simplify: $12 - (8 - |4 - 6| + 2)$

Solution Absolute value symbols may serve as symbols of inclusion. In this problem we find the absolute value of $4 - 6$ as the first step of simplifying within the parentheses.

$12 - (8 - 2 + 2)$	found absolute value of $4 - 6$
$12 - (8)$	simplified within parentheses
4	subtracted

Division bar As we noted in Lesson 52, a division bar can serve as a symbol of inclusion. We simplify above and below the division bar before we divide. We follow the order of operations within the symbol of inclusion.

Example 3 Simplify: $\dfrac{4 + 5 \times 6 - 7}{10 - (9 - 8)}$

Solution We simplify above and below the bar before we divide. Above the bar we multiply first. Below the bar we simplify within the parentheses first. This gives us

$$\frac{4 + 30 - 7}{10 - (1)}$$

We continue by simplifying above and below the division bar.

$$\frac{27}{9}$$

Now we divide and get

$$\mathbf{3}$$

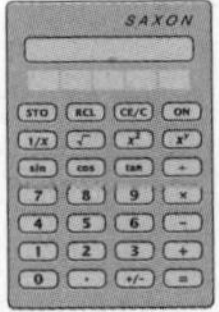

Some calculators with parenthesis keys are capable of dealing with many levels of parentheses (parentheses within parentheses within parentheses). When performing calculations such as the one in example 1, we press the "open parenthesis" key, (, for each opening parenthesis, bracket, or brace. We press the "close parenthesis" key,), for each closing parenthesis, bracket, or brace. For the problem in example 1, the keystrokes are

5 0 − (2 0 + (1 0 − 5)) =

To perform calculations such as the one in example 3 using a calculator, we follow one of these two procedures:

1. We perform the calculations above the bar and record the result. Then we perform the calculations below the bar and record the result. Finally, we perform the division using the two recorded numbers.
2. To perform the calculation with one uninterrupted sequence of keystrokes, we picture the problem like this:

$$\frac{4 + 5 \times 6 - 7}{[10 - (9 - 8)]} =$$

We press the equals key after the 7 to complete the calculations above the bar. Then we press ÷ for the division bar. We place all the operations below the division bar within a set of parentheses so that the denominator is handled by the calculator as though it were one number.

If you have a calculator with parenthesis keys and algebraic logic, perform these calculations and note the display at the indicated location in the sequence of keystrokes.

4 + 5 × 6 − 7 =

The number 27 should be displayed. This number is the numerator.

÷ (1 0 − (9 − 8)) =

The number 3 should be displayed. This number is the quotient.

LESSON PRACTICE

Practice set Simplify:

a. $30 - [40 - (10 - 2)]$

b. $100 - 3[2(6 - 2)]$

c. $\dfrac{10 + 9 \cdot 8 - 7}{6 \cdot 5 - 4 - 3 + 2}$

d. $\dfrac{1 + 2(3 + 4) - 5}{10 - 9(8 - 7)}$

e. $12 + 3(8 - |-2|)$

MIXED PRACTICE

Problem set

1. Jennifer and Jason each earn \$6 per hour doing yard work. On one job Jennifer worked 3 hours, and Jason worked $2\frac{1}{2}$ hours. Altogether, how much money were they paid?
(28)

2. When Jim is resting, his heart beats 70 times per minute. When Jim is jogging, his heart beats 150 times per minute. During a half hour of jogging, Jim's heart beats how many more times than it would if he were resting?
(28, 53)

3. Use a ratio box to solve this problem. The ratio of brachiopods to trilobites in the fossil find was 2 to 9. If 720 trilobites were found, how many brachiopods were found?
(54)

4. During the first 5 days of the journey, the wagon train averaged 18 miles per day. During the next 2 days the wagon train traveled 16 miles and 21 miles. If the total journey is 1017 miles, how much farther does the wagon train have to travel?
(55)

5. During one day of the journey, the wagon train descended from an elevation of 2850 feet to a spot on the desert floor 160 feet below sea level. What was the net descent of the wagon train during the day?
(59)

6. The average distance from Earth to the Sun is 1.496×10^8 km. Use words to write that distance.
(51)

7. Diagram this statement. Then answer the questions that follow.
(22, 48)

   Twelve of the 40 cars pulled by the locomotive were tankers.

   (a) What fraction of the cars were tankers?

   (b) What percent of the cars were not tankers?

8. The top speed of Jamaal's pet snail is 2×10^{-3} mile per hour. Use words to write the snail's top speed.
(57)

9. Use a unit multiplier to convert 1.5 km to m.
(50)

10. Divide 4.36 by 0.012 and write the answer with a bar over the repetend.
(45)

11. (59) Show this addition problem on a number line:

$$(-3) + (+5) + (-2)$$

12. (48) Complete the table.

FRACTION	DECIMAL	PERCENT
(a)	(b)	33%
$\frac{1}{3}$	(c)	(d)

13. (58) Describe the rule of this function.

IN	FUNCTION	OUT
3 →		→ 1
12 →		→ 4
6 →		→ 2
15 →		→ 5

14. (36) What is the probability of drawing a red face card when drawing one card from a normal deck?

In the figure below, $AB = AD = BD = CD = 5$ cm. The measure of angle ABC is 90°. Refer to the figure for problems 15–17.

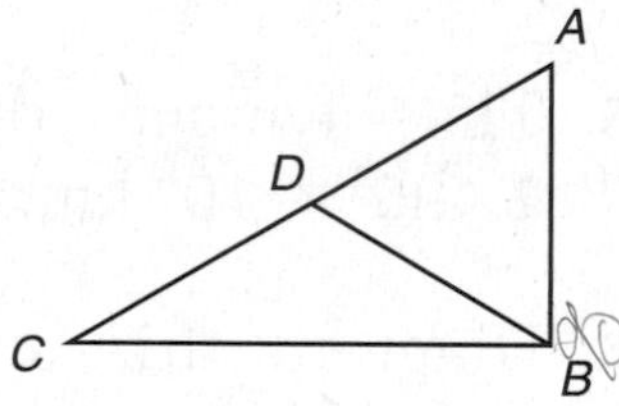

15. (62) (a) Classify ΔBCD by its sides.

(b) What is the perimeter of the equilateral triangle?

(c) Which triangle is a right triangle?

16. (40) Find the measure of each of the following angles:

(a) $\angle BAC$ (b) $\angle ADB$ (c) $\angle BDC$

(d) $\angle DBA$ (e) $\angle DBC$ (f) $\angle DCB$

17. (36) What is the ratio of the length of the shortest side of ΔABC to the length of the longest side?

Solve:

18. (30) $\frac{5}{18} = x + \frac{1}{12}$

19. (45) $2 = 0.4p$

Simplify:

20. (63) $3[24 - (8 + 3 \cdot 2)] - \frac{6 + 4}{|-2|}$

21. (52) $3^3 - \sqrt{3^2 + 4^2}$

22. (56)

	1 week	2 days	7 hr
−		5 days	9 hr

23. (50) $\frac{20 \text{ mi}}{1 \text{ gal}} \cdot \frac{1 \text{ gal}}{4 \text{ qt}}$

24. (30) $4\frac{2}{3} + 3\frac{5}{6} + 2\frac{5}{9}$

25. (26) $12\frac{1}{2} \cdot 4\frac{4}{5} \cdot 3\frac{1}{3}$

26. (26, 30) $6\frac{1}{3} - \left(1\frac{2}{3} \div 3\right)$

27. (52) Evaluate: $x^2 + 2xy + y^2$ if $x = 3$ and $y = 4$

28. (58, 62) Draw an isosceles triangle that is not equilateral, and show its line of symmetry.

29. (61) The coordinates of the four vertices of a parallelogram are (0, 0), (4, 0), (1, –3), and (–3, –3).

(a) Graph the parallelogram.

(b) Find the area of the parallelogram.

(c) What is the measure of each acute angle of the parallelogram?

30. (3) Three identical boxes are balanced on one side of a scale by a 750-g mass on the other side of the scale. What is the mass of each box?

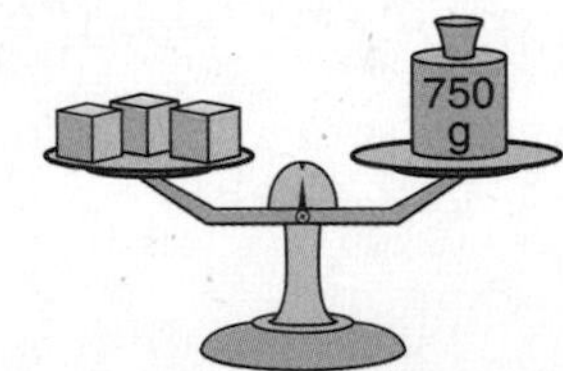

LESSON

64 Adding Signed Numbers

WARM-UP

Facts Practice: + − × ÷ Mixed Numbers (Test N)

Mental Math:

a. 3.6 × 50

b. 7.5 × 10^2

c. $4x - 5 = 35$

d. Convert 20 cm to mm.

e. $\sqrt{9 + 16}$

f. $\frac{5}{9}$ of $1.80

g. 1.5 + 1, × 2, + 3, ÷ 4, − 1.5

Problem Solving:

When all the cards from a 52-card deck are dealt to three players, each player receives 17 cards, and there is one extra card. Dean invented a new deck of cards so that any number of players up to 6 can play and there will be no extra cards. How many cards are in Dean's deck if the number is less than 100?

NEW CONCEPT

From our practice on the number line, we have seen that when we add two negative numbers, the sum is a negative number. When we add two positive numbers, the sum is a positive number.

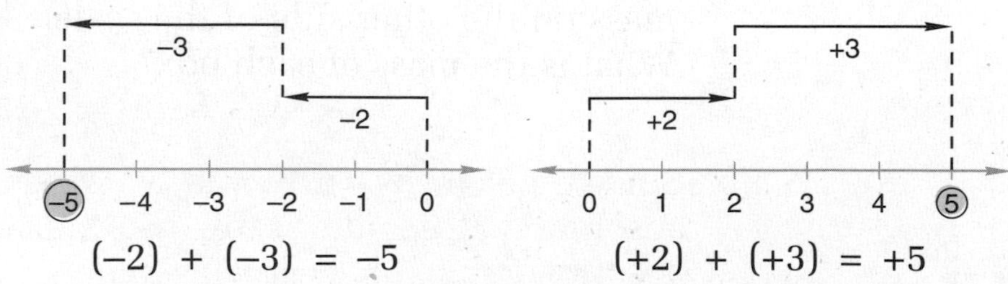

We have also seen that when we add a positive number and a negative number, the sum is positive, negative, or zero depending upon which, if either, of the numbers has the greater absolute value.

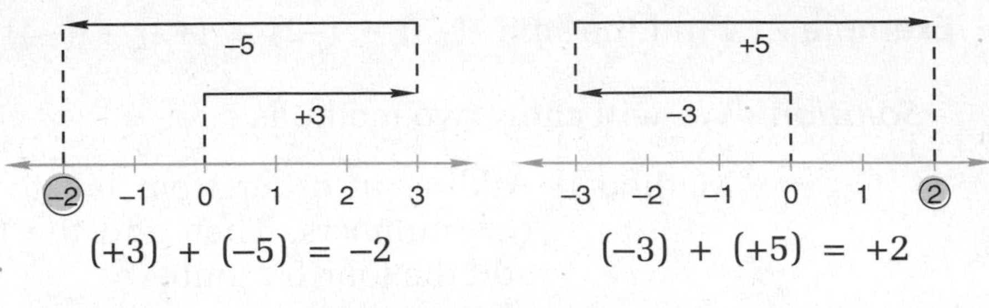

(+3) + (–5) = –2 (–3) + (+5) = +2

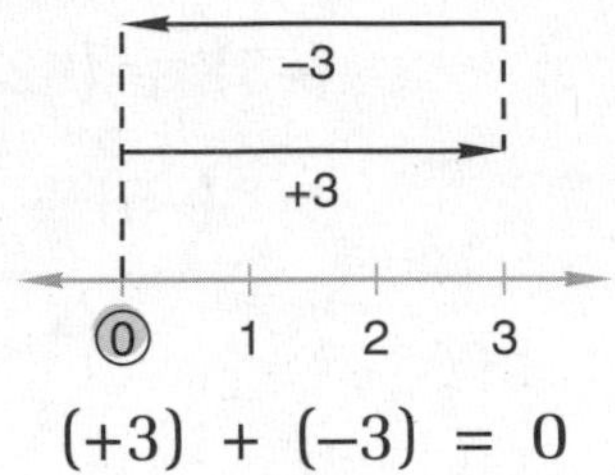

(+3) + (–3) = 0

We can summarize these observations with the following statements.

1. The sum of two numbers with the same sign has an absolute value equal to the sum of their absolute values. Its sign is the same as the sign of the numbers.
2. The sum of two numbers with opposite signs has an absolute value equal to the difference of their absolute values. Its sign is the same as the sign of the number with the greater absolute value.
3. The sum of two opposites is zero.

We can use these observations to help us add signed numbers without drawing a number line.

Example 1 Find each sum:

(a) (–54) + (–78) (b) (+45) + (–67) (c) (–92) + (+92)

Solution (a) Since the signs are the same, we add the absolute values and use the same sign for the sum.

(–54) + (–78) = **–132**

(b) Since the signs are different, we find the difference of the absolute values and keep the sign of –67 because its absolute value, 67, is greater than 45.

(+45) + (–67) = **–22**

(c) The sum of two opposites is zero, a number which has no sign.

(–92) + (+92) = **0**

Example 2 Find the sum: $(-3) + (-2) + (+7) + (-4)$

Solution We will show two methods.

Method 1: Adding in order from left to right, add the first two numbers. Then add the third number. Then add the fourth number.

$(-3) + (-2) + (+7) + (-4)$	problem
$(-5) + (+7) + (-4)$	added −3 and −2
$(+2) + (-4)$	added −5 and +7
$\mathbf{-2}$	added +2 and −4

Method 2: Employing the commutative and associative properties, rearrange the terms and add all numbers with the same sign first.

$(-3) + (-2) + (-4) + (+7)$	rearranged
$(-9) + (+7)$	added
$\mathbf{-2}$	added

Example 3 Find each sum:

(a) $\left(-2\frac{1}{2}\right) + \left(-3\frac{1}{3}\right)$ (b) $(+4.3) + (-7.24)$

Solution These numbers are not integers, but the method for adding these signed numbers is the same as the method for adding integers.

(a) The signs are both negative. We add the absolute values and keep the same sign.

$$\left(-2\frac{1}{2}\right) + \left(-3\frac{1}{3}\right) = \mathbf{-5\frac{5}{6}}$$

$$\begin{array}{r} 2\frac{1}{2} = 2\frac{3}{6} \\ +\ 3\frac{1}{3} = 3\frac{2}{6} \\ \hline 5\frac{5}{6} \end{array}$$

(b) The signs are different. We find the difference of the absolute values and keep the sign of −7.24.

$$\begin{array}{r} \overset{6}{\cancel{7}}.\overset{1}{2}4 \\ -\ 4.3 \\ \hline 2.94 \end{array}$$

$$(+4.3) + (-7.24) = \mathbf{-2.94}$$

Example 4 On one stock trade Tim lost $450. On a second trade Tim gained $800. What was the net result of the two trades?

Solution A loss may be represented by a negative number and a gain by a positive number. So the results of the two trades may be expressed this way:

$$(-450) + (+800)$$

The sum represents the net result of the two trades. The sum is +350, which represents **a gain of $350.**

LESSON PRACTICE

Practice set* Find each sum:

a. $(-56) + (+96)$

b. $(-28) + (-145)$

c. $(-5) + (+7) + (+9) + (-3)$

d. $(-3) + (-8) + (+15)$

e. $(-12) + (-9) + (+16)$

f. $(+12) + (-18) + (+6)$

g. $\left(-3\frac{5}{6}\right) + \left(+5\frac{1}{3}\right)$

h. $(-1.6) + (-11.47)$

i. On three separate stock trades Dawn gained $250, lost $300, and gained $525. Write an expression that shows the results of each trade. Then find the net result of the trades.

MIXED PRACTICE

Problem set

1. (51) Two trillion is how much more than seven hundred fifty billion? Write the answer in scientific notation.

2. (28) The taxi cost $2.25 for the first mile plus 15¢ for each additional tenth of a mile. For a 5.2-mile trip, Eric paid $10 and told the driver to keep the change as a tip. How much was the driver's tip?

3. (44) Mae-Ying wanted to buy packages of crackers and cheese from the vending machine. Each package cost 35¢. Mae-Ying had 5 quarters, 3 dimes, and 2 nickels. How many packages of crackers and cheese could she buy?

4. The two prime numbers p and m are between 50 and 60.
(21) Their difference is 6. What is their sum?

5. What is the mean of 1.74, 2.8, 3.4, 0.96, 2, and 1.22?
(28)

6. Diagram this statement. Then answer the questions
(22, 48) that follow.

The viceroy conscripted two fifths of the 1200 serfs in the province.

(a) How many of the serfs in the province were conscripted?

(b) What percent of the serfs in the province were not conscripted?

7. Write an equation to solve this problem:
(60)

What number is $\frac{5}{9}$ of 100?

8. (a) The temperature at the center of the sun is about
(51, 57) 1.6×10^7 degrees Celsius. Use words to write that temperature.

(b) A red blood cell is about 7×10^{-6} meter in diameter. Use words to write that length.

9. Compare:
(57)

(a) $1.6 \times 10^7 \bigcirc 7 \times 10^{-6}$

(b) $7 \times 10^{-6} \bigcirc 0$

(c) $2^{-3} \bigcirc 2^{-2}$

10. Divide 456 by 28 and write the answer
(44)

(a) as a mixed number.

(b) as a decimal rounded to two decimal places.

(c) rounded to the nearest whole number.

11. Find each sum:
(64)

(a) $(-63) + (-14)$

(b) $(-16) + (+20) + (-32)$

12. On two separate stock trades Josefina lost \$327 and gained \$280. What was the net result of the two trades?
(64)

13. The figure shows an equilateral triangle inscribed in a circle.
(Inv. 2, 62)

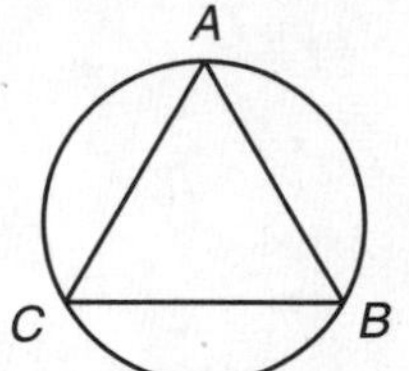

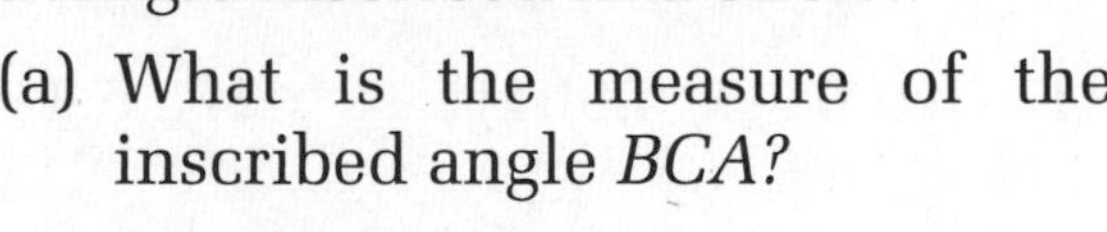

(a) What is the measure of the inscribed angle *BCA?*

(b) Select a chord of this circle, and state whether the chord is longer or shorter than the diameter of the circle and why.

14. Evaluate: $x + xy$ if $x = \frac{2}{3}$ and $y = \frac{3}{4}$
(52)

Refer to the hexagon below to answer problems 15 and 16. Dimensions are in meters. All angles are right angles.

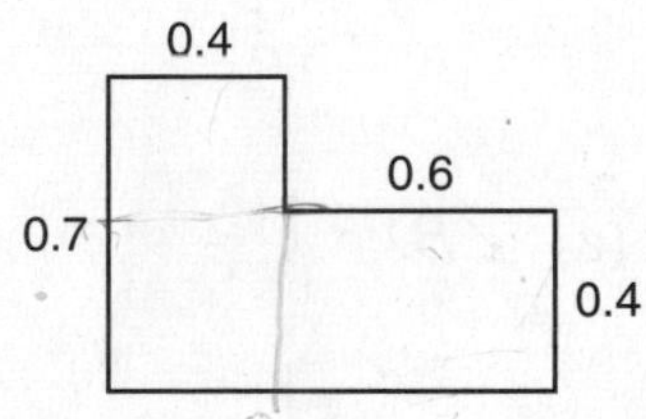

15. What is the perimeter of the hexagon?
(19)

16. What is the area of the hexagon?
(37)

17. The product of x and 12 is 84. The product of y and 12 is 48. What is the product of x and y?
(41)

18. The center of a circle with a radius of three units is (1, 1). Which of these points is on the circle?
(Inv. 3)

A. (4, 4) B. (−2, 1) C. (−4, 1) D. (3, 0)

Solve:

19. $\frac{4}{9} = y - \frac{2}{9}$
(15)

20. $25x = 10$
(44)

Simplify:

21. (63) $\dfrac{3^2 + 4^2}{\sqrt{3^2 + 4^2}}$

22. (26) $2\frac{4}{5} \div \left(6 \div 2\frac{1}{2}\right)$

23. (57, 63) $100 - [20 + 5(4) + 3(2 + 4^0)]$

24. (49)
$$\begin{array}{r} 5 \text{ gal } 2 \text{ qt } 1 \text{ pt } 7 \text{ oz} \\ + \ 1 \text{ gal } 1 \text{ qt } 1 \text{ pt } 9 \text{ oz} \\ \hline \end{array}$$

25. (26, 30) $\left(1\frac{1}{2}\right)^2 - \left(4 - 2\frac{1}{3}\right)$

26. (35) $0.1 - (0.01 - 0.001)$

27. (45) $5.1 \div (5.1 \div 1.5)$

28. (43) Write $3\frac{1}{5}$ as a decimal number, and subtract it from 4.375.

29. (36) What is the probability of rolling an even prime number with one toss of a die?

30. (61) Figure *ABCD* is a parallelogram. Find the measure of

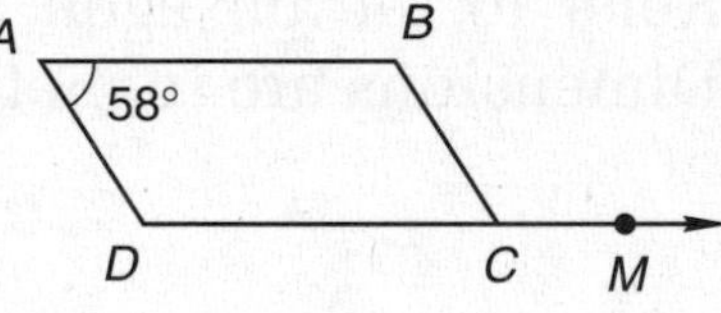

(a) $\angle B$. (b) $\angle BCD$. (c) $\angle BCM$.

LESSON

65 Ratio Problems Involving Totals

WARM-UP

Facts Practice: Metric Conversions (Test M)

Mental Math:

a. 0.42×50

b. 1.25×10^{-1}

c. $\frac{9}{w} = \frac{15}{10}$

d. Convert 0.75 m to mm.

e. $5^3 - 10^2$

f. $\frac{9}{10}$ of \$4.00

g. What is the total cost of a \$20.00 item plus 7% sales tax?

Problem Solving:

Copy this problem and fill in the missing digits:

```
      91½
__)‾‾‾‾‾
    ___
    ==
     __
     ==
      _
```

NEW CONCEPT

Some ratio problems require that we use the total to solve the problem. Consider the following problem:

> *The ratio of boys to girls at the concert was 5 to 4. If there were 180 children at the concert, how many girls were there?*

We begin by making a ratio box. This time we add a third row for the total number of children.

	Ratio	Actual Count
Boys	5	*B*
Girls	4	*G*
Total	9	180

In the ratio column we wrote 5 for boys and 4 for girls, then *added these to get 9 for the total ratio number.* We were given 180 as thc actual count of children. This is a total. We can use two rows from this table to write a proportion. Since we were asked to find the number of girls, we will use the "girls" row.

Because we know both total numbers, we will also use the "total" row. Using these numbers, we solve the proportion.

	Ratio	Actual Count
Boys	5	B
Girls	4	G
Total	9	180

$$\frac{4}{9} = \frac{G}{180}$$
$$9G = 720$$
$$G = 80$$

We find there were 80 girls at the concert. We can use this answer to complete the ratio box.

	Ratio	Actual Count
Boys	5	100
Girls	4	80
Total	9	180

Example The ratio of football players to soccer players in the room was 5 to 7. If the football and soccer players in the room totaled 48, how many were football players?

Solution We use the information in the problem to form a table. We include a row for the total number of players. The total ratio number is 12.

	Ratio	Actual Count
Football players	5	F
Soccer players	7	S
Total players	12	48

$$\frac{5}{12} = \frac{F}{48}$$
$$12F = 240$$
$$F = 20$$

To find the number of football players, we write a proportion from the "football players" row and the "total players" row. We solve the proportion to find that there were **20 football players** in the room. From this information we can complete the ratio box.

	Ratio	Actual Count
Football players	5	20
Soccer players	7	28
Total players	12	48

LESSON PRACTICE

Practice set Solve these problems. Begin by drawing a ratio box.

a. Acrobats and clowns converged on the center ring in the ratio of 3 to 5. If a total of 72 acrobats and clowns performed in the center ring, how many were clowns?

b. The ratio of young men to young women at the ball was 8 to 9. If 240 young men were in attendance, how many young people attended in all?

MIXED PRACTICE

Problem set

1. (46) If 5 pounds of apples cost $2.40, then

(a) what is the price per pound?

(b) what is the cost for 8 pounds of apples?

2. (41) (a) Simplify and compare:

$$(0.3)(0.4) + (0.3)(0.5) \bigcirc 0.3(0.4 + 0.5)$$

(b) What property is illustrated by this comparison?

3. (65) Use a ratio box to solve this problem. The ratio of big fish to little fish in the pond was 4 to 11. If there were 1320 fish in the pond, how many big fish were there?

4. (44, 46) The car traveled 350 miles on 15 gallons of gasoline. The car averaged how many miles per gallon? Round the answer to the nearest tenth.

5. (28) The average of 2 and 4 is 3. What is the average of the reciprocals of 2 and 4?

6. (51) Write 12 billion in scientific notation.

7. (22, 36) Diagram this statement. Then answer the questions that follow.

One sixth of the five dozen eggs were cracked.

(a) How many eggs were not cracked?

(b) What was the ratio of eggs that were cracked to eggs that were not cracked?

(c) What percent of the eggs were cracked?

8. (Inv. 6) (a) Draw segment AB. Draw segment DC parallel to segment AB but not the same length. Draw segments between the endpoints of segments AB and DC to form a quadrilateral.

(b) What type of quadrilateral was formed in (a)?

9. (37) Find the area of each triangle. Dimensions are in centimeters.

(a)

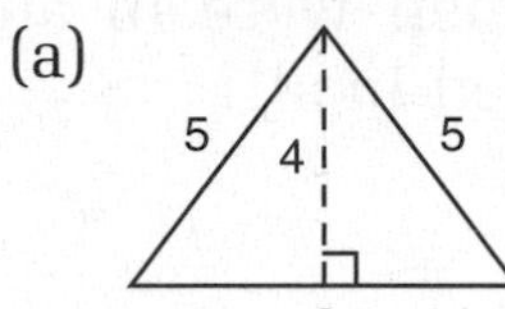

(b)

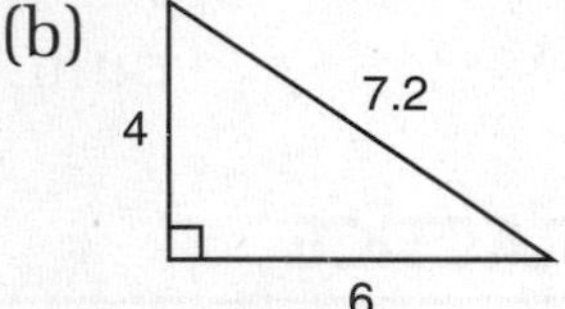

(c)

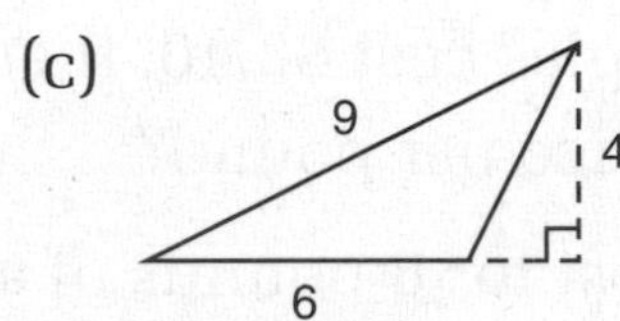

10. (28, 34) What is the average of the two numbers indicated by arrows on the number line below?

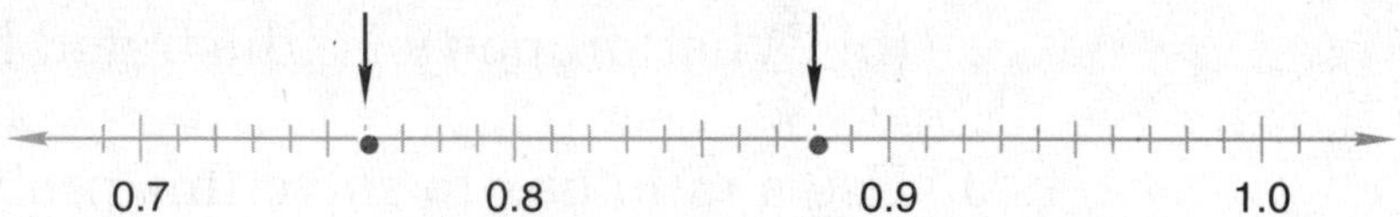

Write equations to solve problems 11 and 12.

11. (60) What number is 75 percent of 64?

12. (46) What is the tax on a \$7.40 item if the sales-tax rate is 8%?

13. (64) Find each sum:

(a) (−3) + (−8)

(b) (+3) + (−8)

(c) (−3) + (+8) + (−5)

14. (Inv. 3) A circle is drawn on a coordinate plane with its center at the origin. One point on the circle is (3, 4). Use a compass and graph paper to graph the circle. Then answer (a) and (b).

(a) What are the coordinates of the points where the circle intersects the x-axis?

(b) What is the diameter of the circle?

15. Use a unit multiplier to convert 0.95 liters to milliliters.
(50)

16. Evaluate: $ab + a + \frac{a}{b}$ if $a = 5$ and $b = 0.2$
(52)

17. How many small blocks were used to build this cube?
(13)

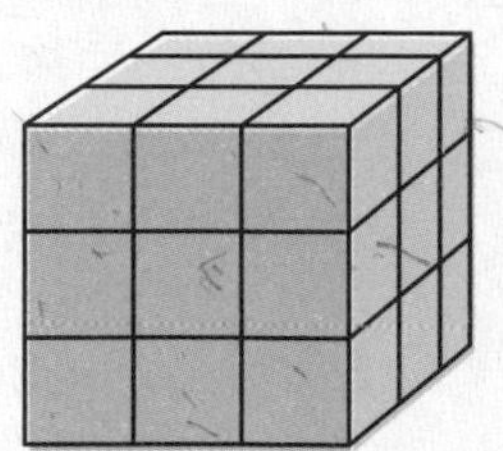

18. Recall that one angle is the complement of another angle if their sum is 90°, and that one angle is the supplement of another if their sum is 180°. In this figure, (a) which angle is a complement of $\angle BOC$ and (b) which angle is a supplement of $\angle BOC$?
(40)

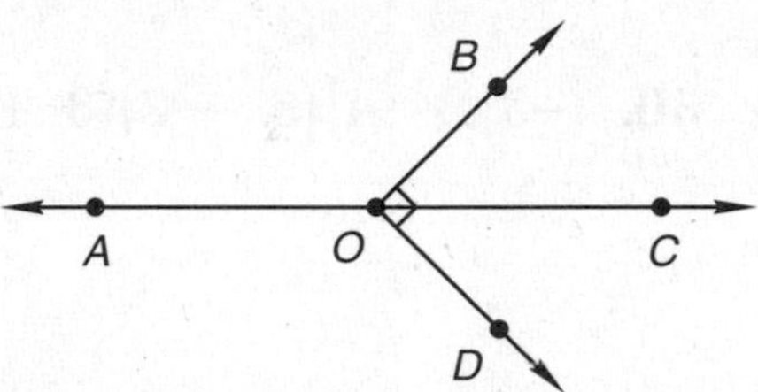

19. Round each number to the nearest whole number to estimate the product of 19.875 and $4\frac{7}{8}$.
(29, 33)

20. Refer to ΔABC at right to answer the following questions:
(58, 62)

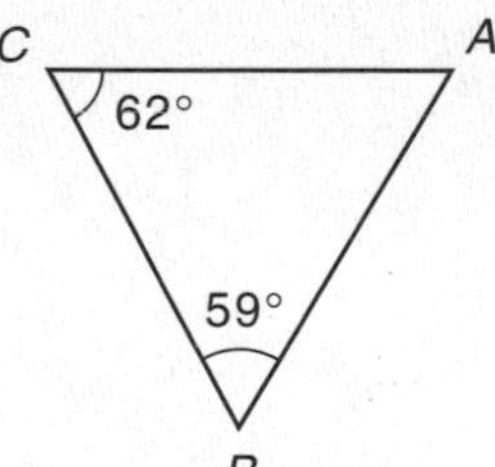

(a) What is the measure of $\angle A$?

(b) Which side of the triangle is the longest side?

(c) Triangle *ABC* is an acute triangle. It is also what other type of triangle?

(d) Triangle *ABC*'s line of symmetry passes through which vertex?

21. (a) Describe how to find the median of this set of 12 scores.
(Inv. 4)

18, 17, 15, 20, 16, 14, 15, 16, 17, 18, 16, 19

(b) What is the median of the set of scores?

22. (62) Answer true or false:

(a) All equilateral triangles are congruent.

(b) All equilateral triangles are similar.

23. (32, 34) The bar was raised from 2.15 meters to 2.2 meters. How many centimeters was the bar raised?

Simplify:

24. (47) $\dfrac{10^3 \cdot 10^3}{10^2}$

25. (56)

$$\begin{array}{rrr} 4 \text{ days} & 5 \text{ hr} & 15 \text{ min} \\ -\ 1 \text{ days} & 7 \text{ hr} & 50 \text{ min} \\ \hline \end{array}$$

26. (45) $4.5 \div (0.4 + 0.5)$

27. (52) $\dfrac{3 + 0.6}{3 - 0.6}$

28. (26) $4\frac{1}{5} \div \left(1\frac{1}{6} \cdot 3\right)$

29. (52) $3^2 + \sqrt{4 \cdot 7 - 3}$

30. (63) $|-3| + 4[(5 - 2)(3 + 1)]$

LESSON

66 Circumference and Pi

WARM-UP

Facts Practice: + – × ÷ Mixed Numbers (Test N)

Mental Math:

a. $3.65 + 1.2 + 2$
b. 1.2×10^{-3}
c. $9y + 3 = 75$
d. Convert 20 decimeters (dm) to meters.
e. $\sqrt{144} + 2^3$
f. 25% of 24
g. Estimate the product of 3.14 and 25.

Problem Solving:

The product of $10 \times 10 \times 10$ is 1000. Find three prime numbers whose product is 1001.

NEW CONCEPT

Recall from Investigation 2 that a **circle** is a smooth curve, every point of which is the same distance from the **center.** The distance from the center to the circle is the **radius.** The plural of radius is **radii.** The distance across a circle through the center is the **diameter.** The distance around a circle is the **circumference.**

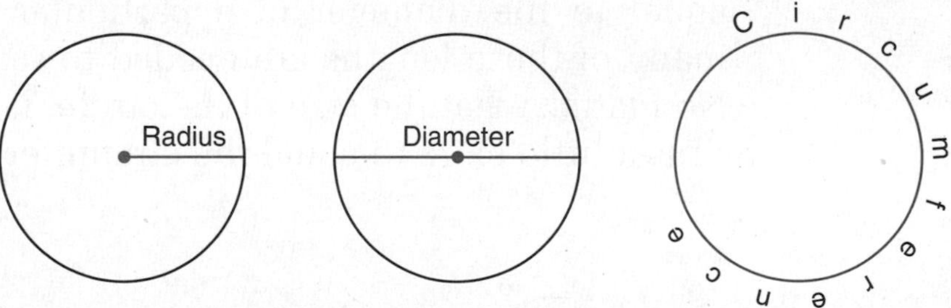

We see that the diameter of a circle is twice the radius of the circle. In the following activity we investigate the relationship between the diameter and the circumference.

Activity: *Investigating Circumference and Diameter*

Materials needed:

- Tape measure (preferably metric)
- Circular objects
- Calculator (optional)

Select a circular object and measure its circumference and its diameter as precisely as you can. To calculate the number of diameters that equal the circumference, divide the circumference by the diameter. Round the quotient to two decimal places. Repeat the activity with another circular object of a different size and record your results in a table similar to the one shown below. Are the ratios of circumference to diameter nearly equal for each object you measured?

Sample Table

Object	Circumference	Diameter	$\frac{\text{Circumference}}{\text{Diameter}}$
Waste basket	94 cm	30 cm	3.13
Plastic cup	22 cm	7 cm	3.14

How many diameters equal a circumference? Mathematicians investigated this question for thousands of years. They found that the answer did not depend on the size of the circle. The circumference of every circle is slightly more than three diameters.

Another way to illustrate this fact is to cut a length of string equal to the diameter of a particular circle and find how many of these lengths are needed to reach around the circle. No matter what the size of the circle, it takes three diameters plus a little extra to equal the circumference.

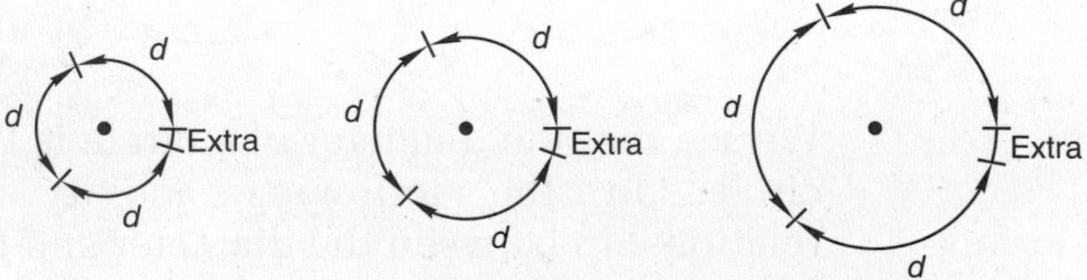

The extra amount needed is about, but not exactly, one seventh of a diameter. Thus the number of diameters needed to equal the circumference of a circle is about

$$3\frac{1}{7} \quad \text{or} \quad \frac{22}{7} \quad \text{or} \quad 3.14$$

Neither $3\frac{1}{7}$ nor 3.14 is the exact number of diameters needed. They are approximations. There is no fraction or decimal number that states the exact number of diameters in a circumference. (Some computers have calculated the number to more than 1 million decimal places.) We use the symbol π, which is the Greek letter **pi** (pronounced like "pie"), to represent this number. Note that π is not a variable. Rather, π is a **constant** because its value does not vary.

The circumference of a circle is π times the diameter of the circle. This idea is expressed by the formula

$$\boldsymbol{C = \pi d}$$

In this formula C stands for circumference and d for diameter. To perform calculations with π, we can use an approximation. The commonly used approximations for π are

$$3.14 \qquad \text{and} \qquad \frac{22}{7}$$

For calculations that require great accuracy, more accurate approximations for π can be used, such as

$$3.14159265359$$

Sometimes π is left as π. Unless directed otherwise, we use 3.14 for π for calculations in this book.

Example 1 The radius of a circle is 10 cm. What is the circumference?

Solution If the radius is 10 cm, the diameter is 20 cm.

$$\begin{aligned} \text{Circumference} &= \pi \cdot \text{diameter} \\ &\approx 3.14 \cdot 20 \text{ cm} \\ &\approx 62.8 \text{ cm} \end{aligned}$$

The circumference is about **62.8 cm.** In the solution we used the symbol $\approx$, which means "approximately equals," because the value for π is not exactly 3.14.

Example 2 Find the circumference of each circle:

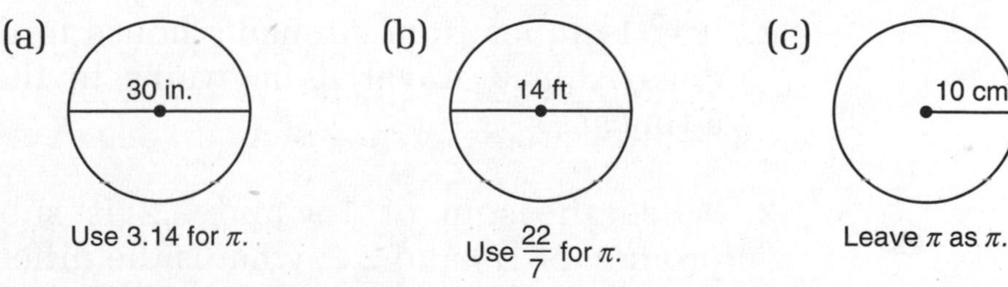

Solution (a) $C = \pi d$

$C \approx 3.14(30 \text{ in.})$

$C \approx \mathbf{94.2 \text{ in.}}$

(b) $C = \pi d$

$C \approx \frac{22}{7}(14 \text{ ft})$

$C \approx \mathbf{44 \text{ ft}}$

(c) $C = \pi d$

$C = \pi(20 \text{ cm})$

$C = \mathbf{20\pi \text{ cm}}$

Note the form of answer (c): first 20 times π, then the unit of measure.

LESSON PRACTICE

Practice set* Find the circumference of each circle:

a.

4 in.

Use 3.14 for π.

b.

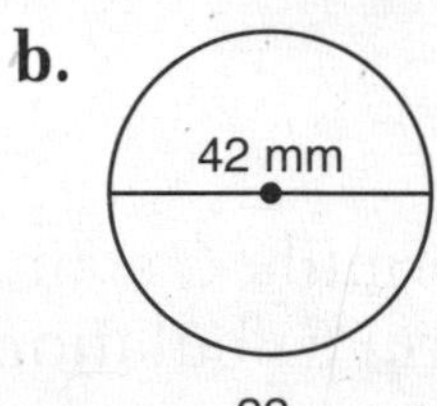

Use $\frac{22}{7}$ for π.

c.

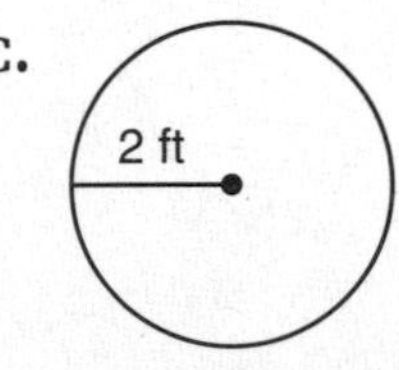

Leave π as π.

d. Sylvia used a compass to draw a circle. If the point of the compass was 3 inches from the point of the pencil, what was the circumference of the circle? (Use 3.14 for π.)

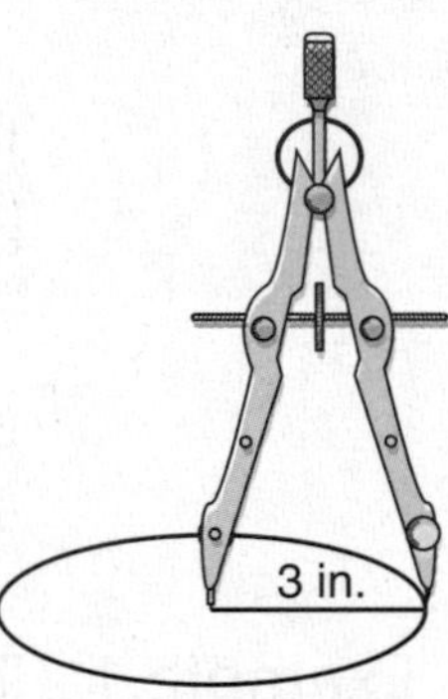

MIXED PRACTICE

Problem set

1. *(38, 60)* According to this circle graph, what percent of Dan's income was spent on items other than food and housing? If his income was $25,000, how much did he spend on food?

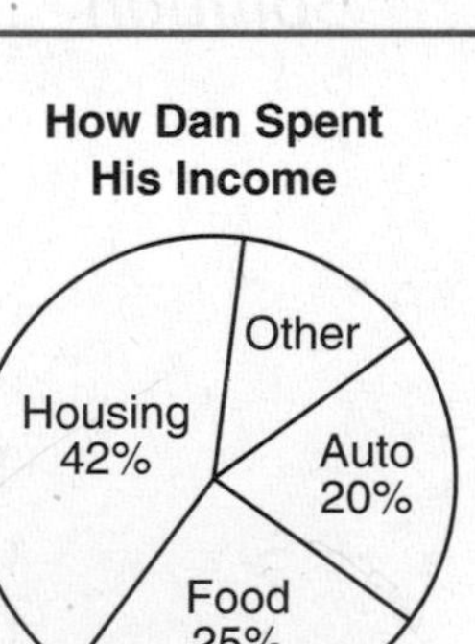

2. *(28)* It is $1\frac{1}{4}$ miles from Ahmad's house to the store. How far does Ahmad travel if he walks to the store and back 5 times?

3. *(35)* When the sum of 1.9 and 2.2 is subtracted from the product of 1.9 and 2.2, what is the difference?

4. (65) Use a ratio box to solve this problem. There was a total of 520 dimes and quarters in the soda machine. If the ratio of dimes to quarters was 5 to 8, how many dimes were there?

5. (51) Saturn's average distance from the Sun is about 900 million miles. Write that distance in scientific notation.

6. (22, 48) Diagram this statement. Then answer the questions that follow.

Three tenths of the 400 acres were planted with alfalfa.

(a) What percent of the land was planted with alfalfa?

(b) How many of the 400 acres were not planted with alfalfa?

7. (36, 48) Twelve of the 30 watermelons were ripe.

(a) What fraction of the watermelons were ripe?

(b) What percent of the watermelons were ripe?

(c) If one of the watermelons is picked at random, which is more likely: that the watermelon is ripe or is not ripe? Why?

8. (66) Find the circumference of each circle:

(a)

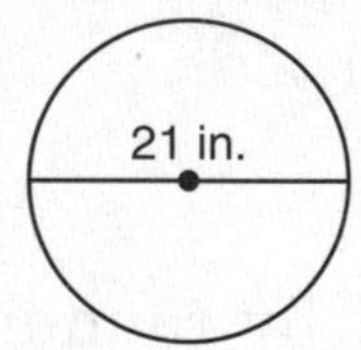

Use 3.14 for π.

(b)

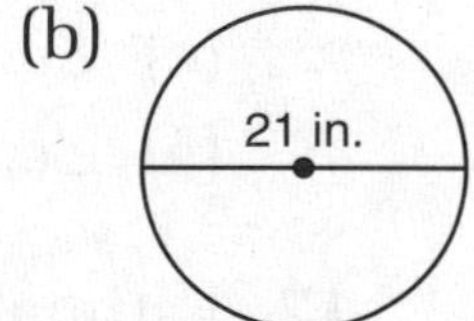

Use $\frac{22}{7}$ for π.

9. (37, 61) Refer to the figure at right to answer (a)–(c). Dimensions are in centimeters.

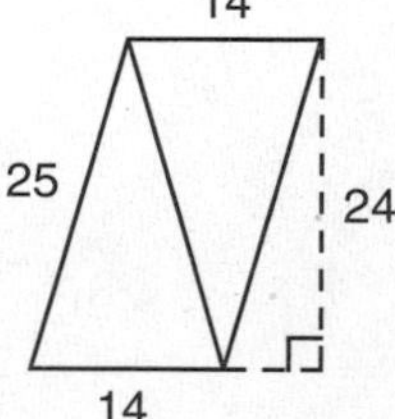

(a) What is the area of the parallelogram?

(b) The two triangles are congruent. What is the area of one of the triangles?

(c) Each triangle is isosceles. What is the perimeter of one of the triangles?

10. (51) Write 32.5 billion in scientific notation.

Write equations to solve problems 11 and 12.

11. (60) What number is 90 percent of 3500?

12. (60) What number is $\frac{5}{6}$ of $2\frac{2}{5}$?

13. (48) Complete the table.

FRACTION	DECIMAL	PERCENT
(a)	0.45	(b)
(c)	(d)	7.5% or $7\frac{1}{2}$%

14. (64) Find each sum:

(a) (5) + (–4) + (6) + (–1)

(b) 3 + (–5) + (+4) + (–2)

15. (50) Use a unit multiplier to convert 1.4 kilograms to grams.

16. (40, 61) In the figure at right, two sides of a parallelogram are extended to form two sides of a right triangle. The measure of ∠*M* is 35°. Find the measure of

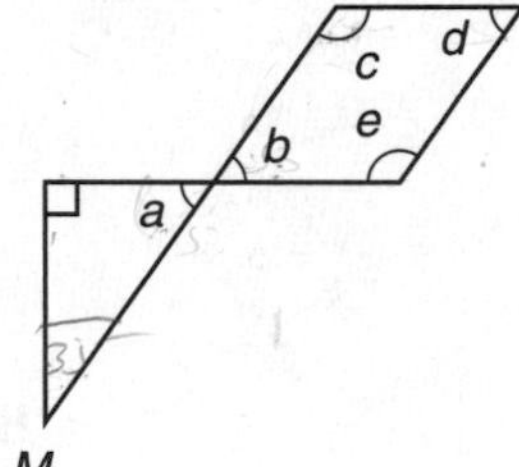

(a) ∠*a*. (b) ∠*b*.

(c) ∠*c*. (d) ∠*d*.

(e) ∠*e*.

17. (29) Estimate this product by rounding each number to one nonzero digit before multiplying:

$$(2876)(513)(18)$$

18. (Inv. 3) The coordinates of the vertices of square *ABCD* are (2, 2), (2, –2), (–2, –2), and (–2, 2). The coordinates of the vertices of square *EFGH* are (2, 0), (0, –2), (–2, 0), and (0, 2). Draw both squares on the same coordinate plane and answer (a)–(d).

(a) What is the area of square *ABCD?*

(b) What is the length of one side of square *ABCD?*

(c) Counting two half squares on the grid as one square unit, what is the area of square *EFGH?*

(d) Remembering that the length of the side of a square is the square root of its area, what is the length of one side of square *EFGH?*

Solve:

19. (39) $\frac{0.9}{1.5} = \frac{12}{n}$

20. (30) $\frac{11}{24} + w = \frac{11}{12}$

Simplify:

21. (57) $2^1 - 2^0 - 2^{-1}$

22. (49)
$$\begin{array}{r} 4 \text{ lb } 12 \text{ oz} \\ + \ 1 \text{ lb } \ \ 7 \text{ oz} \\ \hline \end{array}$$

23. (50) $\frac{3 \text{ ft}}{1 \text{ yd}} \cdot \frac{12 \text{ in.}}{1 \text{ ft}}$

24. (45) $16 \div (0.8 \div 0.04)$

25. (63) $0.4[0.5 - (0.6)(0.7)]$

26. (26) $\frac{3}{8} \cdot 1\frac{2}{3} \cdot 4 \div 1\frac{2}{3}$

27. (63) $30 - 5[4 + (3)(2) - 5]$

28. (13) Write a word problem for this division: $2.88 ÷ 12

29. (3) Two identical boxes balance a 9-ounce weight. What is the weight of each box?

30. (Inv. 2, 62) Refer to the circle with center at point *M* to answer (a)–(c).

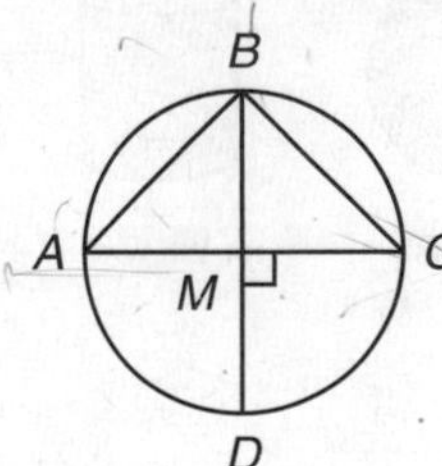

(a) Name two chords that are not diameters.

(b) Classify ΔAMB by its sides.

(c) What is the measure of inscribed angle *ABC*?

LESSON

67 Geometric Solids

WARM-UP

Facts Practice: Metric Conversions (Test M)

Mental Math:

a. $43.6 - 10$

b. 3.85×10^3

c. $\frac{5}{10} = \frac{2.5}{m}$

d. Convert 20 m to decimeters (dm).

e. $10^3 \div 10$

f. 75% of 24

g. A mental calculation technique for multiplying is to double one factor and halve the other factor. The product is the same. Use this technique to multiply 45 and 16.

$$\begin{array}{r} 35 \\ \times\ 14 \\ \hline 490 \end{array} \xrightarrow[\div 2]{\times 2} \begin{array}{r} 70 \\ \times\ 7 \\ \hline 490 \end{array}$$

Problem Solving:

Tom reads 5 pages in 4 minutes, and Jerry reads 4 pages in 5 minutes. If they both begin reading 200-page books at the same time, then Tom will finish how many minutes before Jerry?

NEW CONCEPT

Geometric solids are shapes that take up space. Below we show a few geometric solids.

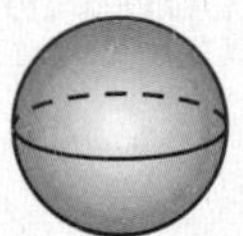
Sphere

Cylinder

Cone

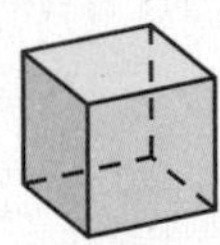
Cube

Triangular prism

Pyramid

Polyhedrons

Some geometric solids, such as spheres, cylinders, and cones, have one or more curved surfaces. If a solid has only flat surfaces that are polygons, the solid is called a **polyhedron.** Cubes, triangular prisms, and pyramids are examples of polyhedrons.

When describing a polyhedron, we may refer to its faces, edges, or vertices. A **face** is one of the flat surfaces. An **edge** is formed where two faces meet. A **vertex** (plural, **vertices**) is formed where three or more edges meet.

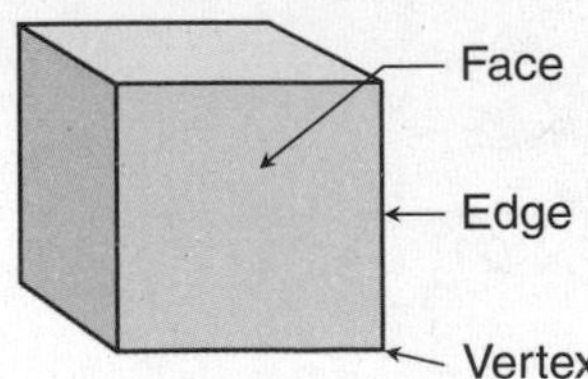

A **prism** is a special kind of polyhedron. A prism has a polygon of a constant size "running through" the prism that appears at opposite faces of the prism and determines the name of the prism. For example, the opposite faces of this prism are congruent triangles; thus this prism is called a **triangular prism.**

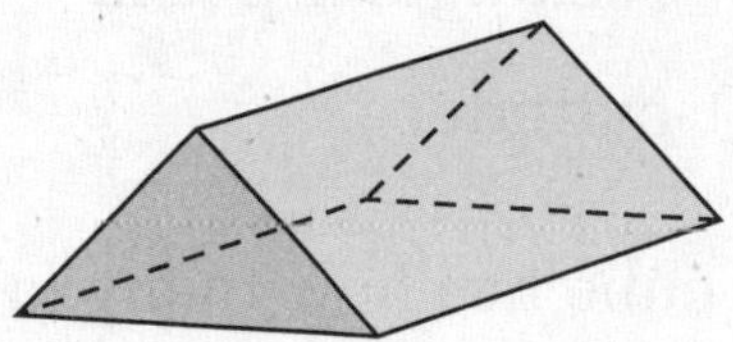

Notice that if we cut through this triangular prism perpendicular to the base, we would see the same size triangle at the cut.

To draw a prism, we draw two identical and parallel polygons, as shown below. Then we draw segments connecting corresponding vertices. We use dashes to indicate edges hidden from view.

Rectangular prism: We draw two congruent rectangles.

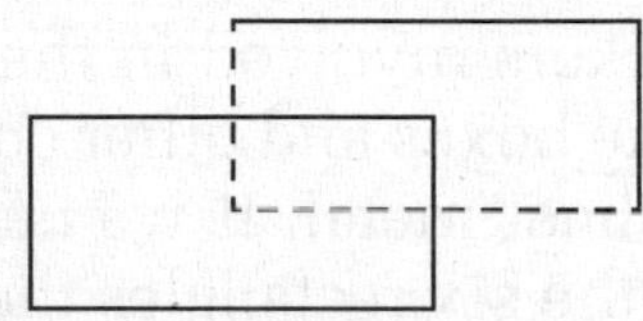

Then we connect the corresponding vertices (using dashes for hidden edges).

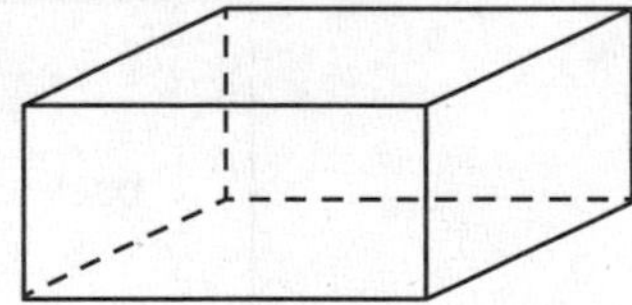

Triangular prism: We draw two congruent triangles.

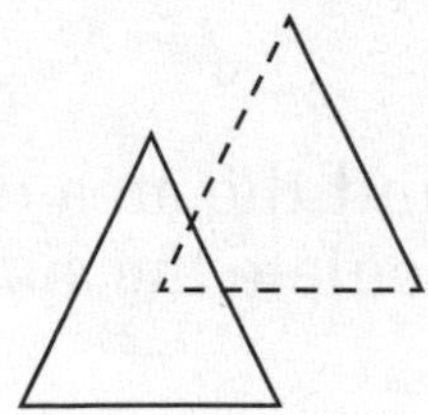

We connect corresponding vertices.

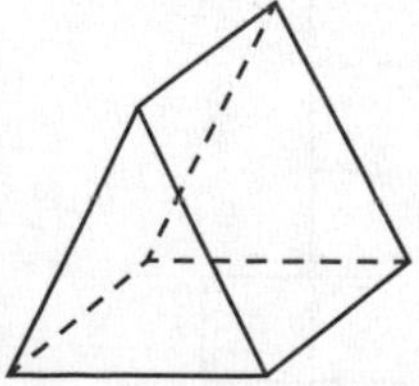

Example 1 Use the name of a geometric solid to describe the shape of each object:

(a) basketball (b) shoe box (c) can of beans

Solution (a) **sphere**

(b) **rectangular prism**

(c) **cylinder**

Example 2 A cube has how many (a) faces, (b) edges, and (c) vertices?

Solution (a) **6 faces** (b) **12 edges** (c) **8 vertices**

Example 3 Draw a cube.

Solution A cube is a special kind of rectangular prism. All faces are squares.

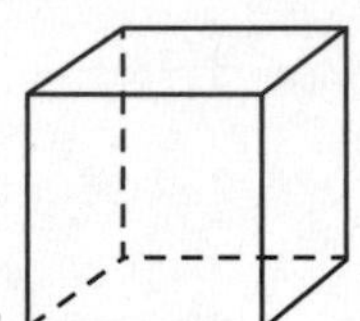

Workers involved in the manufacturing of packaging materials make boxes and other containers out of flat sheets of cardboard or sheet metal. If we cut apart a cereal box and unfold it, we see the six rectangles that form the faces of the box.

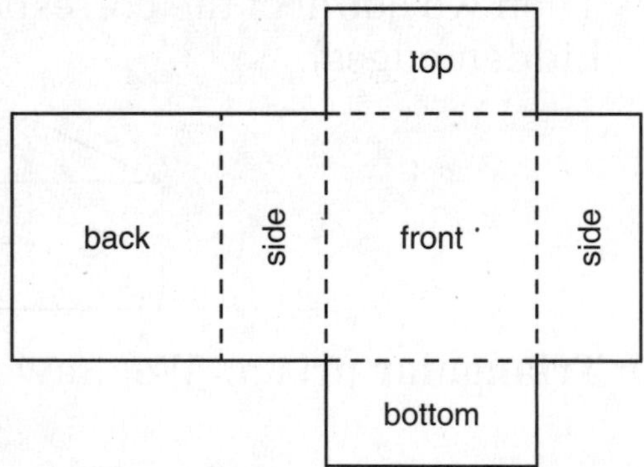

If we find the area of each rectangle and add those areas, we can calculate the **surface area** of the cereal box.

Example 4 Which of these patterns will not fold to form a cube?

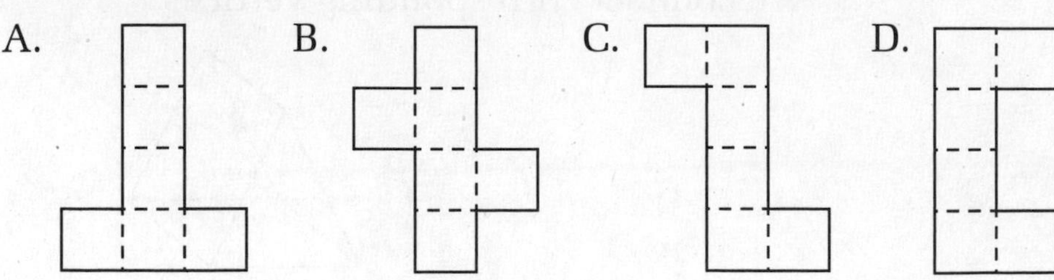

Solution **Pattern D** will not fold into a cube.

Example 5 If each edge of a cube is 5 cm, what is the surface area (the combined area of all of the faces) of the cube?

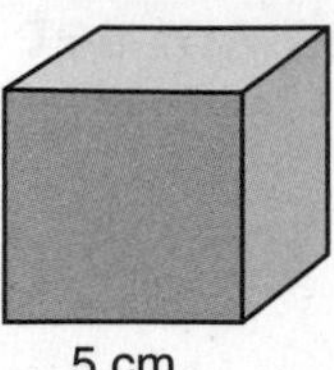

Solution A cube has six congruent square faces. Each face of this cube is 5 cm by 5 cm. So the area of one face is 25 cm^2, and the area of all six faces is

$$6 \times 25\ \text{cm}^2 = \mathbf{150\ cm^2}$$

LESSON PRACTICE

Practice set Use the name of a geometric solid to describe each shape:

a. 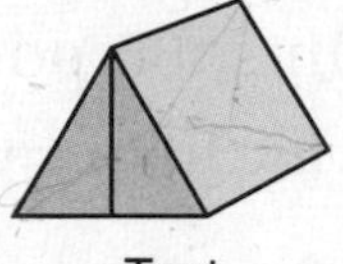
Tent

b. 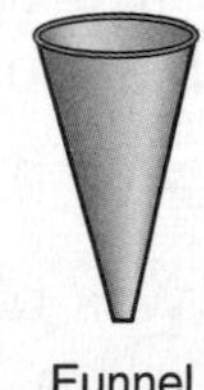
Funnel

c. 

Box

A triangular prism has how many of each of the following?

d. Faces **e.** Edges **f.** Vertices

Draw a representation of each shape. (Refer to the representations at the beginning of this lesson.)

g. Sphere **h.** Rectangular prism

i. Cylinder

j. What three-dimensional figure could be formed by folding this pattern?

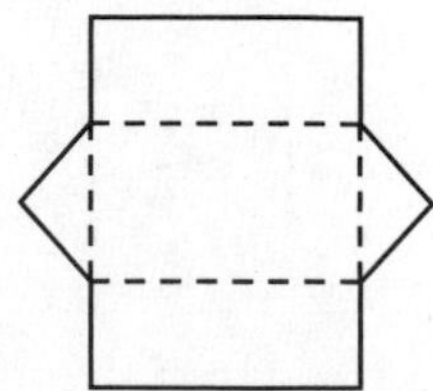

k. Calculate the surface area of a cube whose edges are 3 cm long.

MIXED PRACTICE

Problem set

1. The bag contains 20 red marbles, 30 white marbles, and 40 blue marbles.
(36)
   (a) What is the ratio of red to blue marbles?
   (b) What is the ratio of white to red marbles?
   (c) If one marble is drawn from the bag, what is the probability that the marble will not be white?

2. When the product of $\frac{1}{3}$ and $\frac{1}{2}$ is subtracted from the sum of $\frac{1}{3}$ and $\frac{1}{2}$, what is the difference?
(30)

3. With the baby in his arms, Papa weighed 180 pounds. Without the baby in his arms, Papa weighed $165\frac{1}{2}$ pounds. How much did the baby weigh?
(12, 23)

4. On his first 5 tests, Ché averaged 92 points. On his next 3 tests, Ché scored 94 points, 85 points, and 85 points.
(28, 55)
   (a) What was his average for his last 3 tests?
   (b) What was his average for all 8 tests?

5. Use a ratio box to solve this problem. The jeweler's tray was filled with diamonds and rubies in the ratio of 5 to 2. If 210 gems filled the tray, how many were diamonds?
(65)

6. Diagram this statement. Then answer the questions that follow.
(22, 48)

   Four fifths of the 360 dolls were sold during November.

   (a) How many of the dolls were sold during November?
   (b) What percent of the dolls were not sold during November?

7. The three-dimensional figure that can be formed by folding this pattern has how many
(67)

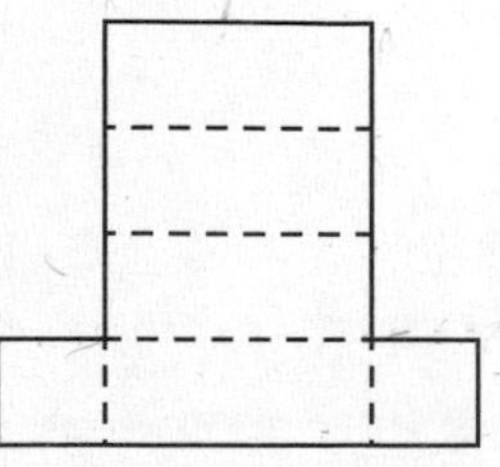

   (a) edges?
   (b) faces?
   (c) vertices?

8. (58, 62) Refer to the triangles below to answer (a)–(d). Dimensions are in meters.

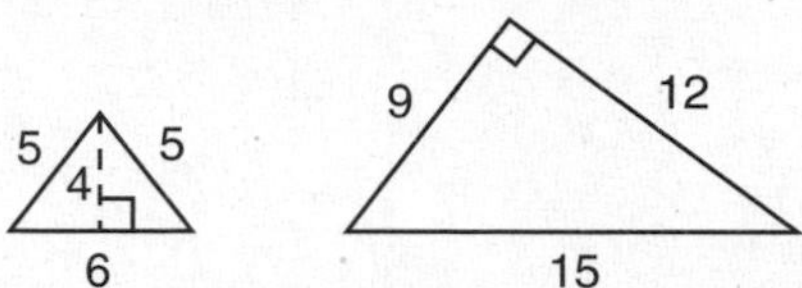

(a) What is the area of the scalene triangle?

(b) What is the perimeter of the isosceles triangle?

(c) If one acute angle of the right triangle measures 37°, then the other acute angle measures how many degrees?

(d) Which of the two triangles is not symmetrical?

9. (28, 34) What is the average of the two numbers marked by arrows on the number line below?

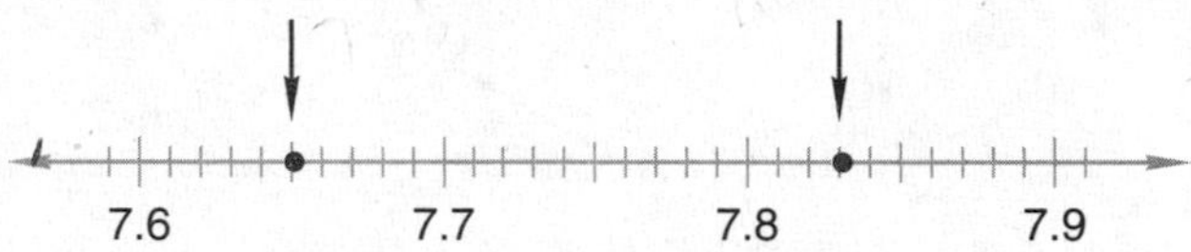

10. (57) Write twenty-five ten-thousandths in scientific notation.

Write equations to solve problems 11 and 12.

11. (60) What number is 24 percent of 75?

12. (60) What number is 120% of 12?

13. (64) Find each sum:

(a) $(-2) + (-3) + (-4)$ (b) $(+2) + (-3) + (+4)$

14. (48) Complete the table.

FRACTION	DECIMAL	PERCENT
(a)	(b)	4%
$\frac{7}{8}$	(c)	(d)

15. (50) Use a unit multiplier to convert 700 mm to cm.

16. (64) In three separate stock trades Dale lost \$560, gained \$850, and lost \$280. What was the net result of the three trades?

17. Describe the rule of the function, and find the missing number.
(58)

In	FUNCTION	Out
7	→	49
0	→	0
11	→	77
1	→	□

18. Round 7856.427
(33)
(a) to the nearest hundredth.
(b) to the nearest hundred.

19. The diameter of Debby's bicycle tire is 24 inches. What is the circumference of the tire to the nearest inch?
(66)

20. Consider angles A, B, C, and D below.
(40)

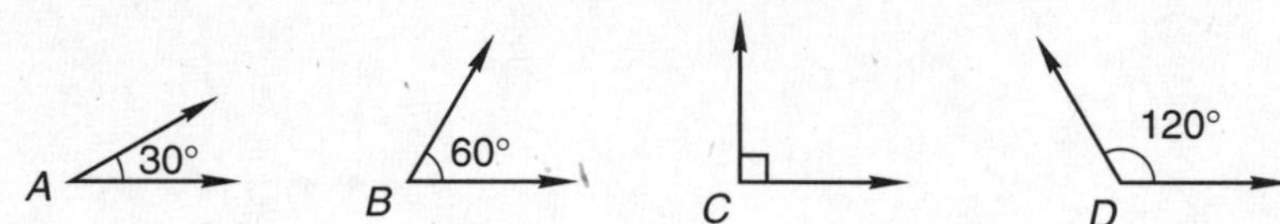

(a) Which two angles are complementary?
(b) Which two angles are supplementary?

21. (a) Show two ways to simplify 2(5 ft + 3 ft).
(41)
(b) Which property is illustrated in (a)?

22. Solve: $\frac{2.5}{w} = \frac{15}{12}$
(39)

Simplify:

23. $9 + 8\{7 \cdot 6 - 5[4 + (3 - 2 \cdot 1)]\}$
(63)

24. 1 yd − 1 ft 3 in.
(56)

25. $6.4 - (0.6 - 0.04)$
(35)

26. $\frac{3 + 0.6}{(3)(0.6)}$
(52)

27. $1\frac{2}{3} + 3\frac{1}{4} - 1\frac{5}{6}$
(30)

28. $\frac{3}{5} \div 3\frac{1}{5} \cdot 5\frac{1}{3} \cdot |-1|$
(26, 59)

29. $3\frac{3}{4} \div \left(3 \div 1\frac{2}{3}\right)$
(26)

30. $5^2 - \sqrt{4^2} + 2^{-2}$
(52, 57)

LESSON 68 Algebraic Addition

WARM-UP

Facts Practice: + − × ÷ Mixed Numbers (Test N)

Mental Math:

a. $0.75 + 0.5$

b. $\sqrt{1} - \left(\frac{1}{2}\right)^2$

c. $4w - 1 = 35$

d. 12×2.5 (halve, double)

e. 20 dm to cm

f. $33\frac{1}{3}\%$ of 24

g. Find the perimeter and area of a rectangle that is 2 m long and 1.5 m wide.

Problem Solving:

How many different triangles are in this figure?

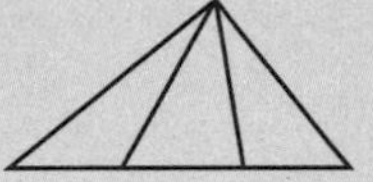

NEW CONCEPT

Recall that the graphs of −3 and 3 are the same distance from zero on the number line. The graphs are on the opposite sides of zero.

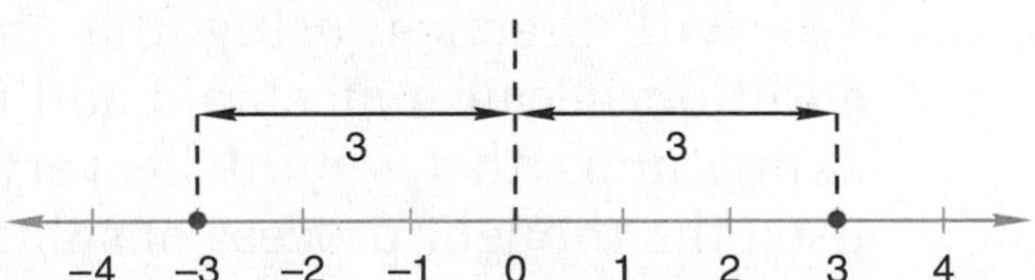

This is why we say that 3 and −3 are the opposites of each other.

3 is the opposite of −3

−3 is the opposite of 3

We can read −3 as "the opposite of 3." Furthermore, −(−3) can be read as "the opposite of the opposite of 3." This means that −(−3) is another way to write 3.

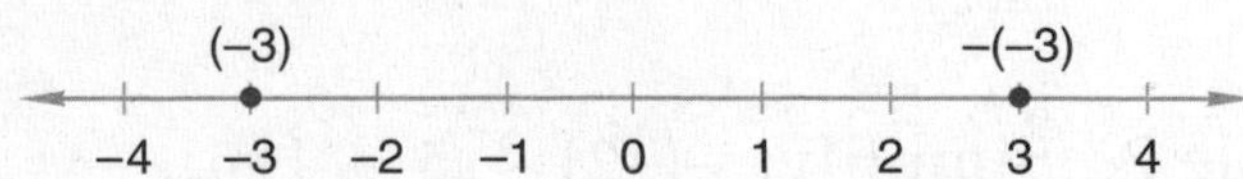

There are two ways to simplify the expression $7 - 3$. The first way is to let the minus sign signify subtraction. When we subtract 3 from 7, the answer is 4.

$$7 - 3 = 4$$

The second way is to use the thought process of **algebraic addition.** To use algebraic addition, we let the minus sign mean that –3 is a negative number and we treat the problem as an addition problem.

$$7 + (-3) = 4$$

Notice that we get the same answer both ways. The only difference is in the way we think about the problem.

We can also use algebraic addition to simplify this expression:

$$7 - (-3)$$

We use an addition thought and think that 7 is added to –(–3). This is what we think:

$$7 + [-(-3)]$$

But the opposite of –3 is 3, so we can write

$$7 + [3] = 10$$

We will practice using the thought process of algebraic addition because algebraic addition can be used to simplify expressions that would be very difficult to simplify if we used the thought process of subtraction.

Example 1 Simplify: $-3 - (-2)$

Solution We think addition. We think we are to *add* –3 and –(–2). This is what we think:

$$(-3) + [-(-2)]$$

The opposite of –2 is 2 itself. So we have

$$(-3) + [2] = \mathbf{-1}$$

Example 2 Simplify: $-(-2) - 5 - (+6)$

Solution We see three numbers. We think *addition,* so we have

$$[-(-2)] + (-5) + [-(+6)]$$

We simplify the first and third numbers and get

$$[+2] + (-5) + [-6] = \mathbf{-9}$$

Note that this time we write 2 as +2. Either 2 or +2 may be used.

LESSON PRACTICE

Practice set* Use algebraic addition to find these sums:

a. (−3) − (+2) **b.** (−3) − (−2)

c. (+3) − (2) **d.** (−3) − (+2) − (−4)

e. (−8) + (−3) − (+2) **f.** (−8) − (+3) + (−2)

MIXED PRACTICE

Problem set

1. (12) The mass of the beaker and the liquid was 1037 g. The mass of the empty beaker was 350 g. What was the mass of the liquid?

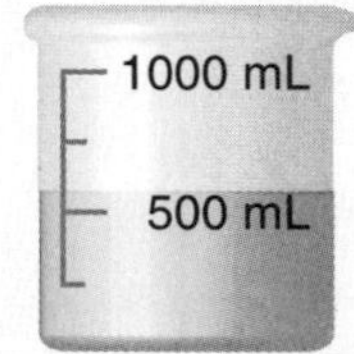

2. (54) Use a ratio box to solve this problem. Adriana's soccer ball is covered with a pattern of pentagons and hexagons in the ratio of 3 to 5. If there are 12 pentagons, how many hexagons are in the pattern?

3. (25, 30) When the sum of $\frac{1}{4}$ and $\frac{1}{2}$ is divided by the product of $\frac{1}{4}$ and $\frac{1}{2}$, what is the quotient?

4. (46) Pens were on sale 4 for $1.24.

(a) What was the price per pen?

(b) How much would 100 pens cost?

5. (46) Christy rode her bike 60 miles in 5 hours.

(a) What was her average speed in miles per hour?

(b) What was the average number of minutes it took to ride each mile?

6. Sound travels about 331 meters per second in air. About how many seconds does it take sound to travel a kilometer?
(32, 53)

7. The following scores were made on a test:
(Inv. 4)

72, 80, 84, 88, 100, 88, 76

(a) Which score was made most often?

(b) What is the median of the scores?

(c) What is the mean of the scores?

8. What is the average of the two numbers marked by arrows on the number line below?
(28, 34)

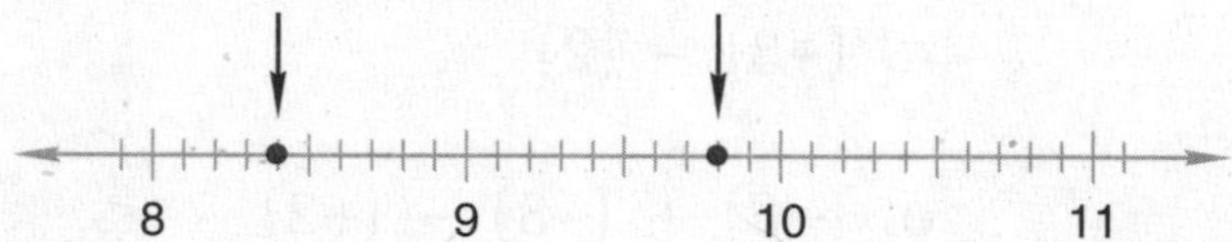

9. This rectangular shape is two cubes tall and two cubes deep.
(67)

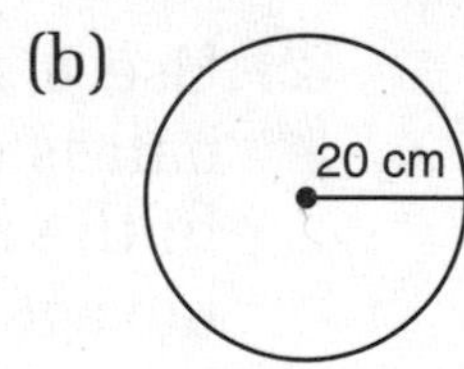

(a) How many cubes were used to build this shape?

(b) What is the name of this shape?

10. Find the circumference of each circle:
(66)

(a) 40 cm

Use 3.14 for π.

(b) 20 cm

Leave π as π.

11. The coordinates of the vertices of ΔABC are A (1, −1), B (−3, −1), and C (1, 3). Draw the triangle and answer these questions:
(58, 62)

(a) What type of triangle is ΔABC classified by angles?

(b) What type of triangle is ΔABC classified by sides?

(c) Triangle ABC's one line of symmetry passes through which vertex?

(d) What is the measure of $\angle B$?

(e) What is the area of ΔABC?

12. (51) Multiply twenty thousand by thirty thousand, and write the product in scientific notation.

Write equations to solve problems 13 and 14.

13. (60) What number is 75 percent of 400?

14. (60) What number is 150% of $1\frac{1}{2}$?

15. (68) Simplify:

(a) $(-4) - (-6)$ (b) $(-4) - (+6)$

(c) $(-6) - (-4)$ (d) $(+6) - (-4)$

16. (67) Find the surface area of a cube that has edges 4 inches long.

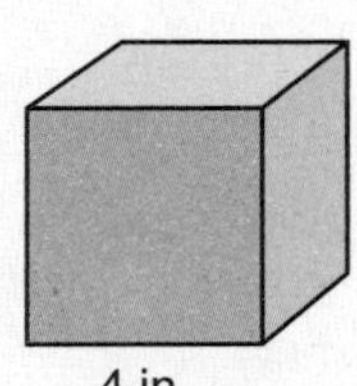

17. (48) Complete the table.

FRACTION	DECIMAL	PERCENT
$\frac{3}{25}$	(a)	(b)
(c)	(d)	120%

18. (52) Evaluate: $x^2 + 2xy + y^2$ if $x = 4$ and $y = 5$

19. (67) Use the name of a geometric solid to describe each object:

(a) 

(b) 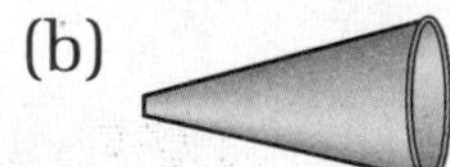

(c) 

20. (40, 61) In this figure parallelogram *ABCD* is divided by a diagonal into two congruent triangles. Angle *DCA* and $\angle BAC$ have equal measures and are complementary. Find the measure of

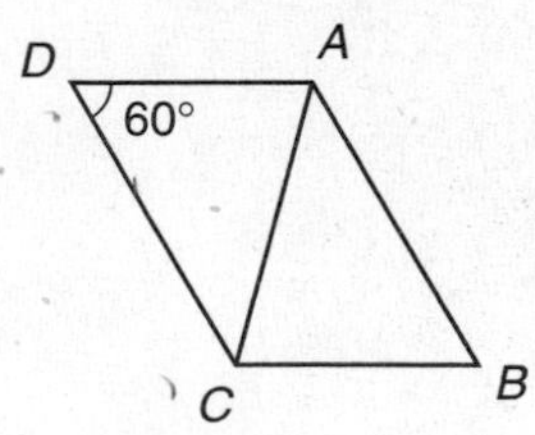

(a) $\angle DCA$. (b) $\angle DAC$.

(c) $\angle CAB$. (d) $\angle ABC$.

(e) $\angle BCA$. (f) $\angle BCD$.

21. (13) Write a word problem for this division: \$3.00 ÷ \$0.25

Solve:

22. (39) $\frac{4}{c} = \frac{3}{7\frac{1}{2}}$

23. (35) $(1.5)^2 = 15w$

Simplify:

24. (56) 1 gal − 1 qt 1 pt 1 oz

25. (45) $16 \div (0.04 \div 0.8)$

26. (63) $10 - [0.1 - (0.01)(0.1)]$

27. (30, 52) $\frac{5}{8} + \frac{2}{3} \cdot \frac{3}{4} - \frac{3}{4}$

28. (26) $4\frac{1}{2} \cdot 3\frac{3}{4} \div 1\frac{2}{3}$

29. (52) $\sqrt{5^2 - 2^4}$

30. (63) $3 + 6[10 - (3 \cdot 4 - 5)]$

LESSON

69 More on Scientific Notation

WARM-UP

Facts Practice: Metric Conversions (Test M)

Mental Math:

a. $4 - 1.5$

b. 75×10^{-3}

c. $\frac{x}{4} = \frac{1.5}{3}$

d. 18×35 (halve, double)

e. 20 cm to dm

f. $66\frac{2}{3}\%$ of 24

g. 5^2, $\times$ 3, $-$ 3, $\div$ 8, $\sqrt{}$, $\times$ 7, $-$ 1, $\div$ 4, $\times$ 10, $-$ 1, $\sqrt{}$, $\div$ 2

Problem Solving:

On a balanced scale are four identical cubes and a 12-ounce weight distributed as shown. What is the weight of each cube?

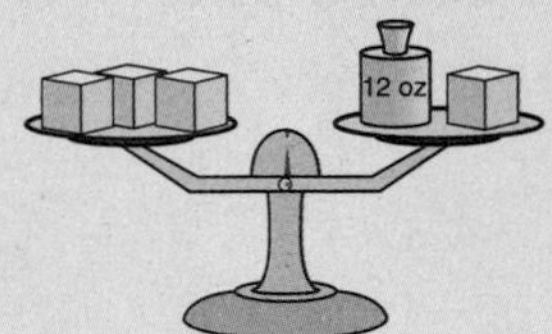

NEW CONCEPT

When we write a number in scientific notation, we usually put the decimal point just to the right of the first digit that is not zero. To write

$$4600 \times 10^5$$

in scientific notation, we use two steps. First we write 4600 in scientific notation. In place of 4600 we write 4.6×10^3. Now we have

$$4.6 \times 10^3 \times 10^5$$

For the second step we change the two powers of 10 into one power of 10. We recall that 10^3 means the decimal point is 3 places to the right and that 10^5 means the decimal point is 5 places to the right. Since 3 places to the right and 5 places to the right is 8 places to the right, the power of 10 is 8.

$$4.6 \times 10^8$$

Example 1 Write 25×10^{-5} in scientific notation.

Solution First we write 25 in scientific notation.

$$2.5 \times 10^1 \times 10^{-5}$$

Then we combine the powers of 10 by remembering that 1 place to the right and 5 places to the left equals 4 places to the left.

$$\mathbf{2.5 \times 10^{-4}}$$

Example 2 Write 0.25×10^4 in scientific notation.

Solution First we write 0.25 in scientific notation.

$$2.5 \times 10^{-1} \times 10^4$$

Since 1 place to the left and 4 places to the right equals 3 places to the right, we can write

$$\mathbf{2.5 \times 10^3}$$

With practice you will soon be able to perform these exercises mentally.

LESSON PRACTICE

Practice set* Write each number in scientific notation:

a. 0.16×10^6

b. 24×10^{-7}

c. 30×10^5

d. 0.75×10^{-8}

e. 14.4×10^8

f. 12.4×10^{-5}

MIXED PRACTICE

Problem set

1. *(Inv. 4)* The following is a list of scores Yori received in a diving competition:

7.0	6.5	6.5	7.4	7.0	6.5	6.0

(a) Which score was received the most often?

(b) What is the median of the scores?

(c) What is the mean of the scores?

(d) What is the range of the scores?

2. *(65)* Use a ratio box to solve this problem. The team won 15 games and lost the rest. If the team's win-loss ratio was 5 to 3, how many games were played?

3. *(53)* Brian swam 4 laps in 6 minutes. At that rate, how many minutes will it take Brian to swim 10 laps?

4. *(69)* Write each number in scientific notation:

(a) 15×10^5

(b) 0.15×10^5

5. Refer to the following statement to answer (a)–(c):
(36, 60)

The survey found that only 2 out of 5 Lilliputians believe in giants.

(a) According to the survey, what fraction of the Lilliputians do not believe in giants?

(b) If 60 Lilliputians were selected for the survey, how many of them would believe in giants?

(c) What is the probability that a randomly selected Lilliputian who participated in the survey would believe in giants?

6. The diameter of the tree stump was 40 cm. Find the circumference of the tree stump to the nearest centimeter.
(66)

7. Use the name of a geometric solid to describe the shape of these objects:
(67)

(a) volleyball (b) water pipe (c) tepee

8. (a) What is the perimeter of the equilateral triangle at right?
(58, 62)

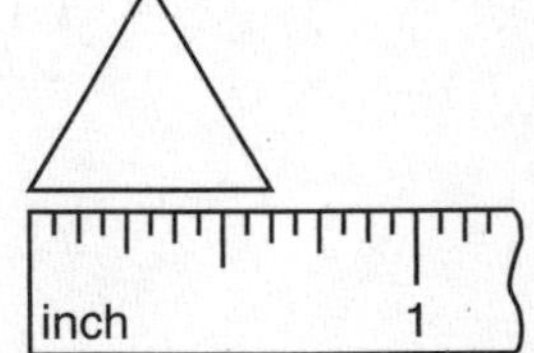

(b) What is the measure of each of its angles?

(c) Trace the triangle on your paper, and show its lines of symmetry.

9. Simplify:
(68)

(a) $(-4) + (-5) - (-6)$

(b) $(-2) + (-3) - (-4) - (+5)$

10. Find the circumference of each circle:
(66)

(a)

7 cm

Use 3.14 for π.

(b)

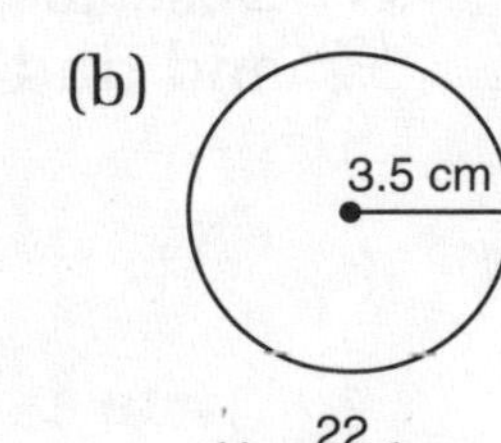

Use $\frac{22}{7}$ for π.

11. (37) Refer to the figure to answer (a)–(c). Dimensions are in millimeters. Corners that look square are square.

12, 6, 10, 9

(a) What is the area of the hexagon?

(b) What is the area of the shaded triangle?

(c) What fraction of the hexagon is shaded?

Write equations to solve problems 12 and 13.

12. (60) What number is 50 percent of 200?

13. (60) What number is 250% of 4.2?

14. (48) Complete the table.

FRACTION	DECIMAL	PERCENT
$\frac{3}{20}$	(a)	(b)
(c)	(d)	150%

15. (40) Refer to this figure to answer (a)–(c):

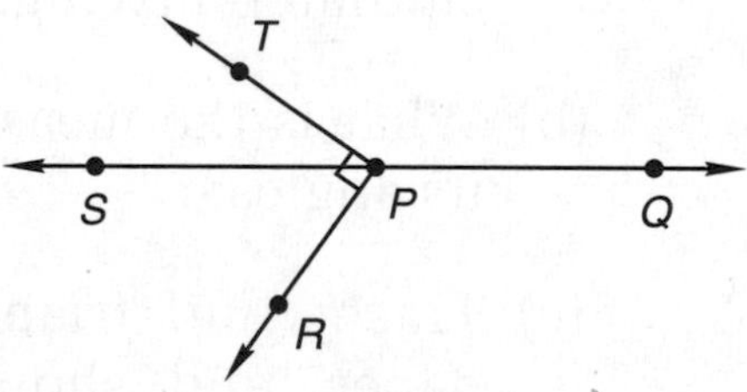

(a) Which angle is supplementary to $\angle SPT$?

(b) Which angle is complementary to $\angle SPT$?

(c) If $\angle QPR$ measures 125°, what is the measure of $\angle QPT$?

16. (52, 57) Evaluate: $a^2 - \sqrt{a} + ab - a^0$ if $a = 4$ and $b = 0.5$

17. (58) Describe the rule of this function, and find the missing number.

IN	FUNCTION	OUT
8	→	15
6	→	11
10	→	19
4	→	□

18. Divide 144 by 11 and write the answer
(44)
(a) as a decimal with a bar over the repetend.

(b) rounded to the nearest whole number.

19. Anders used this formula to convert from degrees Celsius to degrees Fahrenheit:
(41)

$$°F = 1.8°C + 32$$

If the Celsius temperature (°C) is 20°C, what is the Fahrenheit temperature (°F)?

20. The prime number 19 is the average of which two different prime numbers?
(21, 28)

Solve:

21. $t + \frac{5}{8} = \frac{15}{16}$
(30)

22. $\frac{a}{8} = \frac{3\frac{1}{2}}{2}$
(39)

Estimate each answer to the nearest whole number. Then perform the calculation.

23. $\left(3\frac{3}{4} \div 1\frac{2}{3}\right) \cdot 3$
(26)

24. $4\frac{1}{2} + \left(5\frac{1}{6} \div 1\frac{1}{3}\right)$
(26, 30)

Simplify:

25.
(49)

$$\begin{array}{r} 5 \text{ ft } 7 \text{ in.} \\ + \; 6 \text{ ft } 8 \text{ in.} \\ \hline \end{array}$$

26. $\frac{350 \text{ m}}{1 \text{ s}} \cdot \frac{60 \text{ s}}{1 \text{ min}} \cdot \frac{1 \text{ km}}{1000 \text{ m}}$
(50)

27. $6 - (0.5 \div 4)$
(35)

28. $7.50 ÷ 0.075
(45)

29. Use prime factorization to reduce $\frac{432}{675}$.
(24)

30. (a) Convert $2\frac{1}{4}$ to a decimal and add 0.15.
(43)

(b) Convert 6.5 to a mixed number and add $\frac{5}{6}$.

LESSON

70 Volume

WARM-UP

Facts Practice: + – × ÷ Mixed Numbers (Test N)

Mental Math:

a. 4.8 + 3 + 0.3
b. 25^2
c. $5m - 3 = 27$
d. \$4.80 × 50 (halve, double)
e. 20 dm to mm
f. 60% of 25
g. Find the perimeter and area of a square that has sides 0.5 m long.

Problem Solving:

Copy this problem and fill in the missing digits:

$$\begin{array}{r} \_\,\_ \\ \times\ \ \_ \\ \hline 679 \end{array}$$

NEW CONCEPT

Recall from Lesson 67 that geometric solids are shapes that take up space. We use the word **volume** to describe the space occupied by a shape. To measure volume, we use units that occupy space. The units that we use to measure volume are cubes of certain sizes. We can use sugar cubes to help us think of volume.

Example 1 This rectangular prism was constructed of sugar cubes. Its volume is how many cubes?

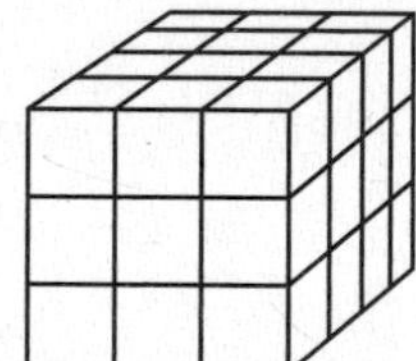

Solution To find the volume of the prism, we calculate the number of cubes it contains. We see that there are 3 layers of cubes. Each layer contains 3 rows of cubes with 4 cubes in each row, or 12 cubes. Three layers with 12 cubes in each layer means that the volume of the prism is **36 cubes.**

Volumes are measured by using cubes of a standard size. A cube whose edges are 1 centimeter long has a volume of 1 cubic centimeter, which we abbreviate by writing 1 cm^3.

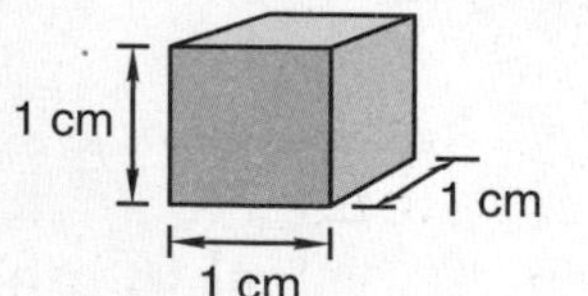

1 cubic centimeter = 1 cm^3

Similarly, if each of the edges is 1 foot long, the volume is 1 cubic foot. If each of the edges is 1 meter long, the volume is 1 cubic meter.

$$1 \text{ cubic foot} = 1 \text{ ft}^3 \qquad 1 \text{ cubic meter} = 1 \text{ m}^3$$

To calculate the volume of a solid, we can imagine constructing the solid out of sugar cubes of the same size. We would begin by constructing the base and then building up the layers to the specified height.

Example 2 Find the number of 1-cm cubes that can be placed inside a rectangular box with the dimensions shown.

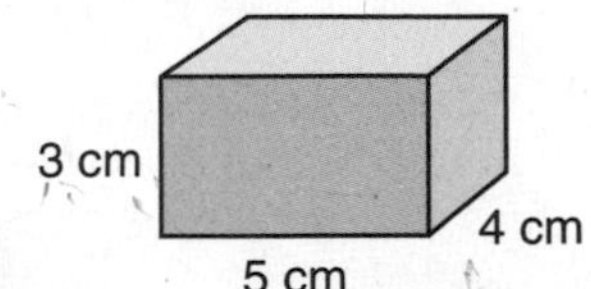

Solution The base of the box is 5 cm by 4 cm, so we can place 5 rows of 4 cubes on the base. Thus there are 20 cubes on the first layer.

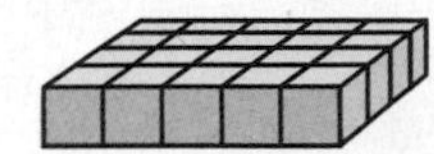

Since the box is 3 cm tall, we can fit 3 layers of cubes in the box.

$$\frac{20 \text{ cubes}}{1 \text{ layer}} \times 3 \text{ layers} = 60 \text{ cubes}$$

We find that **60 1-cm cubes** can be placed in the box.

Example 3 What is the volume of this cube? Dimensions are in inches.

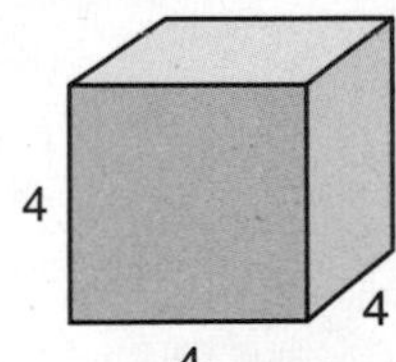

Solution The base is 4 in. by 4 in. Thus 16 cubes can be placed on the base.

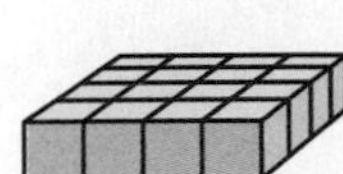

Since the big cube is 4 in. tall, there are 4 layers of small cubes.

$$\frac{16 \text{ cubes}}{1 \text{ layer}} \times 4 \text{ layers} = 64 \text{ cubes}$$

Each small cube has a volume of 1 cubic inch. Thus the volume of the big cube is **64 cubic inches (64 in.^{3})**.

LESSON PRACTICE

Practice set a. This rectangular prism was constructed of sugar cubes. Its volume is how many sugar cubes?

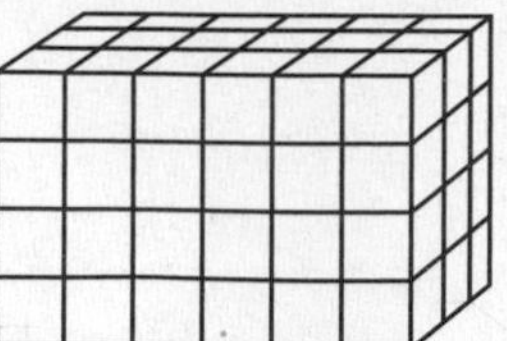

b. Find the number of 1-cm cubes that can be placed inside a box with dimensions as illustrated.

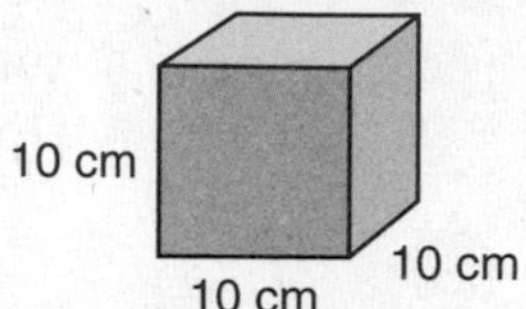

c. What is the volume of this rectangular prism? Dimensions are in feet.

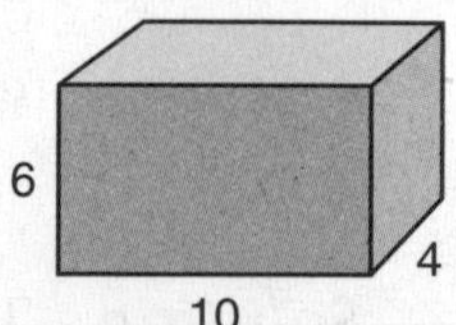

d. Estimate the volume (in cubic yards) of a room in your home. Then use a yardstick to measure the length, width, and height of the room to the nearest yard, and calculate the volume of the room.

MIXED PRACTICE

Problem set

1. It was 38 kilometers from the encampment to the castle. Milton galloped to the castle and cantered back. If the round trip took 4 hours, what was his average speed in kilometers per hour?
(53)

2. The vertices of two angles of a triangle are (3, 1) and (0, −4). The y-axis is a line of symmetry of the triangle.
(37, 58)
   (a) What are the coordinates of the third vertex of the triangle?
   (b) What is the area of the triangle?

3. Using a tape measure, Gretchen found that the circumference of the great oak was 600 cm. Using 3 in place of π, she estimated that the tree's diameter was 200 cm. Was her estimate for the diameter a little too large or a little too small? Why?
(66)

4. Grapes were priced at 3 pounds for \$1.29.
(46)
   (a) What was the price per pound?
   (b) How much would 10 pounds of grapes cost?

5. (35) If the product of nine tenths and eight tenths is subtracted from the sum of seven tenths and six tenths, what is the difference?

6. (22, 48) Three fourths of the batter's 188 hits were singles.

(a) How many of the batter's hits were singles?

(b) What percent of the batter's hits were not singles?

7. (8) On an inch ruler, which mark is halfway between the $1\frac{1}{2}$-inch mark and the 3-inch mark?

8. (70) Find the number of 1-cm cubes that can be placed in the box at right.

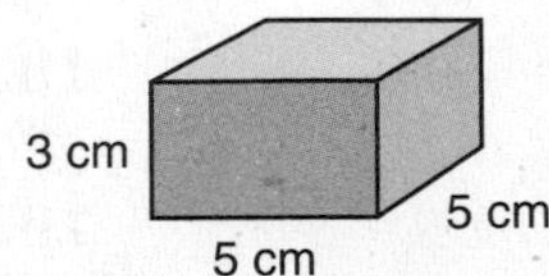

9. (66) Find the circumference of each circle:

(a)

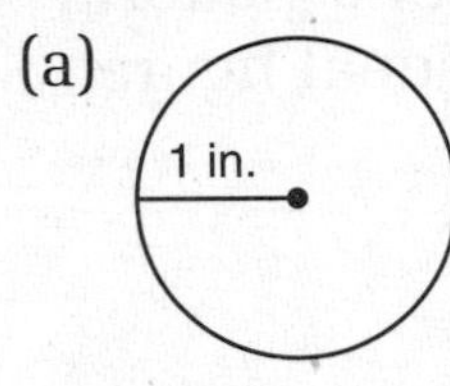

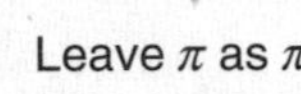

Leave π as π.

(b)

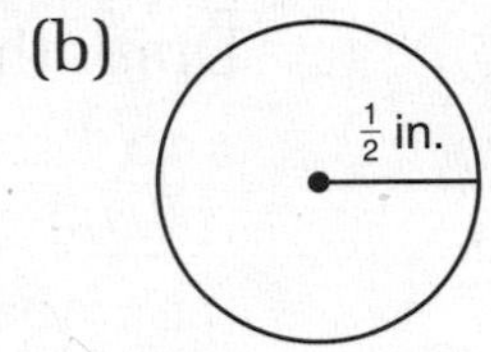

Use 3.14 for π.

10. (69) Write each number in scientific notation:

(a) 12×10^{-6} (b) 0.12×10^{-6}

11. (28, 34) What is the average of the three numbers marked by arrows on the number line below?

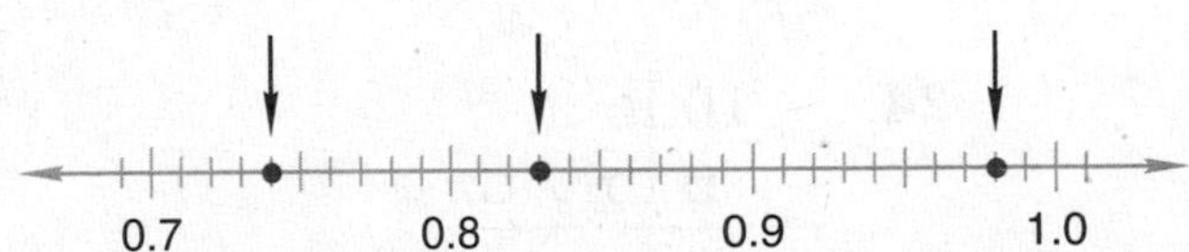

12. (50) Use a unit multiplier to convert 1.25 kilograms to grams.

13. (47, 57) Find each missing exponent:

(a) $2^6 \cdot 2^3 = 2^{\square}$ (b) $2^6 \div 2^3 = 2^{\square}$

(c) $2^3 \div 2^6 = 2^{\square}$ (d) $2^6 \div 2^6 = 2^{\square}$

14. (60) Write an equation to solve this problem:

What number is $\frac{1}{6}$ of 100?

15. Complete the table.
(48)

FRACTION	DECIMAL	PERCENT
(a)	(b)	14%
$\frac{5}{6}$	(c)	(d)

16. Simplify:
(68)
(a) $(-6) - (-4) + (+2)$

(b) $(-5) + (-2) - (-7) - (+9)$

17. Evaluate: $ab - (a - b)$ if $a = 0.4$ and $b = 0.3$
(52)

18. Round $29{,}374.\overline{65}$ to the nearest whole number.
(42)

19. Estimate the product of 6.085 and $7\frac{15}{16}$.
(29, 33)

20. What three-dimensional figure can be formed by folding this pattern? Draw the three-dimensional figure.
(67)

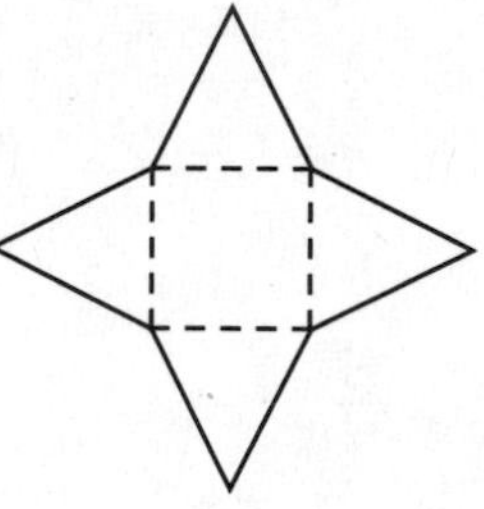

21. What is the surface area of a cube with edges 2 ft long?
(67)

Solve:

22. $4.3 = x - 0.8$
(35)

23. $\frac{2}{d} = \frac{1.2}{1.5}$
(39)

Simplify:

24.
(56)
$$\begin{array}{r} 10 \text{ lb} \\ -\ 6 \text{ lb } 7 \text{ oz} \\ \hline \end{array}$$

25. $\frac{\$5.25}{1 \text{ hr}} \cdot \frac{8 \text{ hr}}{1 \text{ day}} \cdot \frac{5 \text{ days}}{1 \text{ week}}$
(53)

26. $3\frac{3}{4} \div \left(1\frac{2}{3} \cdot 3\right)$
(26)

27. $4\frac{1}{2} + 5\frac{1}{6} - 1\frac{1}{3}$
(30)

28. $(0.06 \div 5) \div 0.004$
(45)

29. Write $9\frac{1}{2}$ as a decimal number, and multiply it by 9.2.
(35, 43)

30. (a) What is the total price of a \$15 meal including 6% sales tax?
(46)

(b) A 15% tip on a \$15 meal would be how much money?

INVESTIGATION 7

Focus on

Balanced Equations

Since Lesson 3 we have solved equations informally by using various strategies for finding the missing number. In this investigation we will practice a more formal, algebraic method for solving equations. To help us see equations from this new perspective, we will use a balance scale as a visual aid.

Equations are sometimes called **balanced equations** because the two sides of the equation "balance" each other. A balance scale can be used as a model of an equation. We replace the equal sign with a balanced scale. The left and right sides of the equation are placed on the left and right trays of the balance. For example, $x + 12 = 33$ becomes

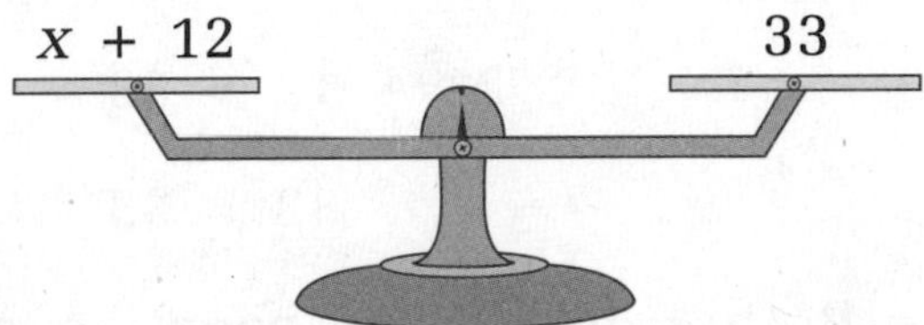

Using a balance-scale model we think of how to get the unknown number, in this case x, alone on one side of the scale. Using our example, we could remove 12 (subtract 12) from the left side of the scale. However, if we did that, the scale would no longer be balanced. So we make this rule for ourselves.

> Whatever operation we perform on one side of an equation, we also perform on the other side of the equation to maintain a balanced equation.

We see that there are two steps to the process.

Step 1: Select the operation that will isolate the variable.

Step 2: Perform the selected operation on both sides of the equation.

In our example we select "subtract 12" as the operation required to isolate x (to "get x alone"). Then we perform this operation on both sides of the equation.

Select operation:

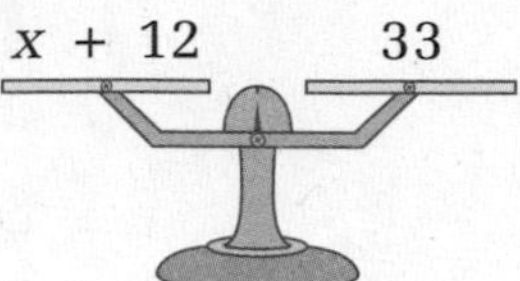

To isolate x, *subtract 12.*

Perform operation:

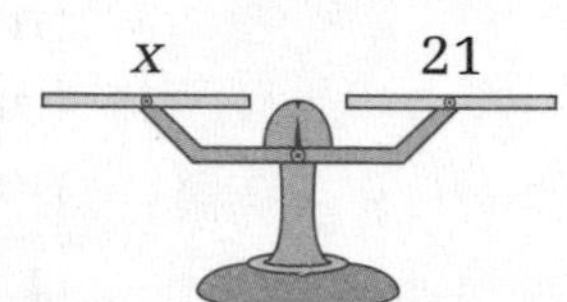

To keep the scale balanced, subtract 12 *from both sides of the equation.*

After subtracting 12 from both sides of the equation, x is isolated on one side of the scale, and 21 balances x on the other side of the scale. This shows that $x = 21$. We check our solution by replacing x with 21 in the original equation.

$x + 12 = 33$	original equation
$21 + 12 = 33$	replaced x with 21
$33 = 33$	simplified left side

Both sides of the equation equal 33. This shows that the solution, $x = 21$, is correct.

Now we will illustrate a second equation, $45 = x + 18$, with a balance-scale model.

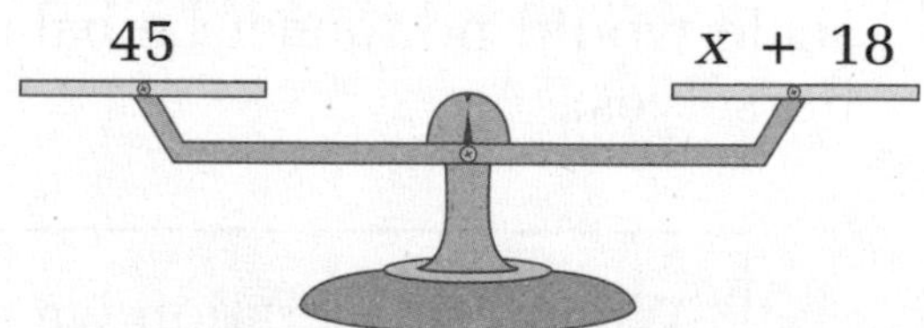

This time the unknown number is on the right side of the balance scale, added to 18.

1. Select the operation that will isolate the variable, and write that operation on your paper.

2. Describe how to perform the operation and keep a balanced scale.

3. Describe what will remain on the left and right side of the balance scale after the operation is performed.

We show the line-by-line solution of the equation below.

$45 = x + 18$	original equation
$45 - 18 = x + 18 - 18$	subtracted 18 from both sides
$27 = x + 0$	simplified both sides
$27 = x$	$x + 0 = x$

We check the solution by replacing x with 27 in the original equation.

$45 = x + 18$	original equation
$45 = 27 + 18$	replaced x with 27
$45 = 45$	simplified right side

By checking the solution in the original equation, we see that the solution is correct. Now we revisit the equation to illustrate one more idea.

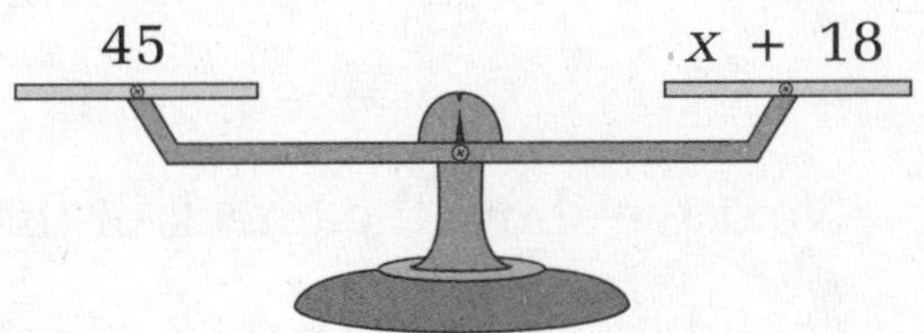

4. Suppose the contents of the two trays of the balance scale were switched. That is, $x + 18$ was moved to the left side, and 45 was moved to the right side. Would the scale still be balanced? Write what the equation would be.

Now we will consider an equation that involves multiplication rather than addition.

$$2x = 132$$

Since $2x$ means two x's $(x + x)$, we can show this equation on a balance scale two ways.

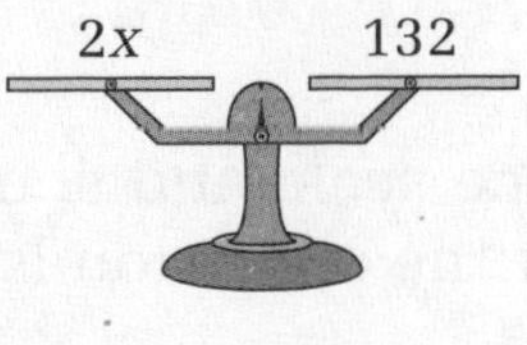

$2x$ 132 or $x + x$ 132

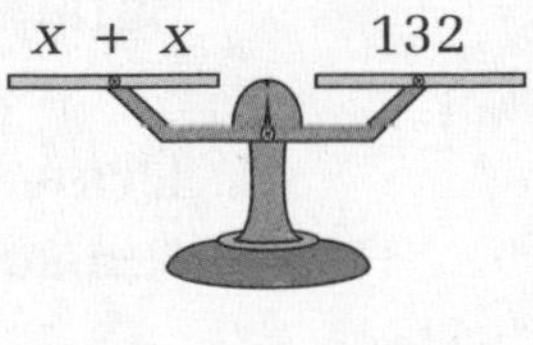

Our goal is to isolate x. We must perform the operations necessary to get one x alone on one side of the scale. We do not subtract 2, because 2 is not added to x. We do not subtract x, because there is no x to subtract from the other side of the equation. To isolate x in this equation, we *divide by 2.* To keep the equation balanced, we *divide both sides by 2.*

Select operation:

To isolate x, *divide by 2.*

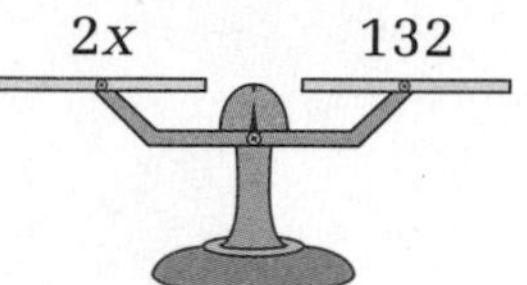

Perform operation:

To keep the equation balanced, divide *both sides by 2.*

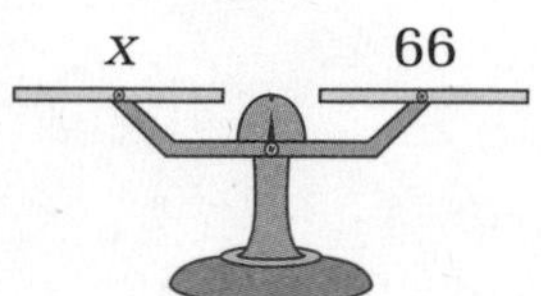

Here we show the line-by-line solution of this equation.

$2x = 132$	original equation
$\dfrac{2x}{2} = \dfrac{132}{2}$	divided both sides by 2
$1x = 66$	simplified both sides
$x = 66$	$1x = x$

Next we show the check of the solution.

$2x = 132$	original equation
$2(66) = 132$	replaced x with 66
$132 = 132$	simplified left side

This check shows that the solution, $x = 66$, is correct.

5. Draw a balance-scale model for the equation $3x = 132$.

6. Select the operation that will isolate the variable, and write that operation on your paper.

7. Describe how to perform the operation and keep a balanced scale.

8. Draw a balance scale and show what is on both sides of the scale after the operation is performed.

9. Write the line-by-line solution of the equation.

10. Show the check of the solution.

Most students choose to solve the equation $3x = 132$ by dividing both sides of the equation by 3. There is another operation that could be selected that is often useful, which we will describe next. First note that the number multiplying the variable, in this case 3, is called the **coefficient** of x. Instead of dividing by the coefficient of x, we could choose to **multiply by the reciprocal** of the coefficient. In this case we could multiply by $\frac{1}{3}$.

$$3x = 132$$

$$\frac{1}{3} \cdot 3x = \frac{1}{3} \cdot 132$$

$$1x = \frac{132}{3}$$

$$x = 44$$

When solving equations with whole number or decimal number coefficients, it is usually easier to think about dividing by the coefficient. However, when solving equations with fractional coefficients, it is usually easier to multiply by the reciprocal of the coefficient. Refer to the following equation for problems 11–14:

$$\frac{3}{4}x = \frac{9}{10}$$

11. Select the operation that will result in $\frac{3}{4}x$ becoming $1x$ in the equation.

12. Describe how to perform the operation and keep the equation balanced.

13. Write a line-by-line solution of the equation.

14. Show the check of the solution.

We find that the solution to the equation is $\frac{6}{5}$ $\left(\text{or } 1\frac{1}{5}\right)$. In arithmetic we usually convert improper fractions to mixed numbers. In algebra we usually leave improper fractions in improper form unless the problem states or implies that a mixed number answer is preferable.

For each of the following equations, (a) state the operation you select to isolate the variable, (b) describe how to perform the operation and keep the equation balanced, (c) write a line-by-line solution of the equation, and (d) show the check of the solution.

15. $x + 2.5 = 7$

16. $3.6 = y + 2$

17. $4w = 132$

18. $1.2m = 1.32$

19. $x + \frac{3}{4} = \frac{5}{6}$

20. $\frac{3}{4}x = \frac{5}{6}$

21. Make up an addition equation with decimal numbers. Solve and check it.

22. Make up a multiplication equation with a fractional coefficient. Solve and check it.

LESSON

71 Finding the Whole Group When a Fraction Is Known

WARM-UP

Facts Practice: Classifying Quadrilaterals and Triangles (Test O)

Mental Math:

a. $(-3) + (-12)$

b. 4.5×10^{-3}

c. $\frac{w}{100} = \frac{24}{30}$

d. $12 \times 2\frac{1}{2}$ (halve, double)

e. 50 cm to m

f. 75% of \$36

g. What is the total cost of a \$30 item plus 8% sales tax?

Problem Solving:

Bry has three different shirts and three different ties he can wear with each shirt. How many different shirt-tie combinations can Bry wear? If the shirts are designated A, B, and C, and the ties 1, 2, and 3, one combination is A1. List all the possible combinations.

NEW CONCEPT

Drawing diagrams of fraction problems can help us understand problems such as the following:

> *Three fifths of the fish in the pond are bluegills. If there are 45 bluegills in the pond, how many fish are in the pond?*

The 45 bluegills are 3 of the 5 parts. We divide 45 by 3 and find there are 15 fish in each part. Since there are 15 fish in each of the 5 parts, there are 75 fish in all.

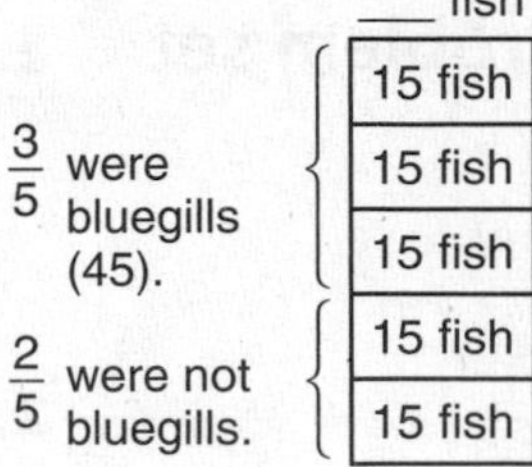

Example 1 When Juan finished page 51, he was $\frac{3}{8}$ of the way through his book. His book had how many pages?

Solution Juan read 51 pages. This is 3 of 8 parts of the book. Since 51 ÷ 3 is 17, each part is 17 pages. Thus the whole book (8 parts) totals 8 × 17, which is **136 pages.**

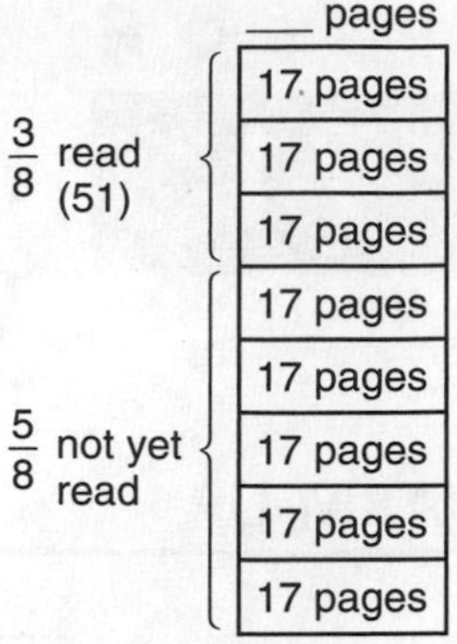

Example 2 As Sakura went from room to room she found that $\frac{3}{5}$ of the lights were on and that 30 lights were off. How many lights were on?

Solution Since $\frac{3}{5}$ of the lights were on, $\frac{2}{5}$ of the lights were off. Because $\frac{2}{5}$ of the lights was 30 lights, each fifth was 15 lights. Thus **45 lights** were on.

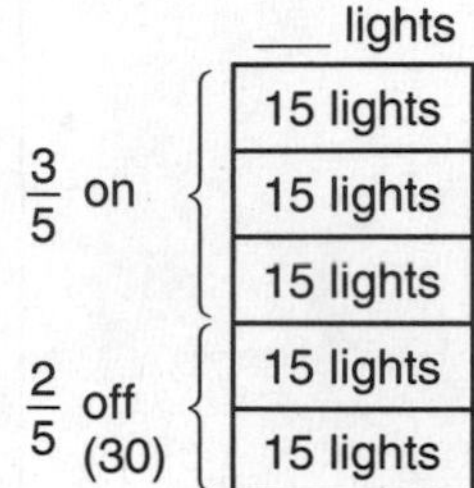

LESSON PRACTICE

Practice set Draw a diagram to solve each problem:

a. Three fifths of the babies in the nursery are boys. If there are 15 boys in the nursery, how many babies are there in all?

b. Five eighths of the clowns had happy faces. If 15 clowns did not have happy faces, how many clowns were there in all?

c. Vincent was chagrined when he looked at the clock, for in $\frac{3}{4}$ of an hour he had only answered 12 test questions. At that rate, how many test questions would Vincent answer in an hour?

MIXED PRACTICE

Problem set

1. *(32, 53)* Nine seconds elapsed from the time Abigail saw the lightning until she heard the thunder. The lightning was about how many kilometers from Abigail? (Sound travels about 331 meters per second in air.)

2. *(28, 34)* What is the average of the three numbers marked by arrows on the number line below?

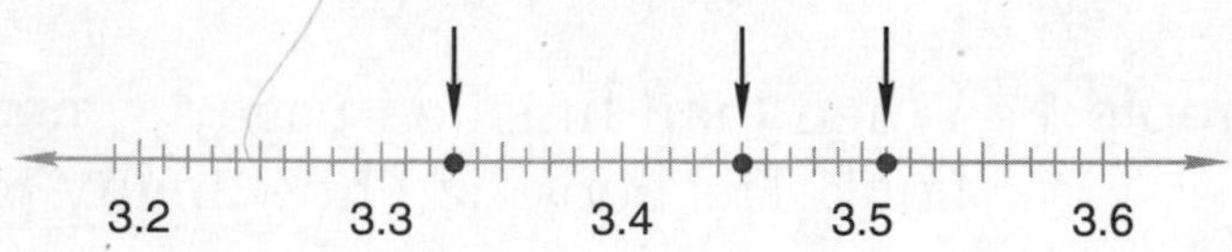

3. *(55)* On his first 2 tests Nate's average score was 80 percent. On his next 3 tests Nate's average score was 90 percent. What was his average score for all 5 tests?

4. *(51)* Twenty billion is how much more than nine billion? Write the answer in scientific notation.

5. *(21)* What is the sum of the first five prime numbers?

6. *(65)* Use a ratio box to solve this problem. The ratio of new ones to used ones in the box was 4 to 7. In all there were 242 new ones and used ones in the box. How many new ones were in the box?

7. *(71)* Diagram this statement. Then answer the questions that follow.

When Rosario finished page 78, she was $\frac{3}{5}$ of the way through her book.

(a) How many pages are in her book?

(b) How many pages does she have left to read?

8. *(70)* Find the number of 1-inch cubes that can be placed in this box. Dimensions are in inches.

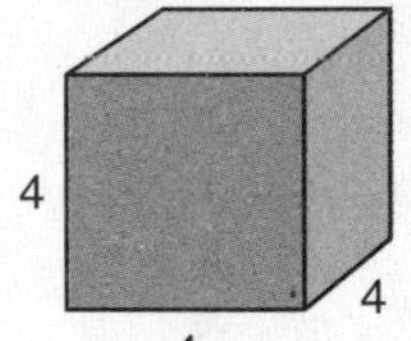

9. *(67)* What is the surface area of the box in problem 8?

10. *(66)* Find the circumference of each circle:

(a)

28 cm

Use 3.14 for π.

(b)

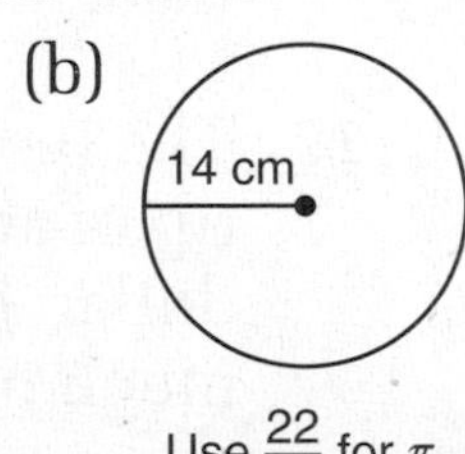

Use $\frac{22}{7}$ for π.

11. *(69)* Write each number in scientific notation:

(a) 25×10^{6} (b) 25×10^{-6}

12. *(48)* Complete the table.

FRACTION	DECIMAL	PERCENT
(a)	0.1	(b)
(c)	(d)	0.5%

13. Write an equation to solve each problem:
(60)

(a) What number is 35% of 80?

(b) Three fourths of 24 is what number?

14. Describe the rule of this function, and find the missing number.
(58)

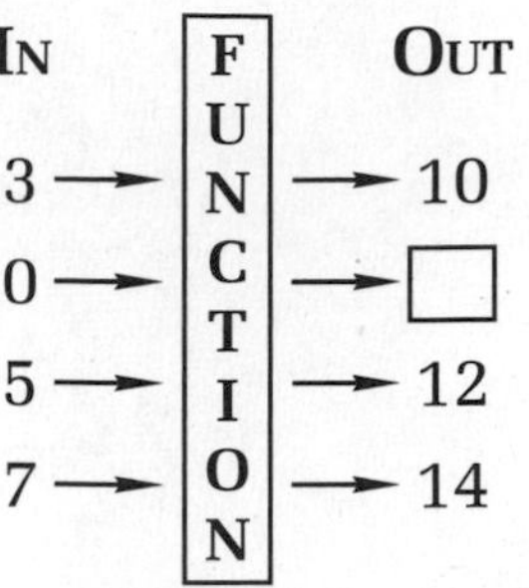

15. Draw a rectangular prism. A rectangular prism has how many vertices?
(67)

16. Figure *ABCD* is a trapezoid. Dimensions are in centimeters.
(37, 62)

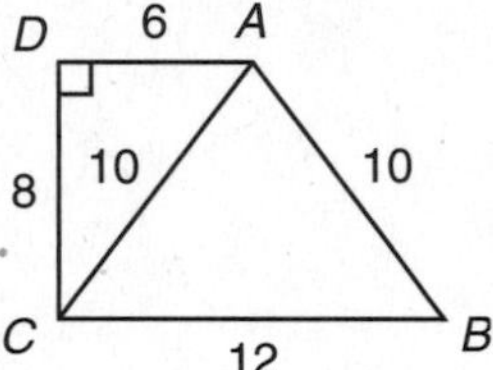

(a) Find the perimeter of the trapezoid.

(b) Find the area of the right triangle.

(c) Find the area of the isosceles triangle.

(d) Combine the areas of the triangles to find the area of the trapezoid.

17. The restaurant bill was \$16.50. Marcia planned to leave a tip of about 15%. After paying for the meal, she had a few dollar bills and quarters left in her purse. About how much money should she leave for the tip?
(46)

18. Frances had several small boxes of raisins. She counted the number of raisins in each box and tallied the results. Make a box-and-whisker plot from this information.
(Inv. 4)

Raisins per Box

Number of Raisins	Tally of Boxes
26	||
27	|||
28	||||
29	|||
30	||
31	|

19. (Inv. 3) The coordinates of three vertices of triangle ABC are A (0, 6), B (8, –2), and C (–9, –3). Graph the triangle. Then use a protractor to find the measures of $\angle A$, $\angle B$, and $\angle C$ to the nearest degree.

20. (52) Evaluate: $y - xy$ if $x = 0.1$ and $y = 0.01$

For problems 21 and 22, solve and check each equation. Show each step.

21. (Inv. 7) $m + 5.75 = 26.4$

22. (Inv. 7) $\frac{3}{4}x = 48$

23. (Inv. 6) What is the name of a parallelogram whose sides are equal in length but whose angles are not necessarily right angles?

Simplify:

24. (63) $\dfrac{4^2 + \{20 - 2[6 - (5 - 2)]\}}{\sqrt{36}}$

25. (56)

$$\begin{array}{r} 1 \text{ yd} \\ - \quad 1 \text{ ft } 1 \text{ in.} \\ \hline \end{array}$$

26. (50) $3.5 \text{ hr} \cdot \dfrac{60 \text{ min}}{1 \text{ hr}} \cdot \dfrac{60 \text{ s}}{1 \text{ min}}$

27. (26) $6\frac{2}{3} \div \left(4\frac{1}{2} \cdot 2\frac{2}{3}\right)$

28. (30) $7\frac{1}{2} - 5\frac{1}{6} + 1\frac{1}{3}$

29. (68) (a) $(-5) + (-6) - |-7|$

(b) $(-15) - (-24) - (+8)$

30. (30, 43) Write 1.5 as a mixed number, and subtract it from $2\frac{2}{3}$.

LESSON

72 Implied Ratios

WARM-UP

Facts Practice: $+ - \times \div$ Mixed Numbers (Test N)

Mental Math:

a. $(-10) + (+17)$

b. $\left(\frac{2}{3}\right)^2$

c. $6x + 2 = 32$

d. What decimal is 10% of 36?

e. 500 g to kg

f. $33\frac{1}{3}\%$ of \$36

g. Find 15% of \$30 by finding 10% of \$30 plus half of 10% of \$30.

Problem Solving:

On a balanced scale are a 6-oz weight, an 18-oz weight, and four identical blocks marked X, which are distributed as shown. What is the weight of each block marked X? Write an equation illustrated by this balanced scale.

NEW CONCEPT

Many rate problems can be solved by completing a proportion. Consider the following problem:

If 12 books weigh 20 pounds, how much would 30 books weigh?

We will illustrate two methods for solving this problem. First we will use the rate method. Since 12 books weigh 20 pounds, we can write two rates.

$$\text{(a)}\ \frac{12 \text{ books}}{20 \text{ pounds}} \qquad \text{(b)}\ \frac{20 \text{ pounds}}{12 \text{ books}}$$

To find the weight of 30 books, we could multiply 30 books by rate (b).

$$30 \cancel{\text{books}} \times \frac{20 \text{ pounds}}{12 \cancel{\text{books}}} = 50 \text{ pounds}$$

We find that 30 books would weigh 50 pounds.

Now we will solve the same problem by completing a proportion. We will record the information in a ratio box.

Instead of using the words "ratio" and "actual count," we will write "Case 1" and "Case 2." We will use p to stand for pounds.

	Case 1	Case 2
Books	12	30
Pounds	20	p

From the table we write a proportion and solve it.

$$\frac{12}{20} = \frac{30}{p} \quad \text{proportion}$$

$$12p = 20 \cdot 30 \quad \text{cross multiplied}$$

$$\frac{\overset{1}{\cancel{12}}p}{\underset{1}{\cancel{12}}} = \frac{20 \cdot 30}{12} \quad \text{divided by 12}$$

$$p = 50 \quad \text{simplified}$$

We find that 30 books would weigh 50 pounds.

Example 1 If 5 pounds of grapes cost \$1.20, how much would 12 pounds of grapes cost? First estimate the answer; then use a ratio box to solve the problem.

Solution Since 12 pounds is a little more than two times 5 pounds, we estimate the cost would be a little more than two times \$1.20, between \$2.50 and \$3.00.

Now we draw the ratio box. We use d for dollars.

	Case 1	Case 2
Pounds	5	12
Dollars	1.2	d

Now we write the proportion and solve for d.

$$\frac{5}{1.2} = \frac{12}{d} \quad \text{proportion}$$

$$5d = 12(1.2) \quad \text{cross multiplied}$$

$$\frac{\overset{1}{\cancel{5}}d}{\underset{1}{\cancel{5}}} = \frac{12(1.2)}{5} \quad \text{divided by 5}$$

$$d = 2.88 \quad \text{simplified}$$

We find that 12 pounds of grapes cost **\$2.88**.

Example 2 Mrs. C can tie 25 bows in 3 minutes. At that rate, how many bows can she tie in 1 hour? Work the problem using (a) rates and (b) a ratio box.

Solution We can use either minutes or hours but not both. *The units must be the same in both cases.* Since there are 60 minutes in 1 hour, we will use 60 minutes instead of 1 hour.

(a) $60 \cancel{\text{min}} \times \dfrac{25 \text{ bows}}{3 \cancel{\text{min}}} = 500 \text{ bows}$

Mrs. C can tie **500 bows** in one hour.

(b)

	Case 1	Case 2
Bows	25	b
Minutes	3	60

Next we write the proportion, cross multiply, and solve by dividing by 3.

$$\frac{25}{3} = \frac{b}{60} \quad \text{proportion}$$

$$25 \cdot 60 = 3b \quad \text{cross multiplied}$$

$$\frac{25 \cdot 60}{3} = \frac{\overset{1}{\cancel{3}}b}{\underset{1}{\cancel{3}}} \quad \text{divided by 3}$$

$$b = 500 \quad \text{simplified}$$

Again we see that Mrs. C can tie **500 bows** in one hour.

Example 3 Six is to 15 as 9 is to what number?

Solution We can sort the numbers in this question using a case 1–case 2 ratio box.

	Case 1	Case 2
First number	6	9
Second number	15	n

Now we write and solve a proportion.

$$\frac{6}{15} = \frac{9}{n}$$

$$6n = 9 \cdot 15$$

$$\frac{\overset{1}{\cancel{6}}n}{\underset{1}{\cancel{6}}} = \frac{9 \cdot 15}{6}$$

$$n = \mathbf{22\tfrac{1}{2}}$$

LESSON PRACTICE

Practice set

a. Kevin rode 30 km in 2 hours. At that rate, how long would it take him to ride 75 km? Estimate an answer. Then use a case 1–case 2 ratio box to solve this problem.

b. If 6 bales are needed to feed 40 head of cattle, how many bales are needed to feed 50 head of cattle? Use the rate method and then use a ratio box to solve this problem.

c. Five is to 15 as 9 is to what number?

MIXED PRACTICE

Problem set

1. Napoleon Bonaparte was born in 1769 and died in 1821. For how many years did he live? (12)

2. In her first 4 games Jill averaged 4 points per game. In her next 6 games Jill averaged 9 points per game. What was her average number of points per game after 10 games? (55)

3. Use a unit multiplier to convert 2.5 liters to milliliters. (50)

4. If the product of $\frac{1}{2}$ and $\frac{2}{5}$ is subtracted from the sum of $\frac{1}{2}$ and $\frac{2}{5}$, what is the difference? (30)

5. Use a ratio box to solve this problem. The ratio of carnivores to herbivores in the jungle was 2 to 7. If there were 126 carnivores in the jungle, how many herbivores were there? (54)

6. Use a ratio box to solve this problem. If 4 books weigh 9 pounds, how many pounds would 14 books weigh? (72)

7. Write an equation to solve each problem: (60)

   (a) Two fifths of 60 is what number?

   (b) How much money is 75% of $24?

8. The diameter of a bicycle tire is 20 in. Find the distance around the tire to the nearest inch. (Use 3.14 for π.) (66)

9. (71) Diagram this statement. Then answer the questions that follow.

Kayla received 150 votes. This was two thirds of the votes cast.

(a) How many votes were cast?

(b) How many votes were not for Kayla?

10. (70) The volume of a block of ice with the dimensions shown is equal to how many 1-by-1-by-1-inch ice cubes?

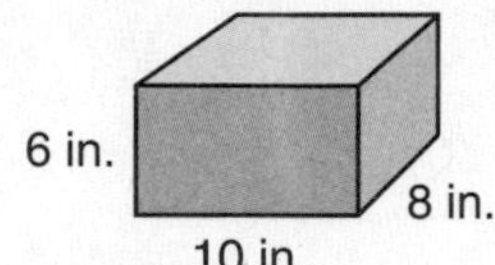

11. (67) Find the area of each of the six surfaces of the block of ice shown in problem 10. Then add the areas to find the total surface area of the block of ice.

12. (69) Write each number in scientific notation:

(a) 0.6×10^6 (b) 0.6×10^{-6}

13. (28, 34) What is the average of the three numbers marked by arrows on the number line below?

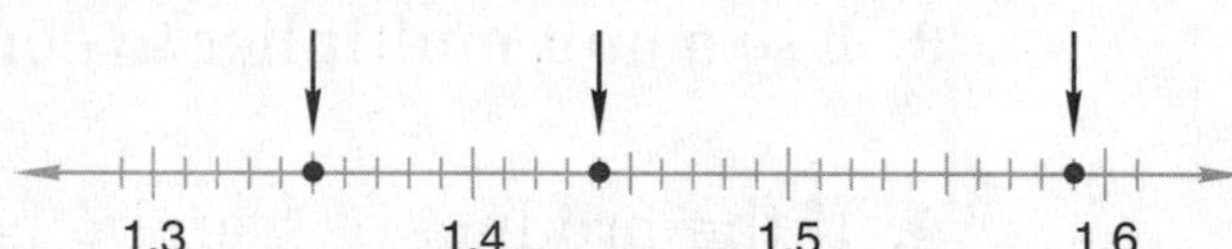

14. (48) Complete the table.

FRACTION	DECIMAL	PERCENT
$\frac{3}{5}$	(a)	(b)
(c)	(d)	2.5%

15. (21) (a) Write the prime factorization of 8100 using exponents.

(b) Find $\sqrt{8100}$.

16. (37, 61) (a) Find the area of the parallelogram.

(b) Find the area of the shaded triangle.

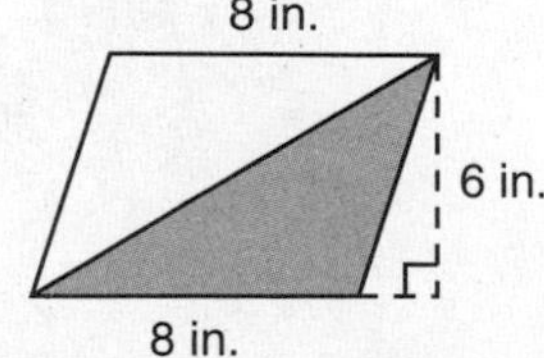

(c) If each acute angle of the parallelogram measures 72°, what is the measure of each obtuse angle of the parallelogram?

17. Name each geometric solid and tell why only one of the figures is a polyhedron:
(67)

(a) 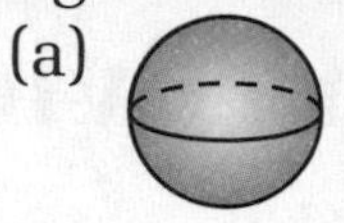

(b) 

(c) 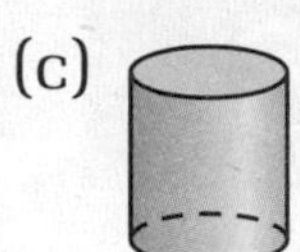

18. Find the circumference of each circle:
(66)

(a)

30 mm

Leave π as π.

(b) 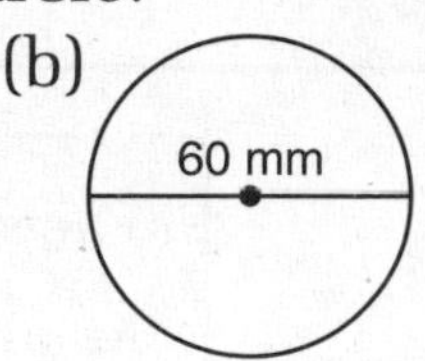

Use 3.14 for π.

19. Compare: $\frac{2}{3} \bigcirc 0.667$
(43)

20. The dots on this number line represent all integers greater than or equal to −3 that are also less than or equal to +3.
(4)

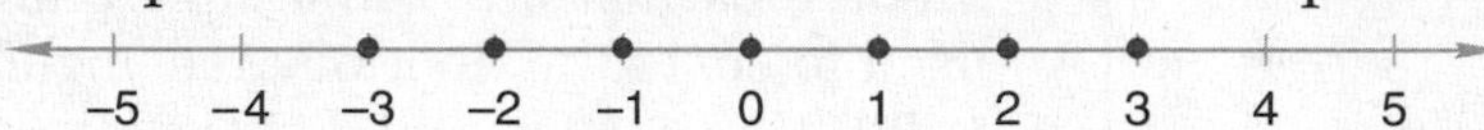

Draw a number line on your paper, and graph all integers greater than −4 that are not positive.

21. For $x = 5$ and $y = 4$, evaluate:
(52, 57)

(a) $x^2 - y^2$

(b) $x^0 - y^{-1}$

For problems 22 and 23, solve and check each equation. Show each step.

22. $m - \frac{2}{3} = 1\frac{3}{4}$
(Inv. 7)

23. $\frac{2}{3}w = 24$
(Inv. 7)

Simplify:

24. $\dfrac{[30 - 4(5 - 2)] + 5(3^3 - 5^2)}{\sqrt{9} + \sqrt{16}}$
(63)

25.
(56)

$$\begin{array}{r} 2 \text{ gal } 1 \text{ qt} \\ -\ 1 \text{ gal } 1 \text{ qt } 1 \text{ pt} \\ \hline \end{array}$$

26. $\frac{1}{2} \text{ mi} \cdot \frac{5280 \text{ ft}}{1 \text{ mi}} \cdot \frac{1 \text{ yd}}{3 \text{ ft}}$
(50)

27. $\left(2\frac{1}{2}\right)^2 \div \left(4\frac{1}{2} \cdot 6\frac{2}{3}\right)$
(26)

28. $7\frac{1}{2} - \left(5\frac{1}{6} + 1\frac{1}{3}\right)$
(30)

29. (a) $(-7) + |+5| + (-9)$
(68)

(b) $(16) + (-24) - (-18)$

30. Write the sum of $5\frac{1}{4}$ and 1.9 as a decimal.
(35, 43)

LESSON

73 Multiplying and Dividing Signed Numbers

WARM-UP

Facts Practice: Classifying Quadrilaterals and Triangles (Test O)

Mental Math:

a. $(+15) + (-25)$

b. 8.75×10^3

c. $\frac{12}{x} = \frac{2.5}{7.5}$

d. $3\frac{1}{2} \times 18$ (double, halve)

e. 500 mL to L

f. $66\frac{2}{3}\%$ of \$36

g. Estimate 15% of 39 by finding 10% of 40 plus half of 10% of 40.

Problem Solving:

Floor tiles that are 1-foot square and come 20 to a box will be used to cover the floor of a rectangular workroom. The workroom is 20 ft 6 in. long and 14 ft 6 in. wide. If all cut-off portions of tiles may be used, how many boxes of tile are needed? If only one portion from each cut tile may be used, how many boxes are needed?

NEW CONCEPT

We can develop the rules for multiplying and dividing signed numbers if we remember that multiplication is a shorthand notation for repeated addition. We remember that 2 times 3 means 3 + 3 and that 2 times –3 means (–3) + (–3), so

$$2(3) = 6 \quad \text{and} \quad 2(-3) = -6$$

We remember that division undoes multiplication, so the following statements must be true:

$$2(3) = 6; \text{ therefore, } \frac{6}{2} = 3 \text{ and } \frac{6}{3} = 2$$

Likewise,

$$2(-3) = -6; \text{ therefore, } \frac{-6}{2} = -3 \text{ and } \frac{-6}{-3} = +2$$

We use these examples to illustrate that when we multiply or divide two positive numbers, the answer is a positive number. Also, when we multiply or divide two numbers whose signs are different, the answer is a negative number.

But what happens if we multiply two negative numbers? Since 2 times −3 equals −6

$$2(-3) = -6$$

then the *opposite of* 2 times −3 equals the *opposite of* −6, which is 6.

$$(-2)(-3) = +6$$

And since division undoes multiplication, these division problems must also be true:

$$\frac{+6}{-2} = -3 \quad \text{and} \quad \frac{+6}{-3} = -2$$

These conclusions give us the rules for multiplying and dividing signed numbers.

RULES FOR MULTIPLYING AND DIVIDING SIGNED NUMBERS

1. If the two numbers in the multiplication or division problem have the same sign, the answer is a positive number.
2. If the two numbers in the multiplication or division problem have different signs, the answer is a negative number.

Here are some examples:

MULTIPLICATION	DIVISION
$(+6)(+2) = +12$	$\frac{+6}{+2} = +3$
$(-6)(-2) = +12$	$\frac{-6}{-2} = +3$
$(-6)(+2) = -12$	$\frac{-6}{+2} = -3$
$(+6)(-2) = -12$	$\frac{+6}{-2} = -3$

Example Divide or multiply:

(a) $\frac{-12}{+4}$ (b) $\frac{-12}{-3}$ (c) (6)(−3) (d) (−6)(−4)

Solution We divide or multiply as indicated. If both signs are the same, the answer is positive. If one sign is positive and the other is negative, the answer is negative. Showing the positive sign is permitted but not necessary.

(a) **−3** (b) **+4** (c) **−18** (d) **+24**

LESSON PRACTICE

Practice set Divide or multiply:

a. (−7)(3) **b.** (+4)(−8) **c.** (8)(+5)

d. (−8)(−3) **e.** $\frac{25}{-5}$ **f.** $\frac{-27}{-3}$

g. $\frac{-28}{4}$ **h.** $\frac{+30}{6}$ **i.** $\frac{+45}{-3}$

MIXED PRACTICE

Problem set

1. (72) Use a ratio box to solve this problem. If Elena can wrap 12 packages in 5 minutes, how many packages can she wrap in 1 hour?

2. (55) Lydia walked 30 minutes a day for 5 days. The next 3 days she walked an average of 46 minutes per day. What was the average amount of time she spent walking per day during those 8 days?

3. (35, 45) If the sum of 0.2 and 0.5 is divided by the product of 0.2 and 0.5, what is the quotient?

4. (50) Use a unit multiplier to convert 23 cm to mm.

5. (65) Use a ratio box to solve this problem. The ratio of paperback books to hardbound books in the town library was 3 to 11. If there were 9240 hardbound books in the library, how many books were there in all?

6. Write each number in scientific notation:
(69)
(a) 24×10^{-5} (b) 24×10^{7}

7. Diagram this statement. Then answer the questions
(71) that follow.

The 30 true-false questions amounted to $\frac{1}{4}$ of the test questions.

(a) How many questions were on the test?

(b) How many of the questions were not true-false?

8. Write an equation to solve each problem:
(60)
(a) Five ninths of 45 is what number?

(b) What number is 80% of 760?

9. Divide or multiply:
(73)

(a) $\frac{-36}{9}$ (b) $\frac{-36}{-6}$

(c) $9(-3)$ (d) $(+8)(+7)$

10. The face of the spinner is divided
(21, 36) into eighths. If the spinner is spun once, what is the probability that it will stop on a composite number?

11. The x-axis is a line of symmetry for ΔRST. The coordinates
(58, 62) of R and S are (6, 0) and (−2, −2), respectively.

(a) What are the coordinates of T?

(b) What type of triangle is ΔRST classified by sides?

(c) If the measure of $\angle R$ is approximately 28°, then what is the approximate measure of $\angle S$?

12. Describe how to determine whether the product of two
(73) signed numbers is positive or negative.

13. Find the number of 1-ft cubes that will fit inside a closet with dimensions as shown.
(70)

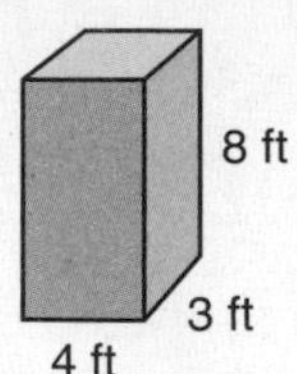

14. Find the circumference of each circle:
(66)

(a)

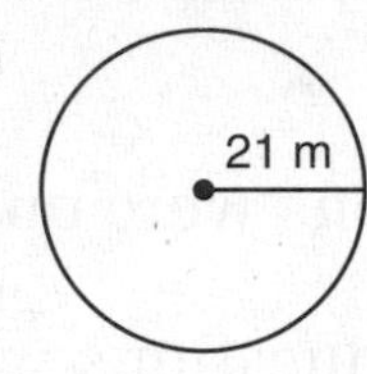

Use 3.14 for π.

(b)

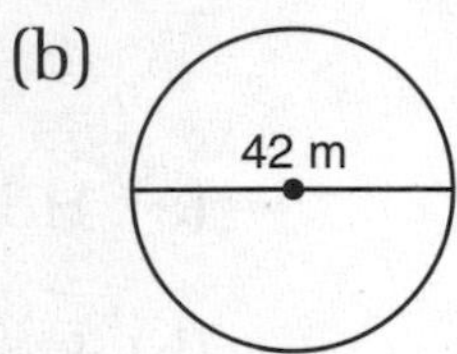

Use $\frac{22}{7}$ for π.

15. Complete the table.
(48)

FRACTION	DECIMAL	PERCENT
(a)	2.5	(b)
(c)	(d)	0.2%

Classify each triangle in problems 16–18 as acute, right, or obtuse. Also classify each triangle as equilateral, isosceles, or scalene. Then find the area of each triangle. Dimensions are in centimeters.

16. (37, 62)

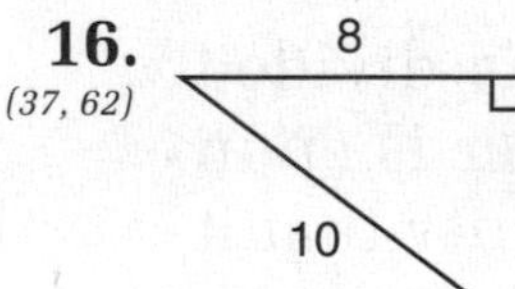

17. (37, 62)

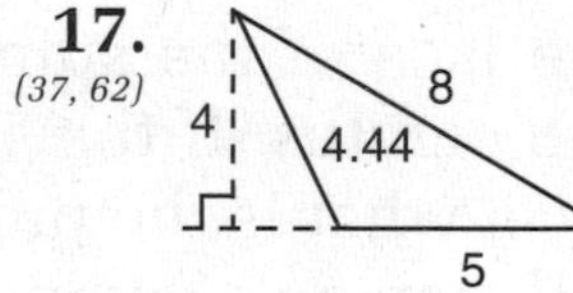

18. (37, 62)

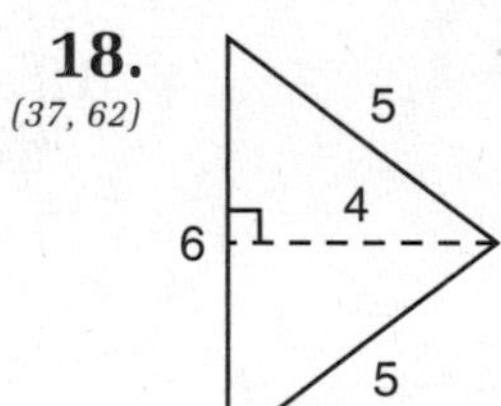

19. Name each three-dimensional figure:
(67)

(a)

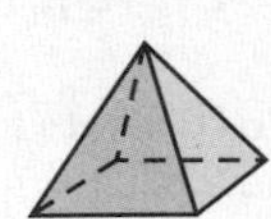

(b)

(c)

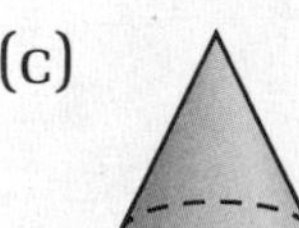

20. Compare: $\frac{2}{3}$ of 96 ◯ $\frac{5}{6}$ of 84
(60)

21. Evaluate: $ab - (a - b)$ if $a = \frac{5}{6}$ and $b = \frac{3}{4}$
(52)

For problems 22–24, solve and check each equation. Show each step.

22. $\frac{3}{5}w = 15$
(Inv. 7)

23. $b - 1.6 = (0.4)^2$
(Inv. 7)

24. $20w = 5.6$
(Inv. 7)

Simplify:

25.
(49)
$$\begin{array}{r} 2\text{ yd } 1\text{ ft } 7\text{ in.} \\ +\ 1\text{ yd } 2\text{ ft } 8\text{ in.} \\ \hline \end{array}$$

26. $0.5\text{ m} \cdot \frac{100\text{ cm}}{1\text{ m}} \cdot \frac{10\text{ mm}}{1\text{ cm}}$
(50)

27. $12\frac{1}{2} \cdot 4\frac{1}{5} \cdot 2\frac{2}{3}$
(26)

28. $7\frac{1}{2} \div \left(6\frac{2}{3} \cdot 1\frac{1}{5}\right)$
(26)

29. (a) $(-8) + (-7) - (-15)$
(68)

(b) $(-15) + (+11) - |+24|$

30. Find the product of 2.25 and $1\frac{1}{3}$.
(26, 43)

LESSON

74 Fractional Part of a Number, Part 2

WARM-UP

Facts Practice: + − × ÷ Mixed Numbers (Test N)

Mental Math:

a. $(-8) - (-12)$

b. 45×10^{-3}

c. $7w + 1 = 50$

d. Estimate 15% tip on a $19.81 bill.

e. 400 m to km

f. 80% of $25

g. 10% of 80, × 2, $\sqrt{}$, × 7, − 1, ÷ 3, $\sqrt{}$, × 12, $\sqrt{}$, ÷ 6

Problem Solving:

Sarita rode the Gravitron at the fair. The cylindrical chamber spins, forcing riders against the wall while the floor drops away. If the chamber is 30 feet in diameter and if it spins around 30 times during a ride, how far do the riders travel? Do riders travel more or less than $\frac{1}{2}$ mile?

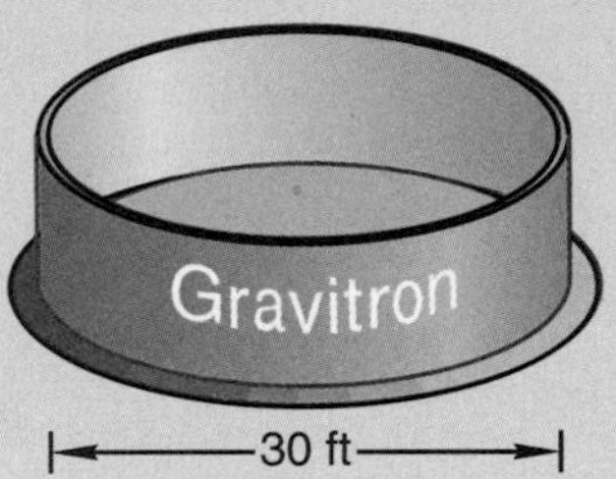

NEW CONCEPT

In some fractional-part-of-a-number problems, the fraction is unknown. In other fractional-part-of-a-number problems, the total is unknown. As discussed in Lesson 60, we can translate these problems into equations by replacing the word "of" with a multiplication sign and the word "is" with an equal sign.

Example 1 What fraction of 56 is 42?

Solution We translate this statement directly into an equation by replacing "what fraction" with W_F, "of" with a multiplication symbol, and "is" with an equal sign.

What fraction of 56 is 42?   question

↓ ↓ ↓ ↓ ↓

$W_F \times 56 = 42$ equation

To solve, we divide both sides by 56.

$$\frac{W_F \times 56}{56} = \frac{42}{56} \quad \text{divided by 56}$$

$$W_F = \frac{3}{4} \quad \text{simplified}$$

If the question had been "What decimal part of 56 is 42?" the procedure would have been the same, but as the last step we would have written $\frac{3}{4}$ as the decimal number 0.75.

$$W_D = 0.75$$

Example 2 Seventy-five is what decimal part of 20?

Solution We make a direct translation.

Seventy-five	is	what decimal part	of	20?	question
↓	↓	↓	↓	↓	
75	=	W_D	×	20	equation

To solve, we divide both sides by 20.

$$\frac{75}{20} = \frac{W_D \times 20}{20} \qquad \text{divided by 20}$$

$$W_D = \mathbf{3.75} \qquad \text{simplified}$$

If the question had asked "what fractional part," we would have written the answer as a fraction or as a mixed number.

$$\frac{75}{20} = W_F \qquad \text{fraction}$$

$$\frac{15}{4} = W_F \qquad \text{reduced}$$

$$3\frac{3}{4} = W_F \qquad \text{mixed number}$$

Example 3 Three fourths of what number is 60?

Solution In this problem the total is the unknown. But we can still translate directly from the question to an equation.

Three fourths	of	what number	is	60?	question
↓	↓	↓	↓	↓	
$\frac{3}{4}$	×	W_N	=	60	equation

To solve, we multiply both sides by $\frac{4}{3}$.

$$\frac{4}{3} \times \frac{3}{4} \times W_N = 60 \times \frac{4}{3} \qquad \text{multiplied by } \frac{4}{3}$$

$$W_N = \mathbf{80} \qquad \text{simplified}$$

Had the question been phrased using 0.75 instead of $\frac{3}{4}$, the procedure would have been similar.

Seventy-five hundredths	of	what number	is	60?	question
↓	↓	↓	↓	↓	
0.75	×	W_N	=	60	equation

To solve, we can divide both sides by 0.75.

$$\frac{0.75 \times W_N}{0.75} = \frac{60}{0.75} \quad \text{divided by 0.75}$$

$$W_N = 80 \quad \text{simplified}$$

LESSON PRACTICE

Practice set Translate each statement into an equation and solve:

a. What fraction of 130 is 80?

b. Seventy-five is what decimal part of 300?

c. Eighty is 0.4 of what number?

d. Sixty is $\frac{5}{6}$ of what number?

e. Sixty is what fraction of 90?

f. What decimal part of 80 is 60?

g. Forty is 0.08 of what number?

h. Six fifths of what number is 60?

MIXED PRACTICE

Problem set

1. (55) During the first 3 days of the week, Mike read an average of 28 pages per day. During the next 4 days, Mike averaged 42 pages per day. For the whole week, Mike read an average of how many pages per day?

2. (46) Twelve ounces of Brand X costs $1.14. Sixteen ounces of Brand Y costs $1.28. Brand X costs how much more per ounce than Brand Y?

3. Use a unit multiplier to convert $4\frac{1}{2}$ feet to inches.
(50)

4. Use a ratio box to solve this problem. The ratio of black-and-white photographs to color photographs at the exhibit was 2 to 3. If 18 of the photographs were color, how many photographs were at the exhibit?
(65)

5. Use a ratio box to solve this problem. If 5 pounds of apples costs \$1.40, how much would 8 pounds of apples cost?
(72)

6. Diagram this statement. Then answer the questions that follow.
(22, 36)

Five sixths of the 300 triathletes completed the course.

(a) How many triathletes completed the course?

(b) What was the ratio of triathletes who completed the course to those who did not complete the course?

Write equations to solve problems 7–11.

7. Fifteen is $\frac{3}{8}$ of what number?
(74)

8. Seventy is what decimal part of 200?
(74)

9. Two fifths of what number is 120?
(74)

10. The store made a 60% profit on the \$180 selling price of the coat. What was the store's profit?
(60)

11. The shoe salesperson received a 20% commission on the sale of a \$35 pair of shoes. What was the salesperson's commission?
(60)

12. (a) What is the volume of this cube?
(67, 70)

(b) What is its surface area?

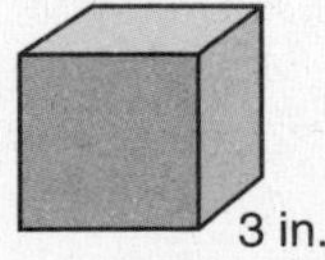

13. Find the circumference of each circle:
(66)

(a)

14 m

Use $\frac{22}{7}$ for π.

(b)

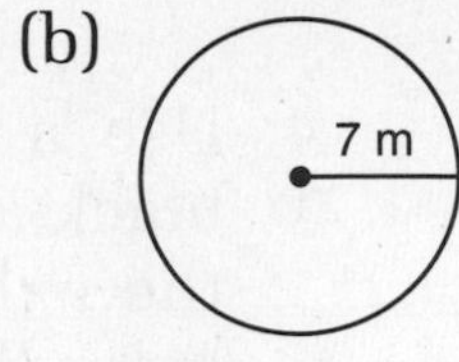

Leave π as π.

14. Complete the table.
(48)

FRACTION	DECIMAL	PERCENT
$3\frac{1}{2}$	(a)	(b)
(c)	(d)	35%

15. Describe the rule of this function, and find the missing numbers.
(58)

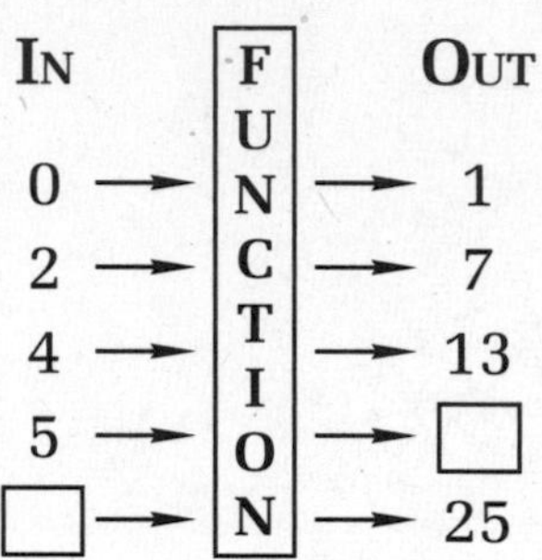

16. Write 425 million in scientific notation.
(51)

Refer to the figure below to answer problems 17 and 18. In the figure, $\overline{AE} \parallel \overline{BD}$, $\overline{AB} \parallel \overline{EC}$, and $EC = ED$.

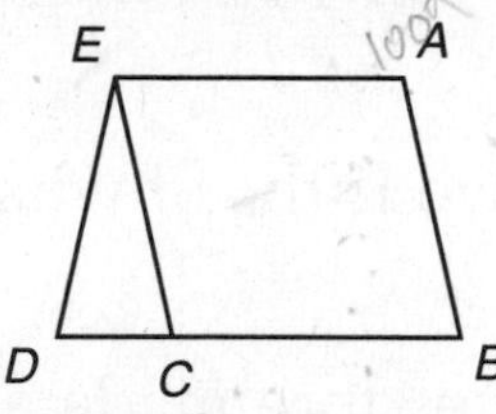

17. (a) What type of quadrilateral is figure *ABCE*?
(Inv. 6, 62)

(b) What type of quadrilateral is figure *ABDE*?

(c) What type of triangle is ΔECD classified by sides?

18. If the measure of $\angle A$ is 100°, what is the measure of
(40, 61)

(a) $\angle ABC$? (b) $\angle BCE$? (c) $\angle ECD$?

(d) $\angle EDC$? (e) $\angle DEC$? (f) $\angle DEA$?

19. Arrange these numbers in order from least to greatest:
(33)

0.013, 0.1023, 0.0103, 0.021

20. Evaluate: $(m + n) - mn$ if $m = 1\frac{1}{2}$ and $n = 2\frac{2}{3}$
(52)

For problems 21–24, solve and check each equation. Show each step.

21. $p + 3\frac{1}{5} = 7\frac{1}{2}$ (Inv. 7)

22. $3n = 0.138$ (Inv. 7)

23. $n - 0.36 = 4.8$ (Inv. 7)

24. $\frac{2}{3}x = \frac{8}{9}$ (Inv. 7)

Simplify:

25. $\sqrt{49} + \{5[3^2 - (2^3 - \sqrt{25})] - 5^2\}$
(63)

26.
(56)

$$\begin{array}{r} 4\text{ hr } \ 5\text{ min } \ 15\text{ s} \\ -\ 1\text{ hr } \ 15\text{ min } \ 30\text{ s} \\ \hline \end{array}$$

27. (a) $(-9) + (-11) - (+14)$ (b) $(26) + (-43) - |-36|$
(68)

28. (a) $(-3)(12)$ (b) $(-3)(-12)$
(73)

(c) $\frac{-12}{3}$ (d) $\frac{-12}{-3}$

29. Write the sum of $8\frac{1}{3}$ and 7.5 as a mixed number.
(30, 43)

30. Florence is facing north. If she turns 180°, which direction will she be facing?
(17)

LESSON

75 Area of a Complex Figure • Area of a Trapezoid

WARM-UP

Facts Practice: Classifying Quadrilaterals and Triangles (Test O)

Mental Math:

a. $(-15) - (+20)$ **b.** 15^2

c. $\frac{30}{40} = \frac{g}{12}$ **d.** $25 \times \$2.40$ (double, halve)(twice)

e. 250 mg to g **f.** 10% of \$35

g. What decimal is half of 10% of 36?

Problem Solving:

Find the missing fraction: $\frac{5}{6} \times \frac{3}{10} \times \frac{16}{3} \times \frac{?}{?} = 1$

NEW CONCEPTS

Area of a complex figure We have practiced finding the areas of figures that can be divided into two or more rectangles. In this lesson we will begin finding the areas of figures that include triangular regions as well.

Example 1 Find the area of this figure. Corners that look square are square. Dimensions are in millimeters.

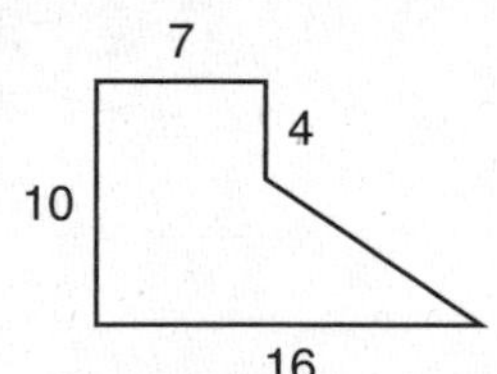

Solution We divide the figure into smaller polygons, the measurements of which we can be certain. In this case we draw dashes that divide the figure into a rectangle and a triangle.

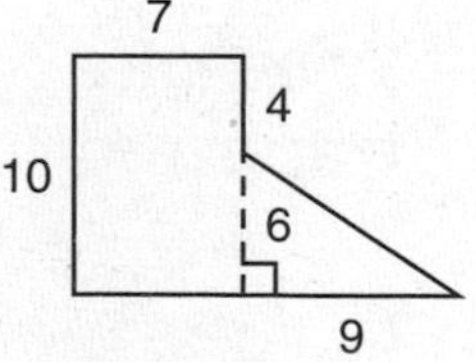

$$\begin{aligned} &\text{Area of rectangle} = 7 \times 10 = 70 \text{ mm}^2 \\ +\ &\text{Area of triangle} = \frac{6 \times 9}{2} = 27 \text{ mm}^2 \\ \hline &\text{Total area} = \mathbf{97\ mm^2} \end{aligned}$$

When dividing figures, we must avoid assumptions based on appearances. Although it may appear that the figure at right is divided into two triangles, the larger "triangle" is actually a quadrilateral. The slanted "segment" bends where the solid and dashed segments intersect. The assumption that the figure is divided into two triangles leads to an incorrect calculation for the area of the figure.

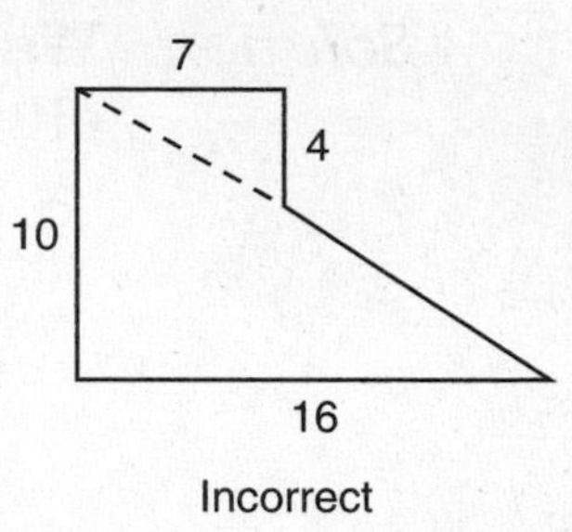

Incorrect

Example 2 Find the area of this figure. Corners that look square are square. Dimensions are in centimeters.

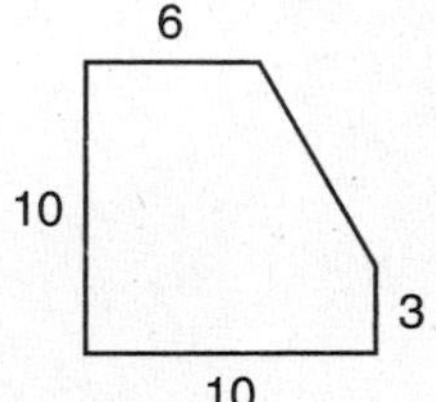

Solution There are many ways to divide this figure.

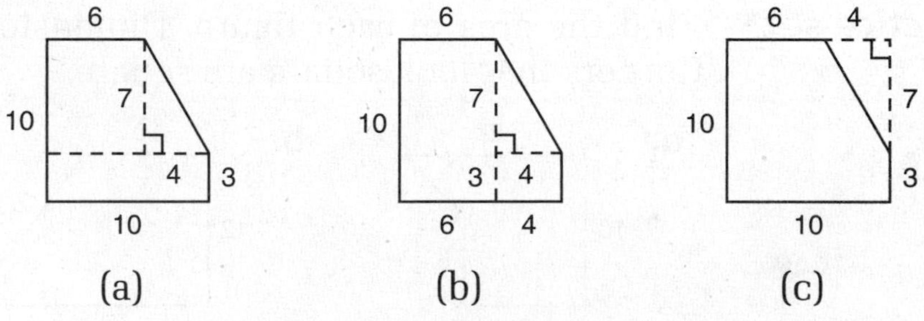

(a) (b) (c)

We decide to use (c). We will find the area of the big rectangle and subtract from it the area of the triangle.

$$\text{Area of rectangle} = 10 \times 10 = 100 \text{ cm}^2$$
$$-\ \text{Area of triangle} = \frac{4 \times 7}{2} = 14 \text{ cm}^2$$
$$\text{Area of figure} = \mathbf{86\ cm^2}$$

Area of a trapezoid Recall that a quadrilateral with just one pair of parallel sides is a trapezoid. One way to find the area of a trapezoid is to divide the trapezoid into two triangular regions and find the combined area of the triangles.

Example 3 Find the area of this trapezoid. Dimensions are in centimeters.

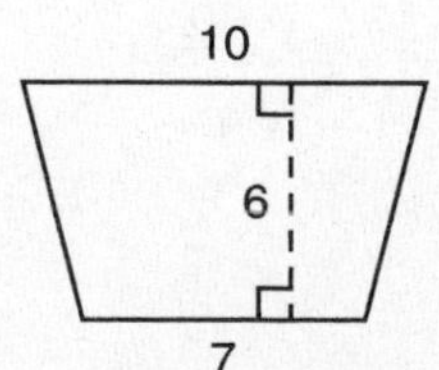

Solution We can divide the trapezoid into two triangles by drawing either diagonal. We show both ways.

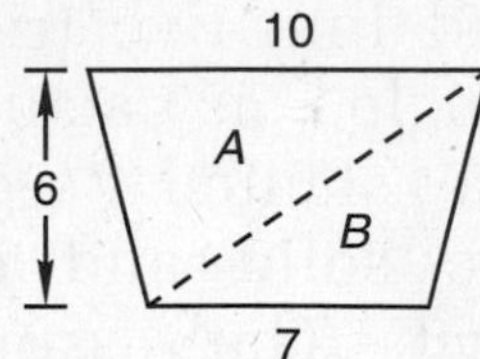

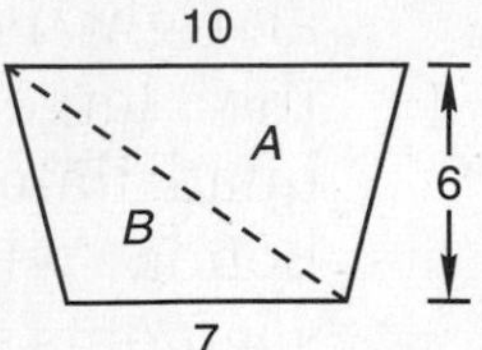

Both figures have two triangles that we have labeled A and B. The height of each triangle is 6 cm.

$$\begin{aligned} \text{Area of triangle } A &= \frac{10 \times 6}{2} = 30 \text{ cm}^2 \\ + \text{ Area of triangle } B &= \frac{7 \times 6}{2} = 21 \text{ cm}^2 \\ \hline \text{Total area} &= \mathbf{51\ cm^2} \end{aligned}$$

LESSON PRACTICE

Practice set* Find the area of each figure. Dimensions are in centimeters. Corners that look square are square.

a.

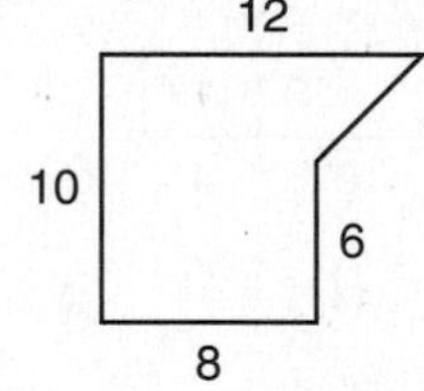

b.

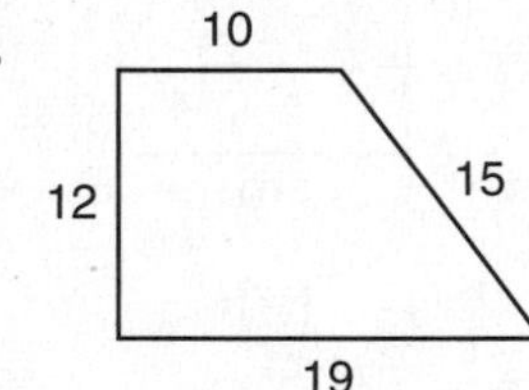

c.

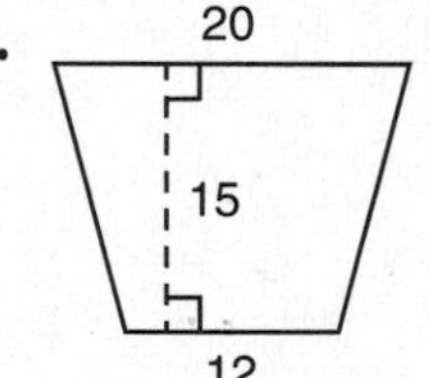

MIXED PRACTICE

Problem set

1. Pablo ran an 8-lap race. For the first 5 laps he averaged 72 seconds per lap. For the rest of the race he averaged 80 seconds per lap. What was his average time per lap for the entire race? *(55)*

2. If 30 ounces of cereal costs \$2.49, what is the cost per ounce? *(46)*

3. One thousand, five hundred meters is how many kilometers? *(32)*

4. The sum of $\frac{1}{2}$ and $\frac{3}{5}$ is how much greater than the product of $\frac{1}{2}$ and $\frac{3}{5}$? *(30)*

5. The ratio of Maela's age to Chelsea's age is 3 to 2. If Maela
(54) is 60 years old, she is how many years older than Chelsea?

6. Compare: $12.5 \times 10^{-4} \bigcirc 1.25 \times 10^{-3}$
(57, 69)

7. Use a ratio box to solve this problem. Martha rode 40 miles
(72) in 3 hours. At this rate, how long would it take Martha to ride 100 miles?

8. Diagram this statement. Then answer the questions
(22) that follow.

Two fifths of the library's 21,000 books were checked out over the course of the year.

(a) How many books were checked out?

(b) How many books were not checked out?

Write equations to solve problems 9–12.

9. Sixty is $\frac{5}{12}$ of what number?
(74)

10. Seventy percent of \$35.00 is how much money?
(60)

11. Thirty-five is what fraction of 80?
(74)

12. Fifty-six is what decimal part of 70?
(74)

13. Simplify:
(73)

(a) $\dfrac{-120}{4}$ (b) $(-12)(11)$

(c) $\dfrac{-120}{-5}$ (d) $12(+20)$

14. Find the volume of this rectangular
(70) prism. Dimensions are in centimeters.

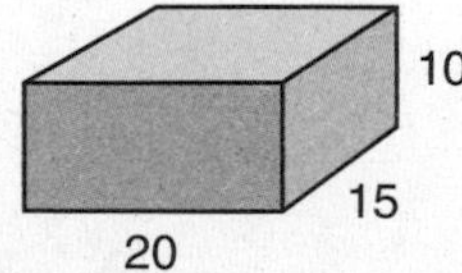

15. The diameter of the plate was 11 inches. Find its
(66) circumference to the nearest half inch.

16. Find the area of the trapezoid at
(75) right. Dimensions are in inches.

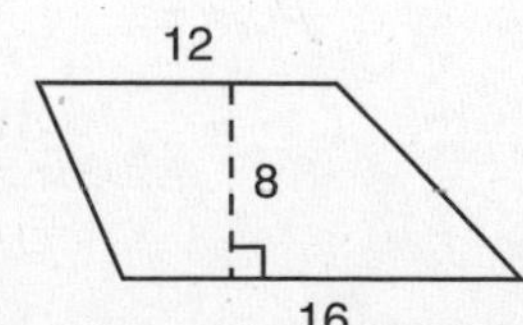

17. (75) A corner was trimmed from a square sheet of paper to make the shape shown. Dimensions are in centimeters.

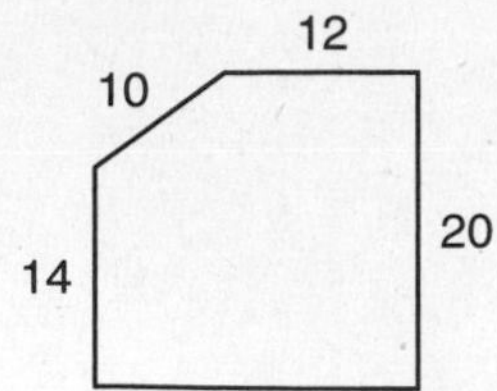

(a) What was the length of each side of the square paper before the corner was trimmed?

(b) Find the perimeter of the figure.

(c) Find the area of the figure.

18. (48) Complete the table.

FRACTION	DECIMAL	PERCENT
(a)	(b)	125%
$\frac{1}{8}$	(c)	(d)

19. (46) The taxicab bill was \$12.50. Sasha tipped the driver 20%. Altogether, how much money did Sasha pay the driver?

20. (52) Evaluate: $x^3 - xy - \dfrac{x}{y}$ if $x = 2$ and $y = 0.5$

For problems 21–23, solve and check each equation. Show each step.

21. (Inv. 7) $\frac{5}{8}x = 40$

22. (Inv. 7) $1.2w = 26.4$

23. (Inv. 7) $y + 3.6 = 8.47$

Simplify:

24. (63) $9^2 - [3^3 - (9 \cdot 3 - \sqrt{9})]$

25. (56)
$$\begin{array}{r} 2\text{ hr } 48\text{ min } 20\text{ s} \\ -\ 1\text{ hr } 23\text{ min } 48\text{ s} \\ \hline \end{array}$$

26. (50) $100\text{ yd} \cdot \dfrac{3\text{ ft}}{1\text{ yd}} \cdot \dfrac{12\text{ in.}}{1\text{ ft}}$

27. (26) $5\frac{1}{3} \cdot \left(3 \div 1\frac{1}{3}\right)$

28. (30) $3\frac{1}{5} + 2\frac{1}{2} - 1\frac{1}{4}$

29. (68) (a) $(-26) + (-15) - (-40)$

(b) $(-5) + (-4) - (-3) - (+2)$

30. (47, 57) Find each missing exponent:

(a) $5^5 \cdot 5^2 = 5^{\square}$

(b) $5^5 \div 5^2 = 5^{\square}$

(c) $5^2 \div 5^2 = 5^{\square}$

(d) $5^2 \div 5^5 = 5^{\square}$

LESSON

76 Complex Fractions

WARM-UP

Facts Practice: + − × ÷ Integers (Test P)

Mental Math:

a. (+6) + (−18)
b. 6.25×10^{-2}
c. $9a - 4 = 32$
d. 12 is $\frac{2}{3}$ of what number?
e. 5 mm to cm
f. What is $\frac{2}{3}$ of 12?
g. What is the total cost of a $25 video game plus 6% sales tax?

Problem Solving:

Kim hit a target like the one shown 6 times, earning a total score of 20. Find two sets of scores Kim could have earned. Shell earned a total score of exactly 20 in the fewest possible number of attempts. How many attempts did Shell make?

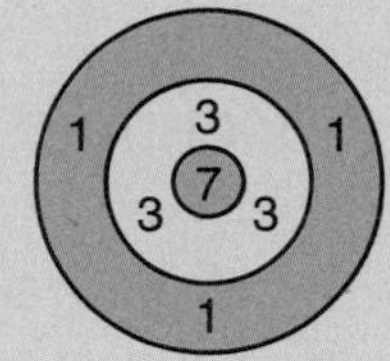

NEW CONCEPT

A **complex fraction** is a fraction that contains one or more fractions in the numerator or denominator. Each of the following is a complex fraction:

$$\frac{\frac{3}{5}}{\frac{2}{3}} \qquad \frac{25\frac{2}{3}}{100} \qquad \frac{15}{7\frac{1}{3}} \qquad \frac{\frac{a}{b}}{\frac{b}{c}}$$

To simplify a complex fraction, we multiply the fraction by a fraction name for 1 that makes the denominator 1.

Example 1 Simplify: $\dfrac{\frac{3}{5}}{\frac{2}{3}}$

Solution We focus our attention on the denominator of the complex fraction, which is $\frac{2}{3}$. We multiply the denominator by the reciprocal of $\frac{2}{3}$, which is $\frac{3}{2}$, so that the new denominator is 1. We also multiply the numerator by $\frac{3}{2}$.

$$\frac{\frac{3}{5}}{\frac{2}{3}} \times \frac{\frac{3}{2}}{\frac{3}{2}} = \frac{\frac{9}{10}}{1} = \mathbf{\frac{9}{10}}$$

We multiplied the complex fraction by a complex name for 1 to change the denominator to 1. Since $\frac{9}{10}$ divided by 1 is $\frac{9}{10}$, the complex fraction simplifies to $\frac{9}{10}$.

An alternative method for simplifying some complex fractions is to treat the fraction as a division problem. We can change the format of the division problem to a more familiar form.

$$\begin{matrix}\text{dividend} \\ \text{divisor}\end{matrix}\ \frac{\frac{3}{5}}{\frac{2}{3}} \longrightarrow \frac{3}{5} \div \frac{2}{3}$$

Then we simplify the division using the method described in Lesson 25.

$$\frac{3}{5} \div \frac{2}{3} = \frac{3}{5} \cdot \frac{3}{2} = \mathbf{\frac{9}{10}}$$

Example 2 Simplify: $\dfrac{25\frac{2}{3}}{100}$

Solution First we write both numerator and denominator as fractions.

$$\frac{\frac{77}{3}}{\frac{100}{1}}$$

Now we multiply the numerator and the denominator by $\frac{1}{100}$.

$$\frac{\frac{77}{3}}{\frac{100}{1}} \cdot \frac{\frac{1}{100}}{\frac{1}{100}} = \frac{\frac{77}{300}}{1} = \mathbf{\frac{77}{300}}$$

Example 3 Simplify: $\dfrac{15}{7\frac{1}{3}}$

Solution We begin by writing both numerator and denominator as improper fractions.

$$\frac{\frac{15}{1}}{\frac{22}{3}}$$

Now we multiply the numerator and the denominator by $\frac{3}{22}$.

$$\frac{\frac{15}{1}}{\frac{22}{3}} \cdot \frac{\frac{3}{22}}{\frac{3}{22}} = \frac{\frac{45}{22}}{1} = \mathbf{2\frac{1}{22}}$$

Example 4 Change $83\frac{1}{3}$ percent to a fraction and simplify:

Solution A percent is a fraction that has a denominator of 100. Thus $83\frac{1}{3}\%$ is

$$\frac{83\frac{1}{3}}{100}$$

Next we write both numerator and denominator as fractions.

$$\frac{\frac{250}{3}}{\frac{100}{1}}$$

Now we multiply the numerator and the denominator by $\frac{1}{100}$.

$$\frac{\frac{250}{3}}{\frac{100}{1}} \cdot \frac{\frac{1}{100}}{\frac{1}{100}} = \frac{\frac{250}{300}}{1} = \mathbf{\frac{5}{6}}$$

LESSON PRACTICE

Practice set Simplify:

a. $\dfrac{37\frac{1}{2}}{100}$ **b.** $\dfrac{12}{\frac{5}{6}}$ **c.** $\dfrac{\frac{2}{5}}{\frac{2}{3}}$

Change each percent to a fraction and simplify:

d. $66\frac{2}{3}\%$ **e.** $8\frac{1}{3}\%$ **f.** $4\frac{1}{6}\%$

MIXED PRACTICE

Problem set

1. (46) Nestor finished a 42-kilometer bicycle race in 1 hour 45 minutes $\left(1\frac{3}{4}\text{ hr}\right)$. What was his average speed in kilometers per hour?

2. (28) Akemi's scores in the diving competition were 7.9, 8.3, 8.1, 7.8, 8.4, 8.1, and 8.2. The highest and lowest scores were not counted. What was the average of the remaining scores?

3. (65) Use a ratio box to solve this problem. The ratio of good guys to bad guys in the movie was 2 to 5. If there were 35 guys in the movie, how many of them were good?

4. Use a unit multiplier to convert 3.5 grams to milligrams.
(50)

5. Change $16\frac{2}{3}$ percent to a fraction and simplify.
(76)

6. Davie is facing north. If he turns 90° in a clockwise direction, what direction will he be facing?
(17)

7. One sixth of the rock's mass was quartz. If the mass of the rock was 144 grams, what was the mass of the quartz in the rock?
(60)

8. For $a = 2$, evaluate
(41, 57)

(a) $\sqrt{2a^3}$ (b) $a^{-1} \cdot a^{-2}$

9. Simplify:
(73)

(a) $\frac{-60}{-12}$ (b) $(-8)(6)$

(c) $\frac{40}{-8}$ (d) $(-5)(-15)$

10. What is the circumference of the circle shown?
(66)

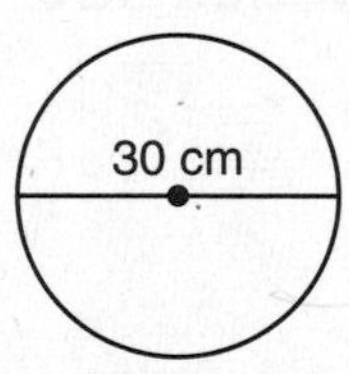

Leave π as π.

11. The figure at right is a pyramid with a square base. Copy the figure and find the number of its (a) faces, (b) edges, and (c) vertices.
(67)

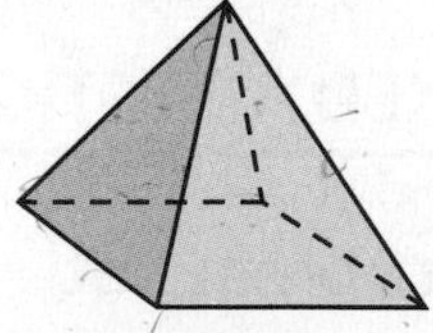

Write equations to solve problems 12–15.

12. What number is 10 percent of \$37.50?
(60)

13. What number is $\frac{5}{8}$ of 72?
(60)

14. Twenty-five is what fraction of 60?
(74)

15. Sixty is what decimal part of 80?
(74)

16. In this figure $AC = AB$. Angles DCA and ACB are supplementary. Find the measure of
(62)

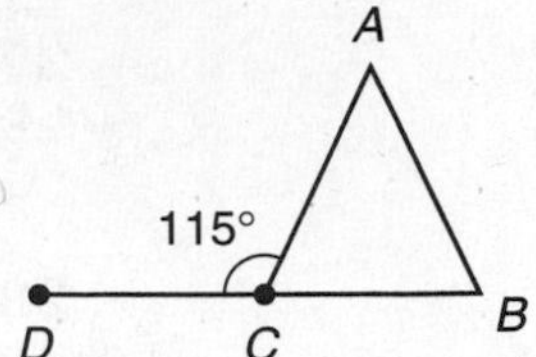

(a) $\angle ACB$.

(b) $\angle ABC$.

(c) $\angle CAB$.

17. Complete the table.
(48)

FRACTION	DECIMAL	PERCENT
$\frac{5}{6}$	(a)	(b)
(c)	(d)	0.1%

18. A square sheet of paper with an area of 81 in.^{2} has a corner cut off, forming a pentagon as shown.
(75)

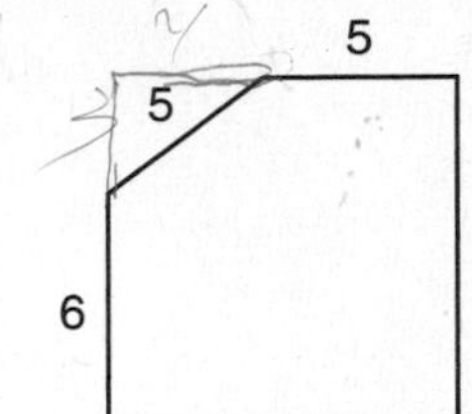

(a) What is the perimeter of the pentagon?

(b) What is the area of the pentagon?

19. What type of parallelogram has four congruent angles but not necessarily four congruent sides?
(Inv. 6)

20. When water increases in temperature from its freezing point to its boiling point, the reading on a thermometer increases from 0°C to 100°C and from 32°F to 212°F. The temperature halfway between 0°C and 100°C is 50°C. What temperature is halfway between 32°F and 212°F?
(28)

For problems 21–24, solve and check each equation. Show each step.

21. $x - 25 = 96$
(Inv. 7)

22. $\frac{2}{3}m = 12$
(Inv. 7)

23. $2.5p = 6.25$
(Inv. 7)

24. $10 = f + 3\frac{1}{3}$
(Inv. 7)

Simplify:

25. $\sqrt{13^2 - 5^2}$
(20)

26. 1 ton − 400 lb
(16)

27. $3\frac{3}{4} \times 4\frac{1}{6} \times (0.4)^2$ (fraction answer)
(26, 43)

28. $3\frac{1}{8} + 6.7 + 8\frac{1}{4}$ (decimal answer)
(35, 43)

29. (a) $(-3) + (-5) - (-3) - |+5|$
(68)
(b) $(-73) + (-24) - (-50)$

30. Before dividing, determine whether the quotient is greater than or less than 1 and state why. Then perform the calculation.
(76)

$$\frac{\frac{5}{6}}{\frac{2}{3}}$$

LESSON

77 Percent of a Number, Part 2

WARM-UP

Facts Practice: + − × ÷ Integers (Test P)

Mental Math:

a. (+12) − (−18) **b.** 4×10^6

c. $\frac{100}{150} = \frac{30}{a}$ **d.** Estimate 15% of $61.

e. 25 cm to m **f.** 12 is $\frac{3}{4}$ of n.

g. 10% of 50, × 6, + 2, ÷ 4, × 2, √, × 9, √, × 7, ÷ 2

Problem Solving:

On a balanced scale are a 25-g mass, a 100-g mass, and five identical blocks marked X, which are distributed as shown. What is the mass of each block marked X? Write an equation illustrated by this balanced scale.

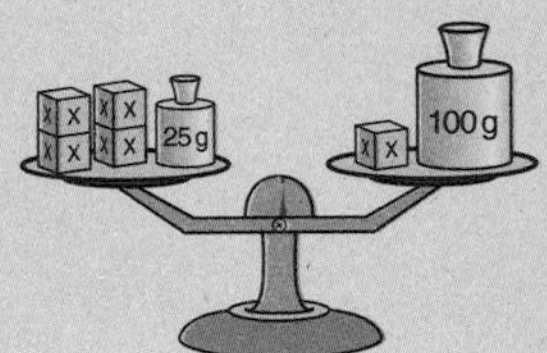

NEW CONCEPT

In Lesson 74 we practiced fractional-part problems involving fractions and decimals. In this lesson we will practice similar problems involving percents. First we translate the problem into an equation; then we solve the equation.

Example 1 What percent of 40 is 25?

Solution We translate the question to an equation and solve.

What percent of 40 is 25? question

↓ ↓ ↓ ↓ ↓

$W_P \times 40 = 25$ equation

To solve we divide both sides of the equation by 40.

$$\frac{W_P \times \overset{1}{\cancel{40}}}{\underset{1}{\cancel{40}}} = \frac{25}{40} \qquad \text{divided by 40}$$

$$W_P = \frac{5}{8} \qquad \text{simplified}$$

Since the question asked "what percent" and not "what fraction," we convert the fraction $\frac{5}{8}$ to a percent.

$$\frac{5}{8} \times 100\% = \mathbf{62\frac{1}{2}\%} \qquad \text{converted to a percent}$$

So $62\frac{1}{2}\%$ (or 62.5%) of 40 is 25.

Example 2 What percent of \$3.50 is \$0.28?

Solution We translate and solve.

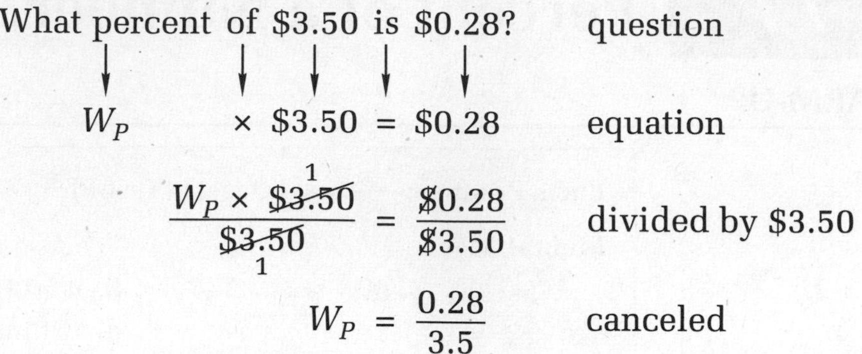

$$\frac{W_P \times \cancel{\$3.50}^{1}}{\cancel{\$3.50}_{1}} = \frac{\$0.28}{\$3.50} \qquad \text{divided by \$3.50}$$

$$W_P = \frac{0.28}{3.5} \qquad \text{canceled}$$

We perform the decimal division.

$$W_P = \frac{0.28}{3.5} = \frac{2.8}{35} = 0.08 \qquad \text{divided}$$

This is a decimal answer. The question asked for a percent answer so we convert the decimal 0.08 to 8%.

$$W_P = \mathbf{8\%} \qquad \text{converted to a percent}$$

Example 3 Seventy-five percent of what number is 600?

Solution We translate the question to an equation and solve. We can translate 75% to a fraction or to a decimal. We choose a fraction for this example.

Seventy-five percent of what number is 600? question

$$\frac{75}{100} \times W_N = 600 \qquad \text{equation}$$

To solve, we multiply both sides by 100 over 75.

$$\frac{\cancel{100}^{1}}{\cancel{75}_{1}} \cdot \frac{\cancel{75}^{1}}{\cancel{100}_{1}} \times W_N = 600 \cdot \frac{100}{75} \qquad \text{multiplied by } \frac{100}{75}$$

$$W_N = \mathbf{800} \qquad \text{simplified}$$

We could have used the fraction $\frac{3}{4}$ for $\frac{75}{100}$ with the same result.

Example 4 Fifty is what percent of 40?

Solution Since 50 is more than 40, the answer will be greater than 100%. We translate to an equation and solve.

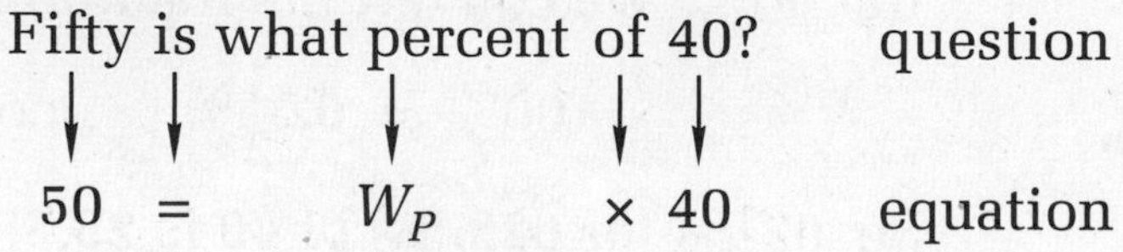

We divide both sides by 40.

$$\frac{50}{40} = \frac{W_P \times \overset{1}{\cancel{40}}}{\underset{1}{\cancel{40}}} \qquad \text{divided by 40}$$

$$\frac{5}{4} = W_P \qquad \text{simplified}$$

We convert $\frac{5}{4}$ to a percent.

$$W_P = \frac{5}{4} \times 100\% = \mathbf{125\%} \qquad \text{converted to a percent}$$

Example 5 Sixty is 150 percent of what number?

Solution We translate by writing 150% as either a decimal or a fraction. We will use the decimal form here.

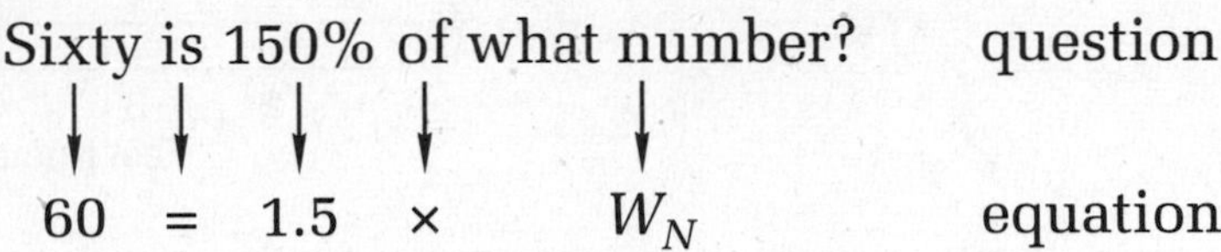

We divide both sides of the equation by 1.5.

$$\frac{60}{1.5} = \frac{\overset{1}{\cancel{1.5}} \times W_N}{\underset{1}{\cancel{1.5}}} \qquad \text{divided by 1.5}$$

$$\mathbf{40} = W_N \qquad \text{simplified}$$

LESSON PRACTICE

Practice set*

a. Twenty-four is what percent of 40?

b. What percent of 6 is 2?

c. Fifteen percent of what number is 45?

d. What percent of 4 is 6?

e. Twenty-four is 120% of what number?

f. Rework example 5, writing 150% as a fraction instead of as a decimal.

g. What percent of \$5.00 is \$0.35?

MIXED PRACTICE

Problem set

1. (65) Use a ratio box to solve this problem. Tammy saved nickels and pennies in a jar. The ratio of nickels to pennies was 2 to 5. If there were 70 nickels in the jar, how many coins were there in all?

Refer to the line graph below to answer problems 2–4.

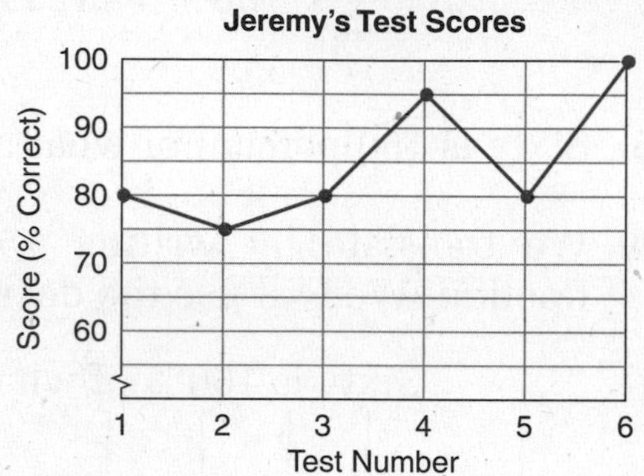

2. (38, 60) If there were 50 questions on Test 1, how many questions did Jeremy answer correctly?

3. (28, 38) What was Jeremy's average score? (What was the mean of the scores?)

4. (38, Inv. 4) (a) Which score did Jeremy make most often? (What was the mode of the scores?)

(b) What was the difference between his highest score and his lowest score? (What was the range of the scores?)

5. (67) Name the shape of each object:

(a) a marble

(b) a length of pipe

(c) a box of tissues

6. (72) Use a case 1–case 2 ratio box to solve this problem. One hundred inches equals 254 centimeters. How many centimeters equals 250 inches?

7. Diagram this statement. Then answer the questions
(36, 71) that follow.

Three fifths of those present agreed, but the remaining 12 disagreed.

(a) What fraction of those present disagreed?

(b) How many were present?

(c) How many of those present agreed?

(d) What was the ratio of those who agreed to those who disagreed?

Write equations to solve problems 8–11.

8. Forty is $\frac{4}{25}$ of what number?
(74)

9. Twenty-four percent of 10,000 is what number?
(60)

10. Twelve percent of what number is 240?
(77)

11. Twenty is what percent of 25?
(77)

12. Simplify:
(73)

(a) $25(-5)$ (b) $-15(-5)$

(c) $\frac{-250}{-5}$ (d) $\frac{-225}{15}$

13. Complete the table.
(48)

FRACTION	DECIMAL	PERCENT
(a)	0.2	(b)
(c)	(d)	2%

14. What is the total price of a $21 item including 7.5% sales
(46) tax?

15. Simplify:
(76)

(a) $\frac{14\frac{2}{7}}{100}$ (b) $\frac{60}{\frac{2}{3}}$

16. Find the area of this symmetrical
(75) figure. Dimensions are in feet. Corners that look square are square.

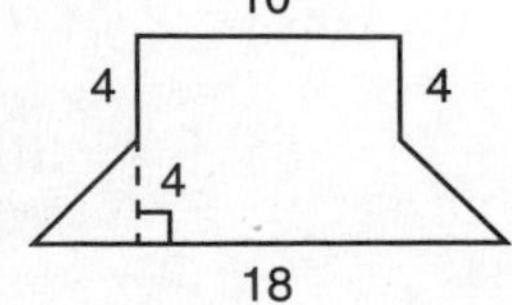

17. Draw a cube with edges 2 cm long.
(67, 70)
(a) What is the volume of the cube?

(b) Describe how to find the surface area of a cube.

18. Write 12 billion in scientific notation.
(51)

19. Find the circumference of each circle:
(66)

(a)

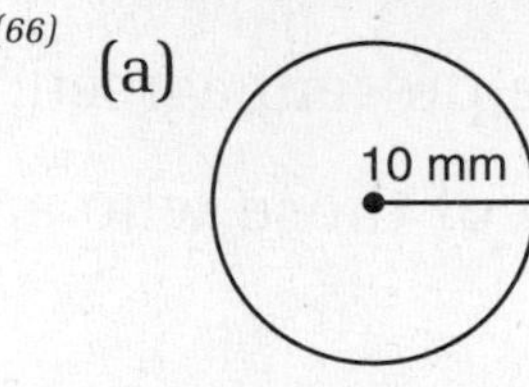

Leave π as π.

(b)

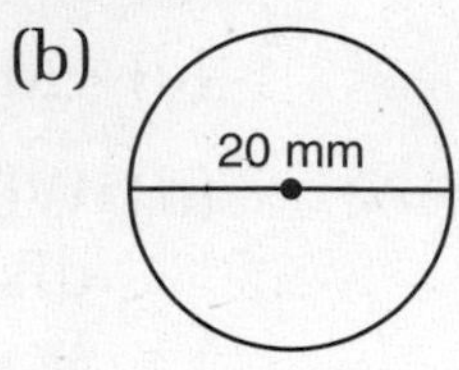

Use 3.14 for π.

For problems 20–22, solve and check each equation. Show each step.

20. $3x = 26.7$
(Inv. 7)

21. $y - 3\frac{1}{3} = 7$
(Inv. 7)

22. $\frac{2}{3}x = 48$
(Inv. 7)

23. The rule of this function has two steps. Describe the rule of the function, and find the missing numbers.
(58)

IN		FUNCTION		OUT
3	→		→	□
1	→		→	5
2	→		→	9
4	→		→	17
□	→		→	1

Simplify:

24. $5^2 - \{2^3 + 3[4^2 - (4)(\sqrt{9})]\}$
(63)

25.
(49)
$$\begin{array}{r} 4 \text{ gal } 3 \text{ qt } 1 \text{ pt} \\ +\ 1 \text{ gal } 2 \text{ qt } 1 \text{ pt} \\ \hline \end{array}$$

26. $1 \text{ ft}^2 \cdot \frac{12 \text{ in.}}{1 \text{ ft}} \cdot \frac{12 \text{ in.}}{1 \text{ ft}}$
(50)

27. $5\frac{1}{3} \div \left(1\frac{1}{3} \div 3\right)$
(26)

28. $3\frac{1}{5} - 2\frac{1}{2} + 1\frac{1}{4}$
(30)

29. $3\frac{1}{3} \div 2.5$ (mixed-number answer)
(43)

30. (a) $(-3) + (-4) - (+5)$
(68)

(b) $(-6) - (-16) - (+30)$

LESSON

78 Graphing Inequalities

WARM-UP

Facts Practice: Classifying Quadrilaterals and Triangles (Test O)

Mental Math:

a. (–8) – (–16)
b. $1^0 + 1^2$
c. $10p + 3 = 63$
d. 5% of $640.00 (double, halve)
e. 750 g to kg
f. 12 is $\frac{1}{6}$ of m.
g. Estimate a 15% tip on a $31.49 bill.

Problem Solving:

When you point a vertex of a dot cube toward you, the dots on three faces are visible. A cube has 8 vertices, so 8 combinations of three faces can be seen by turning a dot cube. List the total number of dots that can be seen in each of the 8 positions.

NEW CONCEPT

We have used the symbols >, <, and = to compare two numbers. In this lesson we will introduce the symbols ≥ and ≤. We will also practice graphing on the number line.

The symbols ≥ and ≤ combine the greater than/less than symbols with the equal sign. Thus, the symbol

$$\geq$$

is read, "greater than or equal to." The symbol

$$\leq$$

is read, "less than or equal to."

To graph a number on the number line, we draw a dot at the point that represents the number. Thus when we graph 4 on the number line, it looks like this:

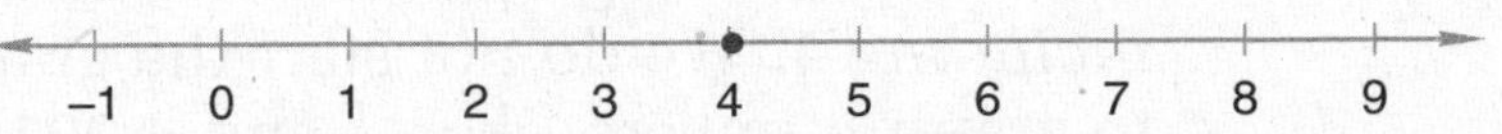

This time we will graph *all the numbers that are greater than or equal to 4.* We might think the graph should look like this:

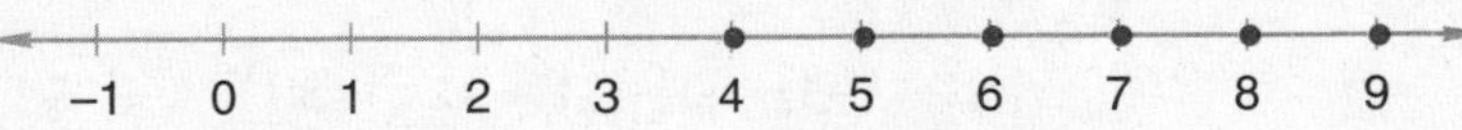

It is true that all the dots mark points that represent numbers greater than or equal to 4. However, we did not graph *all* the numbers that are greater than 4. For instance, we did not graph 10, 11, 12, and so on. Also, we did not graph numbers like $4\frac{1}{2}$, $5\frac{1}{3}$, $\sqrt{29}$, or 2π. If we were to graph all these numbers, the dots would be so close together that we would end up with a ray that goes on and on. Thus we graph all the numbers greater than or equal to 4 like this:

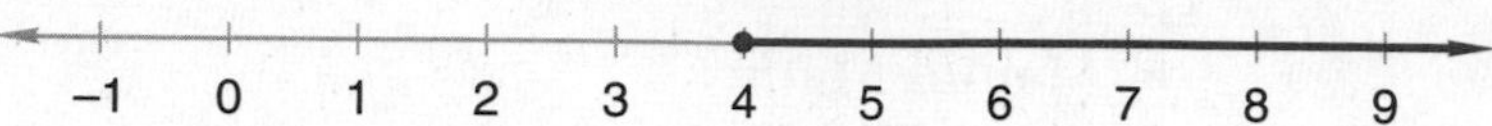

The large dot marks the 4. The colored ray marks the numbers greater than 4. The ray's arrowhead shows that the ray continues without end.

Expressions such as the following are called **inequalities:**

(a) $x \leq 4$ (b) $x > 4$

We read (a) as "x is less than or equal to 4." We read (b) as "x is greater than 4."

We can graph inequalities on the number line by graphing all the numbers that make the inequality a true statement.

Example 1 Graph on a number line: $x \leq 4$

Solution We are told to graph all numbers that are less than or equal to 4. We draw a dot at the point that represents 4, and then we shade all the points to the left of the dot. The arrowhead shows that the ray continues without end.

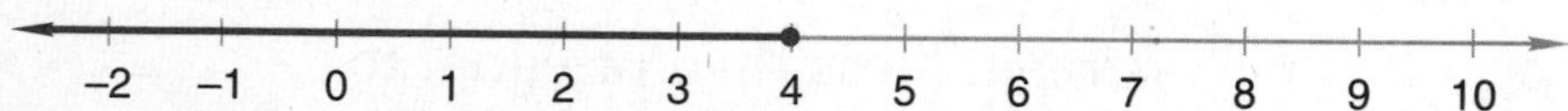

Example 2 Graph on a number line: $x > 4$

Solution We are told to graph all numbers greater than 4 *but not including* 4. We do not start the graph at 5, because we need to graph numbers like $4\frac{1}{2}$ and 4.001. To show that the graph does not include 4, *we draw an empty circle* at 4. Then we shade the portion of the number line to the right of the circle.

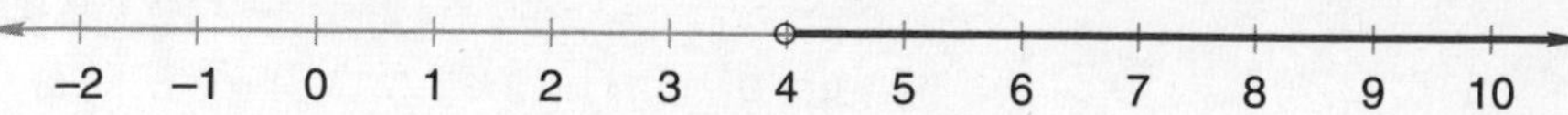

LESSON PRACTICE

Practice set

a. On a number line, graph all the numbers less than 2.

b. On a number line, graph all the numbers greater than or equal to 1.

Graph each inequality on a number line:

c. $x \leq -1$ **d.** $x > -1$

MIXED PRACTICE

Problem set

1. (72) Use a ratio box to solve this problem. If 4 cartons are needed to feed 30 hungry children, how many cartons are needed to feed 75 hungry children?

2. (55) Gabriel's average score after four tests was 88. What score must Gabriel average on the next two tests to have a six-test average of 90?

3. (30) If the sum of $\frac{2}{3}$ and $\frac{3}{4}$ is divided by the product of $\frac{2}{3}$ and $\frac{3}{4}$, what is the quotient?

4. (54) Use a ratio box to solve this problem. The ratio of monocotyledons to dicotyledons in the nursery was 3 to 4. If there were 84 dicotyledons in the nursery, how many monocotyledons were there?

5. (66) The diameter of a nickel is 21 millimeters. Find the circumference of a nickel to the nearest millimeter.

6. (78) Graph each inequality on a separate number line:

(a) $x > 2$ (b) $x \leq 1$

7. (50) Use a unit multiplier to convert 1.5 kg to g.

8. (65) Five sixths of the 30 people who participated in the taste test preferred Brand X. The rest preferred Brand Y.

(a) How many more people preferred Brand X than preferred Brand Y?

(b) What was the ratio of the number of people who preferred Brand Y to the number who preferred Brand X?

Write equations to solve problems 9–12.

9. Forty-two is seven tenths of what number?
(74)

10. One hundred fifty percent of what number is 600?
(77)

11. Forty percent of 50 is what number?
(60)

12. Forty is what percent of 50?
(77)

13. (a) Write 1.5×10^{-3} in standard form.
(57, 69)

(b) Write 25×10^{6} in scientific notation.

14. Simplify:
(73)

(a) $\frac{-45}{9}$ (b) $\frac{-450}{15}$

(c) $15(-20)$ (d) $-15(-12)$

15. Complete the table.
(48)

FRACTION	DECIMAL	PERCENT
(a)	(b)	50%
$\frac{1}{12}$	(c)	(d)

16. Simplify: $\frac{83\frac{1}{3}}{100}$
(76)

17. Find the area of this trapezoid. Dimensions are in millimeters.
(75)

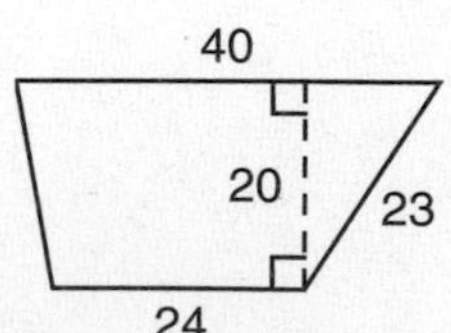

18. A box of tissues is 24 cm long, 12 cm wide, and 10 cm tall. Draw the box and find its volume.
(70)

19. Draw the box in problem 18 as if it were cut open and unfolded so that the six faces are lying flat.
(67)

20. Describe the rule of this function, and find the missing numbers.
(58)

IN		FUNCTION		OUT
2	→		→	9
3	→		→	14
4	→		→	☐
5	→		→	24
☐	→		→	4

21. A merchant sold an item for \$18.50. If 30% of the selling price was profit, how much profit did the merchant make on the sale?
(60)

For problems 22–24, solve and check each equation. Show each step.

22. $m + 8.7 = 10.25$
(Inv. 7)

23. $\frac{4}{3}w = 36$
(Inv. 7)

24. $0.7y = 48.3$
(Inv. 7)

Simplify:

25. $\{4^2 + 10[2^3 - (3)(\sqrt{4})]\} - \sqrt{36}$
(63)

26. $|5 - 3| - |3 - 5|$
(59)

27. $1\text{ m}^2 \cdot \frac{100\text{ cm}}{1\text{ m}} \cdot \frac{100\text{ cm}}{1\text{ m}}$
(50)

28. $7\frac{1}{2} \cdot 3 \cdot \left(\frac{2}{3}\right)^2$
(26)

29. $3\frac{1}{5} - \left(2\frac{1}{2} - 1\frac{1}{4}\right)$
(30)

30. (a) $(-10) - (-8) - (+6)$
(68)

(b) $(+10) + (-20) - (-30)$

LESSON

79 Insufficient Information • Quantitative Comparisons

WARM-UP

Facts Practice: + – × ÷ Integers (Test P)

Mental Math:

a. (–25) + (–15)

b. 3.75×10^3

c. $\frac{c}{100} = \frac{25}{10}$

d. Estimate 15% of $11.95.

e. 1200 mL to L

f. 20 is $\frac{4}{5}$ of n.

g. Square 5, × 2, – 1, $\sqrt{\ }$, × 8, – 1, ÷ 5, × 3, – 1, ÷ 4, × 9, + 3, ÷ 3

Problem Solving:

Emily earned an A, a B, and a C on tests in math, science, and history, although the A was not in math, the B was not in science, and the C was not in history. If the lowest grade was not in math, what test grade did Emily earn in each subject? Make a table to show your work.

NEW CONCEPTS

Insufficient information

Sometimes we encounter problems for which there is **insufficient information** (not enough information) to determine the answer. The following problem has insufficient information to answer the question:

A 10-pound bag of potatoes costs $1.49. What is the average price of each potato?

Since we do not know the number of potatoes in the bag, we do not have enough information to find the average price of each potato.

We will practice recognizing problems with insufficient information as we answer quantitative comparison problems like the examples in this lesson.

Quantitative comparisons We have practiced comparing numbers using the symbols >, <, and =. Beginning in this lesson, we will consider comparison problems in which the information provided is insufficient to make the comparison.

Example 1 The numbers x and y are whole numbers. Compare:

$$x \bigcirc y$$

Solution We are told that x and y are whole numbers, but we are not given enough information to determine which is greater or whether x and y are equal. Since we do not have enough information to make the comparison, we write **insufficient information** as our answer.

Example 2 The number x is positive and y is negative. Compare:

$$x \bigcirc y$$

Solution We are not given enough information to determine what each number is. However, we are given enough information to make the comparison. Any positive number is greater than any negative number. Thus the answer is

$$\boldsymbol{x > y}$$

Example 3 Compare: $a \bigcirc b$ if $a - b = 0$

Solution The equation does not provide enough information to determine the value of either number. However, since their difference is zero, the two numbers must be equal.

$$\boldsymbol{a = b}$$

LESSON PRACTICE

Practice set For each comparison, write >, <, =, or "insufficient information."

a. Compare: $x \bigcirc y$ if $x - y = 1$

b. Compare: $m \bigcirc n$ if $\frac{m}{n} = 1$

c. Compare: $a \bigcirc b$ if $a \cdot b = 1$

d. The number x is not positive, and y is not negative.

Compare: $x \bigcirc y$

MIXED PRACTICE

Problem set

1. (55) Videocassettes were placed on 4 shelves with an average of 33.5 cassettes per shelf. If the cassettes were regrouped on 5 shelves, what would be the average number of cassettes on each shelf?

2. (46) Nelda drove 315 kilometers and used 35 liters of gasoline. Her car averaged how many kilometers per liter of gas?

3. (65) The ratio of winners to losers was 7 to 5. If the total number of winners and losers was 1260, how many more winners were there than losers?

4. (69) Write each number in scientific notation:

(a) 37.5×10^{-6} (b) 37.5×10^{6}

5. (79) Compare: $x \bigcirc y$ if $\frac{y}{x} = 2$

6. (78) Graph each inequality on a separate number line:

(a) $x < 1$ (b) $x \geq -1$

7. (72) Use a case 1–case 2 ratio box to solve this problem. Four inches of snow fell in 3 hours. At that rate, how long would it take for 1 foot of snow to fall?

8. (71) Diagram this statement. Then answer the questions that follow.

Twelve of Ian's pencils were sharpened. This was $\frac{3}{8}$ of Ian's pencils.

(a) How many of Ian's pencils were not sharpened?

(b) What percent of Ian's pencils were not sharpened?

Write equations to solve problems 9–12.

9. (77) Thirty-five is 70% of what number?

10. (77) What percent of 20 is 17?

11. (77) What percent of 20 is 25?

12. (77) Three hundred sixty is 75 percent of what number?

13. Simplify:
(73)

(a) $\dfrac{144}{-8}$ (b) $\dfrac{-144}{+6}$

(c) $-12(12)$ (d) $-16(-9)$

14. Complete the table.
(48)

FRACTION	DECIMAL	PERCENT
$\frac{1}{25}$	(a)	(b)
(c)	(d)	8%

15. At the Citrus used car lot, a salesperson is paid a commission of 5% of the sale price for every car he or she sells. If a salesperson sells a car for $4500, how much would he or she be paid as a commission?
(60)

16. Simplify: $\dfrac{62\frac{1}{2}}{100}$
(76)

17. A square sheet of paper with an area of 100 in.^{2} has a corner cut off, as shown in the figure below.
(75)

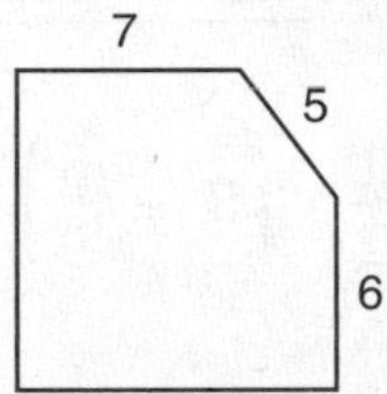

(a) What is the perimeter of the shape?

(b) What is the area of the shape?

18. In the figure at right, each small cube has a volume of 1 cubic centimeter.
(67, 70)

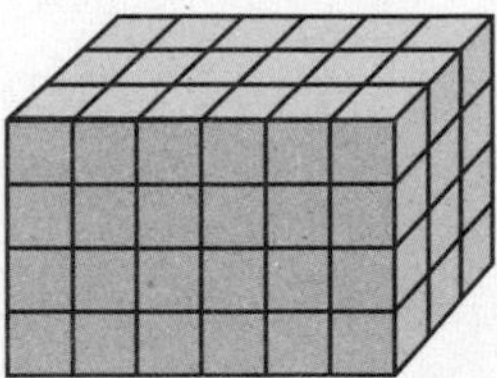

(a) What is the volume of this rectangular prism?

(b) What is the total surface area of the rectangular prism?

19. Find the circumference of each circle:
(66)

(a)

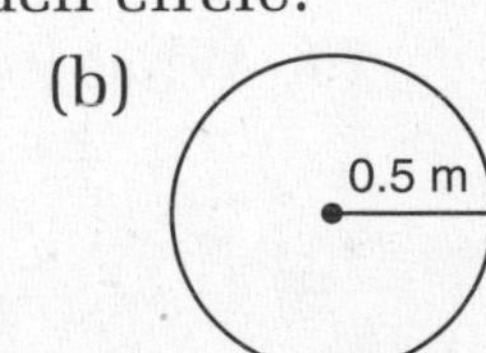

Use 3.14 for π.

(b)
0.5 m

Leave π as π.

20. (62) Identify each triangle as acute, right, or obtuse. Then identify each triangle as equilateral, isosceles, or scalene.

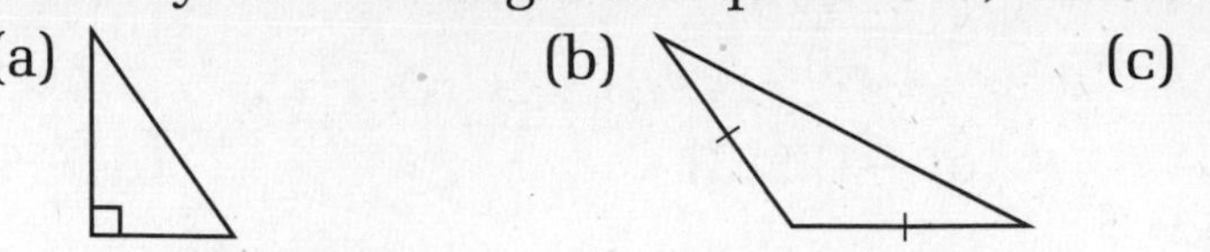

For problems 21–23, solve and check each equation. Show each step.

21. (Inv. 7) $1.2x = 2.88$

22. (Inv. 7) $3\frac{1}{3} = x + \frac{5}{6}$

23. (Inv. 7) $\frac{3}{2}w = \frac{9}{10}$

Simplify:

24. (63) $\frac{\sqrt{100} + 5[3^3 - 2(3^2 + 3)]}{5}$

25. (56)
$$\begin{array}{r} 3\text{ hr } 15\text{ min } 24\text{ s} \\ -\ 2\text{ hr } 45\text{ min } 30\text{ s} \\ \hline \end{array}$$

26. (50) $1\text{ yd}^2 \cdot \frac{3\text{ ft}}{1\text{ yd}} \cdot \frac{3\text{ ft}}{1\text{ yd}}$

27. (26) $7\frac{1}{2} \cdot \left(3 \div \frac{5}{9}\right)$

28. (30) $4\frac{5}{6} + 3\frac{1}{3} + 7\frac{1}{4}$

29. (35, 43) $3\frac{3}{4} \div 1.5$ (decimal answer)

30. (68) (a) $(-10) - (+20) - (-30)$

(b) $(-10) - |(-20) - (+30)|$

LESSON

80 Transformations

WARM-UP

Facts Practice: Classifying Quadrilaterals and Triangles (Test O)

Mental Math:

a. $(-30) - (+45)$ b. 40^2

c. $5q - 4 = 36$ d. 15 is $\frac{3}{5}$ of n.

e. 1500 m to km f. What is $\frac{3}{5}$ of 15?

g. Find the perimeter and area of a square with sides 1.5 m long.

Problem Solving:

Simplify the first three terms of the following sequence. Then write and simplify the next two terms.

$$\sqrt{1^3}, \sqrt{1^3 + 2^3}, \sqrt{1^3 + 2^3 + 3^3}, \ldots$$

NEW CONCEPT

Recall that two figures are congruent if they are the same shape and size. These two triangles are congruent, but they are not in the same position:

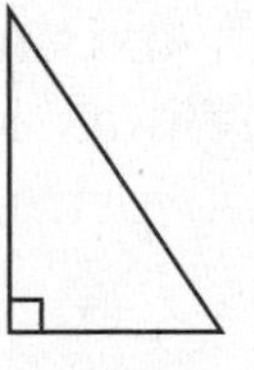

I

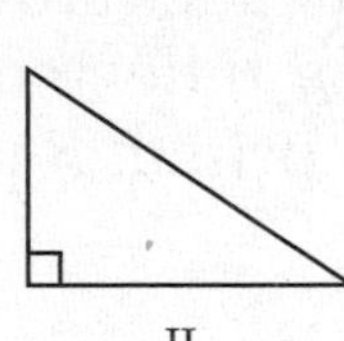

II

We can use three types of position change to move triangle I to the position of triangle II. One change of position is to "flip" triangle I over as though flipping a transparent piece of paper.

I

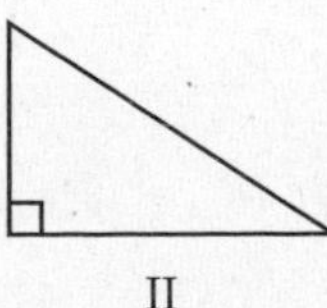

II

A second change of position is to "slide" triangle I to the right.

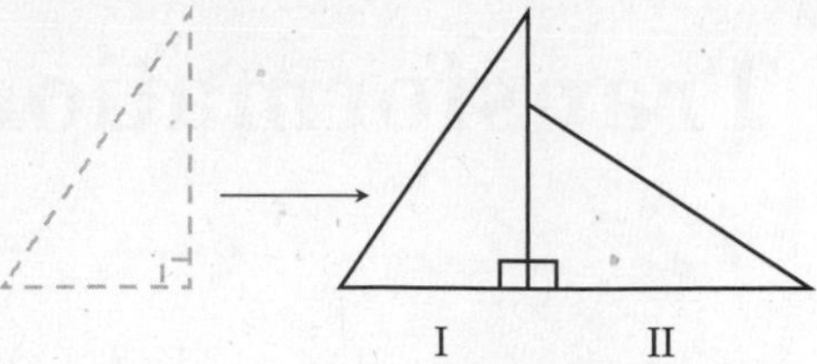

The third change of position is to "turn" triangle I 90° clockwise.

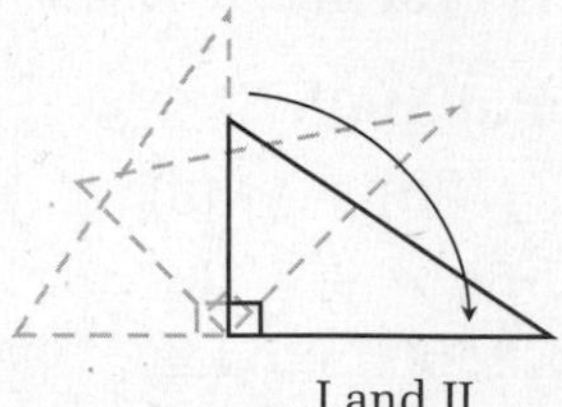

These "flips, slides, and turns" are called **transformations** and have special names, which are listed in this table.

Transformations

Movement	Name
flip	reflection
slide	translation
turn	rotation

A **reflection** of a figure in a line (a "flip") produces a mirror image of the figure that is reflected.

If we reflect ΔABC in the y-axis, the reflection of every point of ΔABC appears on the opposite side of the y-axis the same distance from the y-axis as the original point. We can refer to the reflected triangle as $\Delta A'B'C'$, which we read as "triangle A prime, B prime, C prime."

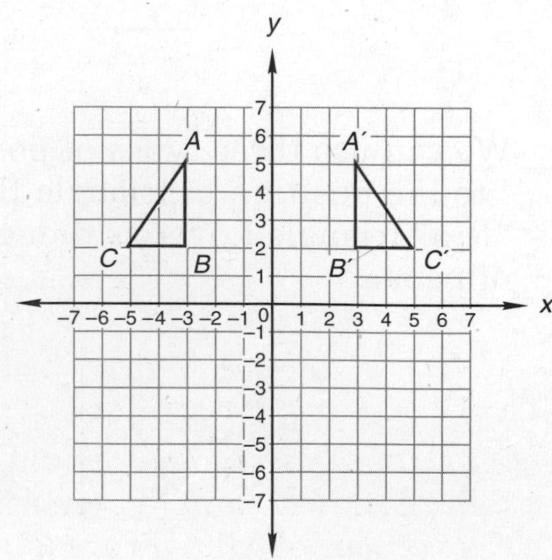

If we then reflect $\Delta A'B'C'$ in the x-axis, we see $\Delta A''B''C''$ ("triangle A double prime, B double prime, C double prime") in the fourth quadrant.

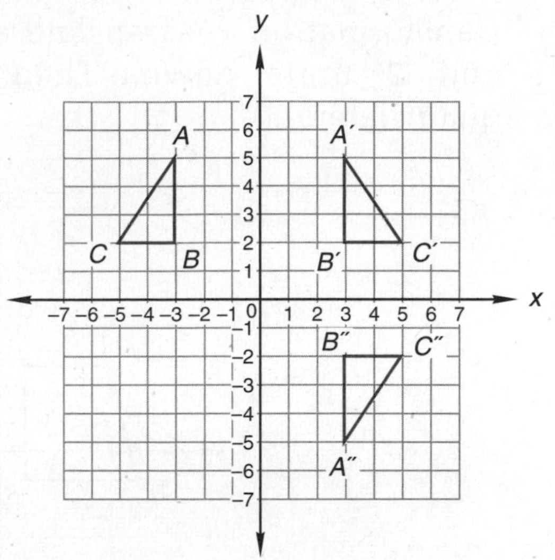

Example 1 The coordinates of the vertices of ΔRST are R (4, 3), S (4, 1), and T (1, 1). Draw ΔRST and its reflection in the x-axis, $\Delta R'S'T'$. What are the coordinates of the vertices of $\Delta R'S'T'$?

Solution We graph the vertices of ΔRST and draw the triangle. The reflection of every point of ΔRST in the x-axis appears on the opposite side of the x-axis the same distance from the x-axis as the original point. We locate the reflected vertices and draw $\Delta R'S'T'$.

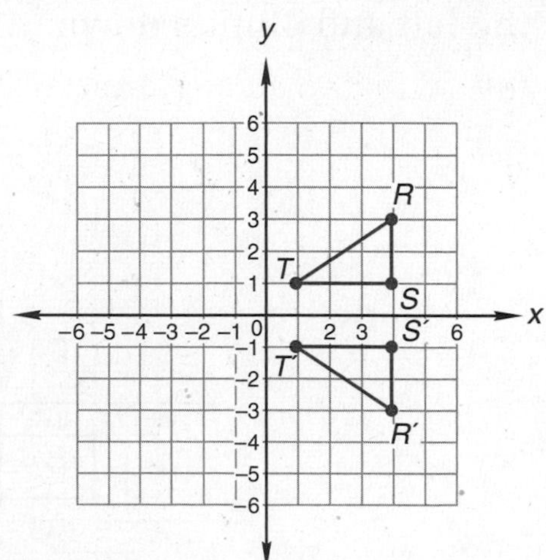

Note that if a segment were drawn between a point and its reflection, the segment would be perpendicular to the line of reflection, which in this case is the x-axis. The coordinates of the vertices of $\Delta R'S'T'$ are **R' (4, –3), S' (4, –1),** and **T' (1, –1).**

A **translation** "slides" a figure to a new position without turning or flipping the figure. If we translate quadrilateral *JKLM* 6 units to the right and 2 units down, quadrilateral *J′K′L′M′* appears in the position shown. To perform the transformation we translate each vertex 6 units to the right and 2 units down. Then we draw the sides of the quadrilateral.

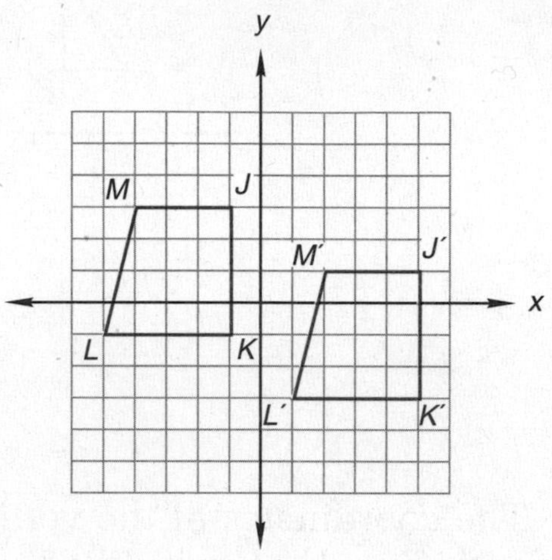

Example 2 The coordinates of the vertices of rectangle *ABCD* are *A* (4, 3), *B* (4, 1), *C* (1, 1), and *D* (1, 3). Draw ▭*ABCD* and its image, ▭*A′B′C′D′*, translated to the left 5 units and down 4 units. What are the coordinates of the vertices of ▭*A′B′C′D′*?

Solution We graph the vertices of ▭*ABCD* and draw the rectangle. Then we graph its image by translating each vertex 5 units to the left and 4 units down.

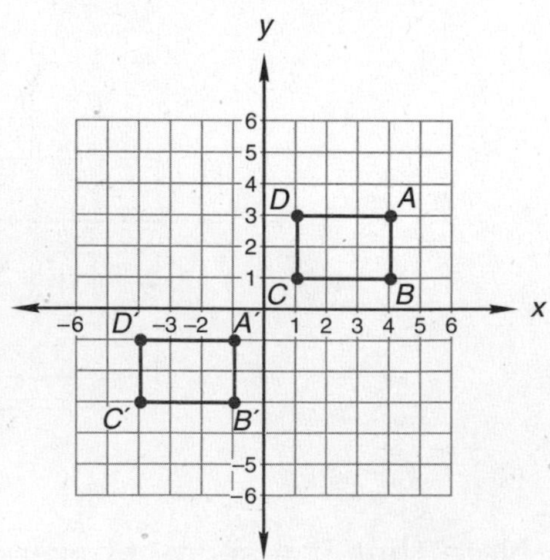

We find that the coordinates of the vertices of ▭*A′B′C′D′* are **A′ (–1, –1), B′ (–1, –3), C′ (–4, –3),** and **D′ (–4, –1).**

A **rotation** of a figure "turns" the figure about a specified point called the *center of rotation.* At the beginning of this lesson we rotated triangle I 90° clockwise. The center of rotation was the vertex of the right angle. In the illustration below, triangle ABC is rotated 180° about the origin.

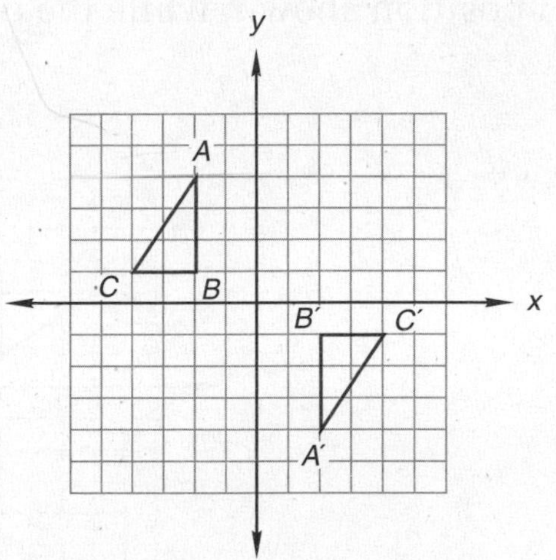

One way to view the effect of a rotation of a figure is to trace the figure on a piece of transparency film. Then place the point of a pencil on the center of rotation and turn the transparency film through the described rotation.

Example 3 The coordinates of the vertices of ΔPQR are P (3, 4), Q (3, 1), and R (1, 1). Draw ΔPQR and also draw its image, $\Delta P'Q'R'$, after a counterclockwise rotation of 90° about the origin. What are the coordinates of the vertices of $\Delta P'Q'R'$?

Solution We graph the vertices of ΔPQR and draw the triangle.

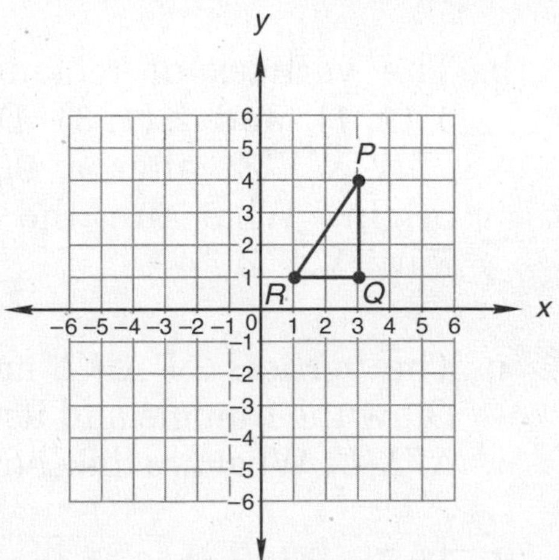

Then we place a piece of transparency film over the coordinate plane and trace the triangle. We also place a mark on the transparency aligned with the x-axis. This mark will align with the y-axis after the transparency is rotated 90°.

After tracing the triangle on the transparency, we place the point of a pencil on the film over the origin, which is the center of rotation in this example. While keeping the graph paper still, we rotate the film 90° (one-quarter turn) counterclockwise. The image of the triangle rotates to the position shown, while the original triangle remains in place.

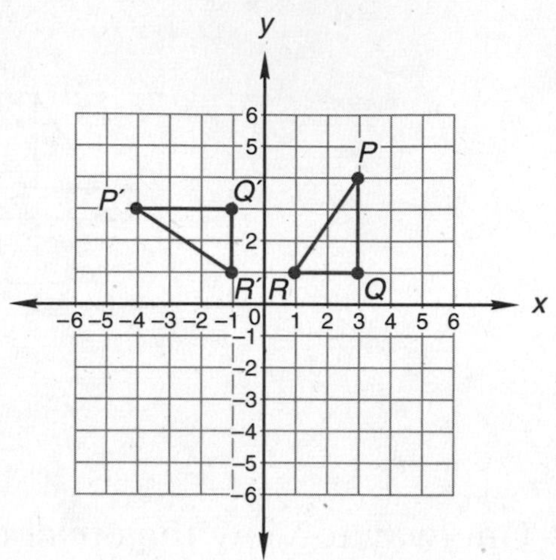

We name the rotated image $\Delta P'Q'R'$ and through the transparency see that the coordinates of the vertices are **P' (−4, 3), Q' (−1, 3),** and **R' (−1, 1).**

LESSON PRACTICE

Practice set

a. Perform each of the examples in this lesson if you have not already done so.

b. The vertices of rectangle $WXYZ$ are W (4, 3), X (4, 1), Y (1, 1), and Z (1, 3). Draw the rectangle and its image $\square W'X'Y'Z'$ after a 90° clockwise rotation about the origin. What are the coordinates of the vertices of $\square W'X'Y'Z'$?

c. The vertices of ΔJKL are J (1, −1), K (3, −2), and L (1, −3). Draw the triangle and its image after reflection in the y-axis, $\Delta J'K'L'$. What are the coordinates of the vertices of $\Delta J'K'L'$?

d. The coordinates of the vertices of $\varnothing PQRS$ are P (0, 3), Q (−1, 1), R (−4, 1), and S (−3, 3). Draw $\varnothing PQRS$ and its image $\varnothing P'Q'R'S'$ translated 6 units to the right and 3 units down. What are the coordinates of the vertices of $\varnothing P'Q'R'S'$?

MIXED PRACTICE

Problem set

1. Tina mowed lawns for 4 hours and earned \$7.00 per hour. Then she washed windows for 3 hours and earned \$6.30 per hour. What was Tina's average hourly pay for the 7-hour period?
(55)

2. Evaluate: $x + (x^2 - xy) - y$ if $x = 4$ and $y = 3$
(52)

3. Compare: $a \bigcirc b$ if $ab = 2$
(79)

4. Use a ratio box to solve this problem. When Nia cleaned her room, she found that the ratio of clean clothes to dirty clothes was 2 to 3. If 30 articles of clothing were discovered, how many were clean?
(65)

5. The diameter of a half-dollar is 3 centimeters. Find the circumference of a half-dollar to the nearest millimeter.
(32, 66)

6. Use a unit multiplier to convert $1\frac{1}{2}$ quarts to pints.
(50)

7. Graph each inequality on a separate number line:
(78)
(a) $x > -2$
(b) $x \leq 0$

8. Use a ratio box to solve this problem. In 25 minutes, 400 customers entered the attraction. At this rate, how many customers would enter the attraction in 1 hour?
(72)

9. Diagram this statement. Then answer the questions that follow.
(71)

Nathan found that it was 18 inches from his knee joint to his hip joint. This was $\frac{1}{4}$ of his total height.

(a) What was Nathan's total height in inches?

(b) What was Nathan's total height in feet?

Write equations to solve problems 10–13.

10. Six hundred is $\frac{5}{9}$ of what number?
(74)

11. Two hundred eighty is what percent of 400?
(77)

12. What number is 4 percent of 400?
(60)

13. Sixty is 60 percent of what number?
(77)

14. Simplify:
(73)

(a) $\frac{600}{-15}$ (b) $\frac{-600}{-12}$

(c) $20(-30)$ (d) $+15(40)$

15. Anil is paid a commission equal to 6% of the price of each appliance he sells. If Anil sells a refrigerator for $850, what is Anil's commission on the sale?
(60)

16. Complete the table.
(48)

FRACTION	DECIMAL	PERCENT
(a)	0.3	(b)
$\frac{5}{12}$	(c)	(d)

17. Write each number in scientific notation:
(69)

(a) 30×10^6 (b) 30×10^{-6}

18. Find the area of the trapezoid shown. Dimensions are in meters.
(75)

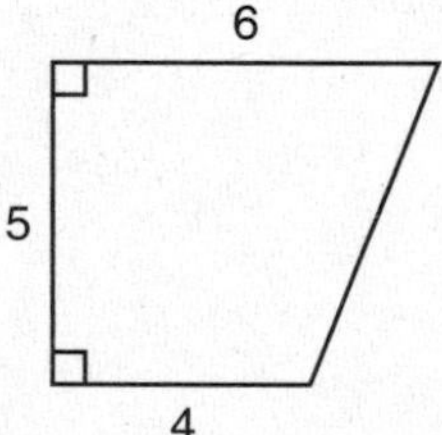

19. Each edge of a cube measures 5 inches.
(67, 70)

(a) What is the volume of the cube?

(b) What is the surface area of the cube?

20. In a bag are 100 marbles: 10 red, 20 white, 30 blue, and 40 green. If one marble is drawn from the bag, what is the probability that the marble will not be red, white, or blue?
(36)

For problems 21–23, solve and check each equation. Show each step.

21. $17a = 408$
(Inv. 7)

22. $\frac{3}{8}m = 48$
(Inv. 7)

23. $1.4 = x - 0.41$
(Inv. 7)

Simplify:

24. $\dfrac{2^3 + 4 \cdot 5 - 2 \cdot 3^2}{\sqrt{25} \cdot \sqrt{4}}$
(52)

25. $7\frac{1}{7} \times 1.4$
(43)

26.
(56)

$$\begin{array}{r} 10 \text{ lb} \quad 6 \text{ oz} \\ -\ \ 7 \text{ lb} \ 11 \text{ oz} \\ \hline \end{array}$$

27. $1 \text{ cm}^2 \cdot \dfrac{10 \text{ mm}}{1 \text{ cm}} \cdot \dfrac{10 \text{ mm}}{1 \text{ cm}}$
(50)

28. $7\frac{1}{2} \div \left(3 \cdot \frac{5}{9}\right)$
(26)

29. $2^{-4} + 4^{-2}$
(30, 57)

30. Triangle ABC with vertices at $A\,(0, 2)$, $B\,(2, 2)$, and $C\,(2, 0)$ is reflected in the x-axis. Draw ΔABC and its image $\Delta A'B'C'$.
(80)

INVESTIGATION 8

Focus on

Using a Compass and Straightedge, Part 2

In Investigation 2 we used a compass to draw circles, and we used a compass and straightedge to inscribe a regular hexagon and a regular triangle in a circle. In this investigation we will use a compass and straightedge to **bisect** (divide in half) a line segment and an angle. We will also inscribe a square and a regular octagon in a circle.

Materials needed:

- Compass
- Ruler or straightedge
- Protractor

Bisecting a line segment Use a metric ruler to draw a segment 6 cm long. Label the endpoints *A* and *C*.

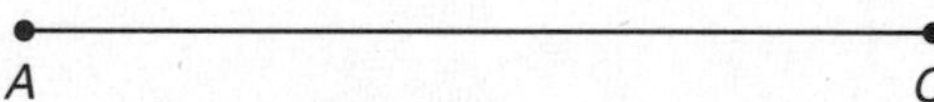

Next open a compass so that the distance between the pivot point and pencil point is more than half the length of the segment to be bisected (in this case, more than 3 cm). You will swing arcs from both endpoints of the segment, so do not change the compass radius once you have it set. Place the pivot point of the compass on one endpoint of the segment, and make a curve by swinging an arc on both sides of the segment as shown.

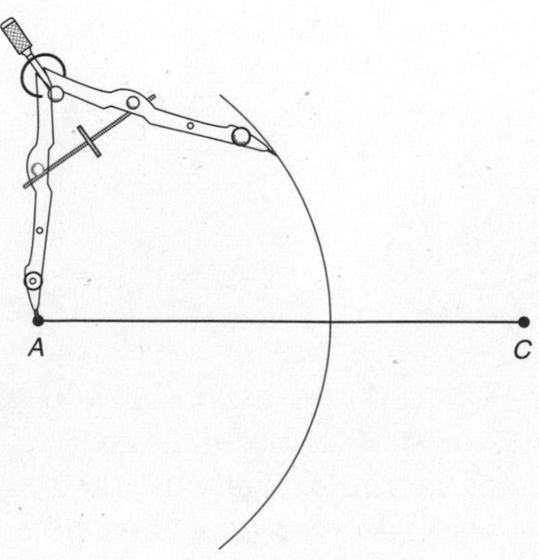

Then move the pivot point of the compass to the other endpoint of the segment, and, without resetting the compass, swing an arc that intersects the other arc on both sides of the segment. Draw a line through the two points where the arcs intersect to divide the original segment into two parts. Label the point where the line intersects the segment point *B*.

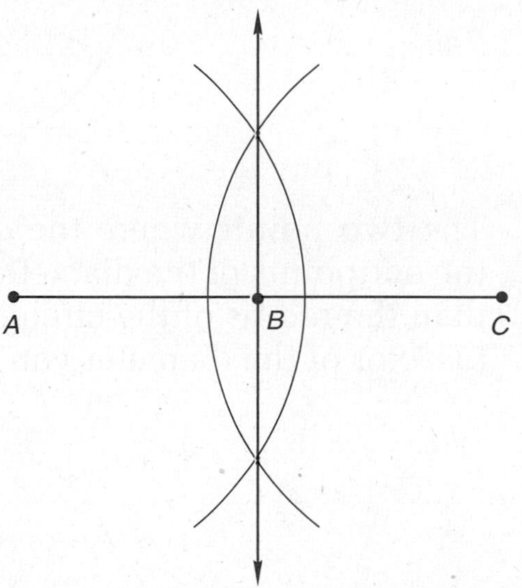

1. Use a metric ruler to find *AB* and *BC*.

2. Where the line and segment intersect, four angles are formed. What is the measure of each angle?

Using a compass and straightedge to create geometric figures is called **construction.** You just constructed the **perpendicular bisector** of a segment.

3. Why is the line you constructed called the perpendicular bisector of the segment?

Inscribing a square in a circle We can use a perpendicular bisector to help us inscribe a square in a circle. Draw a dot on your paper to be the center of a circle. Set the distance between the points of your compass to 2 cm. Then place the pivot point of the compass on the dot and draw a circle. Use a straightedge to draw a diameter of the circle.

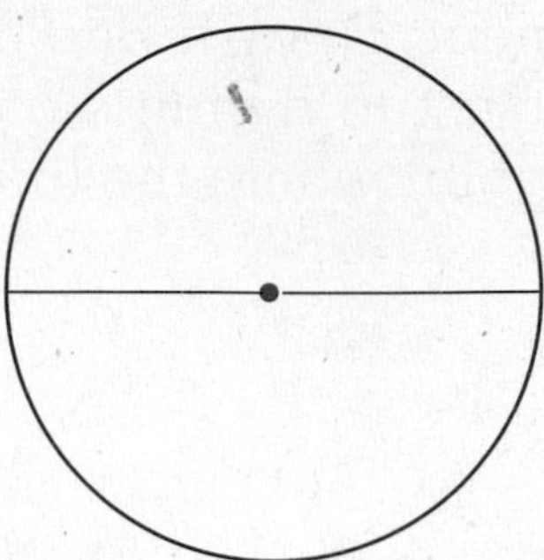

The two points where the diameter intersects the circle are the endpoints of the diameter. Open the compass a little more than the radius of the circle, and construct the perpendicular bisector of the diameter you drew.

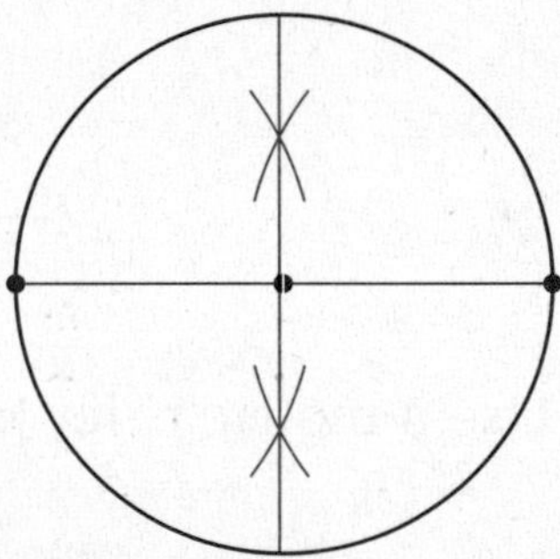

Make the perpendicular bisector another diameter of the circle. The two diameters divide the circle into quarters. Draw chords between the points on the circle that are the endpoints of the two diameters.

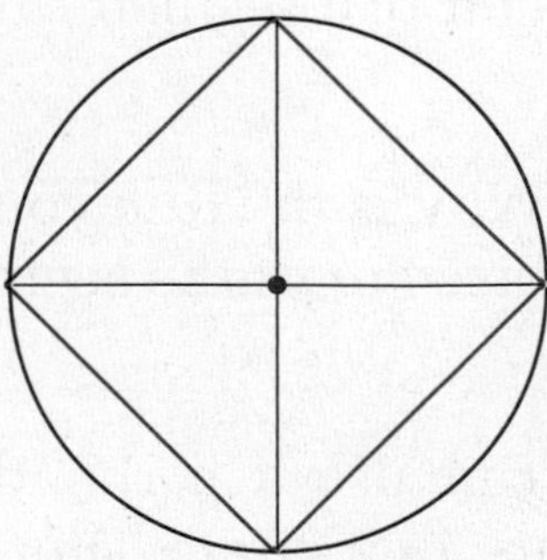

You have inscribed a square in a circle.

4. Each angle of the square is an inscribed angle of the circle. What is the measure of each angle of the square?

5. Within the square inscribed in the circle are four small right triangles. Two sides of each small triangle are radii of the circle. If the radius of the circle is 2 cm, then
   (a) what is the area of each small right triangle?
   (b) how can we find the area of the inscribed square?

Bisecting an angle Use a straightedge to draw an angle. With the pivot point of the compass on the vertex of the angle, draw an arc that intersects the sides of the angle. For reference call these points *R* and *S*, and label the vertex *V*.

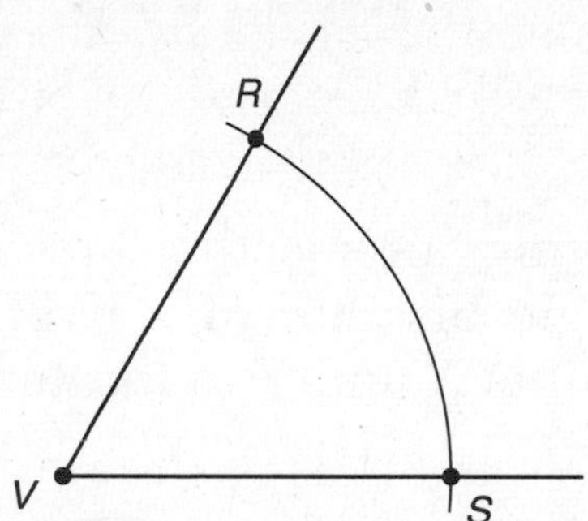

Set the compass so that it is open more than half the distance between *R* and *S*. With the pivot point on *R*, swing an arc. Then, with the pivot point on *S*, swing another arc that intersects the one centered at *R*. Label the point of intersection *T*.

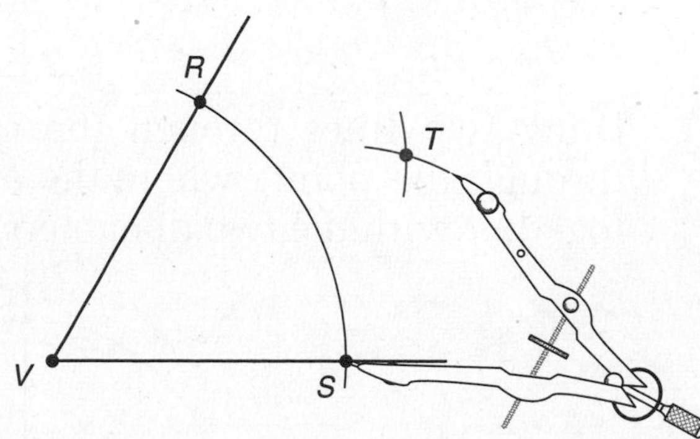

Using a straightedge, draw a ray from the vertex *V* through point *T*. Ray *VT* divides $\angle RVS$ into two congruent angles.

6. Use a protractor to measure the original angle you drew and the two smaller angles formed when you constructed the ray. Record all three angle measures for your answer.

In this activity you constructed an **angle bisector.**

7. Why is the ray called an angle bisector?

Inscribing a regular octagon in a circle Draw a circle and a diameter of the circle. Then construct a diameter that is a perpendicular bisector of the first diameter. Your work should look like the circle shown below. We have labeled points *M, X, Y,* and *Z* for reference.

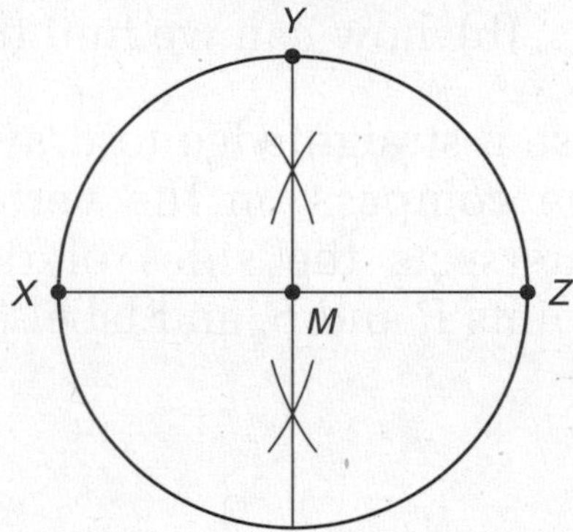

Swing intersecting arcs from points *Y* and *Z* to locate the angle bisector of $\angle YMZ$. Also swing intersecting arcs from points *X* and *Y* to locate the angle bisector of $\angle XMY$.

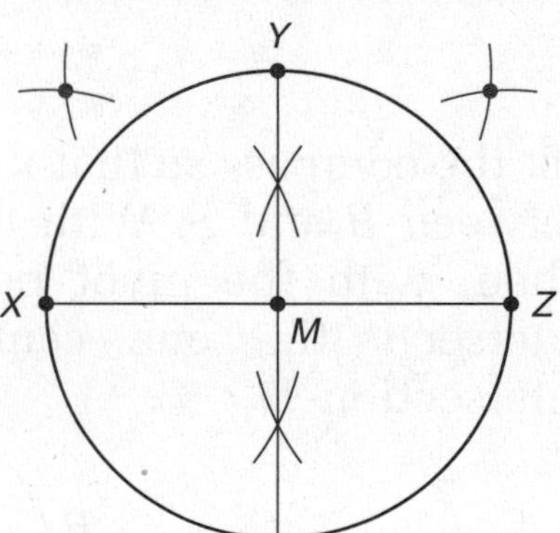

Draw two lines through the center of the circle that pass through the points where the arcs intersect. These two lines together with the two diameters divide the circle into eighths.

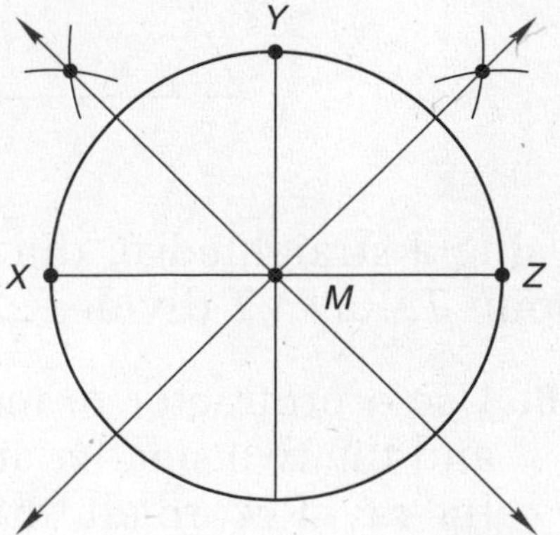

8. What is the measure of each small central angle that is formed?

There are 8 points of intersection around the circle. Draw chords from point to point around the circle to inscribe a regular octagon in the circle.

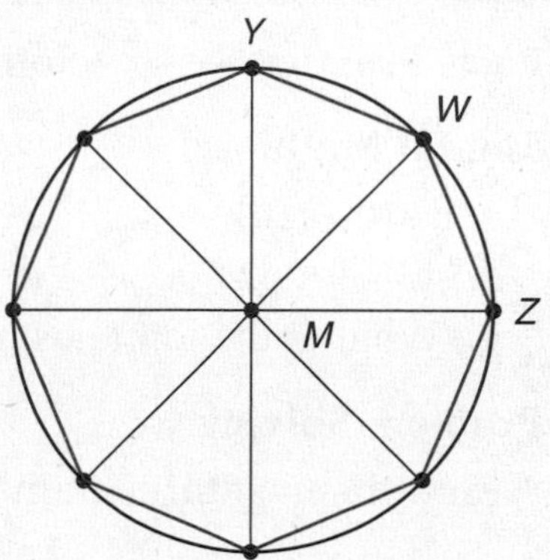

Refer to the inscribed octagon to answer problems 9–12.

9. Segments MY, MW, and MZ are radii of the same circle. Classify ΔYMW and ΔWMZ by sides.

10. What is the measure of the following angles?
 (a) $\angle YMW$ (b) $\angle WMZ$

11. What is the measure of each of the following angles?
 (a) $\angle MYW$ (b) $\angle MWY$
 (c) $\angle MWZ$ (d) $\angle MZW$

12. (a) What is the measure of $\angle YWZ$?
 (b) What is the measure of each inscribed angle formed by the sides of the octagon?

LESSON

81 Using Proportions to Solve Percent Problems

WARM-UP

Facts Practice: + − × ÷ Integers (Test P)

Mental Math:

a. (−10)(−10) b. 12×10^{-4} c. $\frac{40}{t} = \frac{6}{9}$
d. 15% of $24.00 e. 25 mm to cm f. 24 is $\frac{2}{3}$ of n.
g. What is the total cost of a $25 item plus 8% sales tax?

Problem Solving:

Find the next three numbers in this sequence:

1, 1, 2, 3, 5, 8, 13, ...

NEW CONCEPT

A percent is a ratio in which 100 represents the total number in the group. Thus percent problems can be solved using the same method we use to solve ratio problems. Consider the following problem:

Thirty percent of the herd stood in the shade. If 21 cows did not stand in the shade, how many cows were in the herd?

The problem is about two parts of a whole herd. One part of the herd stood in the shade; the other part of the herd did not stand in the shade. The whole herd is 100 percent. The part that stood in the shade was 30 percent. Thus 70 percent did not stand in the shade. We record these numbers in a ratio box just as we do with ratio problems.

	Percent	Actual Count
Stood in the shade	30	
Did not stand in the shade	70	
Whole herd	100	

As we read the problem, we find an actual count as well. There were 21 cows that did not stand in the shade. We record 21 in the appropriate place on the table and use letters in the remaining places.

	Percent	Actual Count
Stood in the shade	30	S
Did not stand in the shade	70	21
Whole herd	100	W

We use two of the rows in the table to write a proportion that we can use to solve the problem. **We always use the row in which both numbers are known.** Since the problem asks for the total number of cows in the herd, we also use the third row.

	Percent	Actual Count
Stood in the shade	30	S
Did not stand in the shade	70	21
Whole herd	100	W

$$\frac{70}{100} = \frac{21}{W}$$

$$70W = 2100$$

$$W = 30$$

By solving the proportion, we find there were 30 cows in the herd.

Example 1 Last week, forty percent of the restaurant's customers ordered chicken. If 480 customers did not order chicken, how many of the customers ordered chicken?

Solution We solve this problem just as we solve a ratio problem. We use the percents to fill the ratio column of the table. Together, all of last week's customers are 100 percent. The part that ordered chicken is 40 percent. Therefore, the part that did not order chicken is 60 percent. The number 480 is the actual count of the customers who did not order chicken. We write these numbers in the table.

	Percent	Actual Count
Ordered chicken	40	C
Did not order chicken	60	480
Total	100	T

Now we use the table to write a proportion. Since we know both numbers in the second row, we will use that row in the proportion. Since the problem asks us to find the actual count of customers who ordered chicken, we also use the first row of the table in the proportion.

	Percent	Actual Count
Ordered chicken	40	C
Did not order chicken	60	480
Total	100	T

$$\frac{40}{60} = \frac{C}{480}$$

$$60C = 19{,}200$$

$$C = 320$$

We find that **320 customers** ordered chicken last week.

Example 2 In the town of Centerville, 261 of the 300 working people do not carpool.

(a) What percent of the people carpool?

(b) According to census data, approximately 13 percent of the working people in the United States carpool to work. Do the carpooling statistics in Centerville reflect the national carpooling statistics?

Solution (a) We make a ratio box and write in the numbers. The total number of working people in Centerville is 300, so 39 people carpool.

	Percent	Actual Count
Carpool	P_C	39
Do not carpool	P_N	261
Total	100	300

We use P_C to stand for the percent who carpool to work. We use the "carpool" row and the "total" row to write the proportion.

	Percent	Actual Count
Carpool	P_C	39
Do not carpool	P_N	261
Total	100	300

$$\frac{P_C}{100} = \frac{39}{300}$$

$$300P_C = 3900$$

$$P_C = 13$$

We find that **13 percent** of the people who work in Centerville carpool to work.

(b) Since census data show that approximately 13 percent of United States workers carpool to work, we find that Centerville's statistics **do reflect the national statistics.**

Example 3 After a difficult spelling test of 40 four-syllable words, Chris was relieved to learn that he had misspelled only 6 words. What percent of the words did Chris spell correctly?

Solution We notice that Chris misspelled a little more than one tenth of the words, so the answer should be a bit less than 90%.

We record the given information in a ratio box. Since 6 of the 40 words were misspelled, we calculate that 34 words were spelled correctly.

	Percent	Actual Count
Correct	P_C	34
Misspelled	P_M	6
Total	100	40

We want to know the percent correct (P_C), so we use the "correct" row and the "total" row to write the proportion.

$$\frac{P_C}{100} = \frac{34}{40}$$

$$40P_C = 3400$$

$$P_C = 85$$

Chris spelled **85%** of the words correctly.

LESSON PRACTICE

Practice set Estimate each answer. Then use a ratio box to solve each problem.

a. Twenty-one of the 70 acres were planted with alfalfa. What percent of the acres was not planted in alfalfa?

b. Lori still has 60% of the book to read. If she has read 120 pages, how many pages does she still have to read?

c. Dewayne missed four of the 30 problems on the problem set. What percent of the problems did Dewayne answer correctly?

MIXED PRACTICE

Problem set

1. (80) The coordinates of the vertices of ΔABC are $A\,(2, -1)$, $B\,(5, -1)$, and $C\,(5, -3)$. Draw the triangle and its image $\Delta A'B'C'$ reflected in the x-axis. What are the coordinates of the vertices of $\Delta A'B'C'$?

Use the information below to answer problems 2 and 3.

> *On his first 15 tests, Rory earned the following scores: 70, 85, 80, 85, 90, 80, 85, 80, 90, 95, 85, 90, 100, 85, 90.*

2. (Inv. 4) (a) What was Rory's average (mean) test score?

(b) If Rory's scores were arranged in order from lowest to highest, which would be the middle score? (What is the median of the scores?)

3. (Inv. 4) (a) Which score did Rory earn most often? (What is the mode of the scores?)

(b) What was the difference between Rory's highest score and his lowest score? (What is the range of the scores?)

4. (56) Danny is 6'1" (6 ft 1 in.) tall. His sister is 5'6$\frac{1}{2}$" tall. Danny is how many inches taller than his sister?

5. (72) Use a ratio box to solve this problem. Carmen bought 5 pencils for 75¢. At this rate, how much would she pay for a dozen pencils?

6. (78) Graph each inequality on a separate number line:

(a) $x < 4$ (b) $x \geq -2$

7. (36, 71) Diagram this statement. Then answer the questions that follow.

> *Gilbert answered 48 questions correctly. This was $\frac{4}{5}$ of the questions on the test.*

(a) How many questions were on the test?

(b) What was the ratio of Gilbert's correct answers to his incorrect answers?

8. If point B is located halfway between points A and C,
(8) what is the length of segment AB?

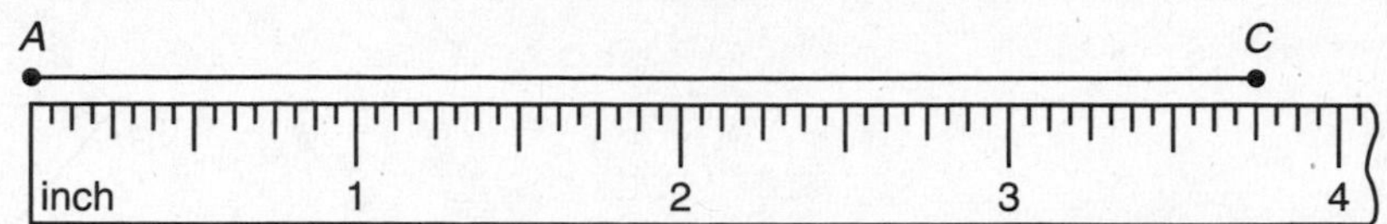

9. Use a ratio box to solve this problem. The ratio of gleeps
(65) to bobbles was 9 to 5. If the total number of gleeps and bobbles was 2800, how many gleeps were there?

10. If $x = 9$, what does $x^2 + \sqrt{x}$ equal?
(41)

11. Compare: $m \bigcirc n$ if $\frac{m}{n} = 0.5$
(79)

12. Complete the table.
(48)

FRACTION	DECIMAL	PERCENT
$2\frac{1}{4}$	(a)	(b)
(c)	(d)	$2\frac{1}{4}\%$

Write equations to solve problems 13 and 14.

13. The store owner makes a profit of 40% of the selling price
(60) of an item. If an item sells for \$12, how much profit does the store owner make?

14. Fifty percent of what number is 0.4?
(77)

15. Simplify: $\dfrac{16\frac{2}{3}}{100}$
(76)

Use ratio boxes to solve problems 16 and 17.

16. Esmeralda correctly answered 21 of the 25 questions.
(81) What percent of the questions did she answer correctly?

17. Twenty percent of the 4000 acres were left fallow. How
(81) many acres were not left fallow?

18. (40) If the measure of $\angle ABC$ is 140°, then

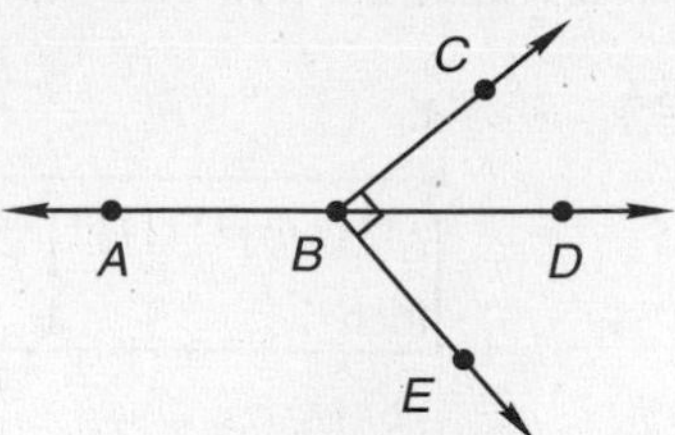

(a) what is the measure of $\angle CBD$? How do you know?

(b) what is the measure of $\angle DBE$? How do you know?

(c) what is the measure of $\angle EBA$? How do you know?

(d) what is the sum of the measures of $\angle ABC$, $\angle CBD$, $\angle DBE$, and $\angle EBA$?

19. (24) Write the prime factorization of the two terms of this fraction. Then reduce the fraction.

$$\frac{3000}{6300}$$

20. (58, 75) (a) Find the area of this isosceles trapezoid. Dimensions are in inches.

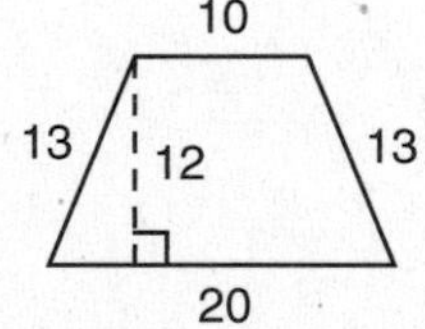

(b) Trace the trapezoid on your paper. Then draw its line of symmetry.

21. (58) Describe the rule of this function, and find the missing numbers.

IN	FUNCTION	OUT
4	→ →	□
−1	→ →	−3
6	→ →	11
3	→ →	5
□	→ →	−1

22. (69) Write each number in scientific notation:

(a) 56×10^{7} (b) 56×10^{-7}

For problems 23 and 24, solve and check the equation. Show each step.

23. $5x = 16.5$
(Inv. 7)

24. $3\frac{1}{2} + a = 5\frac{3}{8}$
(Inv. 7)

Simplify:

25. $3^2 + 5[6 - (10 - 2^3)]$
(63)

26. $\sqrt{2^2 \cdot 3^4 \cdot 5^2}$
(52)

27. $2\frac{2}{3} \times 4.5 \div 6$ (fraction answer)
(43)

28. $\left(3\frac{1}{2}\right)^2 - (5 - 3.4)$ (decimal answer)
(26, 43)

29. (a) $(-12)(-9)$ (b) $(-3)(25)$
(73)

(c) $\frac{-100}{5}$ (d) $\frac{25}{-5}$

30. (a) $(-3) + |-4| - (-5)$
(68)

(b) $(-18) - (+20) + (-7)$

LESSON

82 Area of a Circle

WARM-UP

Facts Practice: Percent-Decimal-Fraction Equivalents (Test Q)

Mental Math:

a. $(-6) - (-24)$ b. 10^{-2} c. $8n + 6 = 78$
d. 3.14×10 ft e. 150 cm to m f. 24 is $\frac{3}{4}$ of n.
g. 25% of 24, × 5, − 2, ÷ 2, + 1, ÷ 3, × 7, + 1, $\sqrt{\ }$, × 10

Problem Solving:

Six identical blocks marked W, a 24-oz weight, and a 60-oz weight are distributed on a balanced scale as shown. What is the weight of each block marked W? Write an equation illustrated by this balanced scale.

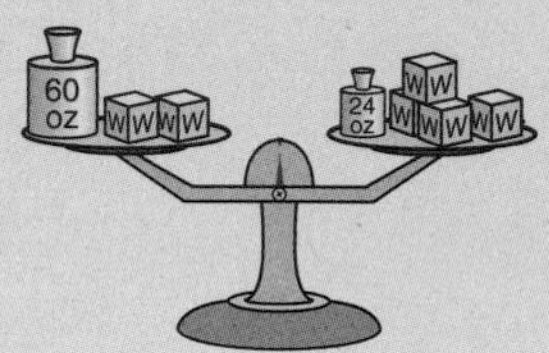

NEW CONCEPT

We can find the areas of some polygons by multiplying two perpendicular dimensions.

- We find the area of a rectangle by multiplying the length by the width.

$$A = lw$$

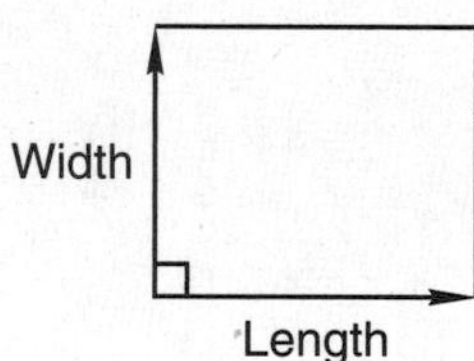

- We find the area of a parallelogram by multiplying the base by the height.

$$A = bh$$

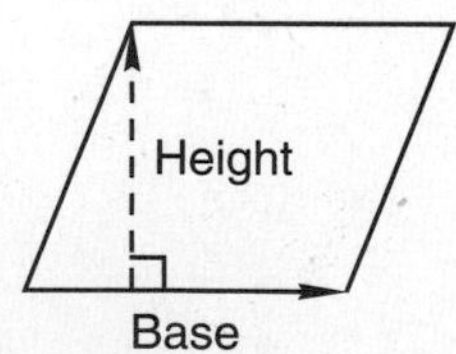

- We find the area of a triangle by multiplying the base by the height (which gives us the area of a parallelogram) and then dividing by 2.

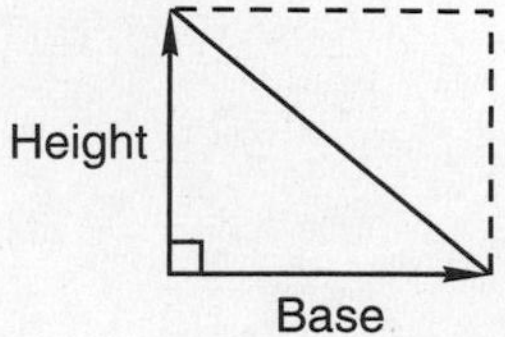

$$A = \frac{bh}{2} \quad \text{or} \quad A = \frac{1}{2}bh$$

To find the area of a circle, we again begin by multiplying two perpendicular dimensions. We multiply the radius by the radius. This gives us the area of a square built on the radius.

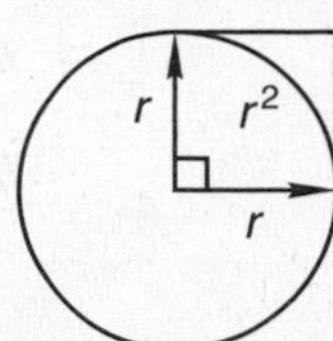

If the radius of the circle is 3, the area of the square is 3^2, which is 9. If the radius of the circle is r, the area of the square is r^2. We see that the area of the circle is less than the area of four of these squares.

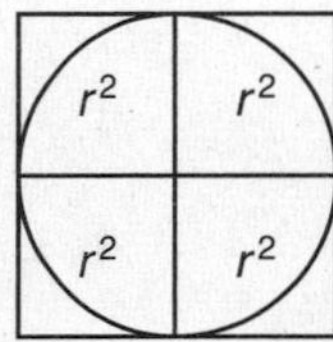

However, the area of the circle is more than the area of three squares.

The number of squares whose area exactly equals the area of the circle is between 3 and 4. The exact number is π. Thus, to find the area of the circle, we first find the area of the square built on the radius; then we multiply that area by π. This is summarized by the equation

$$\boldsymbol{A = \pi r^2}$$

Example Find the area of each circle:

(a)

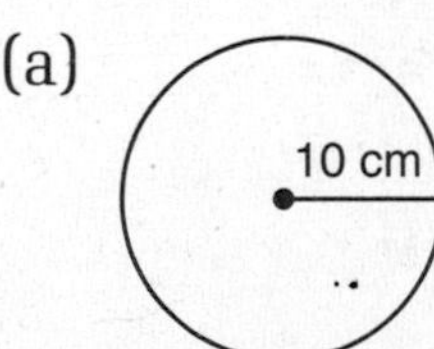

Use 3.14 for π.

(b)

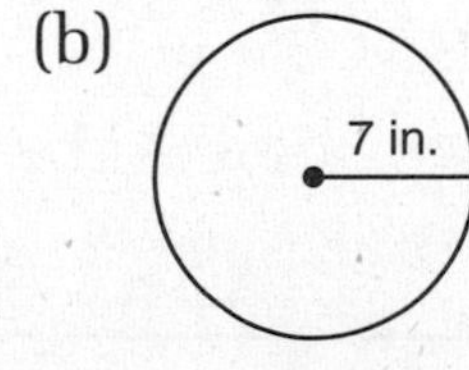

Use $\frac{22}{7}$ for π.

(c)

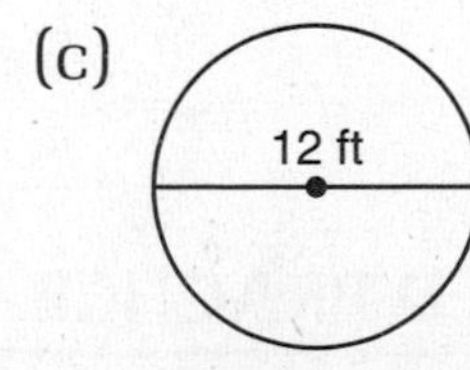

Leave π as π.

Solution (a) The area of a square built on the radius is 100 cm². We multiply this by π.

10 cm | 100 cm² | 10 cm

$$A = \pi r^2$$

$$A \approx (3.14)(100 \text{ cm}^2)$$

$$A \approx \mathbf{314\ cm^2}$$

(b) The area of a square built on the radius is 49 in.^{2}. We multiply this by π.

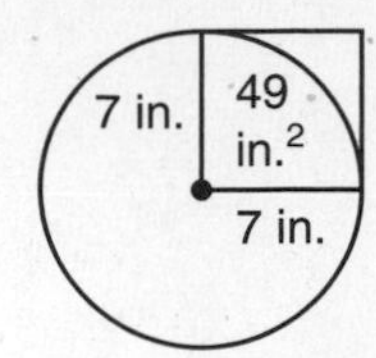

$$A = \pi r^2$$

$$A \approx \frac{22}{\cancel{7}_1} \cdot \cancel{49}^{7} \text{ in.}^2$$

$$A \approx \mathbf{154\ in.^2}$$

(c) Since the diameter is 12 ft, the radius is 6 ft. The area of a square built on the radius is 36 ft^2. We multiply this by π.

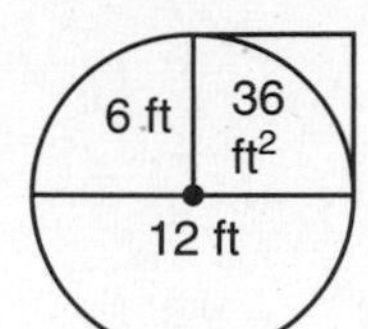

$$A = \pi r^2$$

$$A = \pi \cdot 36 \text{ ft}^2$$

$$A = \mathbf{36\pi\ ft^2}$$

LESSON PRACTICE

Practice set* **a.** Using 3.14 for π, calculate to the nearest square foot the area of circle (c) in this lesson's example.

Find the area of each circle:

b.

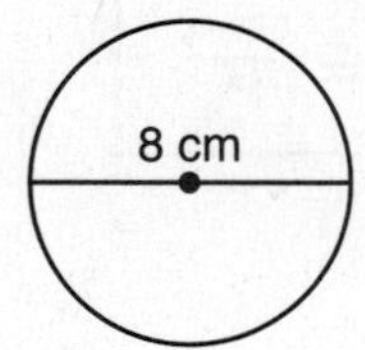

Use 3.14 for π.

c.

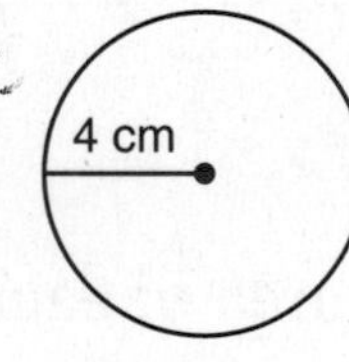

Leave π as π.

d.

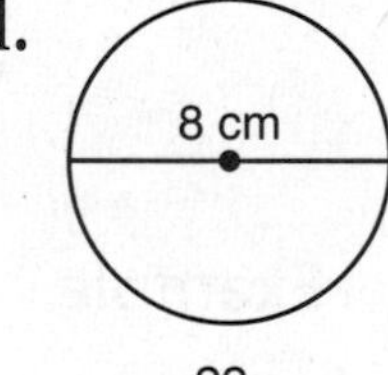

Use $\frac{22}{7}$ for π.

MIXED PRACTICE

Problem set

1. (70) Find the volume of this rectangular prism. Dimensions are in feet.

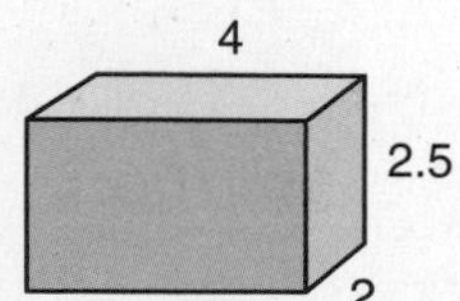

2. (28) The heights of the five basketball starters were 6'3", 6'5", 5'11", 6'2", and 6'1". Find the average height of the five starters. (*Hint:* Change all measures to inches before dividing.)

3. (54) Use a ratio box to solve this problem. The patient-doctor ratio at the large hospital was 20 to 1. If there were 48 doctors, how many patients were there?

4. (50) An inch equals 2.54 centimeters. Use a unit multiplier to convert 2.54 centimeters to meters.

5. (78) Graph each inequality on a separate number line:

(a) $x < -2$ (b) $x \geq 0$

6. (72) Use a case 1–case 2 ratio box to solve this problem. Don's heart beats 225 times in 3 minutes. At that rate, how many times will his heart beat in 5 minutes?

7. (36, 71) Diagram this statement. Then answer the questions that follow.

Two fifths of the children in the club were boys. There were 15 girls in the club.

(a) How many children were in the club?

(b) What was the ratio of girls to boys in the club?

8. (79) Compare: $x^2 - y^2 \bigcirc (x + y)(x - y)$ if $x = 5$ and $y = 3$

9. (Inv. 1) What percent of this circle is shaded?

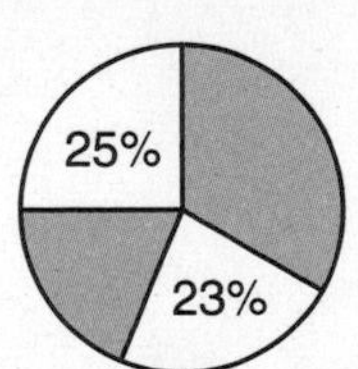

10. (79) Compare: $a \bigcirc b$ if $a - b$ is negative

11. (66) Find the circumference of each circle:

(a)

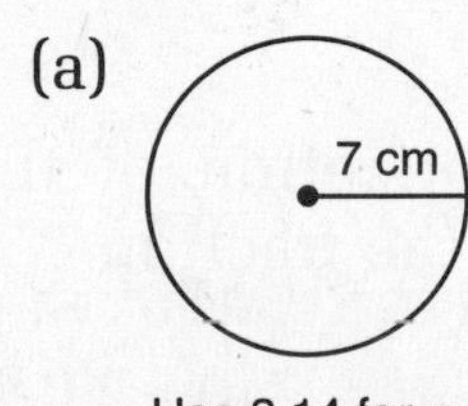

Use 3.14 for π.

(b)

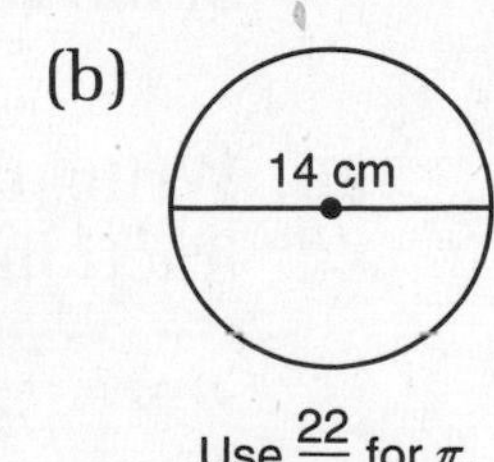

Use $\frac{22}{7}$ for π.

12. Find the area of each circle in problem 11.
(82)

13. Complete the table.
(48)

FRACTION	DECIMAL	PERCENT
(a)	1.6	(b)
(c)	(d)	1.6%

14. Write an equation to solve this problem:
(60)

How much money is 6.4% of $25?

15. Write each number in scientific notation:
(69)

(a) 12×10^5 (b) 12×10^{-5}

Use ratio boxes to solve problems 16 and 17.

16. Sixty-four percent of the students correctly described the process of photosynthesis. If 63 students did not correctly describe the process of photosynthesis, how many students did correctly describe the process?
(81)

17. Ginger still has 40 percent of her book to read. If she has read 180 pages, how many pages does she still have to read?
(81)

18. Find the area of the figure shown. Dimensions are in inches. Corners that look square are square.
(75)

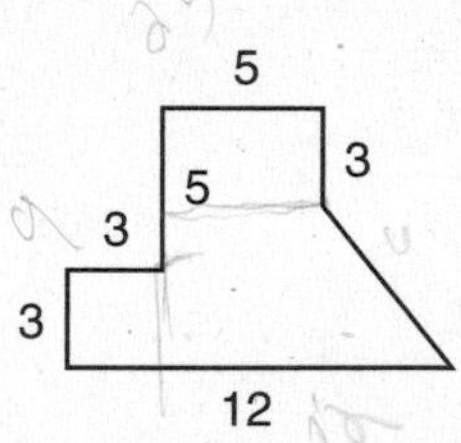

19. The coordinates of the vertices of ΔXYZ are $X(4, 3)$, $Y(4, 1)$, and $Z(1, 1)$. Draw ΔXYZ and its image $\Delta X'Y'Z'$ translated 5 units to the left and 3 units down. What are the coordinates of the vertices of $\Delta X'Y'Z'$?
(80)

20. Write the prime factorization of the two terms of this fraction. Then reduce the fraction.
(24)

$$\frac{240}{816}$$

21. (Inv. 2) The figure illustrates regular hexagon *ABCDEF* inscribed in a circle with center at point *M*.

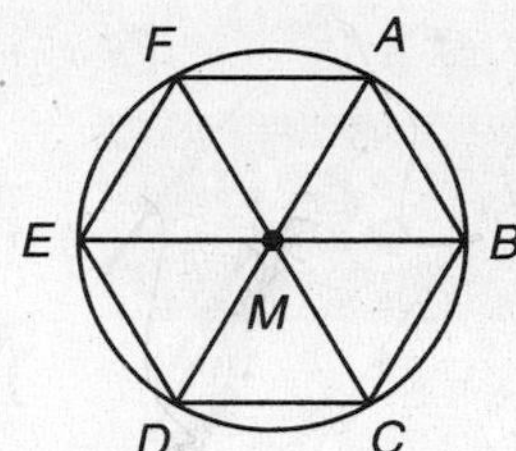

(a) How many illustrated chords are diameters?

(b) How many illustrated chords are not diameters?

(c) What is the measure of central angle *AMB*?

(d) What is the measure of inscribed angle *ABC*?

22. (51) Write 100 million in scientific notation.

For problems 23 and 24, solve and check the equation. Show each step.

23. (Inv. 7) $\frac{3}{4}x = 36$

24. (Inv. 7) $3.2 + a = 3.46$

Simplify:

25. (52) $\frac{\sqrt{3^2 + 4^2}}{5}$

26. (52) $(8 - 3)^2 - (3 - 8)^2$

27. (43, 45) $3\frac{1}{2} \div (7 \div 0.2)$ (decimal answer)

28. (43) $4.5 + 2\frac{2}{3} - 3$ (mixed-number answer)

29. (73) (a) $\frac{(-3)(-4)}{(-2)}$

(b) $(-2)(+3)(-4)$

30. (68) (a) $(-3) + (-4) - (-2)$

(b) $(-20) + (+30) - |-40|$

LESSON

83 Multiplying Powers of 10 • Multiplying Numbers in Scientific Notation

WARM-UP

Facts Practice: Area (Test R)

Mental Math:

a. $(-60) \div (+3)$
b. 6.75×10^6
c. $\frac{100}{150} = \frac{m}{30}$
d. 15% of \$120
e. 500 mg to g
f. 24 is $\frac{3}{8}$ of n.
g. At 60 mph, how far will a car travel in $2\frac{1}{2}$ hours?

Problem Solving:

How many different triangles of any size are in this figure?

NEW CONCEPTS

Multiplying powers of 10

From our earlier work with powers of 10, we remember that

$$10^3 \text{ means } 10 \cdot 10 \cdot 10$$

and

$$10^4 \text{ means } 10 \cdot 10 \cdot 10 \cdot 10$$

Now let us multiply these two powers of 10.

$$\overbrace{10 \cdot 10 \cdot 10}^{10^3} \cdot \overbrace{10 \cdot 10 \cdot 10 \cdot 10}^{10^4}$$

We see that $10^3 \cdot 10^4$ means 7 tens are multiplied. We can write this as 10^7.

$$10^3 \cdot 10^4 = 10^7$$

As we focus our attention on the exponents, we see that

$$3 + 4 = 7$$

The above example illustrates an important rule of mathematics.

When we multiply powers of 10, we add the exponents.

Multiplying numbers in scientific notation To multiply numbers written in scientific notation, we multiply the decimal numbers to find the decimal-number part of the product. Then we multiply the powers of 10 to find the power-of-10 part of the product. We remember that when we multiply powers of 10, we add the exponents.

Example 1 Multiply: $(1.2 \times 10^5)(3 \times 10^7)$

Solution We multiply 1.2 by 3 and get 3.6. Then we multiply 10^5 by 10^7 and get 10^{12}. The product is

$$\mathbf{3.6 \times 10^{12}}$$

Example 2 Multiply: $(4 \times 10^6)(3 \times 10^5)$

Solution We multiply 4 by 3 and get 12. Then we multiply 10^6 by 10^5 and get 10^{11}. The product is

$$12 \times 10^{11}$$

We rewrite this expression in proper scientific notation.

$$(1.2 \times 10^1) \times 10^{11} = \mathbf{1.2 \times 10^{12}}$$

Example 3 Multiply: $(2 \times 10^{-5})(3 \times 10^{-7})$

Solution We multiply 2 by 3 and get 6. To multiply 10^{-5} by 10^{-7}, we add the exponents and get 10^{-12}. Thus the product is

$$\mathbf{6 \times 10^{-12}}$$

Example 4 Multiply: $(5 \times 10^3)(7 \times 10^{-8})$

Solution We multiply 5 by 7 and get 35. We multiply 10^3 by 10^{-8} and get 10^{-5}. The product is

$$35 \times 10^{-5}$$

We rewrite this expression in scientific notation.

$$(3.5 \times 10^1) \times 10^{-5} = \mathbf{3.5 \times 10^{-4}}$$

LESSON PRACTICE

Practice set* Multiply. Write each product in scientific notation.

a. $(4.2 \times 10^6)(1.4 \times 10^3)$

b. $(5 \times 10^5)(3 \times 10^7)$

c. $(4 \times 10^{-3})(2.1 \times 10^{-7})$

d. $(6 \times 10^{-2})(7 \times 10^{-5})$

MIXED PRACTICE

Problem set

1. The 16-ounce box costs \$1.12. The 24-ounce box costs \$1.32. The smaller box costs how much more per ounce than the larger box? *(46)*

2. Use a ratio box to solve this problem. The ratio of good apples to bad apples in the basket was 5 to 2. If there were 70 apples in the basket, how many of them were good? *(65)*

3. Jan's average score after fifteen tests was 82. Her average score on the next five tests was 90. What was her average score for all twenty tests? *(55)*

4. Hakim earns \$6 per hour at a part-time job. How much does he earn if he works for 2 hours 30 minutes? *(53)*

5. Use a unit multiplier to convert 24 shillings to pence (1 shilling = 12 pence). *(50)*

6. Graph $x \leq -1$ on a number line. *(78)*

7. Use a case 1–case 2 ratio box to solve this problem. Five is to 12 as 20 is to what number? *(72)*

8. If $a = 1.5$, what does $4a + 5$ equal? *(41)*

9. Four fifths of the football team's 30 points were scored on pass plays. How many points did the team score on pass plays? *(22)*

10. Compare: $x(x + y) \bigcirc x^2 + xy$ if x and y are integers *(79)*

11. Find the circumference of each circle: *(66)*

(a)

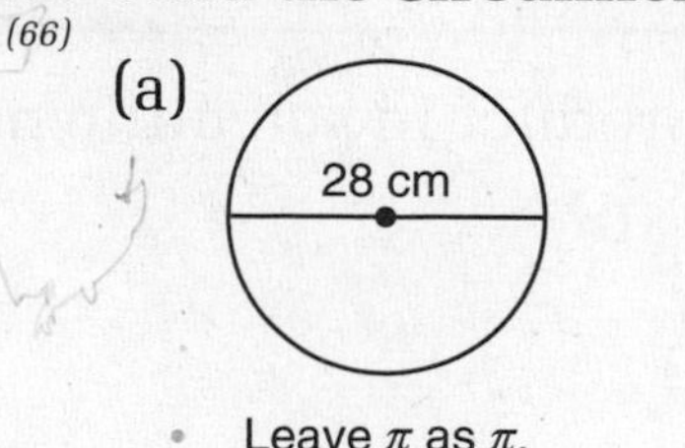

Leave π as π.

(b)

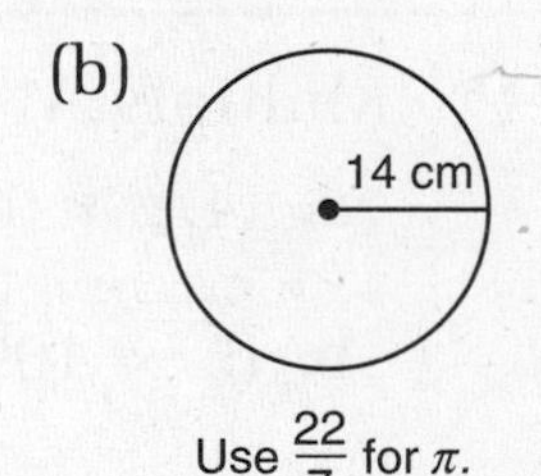

Use $\frac{22}{7}$ for π.

12. Find the area of each circle in problem 11 by using the indicated values for π. *(82)*

13. The edges of a cube are 10 cm long.
(67, 70)

(a) What is the volume of the cube?

(b) What is the surface area of the cube?

14. Complete the table.
(48)

FRACTION	DECIMAL	PERCENT
(a)	(b)	250%
$\frac{7}{12}$	(c)	(d)

15. What is the sales tax on an \$8.50 purchase if the sales-tax rate is $6\frac{1}{2}\%$?
(46)

Use ratio boxes to solve problems 16 and 17.

16. Monifa found that 12 minutes of commercials aired during every hour of prime-time programming. Commercials were shown for what percent of each hour?
(81)

17. Thirty percent of the boats that traveled up the river on Monday were steam-powered. If 42 of the boats that traveled up the river were not steam-powered, how many boats were there in all?
(81)

18. Write the prime factorization of the numerator and denominator of this fraction. Then reduce the fraction.
(24)

$$\frac{420}{630}$$

19. Find the area of the trapezoid at right.
(75)

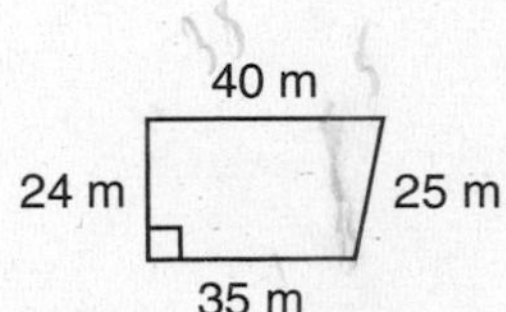

20. In this figure, $\angle A$ and $\angle B$ of $\triangle ABC$ are congruent. The measure of $\angle E$ is 54°. Find the measure of
(40)

(a) $\angle ECD$. (b) $\angle ECB$.

(c) $\angle ACB$. (d) $\angle BAC$.

A D B C E

21. (58) Describe the rule of this function, and find the missing numbers.

IN		FUNCTION		OUT
□	→		→	21
5	→		→	11
7	→		→	15
2	→		→	5
−5	→		→	□

22. (83) Multiply. Write each product in scientific notation.

(a) $(3 \times 10^4)(6 \times 10^5)$ (b) $(1.2 \times 10^{-3})(4 \times 10^{-6})$

For problems 23 and 24, solve and check the equation. Show each step.

23. (Inv. 7) $b - 1\frac{2}{3} = 4\frac{1}{2}$

24. (Inv. 7) $0.4y = 1.44$

Simplify:

25. (52, 57) $2^3 + 2^2 + 2^1 + 2^0 + 2^{-1}$

26. (43) $0.6 \times 3\frac{1}{3} \div 2$

27. (73) (a) $\frac{(-4)(-6)}{(-2)(-3)}$

(b) $(-3)(-4)(-5)$

28. (30) $\frac{5}{24} - \frac{7}{60}$

29. (68) (a) $(-3) + (-4) - (-5)$

(b) $(-15) - (+14) + (+10)$

30. (80) The coordinates of the vertices of ΔPQR are $P(0, 1)$, $Q(0, 0)$, and $R(-2, 0)$. Draw the triangle and its image $\Delta P'Q'R'$ after a 180° clockwise rotation about the origin. What are the coordinates of the vertices of $\Delta P'Q'R'$?

LESSON

84 Algebraic Terms

WARM-UP

Facts Practice: Area (Test R)

Mental Math:

a. $(-12) - (-12)$ **b.** 25^2 **c.** $6m - 10 = 32$

d. 3.14×30 cm **e.** 1.5 cm to mm **f.** 30 is $\frac{5}{6}$ of n.

g. 12×12, $- 4$, $\div 10$, $+ 1$, $\times 2$, $+ 3$, $\div 3$, $\times 5$, $- 1$, $\div 6$, $\sqrt{\ }$

Problem Solving:

A rectangular tablecloth was draped over a rectangular table. Eight inches of cloth hung over the left edge of the table, 3 inches over the back, 4 inches over the right edge, and 7 inches over the front. In which directions (L, B, R, and F) and by how many inches should the tablecloth be shifted so that equal amounts of cloth hang over opposite edges of the table?

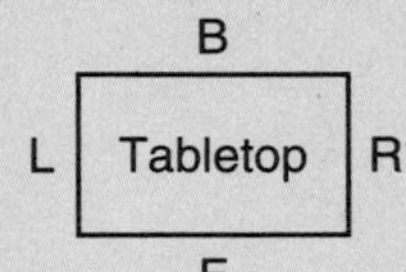

NEW CONCEPT

We have used the word **term** in arithmetic to refer to the numerator or denominator of a fraction. For example, we reduce a fraction to its lowest terms. In algebra *term* refers to a part of an algebraic expression or equation. An algebraic expression may contain one, two, three, or more terms.

Some Algebraic Expressions

Type of Expression	Number of Terms	Example
monomial	1	$-2x$
binomial	2	$a^2 - 4b^2$
trinomial	3	$3x^2 - x - 4$

Terms are separated from one another in an expression by plus or minus signs that are not within symbols of inclusion.

Here we have separated the terms of the binomial and trinomial examples with slashes:

$$a^2 \ / - 4b^2 \qquad 3x^2 \ / - x \ / - 4$$

Each term contains a signed number and may contain one or more variables (letters). Sometimes the signed-number part is understood and not written. For instance, the understood signed-number part of a^2 is +1 since $a^2 = +1a^2$. Likewise, the term $-x$ is understood to mean $-1x$. **When a term is written without a number, it is understood that the number is 1. When a term is written without a sign, it is understood that the sign is positive.** It is not necessary for a term to contain a variable. The third term of the trinomial above is −4. A term that does not contain a variable is often called a **constant term,** because its value never changes.

Constant terms can be combined by algebraic addition.

$$3x + 3 - 1 = 3x + 2 \qquad \text{added +3 and −1}$$

Variable terms can also be combined by algebraic addition if they are **like terms.** Like terms have identical variable parts. That is, the same variables with the same exponents appear in the terms. The terms $-3xy$ and $+xy$ are both xy terms. They are like terms and can be combined by algebraically adding the signed-number part of the terms.

$$-3xy + xy = -2xy$$

The signed number part of $+xy$ is +1. We get $-2xy$ by adding $-3xy$ and $+1xy$.

Example 1 Collect like terms in this algebraic expression.

$$3x + y + x - y$$

Solution There are four terms in this expression. There are two x terms and two y terms. We can use the commutative property to rearrange the terms.

$$3x + x + y - y$$

Adding $+3x$ and $+1x$ we get $+4x$. Then adding $+1y$ and $-1y$ we get $0y$, which is 0.

$$3x + x + y - y$$

$$4x + 0$$

$$\mathbf{4x}$$

Example 2 Collect like terms in this algebraic expression:

$$3x + 2x^2 + 4 + x^2 - x - 1$$

Solution In this expression there are three kinds of terms: x^2 terms, x terms, and constant terms. Using the commutative property we arrange them to put like terms next to each other.

$$2x^2 + x^2 + 3x - x + 4 - 1$$

Now we collect like terms.

$$2x^2 + x^2 + 3x - x + 4 - 1$$

$$\mathbf{3x^2 + 2x + 3}$$

Notice that x^2 terms and x terms are not like terms and cannot be combined by addition. There are other possible arrangements of the collected terms, such as the following:

$$2x + 3x^2 + 3$$

Customarily, however, we arrange terms in descending order of exponents so that the term with the largest exponent is on the left and the constant term is on the right. An expression written without a constant term is understood to have zero as a constant term.

LESSON PRACTICE

Practice set Describe each of these expressions as a monomial, a binomial, or a trinomial:

a. $x^2 - y^2$

b. $3x^2 - 2x - 1$

c. $-2x^3yz^2$

d. $-2x^2y - 4xy^2$

Collect like terms:

e. $3a + 2a^2 - a + a^2$

f. $5xy - x + xy - 2x$

g. $3 + x^2 + x - 5 + 2x^2$

h. $3\pi + 1.4 - \pi + 2.8$

MIXED PRACTICE

Problem set

1. (32) An increase in temperature of 10° on the Celsius scale corresponds to an increase of how many degrees on the Fahrenheit scale?

2. (84) Collect like terms: $2xy + xy - 3x + x$

3. Refer to the graph below to answer (a)–(c).
(Inv. 4)

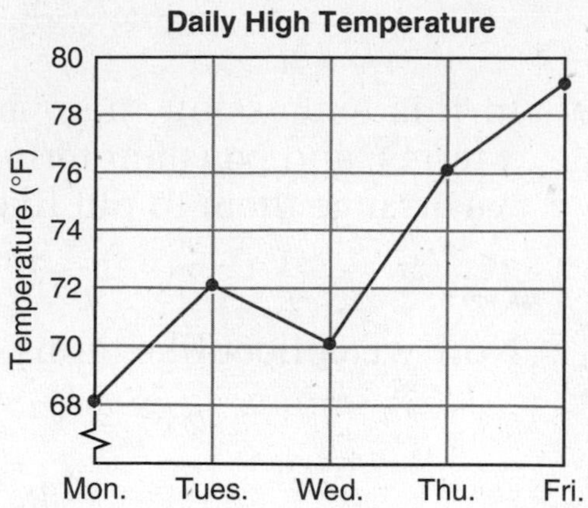

(a) What was the range of the daily high temperatures from Monday to Friday?

(b) Which day had the greatest increase in temperature from the previous day?

(c) Wednesday's high temperature was how much lower than the average high temperature for these 5 days?

4. Frank's scores on ten tests were as follows:
(Inv. 4)

90, 90, 100, 95, 95, 85, 100, 100, 80, 100

For this set of scores, find the (a) mean, (b) median, (c) mode, and (d) range.

5. Use a ratio box to solve this problem. The ratio of rowboats to sailboats in the bay was 3 to 7. If the total number of rowboats and sailboats in the bay was 210, how many sailboats were in the bay?
(65)

6. Recall that the four quadrants of a coordinate plane are numbered 1st, 2nd, 3rd, and 4th in a counterclockwise direction beginning with the upper right quadrant. In which quadrant are the x-coordinates negative and the y-coordinates positive?
(Inv. 3)

7. Write a proportion to solve this problem. If 4 cost \$1.40, how much would 10 cost?
(72)

8. Five eighths of the members supported the treaty, whereas 36 opposed the treaty. How many members supported the treaty?
(71)

9. Evaluate each expression for $x = 5$:
(52)

(a) $x^2 - 2x + 1$ (b) $(x - 1)^2$

10. Compare: $f \bigcirc g$ if $\frac{f}{g} = 1$
(79)

11. (a) Find the circumference of the circle shown.
(66, 82)

(b) Find the area of the circle.

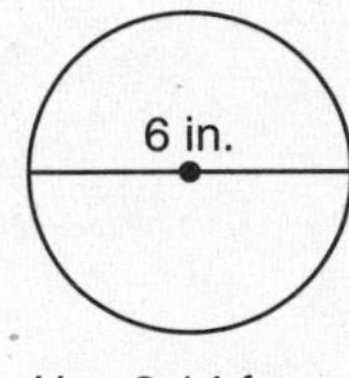

Use 3.14 for π.

12. Use a unit multiplier to convert 4.8 meters to centimeters.
(50)

13. Draw a rectangular prism. A rectangular prism has how many faces?
(67)

14. Complete the table.
(48)

FRACTION	DECIMAL	PERCENT
$1\frac{4}{5}$	(a)	(b)
(c)	(d)	1.8%

15. Write an equation to solve this problem. A merchant priced a product so that 30% of the selling price is profit. If the product sells for $18.00, how much is the merchant's profit?
(60)

16. Simplify: $\dfrac{12\frac{1}{2}}{100}$
(76)

Use ratio boxes to solve problems 17 and 18.

17. When the door was left open, 36 pigeons flew the coop. If this was 40 percent of all pigeons, how many pigeons were originally in the coop?
(81)

18. Sixty percent of the saplings were 3 feet tall or less. If there were 300 saplings in all, how many were more than 3 feet tall?
(81)

19. (19, 75) A square sheet of paper with a perimeter of 48 in. has a corner cut off, forming a pentagon as shown.

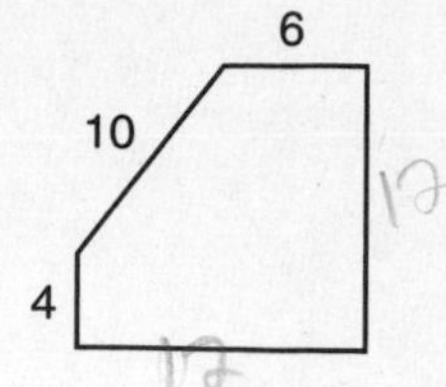

(a) What is the perimeter of the pentagon?

(b) What is the area of the pentagon?

20. (36, Inv. 5) The face of this spinner has been divided into seven sectors, the central angles of which have the following measures:

A 60° B 90° C 45° D 30°

E 75° F 40° G 20°

If the spinner is spun once, what is the probability that it will stop in sector

(a) A? (b) C? (c) E?

21. (2) Describe the rule of this sequence. Then find the next three numbers of the sequence.

1, 3, 7, 15, 31, ...

22. (83) Multiply. Write each product in scientific notation.

(a) $(1.5 \times 10^{-3})(3 \times 10^{6})$ (b) $(3 \times 10^{4})(5 \times 10^{5})$

Find each missing exponent:

23. (57, 83) (a) $10^2 \cdot 10^2 \cdot 10^2 = 10^{\square}$ (b) $\dfrac{10^2}{10^6} = 10^{\square}$

For problems 24 and 25, solve and check the equation. Show each step.

24. (Inv. 7) $b - 4.75 = 5.2$

25. (Inv. 7) $\frac{2}{3}y = 36$

Simplify:

26. (52) $\sqrt{5^2 - 4^2} + 2^3$

27. (32) 1 m − 45 mm

28. (43) $\frac{9}{10} \div 2\frac{1}{4} \cdot 24$ (decimal answer)

29. (73) (a) $\dfrac{(-8)(+6)}{(-3)(+4)}$ (b) $(+3)(-5)(+2)$

30. (68) (a) $(+30) - (-50) - (+20)$

(b) $(-3) - (-4) - (5)$

LESSON

85 Order of Operations with Signed Numbers • Functions, Part 2

WARM-UP

Facts Practice: + − × ÷ Integers (Test P)

Mental Math:

a. (+12)(−6)

b. $(4 \times 10^3)(2 \times 10^6)$

c. $\frac{1}{1.5} = \frac{80}{n}$

d. \$12 is $\frac{1}{4}$ of how much money (m)?

e. 0.8 km to m

f. What is $\frac{1}{4}$ of \$12?

g. Find the perimeter and area of a square with sides 2.5 m long.

Problem Solving:

There are three numbers whose sum is 180. The second number is twice the first number, and the third number is three times the first number (n, $2n$, and $3n$). Find the three numbers. (Try guess and check; then try writing an equation.)

$$\begin{array}{r} \_\_ \\ \_\_ \\ + \ \_\_ \\ \hline 180 \end{array}$$

NEW CONCEPTS

Order of operations with signed numbers To simplify expressions that involve several operations, we perform the operations in a prescribed order. We have practiced simplifying expressions with whole numbers. In this lesson we will begin simplifying expressions that contain both whole numbers and negative numbers.

Example 1 Simplify: $(-2) + (-2)(-2) - \frac{(-2)}{(+2)}$

Solution First we multiply and divide in order from left to right.

$$(-2) + \underbrace{(-2)(-2)} - \underbrace{\frac{(-2)}{(+2)}}$$

$$(-2) + \quad (+4) \quad - \ (-1)$$

Then we add and subtract in order from left to right.

$$\underbrace{\underbrace{(-2) + (+4)}_{(+2)} - (-1)}_{\mathbf{+3}}$$

Mentally separating an expression into its terms can make an expression easier to simplify.

$$(-2) \;/\; + (-2)(-2) \;/\; - \frac{(-2)}{(+2)}$$

First we simplify each term; then we combine the terms.

$$(-2) \;/\; + (-2)(-2) \;/\; - \frac{(-2)}{(+2)}$$

$$-2 \;/\; + 4 \;/\; + 1$$

$$\mathbf{+3}$$

Example 2 Simplify each term. Then combine the terms.

$$-3(2 - 4) - 4(-2)(-3) + \frac{(-3)(-4)}{2}$$

Solution We separate the individual terms with slashes. The slashes precede plus and minus signs that are not enclosed by parentheses or other symbols of inclusion.

$$-3(2 - 4) \;/\; - 4(-2)(-3) \;/\; + \frac{(-3)(-4)}{2}$$

Next we simplify each term.

$$-3(2 - 4) \;/\; - 4(-2)(-3) \;/\; + \frac{(-3)(-4)}{2}$$

$$-3(-2) \;/\; + 8(-3) \;/\; + \frac{12}{2}$$

$$+6 \;/\; - 24 \;/\; + 6$$

Now we combine the simplified terms.

$$+6 - 24 + 6$$

$$-18 + 6$$

$$\mathbf{-12}$$

Example 3 Simplify: $(-2) - [(-3) - (-4)(-5)]$

Solution There are only two terms, −2 and the bracketed quantity. By the order of operations, we simplify within brackets first, multiplying and dividing before adding and subtracting.

$$(-2) \;/\; - [(-3) - (-4)(-5)]$$

$$(-2) \;/\; - [(-3) - (+20)]$$

$$(-2) \;/\; - (-23)$$

$$(-2) \;/\; + 23$$

$$\mathbf{+21}$$

Functions, part 2 We remember that a function is a relationship between two sets of numbers. We have practiced finding missing numbers in functions when some number pairs have been given. For instance, the missing numbers in the functions on the left and the right below are 14 and 7, respectively.

IN	FUNCTION	OUT
0 →		→ 0
3 →		→ 6
4 →		→ 8
7 →		→ □

IN	FUNCTION	OUT
14 →		→ 7
□ →		→ 0
10 →		→ 3
5 →		→ −2

We found the missing numbers by first finding the rule of the function. The rule of the function on the left is *multiply the "in" number by 2 to find the "out" number.* The rule of the function on the right is *subtract 7 from the "in" number to find the "out" number.*

Often the rule of a function is expressed as an equation with x standing for the "in" number and y standing for the "out" number. The equation for the rule of the function on the left is

$$y = 2x$$

The rule of the function on the right is

$$y = x - 7$$

Beginning with this lesson we will practice finding missing numbers in function tables when the rule is given as an equation.

Example 4 Find the missing numbers in the table using the function rule.

$$y = 2x + 1$$

x	y
4	☐
7	☐
0	☐

Solution The letter y stands for the "out" number. The letter x stands for the "in" number. We are given three "in" numbers and are asked to find the "out" number for each by using the rule of the function. The expression $2x + 1$ shows us what to do to find y, the "out" number. It shows us we should multiply the x number by 2 and then add 1. The first x number is 4. We multiply by 2 and add 1.

$y = 2x + 1$

$y = 2(4) + 1$ substituted

$y = 8 + 1$ multiplied

$y = 9$ added

We find that the y number is 9 when x is 4. The next x number is 7. We multiply by 2 and add 1.

$y = 2x + 1$

$y = 2(7) + 1$ substituted

$y = 14 + 1$ multiplied

$y = 15$ added

The third x number is 0. We multiply by 2 and add 1.

$y = 2x + 1$

$y = 2(0) + 1$ substituted

$y = 0 + 1$ multiplied

$y = 1$ added

The missing numbers are **9, 15,** and **1.**

LESSON PRACTICE

Practice set Simplify:

a. $(-3) + (-3)(-3) - \dfrac{(-3)}{(+3)}$ **b.** $(-3) - [(-4) - (-5)(-6)]$

c. $(-2)[(-3) - (-4)(-5)]$ **d.** $(-5) - (-5)(-5) + |-5|$

Find the missing numbers in each table by using the function rule:

e. $y = 3x - 1$

x	y
3	☐
1	☐
0	☐

f. $y = \frac{1}{2}x$

x	y
6	☐
0	☐
☐	4

g. $y = 8 - x$

x	y
7	☐
1	☐
4	☐

h. Jacinta studied a function and found that when x was 1, y was 1; when x was 2, y was 4; and when x was 3, y was 9. Make a table of x, y pairs for this function, and above the table write an equation that expresses the function rule.

MIXED PRACTICE

Problem set

1. Find the (a) mean, (b) median, (c) mode, and (d) range of the following scores:
(Inv. 4)

70, 80, 90, 80, 70, 90, 75, 95, 100, 90

Use ratio boxes to solve problems 2–4:

2. The team's ratio of games won to games played was 3 to 4. If the team played 24 games, how many games did the team fail to win?
(65)

3. Mary was chagrined to find that the ratio of dandelions to marigolds in the garden was 11 to 4. If there were 44 marigolds in the garden, how many dandelions were there?
(54)

4. If sound travels 2 miles in 10 seconds, how far does sound travel in 1 minute?
(72)

5. Use a unit multiplier to convert 0.98 liter to milliliters.
(50)

6. Graph $x > 0$ on a number line.
(78)

7. (71) Diagram this statement. Then answer the questions that follow.

Thirty-five thousand dollars was raised in the charity drive. This was seven tenths of the goal.

(a) The goal of the charity drive was to raise how much money?

(b) The drive fell short of the goal by what percent?

8. (79) Compare: $2a \bigcirc a^2$ if a is a whole number

9. (66, 82) The radius of a circle is 4 meters. Use 3.14 for π to find the

(a) circumference of the circle.

(b) area of the circle.

10. (Inv. 1) What fraction of this circle is shaded?

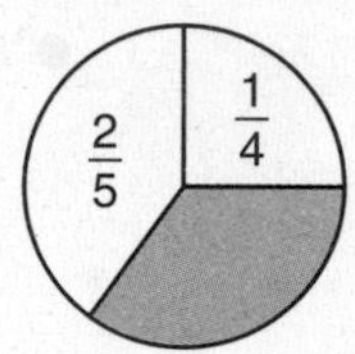

11. (70) A certain rectangular box is 5 in. long, 4 in. wide, and 3 in. high. Draw the box and find its volume.

12. (67) Suppose that the box described in problem 11 is cut open and unfolded. Draw the unfolded pattern and find the surface area.

13. (48) Complete the table.

FRACTION	DECIMAL	PERCENT
$\frac{1}{40}$	(a)	(b)
(c)	(d)	0.25%

14. (60) When the Nelsons sold their house, they paid the realtor a fee of 6% of the selling price. If the house sold for $180,000, how much was the realtor's fee?

15. (21) (a) Write the prime factorization of 17,640 using exponents.

(b) How can you tell by looking at the answer to (a) that $\sqrt{17{,}640}$ is not a whole number?

16. (76) Simplify: $\dfrac{8\frac{1}{3}}{100}$

Use ratio boxes to solve problems 17 and 18.

17. (81) Max was delighted when he found that he had correctly answered 38 of the 40 questions. What percent of the questions had he answered correctly?

18. (81) Before the clowns arrived, only 35 percent of the children had happy faces. If 91 children did not have happy faces, how many children were there in all?

19. (Inv. 6, 61)

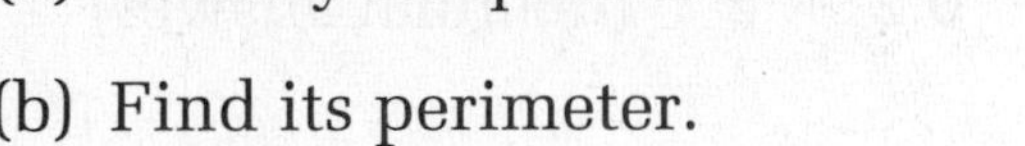

(a) Classify the quadrilateral shown.

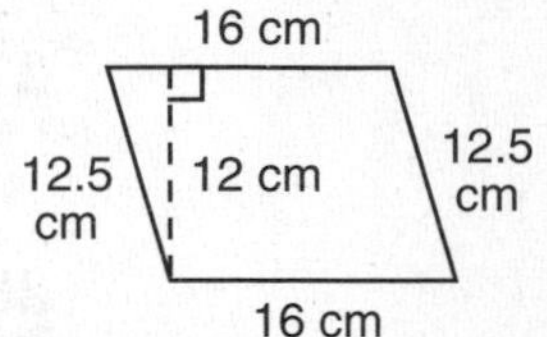

(b) Find its perimeter.

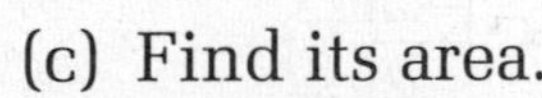

(c) Find its area.

(d) This figure does not have line symmetry, but it does have point symmetry. Trace the figure on your paper, and locate its point of symmetry.

20. (40) Refer to the figure below to answer (a)–(d).

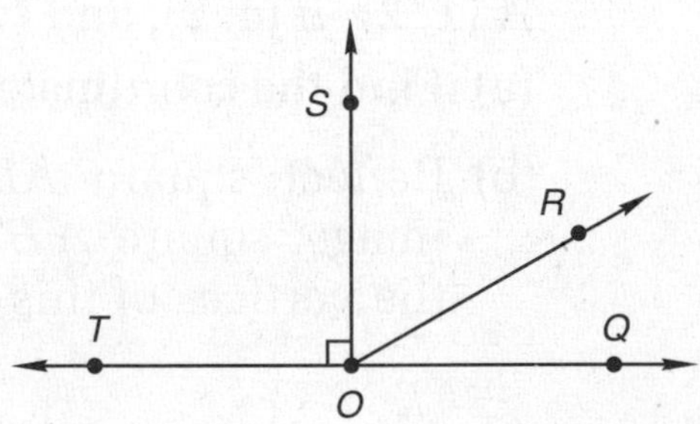

(a) Find m∠TOS.

(b) Find m∠QOT.

(c) Angle QOR is one third of a right angle. Find m∠QOR.

(d) Find m∠TOR.

21. (85) Find the missing numbers in the table by using the function rule.

$y = 2x - 1$

x	y
5	☐
3	☐
1	☐

22. (51, 83) Compare: $(5 \times 10^{-3})(6 \times 10^{8}) \bigcirc (5 \times 10^{8})(6 \times 10^{-3})$

For problems 23 and 24, solve and check the equation. Show each step.

23. $13.2 = 1.2w$
(Inv. 7)

24. $c + \frac{5}{6} = 1\frac{1}{4}$
(Inv. 7)

Simplify:

25. $3\{20 - [6^2 - 3(10 - 4)]\}$
(63)

26.
(56)

$$\begin{array}{r} 3 \text{ hr } 15 \text{ min } 25 \text{ s} \\ -\ 2 \text{ hr } 45 \text{ min } 30 \text{ s} \\ \hline \end{array}$$

27. $2^0 + 0.2 + 2^{-2}$ (decimal answer)
(43, 57)

28. (a) $(-2) + (-2)(+2) - \frac{(-2)}{(-2)}$
(85)

(b) $(-3) - [(-2) - (+4)(-5)]$

29. Collect like terms: $x^2 + 6x - 2x - 12$
(84)

30. The coordinates of three vertices of square *ABCD* are *A* (1, 2), *B* (4, 2), and *C* (4, –1).
(80)

(a) Find the coordinates of *D* and draw the square.

(b) Reflect square *ABCD* in the *y*-axis, and draw its image, square *A′B′C′D′*. What are the coordinates of the vertices of this reflection?

LESSON

86 Number Families

WARM-UP

Facts Practice: Scientific Notation (Test S)

Mental Math:

a. $(-18) + (-40)$

b. $(3 \times 10^{-3})(3 \times 10^{-3})$

c. $7x + 4 = 60$

d. Estimate 15% of \$17.90.

e. 0.2 L to mL

f. \$30 is $\frac{1}{3}$ of m.

g. What is the total cost of a \$200 item plus 7% sales tax?

Problem Solving:

Here are the front, top, and side views of an object. Draw a three-dimensional view of the object from the perspective of the upper right front.

Front

Right Side

Top

NEW CONCEPT

In mathematics we give special names to certain sets of numbers. Some of these sets are the counting numbers, the whole numbers, the integers, and the rational numbers. In this lesson we will review each of these **number families** and discuss how they are related.

- **The Counting Numbers.** Counting numbers are the numbers we say when we count. The first counting number is 1, the next is 2, then 3, and so on.

  Counting numbers: 1, 2, 3, 4, 5, ...

- **The Whole Numbers.** The members of the whole-number family are the counting numbers as well as the number zero.

  Whole numbers: 0, 1, 2, 3, 4, 5, ...

  If we use a dot to mark each of the whole numbers on the number line, the graph looks like this:

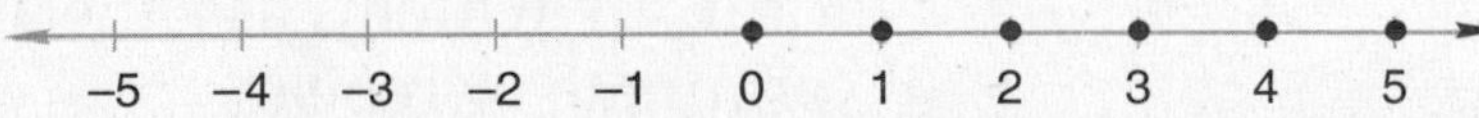

Notice that there are no dots to the left of zero. This is because no whole number is a negative number. Also notice that there are no dots between consecutive whole numbers. Numbers between consecutive whole numbers are not "whole." The arrowhead on the right end of the number line indicates that the whole numbers increase without end.

- **The Integers.** The integer family includes all the whole numbers. It also includes the opposites (negatives) of the positive whole numbers. The list of integers goes on and on in both directions as indicated by the ellipses below.

  Integers: ..., –4, –3, –2, –1, 0, 1, 2, 3, 4, ...

  A graph of the integers looks like this:

  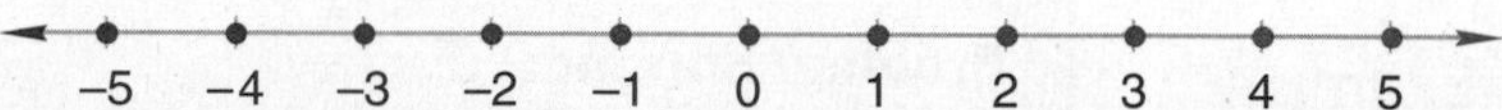

  The arrowheads on both ends of the number line indicate that the set of integers continues without end in both directions. Notice that integers do not include such numbers as $\frac{1}{2}$, $\frac{5}{3}$, and other fractions.

- **The Rational Numbers.** The family of **rational numbers** includes all numbers that can be written as a *ratio* (fraction) of two integers. Here are some examples of rational numbers:

  $$\frac{1}{2} \quad \frac{5}{3} \quad \frac{-3}{2} \quad \frac{-4}{1} \quad \frac{0}{2} \quad \frac{3}{1}$$

  Notice that the family of rational numbers includes all the integers, because every integer can be written as a fraction whose denominator is the number 1. For example, we can write –4 as a fraction by writing

  $$\frac{-4}{1}$$

  The set of rational numbers also includes all the positive and negative mixed numbers, because these numbers can be written as fractions. For example, we can write $4\frac{1}{5}$ as

  $$\frac{21}{5}$$

  Sometimes rational numbers are written in decimal form, in which case the decimal number will either terminate or repeat.

  $$\frac{1}{8} = 0.125 \qquad \frac{5}{6} = 0.8333... = 0.8\overline{3}$$

The diagram below may be helpful in visualizing the relationships between these families of numbers. The diagram shows that the set of rational numbers includes all the other number families described in this lesson.

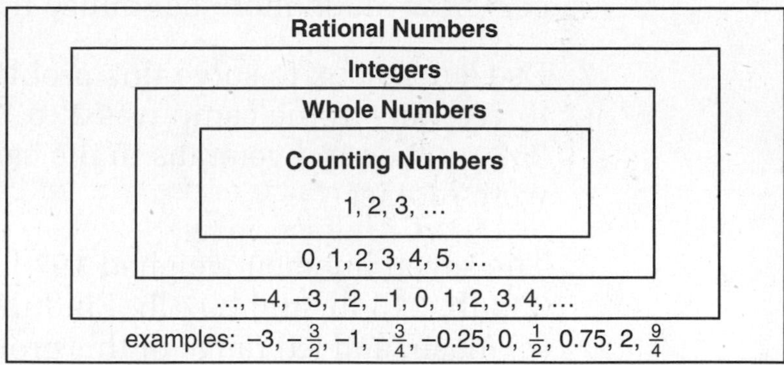

Example 1 Graph the integers that are less than 4.

Solution We draw a number line and mark a dot at every integer that is less than 4. Since the set of integers includes whole numbers, we mark dots at 3, 2, 1, and 0. Since the integers also include the opposites of the positive whole numbers, we continue marking dots at −1, −2, −3, and so on. We then mark an arrowhead on the negative end of the line to indicate that the graph of integers that are less than 4 continues without end.

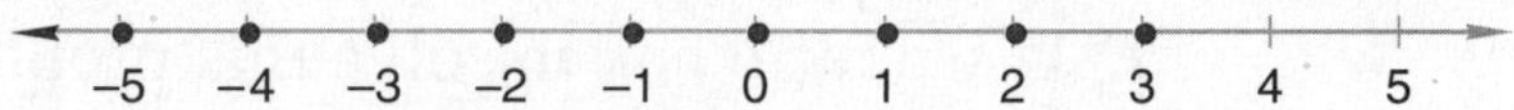

Example 2 Answer true or false:

(a) All whole numbers are integers.

(b) All rational numbers are integers.

Solution (a) **True.** Every whole number is included in the family of integers.

(b) **False.** Although every integer is a rational number, it is not true that every rational number is an integer. Rational numbers such as $\frac{1}{2}$ and $\frac{5}{3}$ are not integers.

LESSON PRACTICE

Practice set

a. Graph the integers that are greater than −4.

b. Graph the whole numbers that are less than 4.

Answer true or false:

c. Every integer is a whole number.

d. Every integer is a rational number.

MIXED PRACTICE

Problem set

1. (46) Heavenly Scent was priced at \$28.50 for 3 ounces, while Eau de Rue cost only \$4.96 for 8 ounces. Heavenly Scent cost how much more per ounce than Eau de Rue?

2. (65) Use a ratio box to solve this problem. The ratio of rookies to veterans in the camp was 2 to 7. Altogether there were 252 rookies and veterans in the camp. How many of them were rookies?

3. (Inv. 4) The seven linemen weighed 197 lb, 213 lb, 246 lb, 205 lb, 238 lb, 213 lb, and 207 lb. Find the (a) mode, (b) median, (c) mean, and (d) range of this group of measures.

4. (50) Use a unit multiplier to convert 12 bushels to pecks (1 bushel = 4 pecks).

5. (46) The Martins drove the car from 7 a.m. to 4 p.m. and traveled 468 miles. Their average speed was how many miles per hour?

6. (86) On a number line, graph the integers that are less than or equal to 3.

7. (72) Use a ratio box to solve this problem. Nine is to 6 as what number is to 30?

8. (22) Nine tenths of the company's 1800 employees attended the company picnic.

(a) How many of the company's employees attended the company picnic?

(b) What percent of the company's employees did not attend the company picnic?

9. (52) Evaluate: $\sqrt{b^2 - 4ac}$ if $a = 1$, $b = 5$, and $c = 4$

10. (79) Compare: $a^2 \bigcirc a$ if a is positive

11. (66, 82) (a) Find the circumference of the circle shown.

(b) Find the area of the circle.

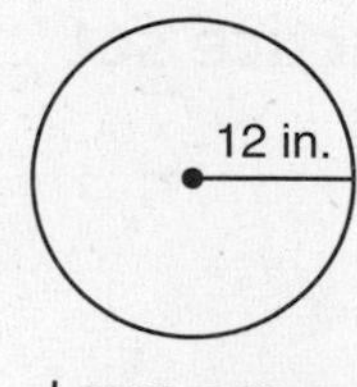

Leave π as π.

12. (47, 57) Find each missing exponent:

(a) $10^8 \cdot 10^{-3} = 10^{\square}$ (b) $10^5 \div 10^8 = 10^{\square}$

13. The figure shown is a triangular prism. Copy the figure on your paper, and find the number of its (a) faces, (b) edges, and (c) vertices.
(67)

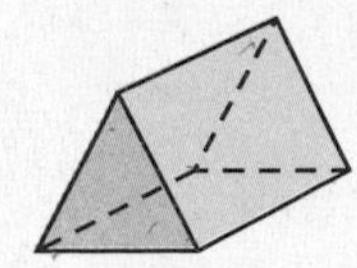

14. Complete the table.
(48)

FRACTION	DECIMAL	PERCENT
(a)	0.9	(b)
$\frac{11}{12}$	(c)	(d)

15. Obi is facing north. If he turns 360° in a clockwise direction, what direction will he be facing?
(17)

Use ratio boxes to solve problems 16 and 17.

16. The sale price of $24 was 60 percent of the regular price. What was the regular price?
(81)

17. Forty-eight corn seeds sprouted. This was 75 percent of the seeds that were planted. How many of the planted seeds did not sprout?
(81)

18. Write an equation to solve this problem:
(77)

Thirty is what percent of 20?

19. (a) Classify the quadrilateral shown.
(Inv. 6, 75)

(b) Find its perimeter.

(c) Find its area.

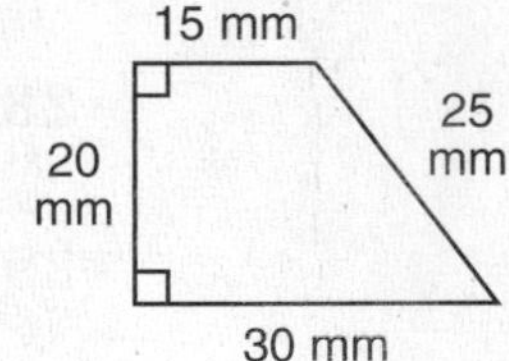

20. Find the measure of each angle.
(17)

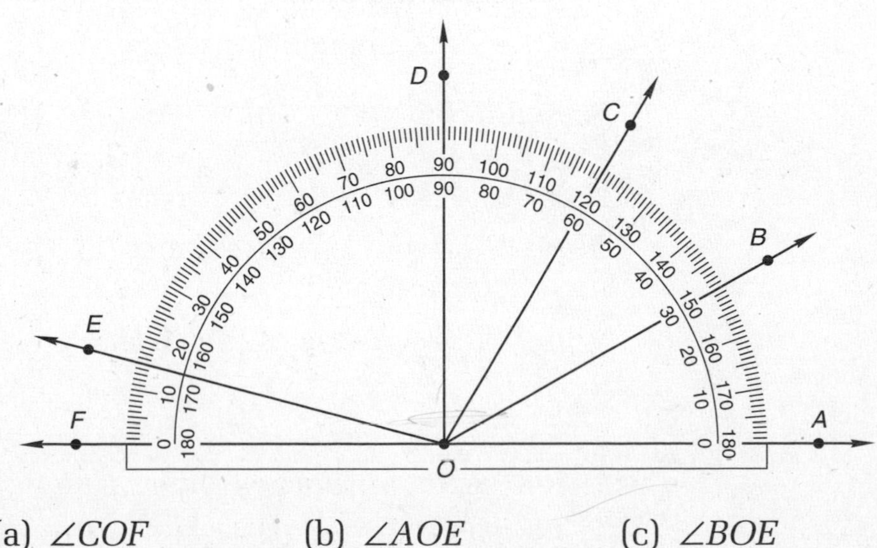

(a) ∠COF (b) ∠AOE (c) ∠BOE

21. Find the missing numbers in the table by using the function rule.
(85)

$y = 3x + 1$

x	y
4	☐
7	☐
0	☐

22. Multiply. Write each product in scientific notation.
(83)

(a) $(1.2 \times 10^5)(1.2 \times 10^{-8})$

(b) $(6 \times 10^{-3})(7 \times 10^{-4})$

For problems 23 and 24, solve and check the equation. Show each step.

23. $56 = \frac{7}{8}w$
(Inv. 7)

24. $4.8 + c = 7.34$
(Inv. 7)

Simplify:

25. $\sqrt{10^2 - 6^2} - \sqrt{10^2 - 8^2}$
(52)

26.
(49)

$$\begin{array}{r} 5 \text{ lb } 9 \text{ oz} \\ + \ 4 \text{ lb } 7 \text{ oz} \\ \hline \end{array}$$

27. $1.4 \div 3\frac{1}{2} \times 10^3$ (decimal answer)
(43, 45)

28. (a) $(-4)(-5) - (-4)(+3)$ (b) $(-2)[(-3) - (-4)(+5)]$
(85)

29. Collect like terms: $x^2 + 3xy + 2x^2 - xy$
(84)

30. The factorization of $6x^2y$ is $2 \cdot 3 \cdot x \cdot x \cdot y$. Write the factorization of $9xy^2$.
(21)

LESSON

87 Multiplying Algebraic Terms

WARM-UP

Facts Practice: Area (Test R)

Mental Math:

a. $(-60) - (-30)$ **b.** $(2 \times 10^5)(4 \times 10^{-3})$

c. $\frac{f}{100} = \frac{10}{25}$ **d.** 3.14×20 ft

e. 750 g to kg **f.** \$50 is $\frac{2}{5}$ of m.

g. $33\frac{1}{3}\%$ of 12, $\times$ 9, $\sqrt{\ }$, $\times$ 8, + 1, $\sqrt{\ }$, $\times$ 3, − 1, ÷ 2, ÷ 2, ÷ 2

Problem Solving:

On a balanced scale, four identical blocks marked M, a 250-g mass, and a 1000-g mass are distributed as shown. Find the mass of each block marked M. Write an equation illustrated by this balanced scale.

NEW CONCEPT

Recall from Lesson 84 that like terms can be added and that adding like terms does not change the variable part of the term.

$$3x + 2x = 5x$$

When we multiply terms, all the factors in the multiplied terms appear in the product.

$$(3x)(2x) = 3 \cdot 2 \cdot x \cdot x$$
$$= 6x^2$$

We multiply the numerical parts of the terms and gather variable factors with exponents. Terms can be multiplied even if they are not like terms.

$$(-2x)(-3y) = 6xy$$

Example 1 Simplify: $(-3x^2y)(2x)(-4xy)$

Solution The minus signs on these terms indicate negative numbers, not subtraction. We will multiply the three terms to make one term. First we list all the factors:

$$(-3) \cdot x \cdot x \cdot y \cdot (+2) \cdot x \cdot (-4) \cdot x \cdot y$$

Using the commutative property, we rearrange the factors as shown below.

$$(-3)(+2)(-4) \cdot x \cdot x \cdot x \cdot x \cdot y \cdot y$$

Using the associative property, we then group the factors by multiplying the numerical factors and gathering the variable factors with exponents.

$$\mathbf{24x^4y^2}$$

Example 2 Simplify: $(-2ab)(a^2b)(3b^3)$

Solution The numerical factors are –2, 1, and 3, and their product is –6. The product of a and a^2 is a^3. The product of b, b, and b^3 is b^5.

$$(-2ab)(a^2b)(3b^3) = \mathbf{-6a^3b^5}$$

LESSON PRACTICE

Practice set Find the following products:

a. $(-3x)(-2xy)$

b. $3x^2(xy^3)$

c. $(2a^2b)(-3ab^2)$

d. $(-5x^2y)(-4x)$

e. $(-xy^2)(xy)(2y)$

f. $(-3m)(-2mn)(m^2n)$

g. $(4wy)(3wx)(-w^2)(x^2y)$

h. $5d(-2df)(-3d^2fg)$

MIXED PRACTICE

Problem set

1. (53) How far will a jet travel in 2 hours 30 minutes if its average speed is 450 miles per hour?

2. (50) Use a unit multiplier to convert 12.5 centimeters to meters.

3. (54) Use a ratio box to solve this problem. If 240 of the 420 children in the audience were girls, what was the ratio of boys to girls in the audience?

4. (28) Geoff and his two brothers are very tall. Geoff's height is 18'3". The heights of his two brothers are 17'10" and 17'11". What is the average height of Geoff and his brother giraffes?

5. (46) The Martins' car traveled 468 miles on 18 gallons of gas. Their car averaged how many miles per gallon?

6. (86) On a number line, graph the whole numbers that are less than or equal to 5.

7. (72) Use a ratio box to solve this problem. The road was steep. Every 100 yards the elevation increased 36 feet. How many feet did the elevation increase in 1500 yards?

8. (61) The quadrilateral in this figure is a parallelogram. Find the measure of

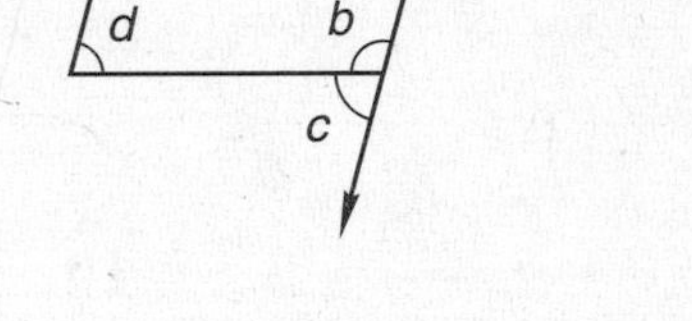

(a) $\angle a$. (b) $\angle b$.

(c) $\angle c$. (d) $\angle d$.

(e) $\angle e$.

9. (41) If $x = -4$ and $y = 3x - 1$, then y equals what number?

10. (66, 82) (a) Find the circumference of the circle at right.

(b) Find the area of the circle.

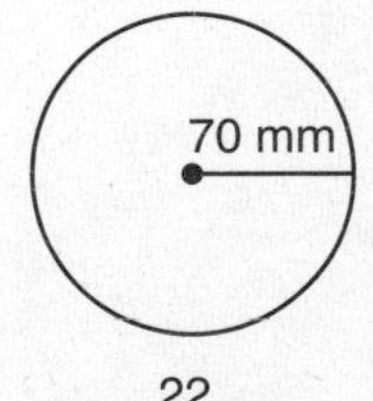

Use $\frac{22}{7}$ for π.

11. (Inv. 3, 61) The coordinates of the vertices of parallelogram $ABCD$ are A (5, 5), B (10, 5), C (5, 0), and D (0, 0).

(a) Find the area of the parallelogram.

(b) Find the measure of each angle of the parallelogram.

12. (70) The shape shown was built of 1-inch cubes. What is the volume of the shape?

13. (48) Complete the table.

FRACTION	DECIMAL	PERCENT
(a)	(b)	$12\frac{1}{2}\%$
$\frac{7}{8}$	(c)	(d)

14. (60) Write an equation to solve this problem:

What number is 25 percent of 4?

Use ratio boxes to solve problems 15 and 16.

15. (81) The sale price of $24 was 80 percent of the regular price. What was the regular price?

16. (81) David had finished 60 percent of the race, but he still had 2000 meters to run. How long was his race?

17. (77) Write an equation to solve this problem:

One hundred is what percent of 80?

18. (75) Find the area of this figure. Dimensions are in centimeters. Corners that look square are square.

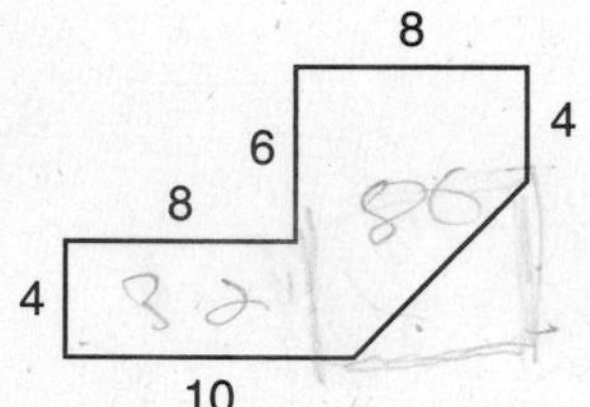

19. (40) In the figure below, angle AOE is a straight angle and $m\angle AOB = m\angle BOC = m\angle COD$.

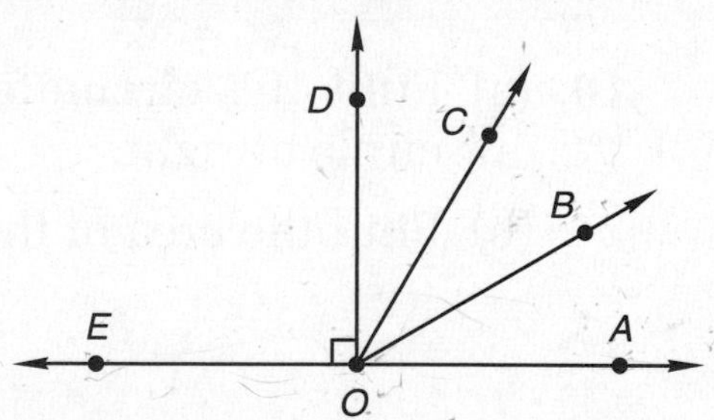

(a) Find $m\angle AOB$.

(b) Find $m\angle AOC$.

(c) Find $m\angle EOC$.

(d) Which angle is the supplement of $\angle EOC$?

20. (76) Simplify: $\dfrac{66\frac{2}{3}}{100}$

21. (85) Find the missing numbers in the table using the function rule.

$y = \dfrac{24}{x}$

x	y
3	
4	
12	

22. Multiply. Write each product in scientific notation.
(83)
(a) $(4 \times 10^{-5})(2.1 \times 10^{-7})$

(b) $(4 \times 10^{5})(6 \times 10^{7})$

For problems 23 and 24, solve and check the equation. Show each step.

23. $d - 8.47 = 9.1$
(Inv. 7)

24. $0.25m = 3.6$
(Inv. 7)

Simplify:

25. $\dfrac{3 + 5.2 - 1}{4 - 3 + 2}$
(52)

26. 1 kg − 75 g
(32)

27. $3.7 + 2\frac{5}{8} + 15$ (decimal answer)
(43)

28. (a) $(-5) - (-2)[(-3) - (+4)]$
(85)

(b) $\dfrac{(-3) + (-3)(+4)}{(+3) + (-4)}$

29. (a) $(3x)(4y)$
(87)

(b) $(6m)(-4m^2n)(-mnp)$

30. Collect like terms: $3ab + a - ab - 2ab + a$
(84)

LESSON

88 Multiple Unit Multipliers • Converting Units of Area

WARM-UP

Facts Practice: Scientific Notation (Test S)

Mental Math:

a. $(-15)(+5)$

b. $(1.5 \times 10^4)(2 \times 10^5)$

c. $3t + 4 = 40$

d. $7\frac{1}{2}\% \times \$200$ (double, halve)

e. 2.54 cm to mm

f. \$1.50 is $\frac{3}{5}$ of m.

g. At 500 mph, how far will an airliner fly in $2\frac{1}{2}$ hours?

Problem Solving:

In the currency of the land a gilder is worth 6 skillings, and a skilling is worth 4 ore. Vincent offered to pay André 10 skillings and 2 ore for the job, but André wanted 2 gilders. André wanted how much more than Vincent's offer?

NEW CONCEPTS

Multiple unit multipliers

We can repeatedly multiply a number by 1 without changing the number.

$$5 \cdot 1 = 5$$

$$5 \cdot 1 \cdot 1 = 5$$

$$5 \cdot 1 \cdot 1 \cdot 1 = 5$$

Since unit multipliers are forms of 1, we can also multiply a measure by several unit multipliers without changing the measure.

$$10 \cancel{\text{yd}} \cdot \frac{3 \cancel{\text{ft}}}{1 \cancel{\text{yd}}} \cdot \frac{12 \text{ in.}}{1 \cancel{\text{ft}}} = 360 \text{ in.}$$

Ten yards is the same distance as 360 inches. We used one unit multiplier to change from yards to feet and a second unit multiplier to change from feet to inches. Of course, we could have changed from yards to inches using the single unit multiplier

$$\frac{36 \text{ in.}}{1 \text{ yd}}$$

In some cases, however, it is easier to prevent mistakes if we use more than one unit multiplier to perform conversions.

Example 1 Use two unit multipliers to convert 5 hours to seconds.

Solution We are changing units from hours to seconds.

hours ⟶ seconds

We will perform the conversion in two steps. We will change from hours to minutes with one unit multiplier and from minutes to seconds with a second unit multiplier. For each step we write a unit multiplier.

hours ⟶ minutes ⟶ seconds

$$5\,\cancel{\text{hr}} \cdot \frac{60\,\cancel{\text{min}}}{1\,\cancel{\text{hr}}} \cdot \frac{60\text{ s}}{1\,\cancel{\text{min}}} = \mathbf{18{,}000\text{ s}}$$

Converting units of area

To convert from one unit of area to another, it is helpful to use two unit multipliers.

Consider this rectangle, which has an area of 6 ft^2:

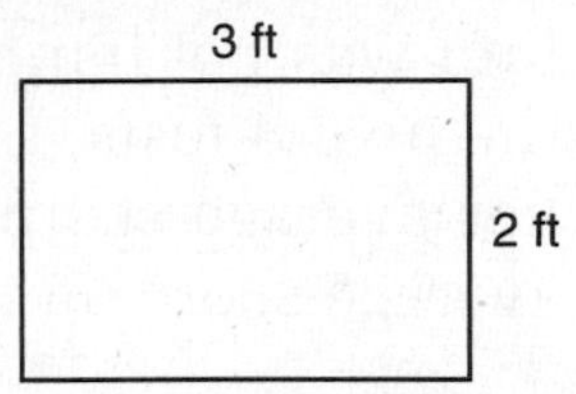

$$3\text{ ft} \cdot 2\text{ ft} = 6\text{ ft}^2$$

Recall that the expression 6 ft^2 means 6 ft · ft. Thus, to convert 6 ft^2 to square inches, we convert from

ft · ft   to   in. · in.

To perform the conversion, we use two unit multipliers.

$$6\text{ ft}^2 = 6\,\cancel{\text{ft}} \cdot \cancel{\text{ft}} \cdot \frac{12\text{ in.}}{1\,\cancel{\text{ft}}} \cdot \frac{12\text{ in.}}{1\,\cancel{\text{ft}}} = 864\text{ in.} \cdot \text{in.} = 864\text{ in.}^2$$

Example 2 Convert 5 yd^2 to square feet.

Solution Since 5 yd^2 means 5 yd · yd, we use two unit multipliers to convert to ft^2.

$$5\,\cancel{\text{yd}} \cdot \cancel{\text{yd}} \cdot \frac{3\text{ ft}}{1\,\cancel{\text{yd}}} \cdot \frac{3\text{ ft}}{1\,\cancel{\text{yd}}} = 45\text{ ft} \cdot \text{ft} = \mathbf{45\text{ ft}^2}$$

Example 3 Convert 1.2 m^2 to square centimeters.

Solution This time, instead of writing 1.2 m^2 as 1.2 m · m, we will simply keep it in mind.

$$1.2\,\cancel{\text{m}^2} \cdot \frac{100\text{ cm}}{1\,\cancel{\text{m}}} \cdot \frac{100\text{ cm}}{1\,\cancel{\text{m}}} = \mathbf{12{,}000\text{ cm}^2}$$

LESSON PRACTICE

Practice set Use two unit multipliers to perform each conversion:

a. 5 yards to inches

b. $1\frac{1}{2}$ hours to seconds

c. 15 yd^2 to square feet

d. 20 cm^2 to square millimeters

MIXED PRACTICE

Problem set

1. (53) Jackson earns $6 per hour at a part-time job. How much does he earn working 3 hours 15 minutes?

2. (55) Lim was not happy with her test average after 6 tests. On the next 4 tests, Lim's average score was 93, which raised her average score for all 10 tests to 84. What was Lim's average score on the first 6 tests?

3. (88) Use two unit multipliers to convert 6 ft^2 to square inches.

4. (54) Use a ratio box to solve this problem. The ratio of woodwinds to brass instruments in the orchestra was 3 to 2. If there were 15 woodwinds, how many brass instruments were there?

5. (86) On a number line, graph the counting numbers that are less than 4.

6. (72) Use a ratio box to solve this problem. Artichokes were on sale 8 for $2. At that price, how much would 3 dozen artichokes cost?

7. (71) Diagram this statement. Then answer the questions that follow.

When Sandra walked through the house, she saw that 18 lights were on and only $\frac{1}{3}$ of the lights were off.

(a) How many lights were off?

(b) What percent of the lights were on?

8. (63) Evaluate: $a - [b - (a - b)]$ if $a = 5$ and $b = 3$

9. (79) Compare: $x \bigcirc y$ if x and y are negative and $\frac{x}{y} = 2$

10. (66, 82) A horse is tied to a stake by a 30-foot rope so that the horse can move about in a circle. Use 3.14 for π to find the

(a) circumference of the circle.

(b) area of the circle.

11. (Inv. 1) What percent of this circle is shaded?

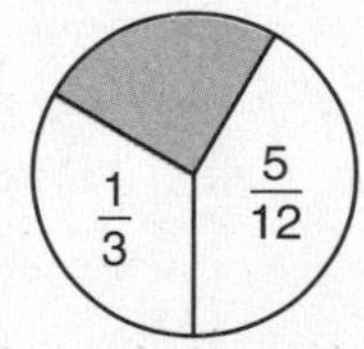

12. (67, 70) Draw a cube with edges 3 cm long.

(a) What is the volume of the cube?

(b) What is the surface area of the cube?

Collect like terms:

13. (84) $2x + 3y - 5 + x - y - 1$

14. (84) $x^2 + 2x - x - 2$

15. (48) Complete the table.

FRACTION	DECIMAL	PERCENT
(a)	0.125	(b)
$\frac{3}{8}$	(c)	(d)

16. (76) Simplify: $\dfrac{60}{1\frac{1}{4}}$

Use ratio boxes to solve problems 17 and 18.

17. (81) The regular price was \$24. The sale price was \$18. The sale price was what percent of the regular price?

18. (81) The auditorium seated 375, but this was enough for only 30 percent of those who wanted a seat. How many wanted a seat but could not get one?

19. (77) Write an equation to solve this problem:

Twenty-four is 25 percent of what number?

20. (Inv. 6, 75) (a) Classify this quadrilateral.

(b) Find its perimeter.

(c) Find its area.

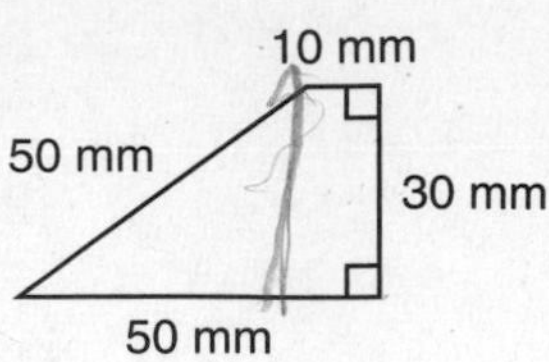

21. (85) Find the missing numbers in the table by using the function rule.

$y = x - 5$

x	y
10	☐
7	☐
5	☐

22. (83) Multiply. Write each product in scientific notation.

(a) $(9 \times 10^{-6})(4 \times 10^{-8})$

(b) $(9 \times 10^{6})(4 \times 10^{8})$

For problems 23 and 24, solve and check the equation. Show each step.

23. (Inv. 7) $8\frac{5}{6} = d - 5\frac{1}{2}$

24. (Inv. 7) $\frac{5}{6}m = 90$

25. (80) Three vertices of rectangle *JKLM* are *J* (−4, 2), *K* (0, 2), and *L* (0, 0).

(a) Find the coordinates of *M* and draw the rectangle.

(b) Translate □*JKLM* 4 units right, 2 down. Draw the translated image □*J′K′L′M′*, and write the coordinates of its vertices.

26. (20, 52) Which of the following does not equal 4^3?

A. 2^6 B. $4 \cdot 4^2$ C. $\frac{4^4}{4}$ D. $4^2 + 4$

27. (43) Find 50% of $\frac{2}{3}$ of 0.12. Write the answer as a decimal.

Simplify:

28. (63) $6\{5 \cdot 4 - 3[6 - (3 - 1)]\}$

29. (85) (a) $\frac{(-3)(-4) - (-3)}{(-3) - (+4)(+3)}$

(b) $(+5) + (-2)[(+3) - (-4)]$

30. (87) (a) $(-2x)(-3x)$

(b) $(ab)(2a^2b)(-3a)$

LESSON

89 Diagonals • Interior Angles • Exterior Angles

WARM-UP

Facts Practice: Area (Test R)

Mental Math:

a. $(-80) \div (-4)$

b. $(2.5 \times 10^{-4})(3 \times 10^{8})$

c. $\frac{8}{g} = \frac{2}{2.5}$

d. Estimate $7\frac{3}{4}\%$ of \$8.29.

e. 1.87 m to cm

f. \$1.00 is $\frac{2}{5}$ of m.

g. 10% of 80, × 3, + 1, $\sqrt{\ }$, × 7, + 1, ÷ 2, ÷ 2, $\sqrt{\ }$, × 10, + 2, ÷ 4

Problem Solving:

Here are the front, top, and side views of an object. Construct this object using 1-inch cubes, or sketch a three-dimensional view. Then calculate the object's volume.

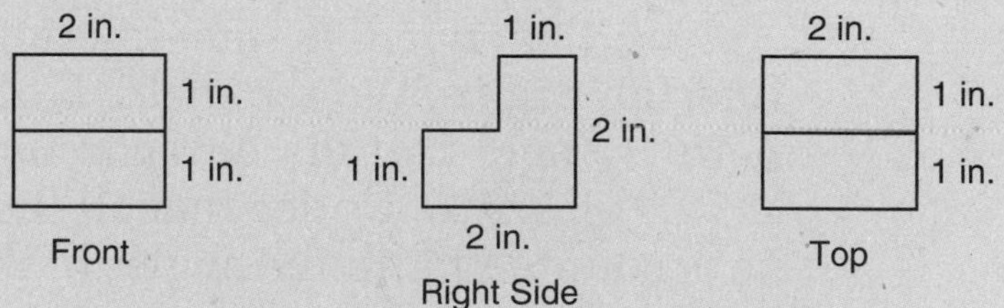

NEW CONCEPTS

Diagonals

Recall that a **diagonal** of a polygon is a line segment that passes through the polygon between two nonadjacent vertices. In the figure below, segment *AC* is a diagonal of quadrilateral *ABCD*.

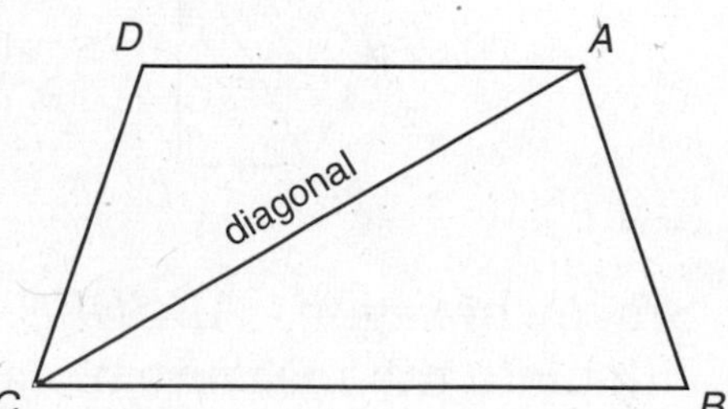

Example 1 From one vertex of regular hexagon *ABCDEF*, how many diagonals can be drawn? (Trace the hexagon and illustrate your answer.)

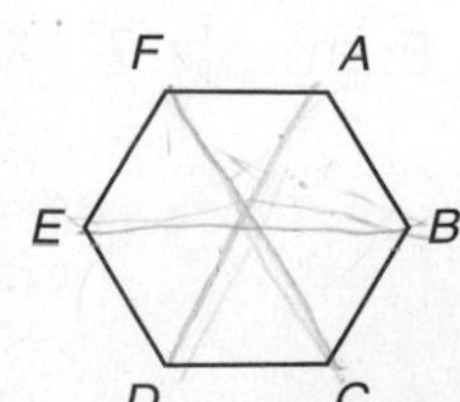

Solution We can select any vertex from which to draw the diagonals. We choose vertex A. Segments AB and AF are sides of the hexagon and are not diagonals. Segments drawn from A to C, D, and E are diagonals. So **3 diagonals** can be drawn.

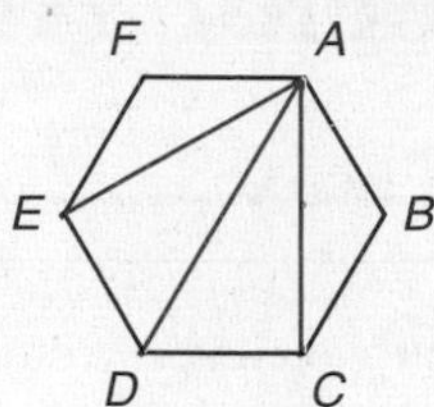

Interior angles Notice in example 1 that the three diagonals from vertex A divide the hexagon into four triangles. We will draw arcs to emphasize each angle of the four triangles.

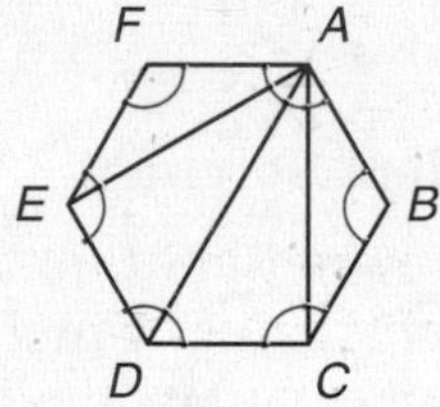

Angles that open to the interior of a polygon are called **interior angles.** We see that $\angle B$ of the hexagon is also $\angle B$ of $\triangle ABC$. Angle C of the hexagon includes $\angle BCA$ of $\triangle ABC$ and $\angle ACD$ of $\triangle ACD$. Notice that all the angles of the triangles can be accounted for in the angles of the hexagon.

Although we may not know the measure of each angle of each triangle, we nevertheless can conclude that the measures of the six angles of a hexagon have the same total as the measures of the angles of four triangles, which is $4 \times 180° = 720°$.

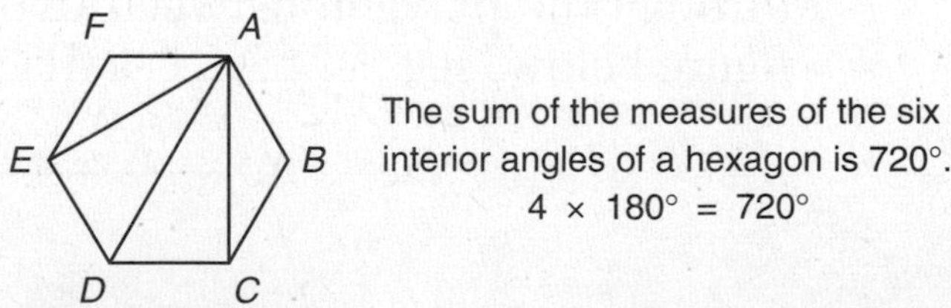

Since hexagon $ABCDEF$ is a regular hexagon, we can calculate the measure of each angle of the hexagon.

Example 2 Maura inscribed a regular hexagon in a circle. Find the measure of each angle of the regular hexagon $ABCDEF$.

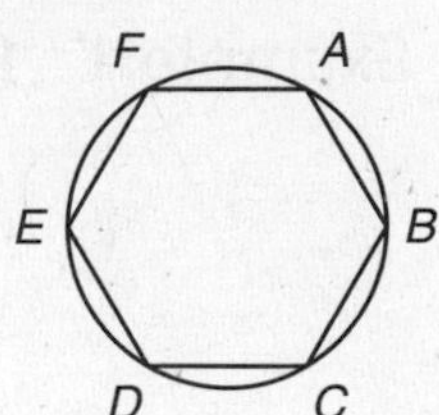

Solution From the explanation above we know that the hexagon can be divided into four triangles. So the sum of the measures of the angles of the hexagon is 4 × 180°, which is 720°. Since the hexagon is regular, the six angles equally share the available 720°. So we divide 720° by 6 to find the measure of each angle.

$$720° \div 6 = 120°$$

We find that each angle of the hexagon measures **120°**.

Example 3 Draw a quadrilateral and one of its diagonals. What is the sum of the measures of the interior angles of the quadrilateral?

Solution We draw a four-sided polygon and a diagonal.

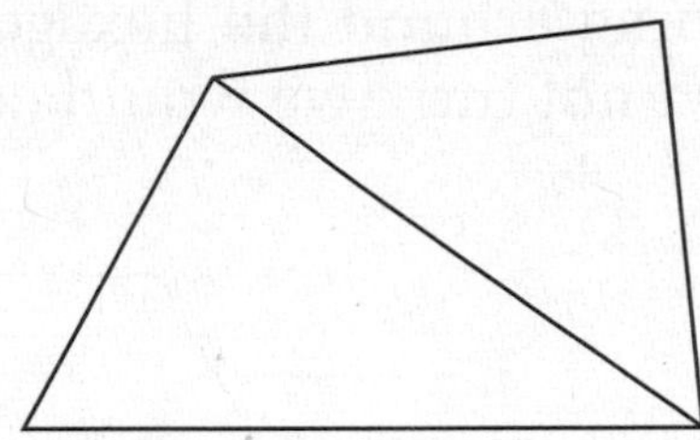

Although we do not know the measure of each angle, we can find the sum of their measures. The sum of the measures of the angles of a triangle is 180°. From the drawing above, we see that the total measure of the angles of the quadrilateral equals the total measure of the angles of two triangles. So the sum of the measures of the interior angles of the quadrilateral is

$$2 \times 180° = \mathbf{360°}$$

Exterior angles In example 2 we found that each interior angle of a regular hexagon measures 120°. By performing the following activity, we can get another perspective on the angles of a polygon.

Activity: *Exterior Angles*

Materials needed:

- A length of string (5 feet or more)
- Chalk
- Masking tape (optional)

If a paved area such as a patio or driveway is available, lay out a regular hexagon on the paved area. This can be done by inscribing a hexagon inside a circle as described in

Investigation 2. Use the string and chalk to sweep out the circle and to mark the vertices and sides of the hexagon. If desired, mark the sides of the hexagon with masking tape.

After the hexagon has been prepared, walk the perimeter of the hexagon while making these observations:

1. The direction you were facing when you started around the hexagon as well as when you finished going around the hexagon after six turns.
2. How much you turned at each "corner" of the hexagon. Did you turn more than, less than, or the same as you would have turned at the corner of a square?

Going around the hexagon, we turned at every corner. If we did not turn, we would continue going straight.

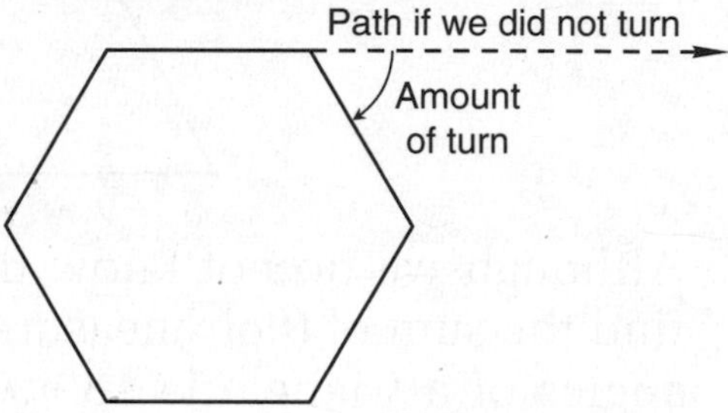

The amount we turned at the corner in order to stay on the hexagon equals the measure of the **exterior angle** of the hexagon at that vertex.

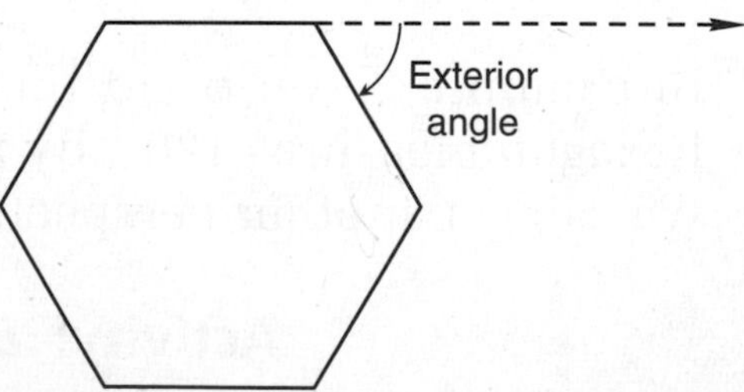

We can calculate the measure of each exterior angle of a regular hexagon by remembering how many turns were required in order to face the same direction as when we started. We remember that we made six small turns. In other words, after six turns we had completed one full turn of 360°.

If all the turns are in the same direction, the sum of the exterior angles of any polygon is 360°.

Example 4 What is the measure of each exterior angle of a regular hexagon?

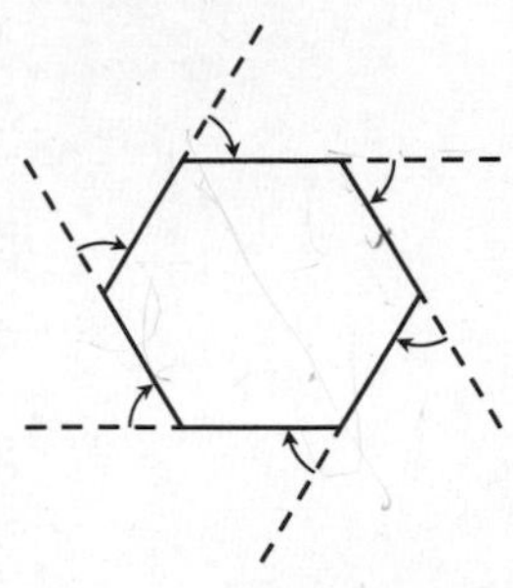

Solution Traveling all the way around the hexagon completes one full turn of 360°. Each exterior angle of a regular hexagon has the same measure, so we can find the measure by dividing 360° by 6.

$$360° \div 6 = 60°$$

We find that each exterior angle of a regular hexagon measures **60°**.

Notice that an interior angle of a polygon and its exterior angle are supplementary, so their combined measures total 180°.

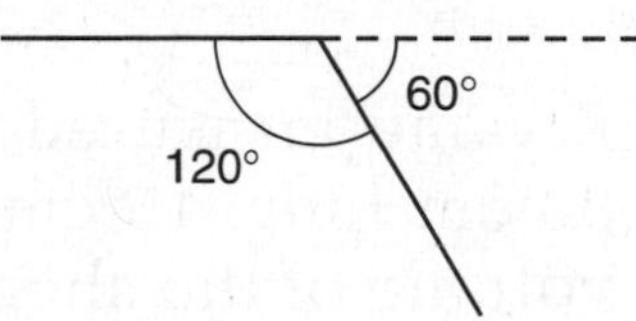

LESSON PRACTICE

Practice set

a. Work examples 1–4 from this lesson if you have not already.

b. Trace this regular pentagon. How many diagonals can be drawn from one vertex? Show your work.

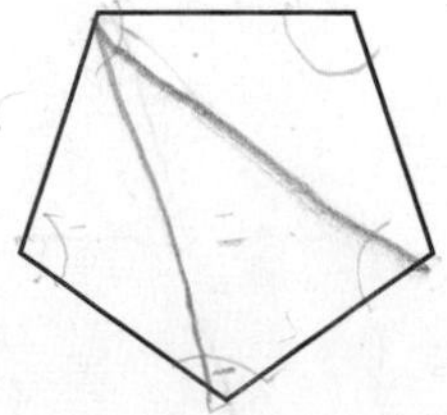

c. The diagonals drawn in problem **b** divide the pentagon into how many triangles?

d. What is the sum of the measures of the five interior angles of a pentagon?

e. What is the measure of each interior angle of a regular pentagon?

f. What is the measure of each exterior angle of a regular pentagon?

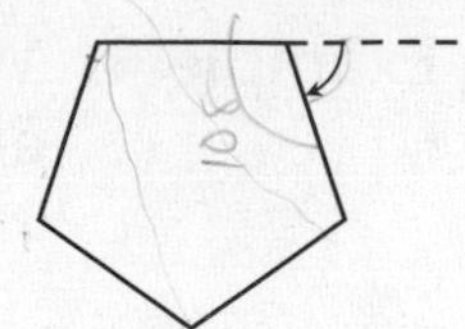

g. What is the sum of the measures of an interior and exterior angle of a regular pentagon?

MIXED PRACTICE

Problem set

1. Use a ratio box to solve this problem. Jason's remote-control car traveled 440 feet in 10 seconds. At that rate, how long would it take the car to travel a mile?
(72)

2. In the forest there were lions, tigers, and bears. The ratio of lions to tigers was 3 to 2. The ratio of tigers to bears was 3 to 4. If there were 18 lions, how many bears were there? Use a ratio box to find how many tigers there were. Then use another ratio box to find the number of bears.
(54)

3. Kwame measured the shoe box and found that it was 30 cm long, 15 cm wide, and 12 cm tall. What was the volume of the shoe box?
(70)

4. A baseball player's batting average is a ratio found by dividing the number of hits by the number of at-bats and writing the result as a decimal number rounded to the nearest thousandth. If Erika had 24 hits in 61 at-bats, what was her batting average?
(44)

5. Use two unit multipliers to convert 18 square feet to square yards.
(88)

6. On a number line, graph the integers greater than –4.
(86)

7. Diagram this statement. Then answer the questions that follow.
(71)

Jimmy bought the shirt for \$12. This was $\frac{3}{4}$ of the regular price.

(a) What was the regular price of the shirt?

(b) Jimmy bought the shirt for what percent of the regular price?

8. Use the figure below to find the measure of each angle.
(40)

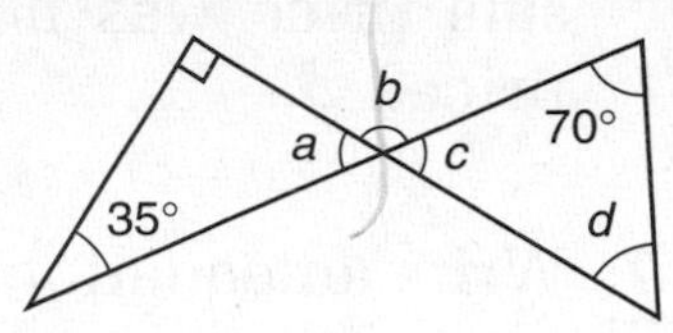

(a) $\angle a$ (b) $\angle b$ (c) $\angle c$ (d) $\angle d$

9. (a) What is the circumference of this circle?.
(66, 82)

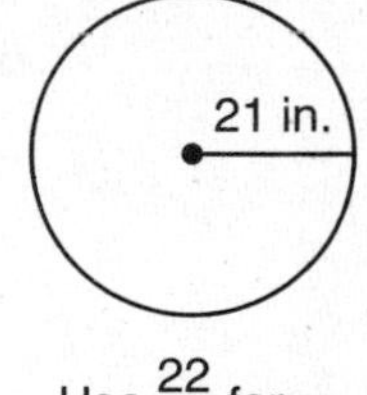

(b) What is the area of this circle?

Use $\frac{22}{7}$ for π.

10. Simplify: $\dfrac{91\frac{2}{3}}{100}$
(76)

11. Evaluate: $\dfrac{ab + a}{a + b}$ if $a = 10$ and $b = 5$
(52)

12. Compare: $a^2 \bigcirc a$ if $a = 0.5$
(79)

13. Complete the table.
(48)

FRACTION	DECIMAL	PERCENT
$\frac{7}{8}$	(a)	(b)
(c)	(d)	875%

14. At three o'clock and at nine o'clock, the hands of a clock form angles equal to $\frac{1}{4}$ of a circle.
(17)

(a) At which two hours do the hands of a clock form angles equal to $\frac{1}{3}$ of a circle?

(b) The angle described in part (a) measures how many degrees?

Use ratio boxes to solve problems 15 and 16.

15. Forty-five percent of the 3000 fast-food customers ordered a hamburger. How many of the customers ordered a hamburger?
(81)

16. (81) The sale price of \$24 was 75% of the regular price. The sale price was how many dollars less than the regular price?

17. (77) Write an equation to solve this problem:

Twenty is what percent of 200?

18. (58, 75) (a) Trace this isosceles trapezoid and draw its line of symmetry.

(b) Find the area of the trapezoid.

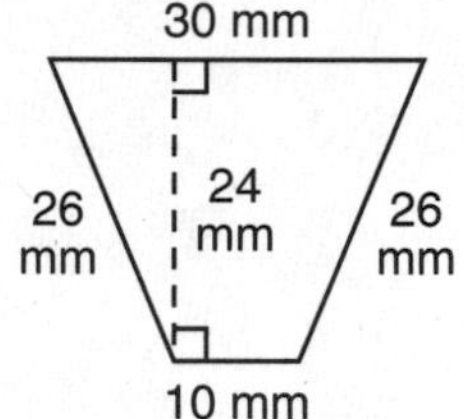

19. (89) What is the measure of each exterior angle of a regular triangle?

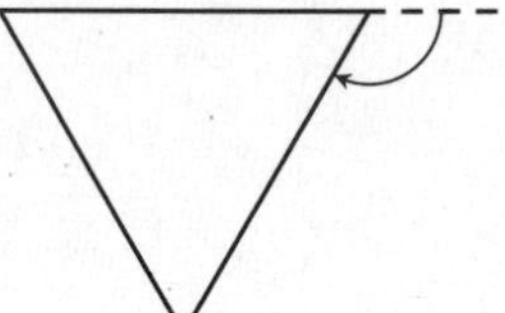

20. (85) Find the missing numbers in the table by using the function rule.

$y = \frac{1}{3}x$

x	y
12	☐
9	☐
☐	6

21. (83) Multiply. Write the product in scientific notation.

$$(1.25 \times 10^{-3})(8 \times 10^{-5})$$

22. (62) The lengths of two sides of an isosceles triangle are 4 cm and 10 cm.

(a) Draw the triangle and find its perimeter.

(b) Can there be more than one answer? Why or why not?

For problems 23 and 24, solve and check the equation. Show each step.

23. (Inv. 7) $\frac{4}{9}p = 72$

24. (Inv. 7) $12.3 = 4.56 + f$

25. Collect like terms: $2x + 3y - 4 + x - 3y - 1$
(84)

Simplify:

26. $\dfrac{9 \cdot 8 - 7 \cdot 6}{6 \cdot 5}$
(63)

27. $3.2 \times 4^{-2} \times 10^2$
(43, 57)

28. $13\frac{1}{3} - \left(4.75 + \frac{3}{4}\right)$ (mixed-number answer)
(43)

29. (a) $\dfrac{(+3) + (-4)(-6)}{(-3) + (-4) - (-6)}$
(85)

(b) $(-5) - (+6)(-2) + (-2)(-3)(-1)$

30. (a) $(3x^2)(2x)$
(87)

(b) $(-2ab)(-3b^2)(-a)$

LESSON

90 Mixed-Number Coefficients • Negative Coefficients

WARM-UP

Facts Practice: Percent-Decimal-Fraction Equivalents (Test Q)

Mental Math:

a. (–50) – (–30)

b. $(4.2 \times 10^{-6})(2 \times 10^{-4})$

c. $4w - 8 = 36$

d. Estimate 15% of $23.89.

e. 800 g to kg

f. $1.00 is $\frac{4}{5}$ of m.

g. A cube with edges 10 inches long has a volume of how many cubic inches?

Problem Solving:

If four people shake hands with one another, we can picture the number of handshakes by drawing four dots (for people) and connecting the dots with segments (for handshakes). Then we count the segments (six). Use this method to count the number of handshakes if five people shake hands with one another.

NEW CONCEPTS

Mixed-number coefficients

We have been solving equations such as

$$\frac{4}{5}x = 7$$

by multiplying both sides of the equation by the reciprocal of the coefficient of x. Here the coefficient of x is $\frac{4}{5}$, so we multiply both sides by the reciprocal of $\frac{4}{5}$, which is $\frac{5}{4}$.

$$\frac{\overset{1}{\cancel{5}}}{\underset{1}{\cancel{4}}} \cdot \frac{\overset{1}{\cancel{4}}}{\underset{1}{\cancel{5}}}x = \frac{5}{4} \cdot 7 \qquad \text{multiplied by } \frac{5}{4}$$

$$x = \frac{35}{4} \qquad \text{simplified}$$

When solving an equation that has a mixed-number coefficient, we convert the mixed number to an improper fraction as the first step. Then we multiply both sides by the reciprocal of the improper fraction.

Example 1 Solve: $3\frac{1}{3}x = 5$

Solution First we write $3\frac{1}{3}$ as an improper fraction.

$$\frac{10}{3}x = 5 \qquad \text{fraction form}$$

Then we multiply both sides of the equation by $\frac{3}{10}$, which is the reciprocal of $\frac{10}{3}$.

$$\frac{\overset{1}{\cancel{3}}}{\underset{1}{\cancel{10}}} \cdot \frac{\overset{1}{\cancel{10}}}{\underset{1}{\cancel{3}}}x = \frac{3}{\underset{2}{\cancel{10}}} \cdot \overset{1}{\cancel{5}} \qquad \text{multiplied by } \frac{3}{10}$$

$$x = \mathbf{\frac{3}{2}} \qquad \text{simplified}$$

In arithmetic we usually convert an improper fraction such as $\frac{3}{2}$ to a mixed number. Recall that in algebra we usually leave improper fractions in fraction form.

Example 2 Solve: $2\frac{1}{2}y = 1\frac{7}{8}$

Solution Since we will be multiplying both sides of the equation by a fraction, we first convert both mixed numbers to improper fractions.

$$\frac{5}{2}y = \frac{15}{8} \qquad \text{fraction form}$$

Then we multiply both sides by $\frac{2}{5}$, which is the reciprocal of $\frac{5}{2}$.

$$\frac{\overset{1}{\cancel{2}}}{\underset{1}{\cancel{5}}} \cdot \frac{\overset{1}{\cancel{5}}}{\underset{1}{\cancel{2}}}y = \frac{\overset{1}{\cancel{2}}}{\underset{1}{\cancel{5}}} \cdot \frac{\overset{3}{\cancel{15}}}{\underset{4}{\cancel{8}}} \qquad \text{multiplied by } \frac{2}{5}$$

$$y = \mathbf{\frac{3}{4}} \qquad \text{simplified}$$

Negative coefficients To solve an equation with a negative coefficient, we multiply (or divide) both sides of the equation by a negative number. The coefficient of x in this equation is negative.

$$-3x = 126$$

To solve this equation, we can either divide both sides by –3 or multiply both sides by $-\frac{1}{3}$. The effect of either method is to make +1 the coefficient of x. We show both ways.

$$-3x = 126 \qquad\qquad -3x = 126$$

$$\frac{-3x}{-3} = \frac{126}{-3} \qquad\qquad \left(-\frac{1}{3}\right)(-3x) = \left(-\frac{1}{3}\right)(126)$$

$$x = -42 \qquad\qquad x = -42$$

Example 3 Solve: $-\frac{2}{3}x = \frac{4}{5}$

Solution We multiply both sides of the equation by the reciprocal of $-\frac{2}{3}$, which is $-\frac{3}{2}$.

$$-\frac{2}{3}x = \frac{4}{5} \qquad \text{equation}$$

$$\left(-\frac{3}{2}\right)\left(-\frac{2}{3}x\right) = \left(-\frac{3}{2}\right)\left(\frac{4}{5}\right) \qquad \text{multiplied by } -\tfrac{3}{2}$$

$$x = \mathbf{-\frac{6}{5}} \qquad \text{simplified}$$

Example 4 Solve: $-5x = 0.24$

Solution We may either multiply both sides by $-\frac{1}{5}$ or divide both sides by –5. Since the right side of the equation is a decimal number, it appears that dividing by –5 will be easier.

$$-5x = 0.24 \qquad \text{equation}$$

$$\frac{-5x}{-5} = \frac{0.24}{-5} \qquad \text{divided by } -5$$

$$x = \mathbf{-0.048} \qquad \text{simplified}$$

LESSON PRACTICE

Practice set Solve:

a. $1\frac{1}{8}x = 36$

b. $3\frac{1}{2}a = 490$

c. $2\frac{3}{4}w = 6\frac{3}{5}$

d. $2\frac{2}{3}y = 1\frac{4}{5}$

e. $-3x = 0.45$

f. $-\frac{3}{4}m = \frac{2}{3}$

g. $-10y = -1.6$

h. $-2\frac{1}{2}w = 3\frac{1}{3}$

MIXED PRACTICE

Problem set

1. (31, 35) The sum of 0.8 and 0.9 is how much greater than the product of 0.8 and 0.9? Use words to write the answer.

2. (Inv. 4) For this set of scores, find the (a) mean, (b) median, (c) mode, and (d) range:

8, 6, 9, 10, 8, 7, 9, 10, 8, 10, 9, 8

3. (46) The 24-ounce container is priced at \$1.20. This container costs how much more per ounce than the 32-ounce container priced at \$1.44?

4. (89) The figure at right is a regular decagon. One of the exterior angles is labeled a, and one of the interior angles is labeled b.

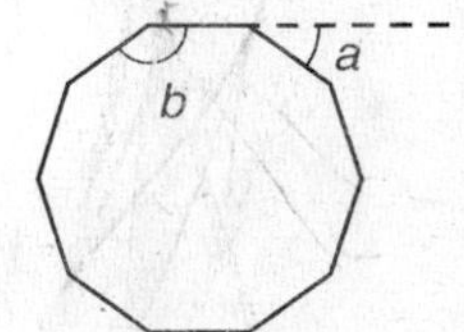

(a) What is the measure of each exterior angle of the decagon?

(b) What is the measure of each interior angle?

5. (84) Collect like terms: $x^2 + 2xy + y^2 + x^2 - y^2$

Use ratio boxes to solve problems 6 and 7.

6. (81) The sale price of \$36 was 90 percent of the regular price. What was the regular price?

7. (81) Seventy-five percent of the citizens voted for Graham. If there were 800 citizens, how many of them did not vote for Graham?

8. (77) Write equations to solve (a) and (b).

(a) Twenty-four is what percent of 30?

(b) Thirty is what percent of 24?

9. (88) Use two unit multipliers to convert 2 ft^2 to square inches.

10. (71) Diagram this statement. Then answer the questions that follow.

Three hundred doctors recommended Brand X. This was $\frac{2}{5}$ of the doctors surveyed.

(a) How many doctors were surveyed?

(b) How many doctors surveyed did not recommend Brand X?

11. (41) If $x = 4.5$ and $y = 2x + 1$, then y equals what number?

12. (79) Compare: $a \bigcirc ab$ if $a < 0$ and $b > 1$

13. (20) If the perimeter of a square is 1 foot, what is the area of the square in square inches?

14. (48) Complete the table.

FRACTION	DECIMAL	PERCENT
(a)	1.75	(b)

15. (46) If the sales-tax rate is 6%, what is the total price of a $325 printer including sales tax?

16. (83) Multiply. Write the product in scientific notation.

$$(6 \times 10^4)(8 \times 10^{-7})$$

17. (67, 70) A cereal box 8 inches long, 3 inches wide, and 12 inches tall is shown.

(a) What is the volume of the box?

(b) What is the surface area of the box?

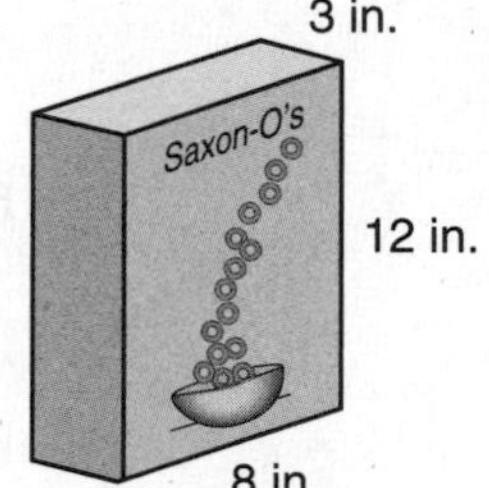

18. (66, 82) (a) Find the circumference of the circle at right.

(b) Find the area of the circle.

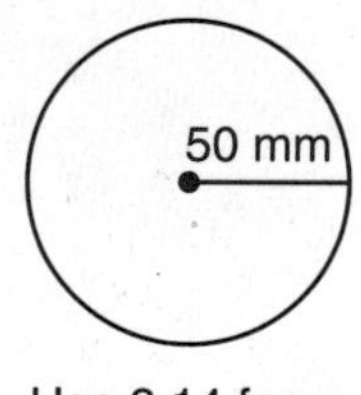

Use 3.14 for π.

19. (86) List the whole numbers that are not counting numbers.

20. (80) The coordinates of three vertices of $\square WXYZ$ are $W(0, 3)$, $X(5, 3)$, and $Y(5, 0)$.

(a) Find the coordinates of Z and draw the rectangle.

(b) Rotate $\square WXYZ$ 90° counterclockwise about the origin, and draw its image $\square W'X'Y'Z'$. Write the coordinates of the vertices.

21. (60) What mixed number is $\frac{2}{3}$ of 20?

22. On a number line graph $x \le 4$.
(78)

For problems 23–25, solve the equation. Show each step.

23. $x + 3.5 = 4.28$ (Inv. 7)

24. $2\frac{2}{3}w = 24$ (90)

25. $-4y = 1.4$ (90)

Simplify:

26. $10^1 + 10^0 + 10^{-1}$ (decimal answer)
(57)

27. $(-2x^2)(-3xy)(-y)$
(87)

28. $\frac{8}{75} - \frac{9}{100}$
(30)

29. (a) $(-3) + (-4)(-5) - (-6)$
(85)

(b) $\frac{(-2)(-4)}{(-4) - (-2)}$

30. Compare for $x = 10$ and $y = 5$:
(79)

$$x^2 - y^2 \bigcirc (x + y)(x - y)$$

INVESTIGATION 9

Focus on

Graphing Functions

Functions can be displayed as graphs on a coordinate plane. To graph a function, we use each (x, y) pair of numbers as coordinates of a point on the plane.

$y = x + 1$

x	y	coordinates
0	1	(0, 1)
1	2	(1, 2)
2	3	(2, 3)
3	4	(3, 4)

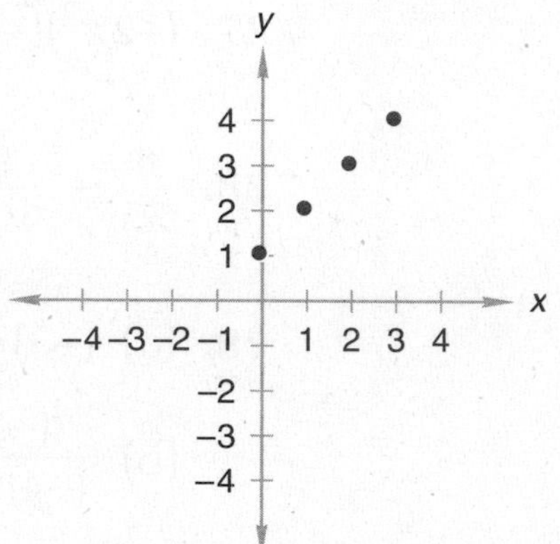

On the coordinate plane above, we graphed four pairs of numbers that satisfy the equation of the function. Although the table lists only four pairs of numbers for the function, the graph of the function includes many other pairs of numbers that satisfy the equation. By extending a line through and beyond the graphed points, we graph all possible pairs of numbers that satisfy the equation. Each point on the graphed line below represents a pair of numbers that satisfies the equation of the function $y = x + 1$.

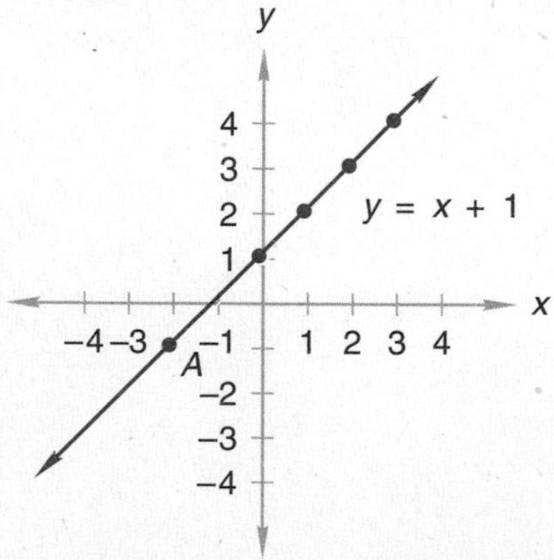

For instance, the (x, y) coordinates of point A are $(-2, -1)$. So $x = -2$ and $y = -1$ should satisfy the equation $y = x + 1$. For confirmation we substitute -2 for x and -1 for y.

$y = x + 1$	equation
$(-1) = (-2) + 1$	substituted
$-1 = -1$	simplified

1. Select another point on the graph of $y = x + 1$ and determine its coordinates. Then replace x and y in the function $y = x + 1$ with the coordinates of the selected point to determine whether the numbers satisfy the function. Show the substitution.

2. This table lists three pairs of numbers that satisfy the function $y = \frac{1}{2}x - 1$. On a coordinate plane, graph all the pairs of numbers that satisfy the function. Then answer (a) and (b).

$y = \frac{1}{2}x - 1$

x	y
0	−1
2	0
4	1

(a) Besides the three (x, y) pairs listed in the table, what is another pair of numbers that satisfies the function?

(b) Can we find coordinates with x negative and y positive that satisfy the function? Why or why not?

The relationship between the length of a side of a square and the perimeter of the square is a function that can be graphed. In the equation we use s for the length of a side and p for the perimeter. We only graph the numerical relationship of the measures, so we do not designate units. It is assumed that the unit used to measure the length of the side is also used to measure the perimeter.

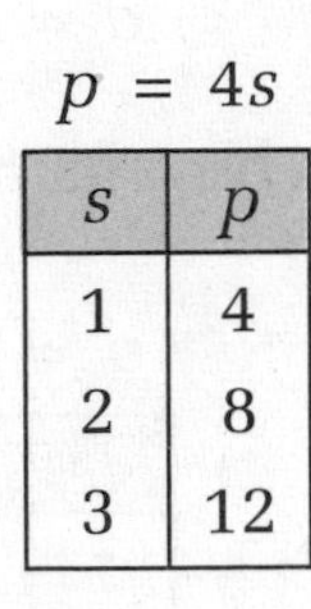

$p = 4s$

s	p
1	4
2	8
3	12

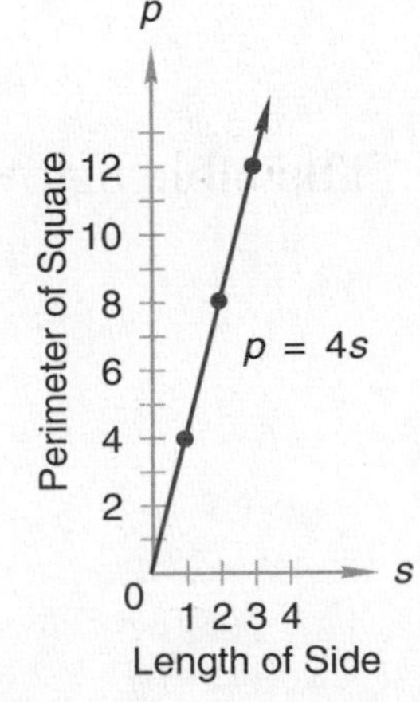

Notice that the x- and y-axes are renamed s and p for this function.

3. The graph of this function is a ray. What are the coordinates of the endpoint of the ray? Why is the graph of this function a ray and not a line?

4. Notice that the graph of the function $p = 4s$ is straight and not curved. Imagine two squares in which the side lengths of the larger square are twice the side lengths of the smaller square. Draw the two squares and find the perimeter of each. What effect does doubling the side length of a square have on the perimeter?

5. The relationship between the length of a side of an equilateral triangle and the perimeter of the triangle is a function that can be graphed. Write an equation for the function using s for the length of a side and p for the perimeter. Next make a table of (s, p) pairs for side lengths 1, 2, and 3. Then draw a rectangular coordinate system with an s-axis and a p-axis, and graph the function.

Rates are functions that can be graphed. Many rates involve time as one of the variables. Speed, for example, is a function of distance and time. Suppose Sam enters a walk-a-thon and walks at a steady rate of 3 miles per hour. The distance (d) in miles that Sam travels is a function of the number of hours (h) that Sam walks. This relationship is expressed in the following equation:

$$d = 3h$$

The table below shows how far Sam walks in 1, 2, and 3 hours.

$d = 3h$

h	d
1	3
2	6
3	9

d = distance in miles
h = time in hours

A graph of the function shows how far Sam walks in any number of hours, including fractions of hours.

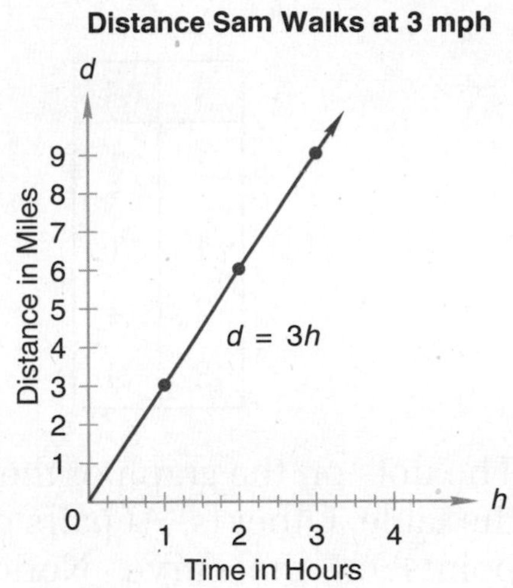

Notice that we labeled both axes of the graph. Also notice that we adjusted the scale of the graph so that each tick mark on the horizontal scale represents one fourth of an hour (15 minutes).

6. Suppose Sam entered a jog-a-thon and was able to jog at a steady pace of 6 miles per hour for 2 hours. Following the pattern described for the walk-a-thon, write an equation for a 6-mile-per-hour rate, and make a table that shows how far Sam had jogged after 1 hour and how far he had jogged after 2 hours. Then draw a graph of the function and label each axis. Let every tick mark on the time axis represent 10 minutes.

7. Refer to the graph drawn in problem 6 to find the distance Sam had jogged in 40 minutes.

8. Refer to the graph drawn in problem 6 to find how long it took Sam to jog 9 miles.

9. Why is the graph of the distance Sam jogged a segment and not a ray?

The graph of a function may be a curve. The relationship between the length of a side of a square and the area of the square is the function graphed below. We use the letters A

and s to represent the area in square units and the side length of the square respectively.

$A = s^2$

s	A
$\frac{1}{2}$	$\frac{1}{4}$
1	1
2	4
3	9

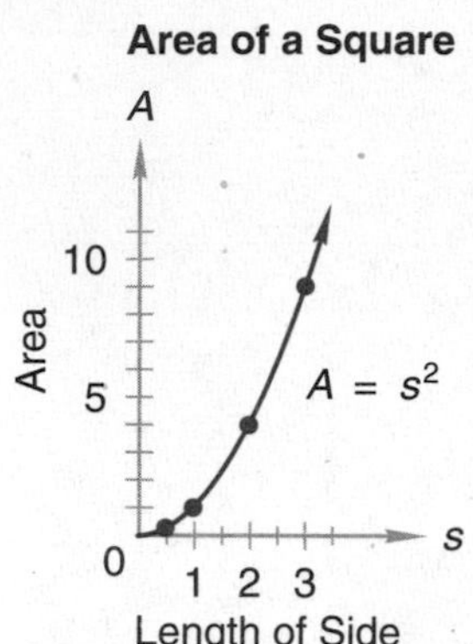

The dots on the graph of the function show the (s, A) pairs from the table. Other (s, A) pairs of numbers are represented by other points on the curve. Notice that the graph of the function becomes steeper as the side of the square becomes longer.

10. Notice that the graph of the function $A = s^2$ is a curve and is not straight. Imagine two squares in which the side lengths of the larger square are twice the side lengths of the smaller square. Draw the two squares, and find the area of each. What effect does doubling the length of a square have on the area of the square?

11. Half of the area of a square is shaded. As the square becomes larger, the area of the shaded region becomes greater, as indicated by the function below in which A represents the area of half of a square and s represents the length of a side of the square. Copy this table and find the missing numbers.

$A = \frac{1}{2}s^2$

s	A
1	$\frac{1}{2}$
2	2
3	☐
4	☐

12. Plot the coordinates from the table in problem 11 on a graph entitled "Area of Half of a Square." Then draw a smooth curve from the origin through the graphed points. Be sure to label the axes of the graph.

Extensions The relationship between the lengths of the edge of a cube and the volume of the cube is a function. Investigate this function through the following activities:

a. Write an equation using e for the length of an edge and V for the volume of a cube. Use the equation to create a table of ordered pairs.

b. Draw a horizontal e-axis and a vertical V-axis. The values of e and V diverge quickly, so space the intervals on the e-axis relatively far apart and reduce the interval length on the V-axis as shown below. Then plot the points from the table created in extension **a,** and draw a smooth curve through the points.

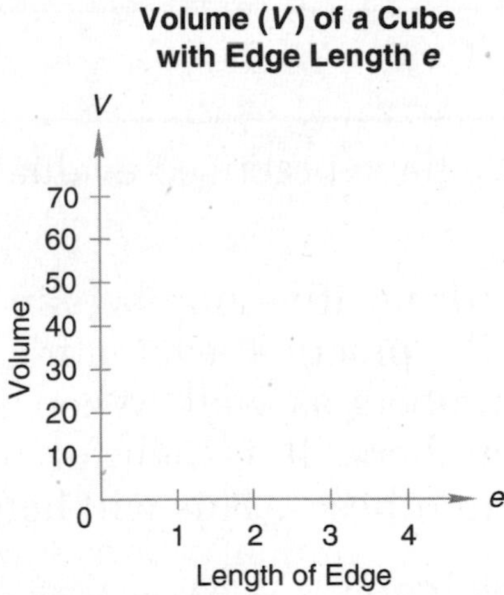

LESSON

91 Evaluations with Signed Numbers • Signed Numbers Without Parentheses

WARM-UP

Facts Practice: Scientific Notation (Test S)

Mental Math:

a. (−84) + (−50)

b. $(1.2 \times 10^3)(1.2 \times 10^3)$

c. $\frac{w}{90} = \frac{80}{120}$

d. $6 \times 2\frac{1}{2}$ (*Think:* $6 \times 2 + 6 \times \frac{1}{2}$)

e. 1.5 L to mL

f. \$20 is $\frac{1}{10}$ of *m*.

g. 50% of 40, + 1, ÷ 3, × 7, + 1, × 2, $\sqrt{\ }$, × 5, − 1, $\sqrt{\ }$, × 4, + 2, ÷ 2

Problem Solving:

Carpeting is sold by the square yard. A room that is 15 feet long and 12 feet wide requires how many square yards of carpeting to cover the floor?

NEW CONCEPTS

Evaluations with signed numbers

We have practiced evaluating expressions such as

$$x - xy - y$$

with positive numbers in place of *x* and *y*. In this lesson we will practice evaluating such expressions with negative numbers as well. When evaluating expressions with signed numbers, it is helpful to first replace each variable with parentheses. This will help prevent making mistakes in signs.

Example 1 Evaluate: $x - xy - y$ if $x = -2$ and $y = -3$

Solution We write parentheses for each variable.

$$(\ \ ) - (\ \ )(\ \ ) - (\ \ ) \qquad \text{parentheses}$$

Now we write the proper numbers within the parentheses.

$$(-2) - (-2)(-3) - (-3) \qquad \text{insert numbers}$$

By the order of operations, we multiply before adding.

$$(-2) - (+6) - (-3) \qquad \text{multiplied}$$

Then we add algebraically from left to right.

$$(-8) - (-3) \qquad \text{added } -2 \text{ and } -6$$

$$\mathbf{-5} \qquad \text{added } -8 \text{ and } +3$$

Signed numbers without parentheses Signed numbers are often written without parentheses. To simplify an expression such as

$$-3 + 4 - 5 - 2$$

we simply add algebraically from left to right.

$$-3 \quad +4 \quad -5 \quad -2 \quad = \quad -6$$

Another way to simplify this expression is to use the commutative and associative properties to rearrange and regroup the terms by their signs.

$$-3 + 4 - 5 - 2$$
$$+4 \underbrace{- 3 - 5 - 2}$$
$$+4 \quad - 10$$

Then we algebraically add the remaining terms.

$$+4 - 10 = -6$$

Using slashes to visually distinguish terms can also help in simplifying an expression, as we show in the next example.

Example 2 Simplify: $-2 + 3(-2) - 2(+4)$

Solution To emphasize the separate terms, we first draw a slash before each plus or minus sign that is not enclosed.

$$-2 \;/\; +3(-2) \;/\; -2(+4)$$

Next we simplify each term.

$$-2 \;/\; +3(-2) \;/\; -2(+4)$$
$$-2 \qquad -6 \qquad -8$$

Then we algebraically add the terms.

$$-2 - 6 - 8 = \mathbf{-16}$$

LESSON PRACTICE

Practice set Evaluate each expression. Write parentheses as the first step.

a. $x + xy - y$ if $x = 3$ and $y = -2$

b. $-m + n - mn$ if $m = -2$ and $n = -5$

Simplify:

c. $-3 + 4 - 5 - 2$

d. $-2 + 3(-4) - 5(-2)$

e. $-3(-2) - 5(2) + 3(-4)$

f. $-4(-3)(-2) - 6(-4)$

MIXED PRACTICE

Problem set

1. On his first six tests Paulo's average score was 86. On his next four tests his average score was 94. What was his average score for his first ten tests?
(55)

2. The mean of these numbers is how much greater than the median?
(Inv. 4)

3, 12, 7, 5, 18, 6, 9, 28

3. The Martins completed the 130-mile trip in $2\frac{1}{2}$ hours. What was their average speed in miles per hour?
(46)

Use ratio boxes to solve problems 4–7.

4. The ratio of laborers to supervisors at the job site was 3 to 5. Of the 120 laborers and supervisors at the job site, how many were laborers?
(65)

5. Vera bought 3 notebooks for \$8.55. At this rate, how much would 5 notebooks cost?
(72)

6. The sale price was 90 percent of the regular price. If the regular price was \$36, what was the sale price?
(60)

7. Forty people came to the party. This was 80 percent of those who were invited. How many were invited?
(81)

8. Write equations to solve (a) and (b).
(77)

(a) Twenty is 40 percent of what number?

(b) Twenty is what percent of 40?

9. Use two unit multipliers to convert 3600 in.^{2} to square feet.
(88)

10. Diagram this statement. Then answer the questions that follow.
(71)

Three fourths of the questions on the test were multiple-choice. There were 60 multiple-choice questions.

(a) How many questions were on the test?

(b) What percent of the questions on the test were not multiple-choice?

11. Evaluate: $x - y - xy$ if $x = -3$ and $y = -2$
(91)

12. (79) Compare: $m \bigcirc n$ if m is an integer and n is a whole number

13. (Inv. 6, 75) (a) Classify this quadrilateral.

15 mm
13 mm
12 mm
20 mm

(b) Find the perimeter of the figure.

(c) Find the area of the figure.

(d) The sum of the measures of the interior angles of a quadrilateral is 360°. If the acute interior angle of the figure measures 75°, what does the obtuse interior angle measure?

14. (2, 41) Which property is illustrated by each equation?

(a) $a + (b + c) = (a + b) + c$

(b) $ab = ba$

(c) $a(b + c) = ab + ac$

15. (62) The lengths of two sides of an isosceles triangle are 5 inches and 1 foot. Draw the triangle and find its perimeter in inches.

16. (83) Multiply. Write the product in scientific notation.

$$(2.4 \times 10^{-4})(5 \times 10^{-7})$$

17. (67) A pyramid with a square base has how many

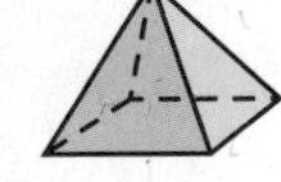

(a) faces?

(b) edges?

(c) vertices?

18. (66, 82) (a) Find the circumference of the circle at right.

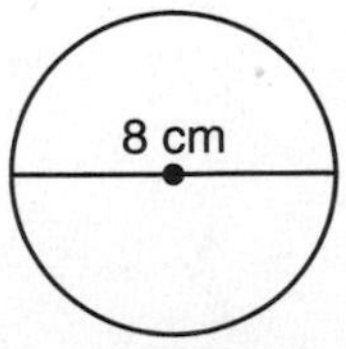

Use 3.14 for π.

(b) Find the area of the circle.

19. (Inv. 9) One yard equals three feet. The table at right shows three pairs of equivalent measures. Plot these points on a coordinate graph, using yards for the horizontal axis and feet for the vertical axis. Then draw a ray through the points to show all pairs of equivalent measures.

yd	ft
1	3
2	6
3	9

20. Refer to the figure below to answer (a)–(d).
(40)

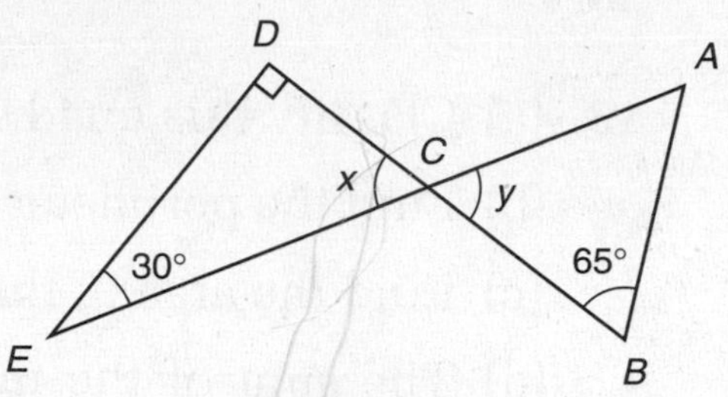

(a) What is $m\angle x$?

(b) What is $m\angle y$?

(c) What is $m\angle A$?

(d) Are the two triangles similar? Why or why not?

21. (a) Add algebraically: $-3x - 3 - x - 1$
(84, 87)

(b) Multiply: $(-3x)(-3)(-x)(-1)$

22. On a number line, graph all integers greater than or equal to −3.
(86)

23. Segment AB is how many millimeters longer than segment BC?
(32)

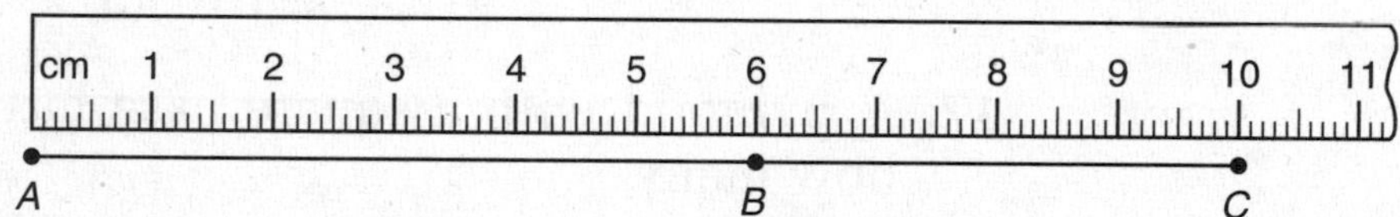

For problems 24–26, solve the equation. Show each step.

24. $5 = y - 4.75$ *(Inv. 7)*

25. $3\frac{1}{3}y = 7\frac{1}{2}$ *(90)*

26. $-9x = 414$ *(90)*

Simplify:

27. $\frac{32 \text{ ft}}{1 \text{ s}} \cdot \frac{60 \text{ s}}{1 \text{ min}}$
(50)

28. $5\frac{1}{3} + 2.5 + \frac{1}{6}$ (mixed-number answer)
(43)

29. $\frac{2\frac{3}{4} + 3.5}{2\frac{1}{2}}$ (decimal answer)
(43, 45)

30. (a) $\frac{(-3) - (-4)(+5)}{(-2)}$
(85, 91)

(b) $-3(+4) - 5(+6) - 7$

LESSON

92 Percent of Change

WARM-UP

Facts Practice: Order of Operations (Test T)

Mental Math:

a. $(-75) - (-50)$

b. $(1.5 \times 10^{-5})(1.5 \times 10^{-5})$

c. $50 = 3m + 2$

d. $6 \times 3\frac{1}{3}$ (*Think:* $6 \times 3 + 6 \times \frac{1}{3}$)

e. 0.3 m to cm

f. \$6 is 10% of m.

g. At \$8.00 per hour, how much money will Kenji earn working $4\frac{1}{2}$ hours?

Problem Solving:

A hexagon can be divided into four triangles by three diagonals drawn from a single vertex. Into how many triangles can a dodecagon be divided by diagonals drawn from one vertex? What is the sum of the measures of the interior angles of a dodecagon?

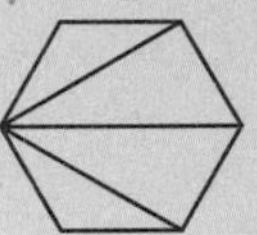

NEW CONCEPT

The percent problems that we have considered before now have used a percent to describe part of a whole. In this lesson we will consider percent problems that use a percent to describe an amount of change. The change may be an increase or a decrease. Adding sales tax to a purchase is an example of an increase. Marking down the price of an item for a sale is an example of a decrease.

INCREASE

original number + amount of change = new number

DECREASE

original number − amount of change = new number

We can use a ratio box to help us with "increase-decrease" problems just as we have with ratio problems and problems about parts of a whole. However, there is a difference in the way we set up the ratio box. When we make a table for a "parts of a whole" problem, the bottom number in the percent column is 100 percent.

	Percent	Actual Count
Part		
Part		
Whole	100	

When we set up a ratio box for an "increase-decrease" problem, we also have three rows. The three rows represent the original number, the amount of change, and the new number. We will use the words ***original, change,*** and ***new*** on the left side of the ratio box. The difference in the setup is where we put 100 percent. Most "increase-decrease" problems consider the original amount to be 100 percent. So the top number in the percent column will be 100 percent.

	Percent	Actual Count
Original	100	
Change		
New		

If the change is an **increase,** we **add** it to the original amount to get the new amount. If the change is a **decrease,** we **subtract** it from the original amount to get the new amount.

Example 1 The county's population increased 15 percent from 1980 to 1990. If the population in 1980 was 120,000, what was the population in 1990?

Solution First we identify the type of problem. The percent describes an amount of change. This is an increase problem. We make a ratio box and write the words "original," "change," and "new" down the side. Since the change was an increase, we write a plus sign in front of "change." In the "percent" column we write 100 percent for the original (1980 population) and 15 percent for the change. We add to get 115 percent for the new (1990 population). In the "actual count" column we write 120,000 for the original population and use the letters C for "change" and N for "new."

	Percent	Actual Count
Original	100	120,000
+ Change	15	C
New	115	N

$$\frac{100}{115} = \frac{120,000}{N}$$

We are asked for the new population. Since we know both numbers in the first row, we use the first and third rows to write the proportion.

$$\frac{100}{115} = \frac{120{,}000}{N}$$

$$100N = 13{,}800{,}000$$

$$N = 138{,}000$$

The county's population in 1990 was **138,000.**

Example 2 The price was reduced 30 percent. If the sale price was $24.50, what was the original price?

Solution First we identify the problem. This is a decrease problem. We make a ratio box and write "original," "change," and "new" down the side, with a minus sign in front of "change." In the percent column we write 100 percent for original, 30 percent for change, and 70 percent for new. The sale price is the new actual count. We are asked to find the original price.

	Percent	Actual Count
Original	100	R
− Change	30	C
New	70	24.50

$$\frac{100}{70} = \frac{R}{24.50}$$

$$70R = 2450$$

$$R = 35$$

The original price was **$35.00.**

Example 3 A merchant bought an item at wholesale for $20 and marked the price up 75% to sell the item at retail. What was the merchant's retail price for the item?

Solution This is an increase problem. We make a table and record the given information.

	Percent	Actual Count
Original (Wholesale)	100	20
+ Change (Markup)	75	M
New (Retail)	175	R

$$\frac{100}{175} = \frac{20}{R}$$

$$100R = 3500$$

$$R = 35$$

The merchant's retail price for the item was **$35.**

LESSON PRACTICE

Practice set Use a ratio box to solve each problem.

a. The regular price was $24.50, but the item was on sale for 30 percent off. What was the sale price?

b. The number of books in John's personal library increased 20% in one year. If there were 60 books in his library this year, how many books were in his library last year?

c. Bikes were on sale for 20 percent off. Tomas bought one for $120. How much money did he save by buying the bike at the sale price instead of at the regular price?

d. The clothing store bought shirts for $15 each and marked up the price 80% to sell the shirts at retail. What was the retail price of each shirt?

MIXED PRACTICE

Problem set

1. (21) The product of the first three prime numbers is how much less than the sum of the next three prime numbers?

2. (55) After five tests Dara's average score was 88. What score must she average on the next two tests to have a seven-test average of 90?

3. (46) Jenna finished a 2-mile race in 15 minutes. What was her average speed in miles per hour?

Use ratio boxes to solve problems 4–7.

4. (65) Forty-five of the 80 children in the club were girls. What was the ratio of boys to girls in the club?

5. (72) Two dozen sparklers cost $3.60. At that rate, how much would 60 sparklers cost?

6. (92) The county's population increased 20 percent from 1980 to 1990. If the county's population in 1980 was 340,000, what was the county's population in 1990?

7. (92) Because of an unexpected cold snap, the cost of tomatoes increased 50% in one month. If the cost after the increase was 96¢ per pound, what was the cost before the increase?

8. *(77)* Write equations to solve (a) and (b).

(a) Sixty is what percent of 75?

(b) Seventy-five is what percent of 60?

9. *(88)* Use two unit multipliers to convert 100 cm^2 to square millimeters.

10. *(71)* Diagram this statement. Then answer the questions that follow.

Five eighths of the trees in the grove were deciduous. There were 160 deciduous trees in the grove.

(a) How many trees were in the grove?

(b) How many of the trees in the grove were not deciduous?

11. *(91)* If $x = -5$ and $y = 3x - 1$, then y equals what number?

12. *(60)* Compare: 30% of 20 ◯ 20% of 30

13. *(58, 75)* (a) Find the area of this isosceles trapezoid.

(b) Trace the figure and draw its line of symmetry.

10 cm
6.5 cm
6 cm
6.5 cm
5 cm

14. *(46, 92)* A merchant bought a stereo at wholesale for $90.00 and marked up the price 75% to sell the stereo at retail.

(a) What was the retail price of the stereo?

(b) If the stereo sells at the retail price, and the sales-tax rate is 6%, what is the total price including sales tax?

15. *(83)* Multiply. Write the product in scientific notation.

$$(8 \times 10^{-5})(3 \times 10^{12})$$

16. *(48)* Complete the table.

FRACTION	DECIMAL	PERCENT
$2\frac{1}{3}$	(a)	(b)
(c)	(d)	$3\frac{1}{3}\%$

17. *(36)* One ace is missing from an otherwise normal deck of cards. If a card is selected at random, what is the probability that the selected card will be an ace?

18. What number is 250 percent of 60?
(60)

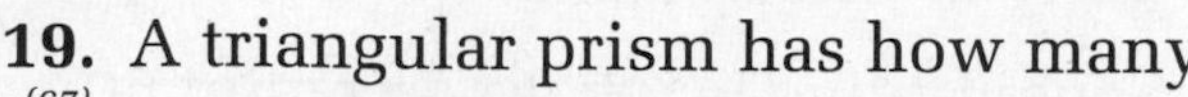

19. A triangular prism has how many
(67)
(a) triangular faces?

(b) rectangular faces?

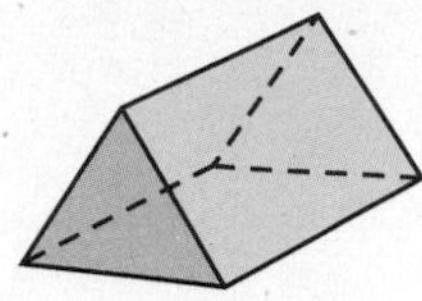

20. John measured the diameter of his bicycle tire and found that it was 24 inches. What is the distance around the tire to the nearest inch? (Use 3.14 for π.)
(66)

21. Find the missing numbers in the table by using the function rule. Plot the (x, y) pairs on a coordinate plane, and then draw a line showing all (x, y) pairs that satisfy the function.
(85, Inv. 9)

$y = 2x + 1$

x	y
0	☐
3	☐
−2	☐

22. The ratio of the measures of two angles was 4 to 5. If the sum of their measures was 180°, what was the measure of the smaller angle?
(65)

23. Simplify:
(84, 87)
(a) $x + y + 3 + x - y - 1$

(b) $(3x)(2x) + (3x)(2)$

24. Draw a pair of parallel lines. Draw a second pair of parallel lines that intersect but are not perpendicular to the first pair. What kind of quadrilateral is formed?
(Inv. 6)

For problems 25–27, solve the equation. Show each step.

25. $3\frac{1}{7}x = 66$
(90)

26. $w - 0.15 = 4.9$
(Inv. 7)

27. $-8y = 600$
(90)

Simplify:

28. $(2 \cdot 3)^2 - 2(3^2)$
(52)

29. $5 - \left(3\frac{1}{3} - 1.5\right)$
(43)

30. (a) $\dfrac{(-8)(-6)(-5)}{(-4)(-3)(-2)}$
(85, 91)

(b) $-6 - 5(-4) - 3(-2)(-1)$

LESSON

93 Two-Step Equations and Inequalities

WARM-UP

Facts Practice: Order of Operations (Test T)

Mental Math:

a. $(-25)(-8)$

b. $(2.5 \times 10^8)(3 \times 10^{-4})$

c. $\frac{100}{x} = \frac{22}{55}$

d. 2 ft^2 equals how many square inches?

e. $8 \times 2\frac{1}{4}$

f. 10% less than 60

g. Estimate the product of 3.14 and 4.9 cm.

Problem Solving:

Michelle's grandfather taught her this method for converting from degrees Celsius to degrees Fahrenheit: "Double the Celsius temperature; subtract 10%; then add 32°." Use this method to convert 20° Celsius to degrees Fahrenheit.

NEW CONCEPT

Since Investigation 7 we have practiced solving one-step balanced equations. In this lesson we will practice solving two-step equations.

This balance scale illustrates a two-step equation:

$$2x + 5 = 35$$

On the left side of the equation are two terms, $2x$ and 5. To solve the equation, we first isolate the variable term by subtracting 5 from (or adding −5 to) both sides of the equation.

$2x + 5 - 5 = 35 - 5$ subtracted 5 from both sides

$2x = 30$ simplified

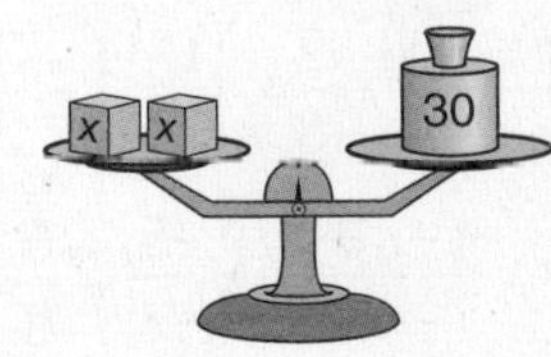

We see that $2x = 30$, so we divide by 2 $\left(\text{or multiply by } \frac{1}{2}\right)$ to find $1x$.

$$\frac{2x}{2} = \frac{30}{2} \quad \text{divided both sides by 2}$$

$$x = 15 \quad \text{simplified}$$

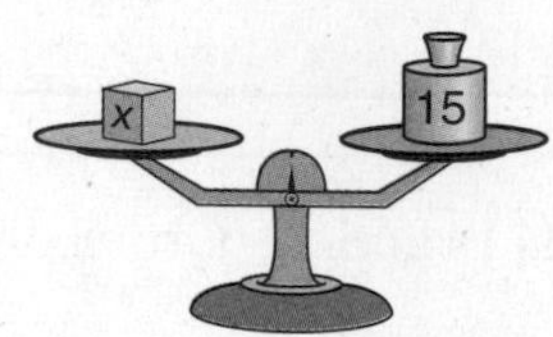

Example 1 Solve this equation. Show all steps.

$$0.4x + 1.2 = 6$$

Solution First we isolate the variable term by subtracting 1.2 from both sides of the equation. Then we divide both sides by 0.4.

$$0.4x + 1.2 = 6 \quad \text{equation}$$

$$0.4x + 1.2 - 1.2 = 6 - 1.2 \quad \text{subtracted 1.2 from both sides}$$

$$0.4x = 4.8 \quad \text{simplified}$$

$$\frac{0.4x}{0.4} = \frac{4.8}{0.4} \quad \text{divided both sides by 0.4}$$

$$x = \mathbf{12} \quad \text{simplified}$$

Example 2 Solve this equation. Show all steps.

$$-\frac{2}{3}x - \frac{1}{2} = \frac{1}{3}$$

Solution We isolate the variable term by adding $\frac{1}{2}$ to both sides of the equation. Then we find $1x$ by multiplying both sides of the equation by $-\frac{3}{2}$.

$$-\frac{2}{3}x - \frac{1}{2} = \frac{1}{3} \quad \text{equation}$$

$$-\frac{2}{3}x - \frac{1}{2} + \frac{1}{2} = \frac{1}{3} + \frac{1}{2} \quad \text{added } \tfrac{1}{2} \text{ to both sides}$$

$$-\frac{2}{3}x = \frac{5}{6} \quad \text{simplified}$$

$$\left(-\frac{3}{2}\right)\left(-\frac{2}{3}x\right) = \left(-\frac{3}{2}\right)\left(\frac{5}{6}\right) \quad \text{multiplied both sides by } -\tfrac{3}{2}$$

$$x = \mathbf{-\frac{5}{4}} \quad \text{simplified}$$

Example 3 Solve: $-15 = 3x + 6$

Solution The variable term is on the right side of the equal sign. We may interchange the entire right side of the equation with the entire left side if we wish (just as we may interchange the contents of one pan of a balance scale with the contents of the other pan). However, we will solve this equation without interchanging the sides of the equation.

$-15 = 3x + 6$	equation
$-15 - 6 = 3x + 6 - 6$	subtracted 6 from both sides
$-21 = 3x$	simplified
$\frac{-21}{3} = \frac{3x}{3}$	divided both sides by 3
$\mathbf{-7 = x}$	simplified

In this lesson we have practiced procedures for solving equations. We may follow similar procedures for solving inequalities in which the variable term is positive.[†] To solve an inequality, we isolate the variable while maintaining the inequality.

Example 4 Solve this inequality and graph its solution: $2x - 5 \geq 1$

Solution We see that the variable term ($2x$) is positive. We begin by adding 5 to both sides of the inequality. Then we divide both sides of the inequality by 2.

$2x - 5 \geq 1$	inequality
$2x - 5 + 5 \geq 1 + 5$	added 5 to both sides
$2x \geq 6$	simplified
$\frac{2x}{2} \geq \frac{6}{2}$	divided both sides by 2
$\mathbf{x \geq 3}$	simplified

We check the solution by replacing x in the original inequality with numbers equal to and greater than 3. We try 3 and 4 below.

$2x - 5 \geq 1$	original inequality
$2(3) - 5 \geq 1$	replaced x with 3
$1 \geq 1$	simplified and checked
$2(4) - 5 \geq 1$	replaced x with 4
$3 \geq 1$	simplified and checked

[†]Procedures for solving inequalities with a negative variable term will be taught in a later course.

Now we graph the solution $x \geq 3$.

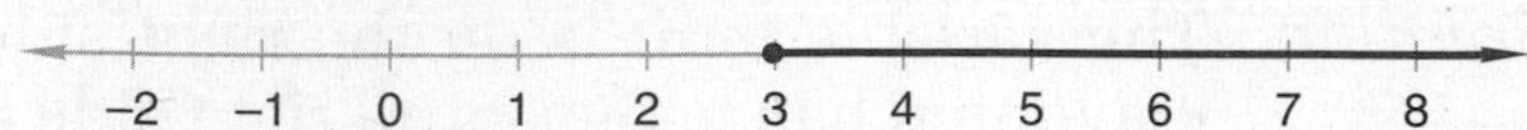

This graph indicates that all numbers greater than or equal to 3 satisfy the original inequality.

LESSON PRACTICE

Practice set* Solve each equation. Show all steps.

a. $8x - 15 = 185$ **b.** $0.2y + 1.5 = 3.7$

c. $\frac{3}{4}m - \frac{1}{3} = \frac{1}{2}$ **d.** $1\frac{1}{2}n + 3\frac{1}{2} = 14$

e. $-6p + 36 = 12$ **f.** $38 = 4w - 26$

g. $-\frac{5}{3}m + 15 = 60$ **h.** $4.5 = 0.6d - 6.3$

Solve these inequalities and graph their solutions:

i. $2x + 5 \geq 1$ **j.** $2x - 5 < 1$

MIXED PRACTICE

Problem set

1. (46) From Sim's house to the lake is 30 kilometers. If he completed the round trip on his bike in 2 hours 30 minutes, what was his average speed in kilometers per hour?

2. (Inv. 4) Find the (a) mean and (b) range for this set of numbers:

3, 9, 7, 5, 10, 4, 5, 8, 5, 4, 8, 40

Use ratio boxes to solve problems 3–5.

3. (36, 65) The ratio of red marbles to blue marbles in a bag of 600 red and blue marbles was 7 to 5.

(a) How many marbles were blue?

(b) If one marble is drawn from the bag, what is the probability that the marble will be blue?

4. (72) The machine could punch out 500 plastic pterodactyls in 20 minutes. At that rate, how many could it punch out in $1\frac{1}{2}$ hours?

5. (92) (a) The price was reduced by 25 percent. If the regular price was $24, what was the sale price?

(b) The price was reduced by 25 percent. If the sale price was $24, what was the regular price?

6. (87) Multiply: $(-3x^2)(2xy)(-x)(3y^2)$

7. (36) Two aces are missing from an otherwise normal deck of cards. If a card is selected at random from the deck, what is the probability that the selected card will be an ace?

8. (88) Use two unit multipliers to convert 7 days to minutes.

9. (22) Diagram this statement. Then answer the questions that follow.

Five ninths of the 45 cars pulled by the locomotive were not cattle cars.

(a) How many cattle cars were pulled by the locomotive?

(b) What percent of the cars pulled by the locomotive were not cattle cars?

10. (33, 48) Compare: $\frac{1}{3} \bigcirc 33\%$

11. (91) Evaluate: $ab - a - b$ if $a = -3$ and $b = -1$

12. (46) Find the total price, including 5% tax, for a meal that includes a $7.95 dish, a 90¢ beverage, and a $2.35 dessert.

13. (19, 75) A corner was cut from a square sheet of paper to form this pentagon. Dimensions are in inches.

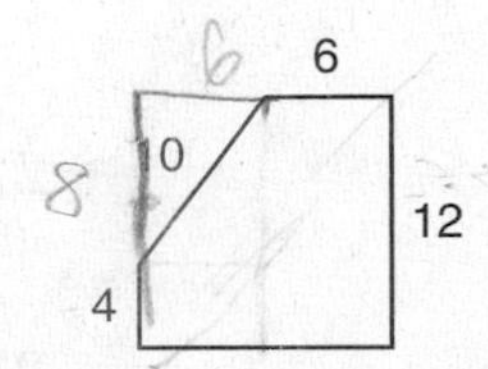

(a) What is the perimeter of the pentagon?

(b) What is the area of the pentagon?

14. (48) Complete the table.

FRACTION	DECIMAL	PERCENT
(a)	0.08	(b)
(c)	(d)	$8\frac{1}{3}\%$

15. (92) A retailer buys a toy for \$3.60 and marks up the price 120%. What is the retail price of the toy?

16. (83) Multiply. Write the product in scientific notation.

$$(8 \times 10^{-3})(6 \times 10^{7})$$

17. (67, 70) Each edge of a cube is 10 cm long.

(a) What is the volume of the cube?

(b) What is the surface area of the cube?

18. (66, 82) (a) What is the area of the circle shown?

(b) What is the circumference of the circle?

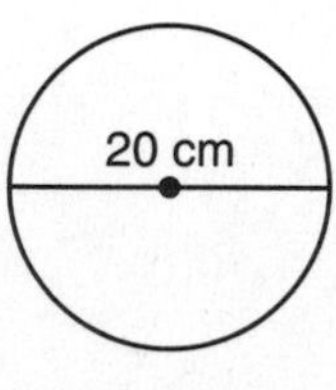

Use 3.14 for π.

19. (84) Collect like terms: $-x + 2x^2 - 1 + x - x^2$

20. (85, Inv. 9) Find the missing numbers in the table by using the function rule. Then graph the function. (Plot the points and draw a line.)

$y = 2x + 3$

x	y
1	☐
0	☐
–2	☐

21. (74) Write an equation to solve this problem:

Sixty is $\frac{3}{8}$ of what number?

22. (93) Solve this inequality and graph its solution: $2x - 5 > -1$

23. (40) Refer to the figure below to answer (a) and (b).

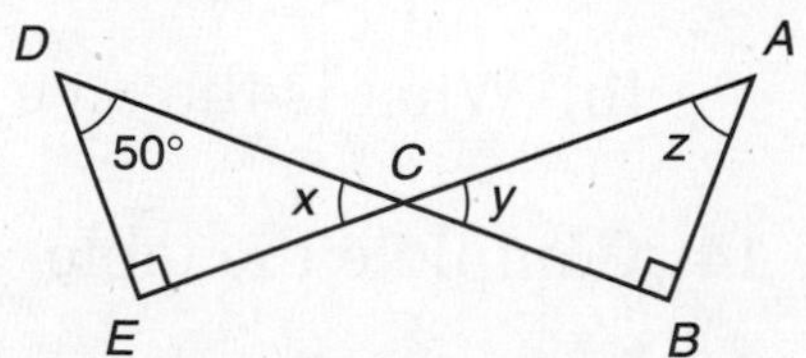

(a) Find the measures of $\angle x$, $\angle y$, and $\angle z$.

(b) Are the two triangles similar? Why or why not?

24. (34) What is the sum of the numbers labeled A and B on the number line below?

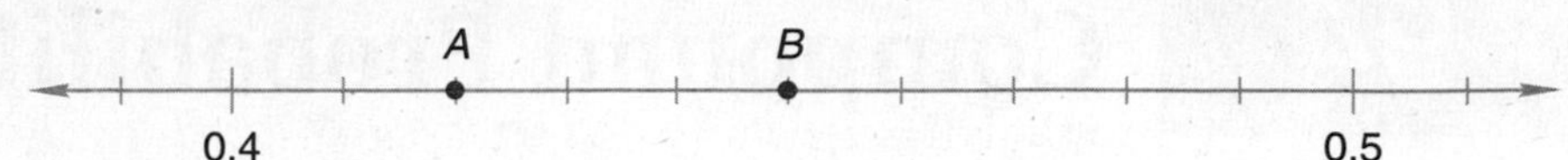

For problems 25–28, solve the equation. Show each step.

25. (93) $3x + 2 = 9$

26. (93) $\frac{2}{3}w + 4 = 14$

27. (93) $0.2y - 1 = 7$

28. (90) $-\frac{2}{3}m = -6$

Simplify:

29. (57, 63) $3(2^3 + \sqrt{16}) - 4^0 - 8 \cdot 2^{-3}$

30. (85, 91) (a) $\frac{(-9)(+6)(-5)}{(-4) - (-1)}$

(b) $-3(4) + 2(3) - 1$

LESSON

94 Compound Probability

WARM-UP

Facts Practice: Scientific Notation (Test S)

Mental Math:

a. $(-144) \div (-6)$ **b.** $(1.5 \times 10^{-8})(4 \times 10^{3})$

c. $5w + 1.5 = 4.5$ **d.** Convert 30°C to degrees Fahrenheit.

e. $6 \times 2\frac{2}{3}$ **f.** 10% more than 50

g. 25% of 40, × 4, + 2, ÷ 6, × 9, + 1, $\sqrt{\ }$, × 3, + 1, $\sqrt{\ }$

Problem Solving:

Six identical blocks marked X, a 1.7-lb weight, and a 4.3-lb weight were balanced on a scale as shown. Write an equation to represent this balanced scale, and find the weight of each block marked X.

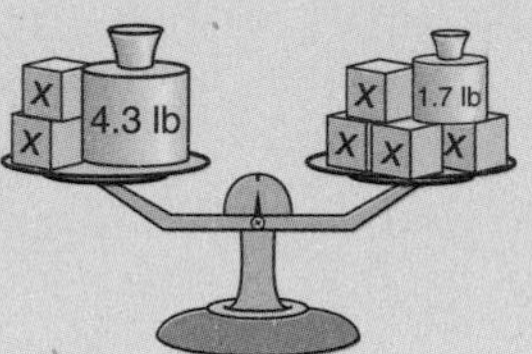

NEW CONCEPT

We know that the probability of getting heads on one toss of a coin is $\frac{1}{2}$. We can state this fact with the following notation in which P(H) stands for "the probability of heads."

$$P(\text{H}) = \frac{1}{2}$$

The probability of getting heads on the second toss of a coin is also $\frac{1}{2}$. So the probability of getting two heads in a row is $\frac{1}{2} \times \frac{1}{2} = \frac{1}{4}$. We can illustrate this fact with a tree diagram. If we toss a coin one time, we can get heads or tails.

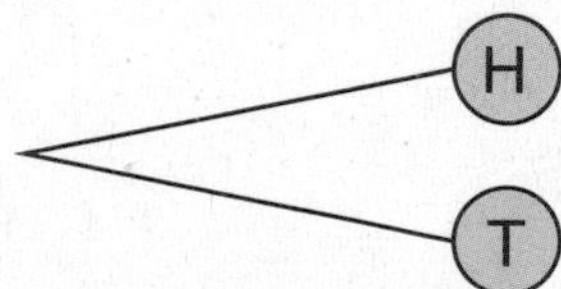

If the first toss came up heads, the second toss could come up either heads or tails.

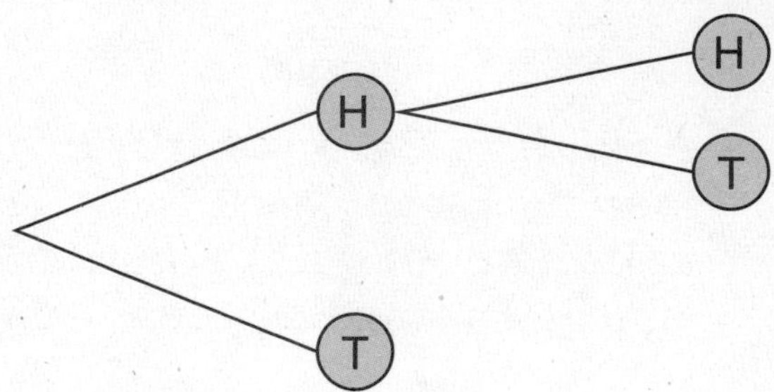

Likewise, if the first toss came up tails, the second toss could come up either heads or tails.

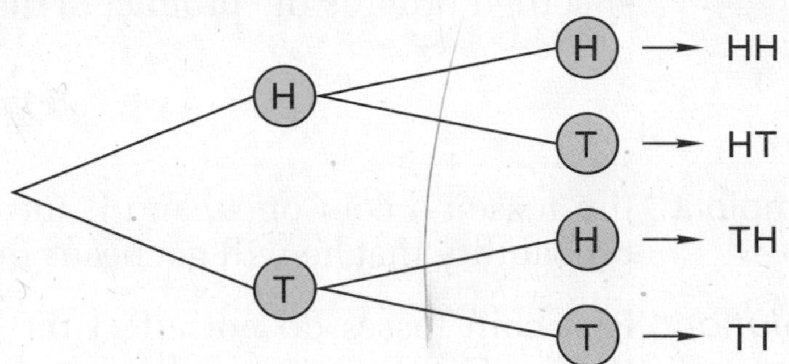

From the tree diagram, we can list the four possible outcomes:

HH HT TH TT

This list of outcomes is called a **sample space.** Since any one of the four outcomes in the sample space is equally likely, the probability of each outcome is one fourth.

Thus, the probability of getting HH is $\frac{1}{4}$, which we can express this way:

$$P(\text{H, H}) = \frac{1}{4}$$

When the outcome of one event does not affect the probability of a second event, the events are **independent.** Each coin toss is an independent event because the outcome of one toss does not affect the outcome of a subsequent toss.

> **The probability of independent events occurring in a specified order is the product of the probabilities of each event.**

Thus, for example,

$P(\text{H, H, T})$ is $\frac{1}{2} \cdot \frac{1}{2} \cdot \frac{1}{2} = \frac{1}{8}$

$P(\text{H, T, T, H})$ is $\frac{1}{2} \cdot \frac{1}{2} \cdot \frac{1}{2} \cdot \frac{1}{2} = \frac{1}{16}$

$P(\text{T, H, T, H, T})$ is $\left(\frac{1}{2}\right)^5 = \frac{1}{32}$

Example 1 The face of this spinner is divided into four congruent sectors. What is the probability of getting a 2 on the first spin and a 1 on the second spin?

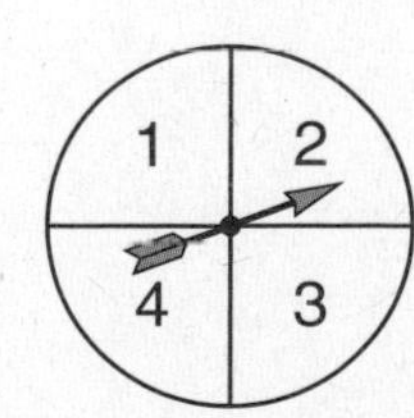

Solution The probability of getting a 2 is $\frac{1}{4}$. The probability of getting a 1 is $\frac{1}{4}$. The probability of independent events occurring in a specified order is the product of the individual probabilities.

$$P(2, 1) = \frac{1}{4} \cdot \frac{1}{4} = \mathbf{\frac{1}{16}}$$

Example 2 Jim tossed a coin once, and it turned up heads. What is the probability that he will get heads on the next toss of the coin?

Solution Past coin tosses do not affect the probability of future coin tosses. So there are only two possible outcomes for the second toss. The coin will turn up either heads or tails. The probability that it will turn up heads is $\mathbf{\frac{1}{2}}$.

Example 3 If a pair of dot cubes is tossed once, what is the probability that the total number rolled will be 12?

Solution Only one combination of two dot-cube faces equals 12, and that is 6 and 6. The probability of one dot cube stopping its roll with 6 on top is $\frac{1}{6}$. So the probability of two dot cubes stopping with 6 on top is $\frac{1}{6} \times \frac{1}{6} = \frac{1}{36}$.

$$P(6, 6) = \left(\frac{1}{6}\right)^2 = \mathbf{\frac{1}{36}}$$

The sample space in the table below shows the 36 possible combinations of rolling two dot cubes. Since only one combination results in a total of 12, the probability of rolling 12 is $\frac{1}{36}$. From the list of outcomes in the sample space, we see that the probability of rolling 11 is $\frac{2}{36}$, which is $\frac{1}{18}$, and the probability of rolling 10 is $\frac{3}{36}$, which is $\frac{1}{12}$.

Outcome of First Dot Cube

Outcome of Second Dot Cube	1	2	3	4	5	6
1	2	3	4	5	6	7
2	3	4	5	6	7	8
3	4	5	6	7	8	9
4	5	6	7	8	9	10
5	6	7	8	9	10	11
6	7	8	9	10	11	12

Example 4 With one toss of a pair of dot cubes, what is the probability of rolling a total greater than 9?

Solution Method 1: From the table showing sample space, we see that there are 36 possible combinations. So 36 is the bottom term of the probability ratio. Also from the table we see that six of the combinations total more than 9 (three total 10, two total 11, one totals 12). So 6 is the top term of the probability ratio.

$$P(>9) = \frac{6}{36} = \mathbf{\frac{1}{6}}$$

Method 2: We regard each roll of a dot cube as an independent event. We think about the combinations that total more than 9. If the first dot cube stops on 1, 2, or 3, there is no way for the total to reach 10, 11, or 12. If the first dot cube stops on 4, the second dot cube must stop on 6. We calculate this probability.

$$P(4, 6) = \frac{1}{6} \cdot \frac{1}{6} = \frac{1}{36}$$

If the first dot cube stops on 5, the second dot cube may stop on 5 or 6 (two favorable outcomes).

$$P(5, 5 \text{ or } 6) = \frac{1}{6} \cdot \frac{2}{6} = \frac{2}{36}$$

If the first dot cube stops on 6, the second dot cube may stop on 4, 5, or 6 (three favorable outcomes).

$$P(6, 4 \text{ or } 5 \text{ or } 6) = \frac{1}{6} \cdot \frac{3}{6} = \frac{3}{36}$$

We add these probabilities to find the total probability of rolling a number greater than 9.

$$P(>9) = \frac{1}{36} + \frac{2}{36} + \frac{3}{36} = \frac{6}{36} = \mathbf{\frac{1}{6}}$$

When the outcome of one event affects the probability of a subsequent event, the events are **dependent.** For instance, if a card is drawn from a deck and not replaced, the probabilities of the draws of any remaining cards are affected by the first draw.

> **The probability of dependent events occurring in a specified order is the product of the first event and the recalculated probabilities of each subsequent event.**

Example 5 From a well-mixed deck of cards, Josefina selected and kept one card, then a second card, then a third card, and finally a fourth card. What is the probability that the four cards Josefina selected are aces?

Solution Since the drawn cards are not replaced, the events are dependent. So the probability of each event must be calculated as though the prior specified events had occurred.

Each card Josefina draws must be an ace. There are 4 chances out of 52 that the first card is an ace. That leaves 3 aces in 51 cards for the second draw. For the third draw there are 2 aces in 50 cards, and for the fourth draw there is 1 ace in 49 cards.

$$P(\text{A, A, A, A}) = \frac{4}{52} \cdot \frac{3}{51} \cdot \frac{2}{50} \cdot \frac{1}{49}$$

We show the calculation below.

$$P(\text{A, A, A, A}) = \frac{\overset{1}{\cancel{4}}}{\underset{13}{\cancel{52}}} \cdot \frac{\overset{1}{\cancel{3}}}{\underset{17}{\cancel{51}}} \cdot \frac{\overset{1}{\cancel{2}}}{\underset{25}{\cancel{50}}} \cdot \frac{1}{49} = \mathbf{\frac{1}{270{,}725}}$$

Many calculators have an **exponent key** such as y^x or x^y that can be used to calculate probabilities.

Suppose you were going to take a ten-question true-false test. Instead of reading the questions, you decide to guess every answer. What is the probability of guessing the correct answer to all ten questions?

The probability of correctly guessing the first answer is $\frac{1}{2}$. The probability of correctly guessing the first two answers is $\frac{1}{2} \cdot \frac{1}{2}$, or $\left(\frac{1}{2}\right)^2$. (Since $\frac{1}{2} \cdot \frac{1}{2} = \frac{1}{4}$, we may also write $\left(\frac{1}{2}\right)^2$ as $\frac{1}{2^2}$.) The probability of correctly guessing the first three answers is $\left(\frac{1}{2}\right)^3$, or $\frac{1}{2^3}$. Thus, the probability of guessing the correct answer to all ten questions is $\left(\frac{1}{2}\right)^{10}$, or $\frac{1}{2^{10}}$. To find 2^{10} on a calculator with a y^x or x^y key, we use these keystrokes:

2 y^x 1 0 =

The number displayed is 1024. Therefore, the probability of correctly guessing all ten true-false answers is

$$\frac{1}{1024}$$

Correctly guessing all ten answers has the same likelihood as tossing heads with a coin ten times in a row.

Probabilities that are extremely unlikely may be displayed by a calculator in scientific notation.

Example 6 Find the probability of correctly guessing the answer to every question on a twenty-question, four-option multiple-choice test.

Solution The probability of correctly guessing the first answer is $\frac{1}{4}$. Since there are 20 questions, the probability of correctly guessing all the answers is $\frac{1}{4^{20}}$. To find 4^{20}, we use these keystrokes:

4 y^x 2 0 =

The number displayed is about 1.1×10^{12}. Therefore, the probability of correctly guessing all 20 answers is

$$\frac{1}{1.1 \times 10^{12}}$$

which is **less than one in a trillion.**

LESSON PRACTICE

Practice set

a. To win the game, Victor needs to roll a total of 9 with one toss of a pair of dot cubes. What is the probability that he will do that on the first try?

b. Draw a tree diagram like the one at the beginning of the lesson to find the sample space of three coin tosses.

c. Jasmine is taking a four-option multiple-choice test. There are two answers she does not know. If she can correctly rule out one option on one question but no options on the other question, what is the probability she will correctly guess both answers?

d. Quentin has a box containing a red marble, a white marble, and a blue marble. Quentin draws a marble, puts it back, and then draws again. Copy and complete the following table to show the sample space for two consecutive draws. The first row has been completed for you. ("R, W" means "red then white.")

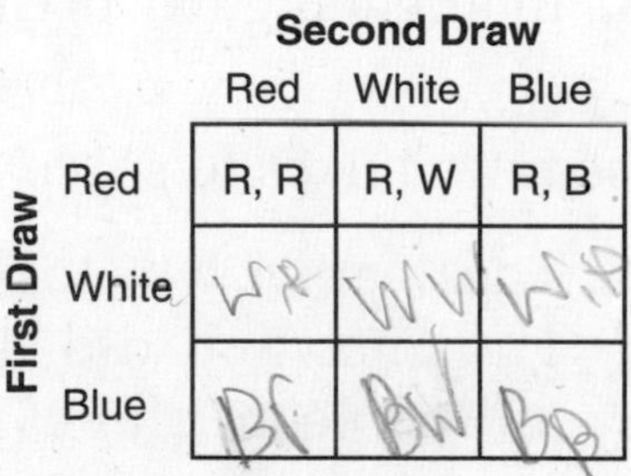

First Draw \ Second Draw	Red	White	Blue
Red	R, R	R, W	R, B
White			
Blue			

e. Quentin draws from the box twice more. This time he did not replace the marble after the first draw. Copy and complete the following table to show the sample space for these two draws. The first row has been completed for you. Notice that the space for "red then red" is blank. This outcome is not possible. If Quentin draws the red marble on the first draw and does not replace it, the red marble will not be in the box for the second draw.

First Draw	Second Draw: Red	White	Blue
Red		R, W	R, B
White			
Blue			

f. Were the two draws in problem **d** dependent or independent events? Were the two draws in problem **e** dependent or independent events?

MIXED PRACTICE

Problem set

1. (51) Twenty-one billion is how much more than 9.8 billion? Write the answer in scientific notation.

2. (53, 55) The train traveled at an average speed of 48 miles per hour for the first 2 hours and at 60 miles per hour for the next 4 hours. What was the train's average speed for the 6-hour trip? (Average speed equals total distance divided by total time.)

3. (46) A 10-pound box of detergent costs $8.40. A 15-pound box costs $10.50. Which box costs the most per pound? How much more per pound does it cost?

4. (36, 67) In a rectangular prism, what is the ratio of faces to edges?

Use ratio boxes to solve problems 5–8.

5. (65) The team's win-loss ratio was 3 to 2. If the team won 12 games and did not tie any games, how many games did the team play?

6. Twenty-four is to 36 as 42 is to what number?
(72)

7. The number that is 20% less than 360 is what percent of 360?
(92)

8. During the sale shirts were marked down 20 percent to \$20. What was the regular price of the shirts (the price before the sale)?
(92)

9. Use two unit multipliers to perform each conversion:
(88)

(a) 12 ft^2 to square inches

(b) 1 kilometer to millimeters

10. Diagram this statement. Then answer the questions that follow.
(71)

The duke conscripted two fifths of the male serfs in his dominion. He conscripted 120 male serfs in all.

(a) How many male serfs were in the duke's dominion?

(b) How many male serfs in his dominion were not conscripted?

11. If a pair of dot cubes is tossed once, what is the probability that the total number rolled will be
(94)

(a) 1? (b) 2? (c) 3?

12. If $y = 4x - 3$ and $x = -2$, then y equals what number?
(91)

13. The perimeter of a certain square is 4 yards. Find the area of the square in square feet.
(16, 20)

14. The sale price of the new car was \$14,500. The sales-tax rate was 6.5 percent.
(46, 60)

(a) What was the sales tax on the car?

(b) What was the total price including tax?

(c) If the commission paid to a salesperson is 2 percent of the sale price, how much is the commission on a \$14,500 sale?

15. Complete the table.
(48)

FRACTION	DECIMAL	PERCENT
(a)	(b)	$66\frac{2}{3}\%$
$1\frac{3}{4}$	(c)	(d)

16. (a) What is 200 percent of \$7.50?
(60, 92)

(b) What is 200 percent more than \$7.50?

17. Multiply. Write the product in scientific notation.
(83)

$$(2 \times 10^8)(8 \times 10^2)$$

18. Robbie stores 1-inch cubes in a box with inside dimensions as shown. How many cubes will fit in this box?
(70)

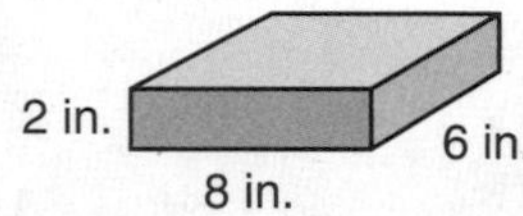

19. The length of each side of the square equals the diameter of the circle. The area of the square is how much greater than the area of the circle?
(82)

Use $\frac{22}{7}$ for π.

20. Divide 7.2 by 0.11 and write the quotient with a bar over the repetend.
(42, 45)

21. Find the missing numbers in the table by using the function rule. Then graph the function.
(85, Inv. 9)

$y = 3x - 2$

x	y
3	☐
0	☐
−1	☐

22. Solve this inequality and graph its solution: $2x - 5 < -1$
(93)

23. In the figure at right, the measure of $\angle AOC$ is half the measure of $\angle AOD$. The measure of $\angle AOB$ is one third the measure of $\angle AOD$.
(40)

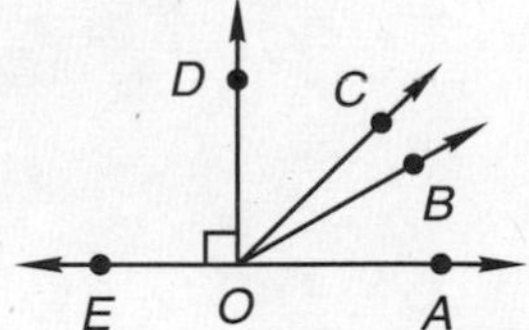

(a) Find m$\angle AOB$.

(b) Find m$\angle EOC$.

24. The length of segment BC is how much less than the length of segment AB?
(8)

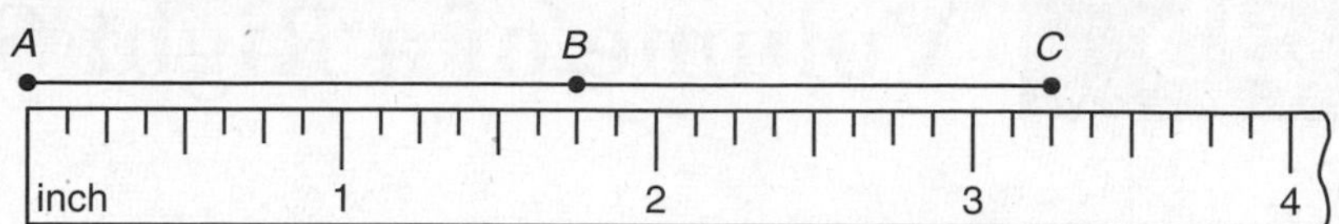

For problems 25 and 26, solve the equation. Show each step.

25. $1.2p + 4 = 28$
(93)

26. $-6\frac{2}{3}m = 1\frac{1}{9}$
(90)

Simplify:

27. (a) $6x^2 + 3x - 2x - 1$
(84, 87)

(b) $(5x)(3x) - (5x)(-4)$

28. (a) $\frac{(-8) - (-6) - (4)}{-3}$
(85, 91)

(b) $-5(-4) - 3(-2) - 1$

29. Evaluate: $b^2 - 4ac$ if $a = -1$, $b = -2$, and $c = 3$
(91)

30. Hugo constructed a sample space of the possible outcomes of flipping a quarter and a dime. He figured the quarter could land either heads up or tails up. For each quarter outcome he listed two outcomes for the flip of the dime. For one possible outcome, heads on both coins, Hugo wrote Q_HD_H. Complete Hugo's list to show all possible outcomes of the experiment.
(94)

LESSON

95 Volume of a Right Solid

WARM-UP

Facts Practice: Order of Operations (Test T)

Mental Math:

a. (72) + (−100)

b. $(2.5 \times 10^6)(2.5 \times 10^6)$

c. $\frac{60}{100} = \frac{y}{1.5}$

d. Convert 25°C to degrees Fahrenheit.

e. $8 \times 2\frac{3}{4}$

f. 50% more than 60

g. 10% of 300 is how much more than 20% of 100?

Problem Solving:

Copy this problem and fill in the missing digits:

```
   _3
×  __
  ___
 ___
 9__1
```

NEW CONCEPT

A **right solid** is a geometric solid whose sides are perpendicular to the base. **The volume of a right solid equals the area of the base times the height.** This rectangular solid is a right solid. It is 5 m long and 2 m wide, so the area of the base is 10 m^2.

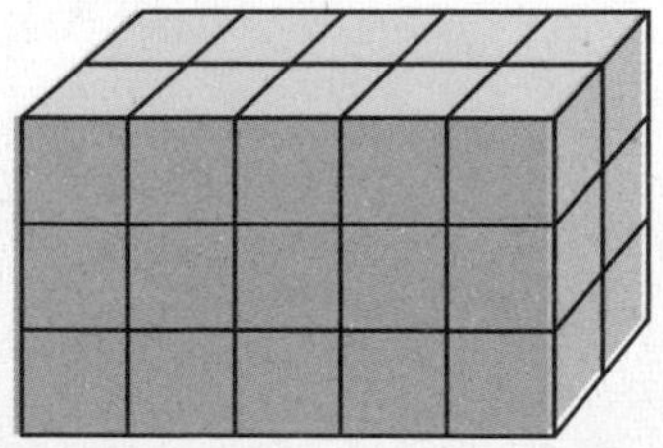

One cube will fit on each square meter of the base, and the cubes are stacked 3 m high, so

$$\begin{aligned} \text{Volume} &= \text{area of the base} \times \text{height} \\ &= 10 \text{ m}^2 \times 3 \text{ m} \\ &= 30 \text{ m}^3 \end{aligned}$$

If the base of the solid is a polygon, the solid is called a **prism.** If the base of a right solid is a circle, the solid is called a **right circular cylinder.**

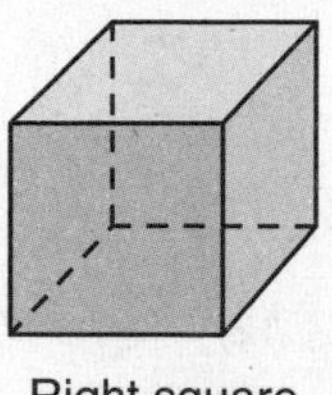
Right square prism

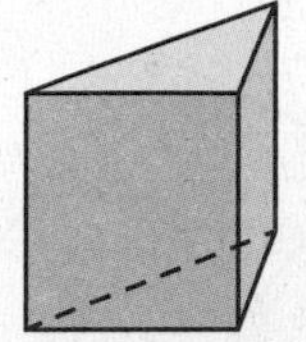
Right triangular prism

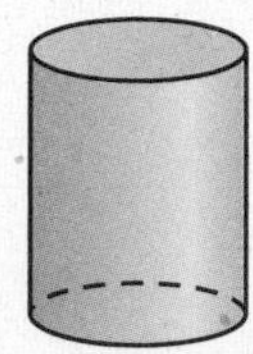
Right circular cylinder

Example 1 Find the volume of the right triangular prism below. Dimensions are in centimeters. We show two views of the prism.

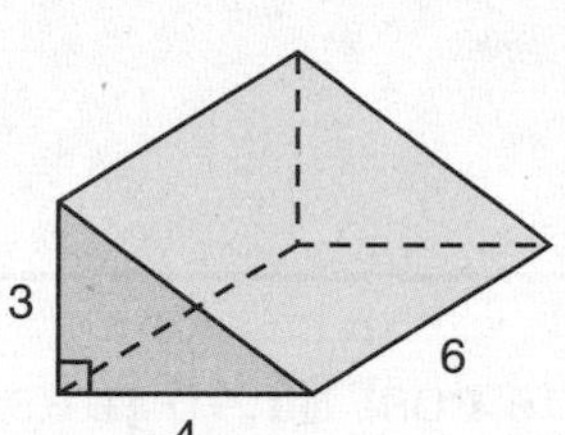

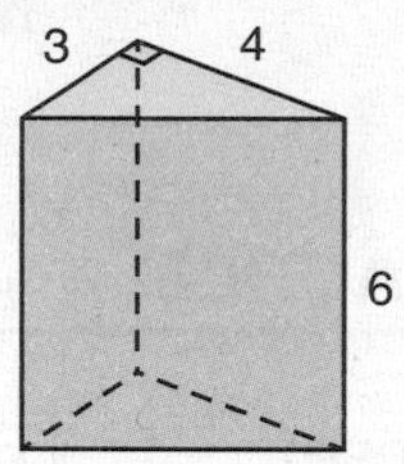

Solution The area of the base is the area of the triangle.

$$\text{Area of base} = \frac{(4\text{ cm})(3\text{ cm})}{2} = 6\text{ cm}^2$$

The volume equals the area of the base times the height.

$$\text{Volume} = (6\text{ cm}^2)(6\text{ cm}) = \mathbf{36\text{ cm}^3}$$

Example 2 The diameter of this right circular cylinder is 20 cm. Its height is 25 cm. What is its volume? Leave π as π.

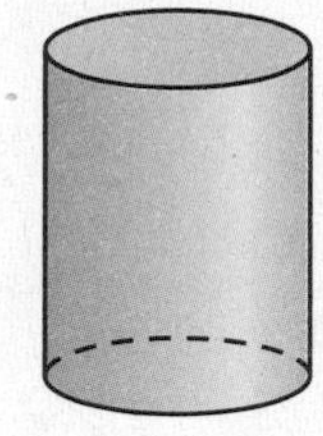

Solution First we find the area of the base. The diameter of the circular base is 20 cm, so the radius is 10 cm.

$$\text{Area of base} = \pi r^2 = \pi(10\text{ cm})^2 = 100\pi\text{ cm}^2$$

The volume equals the area of the base times the height.

$$\text{Volume} = (100\pi\text{ cm}^2)(25\text{ cm}) = \mathbf{2500\pi\text{ cm}^3}$$

LESSON PRACTICE

Practice set Find the volume of each right solid shown. Dimensions are in centimeters.

a. 8, 6, 12

b. 12, 6, 10

c. 10, 6

Leave π as π.

d. 3, 2, 5, 10, 7

e.

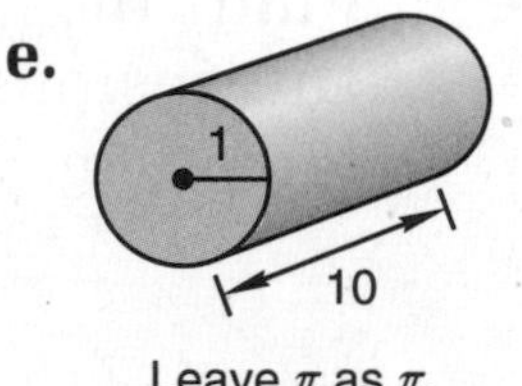

Leave π as π.

MIXED PRACTICE

Problem set

1. The taxi ride cost $1.40 plus 35¢ for each tenth of a mile. What was the average cost per mile for a 4-mile taxi ride?
(55)

2. The table at right shows how many competitors earned certain scores in the event. Create a box-and-whisker plot for these scores.
(Inv. 4)

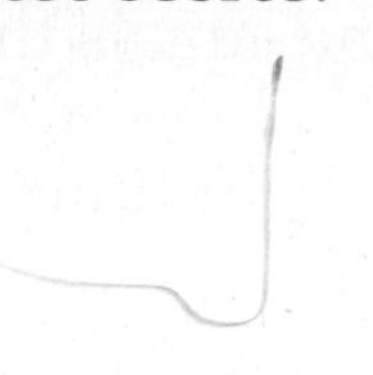

Competitor Scores

Score	Number of Competitors
100	||||
95	𝍸 |
90	𝍸 |||
85	𝍸 ||
80	|||
75	|
70	|

3. The coordinates of the vertices of ΔABC are $A\,(-1, -1)$, $B\,(-1, -4)$, and $C\,(-3, -2)$. The reflection of ΔABC in the axis is its image $\Delta A'B'C'$. Draw both triangles and write the coordinates of the vertices of $\Delta A'B'C'$.
(80)

4. If Jackson is paid $6 per hour, how much will he earn in 4 hours 20 minutes?
(53)

5. (36, 75) In this rectangle, what is the ratio of the shaded area to the unshaded area?

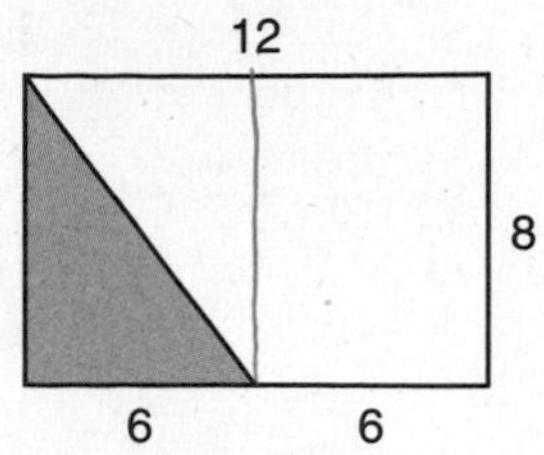

6. (72) If 600 pounds of sand costs \$7.20, what would be the cost of 1 ton of sand at the same price per pound?

7. (92) The cost of production rose 30%. If the new cost is \$3.90 per unit, what was the old cost per unit?

8. (92) If a grocery store marks up cereal 30%, what is the retail price of a large box of cereal that costs the store \$3.90?

9. (88) Use two unit multipliers to convert 1000 mm^2 to square centimeters.

10. (71) Diagram this statement. Then answer the questions that follow.

Three fifths of the Lilliputians believed in giants. The other 60 Lilliputians did not believe in giants.

(a) How many Lilliputians were there?

(b) How many Lilliputians believed in giants?

11. (79) Compare: $a \bigcirc b$ if a is a counting number and b is an integer

12. (91) Evaluate: $m(m + n)$ if $m = -2$ and $n = -3$

13. (94) If a pair of dot cubes is tossed once, what is the probability that the total number rolled will be

(a) 7?

(b) a number less than 7?

14. (95) Find the volume of the triangular prism shown. Dimensions are in millimeters.

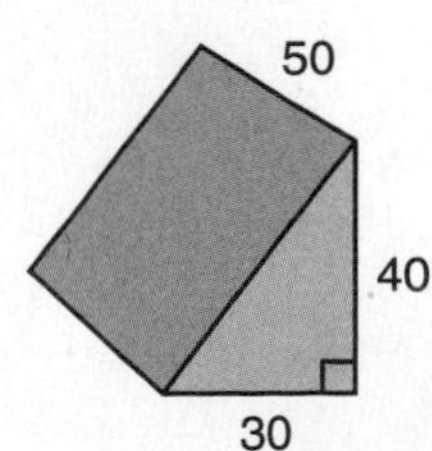

15. (95) The diameter of a soup can is 6 cm. Its height is 10 cm. What is the volume of the soup can? (Use 3.14 for π.)

16. (46) Find the total cost, including 6 percent sales tax, of 3 tacos at \$1.25 each, 2 soft drinks at 95¢ each, and a shake at \$1.30.

17. (94) Esther made a tree diagram for tossing three coins. Copy and complete this diagram, and then make a list of all the possible outcomes for the experiment.

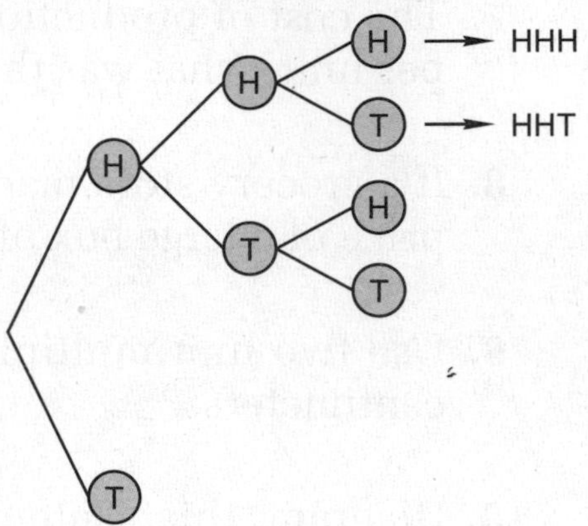

18. (84, 87) Simplify:

(a) $(-2xy)(-2x)(x^2y)$

(b) $6x - 4y + 3 - 6x - 5y - 8$

19. (83) Multiply. Write the product in scientific notation.

$$(8 \times 10^{-6})(4 \times 10^{4})$$

20. (85, Inv. 9) (a) Find the missing numbers in the table by using the function rule. Then graph the function.

(b) At what point does the graph of the function intersect the y-axis?

$y = \frac{1}{2}x + 1$

x	y
6	☐
4	☐
−2	☐

21. (40) Find the measures of the following angles.

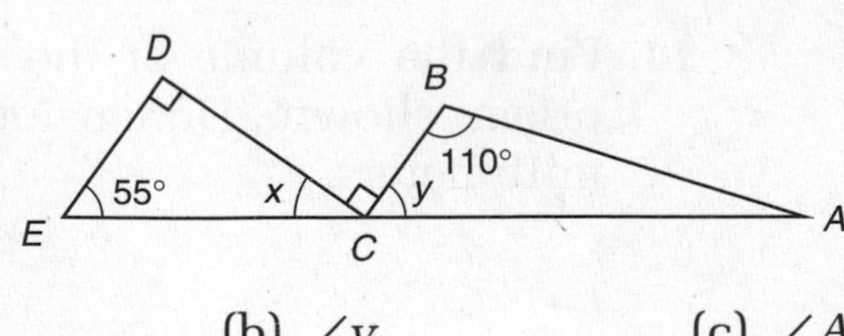

(a) $\angle x$ (b) $\angle y$ (c) $\angle A$

22. (78) On a number line, graph all the negative numbers that are greater than –2.

23. (34) What is the average of the numbers labeled A and B on the number line below?

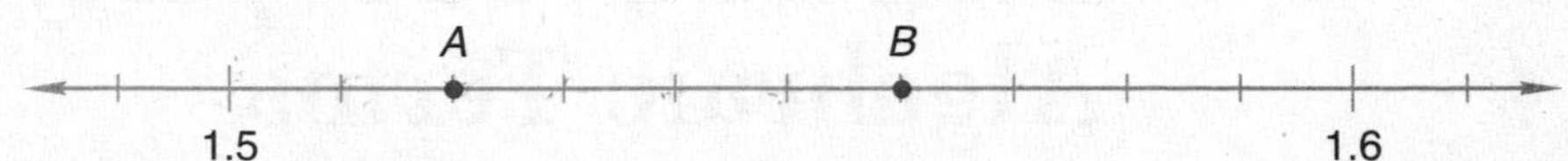

For problems 24 and 25, solve the equation. Show each step.

24. (93) $-5w + 11 = 51$

25. (93) $\frac{4}{3}x - 2 = 14$

26. (93) Solve this inequality and graph its solution:

$$0.9x + 1.2 \leq 3$$

Simplify:

27. (57) $\frac{10^3 \cdot 10^2}{10^5} - 10^{-1}$

28. (63) $\sqrt{1^3 + 2^3} + (1 + 2)^3$

29. (23, 26) $5 - 2\frac{2}{3}\left(1\frac{3}{4}\right)$

30. (85, 91) (a) $\frac{(-10) + (-8) - (-6)}{(-2)(+3)}$

(b) $-8 + 3(-2) - 6$

LESSON

96 Estimating Angle Measures • Distributive Property with Algebraic Terms

WARM-UP

Facts Practice: Two-Step Equations (Test U)

Mental Math:

a. $(-27) - (-50)$

b. $(5 \times 10^5)(2 \times 10^7)$

c. $160 = 80 + 4y$

d. Convert 15°C to degrees Fahrenheit.

e. $9 \times 1\frac{2}{3}$

f. 25% more than \$80

g. Estimate 15% of \$49.75.

Problem Solving:

What is the average of these fractions?

$$\frac{1}{4}, \frac{1}{6}, \frac{1}{12}$$

NEW CONCEPTS

Estimating angle measures

We have practiced reading the measure of an angle from a protractor scale. The ability to measure an angle with a protractor is an important skill. The ability to **estimate** an angle measure is also valuable. In this lesson we will learn a technique for estimating the measure of an angle. We will also practice using a protractor as we check our estimates.

To estimate a measurement, we need a mental image of the units to be used in the measurement. To estimate angle measures, we need a mental image of a degree scale—a mental protractor. We can "build" a mental image of a protractor from a mental image we already have—the face of a clock.

The face of a clock is a full circle, which is 360°, and is divided into 12 numbered divisions that mark the hours. From one numbered division to the next is $\frac{1}{12}$ of a full circle. One twelfth of 360° is 30°. Thus the measure of the angle formed by the hands of a clock at 1 o'clock is 30°, at 2 o'clock is 60°, and at 3 o'clock is 90°. A clock face is further divided into 60 smaller divisions that mark the minutes. From one small division to the next is $\frac{1}{60}$ of a circle. One sixtieth of 360° is 6°.

$$60\overline{)360^\circ}\quad 6^\circ$$

Thus, **from one minute mark to the next on the face of a clock is 6°.**

Here we have drawn an angle on the face of a clock. The vertex of the angle is at the center of the clock. One side of the angle is set at 12, and the other side of the angle is at "8 minutes after."

Since each minute of separation represents 6°, the measure of this angle is 8 × 6°, which is 48°. With some practice we can usually estimate the measure of an angle to within 5° of its actual measure.

Example 1 (a) Estimate the measure of $\angle BOC$ in the figure below.

(b) Use a protractor to find the measure of $\angle BOC$.

(c) By how many degrees did your estimate differ from your measurement?

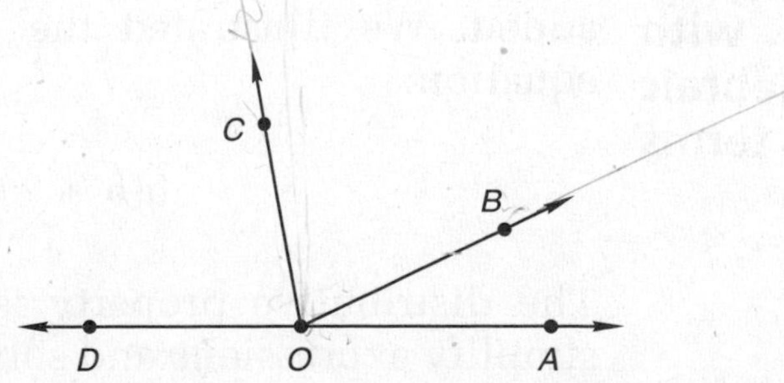

Solution (a) We use a mental image of a clock face on $\angle BOC$ with $\overrightarrow{OC}$ set at 12. Mentally we see that $\overrightarrow{OB}$ falls more than 10 minutes "after." Perhaps it is 12 minutes after. Since $12 \times 6° = 72°$, we estimate that m$\angle BOC$ is **72°**.

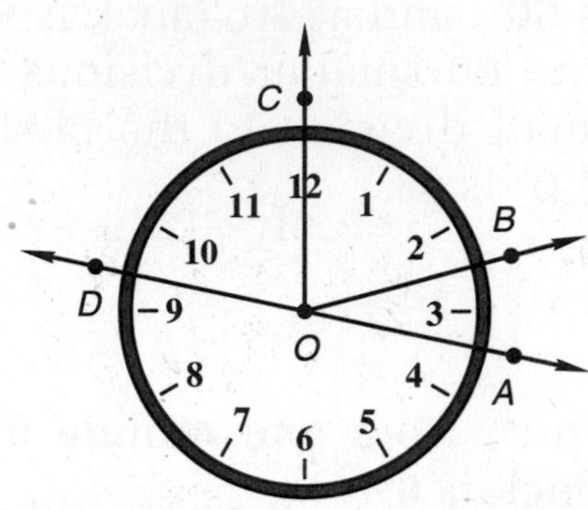

(b) We trace $\angle BOC$ on our paper and extend the sides so that we can use a protractor. We find that m$\angle BOC$ is **75°**.

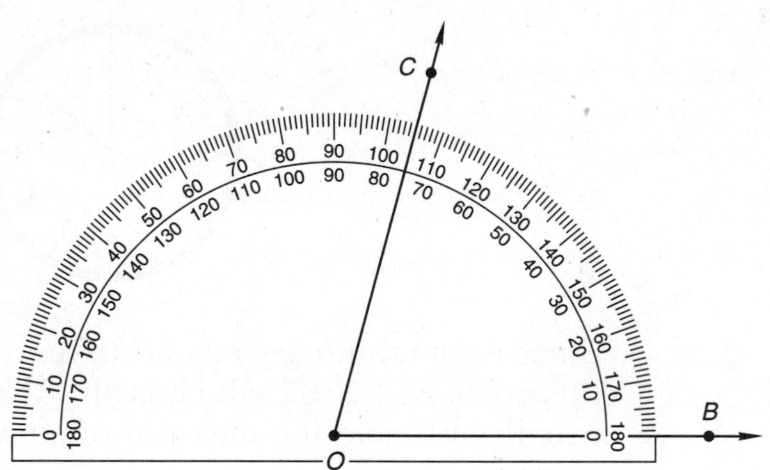

(c) Our estimate of 72° differs from our measurement of 75° by **3°**.

Distributive property with algebraic terms

Recall from Lesson 41 that the distributive property "spreads" multiplication over terms that are algebraically added. We illustrated the distributive property with this equation:

$$a(b + c) = ab + ac$$

The distributive property is frequently used in algebra to simplify expressions and solve equations.

Example 2 Simplify:

(a) $2(x - 3)$ (b) $-2(x + 3)$

(c) $-2(x - 3)$ (d) $x(x - 3)$

Solution (a) We multiply 2 by x, and we multiply 2 by -3.

$$2(x - 3) = \mathbf{2x - 6}$$

(b) We multiply -2 by x, and we multiply -2 by 3.

$$-2(x + 3) = \mathbf{-2x - 6}$$

(c) We multiply -2 by x, and we multiply -2 by -3.

$$-2(x - 3) = \mathbf{-2x + 6}$$

(d) We multiply x by x, and we multiply x by -3.

$$x(x - 3) = \mathbf{x^2 - 3x}$$

Example 3 Simplify: $x^2 + 2x + 3(x - 2)$

Solution We first use the distributive property to clear parentheses. Then we add like terms.

$x^2 + 2x + 3(x - 2)$	expression
$x^2 + 2x + 3x - 6$	distributive property
$\mathbf{x^2 + 5x - 6}$	added $2x$ and $3x$

LESSON PRACTICE

Practice set By counting minute marks on the clock face below, find the measure of each angle:

a. $\angle AOB$ **b.** $\angle AOC$ **c.** $\angle AOD$

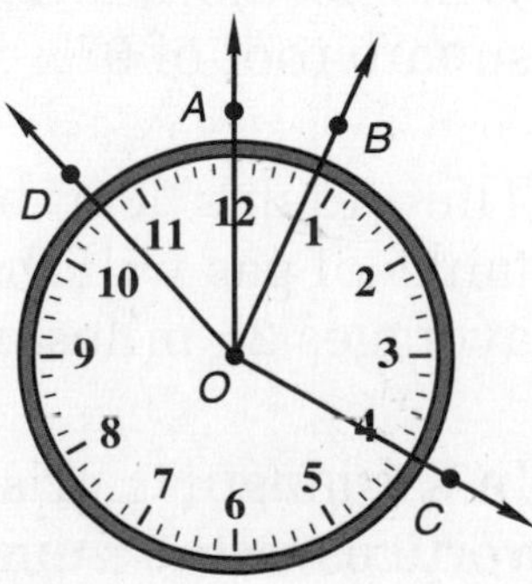

In practice problems **d–g,** estimate the measure of each angle. Then use a protractor to measure each angle. By how many degrees did your estimate differ from your measurement?

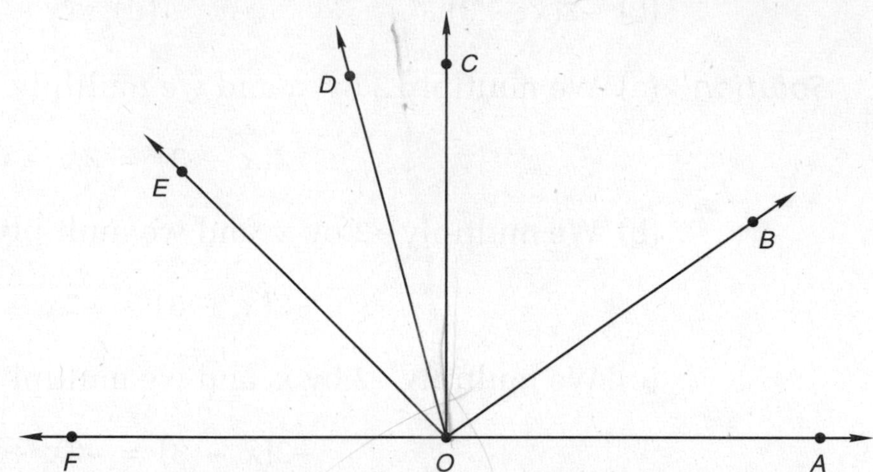

d. $\angle BOC$ **e.** $\angle DOC$

f. $\angle FOE$ **g.** $\angle FOB$

Simplify:

h. $x(x - y)$ **i.** $-3(2x - 1)$

j. $-x(x - 2)$ **k.** $-2(4 - 3x)$

l. $x^2 + 2x - 3(x + 2)$ **m.** $x^2 - 2x - 3(x - 2)$

MIXED PRACTICE

Problem set

1. (55) In May the merchant bought 3 tons of beans at an average price of $280 per ton. In June the merchant bought 5 tons of beans at an average price of $240 per ton. What was the average price of all the beans the merchant bought in May and June?

2. (20) What is the quotient when 9 squared is divided by the square root of 9?

3. (28) The Adams' car has a 16-gallon gas tank. How many tanks of gas will the car use on a 2000-mile trip if the car averages 25 miles per gallon?

4. (36, 67) In a triangular prism, what is the ratio of the number of vertices to the number of edges?

Use ratio boxes to solve problems 5–7.

5. (72) If 12 dollars can be exchanged for 100 yuan, what is the cost in dollars of an item that sells for 475 yuan?

6. (92) Sixty is 20 percent less than what number?

7. (92) The average number of customers per day increased 25 percent during the sale. If the average number of customers per day before the sale was 120, what was the average number of customers per day during the sale?

8. (77) Write equations to solve these problems:

(a) Sixty is what percent of 50?

(b) Fifty is what percent of 60?

9. (88) Use two unit multipliers to convert 1.2 m^2 to square centimeters.

10. (18, 40) Triangle *ABC* is similar to triangle *EDC*.

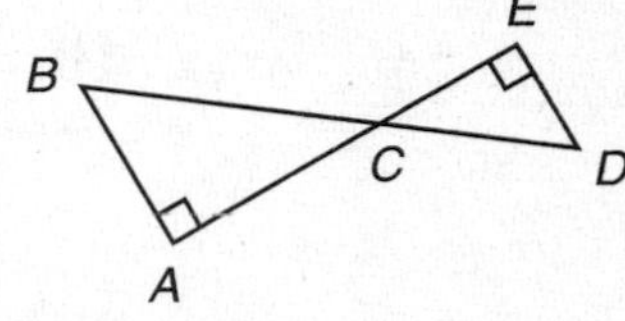

(a) List three pairs of corresponding angles and three pairs of corresponding sides.

(b) If $m\angle ABC = 53°$, what is $m\angle ECD$?

11. (79) Compare: $x + y \bigcirc x - y$ if $y > 0$

12. (91) Evaluate: $c(a + b)$ if $a = -4$, $b = -3$, and $c = -2$

13. (16, 20) The perimeter of a certain square is 1 yard. Find the area of the square in square inches.

14. (94) The face of this spinner is divided into one half and two fourths.

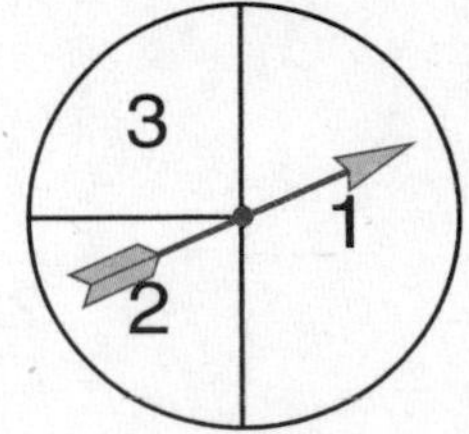

(a) If the spinner is spun two times, what is the probability that it will stop on 3 both times?

(b) If the spinner is spun four times, what is the probability that it will stop on 1 four times in a row?

15. Find the volume of each solid. Dimensions are in centimeters.
(95)

(a)

(b)

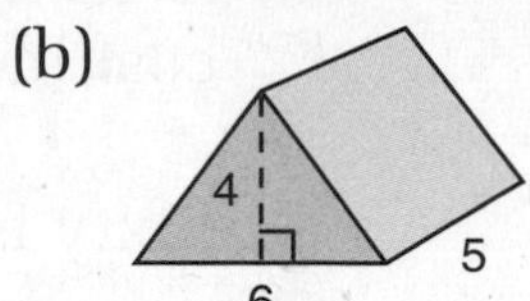

16. Find the total price, including 7 percent sales tax, of 20 square yards of carpeting priced at \$14.50 per square yard.
(46)

17. Complete the table.
(48)

FRACTION	DECIMAL	PERCENT
(a)	(b)	$3\frac{3}{4}\%$

18. Raincoats regularly priced at \$24 were on sale for $33\frac{1}{3}\%$ off.
(60, 92)

(a) What is $33\frac{1}{3}\%$ of \$24?

(b) What is $33\frac{1}{3}\%$ less than \$24?

19. Multiply. Write the product in scientific notation.
(83)

$$(3 \times 10^3)(8 \times 10^{-8})$$

20. (a) Find the circumference of the circle at right.
(66, 82)

(b) Find the area of the circle.

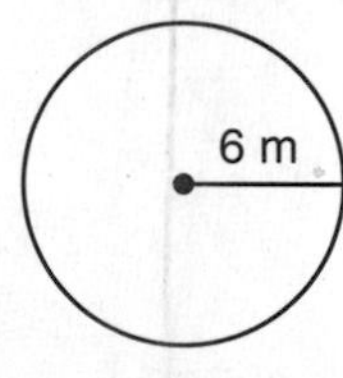

Leave π as π.

21. Use the clock face to estimate the measure of each angle:
(96)

(a) $\angle BOC$

(b) $\angle COA$

(c) $\angle DOA$

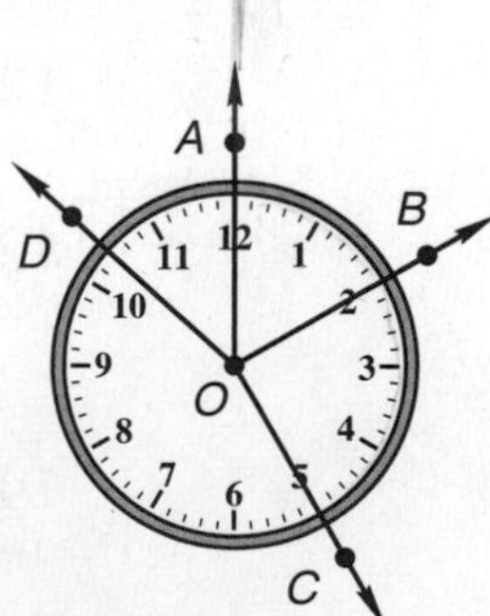

22. (a) What is the measure of each exterior angle of a regular octagon?
(89)

(b) What is the measure of each interior angle of a regular octagon?

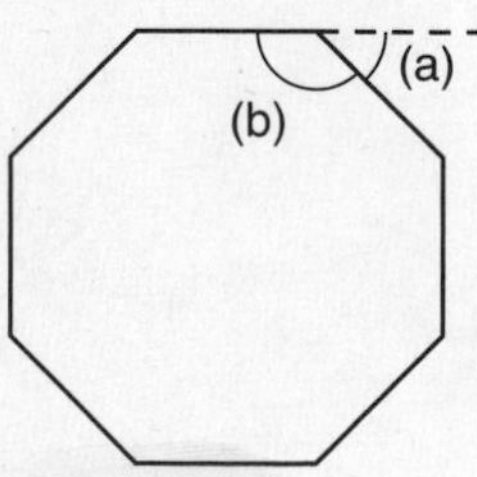

23. Find the coordinates of the vertices of $\square Q'R'S'T'$, which is the image of $\square QRST$ translated 4 units right and 4 units down.
(80)

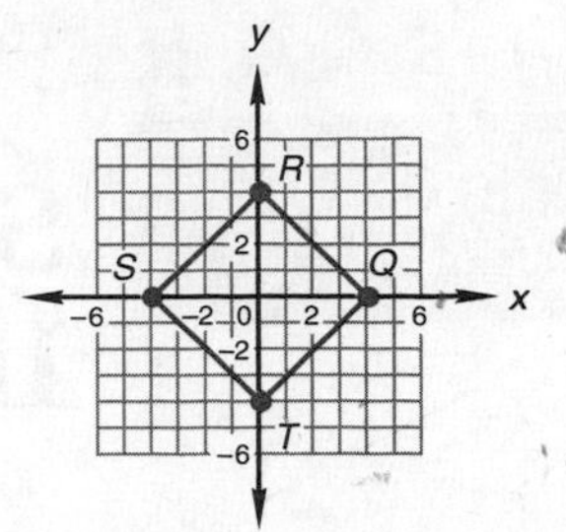

24. Solve this inequality and graph its solution:
(93)

$$0.8x + 1.5 < 4.7$$

For problems 25 and 26, solve the equation. Show each step.

25. $2\frac{1}{2}x - 7 = 13$
(93)

26. $-3x + 8 = -10$
(93)

Simplify:

27. (a) $-3(x - 4)$
(96)

(b) $x(x + y)$

28. (a) $\dfrac{(-4) - (-8)(-3)(-2)}{-2}$
(85)

(b) $(-3)^2 + 3^2$

29. (a) $(-4ab^2)(-3b^2c)(5a)$
(84, 87)

(b) $a^2 + ab - ab - b^2$

30. If the spinner is spun twice, two of the possible outcomes are C,A and A,C. List all the possible outcomes (the sample space) for this experiment.
(94)

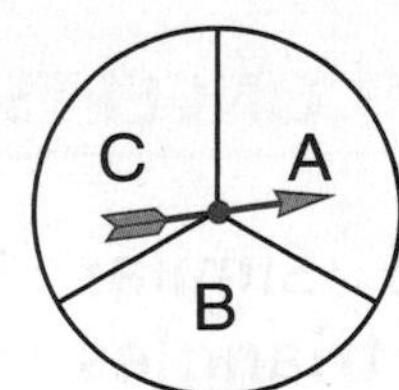

LESSON

97 Similar Triangles • Indirect Measure

WARM-UP

Facts Practice: Order of Operations (Test T)

Mental Math:

a. $(-5)(-5)(-5)$

b. $(5 \times 10^6)(6 \times 10^5)$

c. $\frac{m}{4.5} = \frac{0.6}{3}$

d. Convert 5°C to degrees Fahrenheit.

e. $10 \times 6\frac{1}{2}$

f. 25% less than $80

g. $\sqrt{100}$, × 7, + 2, ÷ 8, × 4, $\sqrt{\ }$, × 5, − 2, ÷ 2, ÷ 2, ÷ 2

Problem Solving:

Find the numbers that complete this table for a hexagon and for an *n*-gon. (An *n*-gon is a polygon with *n*-sides.)

Type of Polygon	Number of Diagonals from One Vertex	Sum of Interior Angles
quadrilateral	1	2 × 180°
pentagon	2	3 × 180°
hexagon		
n-gon		

NEW CONCEPTS

Similar triangles We often use tick marks to indicate that the measures of angles are equal (that the angles are congruent).

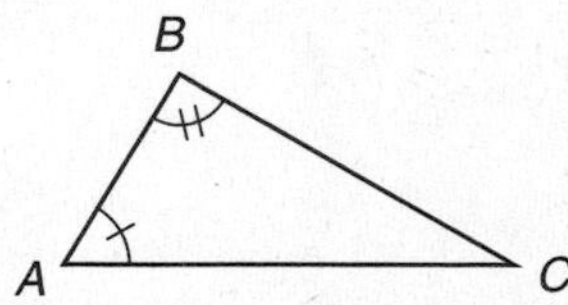

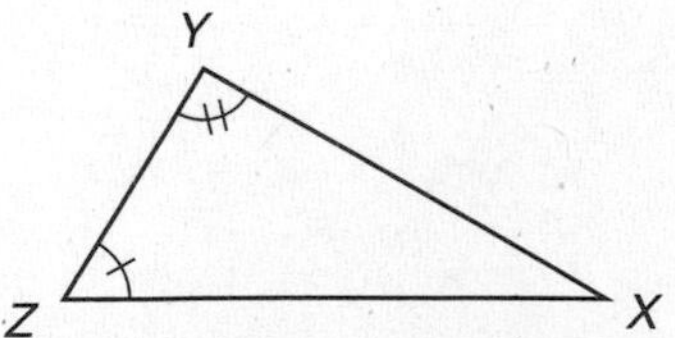

In these figures the single tick marks indicate that angles A and Z have equal measures. The double tick marks indicate that angles B and Y have equal measures. Since the sum of the angle measures of each triangle is 180°, we can conclude that the remaining angles, C and X, also have equal measures.

Recall from Lesson 18 that corresponding angles of similar figures have the same measure. The converse is also true. If three angles in one triangle have the same measures as three angles in another triangle, the triangles are *similar triangles.* So triangles *ABC* and *ZYX* are similar. Also recall that similar triangles have three pairs of corresponding angles and three pairs of corresponding sides. Here we show the corresponding angles and sides for the triangles above:

CORRESPONDING ANGLES	CORRESPONDING SIDES
$\angle A$ and $\angle Z$	$\overline{AB}$ and $\overline{ZY}$
$\angle B$ and $\angle Y$	$\overline{BC}$ and $\overline{YX}$
$\angle C$ and $\angle X$	$\overline{CA}$ and $\overline{XZ}$

In this lesson we will focus our attention on the following characteristic of similar triangles:

The lengths of corresponding sides of similar triangles are proportional.

This means that ratios formed by corresponding sides are equal, as we illustrate with the two triangles below.

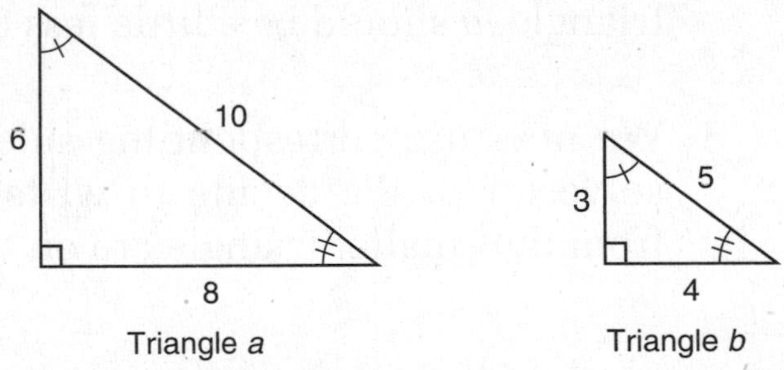

Triangle *a*   Triangle *b*

The lengths of the corresponding sides of triangles *a* and *b* are 6 and 3, 8 and 4, and 10 and 5. These pairs of lengths can be written as equal ratios.

$$\frac{\text{triangle } a}{\text{triangle } b} \quad \frac{6}{3} = \frac{8}{4} = \frac{10}{5}$$

Notice that each of these ratios equals 2. If we choose to put the lengths of the sides of triangle *b* on top, we get three ratios, each equal to $\frac{1}{2}$.

$$\frac{\text{triangle } b}{\text{triangle } a} \quad \frac{3}{6} = \frac{4}{8} = \frac{5}{10}$$

We can write proportions using equal ratios in order to find the lengths of unknown sides of similar triangles.

Example 1 Estimate the length a. Then use a proportion to find a.

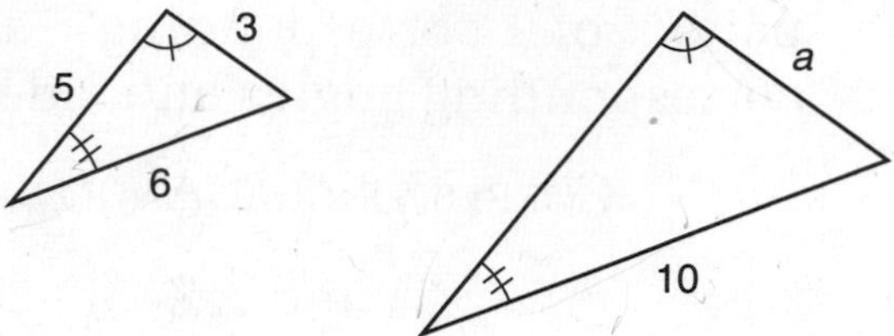

Solution The tick marks indicate two pairs of congruent angles in the triangles. Because the sum of the interior angles of every triangle is 180°, we know that the unmarked pair of angles is also congruent. Three pairs of congruent angles means the two triangles are similar. Therefore, the lengths of corresponding sides are proportional.

The side of length 6 in the smaller triangle corresponds to the side of length 10 in the larger triangle. Thus the side lengths of the larger triangle are not quite double the side lengths of the smaller triangle. Since the side of length a in the larger triangle corresponds to the side of length 3 in the smaller triangle, a should be a little less than 6. We estimate a to be 5.

We now use corresponding sides to write a proportion and solve for a. We decide to write the ratios so that the sides from the smaller triangle are on top.

$$\frac{6}{10} = \frac{3}{a} \quad \text{equal ratios}$$

$$6a = 30 \quad \text{cross multiplied}$$

$$a = \mathbf{5} \quad \text{solved}$$

Indirect measure Sarah looked up and said, "I wonder how tall that tree is." Beth looked down and said, "It's about 25 feet tall." Beth did not *directly* measure the height of the tree. Instead she used her knowledge of proportions to *indirectly* estimate the height of the tree.

The lengths of the shadows cast by two objects are proportional to the heights of the two objects (assuming the objects are in the same general location).

We can separate the objects and their shadows into two "triangles."

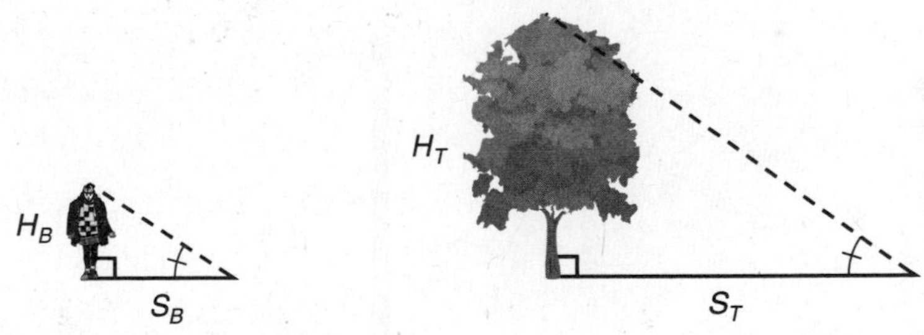

Assuming the ground is flat and level, both the tree and Beth are perpendicular to the ground. The angle of the sun's light is the same for both Beth and the tree. Thus the triangles are similar. The height of Beth (H_B) and the length of Beth's shadow (S_B) are proportional to the height of the tree (H_T) and the length of the tree's shadow (S_T). We can record the relationship in a ratio box.

	Height of object	Length of shadow
Beth	H_B	S_B
Tree	H_T	S_T

How did Beth perform the calculation? We suggest two ways. Knowing her own height (5 ft), she may have estimated the length of her shadow (6 ft) and the length of the shadow of the tree (30 ft). She then could have solved this proportion in which the dimensions are feet:

$$\frac{5}{H_T} = \frac{6}{30}$$

$$6H_T = 5 \cdot 30$$

$$H_T = 25$$

Another way Beth may have estimated the height of the tree is by estimating that the tree's shadow was five times as long as her own shadow. If the tree's shadow was five times as long as her shadow, then the tree's height must have been five times her height.

$$5 \text{ ft} \times 5 = 25 \text{ ft}$$

Example 2 To indirectly measure the height of a light pole on her street, Jenna measured the length of the shadow cast by the pole (24 ft) and the length of the shadow cast by a vertical meterstick (40 cm). About how tall was the light pole?

Solution We sketch the objects and their shadows using the given information.

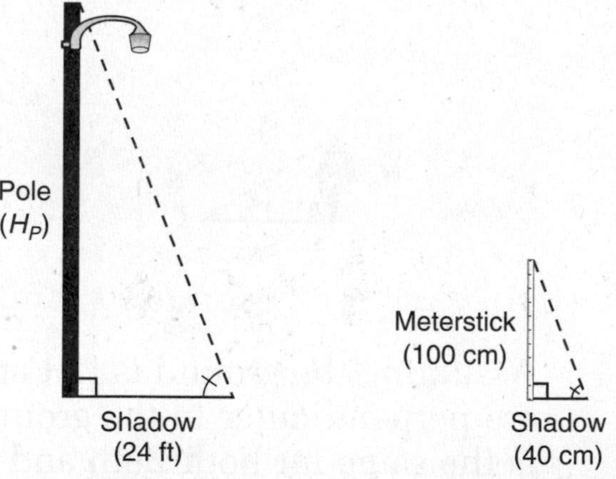

We use a ratio box to record the given information. Then we write and solve a proportion.

	Height of object	Length of shadow
Meterstick	100 cm	40 cm
Pole	H_P	24 ft

$$\frac{100}{H_P} = \frac{40}{24}$$

$$40H_P = 24 \cdot 100$$

$$H_P = 60$$

We find that the height of the light pole was about **60 feet.** Note that the two objects were measured using different units. Exercise caution whenever mixing units in proportions to ensure the solution is expressed in the desired units. You may choose to include units in your calculation, as we have done below.

$$\frac{100 \text{ cm}}{H_P} = \frac{40 \text{ cm}}{24 \text{ ft}}$$

$$40 \text{ cm} \cdot H_P = 24 \text{ ft} \cdot 100 \text{ cm}$$

$$H_P = \frac{\overset{6}{\cancel{24}} \text{ ft} \cdot \overset{10}{\cancel{100}} \cancel{\text{cm}}}{\underset{\underset{1}{\cancel{4}}}{\cancel{40}} \cancel{\text{cm}}}$$

$$H_P = 60 \text{ ft}$$

Example 3 Brad saw that his shadow was the length of about one big step (3 ft). He then walked along the shadow of a nearby flagpole and found that it was eight big steps long. Brad is about 5 ft tall. Which is the best choice for the height of the flagpole?

A. 15 ft B. 24 ft C. 40 ft D. 60 ft

Solution We can estimate the heights of objects by the lengths of their shadows. In this example, Brad, who is about 5 ft tall, cast a shadow about 3 ft long. So Brad is nearly twice as tall as the length of the shadow he cast. The shadow of the flagpole was the length of about eight big steps, or about 24 ft. So the height of the pole is about, but not quite, twice the length of its shadow. Thus, the best choice for the height of the pole is **C. 40 ft.**

A quicker way to choice C is to notice that each step of shadow length corresponds to 5 ft of object height. So a shadow eight steps long was cast by an object whose height is 8 × 5 ft, or 40 ft.

Activity: *Indirect Measure*

Using a yardstick, ruler, and/or tape measure, indirectly measure the height of a tree, building, pole, or other tall object as described in example 2.

LESSON PRACTICE

Practice set a. Identify each pair of corresponding angles and each pair of corresponding sides in these two triangles:

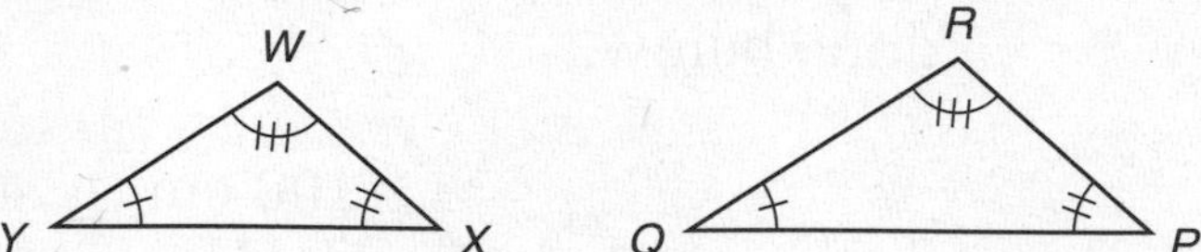

Refer to the figures shown to answer problems **b–e.**

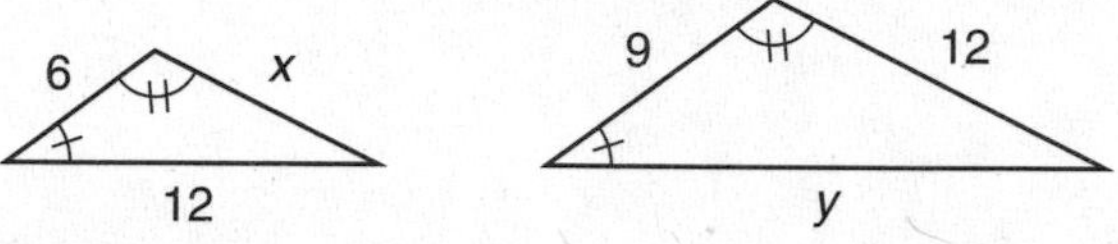

b. Estimate the length x.

c. Find the length x.

d. Estimate the length y.

e. Find the length y.

f. A tree casts a shadow 18 feet long, while a 6-foot pole casts a shadow 9 feet long. How tall is the tree?

g. As Donald stood next to the pole supporting the basketball backboard, he noticed that the shadow of the pole was about twice the length of his own shadow. If Donald is 5 ft 6 in. tall, what is a reasonable estimate of the height of the pole?

MIXED PRACTICE

Problem set

1. (46) Marta gave the clerk \$10 for a CD that cost \$8.95 plus 6 percent tax. How much money should she get back?

2. (51) Three hundred billion is how much less than two trillion? Write the answer in scientific notation.

3. (Inv. 4) Yusuke's history test scores were:

95, 90, 80, 85, 90, 100, 85, 90, 95, 80

Find the (a) median, (b) mode, and (c) range of these scores.

Use ratio boxes to solve problems 4 and 5.

4. (72) Coming down the long hill, Nelson averaged 24 miles per hour. If it took him 5 minutes to come down the hill, how long was the hill?

5. (72) If Nelson traveled 3520 yards in 5 minutes, how far could he travel in 8 minutes at the same rate?

6. (80) Describe the transformation that moves ΔABC to its image $\Delta A'B'C'$.

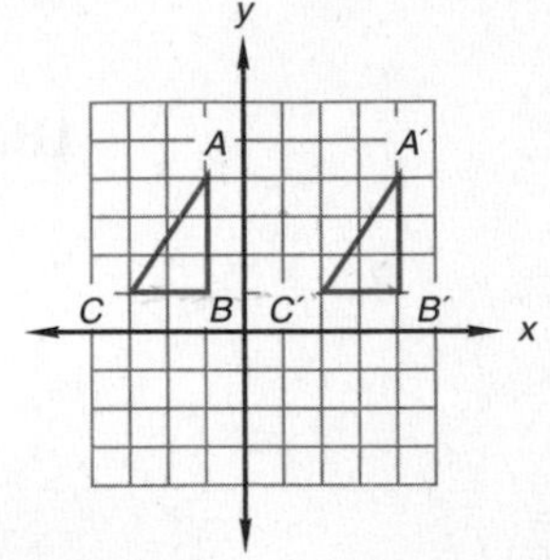

7. (16, 60) Three fourths of a yard is how many inches?

8. (54) Use a ratio box to solve this problem. The ratio of leeks to radishes growing in the garden was 5 to 7. If 420 radishes were growing in the garden, how many leeks were there?

Write equations to solve problems 9–11.

9. (77) Forty is 250 percent of what number?

10. (77) Forty is what percent of 60?

11. (60) What decimal number is 40 percent of 6?

12. (92) Use a ratio box to solve this problem. The tuition increased 10 percent this year. If the tuition this year is $17,600, what was the tuition last year?

13. (28, 34) What is the average of the two numbers marked by arrows on the number line below?

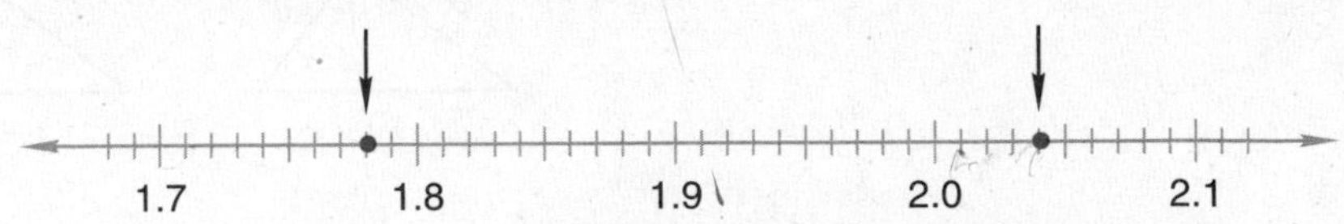

14. (48) Complete the table.

FRACTION	DECIMAL	PERCENT
(a)	3.25	(b)
$\frac{1}{6}$	(c)	(d)

15. Compare: $x + y \bigcirc x - y$ if x is positive and y is negative
(79)

16. Multiply. Write the product in scientific notation.
(83)

$$(5.4 \times 10^8)(6 \times 10^{-4})$$

17. Find the (a) circumference and (b) area of a circle with a radius of 10 millimeters. (Use 3.14 for π.)
(66, 82)

18. Find the area of the trapezoid shown. Dimensions are in feet.
(75)

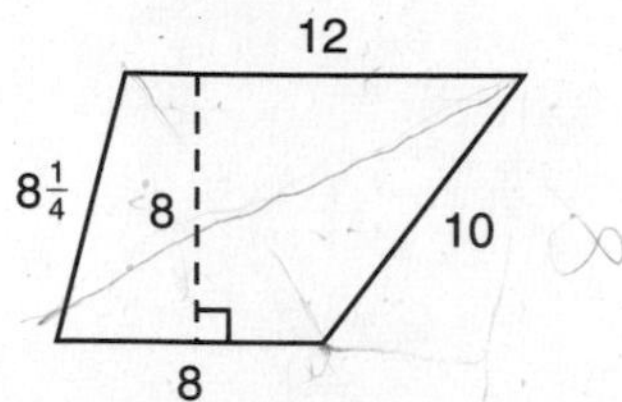

19. Find the volume of each of these right solids. Dimensions are in meters. (Leave π as π.)
(95)

(a)

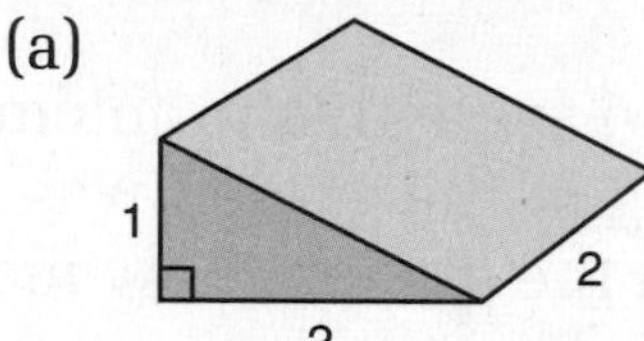

(b)

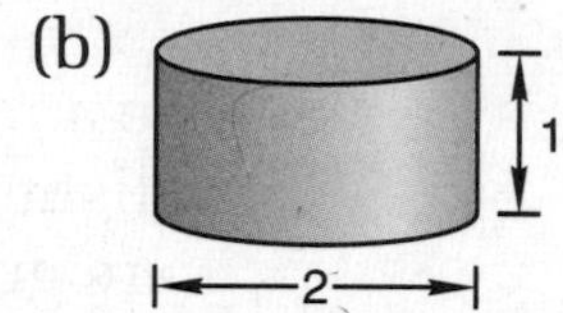

20. Refer to the figure below. What are the measures of the following angles?
(40)

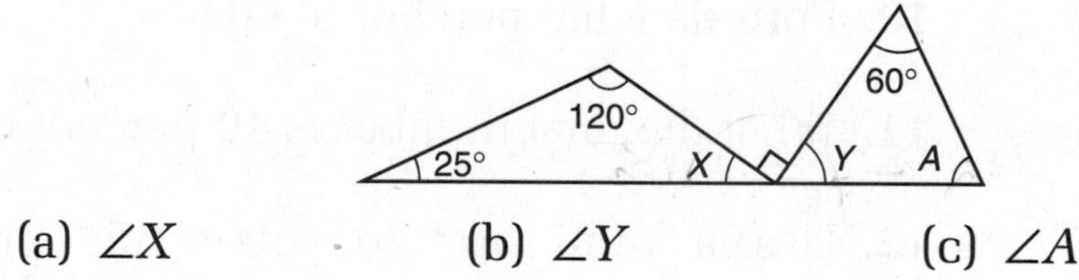

(a) $\angle X$ (b) $\angle Y$ (c) $\angle A$

21. The triangles below are similar. Find x. Dimensions are in centimeters.
(97)

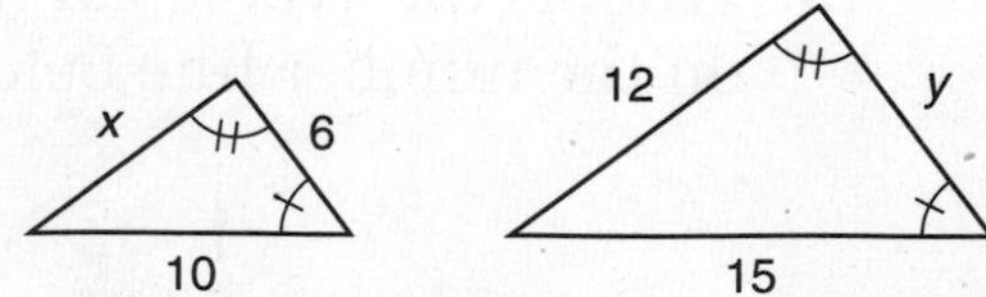

22. How tall is a power pole that casts a 72-foot shadow if a nearby vertical yardstick casts a 48-inch shadow? Draw a diagram to illustrate the problem.
(97)

23. Estimate: $\dfrac{(38{,}470)(607)}{79}$
(29)

For problems 24 and 25, solve the equation. Show each step.

24. $1.2m + 0.12 = 12$
(93)

25. $1\frac{3}{4}y - 2 = 12$
(93)

Simplify:

26. $3x - y + 8 + x + y - 2$
(84)

27. (a) $3(x - y)$
(96)
(b) $x(y - 3)$

28. $3\frac{1}{3} \div \left(4.5 \div 1\frac{1}{8}\right)$
(43)

29. $\dfrac{(-2) - (+3) + (-4)(-3)}{(-2) + (+3) - (+4)}$
(85)

30. Sam placed three cards face down on a table as shown. Then he asked his friends to pick one card and then another. Sarah picked A then B (A,B). Javier picked B then A (B,A). List all the possible outcomes of Sam's experiment.
(94)

LESSON

98 Scale • Scale Factor

WARM-UP

Facts Practice: Two-Step Equations (Test U)

Mental Math:

a. $(-360) \div (8)$

b. $(2.5 \times 10^7)(4 \times 10^{-2})$

c. $2c + 1\frac{1}{2} = 6\frac{1}{2}$

d. Convert 0.02 kg to g.

e. $4 \times 3\frac{3}{4}$

f. $33\frac{1}{3}\%$ more than \$60

g. At 12 mph, how far can Toby ride a bike in 1 hour and 45 minutes?

Problem Solving:

To convert from degrees Celsius to degrees Fahrenheit, Eliana's grandfather said, "Double the Celsius number; subtract 10%; then add 32°." We exercise caution when converting from a Celsius temperature less than 0°. Record the steps for converting from –10°C to degrees Fahrenheit.

NEW CONCEPTS

Scale

In the preceding lesson we discussed similar triangles. Scale models and scale drawings are other examples of similar shapes. Scale models and scale drawings are reduced (or enlarged) renderings of actual objects. As is true of similar triangles, the lengths of corresponding parts of scale models and the objects they represent are proportional.

The **scale** of a model is stated as a ratio. For instance, if a model airplane is $\frac{1}{24}$ the size of the actual airplane, the scale is stated as $\frac{1}{24}$ or 1:24. We can use the given scale to write a proportion to find a measurement on either the model or the actual object. A ratio box helps us put the numbers in the proper places.

Example 1 A model airplane is built with a scale of 1:24. If the wingspan of the model is 18 inches, the wingspan of the actual airplane is how many feet?

Solution The scale indicates that the dimensions of the actual airplane are 24 times the dimensions of the model, so the wingspan is 24 times 18 inches. Note that the model is measured in inches, but the question asks for the wingspan in feet. Thus we will divide the product of 18 and 24 by 12. This is the calculation:

$$\frac{18 \cdot 24}{12}$$

Recall that we can reduce before multiplying.

$$\frac{18 \cdot \overset{2}{\cancel{24}}}{\underset{1}{\cancel{12}}} = 36$$

Another way to approach this problem is to construct a ratio box as we have done with ratio problems. In one column we write the ratio numbers, which are the scale numbers. In the other column we write the measures. The first number of the scale refers to the model. The second number refers to the object. We can use the entries in the ratio box to write a proportion.

	Scale	Measure
Model	1	18
Object	24	*w*

$$\frac{1}{24} = \frac{18}{w}$$

$$w = 432$$

The wingspan of the model was given in inches. Solving the proportion, we find that the full-size wingspan is 432 inches. We are asked for the wingspan in feet, so we convert units from inches to feet.

$$432\ \cancel{\text{in.}} \cdot \frac{1\text{ ft}}{12\ \cancel{\text{in.}}} = 36\text{ ft}$$

We find that the wingspan of the airplane is **36 feet.**

We can graphically portray the relationship between the measures of an object and a scale model. Below we show a graph of the measures of the airplane and its model. Notice that the scale of the model is inches while the scale of the actual airplane is feet. Since the scale is 1:24, one inch on the model corresponds to 24 inches, which is 2 feet, on the actual airplane.

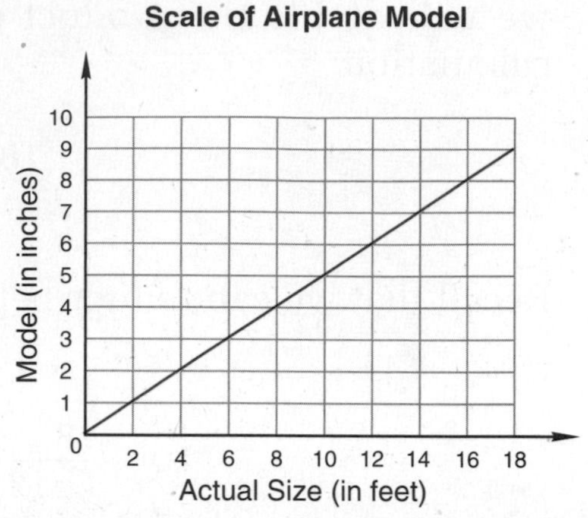

Every point on the graphed line represents a length on the airplane and its corresponding measure on the model.

Example 2 Sofia is molding a model of a car from clay. The scale of the model is 1:36. If the height of the car is 4 feet 6 inches, what should be the height of the model in inches?

Solution First we convert 4 feet 6 inches to inches:

$$4 \text{ feet } 6 \text{ inches} = 4(12) + 6 = 54 \text{ inches}$$

Then we construct a ratio box using 1 and 36 as the ratio numbers, write the proportion, and solve.

	Scale	Measure
Model	1	m
Object	36	54

$$\frac{1}{36} = \frac{m}{54}$$

$$36m = 54$$

$$m = \frac{54}{36} = 1\tfrac{1}{2}$$

The height of the model car should be **$1\frac{1}{2}$ inches.**

Scale factor

We have solved proportions by using cross products. Sometimes a proportion can be solved more quickly by noting the **scale factor.** The scale factor is the number of times larger (or smaller) the terms of one ratio are when compared with the terms of the other ratio. The scale factor in the proportion below is 6 because the terms of the second ratio are 6 times the terms of the first ratio.

$$\frac{3}{4} = \frac{18}{24}$$

Example 3 Solve: $\frac{3}{7} = \frac{15}{n}$

Solution Instead of finding cross products, we note that multiplying the numerator 3 by 5 gives us the other numerator, 15. Thus the scale factor is 5. We use this scale factor to find n.

$$\frac{3}{7} \times \frac{5}{5} = \frac{15}{35}$$

We find that n is **35.**

Example 4 These two triangles are similar. Calculate the scale factor from the smaller triangle to the larger triangle.

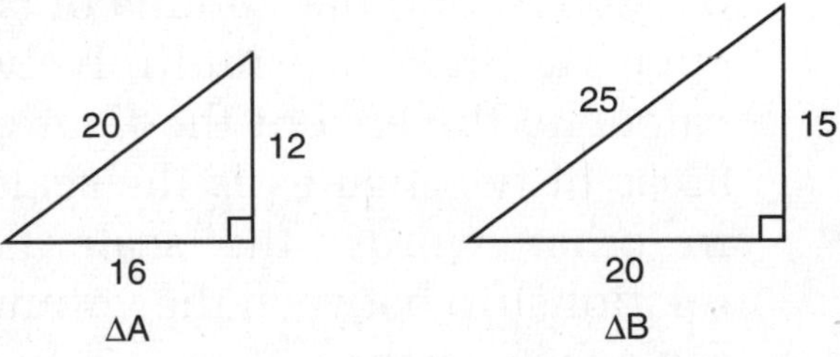

Solution If we multiply the length of one side of the smaller triangle by the scale factor, we get the length of the corresponding side of the larger triangle.

$$\text{Dimension of } \Delta\text{A} \times \text{scale factor} = \text{dimension of } \Delta\text{B}$$

We may select any pair of corresponding sides to calculate the scale factor. Here, we will select the longest sides. We write an equation using f for the scale factor and then solve for f.

$$20f = 25$$

$$f = \frac{25}{20}$$

$$f = \mathbf{\frac{5}{4}} \text{ or } \mathbf{1.25}$$

In this book we will express the scale factor in decimal form unless otherwise directed.

Note that the scale factor refers to the *linear measures* of two similar figures and not to the area or volume measures of the figures. The scale factor from cube A to cube B below is 2 because the linear measures of cube B are twice the corresponding measures of cube A.

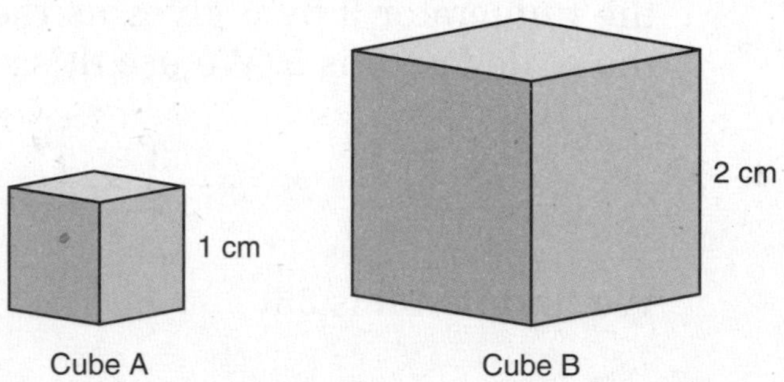

However, the surface area of cube B is 4 times the surface area of cube A, and the volume of cube B is 8 times the volume of cube A. Since we multiply two dimensions of a figure to calculate the area of the figure, the relationship between the areas of two figures is the scale factor times the scale factor; in other words, the scale factor squared. Likewise, the relationship between the volumes of two similar figures is the scale factor cubed.

Example 5 The smaller of two similar rectangular prisms has dimensions of 2 cm by 3 cm by 4 cm. The larger rectangular prism has dimensions of 6 cm by 9 cm by 12 cm.

(a) What is the scale factor from the smaller to the larger rectangular prism?

(b) The area of any face of the larger prism is how many times the area of the corresponding face of the smaller prism?

(c) The volume of the larger solid is how many times the volume of the smaller solid?

Solution Before answering the questions, we draw the two figures.

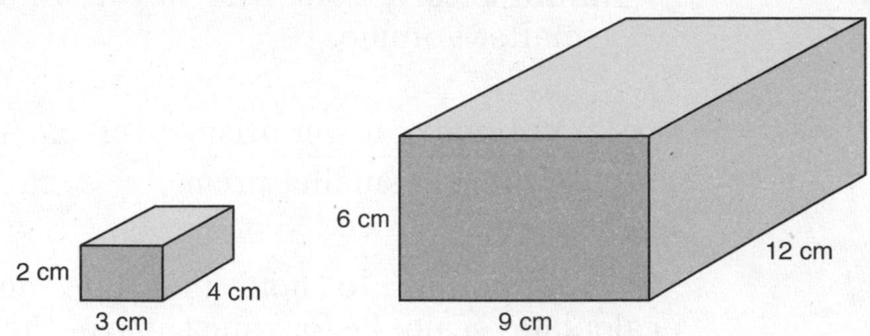

(a) We select any two corresponding linear measures to calculate the scale factor. We choose the 2-cm and the 6-cm measures.

Dimension of smaller × scale factor = dimension of larger

$$2f = 6$$
$$f = 3$$

We find that the scale factor is **3.**

(b) Since the scale factor from the smaller to the larger figure is 3, the area of any face of the larger figure should be $3^2 = 9$ times the area of the corresponding face of the smaller figure. We confirm this relationship by comparing the area of a 6-by-9-cm face of the larger prism with the corresponding 2-by-3-cm face of the smaller prism.

Area of 6-by-9-cm face = 54 cm^2

Area of 2-by-3-cm face = 6 cm^2

We see that the area of the selected face of the larger prism is indeed **9 times** the area of the corresponding face of the smaller prism.

(c) Since the scale factor of the linear dimensions of the two figures is 3, the volume of the larger prism should be $3^3 = 27$ times the volume of the smaller prism. We confirm this relationship by performing the calculations.

Volume of 6-by-9-by-12-cm prism = 648 cm^3

Volume of 2-by-3-by-4-cm prism = 24 cm^3

Dividing 648 cm^3 by 24 cm^3, we find that the larger volume is indeed **27 times** the smaller volume.

$$\frac{648 \text{ cm}^3}{24 \text{ cm}^3} = 27$$

Showing this calculation another way demonstrates more clearly why the larger volume is 3^3 times the smaller volume.

$$\frac{\text{Volume of larger prism}}{\text{Volume of smaller prism}} = \frac{\overset{3}{\cancel{6\text{ cm}}} \cdot \overset{3}{\cancel{9\text{ cm}}} \cdot \overset{3}{\cancel{12\text{ cm}}}}{\underset{1}{\cancel{2\text{ cm}}} \cdot \underset{1}{\cancel{3\text{ cm}}} \cdot \underset{1}{\cancel{4\text{ cm}}}} = 3^3$$

It is important to note that the measurements used to calculate scale factor must have the same units. If the measurements have different units, we should convert before calculating scale factor.

LESSON PRACTICE

Practice set

a. The blueprints were drawn to a scale of 1:24. If a length of a wall on the blueprint was 6 in., what was the length in feet of the wall in the house?

b. Bret is carving a model ship from balsa wood using a scale of 1:36. If the ship is 54 feet long, the model ship should be how many inches long?

Solve by using the scale factor:

c. $\frac{5}{7} = \frac{15}{w}$ **d.** $\frac{x}{3} = \frac{42}{21}$

e. These two rectangles are similar. Calculate the scale factor from the smaller rectangle to the larger rectangle.

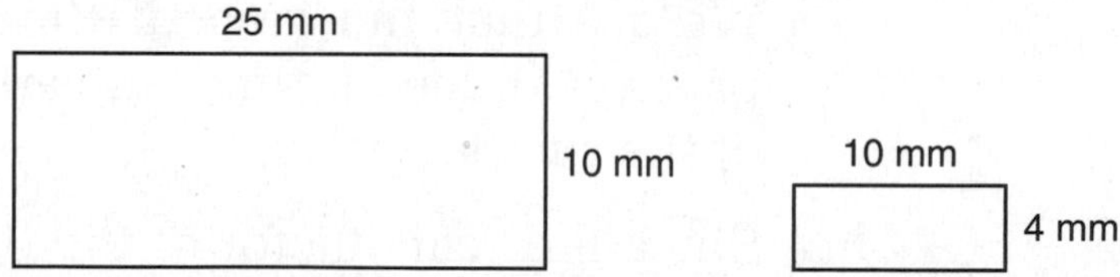

f. The area of the larger rectangle above is how many times the area of the smaller rectangle?

g. The scale of the car model in example 2 is 1:36. This means 1 inch on the model corresponds to 36 inches (that is, 3 feet) on the actual car. On grid paper make a graph that shows this relationship. Make the units of the horizontal axis feet to represent the car. On the vertical axis use inches for the model. Use the graph following example 1 as a pattern.

h. The statue of the standing World War II general was $1\frac{1}{2}$ times life-size. Which is the most reasonable estimate for the height of the statue?

A. 4 ft B. 6 ft C. 9 ft D. 15 ft

MIXED PRACTICE

Problem set

1. (94) If a pair of dot cubes is tossed once, what is the probability that the total number rolled will be

(a) greater than 7?

(b) less than 2?

Use ratio boxes to solve problems 2–4.

2. (92) The regular price of the item was $45, but the item was on sale for 20 percent off. What was the sale price?

3. (72) If 5 dollars equals 40 kroner, what is the cost in dollars of an item priced at 100 kroner?

4. (92) The number of club members increased 25 percent this year. If there are 20 more club members this year than there were last year, how many club members are there this year?

5. (84, 87) Simplify: $(3x)(x) - (x)(2x)$

6. (55) In her first 6 games Ann averaged 10 points per game. In her next 9 games Ann averaged 15 points per game. How many points per game did Ann average during her first 15 games?

7. (46) Ingrid started her trip at 8:30 a.m. with a full tank of gas and an odometer reading of 43,764 miles. When she stopped for gas at 1:30 p.m., the odometer read 44,010 miles.

(a) If it took 12 gallons to fill the tank, her car averaged how many miles per gallon?

(b) Ingrid traveled at an average speed of how many miles per hour?

8. (74) Write an equation to solve this problem. Three fifths of Tom's favorite number is 60. What is Tom's favorite number?

9. (Inv. 3, 80) On a coordinate plane, graph the points (−3, 2), (3, 2), and (−3, −2).

(a) If these points designate three of the vertices of a rectangle, what are the coordinates of the fourth vertex of the rectangle? Draw the rectangle.

(b) Draw the image of the rectangle in (a) after a 90° clockwise rotation about the origin. What are the coordinates of the vertices of the rotated image?

10. (36, 86) What is the ratio of counting numbers to integers in this set of numbers?

$$\{-3, -2, -1, 0, 1, 2\}$$

11. (20, 41) Find a^2 if $\sqrt{a} = 3$.

Write equations to solve problems 12 and 13.

12. (77) Forty is what percent of 250?

13. (77) Forty percent of what number is 60?

14. (92) An antique dealer bought a chair for $40 and sold the chair for 60% more. What was the selling price?

15. (8) Segment *BC* is how much longer than segment *AB*?

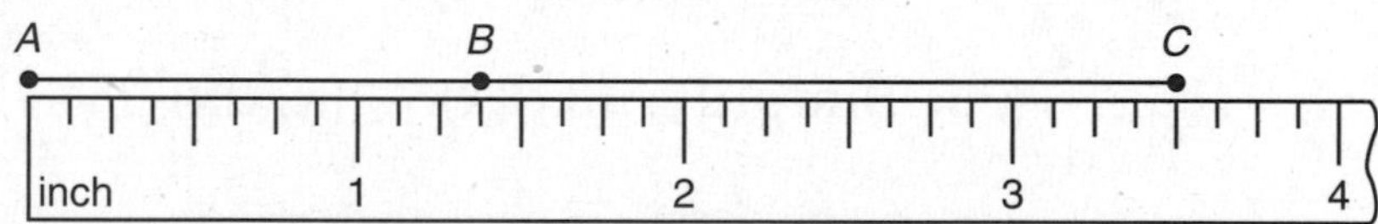

16. (78) Graph on a number line: $x \leq 3$

17. (48) Complete the table.

FRACTION	DECIMAL	PERCENT
(a)	(b)	1.4%

18. (83) Multiply. Write the product in scientific notation.

$$(1.4 \times 10^{-6})(5 \times 10^{4})$$

19. (85, Inv. 9) Find the missing numbers in the table by using the function rule. Then graph the function. Where does the graph of the function intersect the *y*-axis?

$y = -2x$

x	y
3	☐
0	☐
−2	☐

20. Find the (a) circumference and (b) area of a circle that has
(66, 82) a diameter of 2 feet. (Use 3.14 for π.)

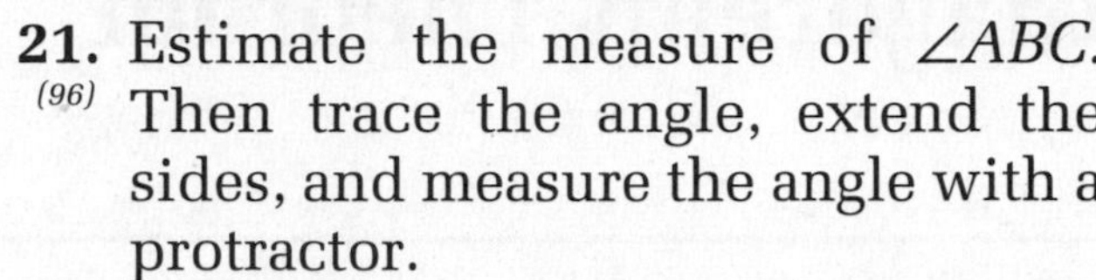

21. Estimate the measure of $\angle ABC$.
(96) Then trace the angle, extend the sides, and measure the angle with a protractor.

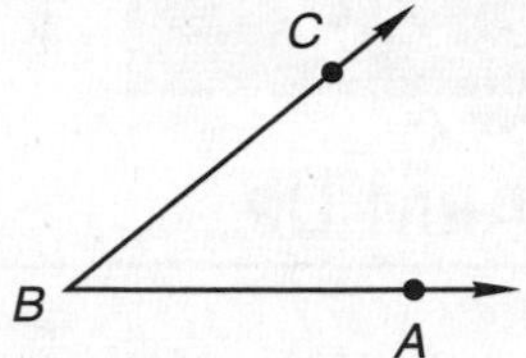

Refer to the figures below to answer problems 22 and 23.

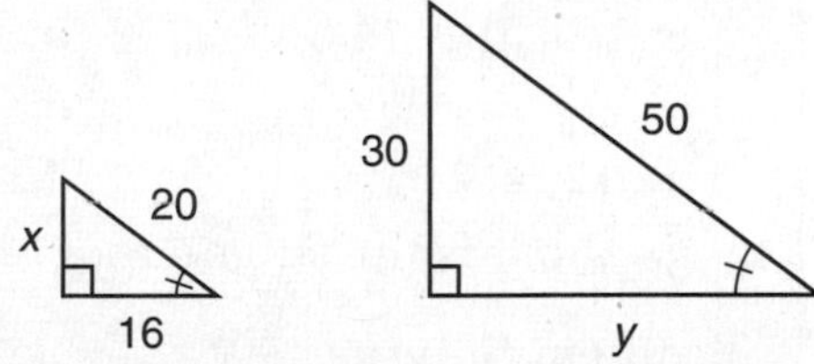

22. Find x and y. Then find the area of the smaller triangle.
(97) Dimensions are in inches.

23. Calculate the scale factor from the smaller triangle to the
(98) larger triangle.

For problems 24 and 25, solve the equation. Show each step.

24. $-\frac{3}{5}m + 8 = 20$
(93)

25. $0.3x - 2.7 = 9$
(93)

Simplify:

26. $\sqrt{5^3 - 5^2}$
(52)

27.
(56)
$$\begin{array}{r} 1 \text{ gal } 1 \text{ qt} \\ - \quad 1 \text{ qt } 1 \text{ pt} \\ \hline \end{array}$$

28. $(0.25)\left(1\frac{1}{4} - 1.2\right)$
(43)

29. $7\frac{1}{3} - \left(1\frac{3}{4} \div 3.5\right)$
(43)

30. $\dfrac{(-2)(3) - (3)(-4)}{(-2)(-3) - (4)}$
(85)

LESSON

99 Pythagorean Theorem

WARM-UP

Facts Practice: Order of Operations (Test T)

Mental Math:

a. $(-1.5) + (4.5)$

b. $(8 \times 10^6)(4 \times 10^4)$

c. $\frac{0.15}{30} = \frac{0.005}{n}$

d. Convert $-15°$C to degrees Fahrenheit.

e. $12 \times 2\frac{1}{3}$

f. $33\frac{1}{3}\%$ less than \$60

g. What is the square root of the sum of 6^2 and 8^2?

Problem Solving:

Here are the front, top, and side views of an object. Construct this object using 1-inch cubes, or draw a three-dimensional view. Then find the volume of the object.

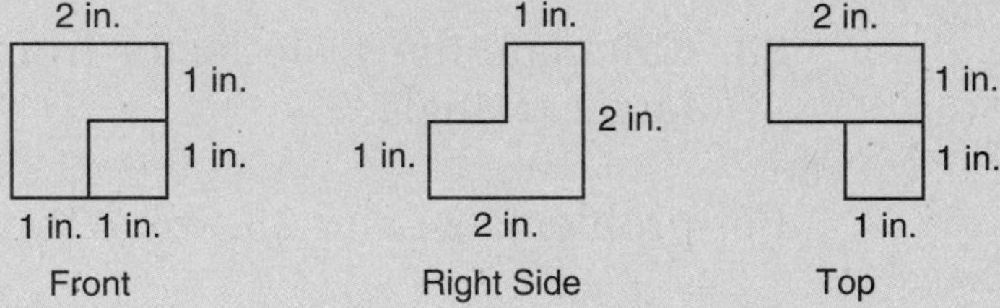

NEW CONCEPT

The longest side of a right triangle is called the **hypotenuse.** The other two sides are called **legs.** Every right triangle has a property that makes right triangles very important in mathematics. **The area of the square drawn on the hypotenuse of a right triangle equals the sum of the areas of the squares drawn on the legs.**

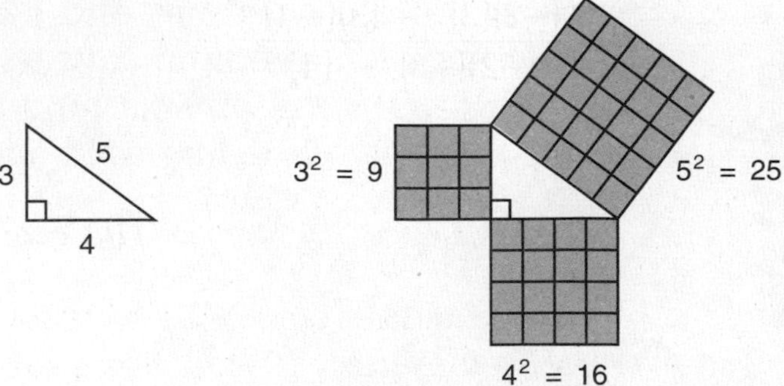

The triangle on the left is a right triangle. On the right we have drawn and shaded a square on each side of the triangle.

We have divided the squares into units and can see that their areas are 9, 16, and 25. Notice that the area of the largest square equals the sum of the areas of the other two squares.

$$25 = 16 + 9$$

This property of right triangles was known to the Egyptians as early as 2000 B.C., but it is named for a Greek mathematician who lived about 550 B.C. The Greek's name was Pythagoras, and the property is called the **Pythagorean theorem.** The Greeks are so proud of Pythagoras that they have issued a postage stamp that illustrates the theorem. Here we show a reproduction of the stamp:

To solve right-triangle problems using the Pythagorean theorem, we will draw the right triangle, as well as squares on each side of the triangle.

Example 1 Copy this triangle. Draw a square on each side. Find the area of each square. Then find *c*.

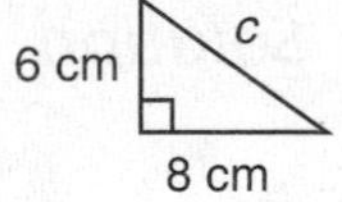

Solution We copy the triangle and draw a square on each side of the triangle as shown.

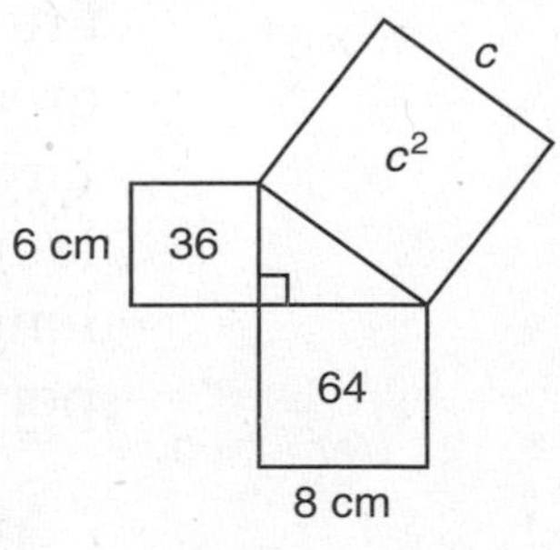

We were given the lengths of the two shorter sides. The areas of the squares on these sides are **36 cm²** and **64 cm².** The Pythagorean theorem says that the sum of the areas of the smaller squares equals the area of the largest square.

$$36\text{ cm}^2 + 64\text{ cm}^2 = \mathbf{100\text{ cm}^2}$$

This means that each side of the largest square must be 10 cm long because $(10\text{ cm})^2$ equals 100 cm^2. Thus

6 cm
10 cm
8 cm

$$c = \mathbf{10\text{ cm}}$$

Example 2 In this triangle, find a. Dimensions are in inches.

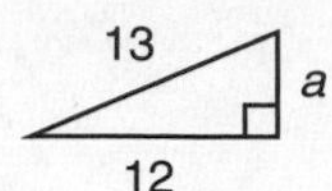

Solution We copy the triangle and draw a square on each side. The area of the largest square is 169 in.^{2}. The areas of the smaller squares are 144 in.^{2} and a^2. By the Pythagorean theorem a^2 plus 144 in.^{2} must equal 169 in.^{2}.

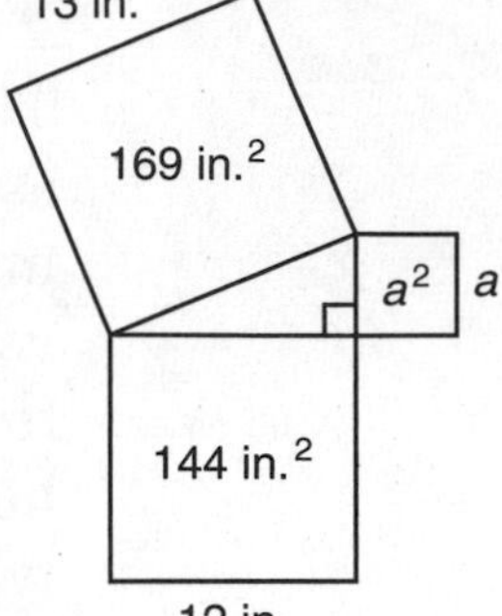

$$a^2 + 144 \text{ in.}^2 = 169 \text{ in.}^2$$

Subtracting 144 in.^{2} from both sides, we see that

$$a^2 = 25 \text{ in.}^2$$

This means that a equals **5 in.**, because (5 in.)^2 is 25 in^2.

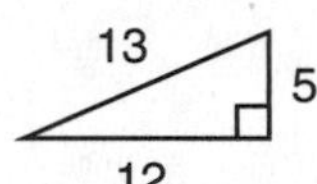

Example 3 Find the perimeter of this triangle. Dimensions are in centimeters.

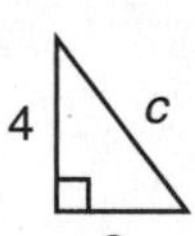

Solution We can draw a square on each side and use the Pythagorean theorem to find c. The areas of the two smaller squares are 16 cm^2 and 9 cm^2. The sum of these areas is 25 cm^2, so the area of the largest square is 25 cm^2. Thus the length c is 5 cm. Now we add the lengths of the sides to find the perimeter.

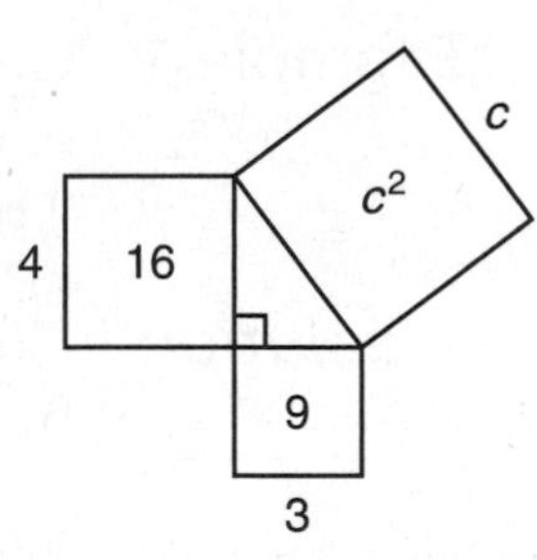

$$\text{Perimeter} = 4 \text{ cm} + 3 \text{ cm} + 5 \text{ cm}$$
$$= \mathbf{12\ cm}$$

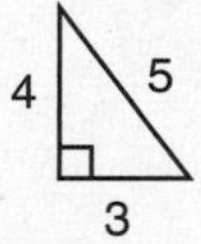

Example 4 Copy this right triangle and draw squares on the sides. Then write an equation that shows the relationship between the areas of the squares.

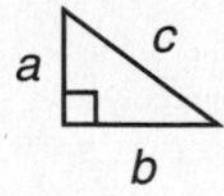

Solution We copy the triangle and draw a square on each side. Squaring the lengths of the sides of the triangle gives us the areas of the squares: a^2, b^2, and c^2. By the Pythagorean theorem, the sum of the areas of the smaller two squares is equal to the area of the largest square.

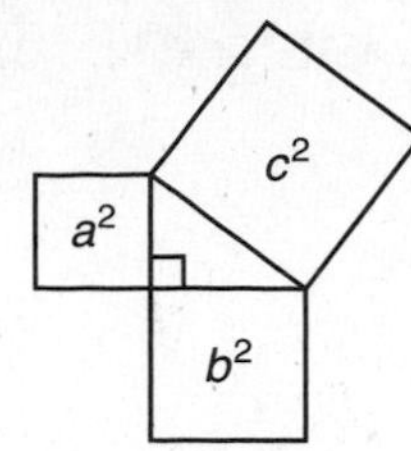

$$a^2 + b^2 = c^2$$

This equation is commonly used to algebraically express the Pythagorean theorem. The sum of the squares of the legs equals the square of the hypotenuse.

LESSON PRACTICE

Practice set Copy the triangles and draw the squares on the sides of the triangles as you work problems **a–c.**

a. Use the Pythagorean theorem to find the length a.

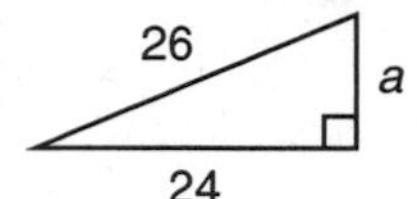

b. Use the Pythagorean theorem to find the length b.

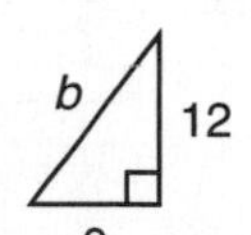

c. Find the perimeter of this triangle. Dimensions are in feet.

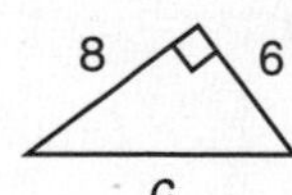

MIXED PRACTICE

Problem set

1. (46) The meal cost \$15. Christie left a tip that was 15 percent of the cost of the meal. How much money did Christie leave for a tip?

2. (57) Twenty-five ten-thousandths is how much greater than twenty millionths? Write the answer in scientific notation.

3. (Inv. 4) Find the (a) mean, (b) median, (c) mode, and (d) range of the number of days in the months of a leap year.

4. (46) The 2-pound box costs \$2.72. The 48-ounce box costs \$3.60. The smaller box costs how much more per ounce than the larger box?

Use ratio boxes to solve problems 5 and 6.

5. (72) If 80 pounds of seed costs \$96, what would be the cost of 300 pounds of seed?

6. (65) The ratio of stalactites to stalagmites in the cavern was 9 to 5. If the total number of stalactites and stalagmites was 1260, how many stalagmites were in the cavern?

7. (16, 60) Five eighths of a pound is how many ounces?

8. (77) Write equations to solve (a) and (b).

(a) Ten percent of what number is 20?

(b) Twenty is what percent of 60?

9. (62) In this figure, central angle *BDC* measures 60°, and inscribed angle *BAC* measures 30°. Angles *ACD* and *BCD* are complementary.

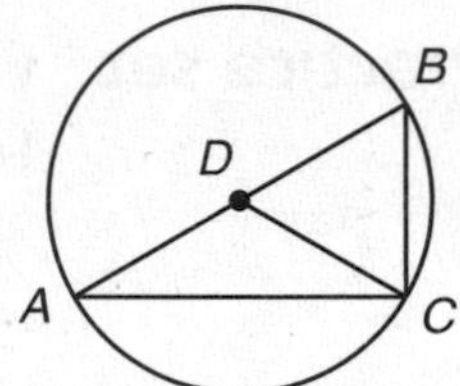

(a) Classify ΔABC by angles.

(b) Classify ΔBCD by sides.

(c) Classify ΔADC by sides.

10. (92) Use a ratio box to solve this problem. The cost of a 10-minute call to Boise decreased by 20%. If the cost before the decrease was \$3.40, what was the cost after the decrease?

11. (92) If an item is on sale for 20% off the regular price, then the sale price is what percent of the regular price?

12. (37) What is the area of the shaded region of this rectangle?

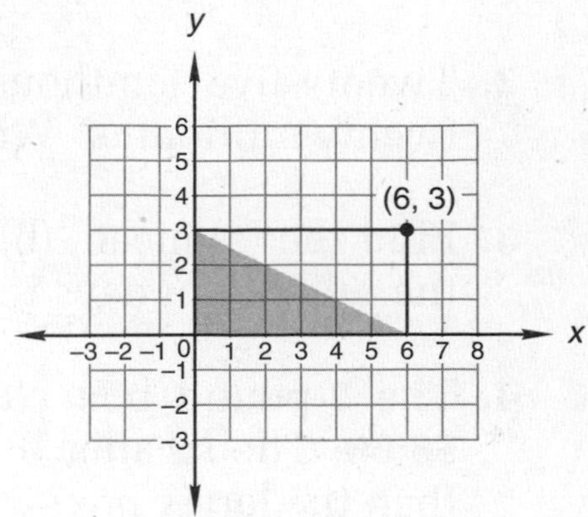

13. (98) Use a ratio box to solve this problem. On a 1:60 scale-model airplane, the wingspan is 8 inches. The wingspan of the actual airplane is how many inches? What is the wingspan of the actual airplane in feet?

14. (48) Complete the table.

FRACTION	DECIMAL	PERCENT
$1\frac{1}{3}$	(a)	(b)
(c)	(d)	$1\frac{1}{3}\%$

15. (84, 87) Simplify:

(a) $(ax^2)(-2ax)(-a^2)$ (b) $\frac{1}{2}\pi + \frac{2}{3}\pi - \pi$

16. (83) Multiply. Write the product in scientific notation.

$$(8.1 \times 10^{-6})(9 \times 10^{10})$$

17. (20, 52) Evaluate: $\sqrt{c^2 - b^2}$ if $c = 15$ and $b = 12$

18. (99) Use the Pythagorean theorem to find c.

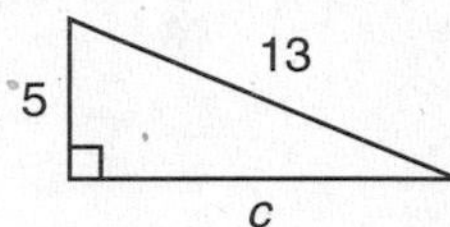

19. (95) Find the volume of this right solid. Dimensions are in centimeters. (Use 3.14 for π.)

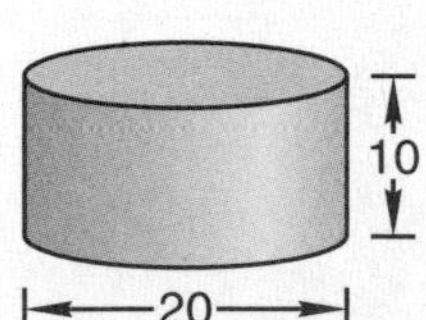

20. (40) Refer to the figure below to find the measures of the following angles.

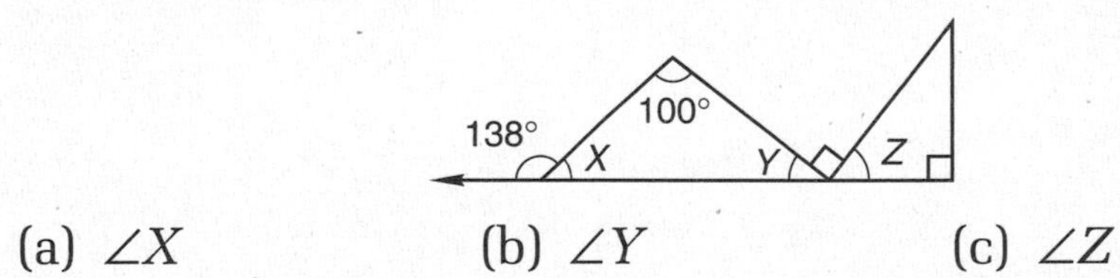

(a) $\angle X$ (b) $\angle Y$ (c) $\angle Z$

21. (97, 98) These triangles are similar. Dimensions are in inches.

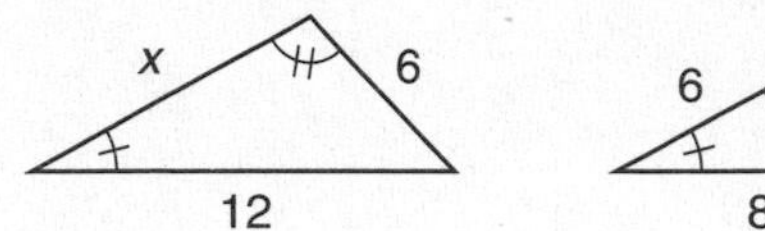

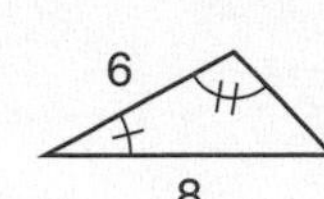

(a) Find x.

(b) What is the scale factor from the smaller triangle to the larger triangle?

(c) The area of the larger triangle is how many times the area of the smaller triangle? *Hint:* Use the scale factor found in (b).

22. Estimate: $\dfrac{(41{,}392)(395)}{81}$
(29)

For problems 23 and 24, solve the equation. Show each step.

23. $4n + 1.64 = 2$
(93)

24. $3\frac{1}{3}x - 1 = 49$
(93)

25. $\dfrac{17}{25} = \dfrac{m}{75}$
(98)

Simplify:

26. $3^3 + 4^2 - \sqrt{225}$
(20, 52)

27. $\sqrt{225} - 15^0 + 10^{-1}$
(52, 57)

28. $\left(3\frac{1}{3}\right)(0.75)(40)$
(43)

29. $\dfrac{-12 - (6)(-3)}{(-12) - (-6) + (3)}$
(85, 91)

30. Using the distributive property, we know that $2(x - 4)$ equals $2x - 8$. Use the distributive property to multiply $3(x - 2)$.
(96)

LESSON

100 Estimating Square Roots • Irrational Numbers

WARM-UP

Facts Practice: Two-Step Equations (Test U)

Mental Math:

a. $(-1.5) - (-7.5)$
b. $(5 \times 10^{-5})(5 \times 10^{-5})$
c. $100 = 5w - 20$
d. Convert −20°C to degrees Fahrenheit.
e. $20 \times 3\frac{3}{4}$
f. $33\frac{1}{3}\%$ less than \$24
g. 25% of 44, × 3, − 1, ÷ 4, × 7, − 1, ÷ 5, × 9, + 1, $\sqrt{\ }$, − 1, $\sqrt{\ }$

Problem Solving:

If two people shake hands, there is one handshake. If three people shake hands, there are three handshakes. From this table can you predict the number of handshakes with 6 people? Draw a diagram to confirm your prediction. (If six people are available, you may wish to act out the story.)

Number in group	2	3	4	5	6
Number of handshakes	1	3	6	10	?

NEW CONCEPTS

Estimating square roots

These counting numbers are perfect squares:

1, 4, 9, 16, 25, 36, 49, 64, ...

Recall that the square root of a perfect square is an integer.

$$\sqrt{25} = 5 \qquad \sqrt{36} = 6$$

The square root of a number that is between two perfect squares is not an integer but can be estimated.

$$\sqrt{29} = ?$$

Since 29 is between the perfect squares 25 and 36, we can conclude that $\sqrt{29}$ is between $\sqrt{25}$ and $\sqrt{36}$.

$$\sqrt{25} = 5 \qquad \sqrt{29} = ? \qquad \sqrt{36} = 6$$

We see that $\sqrt{29}$ is between 5 and 6. On this number line we see that $\sqrt{29}$ is between 5 and 6 but not exactly halfway between.

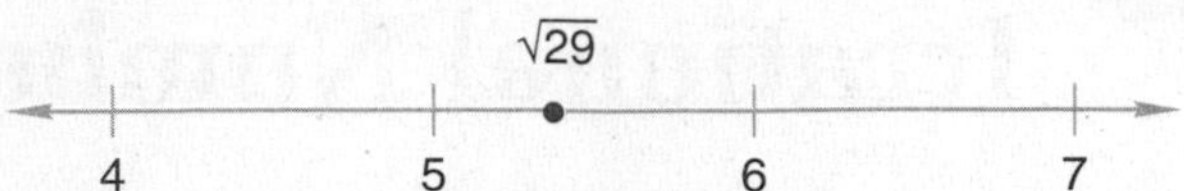

Example 1 Between which two consecutive whole numbers is $\sqrt{200}$?

Solution We remember that $\sqrt{100}$ is 10, so $\sqrt{200}$ is more than 10. We might guess that $\sqrt{200}$ is 20. We check our guess.

$$20 \times 20 = 400 \qquad \text{too large}$$

Our guess is much too large. Next we guess 15.

$$15 \times 15 = 225 \qquad \text{too large}$$

Since 15 is still too large, we try 14.

$$14 \times 14 = 196 \qquad \text{too small}$$

We see that 14 is less than $\sqrt{200}$ and 15 is more than $\sqrt{200}$. So $\sqrt{200}$ is between the consecutive whole numbers **14** and **15.**

Irrational numbers At the beginning of this lesson we found that $\sqrt{29}$ is between 5 and 6. We can refine our estimate by finding a decimal (or fraction) that is closer to $\sqrt{29}$. We try 5.4.

$$5.4 \times 5.4 = 29.16 \qquad \text{too large}$$

Since 5.4 is too large, we try 5.3.

$$5.3 \times 5.3 = 28.09 \qquad \text{too small}$$

We see that $\sqrt{29}$ is between 5.3 and 5.4. We could continue refining our estimate by finding numbers closer to $\sqrt{29}$. However, no matter how many numbers we try, we will not find a decimal (or fraction) that equals $\sqrt{29}$.

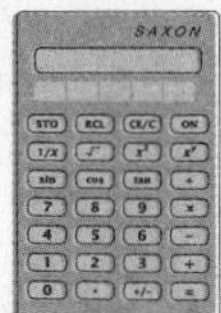

If we use a calculator, we can quickly find a number close to $\sqrt{29}$. If we enter these keystrokes

(depending on the type of calculator)

the number displayed on an 8-digit calculator is

5.3851648

This number is close to $\sqrt{29}$ but does not equal $\sqrt{29}$, as we see in the first step of checking the answer:

$$\begin{array}{r} \overset{6}{} \\ 5.3851648 \\ \times\ 5.3851648 \\ \hline 4 \end{array}$$

Since both factors have seven decimal places, the product has 14 decimal places. However, we see immediately that the product is not 29.00000000000000 because the digit in the fourteenth decimal place is 4.

Actually $\sqrt{29}$ is a number that cannot be exactly expressed as a decimal or fraction and therefore is *not a rational number.* Rather $\sqrt{29}$ is an **irrational number**—a number that cannot be expressed as a ratio of two integers. Nevertheless, $\sqrt{29}$ is a number that has an exact value. For instance, if the legs of this right triangle are exactly 2 cm and 5 cm, we find by the Pythagorean theorem that the length of the hypotenuse is $\sqrt{29}$ cm.

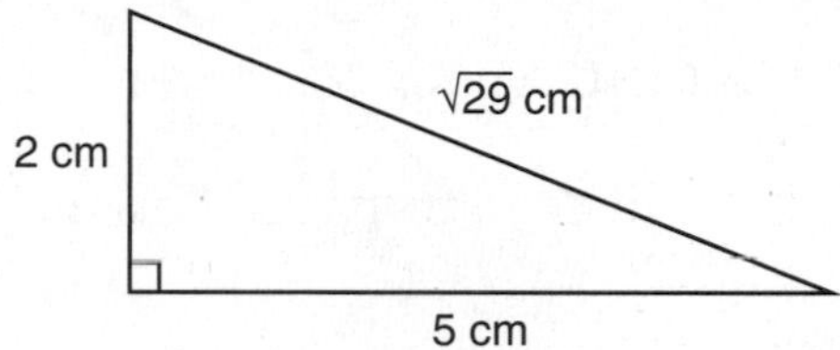

If we measure the hypotenuse with a centimeter ruler, we find that the length is about 5.4 cm, which is an approximation of $\sqrt{29}$ cm.

Other examples of irrational numbers include π (the circumference of a circle with a diameter of 1), $\sqrt{2}$ (the length of the diagonal of a square with sides of 1), and the square roots of counting numbers that are not perfect squares. The irrational numbers, together with the rational numbers, make up the set of **real numbers.**

Real Numbers

Rational Numbers	Irrational Numbers

All of the numbers represented by points on the number line are real numbers and are either rational or irrational.

Example 2 Draw a number line and show the approximate location of the points representing the following real numbers. Then describe each number as rational or irrational.

$$\pi \qquad \sqrt{2} \qquad 2.\overline{3} \qquad -\frac{1}{2}$$

Solution We draw a number line and mark the location of the integers from –1 through 4. We position π ($\approx$ 3.14) between 3 and 4 but closer to 3. Since $\sqrt{2}$ is between $\sqrt{1}$ (= 1) and $\sqrt{4}$ (= 2), we position $\sqrt{2}$ between 1 and 2 but closer to 1. The repeating decimal $2.\overline{3}$ $\left(= 2\frac{1}{3}\right)$ is closer to 2 than to 3. The negative fraction $-\frac{1}{2}$ is halfway between 0 and –1.

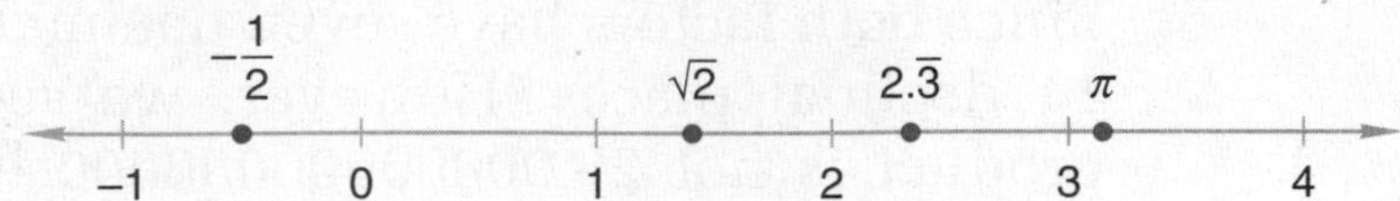

Repeating decimal numbers are rational. Thus both $-\frac{1}{2}$ **and** $\mathbf{2.\overline{3}}$ **are rational, while** $\sqrt{2}$ **and** π **are irrational.**

LESSON PRACTICE

Practice set Each square root below is between which two consecutive whole numbers?

a. $\sqrt{7}$ **b.** $\sqrt{70}$ **c.** $\sqrt{700}$

d. Find *x*:

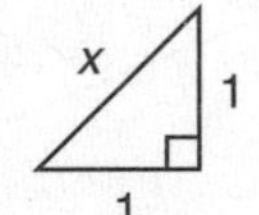

e. Find *y*:

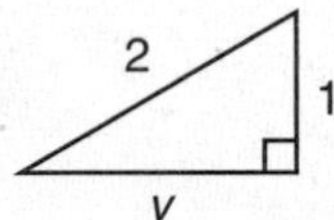

f. Draw a number line and show the approximate location of the points representing these real numbers. Which of them are irrational?

$\sqrt{3}$ $0.\overline{3}$ π $-\frac{1}{3}$

MIXED PRACTICE

Problem set

1. (28) Gabriel paid $20.00 for $2\frac{1}{2}$ pounds of cheese that cost $2.60 per pound and 2 boxes of crackers that cost $1.49 each. How much money should he get back?

2. (21, 94) The face of this spinner is divided into fifths.

(a) What is the probability that the spinner will not stop on a prime number on one spin?

(b) If the spinner is spun twice, what is the probability that it will stop on a prime number both times?

3. What is the average of the first 10 counting numbers?
(28, 86)

4. At an average speed of 50 miles per hour, how long would it take to complete a 375-mile trip?
(53)

Use ratio boxes to solve problems 5–7.

5. The Johnsons traveled 300 kilometers in 4 hours. At that rate, how long will it take them to travel 500 kilometers? Write the answer in hours and minutes.
(72)

6. The ratio of winners to losers in the contest was 1 to 15. If there were 800 contestants, how many winners were there?
(65)

7. The population of the colony decreased 30 percent after the first winter. If the population after the first winter was 350, what was the population before the first winter?
(92)

8. Three fourths of Genevieve's favorite number is 36. What number is one half of Genevieve's favorite number?
(60, 74)

9. Write equations to solve (a) and (b).
(77)

(a) Three hundred is 6 percent of what number?

(b) Twenty is what percent of 10?

10. What is the total price of a $40 item including 6.5% sales tax?
(46)

11. Using the distributive property, we know that $3(x + 3)$ equals $3x + 9$. Use the distributive property to multiply $x(x + 3)$.
(96)

12. The ordered pairs (0, 0), (−2, −4), and (2, 4) designate points that lie on the graph of the equation $y = 2x$. Graph the equation on a coordinate plane, and name another (x, y) pair from the 3rd quadrant that satisfies the equation.
(Inv. 9)

13. (Inv. 9, 98) Nathan used the data in the graph below to mold a scale model of a car from clay. The car is 4 feet tall, and he used the graph to find that the model should be 2 inches tall.

(a) The length of the car's bumper is 5 feet. Use the graph to find the proper length of the model's bumper.

(b) What is the scale factor from the car to the model? Write the scale factor as a fraction.

(c) Nathan's completed model was 7 inches long. Estimate the length of the car.

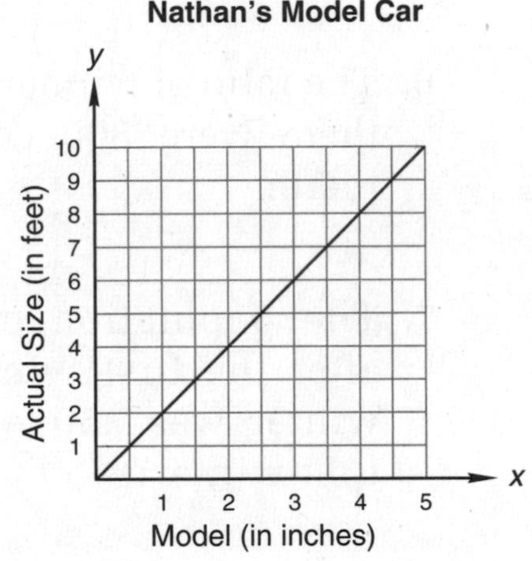

14. (98) The edge of one cube measures 2 cm. The edge of a larger cube measures 6 cm.

(a) What is the scale factor from the smaller cube to the larger cube?

(b) The area of each face of the larger cube is how many times the area of a face of the smaller cube?

(c) The volume of the larger cube is how many times the volume of the smaller cube?

15. (79) Compare: $xy \bigcirc \frac{x}{y}$ if x is positive and y is negative

16. (48) Complete the table.

FRACTION	DECIMAL	PERCENT
(a)	(b)	72%

17. (83) Multiply. Write the product in scientific notation.

$$(4.5 \times 10^6)(6 \times 10^3)$$

18. Each square root is between which two consecutive whole numbers?
(100)

(a) $\sqrt{40}$ (b) $\sqrt{20}$

19. Find the (a) circumference and (b) area of a circle that has a radius of 7 inches. (Use $\frac{22}{7}$ for π.)
(66, 82)

20. Use the Pythagorean theorem to find a. Dimensions are in centimeters.
(99)

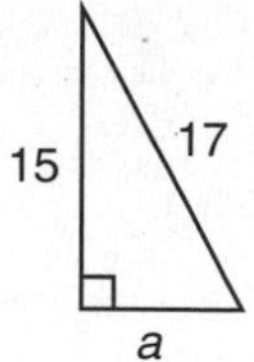

Find the volume of each right solid. Dimensions are in centimeters. (Use 3.14 for π.)

21.
(95)

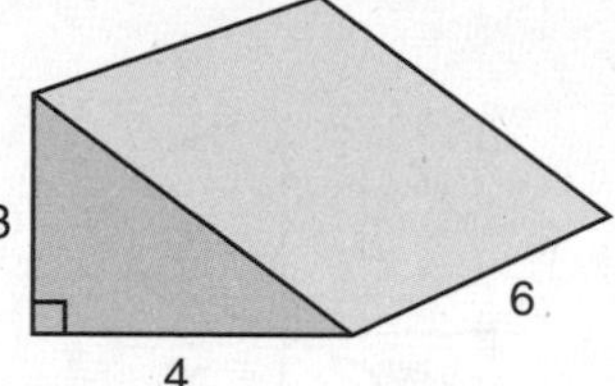

22.
(95)

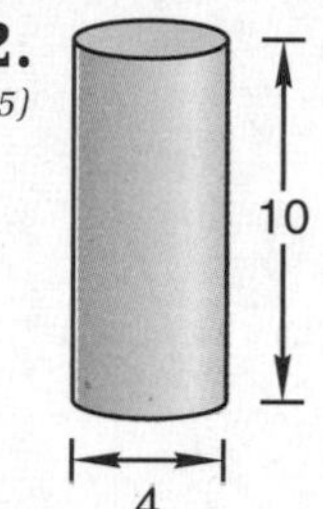

23. In the figure at right, find the measures of angles a, b, and c.
(40)

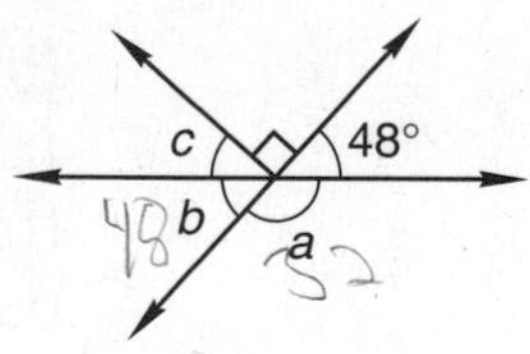

For problems 24 and 25, solve the equation. Show each step.

24. $-4\frac{1}{2}x + 8^0 = 4^3$
(57, 93)

25. $\frac{15}{w} = \frac{45}{3.3}$
(98)

Simplify:

26. $\sqrt{6^2 + 8^2}$
(52)

27. $3\frac{1}{3}\left(7.2 \div \frac{3}{5}\right)$
(43)

28. $8\frac{5}{6} - 2.5 - 1\frac{1}{3}$
(43)

29. $\frac{|-18| - (2)(-3)}{(-3) + (-2) - (-4)}$
(85)

30. Draw a number line and show the approximate locations of 1.5, –0.5, and $\sqrt{5}$.
(100)

INVESTIGATION 10

Focus on

Probability, Chance, and Odds

Probability, chance, and odds are different ways of expressing the likelihood of an event. Recall that **probability** is the ratio of the number of favorable outcomes to the total number of possible outcomes.

$$\text{Probability} = \frac{\text{number of favorable outcomes}}{\text{number of possible outcomes}}$$

Thus the probability that this spinner will end up in region A is $\frac{1}{4}$.

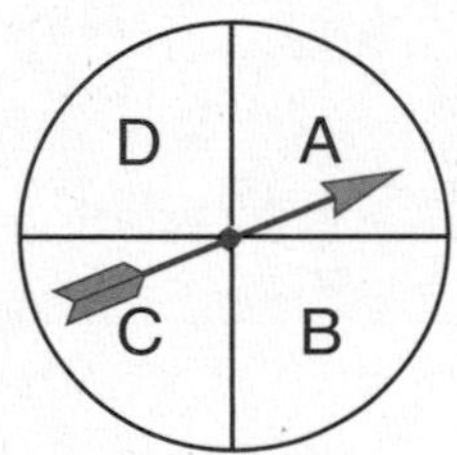

Probability can be expressed in decimal form. Since the fraction $\frac{1}{4}$ is equivalent to the decimal 0.25, the probability that the spinner will stop in region A can be expressed as 0.25. Recall that probabilities range from 0 (certain not to occur) to 1 (certain to occur). It is usually easier to compare probabilities when they are expressed as decimals than when they are expressed as fractions.

Example 1 Which is more likely: rolling a total of 7 with one toss of a pair of dot cubes or drawing a "face card" with one draw from a normal deck of cards?

Solution The probability of rolling a total of 7 with one toss of a pair of dot cubes is $\frac{6}{36}$, which reduces to $\frac{1}{6}$. The probability of drawing a face card is $\frac{12}{52}$, which reduces to $\frac{3}{13}$. We will convert $\frac{1}{6}$ and $\frac{3}{13}$ to decimal form (rounded to two places) and then compare.

$$\text{Probability of rolling a total of 7} = \frac{1}{6} \quad \text{or about } 0.17$$

$$\text{Probability of drawing a face card} = \frac{3}{13} \quad \text{or about } 0.23$$

The greater the probability, the greater the likelihood, so **drawing a face card is more likely than rolling a total of 7.**

1. Express as a decimal the probability that a flipped coin will land heads up.

2. The probability of rolling a total of 12 with one toss of a pair of dot cubes is $\frac{1}{36}$. Express this probability as a decimal rounded to two decimal places.

3. Pedro figures that the probability of drawing an ace from a normal deck of cards is $\frac{1}{13}$, or 0.08. Which number is more precise and why?

Besides expressing probability as a fraction or as a decimal, we can also express probability as a percent. We often use the word **chance** when expressing a probability in percent form. Referring to the spinner at the beginning of this lesson, we could say that the spinner has a 25% chance of stopping on A and a 75% chance of not stopping on A. Chance ranges from 0% (certain not to occur) to 100% (certain to occur).

4. The weather forecast stated that the chance of rain is 40%. What is the chance that it will not rain?

5. What is the chance that a flipped coin will land heads up?

6. The probability of selecting a heart by drawing one card from a normal deck of cards is 0.25. What is the chance that a card drawn from a normal deck of cards will not be a heart?

Another way to describe the likelihood of an event is with **odds.** While probability is the ratio of the number of favorable outcomes to the number of possible outcomes, *odds* is the ratio of the number of favorable outcomes to the number of unfavorable outcomes.

$$\text{Odds} = \text{favorable to unfavorable}$$

Using the spinner example, one outcome is A and three outcomes are not A. Thus the odds of the spinner ending up in region A are

1 to 3

or

1:3

Note that odds are usually expressed by using the word "to," or with a colon, and not by using a division bar. However, odds are reduced as are other ratios.

Example 2 A 20 percent chance of rain was forecast. What are the odds that it will rain?

Solution Since 20% is equivalent to the fraction $\frac{1}{5}$, the forecast means that there is 1 chance in 5 that it will rain. Thus there are 4 chances in 5 that it will not rain. Therefore, the odds that it will rain are **1 to 4.**

Example 3 The odds that a marble drawn from a bag will be red are 3 to 2.

(a) What is the probability that a red marble will be drawn?

(b) What is the chance of drawing a red marble?

Solution If the odds are 3 to 2, we mean

$$\begin{array}{rl} 3 & \text{favorable outcomes} \\ + \; 2 & \text{unfavorable outcomes} \\ \hline 5 & \text{possible outcomes} \end{array}$$

(a) The probability of drawing a red marble is

$$\frac{\text{favorable}}{\text{possible}} = \mathbf{\frac{3}{5}}$$

(b) The chance of drawing a red marble is

$$\frac{3}{5} = \mathbf{60\%}$$

Refer to the following sentence for problems 7–9:

> *One marble will be drawn from a bag containing 3 red marbles, 4 white marbles, and 5 blue marbles.*

7. What is the probability of drawing a white marble?

8. What is the probability of not drawing a red marble?

9. What are the odds of drawing a blue marble?

We can distinguish between **theoretical probability** and **statistical probability.** Theoretical probability can be calculated when the number of favorable outcomes and the number of possible outcomes can be counted without testing. Games of chance are based upon theoretical probability involving actions with cards, dot cubes, spinners, and similar objects with countable outcomes.

Statistical probability is used to determine the likelihood of an event based upon a *record of past experience.* For example, a baseball manager may select a pinch hitter by comparing the batting average of several players. Suppose a baseball player has had 125 hits in 375 at-bats. The manager can calculate the probability that the player will get a hit by dividing the player's number of previous hits by the player's number of previous at-bats.

$$\text{Probability of a hit} = \frac{125 \text{ hits}}{375 \text{ at-bats}} = \frac{1}{3} \approx 0.333$$

In statistical probability, we assume that the number of times an event happens out of the number of times the event could have happened is the probability of the event. Statistical probability is used by the insurance industry, which assesses risk and calculates insurance premiums on the basis of statistical records.

10. During the last three basketball seasons, Darcy made 64 free throws and missed 32 free throws. What is the statistical probability that Darcy will make her next free throw?

11. During clinical trials of the new medication, 15 of the 1000 participants developed unfavorable side effects. If Roger is treated with the new medication, what is the statistical probability that he will develop unfavorable side effects?

12. If drivers under the age of 25 are far more likely to be involved in an auto accident than drivers over the age of 25, then what is probably true about the insurance rates for younger drivers compared to the rates for older drivers?

Knowing the probability of an event helps us *predict* the outcome of repeated activity. For instance, knowing that the theoretical probability of rolling a total of 7 with one toss of a pair of dot cubes is $\frac{1}{6}$, we can predict that over a large number of rolls, about $\frac{1}{6}$ of the rolls will result in a total of 7. If we graph the results of an activity subject to the laws of probability, patterns may emerge that help us visualize the effects of these laws.

Activity: *Experimental Probability*

Materials needed:

- Activity Sheet 7 (available in *Saxon Math 8/7—Homeschool Tests and Worksheets*)
- Pair of dot cubes

Section A of Activity Sheet 7 displays the 36 equally likely outcomes of rolling a pair of dot cubes. Section B is the outline of a bar graph. On the graph draw bars to indicate the theoretical outcome of rolling a pair of dot cubes 36 times.

After completing Section B, roll the dot cubes and record the results for 36 rolls. Record and graph the results of the tosses in Section C.

If the results of the experiment differ from the theoretical outcome, think of possible reasons for this difference and write your reasons in Section D.

Extension Repeat the experiment two more times, and create two additional bar graphs to represent the results of all three experiments. First create a "Theoretical Outcomes" graph, and then create an "Actual Results" graph by combining the results from all three experiments. Does increasing the number of rolls (by counting the rolls in all three experiments) produce results closer to theoretical outcomes than were attained with just 36 rolls?

LESSON

101 Translating Expressions into Equations

WARM-UP

Facts Practice: Percent-Decimal-Fraction Equivalents (Test Q)

Mental Math:[†]

a. 110 (base 2)
b. XXXIV
c. $(-2)(-2)(-2)(-2)$
d. $(5 \times 10^{-5})(6 \times 10^{-6})$
e. $\frac{0.2}{w} = \frac{0.4}{0.12}$
f. Convert –25°C to degrees Fahrenheit.
g. $\frac{3}{4}$ of \$80
h. 25% less than \$80
i. Estimate a 15% tip on a \$29.78 bill.

Problem Solving:

At three o'clock the hands of a clock form a 90° angle. What angle is formed by the hands of a clock two hours after three o'clock?

NEW CONCEPT

An essential skill in mathematics is the ability to translate language, situations, and relationships into mathematical form. Since the earliest lessons of this book we have practiced translating stories into equations that we then solved. In this lesson we will practice translating other common patterns of language into algebraic form. We will also use our knowledge of geometric relationships to write equations to solve geometry problems.

Consider the following examples of mathematical phrases and how they are translated into algebraic form.

Examples of Translations

Phrase	Translation
twice a number	$2n$
five more than a number	$x + 5$
three less than a number	$a - 3$
half of a number	$\frac{1}{2}h$ or $\frac{h}{2}$
the product of a number and seven	$7b$
seventeen is five more than twice a number	$17 = 2n + 5$

[†]In Lessons 101–120, Mental Math problems **a** and **b** review concepts from Appendix Topic A. For each problem, convert the given base 2 number or Roman numeral to a base 10 number. Skip these Warm-up problems if you have not covered Appendix Topic A.

The last translation in the examples above resulted in an equation that can be solved to find the unstated number. We will solve a similar equation in the next example.

Example 1 If five less than twice a number is seventeen, what is the number?

Solution We will use the letter x to represent the unknown as we translate the sentence into an equation.

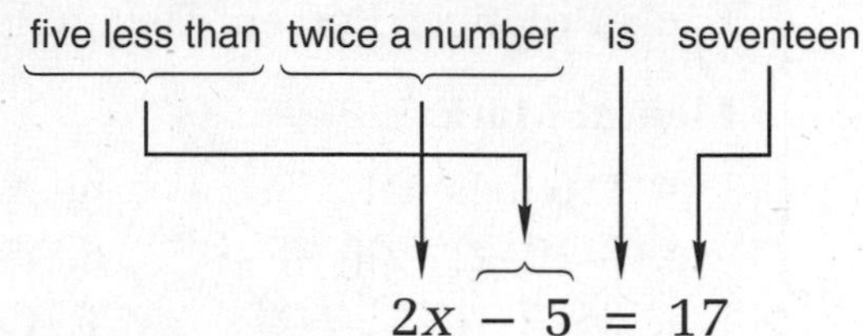

Now we solve the equation.

$2x - 5 = 17$	equation
$2x - 5 + 5 = 17 + 5$	added 5 to both sides
$2x = 22$	simplified
$\frac{2x}{2} = \frac{22}{2}$	divided both sides by 2
$x = 11$	simplified

We find that the number described is **11.**

Five less than twice eleven is seventeen.

We can also translate geometric relationships into algebraic expressions.

Example 2 The angles marked x and $2x$ in this figure are supplementary. What is the measure of the larger angle?

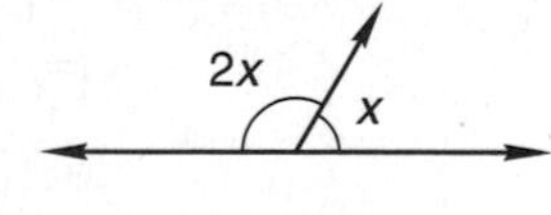

Solution The sum of the angle measures is 180°. We write this relationship as an equation.

$$2x + x = 180°$$

Since $2x + x = 3x$, we can simplify then solve the equation.

$2x + x = 180°$	equation
$3x = 180°$	simplified
$= \frac{180°}{3}$	divided both sides by 3
$x = 60°$	simplified

The solution of the equation is 60°, but 60° is not the answer to the question. We were asked to find the measure of the larger angle, which in the diagram is marked $2x$. Since x is 60°, we find that the larger angle measures $2(60°) = \mathbf{120°}$.

LESSON PRACTICE

Practice set* Write and solve an equation for each of these problems:

a. Six more than the product of a number and three is 30. What is the number?

b. Ten less than half of what number is 30?

c. What is the measure of the smallest angle in this figure?

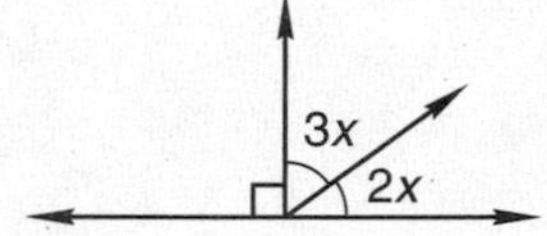

d. Find the measure of each angle of this triangle.

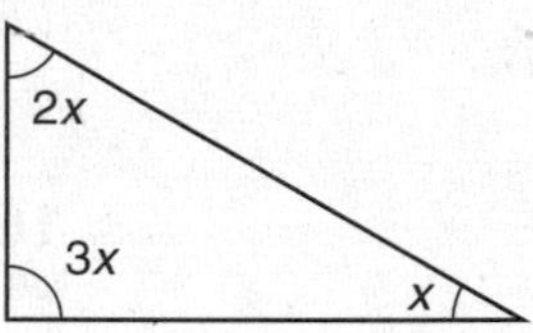

MIXED PRACTICE

Problem set

1. (Inv. 4) The following marks are Katie's 100-meter dash times, in seconds, during track season. Find the (a) median, (b) mode, and (c) range of these times.

12.3, 11.8, 11.9, 11.7, 12.0, 11.9, 12.1, 11.6, 11.8

2. (53) How much money does Jackson earn working 3 hours 45 minutes at \$6 per hour?

Use ratio boxes to solve problems 3–6.

3. (72) The recipe called for 3 cups of flour and 2 eggs to make 6 servings. If 15 cups of flour were used to make more servings, how many eggs should be used?

4. (72) Lester can type 48 words per minute. At that rate, how many words can he type in 90 seconds?

5. (81) Ten salespeople surpassed the quota. This was 40 percent of the sales staff. How many salespeople were on the sales staff?

6. The dress was on sale for 40 percent off the regular price. If the regular price was \$24, what was the sale price?
(92)

7. Use the distributive property to clear parentheses. Then simplify by adding like terms.
(96)

$$3(x - 4) - x$$

8. Use two unit multipliers to convert 3 gallons to pints.
(88)

9. Diagram this statement. Then answer the questions that follow.
(36, 71)

The Trotters won $\frac{5}{6}$ of their games. They won 20 games and lost the rest.

(a) How many games did they play?

(b) What was the Trotters' win-loss ratio?

10. Between which two consecutive whole numbers is $\sqrt{200}$?
(100)

11. Compare: $w \bigcirc m$ if w is 0.5 and m is the reciprocal of w
(79)

12. Find the area of the hexagon shown at right. Dimensions are in centimeters. Corners that look square are square.
(75)

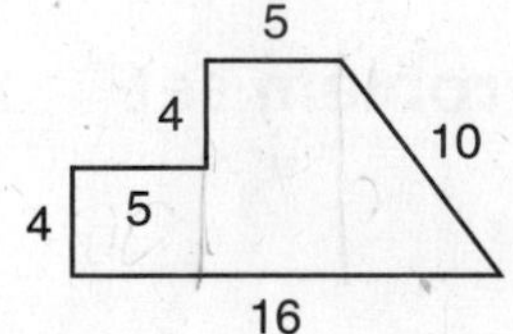

13. Write an equation to solve this problem:
(101)

Three less than the product of six and what number is 45?

14. Multiply. Write the product in scientific notation.
(83)

$$(8 \times 10^8)(4 \times 10^{-2})$$

15. Complete the table.
(48)

Fraction	Decimal	Percent
(a)	0.02	(b)
(c)	(d)	0.2%

16. Find the missing numbers in the table by using the function rule. Then graph the function on a coordinate plane.
(85, Inv. 9)

$y = 2x + 1$

x	y
−1	☐
0	☐
1	☐
2	☐

17. Copy this cube on your paper. Then find its (a) volume and (b) surface area. Dimensions are in inches.
(67, 70)

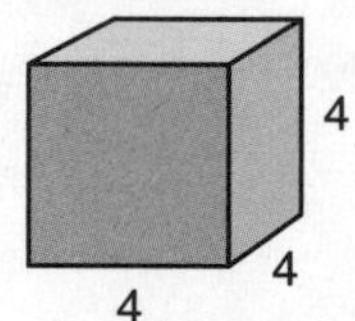

18. (a) Find the circumference of the circle at right.
(66, 82)

(b) Find the area of the circle.

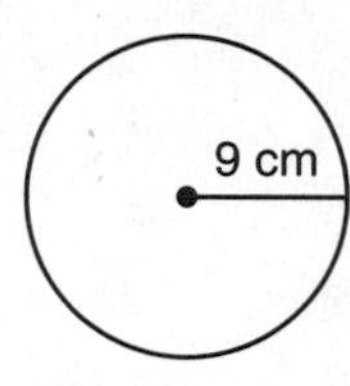

Leave π as π.

19. What are the odds that a tossed coin will land heads up?
(Inv. 10)

20. The two acute angles in this figure are complementary. What are the measures of the two angles?
(101)

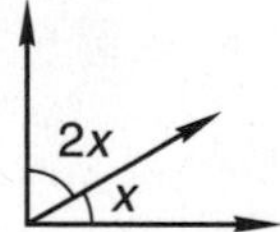

21. Divide 1.23 by 9 and write the quotient
(44)

(a) with a bar over the repetend.

(b) rounded to three decimal places.

22. If BC is 9 cm and AC is 12 cm, then what is AB?
(99)

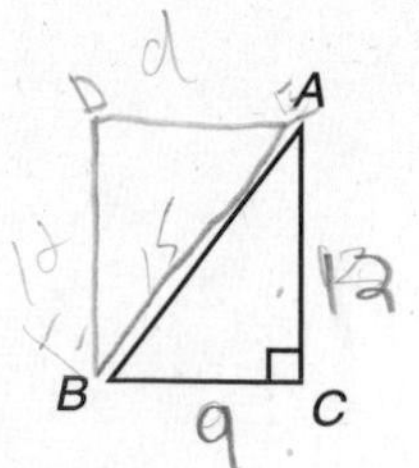

23. If the scale factor from ΔABC in problem 22 to ΔDEF is 2, then
(98)

(a) what is the perimeter of ΔDEF?

(b) what is the area of ΔDEF?

24. In a bag were 6 red marbles and 4 blue marbles. If Lily pulls a marble out of the bag with her left hand and then pulls a marble out with her right hand, what is the probability that the marble in each hand will be blue?
(94)

25. Solve: $3\frac{1}{7}d = 88$
(90)

26. Solve this inequality and graph its solution:
(93)

$$3x + 20 \geq 14$$

Simplify:

27. $5^2 + (3^3 - \sqrt{81})$
(52)

28. $3x + 2(x - 1)$
(96)

29. $\left(4\frac{4}{9}\right)(2.7)\left(1\frac{1}{3}\right)$
(43)

30. $(-2)(-3) - (-4)(-5)$
(85)

LESSON

102 Transversals • Simplifying Equations

WARM-UP

Facts Practice: Two-Step Equations (Test U)

Mental Math:

a. 1011 (base 2)

b. XCIV

c. (–0.25) + (–0.75)

d. $(3 \times 10^{10})(2 \times 10^{-2})$

e. $3x + 2\frac{1}{2} = 10$

f. Convert 500 mL to L.

g. 150% of \$40

h. 150% more than \$40

i. Start with a score, – 5, × 2, + 2, ÷ 4, + 1, $\sqrt{\ }$, × 7, – 1, ÷ 10.

Problem Solving:

An octagon can be divided into six triangles by five diagonals drawn from a single vertex. Into how many triangles can a 22-gon be divided by diagonals drawn from a single vertex? What is the sum of the measures of the interior angles of a 22-gon?

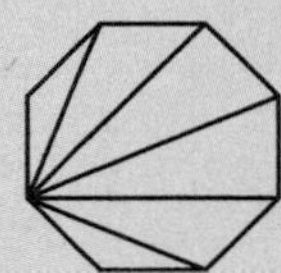

NEW CONCEPTS

Transversals

A **transversal** is a line that intersects one or more other lines in a plane. In this lesson we will pay particular attention to the angles formed when a transversal intersects a pair of parallel lines. Notice the eight angles that are formed.

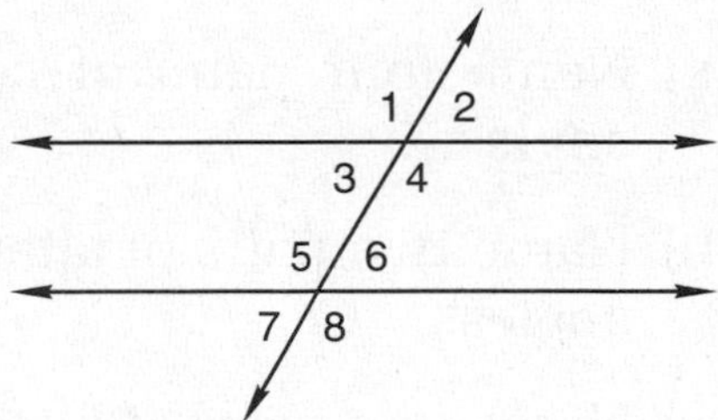

In this figure there are four acute angles numbered 2, 3, 6, and 7, and there are four obtuse angles numbered 1, 4, 5, and 8. All of the acute angles have the same measure, and all of the obtuse angles have the same measure.

Example 1 Transversal t intersects parallel lines l and m so that the measure of $\angle a$ is 105°. Find the measure of angles b–h.

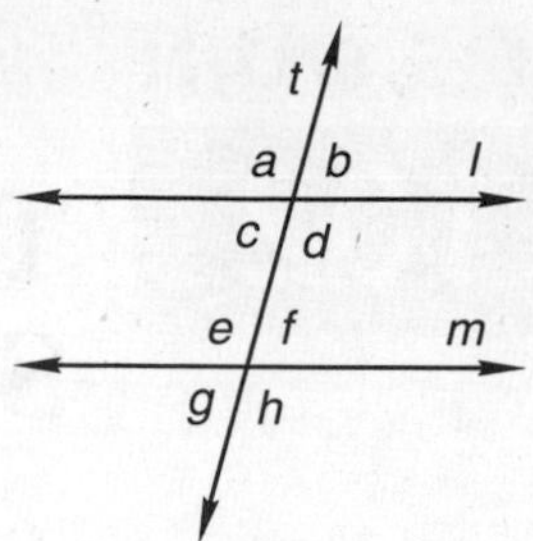

Solution All the obtuse angles have the same measure, so $\angle$***d,*** $\angle$***e,*** and $\angle$***h*** **each measure 105°**. Each of the acute angles is a supplement of an obtuse angle, so each acute angle measures

$$180° - 105° = 75°$$

Thus, $\angle$***b,*** $\angle$***c,*** $\angle$***f,*** and $\angle$***g*** **each measure 75°.**

When a transversal intersects a pair of lines, special pairs of angles are formed. In example 1, $\angle a$ and $\angle e$ are **corresponding angles** because the position of $\angle e$ corresponds to the position of $\angle a$ (to the left of the transversal and above lines m and l). Angle b and $\angle f$, $\angle c$ and $\angle g$, and $\angle d$ and $\angle h$ are three more pairs of corresponding angles in example 1.

Angle a and $\angle h$ in example 1 also form a special pair of angles. They are on alternate sides of the transversal and are outside of (not between) the parallel lines. So $\angle a$ and $\angle h$ are called **alternate exterior angles.** Angle b and $\angle g$ are another pair of alternate exterior angles in example 1.

Angles d and e in example 1 are **alternate interior angles** because they are on alternate sides of the transversal and in the interior of (between) the parallel lines. Angle c and $\angle f$ are another pair of alternate interior angles in example 1.

Example 2 Transversal r intersects parallel lines p and q to form angles 1–8.

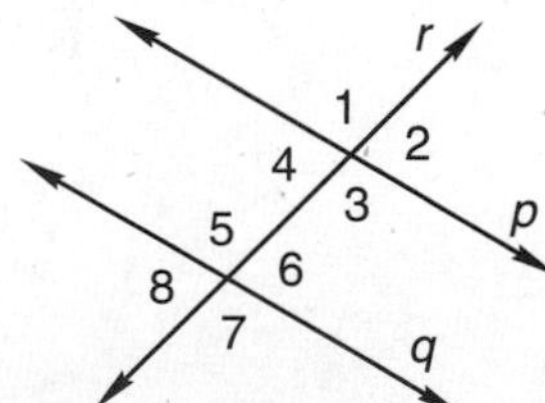

(a) Name four pairs of corresponding angles.

(b) Name two pairs of alternate exterior angles.

(c) Name two pairs of alternate interior angles.

Solution (a) **$\angle 1$ and $\angle 5$, $\angle 2$ and $\angle 6$, $\angle 3$ and $\angle 7$, $\angle 4$ and $\angle 8$**

(b) **$\angle 1$ and $\angle 7$, $\angle 2$ and $\angle 8$**

(c) **$\angle 4$ and $\angle 6$, $\angle 3$ and $\angle 5$**

Simplifying equations One step to solving some equations is to collect like terms. In this equation we can collect variable terms and constant terms as the first step of the solution:

$$\underbrace{3x - x}_{2x} \underbrace{- 5 + 8}_{+3} = 17$$

$$2x + 3 = 17$$

Example 3 Simplify and then solve this equation:

$$3x + 5 - x = 17 + x - x$$

Solution We collect like terms on each side of the equal sign. On the left side, combining $3x$ and $-x$ gives us $2x$. On the right side, combining $+x$ and $-x$ gives us zero.

$3x + 5 - x = 17 + x - x$	equation
$2x + 5 = 17$	simplified

Now we solve the simplified equation.

$2x + 5 = 17$	equation
$2x + 5 - 5 = 17 - 5$	subtracted 5 from both sides
$2x = 12$	simplified
$\frac{2x}{2} = \frac{12}{2}$	divided both sides by 2
$x = \mathbf{6}$	simplified

Example 4 Simplify this equation by removing the variable term from one side of the equation. Then solve the equation.

$$5x - 17 = 2x - 5$$

Solution We see an x-term on both sides of the equal sign. We may remove the x-term from either side. We choose to remove the variable term from the right side. We do this by subtracting $2x$ from both sides of the equation.

$5x - 17 = 2x - 5$	equation
$5x - 17 - 2x = 2x - 5 - 2x$	$2x$ from both sides
$3x - 17 = -5$	simplified

Now we solve the simplified equation.

$3x - 17 = -5$	equation
$3x - 17 + 17 = -5 + 17$	added 17 to both sides
$3x = 12$	simplified
$\frac{3x}{3} = \frac{12}{3}$	divided both sides by 3
$x = \mathbf{4}$	simplified

Example 5 Solve: $3x + 2(x - 4) = 32$

Solution We first apply the distributive property to clear parentheses.

$3x + 2(x - 4) = 32$	equation
$3x + 2x - 8 = 32$	distributive property
$5x - 8 = 32$	added $3x$ and $2x$
$5x = 40$	added 8 to both sides
$x = \mathbf{8}$	divided both sides by 5

LESSON PRACTICE

Practice set* Refer to this figure to answer problems **a–d.**

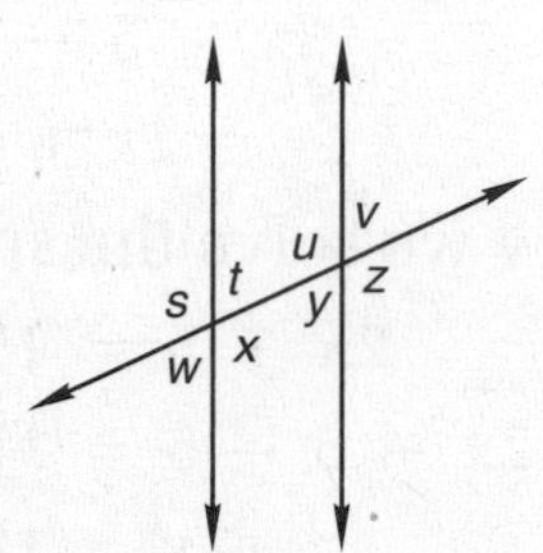

a. Name four pairs of corresponding angles.

b. Name two pairs of alternate interior angles.

c. Name two pairs of alternate exterior angles.

d. If the measure of $\angle w$ is 80°, what is the measure of each of the other angles?

Simplify and solve the following equations:

e. $3w - 10 + w = 90$

f. $x + x + 10 + 2x - 10 = 180$

g. $3y + 5 = y - 25$ **h.** $4n - 5 = 2n + 3$

i. $3x - 2(x - 4) = 32$ **j.** $3x = 2(x - 4)$

MIXED PRACTICE

Problem set

1. (Inv. 10) Home base was 5 spaces away. Jorge nervously tossed the pair of dot cubes.

(a) What is the probability that the dots on the rolled dot cubes will total 5?

(b) What are the odds of Jorge failing to reach home base on this turn?

2. Write an equation to solve this problem:
(101)

Twelve less than the product of what number and three is 36?

3. The figure shows three sides of a regular decagon.
(89)

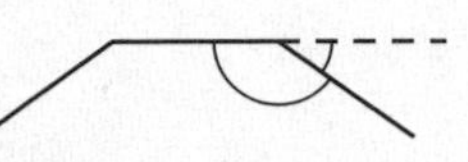

(a) What is the measure of each exterior angle?

(b) What is the measure of each interior angle?

Use ratio boxes to solve problems 4–7.

4. The ratio of youths to adults at the convocation was 3 to 7. If 4500 attended the convocation, how many adults were present?
(65)

5. Every time the knight moved over 2, he moved up 1. If the knight moved over 8, how far did he move up?
(72)

6. Eighty percent of those who were invited came to the party. If 40 people were invited to the party, how many did not come?
(81)

7. The dress was on sale for 60 percent of the regular price. If the sale price was \$24, what was the regular price?
(81)

8. Write an equation to solve this problem:
(101)

Three more than twice a number is −13.

9. The obtuse and acute angles in the figure at right are supplementary. What is the measure of each angle?
(101)

$3x - 25$ $x + 5$

10. Diagram this statement. Then answer the questions that follow.
(71)

Exit polls showed that 7 out of 10 voters cast their ballots for the incumbent. The incumbent received 1400 votes.

(a) How many voters cast their ballots?

(b) What percent of the voters did not vote for the incumbent?

11. Evaluate: $x + xy - xy$ if $x = 3$ and $y = -2$
(91)

12. Compare: $a \bigcirc a - a$ if $a < 0$
(79)

13. If the perimeter of a square is 1 meter, what is the area of the square in square centimeters?
(32)

14. Find the total price, including tax, of a \$12.95 bat, a \$7.85 baseball, and a \$49.50 glove. The tax rate is 7 percent.
(46)

15. Multiply. Write the product in scientific notation.
(83)

$$(3.5 \times 10^5)(3 \times 10^6)$$

16. Lines l and m are parallel and are intersected by transversal q.
(102)

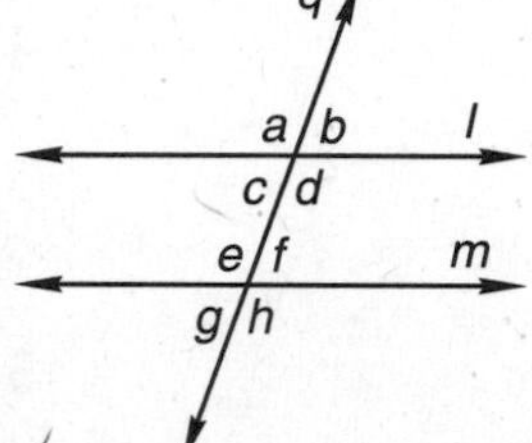

(a) Which angle corresponds to $\angle c$?

(b) Which angle is the alternate interior angle of $\angle e$?

(c) Which angle is the alternate exterior angle of $\angle h$?

(d) If $m\angle a$ is 110°, what is $m\angle f$?

17. (a) What number is 125% of 84?
(60, 92)

(b) What number is 25% more than 84?

18. If the chance of rain is 40%, what are the odds that it will not rain?
(Inv. 10)

19. What is the volume of this rectangular prism? Dimensions are in feet.
(70)

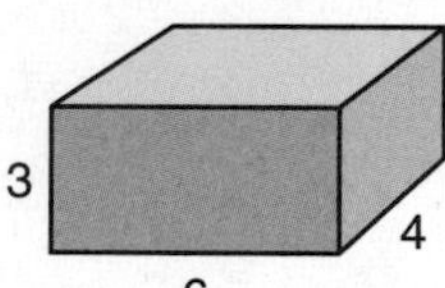

20. (a) Find the circumference of the circle at right.
(66, 82)

(b) Find the area of the circle.

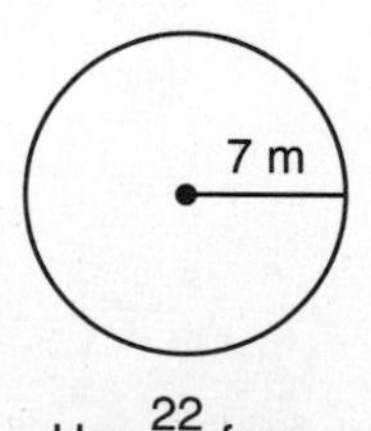

21. Find the missing numbers in the table by using the function rule. Then graph the function on a coordinate plane.
(85, Inv. 9)

$y = 3x$

x	y
2	☐
−1	☐
0	☐

22. Polygon $ZWXY$ is a rectangle. Find the measures of the following angles:
(40)

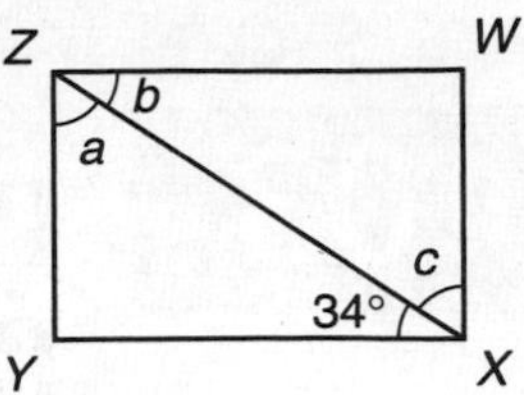

(a) $\angle a$ 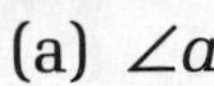 (b) $\angle b$ (c) $\angle c$

23. Graph on a number line: $x \geq -2$
(78)

24. Draw a number line and show the locations of these numbers. Which are rational?
(100)

$$0.4, \tfrac{1}{4}, \sqrt{4}$$

Solve:

25. $3x + x + 3^0 = 49$
(57, 102)

26. $3y + 2 = y + 32$
(102)

27. $x + 2(x + 3) = 36$
(102)

Simplify:

28. (a) $(3x^2y)(-2x)(xy^2)$
(84, 87)

(b) $-3x + 2y - x - y$

29. $\left(4\frac{1}{2}\right)(0.2)(10^2)$
(43)

30. $\dfrac{(-4)(+3)}{(-2)} - (-1)$
(85)

LESSON

103 Powers of Negative Numbers • Dividing Terms

WARM-UP

Facts Practice: Two-Step Equations (Test U)

Mental Math:

a. 11 (base 2)

b. LXXXI

c. $(-2.5) \div (-5)$

d. $(3 \times 10^{-4})(4 \times 10^{-3})$

e. $2y + y = 45$

f. Convert −30°C to degrees Fahrenheit.

g. $33\frac{1}{3}\%$ of \$600

h. $33\frac{1}{3}\%$ less than \$600

i. The expression $(3 \times 10^3)^2$ means $(3 \times 10^3)(3 \times 10^3)$. Write the product in scientific notation.

Problem Solving:

The expression $\sqrt[3]{8}$ means "the cube root of 8." The cube root of 8 is 2 because $2 \cdot 2 \cdot 2 = 8$. Find $\sqrt[3]{64}$. Find $\sqrt[3]{125}$.

NEW CONCEPTS

Powers of negative numbers

One way to multiply three or more signed numbers is to multiply the factors in order from left to right, keeping track of the signs with each step, as we show here:

$(-3)(-4)(+5)(-2)(+3)$	problem
$(+12)(+5)(-2)(+3)$	multiplied $(-3)(-4)$
$(+60)(-2)(+3)$	multiplied $(+12)(+5)$
$(-120)(+3)$	multiplied $(+60)(-2)$
-360	multiplied $(-120)(+3)$

Another way to keep track of signs when multiplying signed numbers is to count the number of negative factors. Notice the pattern in the multiplications below.

$-1 = -1$	odd
$(-1)(-1) = +1$	even
$(-1)(-1)(-1) = -1$	odd
$(-1)(-1)(-1)(-1) = +1$	even
$(-1)(-1)(-1)(-1)(-1) = -1$	odd

When there is an even number of negative factors, the product is positive. When there is an odd number of negative factors, the product is negative.

Note: If any of the multiplied factors are zero, their product is also zero (which has no sign).

Example 1 Find the product: $(+3)(+4)(-5)(-2)(-3)$

Solution There are three negative factors (an odd number), so the product will be a negative number. We multiply and get

$$(+3)(+4)(-5)(-2)(-3) = \mathbf{-360}$$

We did not need to count the number of positive factors, because positive factors do not change the sign of a product.

We remember that an exponent indicates how many times the base is used as a factor.

$$(-3)^4 \text{ means } (-3)(-3)(-3)(-3)$$

Example 2 Simplify: (a) $(-2)^4$ (b) $(-2)^5$

Solution (a) The expression $(-2)^4$ means $(-2)(-2)(-2)(-2)$. There is an even number of negative factors, so the product is a positive number. Since 2^4 is 16, we find that $(-2)^4$ is **+16.**

(b) The expression $(-2)^5$ means $(-2)(-2)(-2)(-2)(-2)$. This time there is an odd number of negative factors, so the product is a negative number. Since $2^5 = 32$, we find that $(-2)^5 = \mathbf{-32.}$

Dividing terms We have used exponents to show the multiplication of algebraic terms.

$$(4a^2b)(ab) = 4a^3b^2$$

We divide terms by removing pairs of factors that equal 1.

$$\frac{4a^3b^2}{ab} = \frac{2 \cdot 2 \cdot \overset{1}{\cancel{a}} \cdot a \cdot a \cdot \overset{1}{\cancel{b}} \cdot b}{\underset{1}{\cancel{a}} \cdot \underset{1}{\cancel{b}}}$$

$$= 4a^2b$$

Example 3 Simplify: $\dfrac{12x^3yz^2}{3x^2y}$

Solution We factor the two terms and remove pairs of factors that equal 1. Then we regroup the remaining factors.

$$\frac{12x^3yz^2}{3x^2y} = \frac{2 \cdot 2 \cdot \overset{1}{\cancel{3}} \cdot \overset{1}{\cancel{x}} \cdot \overset{1}{\cancel{x}} \cdot x \cdot \overset{1}{\cancel{y}} \cdot z \cdot z}{\underset{1}{\cancel{3}} \cdot \underset{1}{\cancel{x}} \cdot \underset{1}{\cancel{x}} \cdot \underset{1}{\cancel{y}}}$$

$$= \mathbf{4xz^2}$$

Example 4 Simplify: $\frac{10a^3bc^2}{8ab^2c}$

Solution We factor the terms and remove common factors. Then we regroup the remaining factors.

$$\frac{10a^3bc^2}{8ab^2c} = \frac{\overset{1}{\cancel{2}} \cdot 5 \cdot \overset{1}{\cancel{a}} \cdot a \cdot a \cdot \overset{1}{\cancel{b}} \cdot \overset{1}{\cancel{c}} \cdot c}{\underset{1}{\cancel{2}} \cdot 2 \cdot 2 \cdot \underset{1}{\cancel{a}} \cdot \underset{1}{\cancel{b}} \cdot b \cdot \underset{1}{\cancel{c}}}$$

$$= \mathbf{\frac{5a^2c}{4b}}$$

LESSON PRACTICE

Practice set Simplify:

a. $(-5)(-4)(-3)(-2)(-1)$

b. $(+5)(-4)(+3)(-2)(+1)$

c. $(-2)^3$

d. $(-3)^4$

e. $(-9)^2$

f. $(-1)^5$

g. $\frac{6a^2b^3c}{3ab}$

h. $\frac{8xy^3z^2}{6x^2y}$

i. $\frac{15mn^2p}{25m^2n^2}$

MIXED PRACTICE

Problem set

1. The dinner bill totaled \$25. Mike left a 15% tip. How much money did Mike leave for a tip?
(46)

2. The table below shows a tally of the scores Maxine earned on this year's math tests. Find the (a) mode and (b) median of her 29 test scores.
(Inv. 4)

Maxine's Test Scores

Score	Number of Tests
100				
95	卌	
90	卌	
85	卌 ||||
80					
70		

3. Draw a box-and-whisker plot for the data presented in problem 2.
(Inv. 4)

4. The plane completed the flight in $2\frac{1}{2}$ hours. If the flight covered 1280 kilometers, what was the plane's average speed in kilometers per hour?
(46)

Use ratio boxes to solve problems 5–9.

5. Jeremy earned \$25 for 4 hours of work. How much would he earn for 7 hours of work at the same rate?
(72)

6. If 40 percent of the lights were on, what was the ratio of lights on to lights off?
(54)

7. Lesley saved \$25 buying the pantsuit at a sale that offered 20 percent off. What was the regular price of the pantsuit?
(92)

8. The merchant bought the item for \$30 and sold it for 60 percent more. How much profit did the merchant make on the item?
(92)

9. (a) The $\frac{1}{20}$ scale model of the rocket stood 54 inches high. What was the height of the actual rocket?
(50, 98)

(b) Find the height of the actual rocket in feet.

10. The volume of the rocket in problem 9 is how many times the volume of the model?
(98)

11. A mile is about eight fifths kilometers. Eight fifths kilometers is how many meters?
(60)

12. Use the Pythagorean theorem to find the length of the longest side of a right triangle whose vertices are (3, 1), (3, −2), and (−1, −2).
(99)

13. Alina has made 35 out of 50 free throws. What is the statistical chance that Alina will make her next free throw?
(Inv. 10)

14. What percent of 25 is 20?
(77)

15. (a) The shaded sector is what fraction of the whole circle?
(Inv. 8)

(b) The unshaded sector is what percent of the circle?

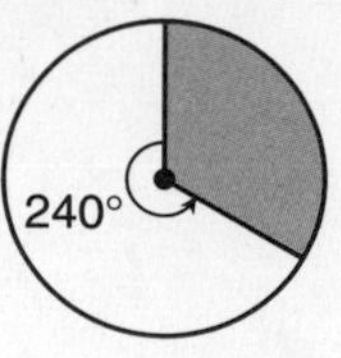

16. Find the missing numbers in the table by using the function rule. Then graph the function on a coordinate plane.
(85, Inv. 9)

$y = -x$

x	y
2	☐
0	☐
–1	☐

17. Write an equation to solve this problem:
(101)

Three less than twice what number is –7?

18. Quadrilateral $ABCD$ is a rectangle. The measure of $\angle ACB$ is 36°. Find the measures of the following angles:
(40)

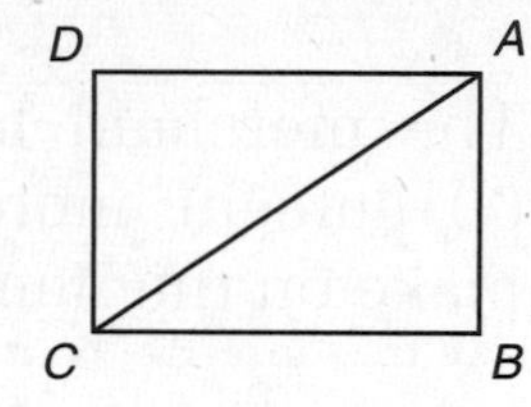

(a) $\angle CAB$ (b) $\angle CAD$ (c) $\angle ACD$

19. The two triangles below are similar.
(97, 98)

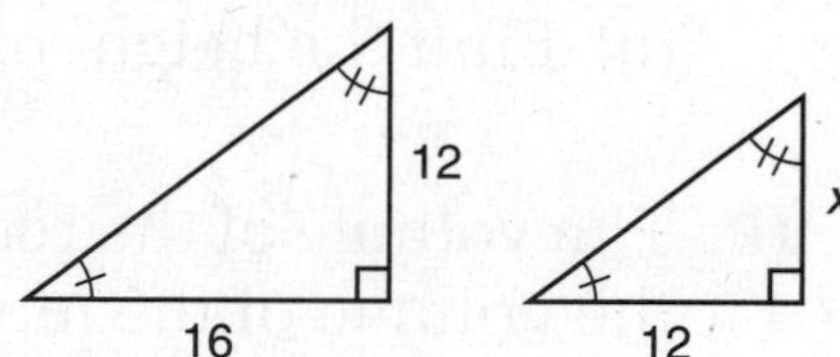

(a) Estimate, then calculate, the length x.

(b) Find the scale factor from the larger triangle to the smaller triangle.

20. Estimate the measure of $\angle AOB$. Then use a protractor to measure the angle.
(96)

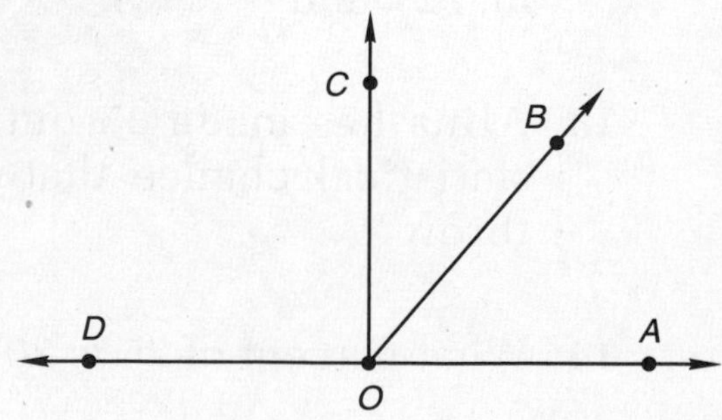

21. Find the (a) circumference and (b) area of a circle with a diameter of 2 feet. (Use 3.14 for π.)
(66, 82)

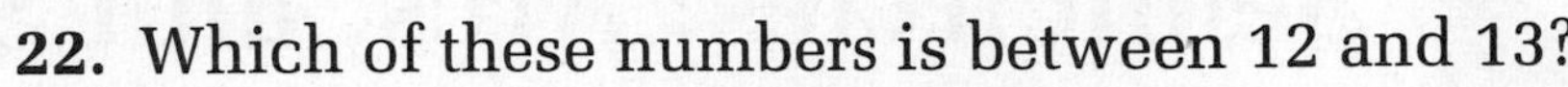

22. Which of these numbers is between 12 and 13?
(100)

A. $\sqrt{13}$ B. $\sqrt{130}$ C. $\sqrt{150}$

23. Find the volume of this right prism.
(95)

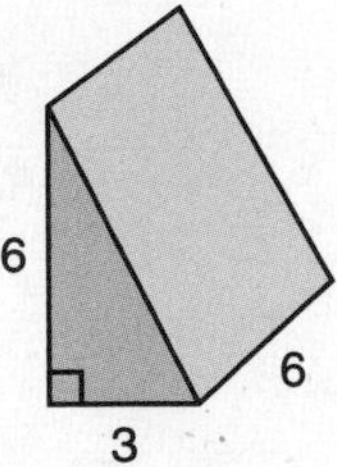

24. Find the volume of this right circular cylinder. (Use 3.14 for π.)
(95)

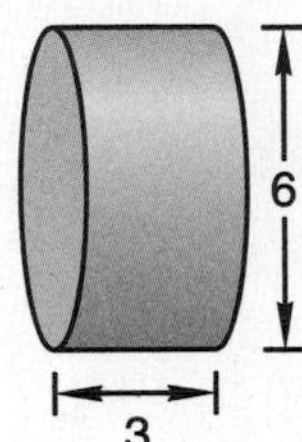

Solve:

25. $3x + x - 5 = 2(x - 2)$
(102)

26. $6\frac{2}{3}f - 5 = 5$
(90, 93)

Simplify:

27. $10\frac{1}{2} \cdot 1\frac{3}{7} \cdot 5^{-2}$
(26, 57)

28. $12.5 - 8\frac{1}{3} + 1\frac{1}{6}$
(43)

29. (a) $\dfrac{(-3)(-2)(-1)}{-|(-3)(+2)|}$ (b) $3^2 - (-3)^2$
(85, 103)

30. (a) $\dfrac{6a^3b^2c}{2abc}$ (b) $\dfrac{8x^2yz^3}{12xy^2z}$
(103)

LESSON 104 Semicircles, Arcs, and Sectors

WARM-UP

Facts Practice: + – × ÷ Algebraic Terms (Test V)

Mental Math:

a. 1111 (base 2)

b. CXC

c. (–0.25) – (–0.75)

d. $(5 \times 10^5)^2$

e. $80 = 4m - 20$

f. Convert 1 sq. yd to sq. ft.

g. 200% of $25

h. 200% more than $25

i. At 12 mph, how far can Sherry skate in 45 minutes?

Problem Solving:

Five identical blocks marked X and a 60-oz weight were balanced on a scale as shown. Write an equation to represent this balanced scale, and find the weight of each block marked X.

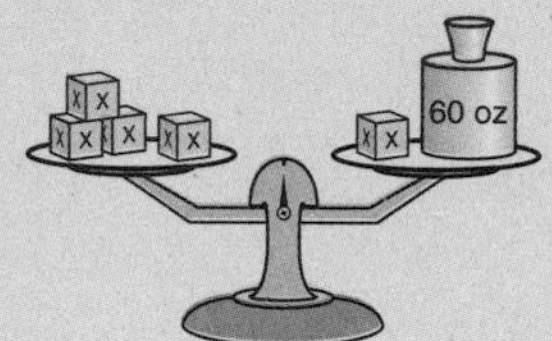

NEW CONCEPT

A **semicircle** is half of a circle. Thus the length of a semicircle is half the circumference of a circle with the same radius. The area enclosed by a semicircle and its diameter is half the area of the full circle.

We will practice finding the lengths of semicircles and the areas they enclose by calculating the perimeters and areas of figures that contain semicircles. We will also find the lengths of arcs and areas of sectors that are not semicircles.

Example 1 Find the perimeter of this figure. Dimensions are in meters.

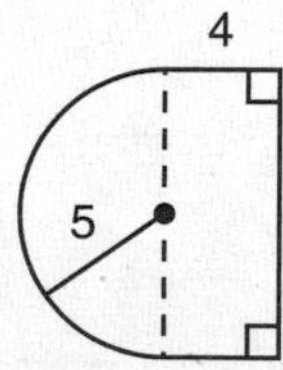

Solution The perimeter of the semicircle is half the perimeter of a circle whose diameter is 10.

$$\text{Perimeter of semicircle} = \frac{\pi d}{2}$$

$$\approx \frac{(3.14)(10\text{ m})}{2}$$

$$\approx 15.7\text{ m}$$

Now we can label all the dimensions on the figure and add them to find the perimeter.

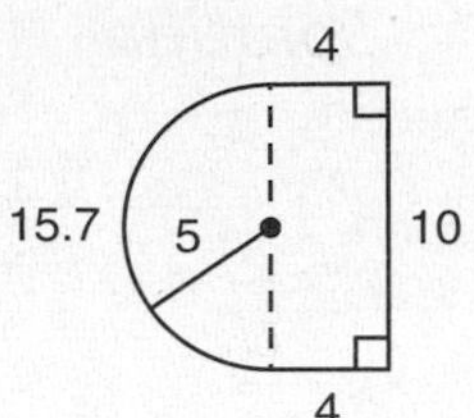

$$\text{Perimeter} \approx 10 \text{ m} + 4 \text{ m} + 15.7 \text{ m} + 4 \text{ m}$$

$$\approx \mathbf{33.7 \text{ m}}$$

Example 2 Find the area of this figure. Dimensions are in meters.

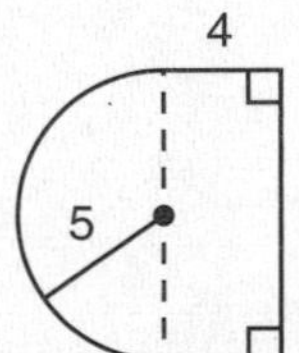

Solution We divide the figure into two parts. Then we find the area of each part.

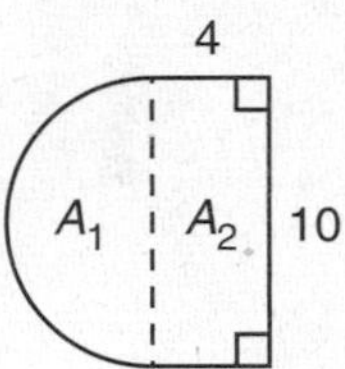

$$A_1 = \frac{\pi r^2}{2} \qquad A_2 = l \times w$$

$$\approx \frac{3.14(25 \text{ m}^2)}{2} \qquad = 4 \text{ m} \times 10 \text{ m}$$

$$\approx 39.25 \text{ m}^2 \qquad = 40 \text{ m}^2$$

The total area of the figure equals $A_1 + A_2$.

$$\text{Total area} = A_1 + A_2$$

$$\approx 39.25 \text{ m}^2 + 40 \text{ m}^2$$

$$\approx \mathbf{79.25 \text{ m}^2}$$

We can calculate the lengths of arcs and the areas of sectors by determining the fraction of a circle represented by the arc or sector.

Example 3 Find the area of the shaded sector of this circle.

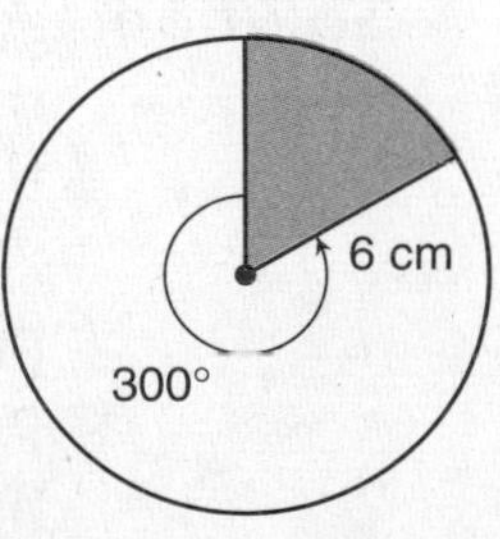

Solution The central angle of the shaded sector measures 60° (the full circle, 360°, minus the given angle, 300°). Since 60° is $\frac{1}{6}$ of a circle $\left(\frac{60}{360} = \frac{1}{6}\right)$, the area of the sector is $\frac{1}{6}$ of the area of the circle.

$$\text{Area of } 60° \text{ sector} = \frac{\pi r^2}{6}$$

$$\approx \frac{3.14(\overset{1}{\cancel{6}} \text{ cm})(6 \text{ cm})}{\underset{1}{\cancel{6}}}$$

$$\approx \mathbf{18.84 \ cm^2}$$

As we discussed in Investigation 2, an arc is part of the circumference of a circle. In the following figure we see two arcs between point *A* and point *B*.

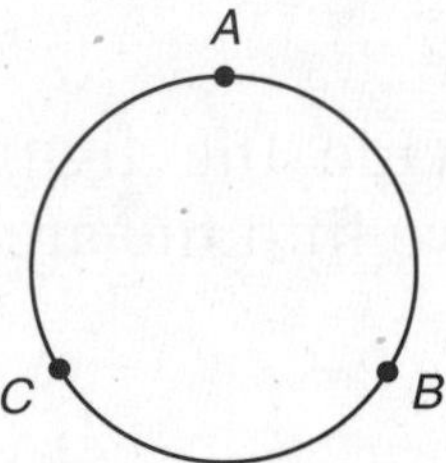

The arc from *A* clockwise to *B* is called a **minor arc** because it is less than a semicircle. We may refer to this arc as arc *AB* (abbreviated $\overset{\frown}{AB}$). The arc from *A* counterclockwise to *B* is called a **major arc** because it is greater than a semicircle. Major arcs are named with three letters. The major arc between point *A* and point *B* may be named arc *ACB*.

We can measure an arc in degrees. The number of degrees in an arc equals the measure of the central angle of the arc. If minor arc *AB* in the figure above measures 120°, then the measure of major arc *ACB* is 240° because the sum of the measures of the major arc and minor arc must be 360°.

Example 4 In this figure, central angle *AOC* measures 70°. What is the measure of major arc *ADC*?

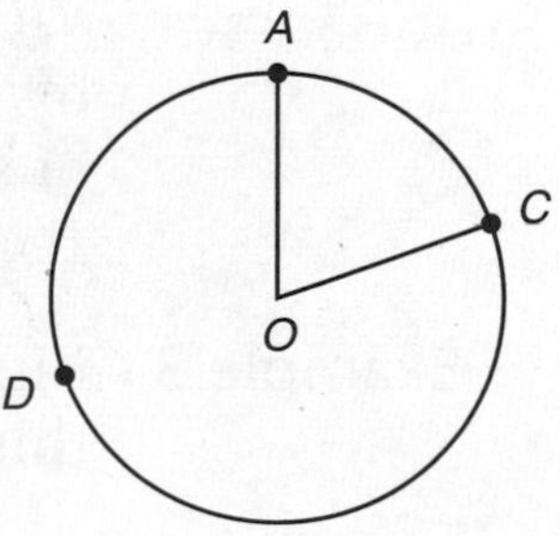

Solution An arc may be described by the measure of its central angle. The arc in the interior of the 70° angle *AOC* is a 70° arc. However, the larger arc from point *A* counterclockwise through point *D* to point *C* measures 360° minus 70°, which is **290°**.

Example 5 A minor arc with a radius of 2, centered at the origin, is drawn from the positive x-axis to the positive y-axis. What is the length of the arc?

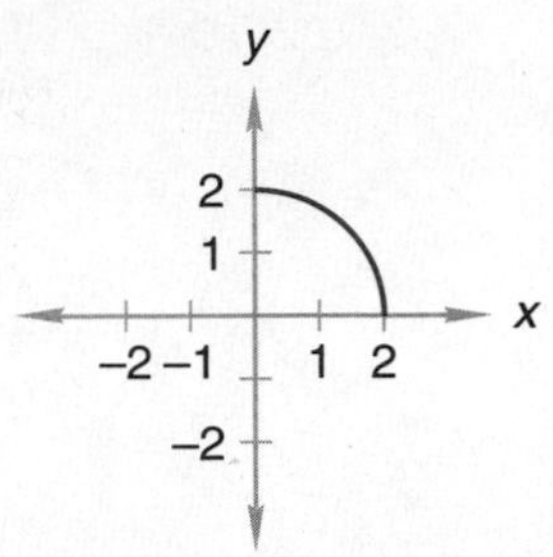

Solution A minor arc is less than 180°. We see that the arc is $\frac{1}{4}$ of a circle, which is 90°. The length of the arc is $\frac{1}{4}$ of the circumference of a circle with a radius of 2 (and a diameter of 4).

$$\text{Length of } 90° \text{ arc} = \frac{\pi d}{4}$$

$$\approx \frac{3.14(\overset{1}{\cancel{4}} \text{ units})}{\underset{1}{\cancel{4}}}$$

$$\approx \mathbf{3.14 \text{ units}}$$

LESSON PRACTICE

Practice set Find the **a.** perimeter and **b.** area of this figure. Dimensions are in centimeters. (Use 3.14 for π.)

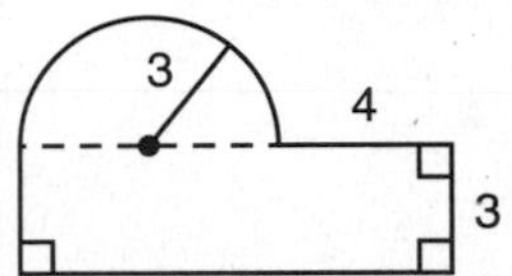

c. Find the area of this 45° sector. (Leave π as π.)

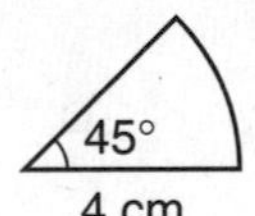

d. Find the perimeter of the figure in problem **c.** (Include the arc and two segments. Use 3.14 for π.)

MIXED PRACTICE

Problem set

1. (60) The merchant sold the item for $12.50. If 40 percent of the selling price was profit, how much money did the merchant earn in profit?

2. (94) With one toss of a pair of dot cubes, what is the probability that the total rolled will be a prime number? (Add the probabilities for each prime-number total.)

3. (55) Keisha's average score for 10 tests is 88. If her lowest score, 70, is not counted, what is her average for the remaining 9 tests?

4. (46) The 36-ounce container cost \$3.42. The 3-pound container cost \$3.84. The smaller container cost how much more per ounce than the larger container?

5. (72) Sean read 18 pages in 30 minutes. If he has finished page 128, how many hours will it take him to finish his 308-page book if he continues reading at the same rate?

6. (74) Matthew was thinking of a certain number. If $\frac{5}{6}$ of the number was 75, what was $\frac{3}{5}$ of the number?

7. (54) Use a ratio box to solve this problem. The ratio of crawfish to tadpoles in the creek was 2 to 21. If there were 1932 tadpoles in the creek, how many crawfish were there?

8. (60, 77) Write equations to solve (a) and (b).

(a) What percent of \$60 is \$45?

(b) How much money is 45 percent of \$60?

In the figure below, $\overline{AD}$ is a diameter and $\overline{CB}$ is a radius of 12 units. Central angle *ACB* measures 60°. Refer to the figure to answer problems 9–11.

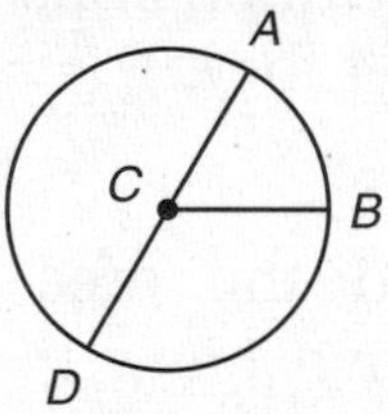

Leave π as π.

9. (66, 82) (a) What is the area of the circle?

(b) What is the circumference of the circle?

10. (104) What is the area of sector *BCD?*

11. (104) (a) How many degrees is the major arc from *B* through *A* to *D* (arc *BAD*)?

(b) How long is arc *BAD?*

12. (Inv. 9) Make a table of ordered pairs for the function $y = 2x - 1$. Use –1, 0, and 1 as x values in the table. Then graph the function on a coordinate plane.

13. (48) Complete the table.

FRACTION	DECIMAL	PERCENT
(a)	(b)	2.2%

14. The graph below shows the distance a car traveled at a certain constant speed.
(Inv. 9)

(a) According to this graph, how far did the car travel in 1 hour 15 minutes?

(b) Estimate the speed of the car in miles per hour as indicated by the graph.

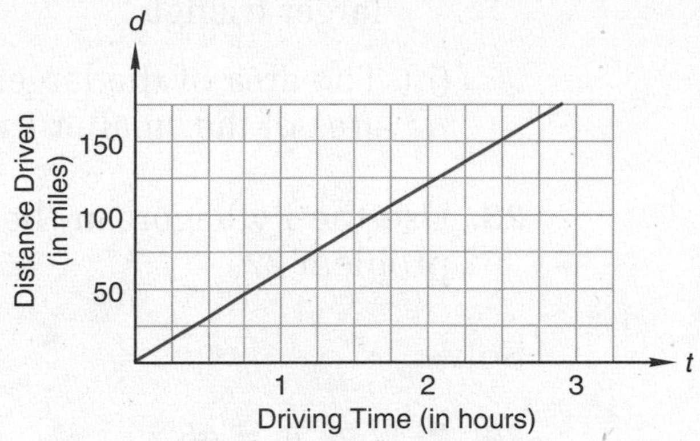

15. Compare: $ab \bigcirc a - b$ if a is positive and b is negative
(79)

16. Multiply. Write the product in scientific notation.
(83)

$$(3.6 \times 10^{-4})(9 \times 10^{8})$$

17. Find the area of the figure at right. Dimensions are in centimeters.
(104)

18. Find the perimeter of the figure in problem 17.
(104)

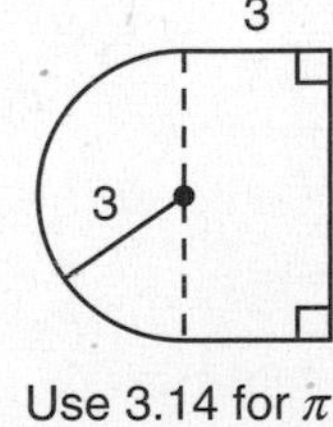

Use 3.14 for π.

19. (a) Find the volume of this solid in cubic inches. Dimensions are in feet.
(67, 70)

(b) Find the surface area of this cube in square feet.

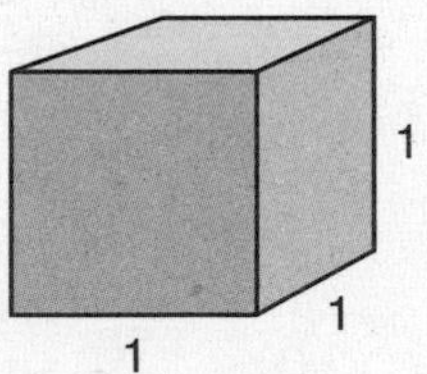

20. What angle is formed by the hands of a clock at 5:00?
(96)

21. Find m∠x in the figure at right.
(40)

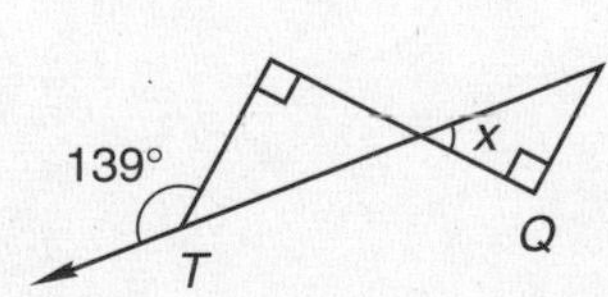

22. The triangles below are similar.
(97, 98)

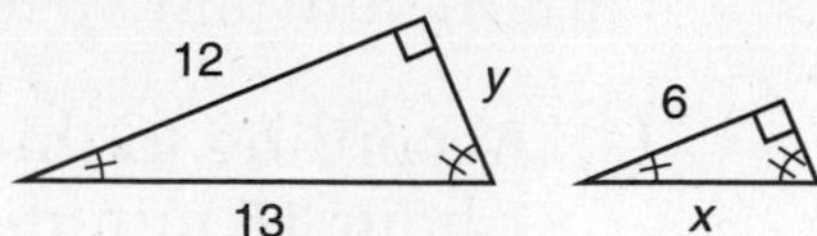

(a) Find x.

(b) Find the scale factor from the smaller triangle to the larger triangle.

(c) The area of the larger triangle is how many times the area of the smaller triangle?

23. Use the Pythagorean theorem to find y in the triangle in problem 22.
(99)

Solve:

24. $2\frac{3}{4}w + 4 = 48$
(90, 93)

25. $2.4n + 1.2n - 0.12 = 7.08$
(102)

Simplify:

26. $\sqrt{(3^2)(10^2)}$
(20)

27. (a) $\dfrac{24x^2y}{8x^3y^2}$ (b) $3x^2 + 2x(x - 1)$
(102, 103)

28. $12.5 - \left(8\frac{1}{3} + 1\frac{1}{6}\right)$
(43)

29. $4\frac{1}{6} \div 3\frac{3}{4} \div 2.5$
(43)

30. (a) $\dfrac{(-3)(4)}{-2} - \dfrac{(-3)(-4)}{-2}$ (b) $\dfrac{(-2)^3}{(-2)^2}$
(85, 103)

LESSON

105 Surface Area of a Right Solid • Surface Area of a Sphere • More on Roots

WARM-UP

Facts Practice: Percent-Decimal-Fraction Equivalents (Test Q)

Mental Math:

a. 10000 (base 2)
b. MCCIX
c. $(-2)^4$
d. $(4 \times 10^{-4})^2$
e. $2w + 3w = 60$
f. Convert –35°C to degrees Fahrenheit.
g. 200% of $50
h. 100% more than $50
i. Square 10, − 1, ÷ 9, × 4, + 1, ÷ 9, × 10, − 1, $\sqrt{\ }$, × 5, + 1, $\sqrt{\ }$, ÷ 3.

Problem Solving:

Shayla constructed this shape using 1-inch cubes. Find the volume of the shape. Then draw two-dimensional views from the front, from the right side, and from the top. (Only the solid line segments should show in your drawings.)

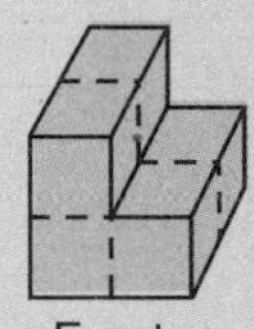

NEW CONCEPTS

Surface area of a right solid

Recall that the total area of the surface of a geometric solid is called the **surface area** of the solid.

The block shown has six rectangular faces. The areas of the top and bottom are equal, the areas of the front and back are equal, and the areas of the left and right sides are equal. We add the areas of these six faces to find the total surface area.

Area of top	= 5 cm × 6 cm =	30 cm²
Area of bottom	= 5 cm × 6 cm =	30 cm²
Area of front	= 3 cm × 6 cm =	18 cm²
Area of back	= 3 cm × 6 cm =	18 cm²
Area of side	= 3 cm × 5 cm =	15 cm²
+ Area of side	= 3 cm × 5 cm =	15 cm²
Total surface area	=	126 cm²

Example 1 Find the surface area of this triangular prism. Dimensions are in centimeters.

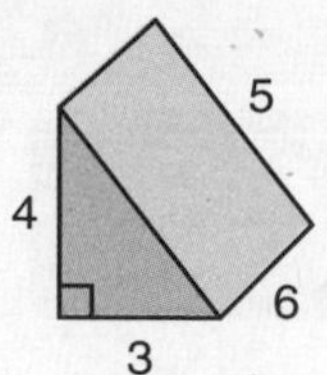

Solution There are two triangular faces and three rectangular faces.

$$\text{Area of triangle} \quad \frac{3\text{ cm} \cdot 4\text{ cm}}{2} = 6\text{ cm}^2$$

$$\text{Area of triangle} \quad \frac{3\text{ cm} \cdot 4\text{ cm}}{2} = 6\text{ cm}^2$$

$$\begin{aligned} \text{Area of rectangle} &= 3\text{ cm} \cdot 6\text{ cm} = 18\text{ cm}^2 \\ \text{Area of rectangle} &= 4\text{ cm} \cdot 6\text{ cm} = 24\text{ cm}^2 \\ + \text{ Area of rectangle} &= 5\text{ cm} \cdot 6\text{ cm} = 30\text{ cm}^2 \\ \hline \text{Total surface area} &= \mathbf{84\text{ cm}^2} \end{aligned}$$

Example 2 (a) What is the area of the label on a soup can with dimensions as shown?

(b) What is the total surface area of the soup can?

Solution (a) If we remove the label from a soup can, we see that it is a rectangle. One dimension of the rectangle is the circumference of the can, and the other dimension is the height of the can. The area of the label is the *lateral surface area* of the soup can.

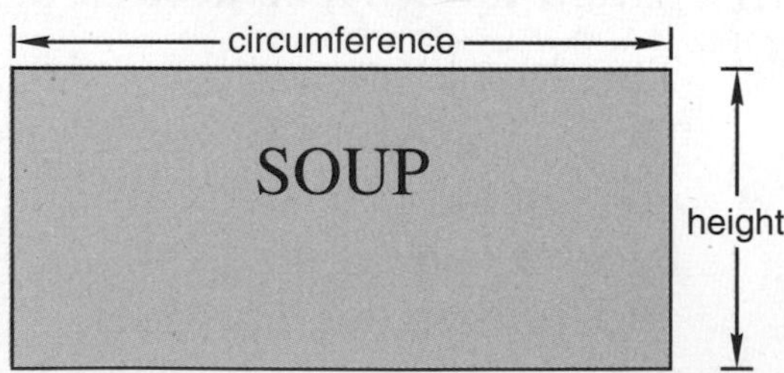

To find the area of the label, we multiply these two dimensions.

$$\begin{aligned} \text{Area} &= \text{circumference} \cdot \text{height} \\ &= \pi d \cdot \text{height} \\ &\approx \frac{22}{\cancel{7}_1} \cdot \overset{1}{\cancel{7}}\text{ cm} \cdot 10\text{ cm} \\ &\approx \mathbf{220\text{ cm}^2} \end{aligned}$$

(b) The total surface area of the can consists of the lateral surface area plus the circular top and bottom of the can. Recall that the area of a circle is found by multiplying the square of the radius by π.

$$A = \pi r^2$$

We use $\frac{22}{7}$ for π and $\frac{7}{2}$ cm (or 3.5 cm) for the radius.

$$A \approx \frac{22}{7}\left(\frac{7}{2}\text{ cm}\right)^2$$

$$A \approx \frac{\overset{11}{\cancel{22}}}{\underset{1}{\cancel{7}}} \cdot \frac{\overset{1}{\cancel{7}}}{\underset{1}{\cancel{2}}} \cdot \frac{7}{2}\text{ cm}^2$$

$$A \approx \frac{77}{2}\text{ cm}^2 \text{ (or 38.5 cm}^2\text{)}$$

We have found the area of one circular surface. However, the can has both a top and a bottom, so we add the areas of the top, bottom, and lateral surface.

Area of top	=	38.5 cm^2
Area of bottom	=	38.5 cm^2
+ Area of lateral surface	=	220.0 cm^2
Total surface area	=	297.0 cm^2

The total surface area of the soup can is **297 cm^2.**

Surface area of a sphere To calculate the surface area of a sphere, we may first calculate the area of the largest cross section of the sphere. Slicing a grapefruit in half provides a visual representation of a cross section of a sphere.

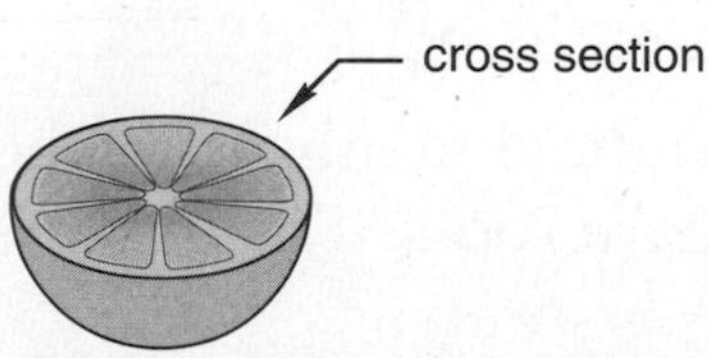

The circle formed by cutting the grapefruit in half is the cross section of the spherical grapefruit. The surface area of the entire sphere is four times the area of this circle. To find the surface area of the sphere, we calculate the area of its largest cross section ($A = \pi r^2$); then we multiply the cross sectional area by four.

Surface area of a sphere = $4\pi r^2$

Example 3 A tennis ball has a diameter of about 6 cm. Find the surface area of the tennis ball to the nearest square centimeter.

Use 3.14 for π.

Solution A tennis ball is spherical. If its diameter is 6 cm, then its radius is 3 cm.

$$\begin{aligned}\text{Surface area} &= 4\pi r^2 \\ &\approx 4(3.14)(3\text{ cm})^2 \\ &\approx 4(3.14)(9\text{ cm}^2) \\ &\approx 113.04\text{ cm}^2\end{aligned}$$

We round the answer to **113 cm²**.

More on roots The perfect square 25 has both positive and negative square roots, 5 and –5.

$$5 \cdot 5 = 25 \qquad (-5)(-5) = 25$$

Thus the equation $x^2 = 25$ has two solutions, 5 and –5.

The positive square root of a number is sometimes called the *principal* square root. So the principal square root of 25 is 5. The radical symbol $\sqrt{\ }$ implies the principal root. So $\sqrt{25}$ is 5 only and does not include –5.

Example 4 What are the two square roots of 5?

Solution The two square roots of 5 are $\mathbf{\sqrt{5}}$ and the opposite of $\sqrt{5}$, which is $\mathbf{-\sqrt{5}}$.

A radical symbol may be used to indicate other roots besides square roots. The expression

$$\sqrt[3]{64}$$

means **cube root** of 64. The small 3 is called the **index** of the root. The cube root of 64 is the number that, when used as a factor three times, yields a product of 64.

$$(?)(?)(?) = 64$$

Thus the cube root of 64 is 4 because

$$4 \cdot 4 \cdot 4 = 64$$

Example 5 Simplify: (a) $\sqrt[3]{1000}$ (b) $\sqrt[3]{-27}$

Solution (a) The cube root of 1000 is **10** because $10 \cdot 10 \cdot 10 = 1000$. Notice that -10 is not a cube root of 1000, because $(-10)(-10)(-10) = -1000$.

(b) The cube root of -27 is $\mathbf{-3}$ because $(-3)(-3)(-3) = -27$.

LESSON PRACTICE

Practice set

a. Find the surface area of this rectangular solid. Dimensions are in meters.

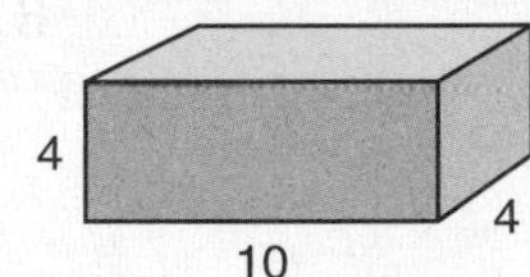

b. Find the surface area of this triangular prism. Dimensions are in inches.

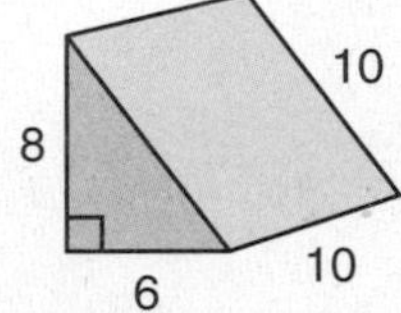

c. Find the area of the label on this can of tuna.

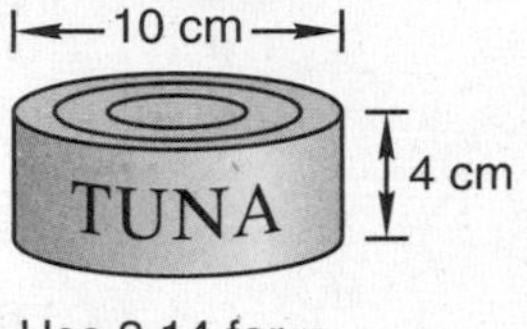

Use 3.14 for π.

d. Find the total surface area of the can.

e. The diameter of a golf ball is about 4 cm. Find the surface area of a golf ball to the nearest square centimeter. (Use 3.14 for π.)

f. What are the two square roots of 16?

Simplify:

g. $\sqrt[3]{125}$ **h.** $\sqrt[3]{-8}$

MIXED PRACTICE

Problem set

1. Twenty billion is how much greater than nine hundred million? Write the answer in scientific notation.
(51)

2. The mean of the following numbers is how much less than the median?
(Inv. 4)

3.2, 4.28, 1.2, 3.1, 1.17

3. Evaluate: $\sqrt{a^2 - b^2}$ if $a = 10$ and $b = 8$
(52)

4. If Glenda is paid at a rate of \$8.50 per hour, how much
(53) will she earn if she works $6\frac{1}{2}$ hours?

Use ratio boxes to solve problems 5–9.

5. If 6 kilograms of flour costs \$2.48, what is the cost of
(72) 45 kilograms of flour?

6. The regular price of the dress was \$30. The dress was on
(92) sale for 25% off.

(a) What was the sale price?

(b) What percent of the regular price was the sale price?

7. The ratio of Whigs to Tories at the assembly was 7 to 3. If
(65) the total number of Whigs and Tories assembled was 210, how many were Tories?

8. The merchant sold the item at a 30-percent discount from
(92) the regular price. If the regular price was \$60, what was the sale price?

9. Brandon is making a model plane at a 1:36 scale. If the
(98) length of the actual plane is 60 feet, how many inches long should he make his model? Begin by converting 60 feet to inches.

10. Transversal r intersects parallel
(102) lines s and t. If the measure of each acute angle is $4x$ and the measure of each obtuse angle is $5x$, then how many degrees is x?

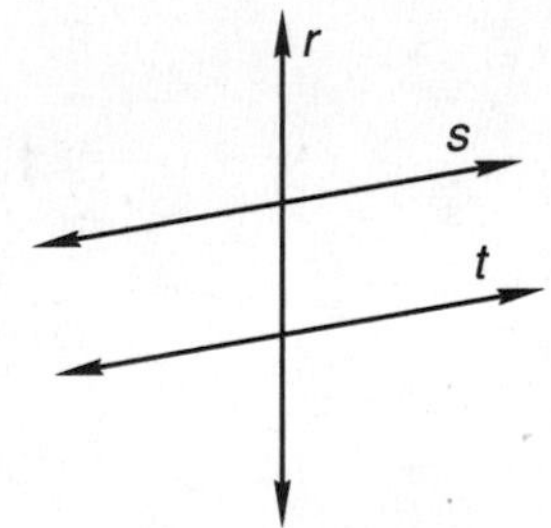

Write equations to solve problems 11–13.

11. What percent of \$60 is \$3?
(77)

12. What fraction is 10 percent of 4?
(60)

13. Twelve less than twice what number is 86?
(101)

14. The coordinates of the vertices of a right triangle are (–2, –2),
(99) (–2, 2), and (1, –2). Find the length of the hypotenuse of this triangle.

15. Compare: $a^3 \bigcirc a^2$ if a is negative
(79, 103)

16. If Carmela draws one card from a normal deck of cards, keeps the card, then draws another card, what is the probability that both cards will be hearts?
(94)

17. Bigler bounced a big ball with a diameter of 20 inches. Using 3.14 for π, find the surface area of the ball.
(105)

18. Multiply. Write the product in scientific notation.
(83)

$$(8 \times 10^{-4})(3.2 \times 10^{-10})$$

19. Find the perimeter of the figure at right. Dimensions are in meters. (Use 3.14 for π.)
(104)

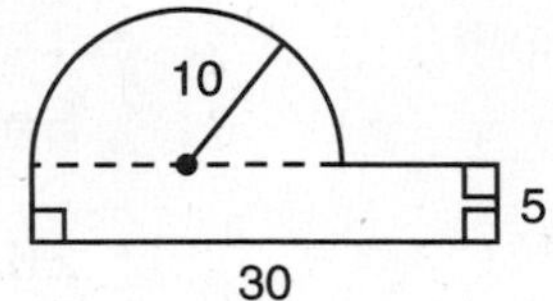

20. Find the missing numbers in the table by using the function rule. Then graph the function on a coordinate plane.
(85, Inv. 9)

$y = -2x - 1$

x	y
3	☐
−2	☐
0	☐

21. Find the (a) volume and (b) surface area of this cube. Dimensions are in millimeters.
(67, 70)

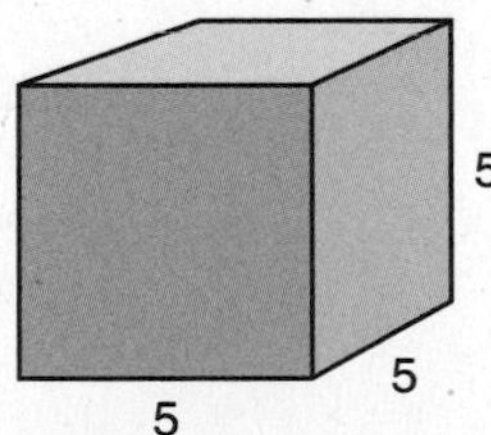

22. Find the volume of this right circular cylinder. Dimensions are in centimeters.
(95)

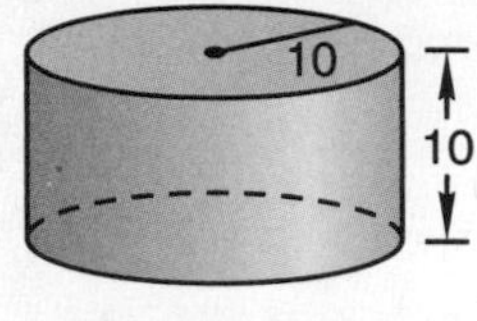

Use 3.14 for π.

23. The total surface area of the cylinder in problem 22 includes the areas of two circles and the curved side. What is the total surface area of the cylinder?
(105)

24. Find m$\angle b$ in the figure at right.
(40)

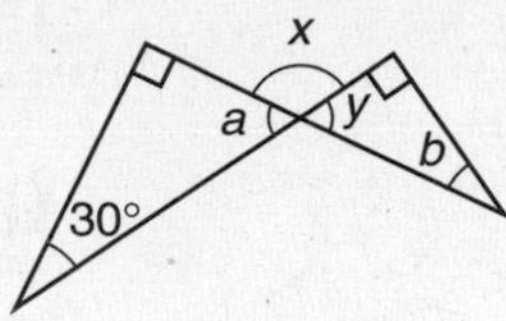

25. The triangles shown are similar.
(97, 98)

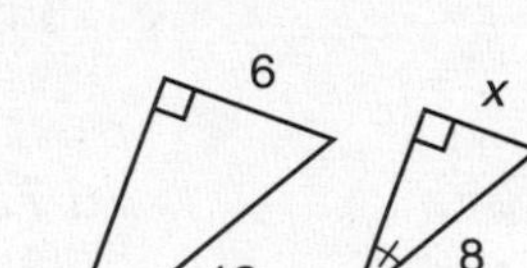

(a) Estimate, then calculate, the length x.

(b) Find the scale factor from the smaller triangle to the larger triangle.

(c) The area of the larger triangle is how many times the area of the smaller triangle?

Solve:

26. $4\frac{1}{2}x + 4 = 48 - x$
(102)

27. $\frac{3.9}{75} = \frac{c}{25}$
(98)

Simplify:

28. $3.2 \div \left(2\frac{1}{2} \div \frac{5}{8}\right)$
(43)

29. (a) $\frac{(2xy)(4x^2y)}{8x^2y}$
(96, 103)

(b) $3(x - 3) - 3$

30. (a) $\frac{(-10)(-4) - (3)(-2)(-1)}{(-4) - (-2)}$
(57, 85, 105)

(b) $(-2)^4 - (-2)^2 + \sqrt[3]{-1} + 2^0$

LESSON

106 Solving Literal Equations • Transforming Formulas

WARM-UP

Facts Practice: + − × ÷ Algebraic Terms (Test V)

Mental Math:

a. 10001 (base 2)
b. MMI
c. $(-5)^3$
d. $(8 \times 10^3)(5 \times 10^{-5})$
e. $\frac{a}{3.6} = \frac{0.9}{1.8}$
f. Convert 2 sq. yd to sq. ft.
g. $66\frac{2}{3}\%$ of \$45
h. $33\frac{1}{3}\%$ less than \$45
i. Estimate a 15% tip on a \$39.67 bill.

Problem Solving:

What is the average of these fractions?

$$\frac{1}{12}, \frac{1}{6}, \frac{1}{4}, \frac{1}{3}, \frac{5}{12}$$

NEW CONCEPTS

Solving literal equations A **literal equation** is an equation that contains letters instead of numbers. We can rearrange (transform) literal equations by using the rules we have learned.

Example 1 Solve for x: $x + a = b$

Solution We solve for x by isolating x on one side of the equation. We do this by adding $-a$ to both sides of the equation.

$$x + a = b \quad \text{equation}$$

$$x + a - a = b - a \quad \text{added } -a \text{ to both sides}$$

$$x = \mathbf{b - a} \quad \text{simplified}$$

Example 2 Solve for x: $ax = b$

Solution To solve for x, we divide both sides of the equation by a.

$$ax = b \quad \text{equation}$$

$$\frac{\overset{1}{\cancel{a}}x}{\underset{1}{\cancel{a}}} = \frac{b}{a} \quad \text{divided by } a$$

$$x = \mathbf{\frac{b}{a}} \quad \text{simplified}$$

Transforming formulas **Formulas** are literal equations that we can use to solve certain kinds of problems. Often it is necessary to change the way a formula is written.

Example 3 Solve for w: $A = lw$

Solution This is a formula for finding the area of a rectangle. We see that w is to the right of the equal sign and is multiplied by l. To undo the multiplication by l, we can divide both sides of the equation by l.

$$A = lw \qquad \text{equation}$$

$$\frac{A}{l} = \frac{\overset{1}{\cancel{l}}w}{\underset{1}{\cancel{l}}} \qquad \text{divided by } l$$

$$w = \frac{A}{l} \qquad \text{simplified}$$

LESSON PRACTICE

Practice set

a. Solve for a: $a + b = c$

b. Solve for w: $wx = y$

c. Solve for y: $y - b = mx$

d. The formula for the area of a parallelogram is

$$A = bh$$

Solve this equation for b.

MIXED PRACTICE

Problem set

1. (46) Otis paid \$20.00 for 3 pairs of socks priced at \$1.85 per pair and a T-shirt priced at \$8.95. The sales tax was 6 percent. How much money should he get back?

2. (Inv. 10) The face of this spinner is divided into twelfths.

(a) If the spinner is spun once, what are the odds that the spinner will land on a one-digit prime number?

(b) If the spinner is spun twice, what is the chance that it will land on an even number both times?

3. *(46)* At $2.80 per pound, the cheddar cheese costs how many cents per ounce?

4. *(55)* Brenda's average score after 6 tests is 90. If her lowest score, 75, is not counted, what is her average score on the remaining 5 tests?

5. *(99)* The ordered pairs (2, 4), (2, –1), and (0, –1) designate the vertices of a right triangle. What is the length of the hypotenuse of the triangle?

Use ratio boxes to solve problems 6–8.

6. *(72)* Justin finished 3 problems in 4 minutes. At that rate, how long will it take him to finish the remaining 27 problems?

7. *(65)* The ratio of residents to visitors in the pool was 2 to 3. If there were 60 people in the pool, how many were visitors?

8. *(92)* The dog groomer's clientele increased 25 percent this year. If the groomer has 80 clients this year, how many clients did the groomer have last year?

9. *(105)* (a) What are the two square roots of 64?

(b) What is the cube root of –64?

10. *(60)* Write an equation to solve this problem:

What number is 225 percent of 40?

11. *(100)* (a) Draw a number line and show the locations of these numbers:

$$|-2|,\ \frac{2}{2},\ \sqrt{2},\ 2^2$$

(b) Which of these numbers are rational numbers?

Write equations to solve problems 12 and 13.

12. *(77)* Sixty-six is $66\frac{2}{3}$ percent of what number?

13. *(77)* Seventy-five percent of what number is 2.4?

14. *(48)* Complete the table.

FRACTION	DECIMAL	PERCENT
(a)	(b)	105%

15. (Inv. 9) Make a table of ordered pairs for the function $y = x - 2$. Then graph the function on a coordinate plane.

16. (42) Divide 6.75 by 81 and write the quotient rounded to three decimal places.

17. (83) Multiply. Write the product in scientific notation.

$$(4.8 \times 10^{-10})(6 \times 10^{-6})$$

18. (91) Evaluate: $x^2 + bx + c$ if $x = -3$, $b = -5$, and $c = 6$

19. (104) Find the area of this figure. Dimensions are in millimeters. Corners that look square are square. (Use 3.14 for π.)

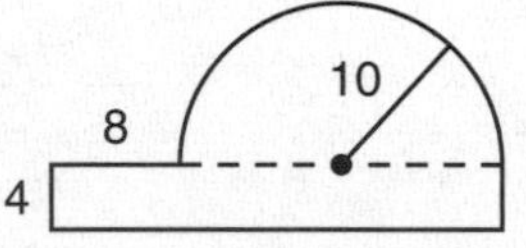

20. (105) Find the surface area of this right triangular prism. Dimensions are in centimeters.

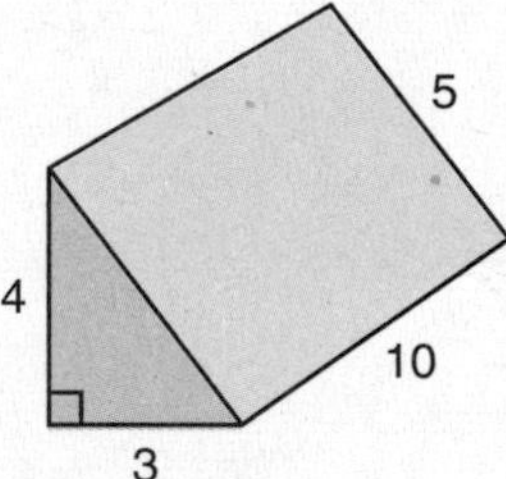

21. (95) Find the volume of this right circular cylinder. Dimensions are in inches. (Use 3.14 for π.)

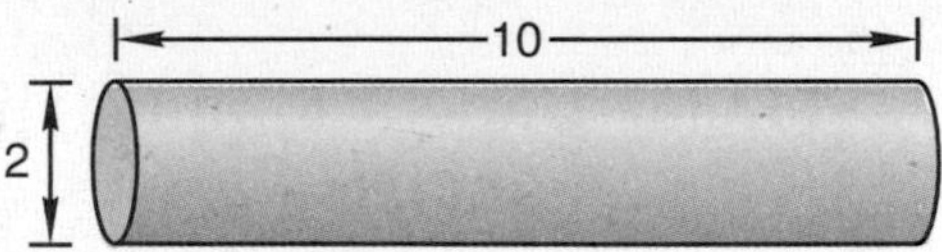

22. (40) Find m$\angle b$ in the figure below.

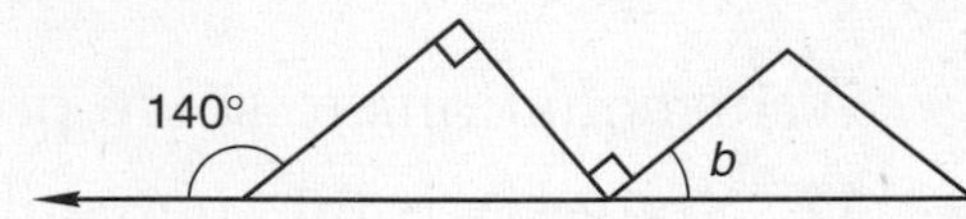

23. (106) (a) Solve for x: $x + c = d$

(b) Solve for n: $an = b$

24. (102) Solve: $6w - 2(4 + w) = w + 7$

25. Solve this inequality and graph its solution:
(93)

$$6x + 8 < 14$$

26. Thirty-seven is five less than the product of what number and three?
(101)

Simplify:

27. $25 - [3^2 + 2(5 - 3)]$
(63)

28. $\dfrac{6x^2 + (5x)(2x)}{4x}$
(103)

29. $4^0 + 3^{-1} + 2^{-2}$
(57)

30. $(-3)(-2)(+4)(-1) + (-3)^2 + \sqrt[3]{-64} - (-2)^3$
(103, 105)

LESSON

107 Slope

WARM-UP

Facts Practice: Percent-Decimal-Fraction Equivalents (Test Q)

Mental Math:

a. 11000 (base 2) **b.** DCCC

c. $(-2.5)(-4)$ **d.** $(2.5 \times 10^6)^2$

e. $2x - 1\frac{1}{2} = 4\frac{1}{2}$ **f.** Convert $-50°$C to degrees Fahrenheit.

g. 75% of \$60 **h.** 75% more than \$60

i. $7 \times 8, - 1, \div 5, \times 3, + 2, \div 5, \times 7, + 1, \times 2, - 1, \div 3, + 3, \sqrt{\ }$

Problem Solving:

In the 3-by-3 square at right, we can find nine 1-by-1 squares, four 2-by-2 squares, and one 3-by-3 square. Find the total number of squares of any size in the 4-by-4 square.

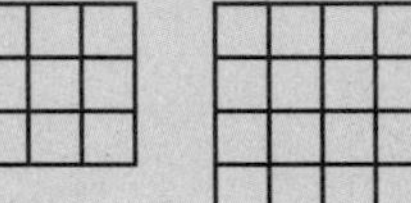

NEW CONCEPT

Below are the graphs of two functions. The graph of the function on the left indicates the number of feet that equal a given number of yards. Changing the number of yards by one changes the number of feet by three. The graph of the function on the right shows the inverse relationship, the number of yards that equal a given number of feet. Changing the number of feet by one changes the number of yards by one third.

Yards to Feet

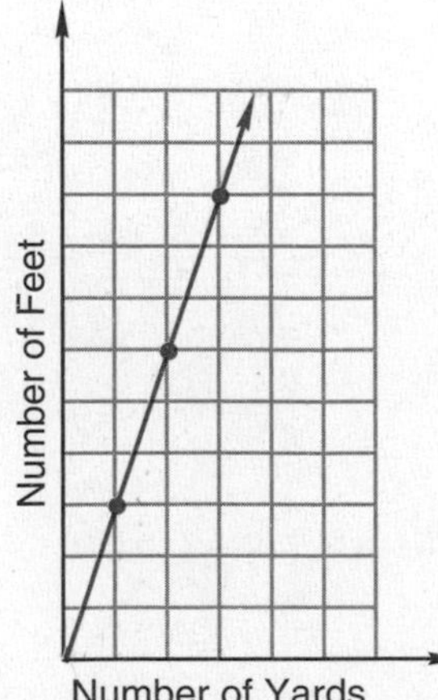

Feet to Yards

Number of Yards

Number of Feet

Notice that the graph of the function on the left has a steep upward slant going from left to right, while the graph of the function on the right also has an upward slant but is not as steep. The "slant" of the graph of a function is called its **slope.** We assign a number to a slope to indicate how steep the slope is and whether the slope is upward or downward. If the slope is upward, the number is positive. If the slope is downward, the number is negative. If the graph is horizontal, the slope is neither positive nor negative; it is zero. If the graph is vertical, the slope cannot be determined.

Example 1 State whether the slope of each line is positive, negative, zero, or cannot be determined.

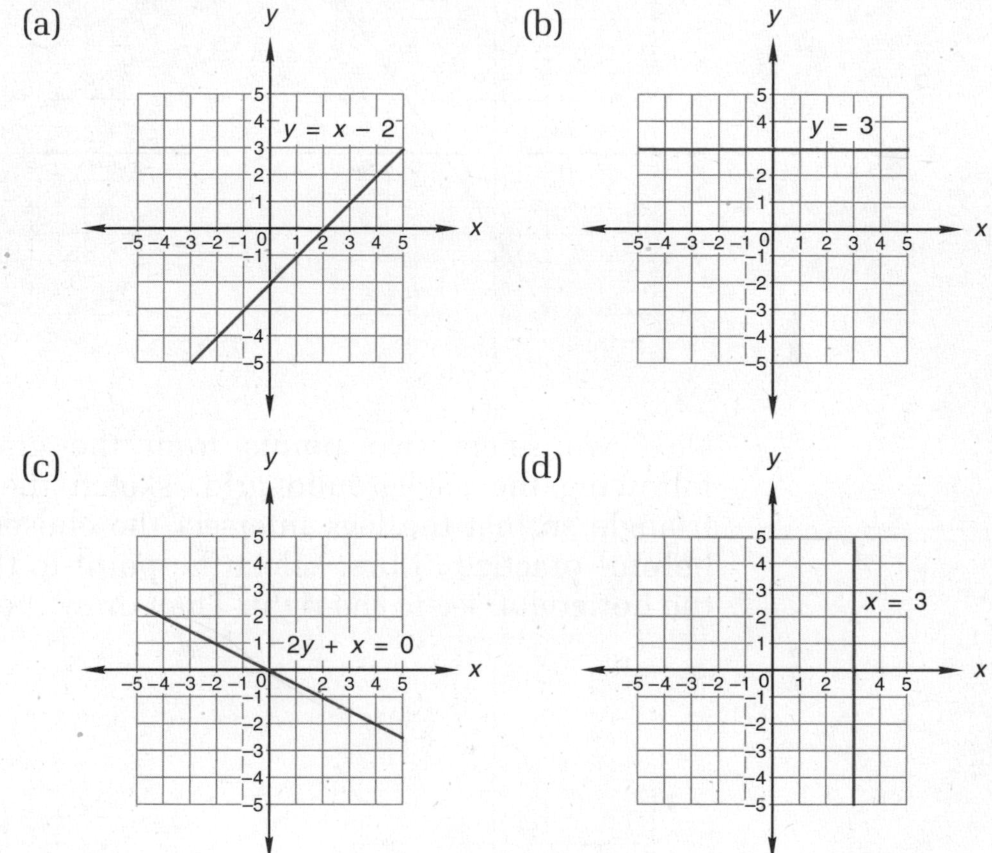

Solution To determine the sign of the slope, follow the graph of the function with your eyes *from left to right* as though you were reading.

(a) From left to right, the graphed line rises, so the slope is **positive.**

(b) From left to right, the graphed line does not rise or fall, so the slope is **zero.**

(c) From left to right, the graphed line slopes downward, so the slope is **negative.**

(d) There is no left to right component of the graphed line, so we cannot determine if the line is rising or falling. The slope is not positive, not negative, and not zero. The slope of a vertical line **cannot be determined.**

To determine the numerical value of the slope of a line, it is helpful to draw a right triangle using the background grid of the coordinate plane and a portion of the graphed line. First we look for points where the graphed line crosses intersections of the grid. We have circled some of these points on the graphs below.

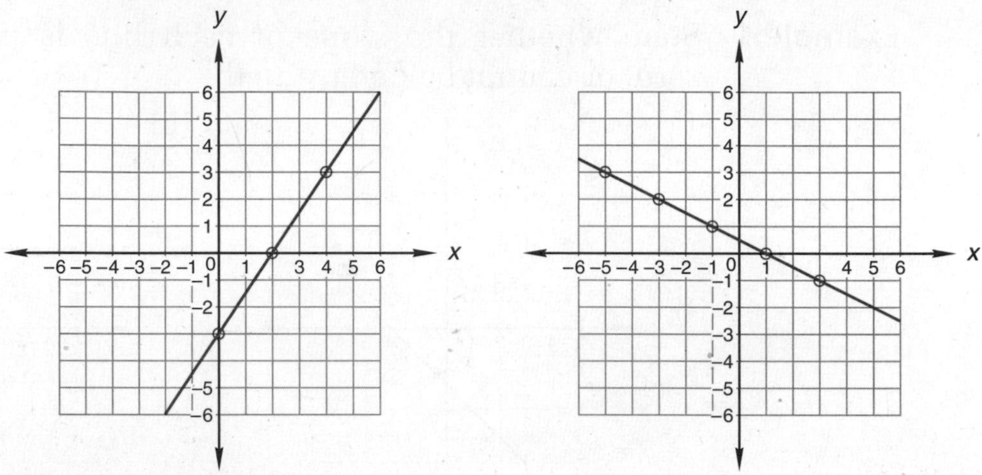

Next we select two points from the graphed line and, following the background grid, sketch the legs of a right triangle so that the legs intersect the chosen points. (It is a helpful practice to first select the point to the left and draw the horizontal leg to the right. Then draw the vertical leg.)

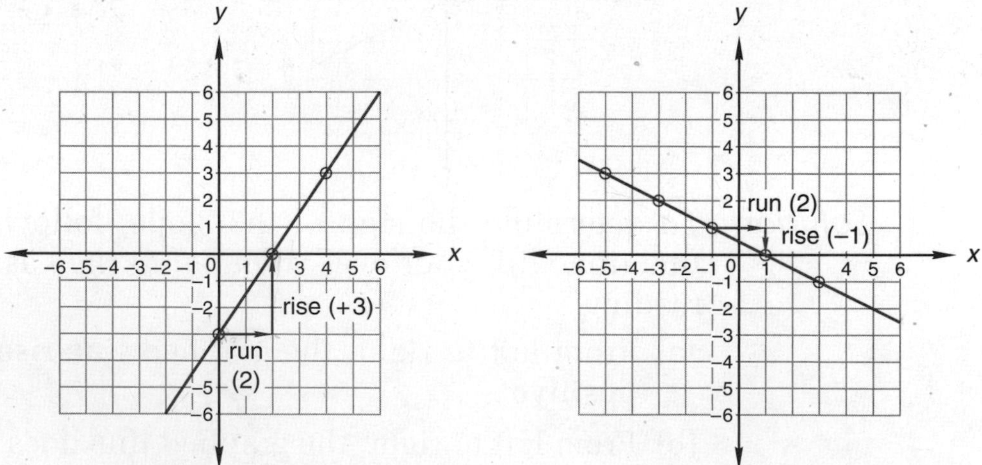

We use the words **run** and **rise** to describe the two legs of the right triangle. The *run* is the length of the horizontal leg, and the *rise* is the length of the vertical leg. We assign a positive sign to the rise if it goes up to meet the graphed line and a negative sign if it goes down to meet the graphed line. In the graph on the left, the run is 2 and the rise is +3. In the graph on the right, the run is 2 and the rise is –1. We use these numbers to write the slope of each graphed line.

So the slopes of the graphed lines are these ratios:

$$\frac{\text{rise}}{\text{run}} = \frac{+3}{2} = \frac{3}{2} \qquad \frac{\text{rise}}{\text{run}} = \frac{-1}{2} = -\frac{1}{2}$$

The slope of a line is the ratio of its rise to its run ("rise over run").

$$\textbf{slope} = \frac{\textbf{rise}}{\textbf{run}}$$

A line whose rise and run have equal values has a slope of 1. A line whose rise has the opposite value of its run has a slope of –1.

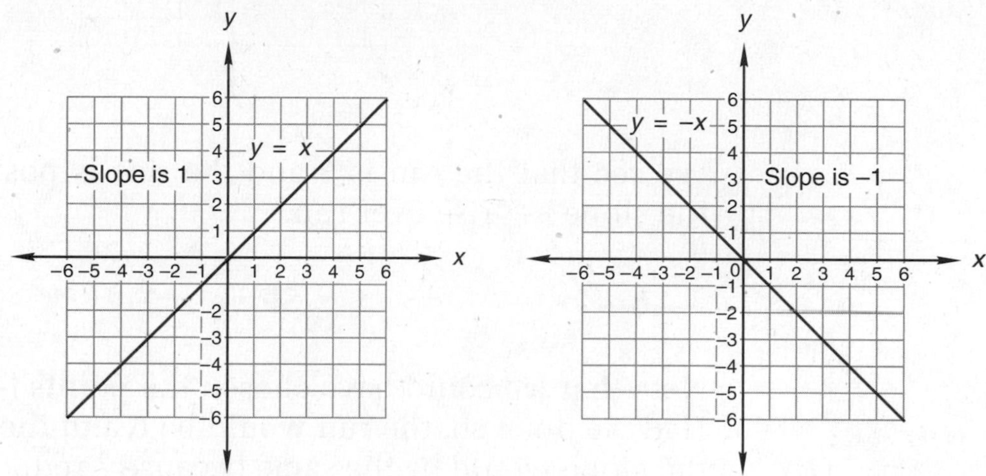

A line that is steeper than the lines above has a slope either greater than 1 or less than –1. A line that is less steep than the lines above has a slope that is between –1 and 1.

Example 2 Find the slope of the graphed line below.

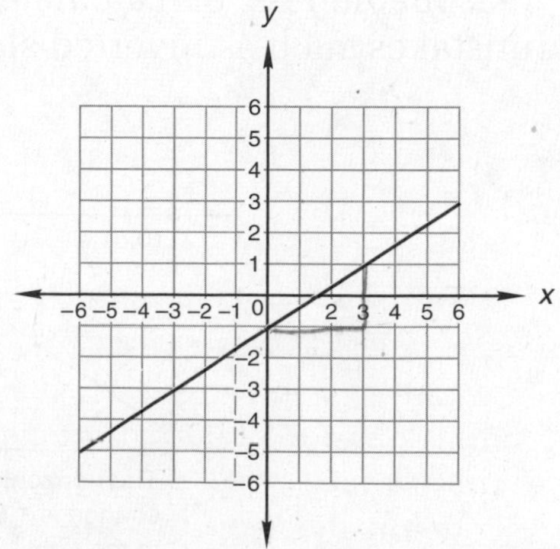

Solution We note that the slope is positive. We locate and select two points where the graphed line passes through intersections of the grid. We choose the points (0, –1) and (3, 1). Starting from the point to the left, (0, –1), we draw the horizontal leg to the right. Then we draw the vertical leg up to (3, 1).

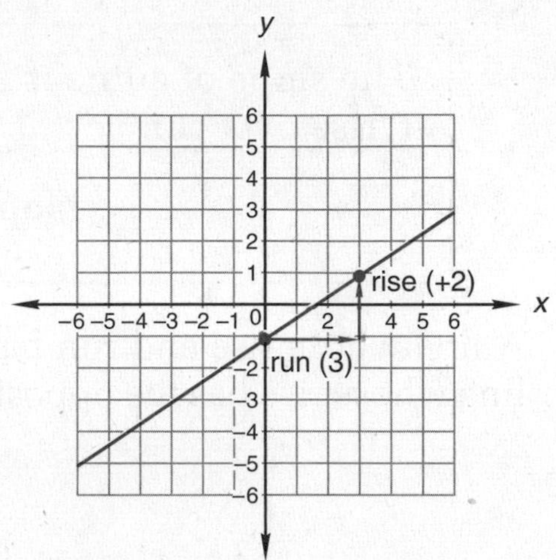

We see that the run is 3 and the rise is positive 2. We write the slope as "rise over run."

$$\text{Slope} = \mathbf{\frac{2}{3}}$$

Note that we could have chosen the points (–3, –3) and (3, 1). Had we done so, the run would be 6 and the rise 4. However, the slope would be the same because $\frac{4}{6}$ reduces to $\frac{2}{3}$.

One way to check the calculation of a slope is to "zoom in" on the graph. When the horizontal change is one unit to the right, the vertical change will equal the slope. To illustrate this, we will zoom in on the square just below and to the right of the origin on this graph. This method is a check for reasonableness of calculated slopes and can help prevent mistakes such as inverted slopes.

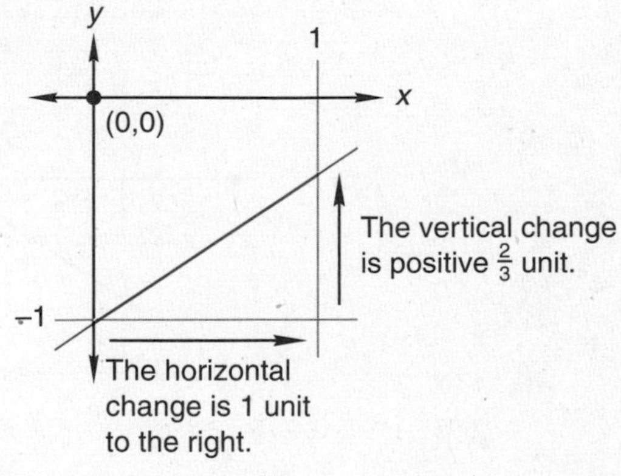

Activity: *Slope*

Materials needed:

- Activity Sheet 8 (available in *Saxon Math 8/7—Homeschool Tests and Worksheets*)

Calculate the slope (rise over run) of each graphed line on the activity sheet by drawing right triangles.

LESSON PRACTICE

Practice set

a. Find the slopes of the "Yards to Feet" and the "Feet to Yards" graphs at the beginning of this lesson.

b. Find the slopes of graphs (a) and (c) in example 1.

c. Mentally calculate the slope of each graphed line below by counting the run and rise rather than by drawing right triangles.

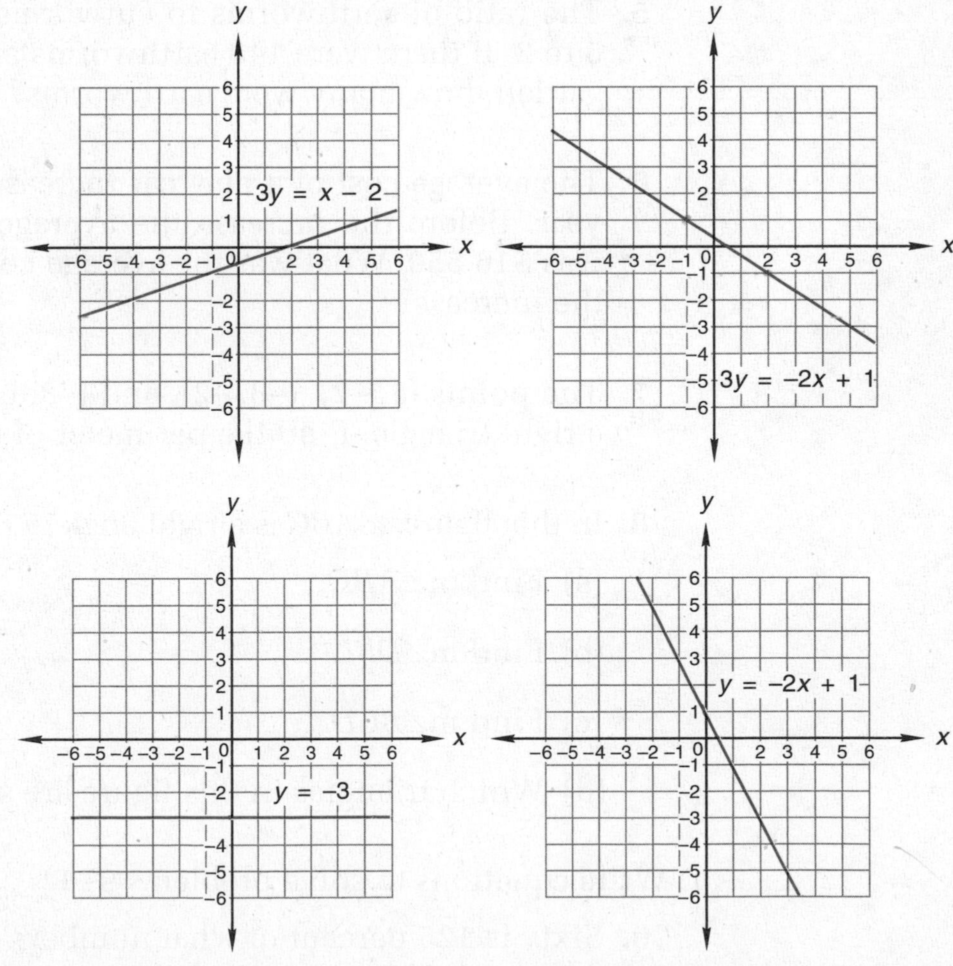

d. For each unit of horizontal change to the right on the graphed lines above, what is the vertical change?

MIXED PRACTICE

Problem set

1. (92) The shirt regularly priced at \$21 was on sale for $\frac{1}{3}$ off. What was the sale price?

2. (51) Nine hundred seventy-five billion is how much less than one trillion? Write the answer in scientific notation.

3. (Inv. 4) What is the (a) range and (b) mode of this set of numbers?

16, 6, 8, 17, 14, 16, 12

Use ratio boxes to solve problems 4–6.

4. (72) Riding her bike from home to the lake, Sonia averaged 18 miles per hour (per 60 minutes). If it took her 40 minutes to reach the lake, how far did she ride?

5. (65) The ratio of earthworms to cutworms in the garden was 5 to 2. If there were 140 earthworms and cutworms in the garden, how many were earthworms?

6. (92) The average cost of a new car increased 8 percent in one year. Before the increase the average cost of a new car was \$16,550. What was the average cost of a new car after the increase?

7. (99) The points (3, –2), (–3, –2), and (–3, 6) are the vertices of a right triangle. Find the perimeter of the triangle.

8. (40) In this figure, $\angle ABC$ is a right angle.

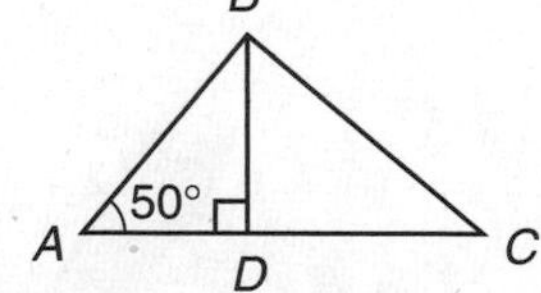

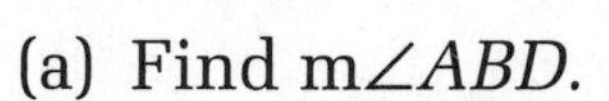
(a) Find m$\angle ABD$.

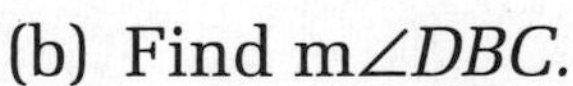
(b) Find m$\angle DBC$.

(c) Find m$\angle BCD$.

(d) Which triangles in this figure are similar?

Write equations to solve problems 9–11.

9. (77) Sixty is 125 percent of what number?

10. (77) Sixty is what percent of 25?

11. Sixty is four more than twice what number?
(101)

12. In a can are 100 marbles: 10 yellow, 20 red, 30 green, and 40 blue.
(94, Inv. 10)

(a) If a marble is drawn from the can, what is the chance that the marble will not be red?

(b) If the first marble is not replaced and a second marble is drawn from the can, what is the probability that both marbles will be yellow?

13. Complete the table.
(48)

FRACTION	DECIMAL	PERCENT
$\frac{5}{6}$	(a)	(b)

14. Compare: $(x - y)^2 \bigcirc (y - x)^2$ if $x > y$
(79)

15. Multiply. Write the product in scientific notation.
(83)

$$(1.8 \times 10^{10})(9 \times 10^{-6})$$

16. (a) Between which two consecutive whole numbers is $\sqrt{600}$?
(100, 105)

(b) What are the two square roots of 10?

17. Find three x, y pairs for the function $y = x + 1$.
(Inv. 9, 107)

(a) Graph these number pairs on a coordinate plane and draw a line through the points.

(b) What is the slope of the graphed line?

18. If the radius of this circle is 6 cm, what is the area of the shaded region?
(104)

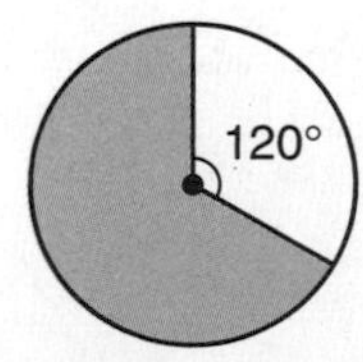

Leave π as π.

19. Find the surface area of this rectangular solid. Dimensions are in inches.
(105)

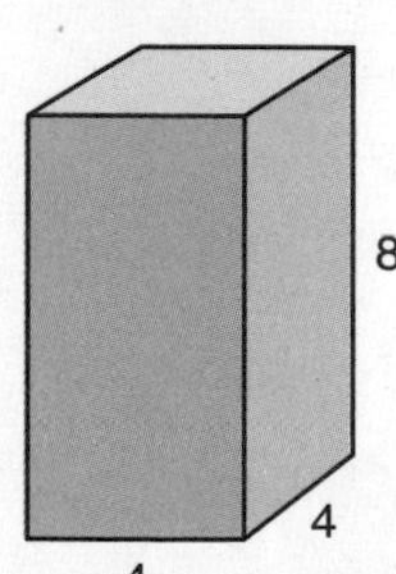

20. (95) Find the volume of this right circular cylinder. Dimensions are in centimeters.

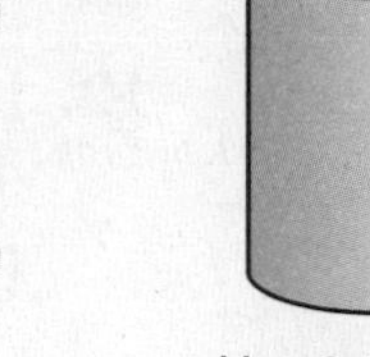

Use 3.14 for π.

21. (105) Find the total surface area of the cylinder in problem 20.

22. (40) The polygon $ABCD$ is a rectangle. Find $m\angle x$.

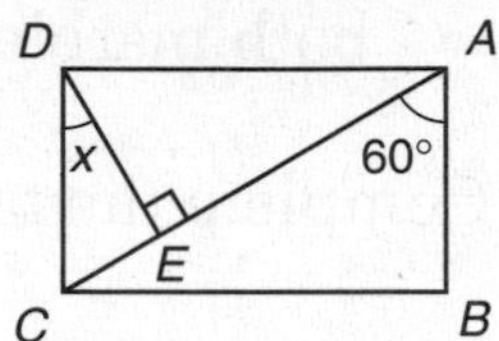

23. (107) Find the slope of the graphed line:

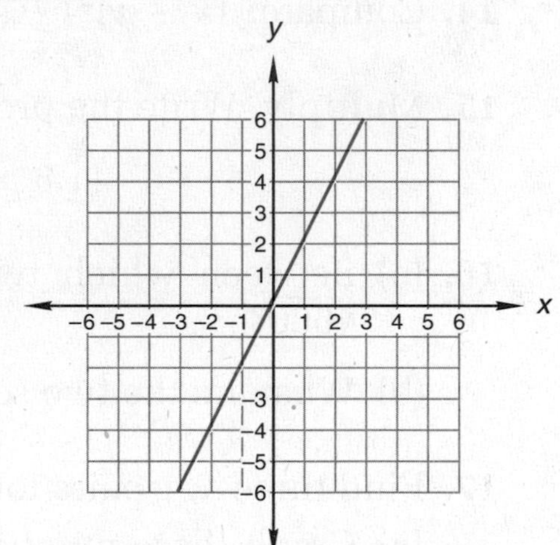

24. (106) Solve for x in each literal equation:

(a) $x - y = z$ (b) $w = xy$

Solve:

25. (98) $\frac{a}{21} = \frac{1.5}{7}$

26. (102) $6x + 5 = 7 + 2x$

Simplify:

27. (63) $62 + 5\{20 - [4^2 + 3(2 - 1)]\}$

28. (103) $\frac{(6x^2y)(2xy)}{4xy^2}$

29. (43) $5\frac{1}{6} + 3.5 - \frac{1}{3}$

30. (85) $\frac{(5)(-3)(2)(-4) + (-2)(-3)}{|-6|}$

LESSON

108 Formulas and Substitution

WARM-UP

Facts Practice: + − × ÷ Algebraic Terms (Test V)

Mental Math:

a. 1110 (base 2)

b. XLV

c. $(-1)^5 + (-1)^6$

d. $(2.5 \times 10^{-5})(4 \times 10^{-3})$

e. $5y - 2y = 24$

f. Convert 3 sq. yd to sq. ft.

g. 150% of \$120

h. \$120 increased 50%

i. At 60 mph, how far can Freddy drive in $3\frac{1}{2}$ hours?

Problem Solving:

Recall that $\sqrt[3]{8}$ means "the cube root of 8" and that $\sqrt[3]{8}$ equals 2. Find $\sqrt[3]{1{,}000{,}000}$.

NEW CONCEPT

A formula is a literal equation that describes a relationship between two or more variables. Formulas are used in mathematics, science, economics, the construction industry, food preparation—anywhere that measurement is used.

To use a formula, we replace the letters in the formula with measures that are known. Then we solve the equation for the measure we wish to find.

Example 1 Use the formula $d = rt$ to find t when d is 36 and r is 9.

Solution This formula describes the relationship between distance (d), rate (r), and time (t). We replace d with 36 and r with 9. Then we solve the equation for t.

$$d = rt \quad \text{formula}$$

$$36 = 9t \quad \text{substituted}$$

$$t = \mathbf{4} \quad \text{divided by 9}$$

Another way to find t is to first solve the formula for t.

$$d = rt \quad \text{formula}$$

$$t = \frac{d}{r} \quad \text{divided by } r$$

Then replace d and r with 36 and 9 and simplify.

$$t = \frac{36}{9} \quad \text{substituted}$$

$$t = \mathbf{4} \quad \text{divided}$$

Example 2 Use the formula $F = 1.8C + 32$ to find F when C is 37.

Solution This formula is used to convert measurements of temperature from degrees Celsius to degrees Fahrenheit. We replace C with 37 and simplify.

$$F = 1.8C + 32 \quad \text{formula}$$

$$F = 1.8(37) + 32 \quad \text{substituted}$$

$$F = 66.6 + 32 \quad \text{multiplied}$$

$$F = \mathbf{98.6} \quad \text{added}$$

Thus, 37 degrees Celsius equals 98.6 degrees Fahrenheit.

LESSON PRACTICE

Practice set

a. Use the formula $A = bh$ to find b when A is 20 and h is 4.

b. Use the formula $A = \frac{1}{2}bh$ to find b when A is 20 and h is 4.

c. Use the formula $F = 1.8C + 32$ to find F when C is −40.

MIXED PRACTICE

Problem set

1. (46) The main course cost $8.35. The beverage cost $1.25. Dessert cost $2.40. Kordell left a tip that was 15 percent of the total price of the meal. How much money did Kordell leave for a tip?

2. (57) Twelve hundred-thousandths is how much greater than twenty millionths? Write the answer in scientific notation.

3. (Inv. 4) Arrange the following numbers in order from least to greatest. Then find the median and the mode of the set of numbers.

   8, 12, 9, 15, 8, 10, 9, 8, 7, 4

4. (94) Two cards will be drawn from a normal deck of 52 cards. The first card will not be replaced before the second card is drawn. What is the probability that both cards will be 5's?

Use ratio boxes to solve problems 5–7.

5. *(72)* Milton can exchange \$200 for 300 Swiss francs. At that rate, how many dollars would a 240-franc Swiss watch cost?

6. *(65)* The jar was filled with red beans and brown beans in the ratio of 5 to 7. If there were 175 red beans in the jar, what was the total number of beans in the jar?

7. *(92)* During the off-season the room rates at the resort were reduced by 35 percent. If the usual rates were \$90 per day, what would be the cost of a 2-day stay during the off-season?

8. *(60)* Three eighths of a ton is how many pounds?

Write equations to solve problems 9–11.

9. *(60)* What number is 2.5 percent of 800?

10. *(77)* Ten percent of what number is \$2500?

11. *(101)* Fifty-six is eight less than twice what number?

12. *(107)* Find the slope of the graphed line:

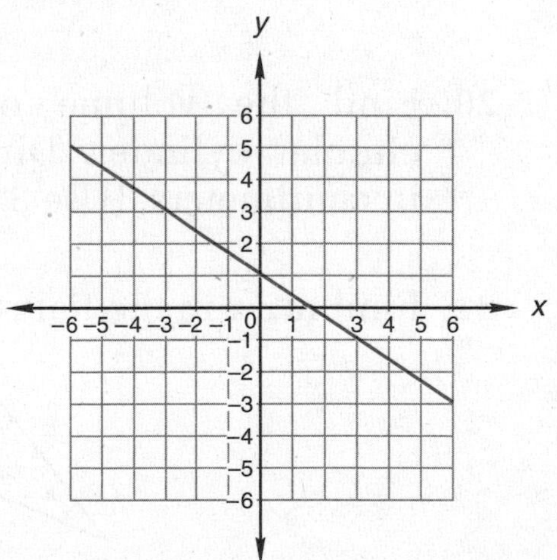

13. *(98)* Liz is drawing a floor plan of her house. On the plan, 1 inch equals 2 feet.

(a) What is the floor area of a room that measures 6 inches by $7\frac{1}{2}$ inches on the plan? Use a ratio box to solve the problem.

(b) One of the walls in Liz's house is 17 feet $9\frac{1}{2}$ inches long. Estimate how long this wall would appear in Liz's floor plan, and explain how you arrived at your estimate.

14. Find the measure of each angle of this triangle by writing and solving an equation.
(101)

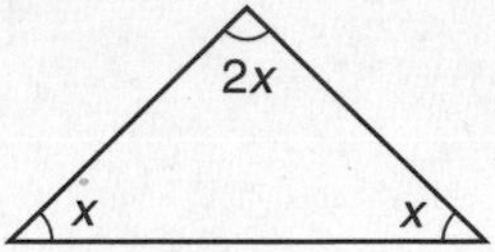

15. Multiply. Write the product in scientific notation.
(83)

$$(2.8 \times 10^5)(8 \times 10^{-8})$$

16. The formula $c = 2.54n$ is used to convert inches (n) to centimeters (c). Find c when n is 12.
(108)

17. Make a table that shows three pairs of numbers that satisfy the function $y = 2x$. Then graph the number pairs on a coordinate plane, and draw a line through the points.
(Inv. 9)

18. Find the perimeter of this figure. Dimensions are in inches. (Use 3.14 for π.)
(104)

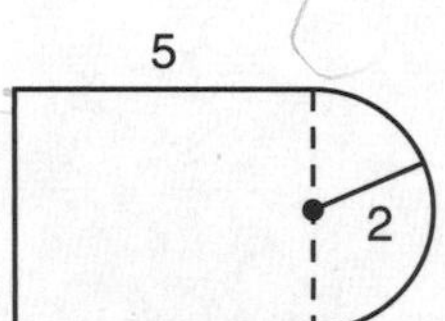

19. Find the surface area of this cube. Dimensions are in inches.
(105)

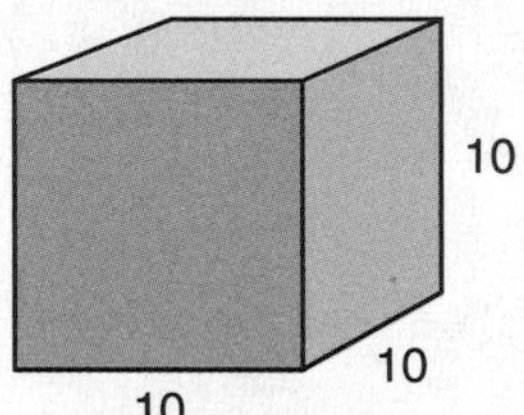

20. Find the volume of this right circular cylinder. Dimensions are in centimeters. (Use 3.14 for π.)
(95)

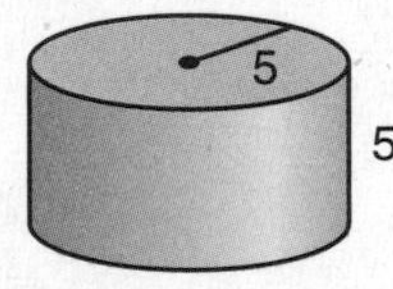

21. Find m∠x in the figure below.
(40)

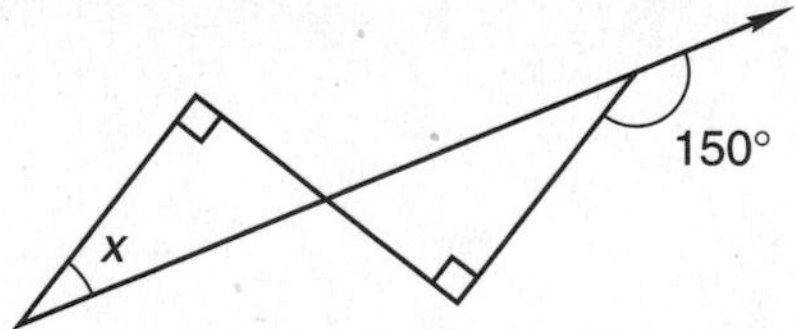

22. These triangles are similar. Dimensions are in centimeters.
(97, 98)

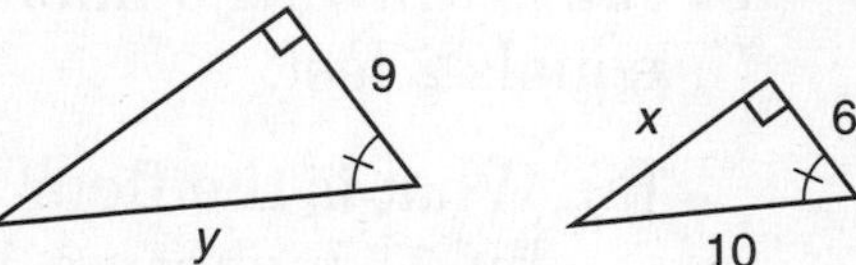

(a) Find y.

(b) Find the scale factor from the smaller to the larger triangle.

(c) The area of the larger triangle is how many times the area of the smaller triangle?

23. Use the Pythagorean theorem to find x in the smaller triangle from problem 22.
(99)

24. Find the surface area of a globe that has a diameter of 10 inches. (Use 3.14 for π.)
(105)

25. Solve: $1\frac{2}{3}x = 32 - x$
(102)

Simplify:

26. $x^2 + x(x + 2)$
(96)

27. $\dfrac{(-4ax)(3xy)}{-6x^2}$
(103)

28. $1.1\{1.1[1.1(1000)]\}$
(63)

29. $3\frac{3}{4} \cdot 2\frac{2}{3} \div 10$
(26)

30. (a) $(-6) - (7)(-4) + \sqrt[3]{125} + \dfrac{(-8)(-9)}{(-3)(-2)}$
(103, 105)

(b) $(-1) + (-1)^2 + (-1)^3 + (-1)^4$

LESSON

109 Equations with Exponents

WARM-UP

Facts Practice: + − × ÷ Algebraic Terms (Test V)

Mental Math:

a. 10010 (base 2)

b. MCMLX

c. $\left(-\frac{1}{2}\right)\left(-\frac{1}{2}\right)$

d. $(1.2 \times 10^{12})^2$

e. $\frac{2.4}{0.6} = \frac{c}{0.25}$

f. Convert 150 cm to m.

g. $12\frac{1}{2}\%$ of \$80

h. $12\frac{1}{2}\%$ less than \$80

i. Find $\frac{1}{3}$ of 60, + 5, × 2, − 1, $\sqrt{}$, × 4, − 1, ÷ 3, square that number, − 1, ÷ 2.

Problem Solving:

Here are the front, top, and side views of an object. Draw a three-dimensional view of the object from the perspective of the upper right front.

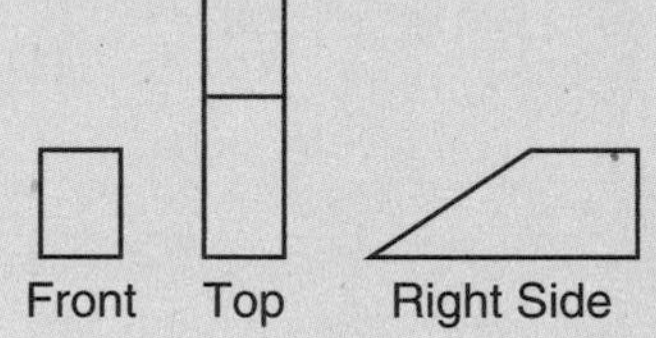

NEW CONCEPT

In the equations we have solved thus far, the variables have had an exponent of 1. You have not seen the exponent, because we usually do not write the exponent when it is 1. In this lesson we will consider equations that have variables with exponents of 2, such as the following equation:

$$3x^2 + 1 = 28$$

Isolating the variable in this equation takes three steps: first we subtract 1 from both sides; next we divide both sides by 3; then we find the square root of both sides. We show the results of each step below.

$3x^2 + 1 = 28$	equation
$3x^2 = 27$	subtracted 1 from both sides
$x^2 = 9$	divided both sides by 3
$x = 3, -3$	found the square root of both sides

Notice that there are two solutions, 3 and –3. Both solutions satisfy the equation, as we show below.

$$3(3)^2 + 1 = 28 \qquad 3(-3)^2 + 1 = 28$$
$$3(9) + 1 = 28 \qquad 3(9) + 1 = 28$$
$$27 + 1 = 28 \qquad 27 + 1 = 28$$
$$28 = 28 \qquad 28 = 28$$

When the variable of an equation has an exponent of 2, we remember to look for two solutions.

Example 1 Solve: $3x^2 - 1 = 47$

Solution There are three steps. We show the results of each step.

$3x^2 - 1 = 47$	equation
$3x^2 = 48$	added 1 to both sides
$x^2 = 16$	divided both sides by 3
$x = \mathbf{4, -4}$	found the square root of both sides

Example 2 Solve: $2x^2 = 10$

Solution We divide both sides by 2. Then we find the square root of both sides.

$2x^2 = 10$	equation
$x^2 = 5$	divided both sides by 2
$x = \mathbf{\sqrt{5}, -\sqrt{5}}$	found the square root of both sides

Since $\sqrt{5}$ is an irrational number, we leave it in radical form. The negative of $\sqrt{5}$ is $-\sqrt{5}$ and not $\sqrt{-5}$.

Example 3 Five less than what number squared is 20?

Solution We translate the question into an equation.

$$n^2 - 5 = 20$$

We solve the equation in two steps.

$n^2 - 5 = 20$	equation
$n^2 = 25$	added 5 to both sides
$n = 5, -5$	found the square root of both sides

There are two numbers that answer the question, **5** and **–5.**

Example 4 In this figure the area of the larger square is 4 square units, which is twice the area of the smaller square. What is the length of each side of the smaller square?

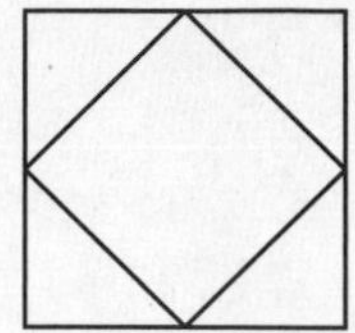

Solution We will use the letter s to stand for the length of each side of the smaller square. So s^2 is the area of the small square. Since the area of the large square (4) is twice the area of the small square, we can write this equation:

$$2s^2 = 4$$

We solve the equation in two steps.

$2s^2 = 4$	equation
$s^2 = 2$	divided both sides by 2
$s = \sqrt{2}, -\sqrt{2}$	found the square root of both sides

Although there are two solutions to the equation, there is only one answer to the question because lengths are positive, not negative. Thus, each side of the smaller square is $\sqrt{2}$ **units.**

Example 5 Solve: $\frac{x}{3} = \frac{12}{x}$

Solution First we cross multiply. Then we find the square root of both sides.

$\frac{x}{3} = \frac{12}{x}$	proportion
$x^2 = 36$	cross multiplied
$x = 6, -6$	found the square root of both sides

There are two solutions to the proportion, **6** and **−6.**

LESSON PRACTICE

Practice set Solve each equation:

a. $3x^2 - 8 = 100$

b. $x^2 + x^2 = 12$

c. Five less than twice what negative number squared is 157?

d. If the product of the square of a positive number and 7 is 21, then what is the number?

e. $\frac{w}{4} = \frac{9}{w}$

MIXED PRACTICE

Problem set

1. (45) What is the quotient when the product of 0.2 and 0.05 is divided by the sum of 0.2 and 0.05?

2. (102) In the figure at right, a transversal intersects two parallel lines.

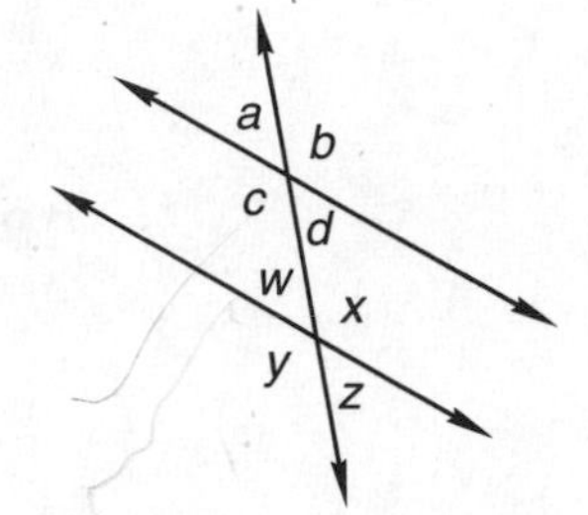

(a) Which angle corresponds to $\angle d$?

(b) Which angle is the alternate interior angle to $\angle d$?

(c) Which angle is the alternate exterior angle to $\angle b$?

(d) If the measure of $\angle a$ is m and the measure of $\angle b$ is $3m$, then each obtuse angle measures how many degrees?

3. (101) Twenty is five more than the product of ten and what decimal number?

4. (88) Use two unit multipliers to convert 1 km^2 to square meters.

5. (36) Santiago has \$5 in quarters and \$5 in dimes. What is the ratio of the number of quarters to the number of dimes?

Use ratio boxes to solve problems 6–8.

6. (72) Jaime ran the first 3000 meters in 9 minutes. At that rate, how long will it take Jaime to run 5000 meters?

7. (92) Sixty is 20 percent more than what number?

8. (92) To attract customers, the merchant reduced all prices by 25 percent. What was the reduced price of an item that cost \$36 before the price reduction?

9. (77) Write an equation to solve this problem:

Sixty is 150 percent of what number?

10. (71) Diagram this statement. Then answer the questions that follow.

Diane kept $\frac{2}{3}$ of her baseball cards and gave the remaining 234 cards to her brother.

(a) How many cards did Diane have before she gave some to her brother?

(b) How many baseball cards did Diane keep?

11. (79) Compare: $a - b \bigcirc b - a$ if $a > b$

12. (94) Warner knew the correct answer to 15 of the 20 true-false questions but guessed on the rest. What is the probability of Warner correctly guessing the answers to all of the remaining true-false questions?

13. (75) Find the area of this trapezoid. Dimensions are in centimeters.

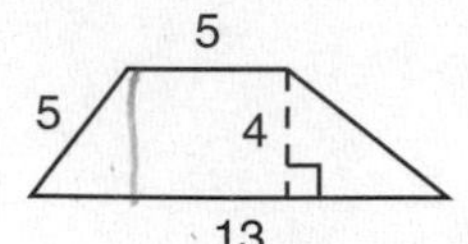

14. (95) Find the volume of this triangular prism. Dimensions are in inches.

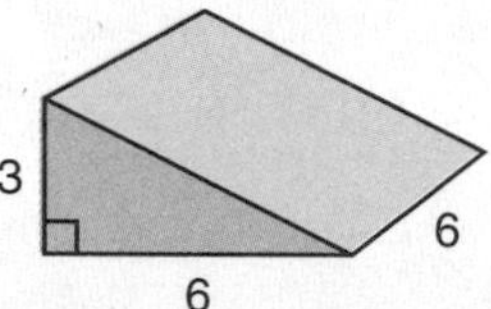

15. (105) A rectangular label is wrapped around a can with the dimensions shown. The label has an area of how many square inches?

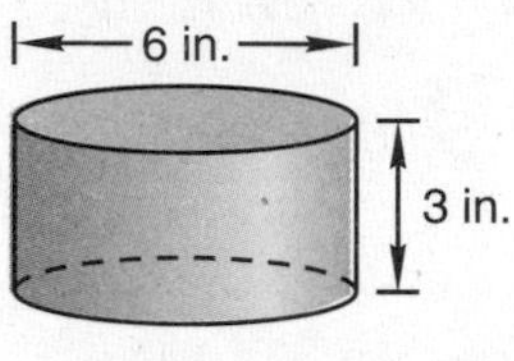

16. (46) The skateboard costs $36. The tax rate is 6.5 percent.

(a) What is the tax on the skateboard?

(b) What is the total price, including tax?

17. (48) Complete the table.

FRACTION	DECIMAL	PERCENT
(a)	(b)	$\frac{1}{2}$%

18. (92) What number is $66\frac{2}{3}$ percent more than 48?

19. (83) Multiply. Write the product in scientific notation.

$$(6 \times 10^{-8})(8 \times 10^{4})$$

20. (Inv. 9, 107) Find the missing numbers in the table by using the function rule. Then graph the function on a coordinate plane. What is the slope of the graphed line?

$y = \frac{2}{3}x - 1$

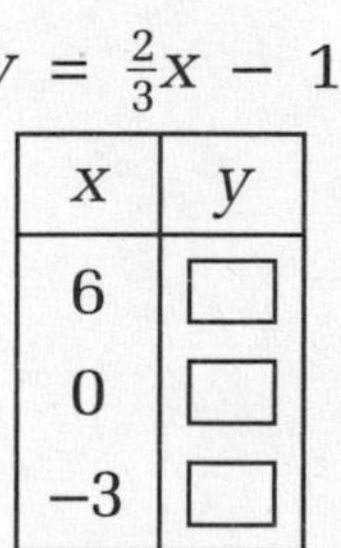

x	y
6	☐
0	☐
−3	☐

21. Use a ratio box to solve this problem. The ratio of the measures of the two acute angles of the right triangle is 7 to 8. What is the measure of the smallest angle of the triangle?
(65)

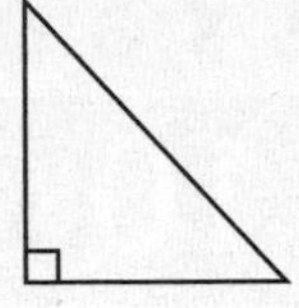

22. The relationship between the measures of four central angles of a circle is shown in this figure. What is the measure of the smallest central angle shown?
(101)

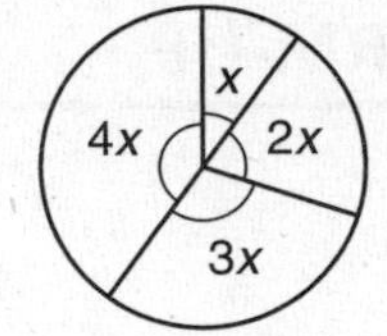

23. We can use the Pythagorean theorem to find the distance between two points on a coordinate plane. To find the distance from point M to point P, we draw a right triangle and use the lengths of the legs to find the length of the hypotenuse. What is the distance from point M to point P?
(99)

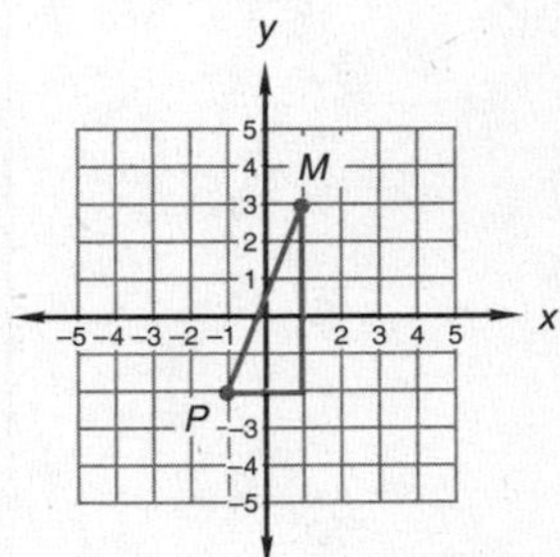

Solve:

24. $3m^2 + 2 = 50$
(109)

25. $7(y - 2) = 4 - 2y$
(102)

Simplify:

26. $\sqrt{144} - (\sqrt{36})(\sqrt{4})$
(20)

27. $x^2y + xy^2 + x(xy - y^2)$
(96)

28. $\left(1\frac{5}{9}\right)(1.5) \div 2\frac{2}{3}$
(43)

29. $9.5 - \left(4\frac{1}{5} - 3.4\right)$
(43)

30. (a) $\dfrac{(-18) + (-12) - (-6)(3)}{-3}$
(57, 91, 105)

(b) $\sqrt[3]{1000} - \sqrt[3]{125}$

(c) $2^2 + 2^1 + 2^0 + 2^{-1}$

LESSON

110 Simple Interest and Compound Interest • Successive Discounts

WARM-UP

Facts Practice: Percent-Decimal-Fraction Equivalents (Test Q)

Mental Math:

a. 11110 (base 2)
b. DCLXXVIII
c. $\left(-\frac{1}{2}\right)^2$
d. $(9 \times 10^6)(6 \times 10^9)$
e. $4w - 1 = 9$
f. Convert 1.5 L to mL.
g. 150% of $60
h. $60 increased 50%
i. Start with the number of minutes in half an hour. Multiply by the number of feet in a yard; add the number of years in a decade; then find the square root of that number. What is the answer?

Problem Solving:

At nine o'clock the hands of a clock form a 90° angle. What angle is formed by the hands of a clock $1\frac{1}{2}$ hours after nine o'clock?

NEW CONCEPTS

Simple interest and compound interest

When you deposit money in a bank, the bank does not simply hold your money for safekeeping. Instead, it spends your money in other places to make more money. For this opportunity the bank pays you a percentage of the money deposited. The amount of money you deposit is called the **principal.** The amount of money the bank pays you is called **interest.**

There is a difference between **simple interest** and **compound interest.** Simple interest is paid on the principal only and not paid on any accumulated interest. For instance, if you deposited $100 in an account that pays 6% simple interest, you would be paid 6% of $100 ($6) each year your $100 was on deposit. If you take your money out after three years, you would have a total of $118.

Simple Interest

$100.00	principal
$6.00	first-year interest
$6.00	second-year interest
+ $6.00	third-year interest
$118.00	total

Most interest-bearing accounts, however, are compound-interest accounts, not simple-interest accounts. In a compound-interest account, interest is paid on accumulated interest as well as on the principal. If you deposited $100 in an account with 6% annual percentage rate, the amount of interest you would be paid each year increases if the earned interest is left in the account. After three years you would have a total of $119.10.

Compound Interest

$100.00	principal
$6.00	first-year interest (6% of $100.00)
$106.00	total after one year
$6.36	second-year interest (6% of $106.00)
$112.36	total after two years
$6.74	third-year interest (6% of $112.36)
$119.10	total after three years

Notice that in three years, $100.00 grows to $118.00 at 6% simple interest, while it grows to $119.10 at 6% compound interest. The difference is not very large in three years, but as this table shows, the difference can become large over time.

Total Value of $100 at 6% Interest

Number of Years	Simple Interest	Compound Interest
3	$118.00	$119.10
10	$160.00	$179.08
20	$220.00	$320.71
30	$280.00	$574.35
40	$340.00	$1028.57
50	$400.00	$1842.02

Example 1 Make a table that shows the value of a $1000 investment growing at 10% compounded annually after 1, 2, 3, 4, and 5 years.

Solution After the first year, \$1000 grows 10% to \$1100. After the second year, the value increases 10% of \$1100 (\$110) to a total of \$1210. We continue the pattern for five years in the table below.

Total Value of \$1000 at 10% Interest

Number of Years	Compound Interest
1	\$1100.00
2	\$1210.00
3	\$1331.00
4	\$1464.10
5	\$1610.51

Notice that the amount of money in the account after one year is 110% of the original deposit of \$1000. This 110% is composed of the starting amount, 100%, plus 10%, which is the interest earned in one year. Likewise, the amount of money in the account the second year is 110% of the amount in the account after one year. To find the amount of money in the account each year, we multiply the previous year's balance by 110% (or the decimal equivalent, which is 1.1).

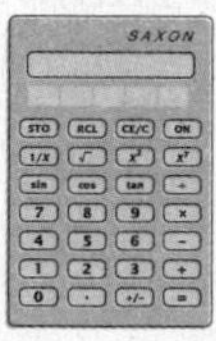

Even with a simple calculator we can calculate compound interest. To perform the calculation in example 1, we could follow this sequence:

$$1000 \times 1.1 \times 1.1 \times 1.1 \times 1.1 \times 1.1 =$$

The circuitry of some calculators permits repeating a calculation by pressing the = key repeatedly.† To make the calculations in example 1, we try this keystroke sequence:

This keystroke sequence first enters 1.1, which is the decimal form of 110% (100% principal plus 10% interest), then the times sign, then 1000 for the \$1000 investment. Pressing the = key once displays

which is the value (\$1100) after one year. Pressing the = key a second time multiplies the displayed number by 1.1, the

†This calculator function varies with make and model of calculator. See instructions for your calculator if the keystroke sequence described in this lesson does not work for you.

first number we entered. The new number displayed is

representing $1210, the value after two years. Each time the = key is pressed, the calculator displays the account value after a successive year. Using this method, we find that the value of the account after 10 years is $2593.74. After 20 years, the value of the account is $6727.50.

Try entering the factors in reverse order and then repeatedly pressing the = key.

We find that different amounts are displayed. This is because reversing the order of the factors causes the displayed number to be multiplied by 1000, not 1.1.

Example 2 Use a calculator to find the value after 12 years of a $2000 investment that earns $7\frac{1}{2}$% interest compounded annually.

Solution The interest rate is $7\frac{1}{2}$%, which is 0.075 in decimal form. We want to find the total value, including the principal. So we multiply the $2000 investment by $107\frac{1}{2}$%, which we enter as 1.075. The keystroke sequence is

We then press the = key 11 more times to find the value after 12 years. We round the final display to the nearest cent, **$4763.56.**

Example 3 Calculate the interest earned on an $8000 deposit in 9 months if the annual interest rate is 6%.

Solution The deposit earns 6% interest in one year, which is

$$0.06 \times \$8000 = \$480$$

In 9 months the deposit earns just $\frac{9}{12}$ of this amount.

$$\frac{9}{12} \times \$480 = \mathbf{\$360}$$

Successive discounts Related to compound interest is **successive discount.** To calculate successive discount, we find a percent of a percent. In the following example we show two methods for finding successive discounts.

Example 4 An appliance store reduced the price of a $400 washing machine 25%. When the washing machine did not sell at the sale price, the store reduced the sale price 20% to its clearance price. What was the clearance price of the washing machine?

Solution One way to find the answer is to first find the sale price and then find the clearance price. We will use a ratio box to find the sale price.

	Percent	Actual Count
Original	100	400
− Change	25	D
New (Sale)	75	S

$$\frac{100}{75} = \frac{400}{S}$$
$$100S = 30{,}000$$
$$S = 300$$

We find that the sale price was $300. The second discount, 20%, was applied to the sale price, not to the original price. So for the next calculation we consider the sale price to be 100% and the clearance price to be what remains after the discount.

	Percent	Actual Count
Original (Sale)	100	300
− Change	20	D
New (Clearance)	80	C

$$\frac{100}{80} = \frac{300}{C}$$
$$100C = 24{,}000$$
$$C = 240$$

We find that the clearance price of the washing machine was **$240.**

Another way to look at this problem is to consider what percent of the original price is represented by the sale price. Since the original price was discounted 25%, the sale price represents 75% of the original price.

Sale price = 75% of the original price

Furthermore, since the sale price was discounted 20%, the clearance price was 80% of the sale price.

Clearance price = 80% of the sale price

So the clearance price was 80% of the sale price, which was 75% of the original price.

$$\begin{aligned}\text{Clearance price} &= 80\% \text{ of } 75\% \text{ of } \$400\\ &= 0.8 \times 0.75 \times \$400\\ &= 0.6 \times \$400\\ &= \mathbf{\$240}\end{aligned}$$

LESSON PRACTICE

Practice set

a. When Mai turned 21, she invested $2000 in an Individual Retirement Account (IRA) that has grown at a rate of 10% compounded annually. If the account continues to grow at that rate, what will be its value when Mai turns 65? (Use the calculator method taught in this lesson.)

b. Hannah deposited $6000 in an account paying 8% interest annually. After 8 months Hannah withdrew the $6000 plus interest. Altogether, how much money did Hannah withdraw? What fraction of a year's interest was earned?

c. A television regularly priced at $300 was placed on sale for 20% off. When the television still did not sell, the sale price was reduced 20% for clearance. What was the clearance price of the television?

MIXED PRACTICE

Problem set

1. (46) Rosita bought 3 paperback books for $5.95 each. The sales-tax rate was 6 percent. If she paid for the purchase with a $20 bill, how much money did she get back?

2. (110) Hector has a coupon for 10% off the price of any item in the store. He decides to buy a shirt regularly priced at $24 that is on sale for 25% off. If he uses his coupon, how much will Hector pay for the shirt before sales tax is applied?

3. (80) Triangle ABC with vertices A (3, 0), B (3, 4), and C (0, 0) is rotated 90° clockwise about the origin to its image $\Delta A'B'C'$. Graph both triangles. What are the coordinates of the vertices of $\Delta A'B'C'$?

4. (72) George burned 100 calories by running 1 mile. At that rate, how many miles would he need to run to burn 350 calories?

5. (72) If a dozen roses costs \$4.90, what is the cost of 30 roses?

6. (104) A semicircle was cut out of a square as shown. What is the perimeter of the resulting figure?

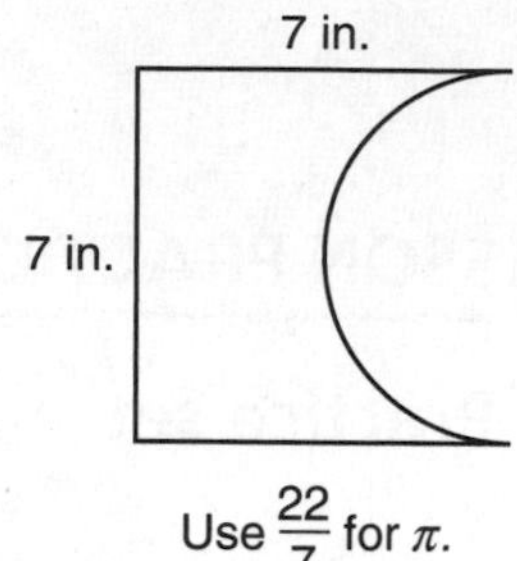

Use $\frac{22}{7}$ for π.

7. (55) The average of four numbers is 8. Three of the numbers are 2, 4, and 6. What is the fourth number?

Write equations to solve problems 8–10.

8. (77) One hundred fifty is what percent of 60?

9. (77) Sixty percent of what number is 150?

10. (109) Six more than the square of what negative number is 150?

11. (99) Graph the points (3, 1) and (−1, −2) on a coordinate plane. Then draw a right triangle, and use the Pythagorean theorem to find the distance between the points.

Use ratio boxes to solve problems 12 and 13.

12. (92) The price of the dress was reduced by 40 percent. If the sale price was \$48, what was the regular price?

13. (98) The car model was built on a 1:36 scale. If the length of the car is 180 inches, how many inches long is the model?

14. (100) The positive square root of 80 is between which two consecutive whole numbers?

15. (Inv. 9, 107) Make a table of ordered pairs for the function $y = -x + 1$. Then graph the function. What is the slope of the graphed line?

16. Solve this inequality and graph its solution:
(93)

$$5x + 12 \geq 2$$

17. Multiply. Write the product in scientific notation.
(83)

$$(6.3 \times 10^7)(9 \times 10^{-3})$$

18. Solve for y: $\frac{1}{2}y = x + 2$
(96, 106)

19. What is the total account value after 3 years on a deposit of $4000 at 9% interest compounded annually?
(110)

20. The triangles below are similar. Dimensions are in inches.
(97, 98)

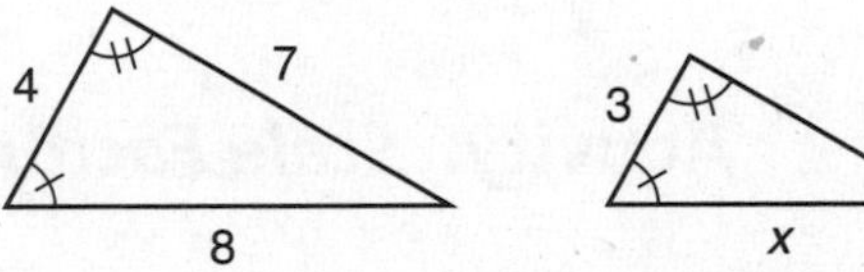

(a) Estimate, then calculate, the length x.

(b) Find the scale factor from the larger to the smaller triangle.

21. Find the volume of this triangular prism. Dimensions are in inches.
(95)

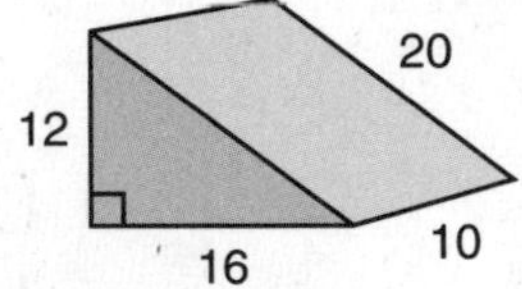

22. Find the total surface area of the triangular prism in problem 21.
(105)

23. Find m∠x in the figure at right.
(40)

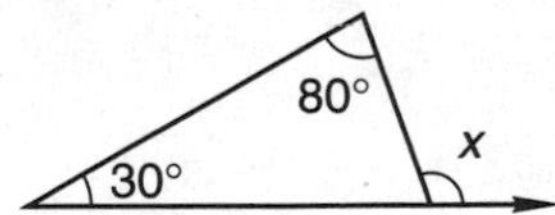

Solve:

24. $\frac{w}{2} = \frac{18}{w}$
(109)

25. $3\frac{1}{3}w^2 - 4 = 26$
(109)

Simplify:

26. $16 - \{27 - 3[8 - (3^2 - 2^3)]\}$
(63)

27. $\frac{(6ab^2)(8ab)}{12a^2b^2}$
(103)

28. $3\frac{1}{3} + 1.5 + 4\frac{5}{6}$
(43)

29. $20 \div \left(3\frac{1}{3} \div 1\frac{1}{5}\right)$
(26)

30. $(-3)^2 + (-2)^3$
(103)

INVESTIGATION 11

Focus on

Scale Factor in Surface Area and Volume

In this investigation we will study the relationship between length, surface area, and volume of three-dimensional shapes. We begin by comparing the measures of cubes of different sizes.

Activity: *Scale Factor in Surface Area and Volume*

Materials needed:

- Activity Sheets 9–11 (available in *Saxon Math 8/7—Homeschool Tests and Worksheets*)
- Scissors
- Tape

Use the materials to build models of four cubes with edges 1 cm, 2 cm, 3 cm, and 4 cm long. Mark, cut, fold, and tape the grid paper so that the *grid is visible* when each model is finished.

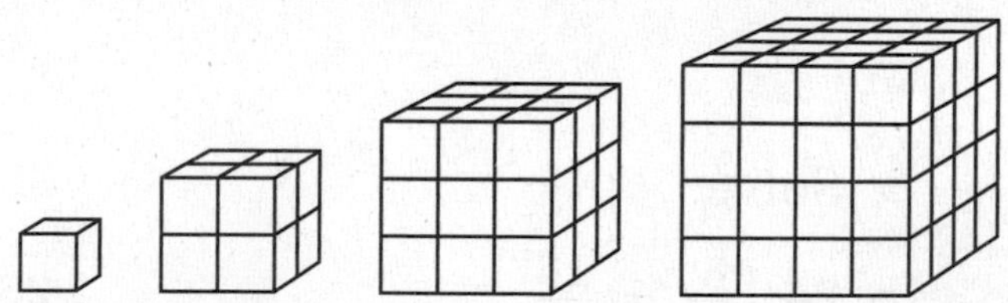

One pattern that folds to form a model of a cube is shown below. Several other patterns also work.

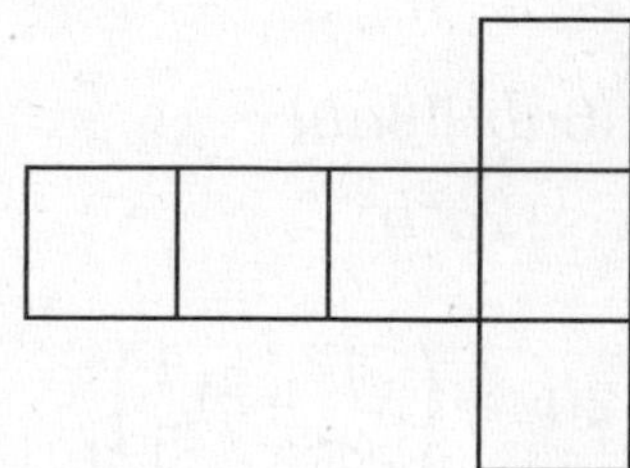

Copy this table on your paper and record the measures for each cube.

Measures of Four Cubes

	1-cm cube	2-cm cube	3-cm cube	4-cm cube
Edge length (cm)				
Surface area (cm^2)				
Volume (cm^3)				

Refer to the table to answer the following questions:

Compare the 2-cm cube to the 1-cm cube.

1. The edge length of the 2-cm cube is how many times the edge length of the 1-cm cube?
2. The surface area of the 2-cm cube is how many times the surface area of the 1-cm cube?
3. The volume of the 2-cm cube is how many times the volume of the 1-cm cube?

Compare the 4-cm cube to the 2-cm cube.

4. The edge length of the 4-cm cube is how many times the edge length of the 2-cm cube?
5. The surface area of the 4-cm cube is how many times the surface area of the 2-cm cube?
6. The volume of the 4-cm cube is how many times the volume of the 2-cm cube?

Compare the 3-cm cube to the 1-cm cube.

7. The edge length of the 3-cm cube is how many times the edge length of the 1-cm cube?
8. The surface area of the 3-cm cube is how many times the surface area of the 1-cm cube?

9. The volume of the 3-cm cube is how many times the volume of the 1-cm cube?

Use the patterns that can be found in answers 1–9 to predict the comparison of a 6-cm cube to a 2-cm cube.

10. The edge length of a 6-cm cube is how many times the edge length of a 2-cm cube?

11. The surface area of a 6-cm cube is how many times the surface area of a 2-cm cube?

12. The volume of a 6-cm cube is how many times the volume of a 2-cm cube?

13. Calculate (a) the surface area and (b) the volume of a 6-cm cube.

14. The calculated surface area of a 6-cm cube is how many times the surface area of a 2-cm cube?

15. The calculated volume of a 6-cm cube is how many times the volume of a 2-cm cube?

In problems 1–6 we compared the measures of a 2-cm cube to a 1-cm cube and the measures of a 4-cm cube to a 2-cm cube. In both sets of comparisons, the scale factors from the smaller cube to the larger cube were calculated.

Scale Factors from Smaller Cube to Larger Cube

Measurement	Scale Factor
Edge length	2
Surface area	$2^2 = 4$
Volume	$2^3 = 8$

Likewise, in problems 7–15 we compared the measures of a 3-cm cube to a 1-cm cube and a 6-cm cube to a 2-cm cube. We calculated the following scale factors:

Scale Factors from Smaller Cube to Larger Cube

Measurement	Scale Factor
Edge length	3
Surface area	$3^2 = 9$
Volume	$3^3 = 27$

Refer to the above description of scale factors to answer problems 16–20.

16. Manuel calculated the scale factors from a 6-cm cube to a 24-cm cube. From the smaller cube to the larger cube, what are the scale factors for (a) edge length, (b) surface area, and (c) volume?

17. Bethany noticed that the scale factor relationships for cubes also applies to spheres. She found the approximate diameters of a table tennis ball $\left(1\frac{1}{2}\text{ in.}\right)$, a baseball (3 in.), and a playground ball (9 in.). Find the scale factor for (a) the volume of the table tennis ball to the volume of the baseball and (b) the surface area of the baseball to the surface area of the playground ball.

18. The photo lab makes 5-by-7-in. enlargements from $2\frac{1}{2}$-by-$3\frac{1}{2}$-in. wallet-size photos. Find the scale factor from the smaller photo to the enlargement for (a) side length and (b) picture area.

19. Rommy wanted to charge the same price per square inch of cheese pizza regardless of the size of the pizza. Since all of Rommy's pizzas were the same thickness, he based his prices on scale factor for area. If he sells a 10-inch diameter cheese pizza for \$10.00, how much should he charge for a 15-inch diameter cheese pizza?

20. The Egyptian archaeologist knew that the scale-factor relationships for cubes also applies to similar pyramids. The archaeologist built a $\frac{1}{100}$ scale model of the Great Pyramid. Each edge of the base of the model was 2.3 meters, while each edge of the base of the Great Pyramid measured 230 meters. From the smaller model to the Great Pyramid, what was the scale factor for (a) the length of corresponding edges, (b) the area of corresponding faces, and (c) the volume of the pyramids?

Notice from the chart that you completed near the beginning of this investigation that as the size of the cube becomes greater, the surface area and volume become much greater. Also notice that the volume increases at a faster rate than the surface area. The ratio of surface area to volume changes as the size of an object changes.

Ratio of Surface Area to Volume of Four Cubes

	1-cm cube	2-cm cube	3-cm cube	4-cm cube
Surface Area to Volume	6 to 1	3 to 1	2 to 1	1.5 to 1

The ratio of surface area to volume affects the size and shape of containers used to package products. The ratio of surface area to volume also affects the world of nature. Consider the relationship between surface area and volume as you answer problems 21–25.

21. Sixty-four 1-cm cubes were arranged to form one large cube. Austin wrapped the large cube with paper and sent the package to Betsy. The volume of the package was 64 cm^3. What was the surface area of the exposed wrapping paper?

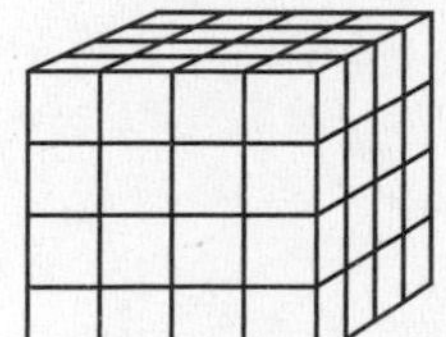

22. When Betsy received the package she divided the contents into eight smaller cubes composed of eight 2-cm cubes. Betsy wrapped the eight packages and sent them on to Charlie. The total volume of the eight packages was still 64 cm^3. What was the total surface area of the exposed wrapping paper of the eight packages?

23. Charlie opened each of the eight packages and wrapped each 1-cm cube. Since there were 64 cubes, the total volume was still 64 cm^3. What was the total surface area of exposed wrapping paper for all 64 packages?

24. After a summer picnic the ice in two large insulated containers was emptied on the ground to melt. A large block of ice in the form of a 6-inch cube fell out of one container. An equal quantity of ice, but in the form of 1-inch cubes, fell scattered out of the other container. Which, if either, do you think will melt sooner, the large block of ice or the small scattered cubes? Explain your answer.

25. If someone does not eat very much, we might say that he or she "eats like a bird." However, birds must eat large amounts, relative to their body weights, in order to maintain their body temperature. Since mammals and birds regulate their own body temperature, there is a limit to how small a mammal or bird may be. Comparing a hawk and a sparrow in the same environment, which of the two might eat a greater percentage of its weight in food every day? Explain your answer.

Extensions

a. Investigate how the weight of a bird and its wingspan are related.

b. Investigate reasons why the largest sea mammals are so much larger than the largest land mammals.

c. Brad's dad is 25% taller than Brad and weighs twice as much. Explain why you think this height-weight relationship may or may not be reasonable.

LESSON

111 Dividing in Scientific Notation

WARM-UP

Facts Practice: $+ - \times \div$ Algebraic Terms (Test V)

Mental Math:

a. 10111 (base 2)

b. CCCXXI

c. $(-0.25) \div (-5)$

d. $(8 \times 10^{-4})^2$

e. $3m + 7m = 600$

f. Convert 1 ft^2 to square inches.

g. $33\frac{1}{3}\%$ of \$150

h. \$150 reduced by $33\frac{1}{3}\%$

i. Estimate 8% tax on a \$198.75 purchase.

Problem Solving:

Carpeting is sold by the square yard. If carpet is priced at \$25 per square yard (including tax and installation), how much would it cost to carpet a meeting room that is 36 feet long and 36 feet wide?

NEW CONCEPT

One unit astronomers use to measure distances within our solar system is the **astronomical unit** (AU). An astronomical unit is the average distance between Earth and the Sun, which is roughly 150,000,000 km (or 93,000,000 mi).

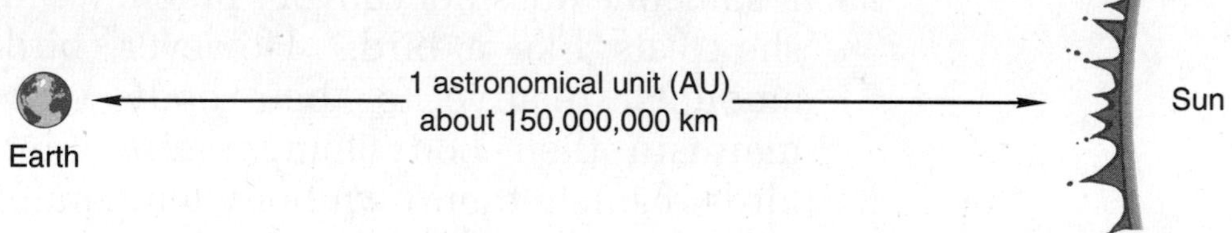

For instance, at a point in Saturn's orbit when it is 1.5 billion kilometers from the Sun, its distance from the Sun is 10 AU.

$$1{,}500{,}000{,}000 \text{ km} \cdot \frac{1 \text{ AU}}{150{,}000{,}000 \text{ km}} = 10 \text{ AU}$$

This means that the distance from Saturn to the Sun is about 10 times the average distance between Earth and the Sun.

When dividing very large or very small numbers, it is helpful to use scientific notation. Here we show the same calculation in scientific notation:

$$\frac{1.5 \times 10^9 \text{ km}}{1.5 \times 10^8 \text{ km/AU}} = 10 \text{ AU}$$

In this lesson we will practice dividing numbers in scientific notation.

Recall that when we multiply numbers in scientific notation, we multiply the powers of 10 by adding their exponents.

$$(6 \times 10^6)(1.5 \times 10^2) = 9 \times 10^8$$

Furthermore, we have this important rule:

> **When we divide numbers written in scientific notation, we divide the powers of 10 by subtracting their exponents.**

$$\frac{6 \times 10^6}{1.5 \times 10^2} = 4 \times 10^4 \longleftarrow (6 - 2 = 4)$$

Example 1 Write each quotient in scientific notation:

(a) $\dfrac{6 \times 10^8}{1.2 \times 10^6}$ (b) $\dfrac{3 \times 10^3}{6 \times 10^6}$ (c) $\dfrac{2 \times 10^{-2}}{8 \times 10^{-8}}$

Solution (a) To find the quotient, we divide 6 by 1.2 and 10^8 by 10^6.

$$12.\overline{)60.} = 5. \qquad 10^8 \div 10^6 = 10^2 \longleftarrow (8 - 6 = 2)$$

The quotient is $\mathbf{5 \times 10^2}$.

(b) We divide 3 by 6 and 10^3 by 10^6.

$$6\overline{)3.0} = 0.5 \qquad 10^3 \div 10^6 = 10^{-3} \longleftarrow (3 - 6 = -3)$$

The quotient, 0.5×10^{-3}, is not in proper form. We write the quotient in scientific notation.

$$\mathbf{5 \times 10^{-4}}$$

(c) We divide 2 by 8 and 10^{-2} by 10^{-8}.

$$8\overline{)2.00} = 0.25 \qquad 10^{-2} \div 10^{-8} = 10^6 \longleftarrow [-2 - (-8) = 6]$$

The quotient, 0.25×10^6, is not in proper form. We write the quotient in scientific notation.

$$\mathbf{2.5 \times 10^5}$$

Example 2 The distance from the Sun to Earth is about 1.5×10^8 km. Light travels at a speed of about 3×10^5 km per second. About how many seconds does it take light to travel from the Sun to Earth?

Solution We divide 1.5×10^8 km by 3×10^5 km/s.

$$\frac{1.5 \times 10^8 \text{ km}}{3 \times 10^5 \text{ km/s}} = 0.5 \times 10^3 \text{ s}$$

We may write the quotient in proper scientific notation, $\mathbf{5 \times 10^2}$ **s.** We may also write the answer in standard form, **500 s.** It takes about 500 seconds for light from the Sun to reach Earth.

LESSON PRACTICE

Practice set* Write each quotient in scientific notation:

a. $\dfrac{3.6 \times 10^9}{2 \times 10^3}$ **b.** $\dfrac{7.5 \times 10^3}{2.5 \times 10^9}$

c. $\dfrac{4.5 \times 10^{-8}}{3 \times 10^{-4}}$ **d.** $\dfrac{6 \times 10^{-4}}{1.5 \times 10^{-8}}$

e. $\dfrac{4 \times 10^{12}}{8 \times 10^4}$ **f.** $\dfrac{1.5 \times 10^4}{3 \times 10^{12}}$

g. $\dfrac{3.6 \times 10^{-8}}{6 \times 10^{-2}}$ **h.** $\dfrac{1.8 \times 10^{-2}}{9 \times 10^{-8}}$

MIXED PRACTICE

Problem set

1. (12) The first Indian-head penny was minted in 1859. The last Indian-head penny was minted in 1909. For how many years were Indian-head pennies minted?

2. (28) The product of y and 15 is 600. What is the sum of y and 15?

3. (36, 54) Thirty percent of those gathered agreed that the king should abdicate his throne. All the rest disagreed.

(a) What fraction of those gathered disagreed?

(b) What was the ratio of those who agreed to those who disagreed?

4. (80) Triangle ABC with vertices A (0, 3), B (0, 0), and C (4, 0) is translated one unit left, one unit down to make the image $\triangle A'B'C'$. What are the coordinates of the vertices of $\triangle A'B'C'$?

5. (a) Write the prime factorization of 1024 using exponents.
(21)
(b) Find $\sqrt{1024}$.

6. A portion of a regular polygon is shown at right. Each interior angle measures 150°.
(89)

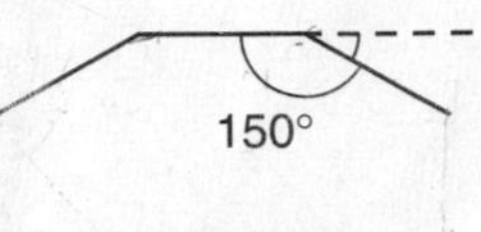

(a) What is the measure of each exterior angle?

(b) The polygon has how many sides?

(c) What is the name for a polygon with this number of sides?

7. The sale price of an item on sale for 40% off is \$48. What was the regular price?
(92)

8. In a bag are 12 marbles: 3 red, 4 white, and 5 blue. One marble is drawn from the bag and not replaced. A second marble is drawn and not replaced. Then a third marble is drawn.
(94)

(a) What is the probability of drawing a red, a white, and a blue marble in that order?

(b) What is the probability of drawing a blue, a white, and a red marble in that order?

9. Write an equation to solve this problem:
(101)

Six more than twice what number is 36?

10. What is the measure of each acute angle of this triangle?
(101)

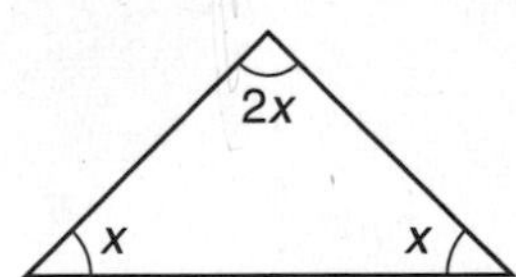

11. Solve for c^2: $c^2 - b^2 = a^2$
(106)

12. In the figure below, if $l \parallel q$ and $m\angle h = 105°$, what is the measure of
(102)

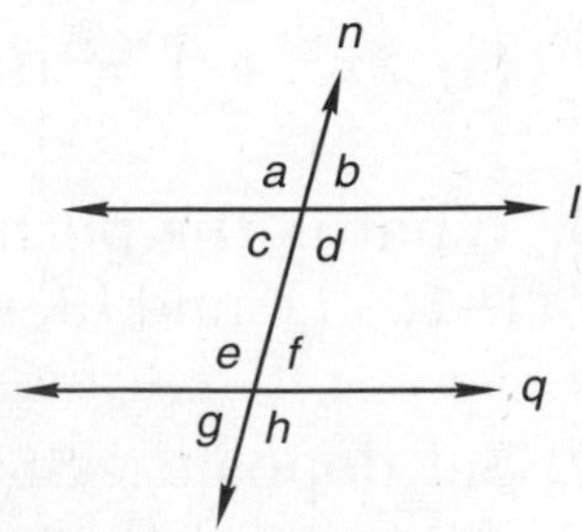

(a) $\angle a$? (b) $\angle b$? (c) $\angle c$? (d) $\angle d$?

13. (108) The formula below may be used to convert temperature measurements from degrees Celsius (C) to degrees Fahrenheit (F). Find F to the nearest degree when C is 17.

$$F = 1.8C + 32$$

14. (104) What is the area of a 45° sector of a circle with a radius of 12 in.? Use 3.14 for π and round the answer to the nearest square inch.

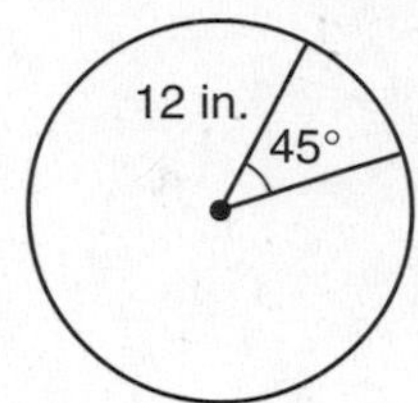

15. (Inv. 9) Make a table of ordered pairs showing three or four solutions for the equation $x + y = 1$. Then graph all possible solutions.

16. (107) Refer to the graph in problem 15 to answer (a) and (b).

(a) What is the slope of the graph of $x + y = 1$?

(b) Where does the graph of $x + y = 1$ intersect the y-axis?

17. (105) The city manager decided to wrap posters around the town's trash cans to encourage people to properly dispose of trash. The illustration shows the dimensions of the trash can. Converting the dimensions to feet and using 3.14 for π, find the number of square feet of paper needed to wrap around each trash can.

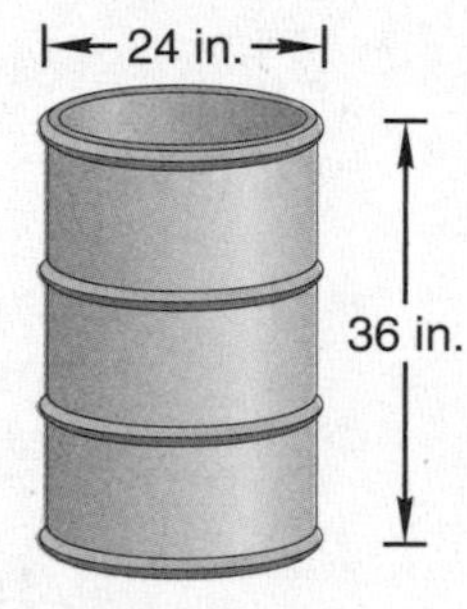

18. (95) The trash can illustrated in problem 17 has the capacity to hold how many cubic feet of trash?

19. (109) Find two solutions to each of these equations:

(a) $2x^2 + 1 = 19$ (b) $2x^2 - 1 = 19$

20. (99) What is the perimeter of a triangle with vertices (–1, 2), (–1, –1), and (3, –1)?

21. (110) Sal deposited \$5000 in an account that paid 5% interest compounded annually. What was the total value of the account after 5 years?

22. The figure at right shows three similar triangles. If AC is 15 cm and BC is 20 cm,
(97, 99)

(a) what is AB?

(b) what is CD?

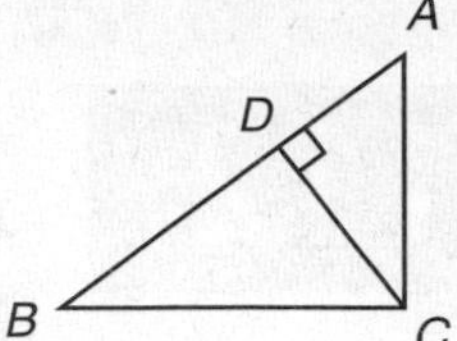

23. Express each quotient in scientific notation:
(111)

(a) $\dfrac{3.6 \times 10^8}{6 \times 10^6}$ (b) $\dfrac{3.6 \times 10^{-8}}{1.2 \times 10^{-6}}$

24. In the figure below, if the measure of $\angle x$ is 140°, what is the measure of $\angle y$?
(40)

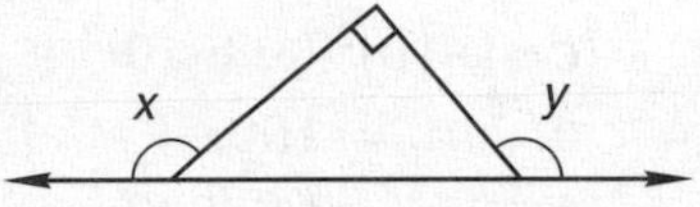

Solve:

25. $5x + 3x = 18 + 2x$
(102)

26. $\dfrac{3.6}{x} = \dfrac{4.5}{0.06}$
(98)

Simplify:

27. (a) $(-1)^6 + (-1)^5$ (b) $(-10)^6 \div (-10)^5$
(103)

28. (a) $\dfrac{(4a^2b)(9ab^2c)}{6abc}$ (b) $x(x - c) + cx$
(96, 103)

29. $(-3) + (+2)(-4) - (-6)(-2) - (-8)$
(85)

30. $\dfrac{3\frac{1}{3} \cdot 1\frac{4}{5} + 1.5}{0.03}$
(43, 45)

LESSON

112 Applications of the Pythagorean Theorem

WARM-UP

Facts Practice: Multiplying and Dividing in Scientific Notation (Test W)

Mental Math:

a. 100000 (base 2)
b. XCIX
c. $(-10)^2 + (-10)^3$
d. $(8 \times 10^6) \div (4 \times 10^3)$
e. $m^2 = 100$
f. Convert 50°C to degrees Fahrenheit.
g. 25% of $2000
h. $2000 increased 25%
i. Start with 2 dozen, + 1, × 4, + 20, ÷ 3, + 2, ÷ 6, × 4, − 3, $\sqrt{\ }$, ÷ 2.

Problem Solving:

Mariabella was $\frac{1}{4}$ of the way through her book. Twenty pages later she was $\frac{1}{3}$ of the way through her book. When she is $\frac{3}{4}$ of the way through the book, how many pages will she have to read to finish the book?

NEW CONCEPT

Workers who construct buildings need to be sure that the structures have square corners. If the corner of a 40-foot-long building is 89° or 91° instead of 90°, the other end of the building will be about 8 inches out of position.

One way construction workers can check whether a building under construction is square is by using a **Pythagorean triplet.** The numbers 3, 4, and 5 satisfy the Pythagorean theorem and are an example of a Pythagorean triplet.

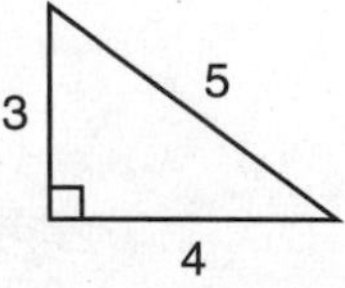

$$3^2 + 4^2 = 5^2$$

Multiples of 3-4-5 are also Pythagorean triplets.

3-4-5

6-8-10

9-12-15

12-16-20

Before pouring a concrete foundation for a building, construction workers build wooden forms to hold the concrete. Then a worker or building inspector can use a Pythagorean triplet to check that the forms make a right angle. First the perpendicular sides are marked at selected lengths.

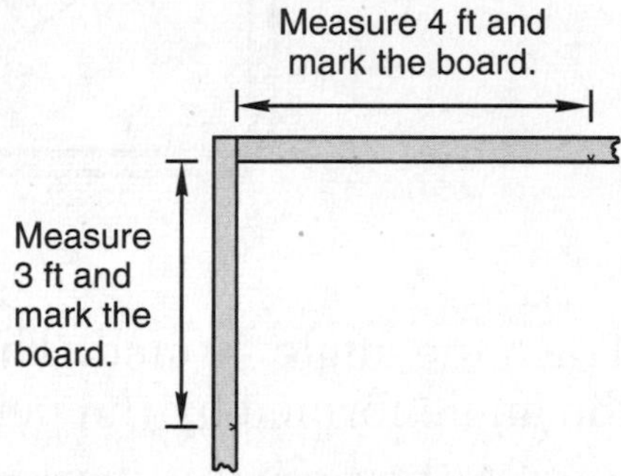

Then the distance between the marks is checked to be sure the three measures are a Pythagorean triplet.

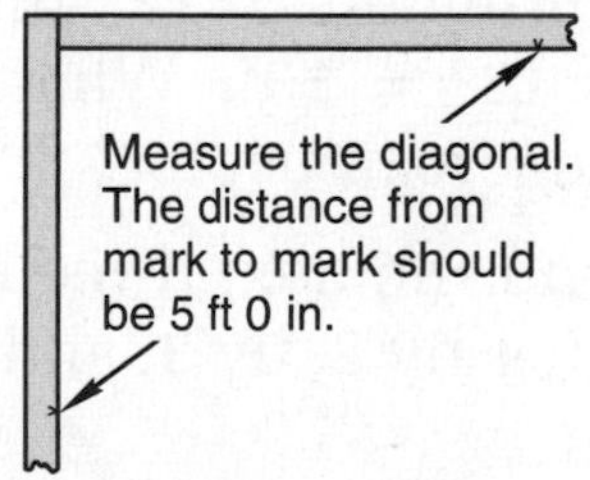

If the three measures are a Pythagorean triplet, the worker can be confident that the corner forms a 90° angle.

Activity: *Application of the Pythagorean Theorem*

Materials needed:

- Two full-length, unsharpened pencils (or other straightedges)
- Ruler
- Protractor

Position two pencils (or straightedges) so that they appear to form a right angle. Mark one pencil 3 inches from the vertex of the angle and the other pencil 4 inches from the vertex. Then measure from mark to mark to see whether the distance between the marks is 5 inches. Adjust the pencils if necessary.

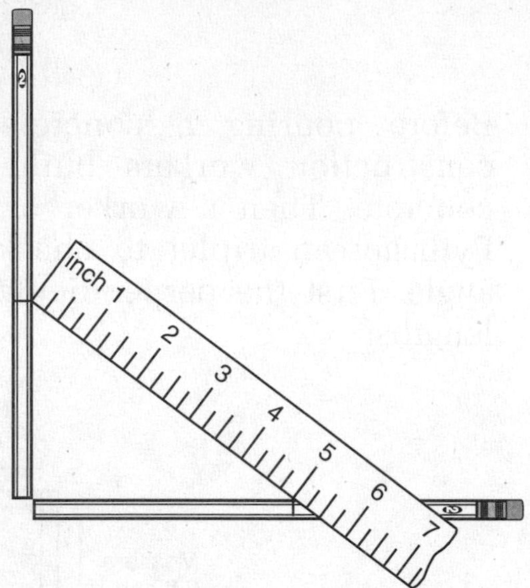

Trace the angle formed. Then use a protractor to confirm that the angle formed by the pencils measures 90°.

Repeat the activity, marking the pencils at 6 cm and 8 cm. The distance between the marks should be 10 cm.

Example 1 The numbers 5, 12, and 13 are a Pythagorean triplet because $5^2 + 12^2 = 13^2$. What are the next three multiples of this Pythagorean triplet?

Solution To find the next three multiples of 5-12-13, we multiply each number by 2, by 3, and by 4.

10-24-26

15-36-39

20-48-52

Example 2 A roof is being built over a 24-ft-wide room. The slope of the roof is 4 in 12. Calculate the length of the rafters needed for the roof. (Include 2 ft for the rafter tail.)

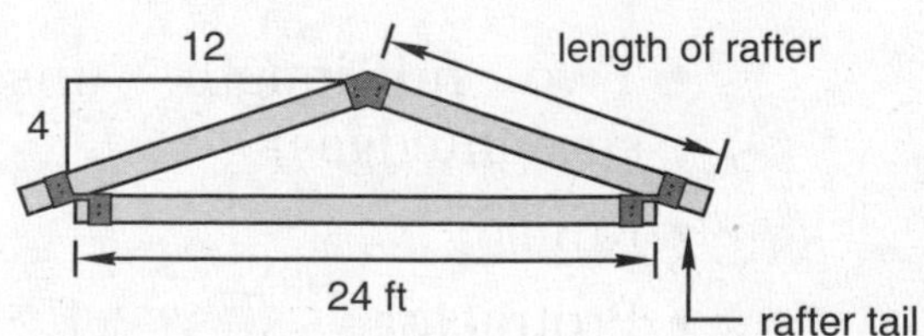

Solution We consider a rafter to be the hypotenuse of a right triangle. The width of the room is 24 ft, but a rafter spans only half the width of the room. So the base of the right triangle is 12 ft. The slope of the roof, 4 in 12, means that for every 12 horizontal units, the roof rises (or falls) 4 vertical units. Thus, since the base of the triangle is 12 ft, its height is 4 ft.

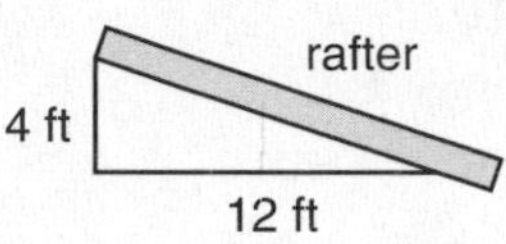

We use the Pythagorean theorem to calculate the hypotenuse.

$$a^2 + b^2 = c^2$$

$$(4 \text{ ft})^2 + (12 \text{ ft})^2 = c^2$$

$$16 \text{ ft}^2 + 144 \text{ ft}^2 = c^2$$

$$160 \text{ ft}^2 = c^2$$

$$\sqrt{160} \text{ ft} = c$$

$$12.65 \text{ ft} \approx c$$

Using a calculator we find that the hypotenuse is about 12.65 feet. We add 2 feet for the rafter tail.

$$12.65 \text{ ft} + 2 \text{ ft} = 14.65 \text{ ft}$$

To convert 0.65 ft to inches, we multiply.

$$0.65 \text{ ft} \times \frac{12 \text{ in.}}{1 \text{ ft}} = 7.8 \text{ in.}$$

We round this up to 8 inches. So the length of each rafter is about **14 ft 8 in.**

Example 3 Serena went to a level field to fly a kite. She let out all 200 ft of string and tied it to a stake. Then she walked out on the field until she was directly under the kite, 150 feet from the stake. About how high was the kite?

Solution We begin by sketching the problem. The length of the kite string is the hypotenuse of a right triangle, and the distance between Serena and the stake is one leg of the triangle. We use the Pythagorean theorem to find the remaining leg, which is the height of the kite.

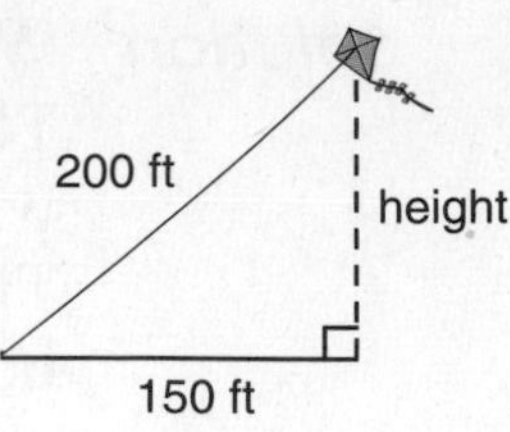

$$a^2 + b^2 = c^2$$

$$a^2 + (150 \text{ ft})^2 = (200 \text{ ft})^2$$

$$a^2 + 22{,}500 \text{ ft}^2 = 40{,}000 \text{ ft}^2$$

$$a^2 = 17{,}500 \text{ ft}^2$$

$$a = \sqrt{17{,}500} \text{ ft}$$

$$a \approx 132 \text{ ft}$$

Using a calculator, we find that the height of the kite was about **132 ft.**

LESSON PRACTICE

Practice set

a. A 12-foot ladder was leaning against a building. The base of the ladder was 5 feet from the building. How high up the side of the building did the ladder reach? Write the answer in feet and inches rounded to the nearest inch.

b. Figure *ABCD* illustrates a rectangular field 400 feet long and 300 feet wide. The path from *A* to *C* is how much shorter than the path from *A* to *B* to *C*?

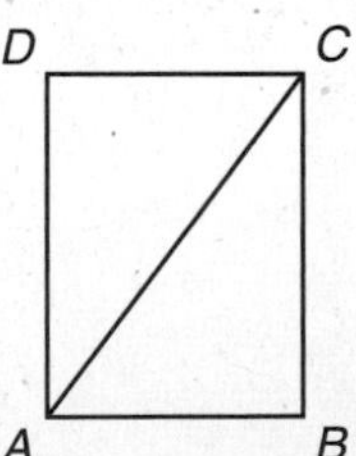

MIXED PRACTICE

Problem set

1. (110) Sherman deposited $3000 in an account paying 8 percent interest compounded annually. He withdrew his money and interest 3 years later. How much did he withdraw?

2. (20, 28) What is the square root of the sum of 3 squared and 4 squared?

3. Find the (a) median and (b) mode of the following quiz scores:
(Inv. 4)

Kristin's Quiz Scores

Score	Number of Quizzes
100	2
95	7
90	6
85	6
80	3
70	3

4. The trucker completed the 840-kilometer haul in 10 hours 30 minutes. What was the trucker's average speed in kilometers per hour?
(46)

Use ratio boxes to solve problems 5–7:

5. Barbara earned \$28 for 6 hours of work. At that rate, how much would she earn for 9 hours of work?
(72)

6. José paid \$48 for a jacket at 25 percent off of the regular price. What was the regular price of the jacket?
(92)

7. Troy bought a baseball card for \$6 and sold it for 25 percent more than he paid for it. How much profit did he make on the sale?
(92)

8. At a yard sale an item marked \$1.00 was reduced 50%. When the item still did not sell, the sale price was reduced 50%. What was the price of the item after the second discount?
(110)

9. If 60% of the children were boys, what was the ratio of boys to girls?
(36)

10. The points (3, 11), (–2, –1), and (–2, 11) are the vertices of a right triangle. Use the Pythagorean theorem to find the length of the hypotenuse of this triangle.
(99)

11. The frame of this kite is formed by two perpendicular pieces of wood whose lengths are shown in inches. A loop of string connects the four ends of the sticks. How long is the string?
(112)

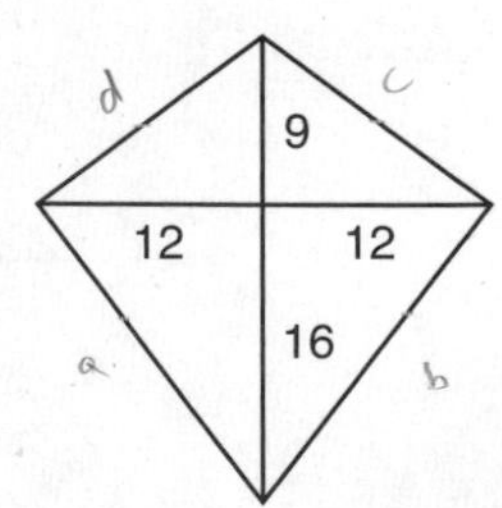

12. (77) What percent of 2.5 is 2?

13. (Inv. 10) What are the odds of having a coin land tails up on 4 consecutive tosses of a coin?

14. (110) How much interest is earned in 6 months on $4000 deposited at 9 percent simple interest?

15. (48) Complete the table.

FRACTION	DECIMAL	PERCENT
$\frac{5}{8}$	(a)	(b)

16. (111) Divide. Write each quotient in scientific notation:

(a) $\frac{5 \times 10^8}{2 \times 10^4}$ (b) $\frac{1.2 \times 10^4}{4 \times 10^8}$

17. (50) Use a unit multiplier to convert 300 kilograms to grams.

18. (106) Solve for t: $d = rt$

19. (Inv. 9) Make a table that shows three pairs of numbers for the function $y = -x$. Then graph the number pairs on a coordinate plane, and draw a line through the points.

20. (104) Find the perimeter of this figure. The arc in the figure is a semicircle. Dimensions are in centimeters. (Use 3.14 for π.)

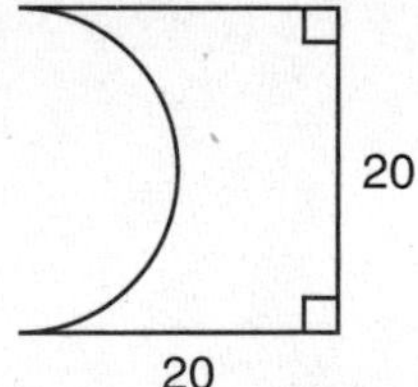

21. (105) Find the surface area of this right triangular prism. Dimensions are in feet.

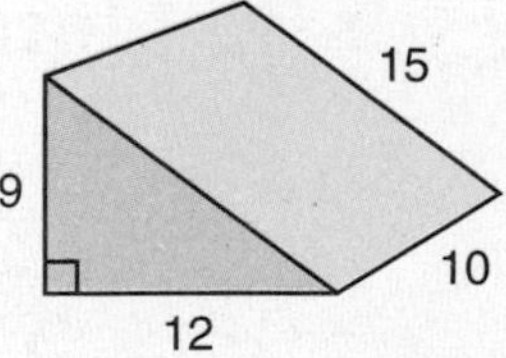

22. (21) (a) Write the prime factorization of 1 trillion using exponents.

(b) Find the positive square root of 1 trillion.

23. (97, 98) The triangles below are similar. Dimensions are in centimeters.

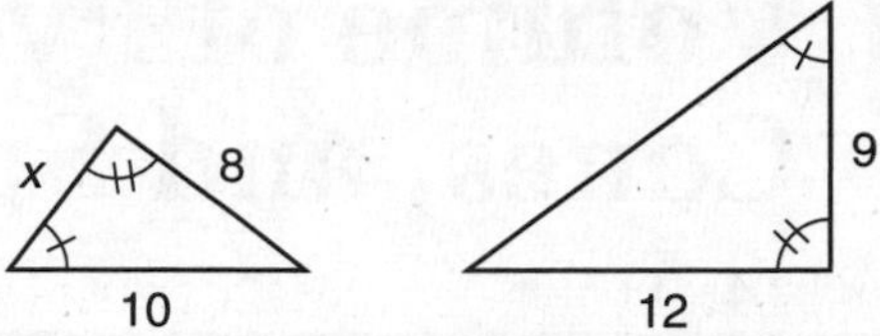

(a) Find x.

(b) Find the scale factor from the smaller to the larger triangle.

Solve:

24. (98) $\dfrac{16}{2.5} = \dfrac{48}{f}$

25. (93) $2\frac{2}{3}x - 3 = 21$

Simplify:

26. (57, 63) $10^2 - [10 - 10(10^0 - 10^{-1})]$

27. (43) $2\frac{3}{4} - \left(1.5 - \frac{1}{6}\right)$

28. (43) $3.5 \div 1\frac{2}{5} \div 3$

29. (85) $|-4| - (-3)(-2)(-1) + \dfrac{(-5)(4)(-3)(2)}{-1}$

30. (105) The large grapefruit was nearly spherical and had a diameter of 14 cm. Using $\frac{22}{7}$ for π, find the approximate surface area of the grapefruit to the nearest hundred square centimeters.

LESSON

113 Volume of Pyramids, Cones, and Spheres

WARM-UP

Facts Practice: + − × ÷ Algebraic Terms (Test V)

Mental Math:

a. 11111 (base 2)
b. MCMXLI
c. $0.75 \div (-3)$
d. $(6 \times 10^6) \div (3 \times 10^{10})$
e. $10m + 1.5 = 7.5$
f. Convert 2 ft^2 to square inches.
g. $66\frac{2}{3}\%$ of \$2400
h. \$2400 reduced by $66\frac{2}{3}\%$
i. At 500 mph, how long will it take a plane to travel 1250 miles?

Problem Solving:

The sum of two numbers is 17, and their product is 60. Use guess and check to find the two numbers.

NEW CONCEPTS

Volume of pyramids

A **pyramid** is a geometric solid that has three or more triangular faces and a base that is a polygon. Each of these figures is a pyramid:

 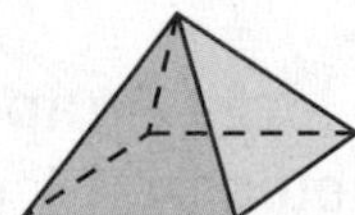 

The volume of a pyramid is $\frac{1}{3}$ the volume of a prism that has the same base and height. Recall that the volume of a prism is equal to the area of its base times its height.

$$\text{Volume of a prism} = \text{area of base} \cdot \text{height}$$

To find the volume of a pyramid, we first find the volume of a prism that has the same base and height. Then we divide the result by 3 $\left(\text{or multiply by } \frac{1}{3}\right)$.

$$\textbf{Volume of a pyramid} = \frac{1}{3} \cdot \textbf{area of base} \cdot \textbf{height}$$

Example 1 The pyramid at right has the same base and height as the cube that contains it. Each edge of the cube is 6 centimeters.

(a) Find the volume of the cube.

(b) Find the volume of the pyramid.

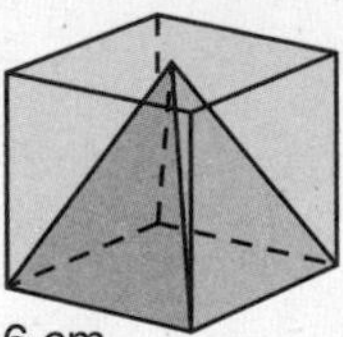

Solution (a) The volume of the cube equals the area of the base times the height.

$$\text{Area of base} = 6 \text{ cm} \times 6 \text{ cm} = 36 \text{ cm}^2$$

$$\begin{aligned}\text{Volume of cube} &= \text{area of base} \cdot \text{height} \\ &= (36 \text{ cm}^2)(6 \text{ cm}) \\ &= \mathbf{216\ cm^3}\end{aligned}$$

(b) The volume of the pyramid is $\frac{1}{3}$ the volume of the cube, so we divide the volume of the cube by 3 $\left(\text{or multiply by } \frac{1}{3}\right)$.

$$\begin{aligned}\text{Volume of pyramid} &= \frac{1}{3}\,(\text{volume of prism}) \\ &= \frac{1}{3}\,(216 \text{ cm}^3) \\ &= \mathbf{72\ cm^3}\end{aligned}$$

Volume of cones The volume of a cone is $\frac{1}{3}$ the volume of a cylinder with the same base and height.

$$\textbf{Volume of a cone} = \frac{1}{3} \cdot \textbf{area of base} \cdot \textbf{height}$$

Example 2 Find the volume of this circular cone. Dimensions are in centimeters. (Use 3.14 for π.)

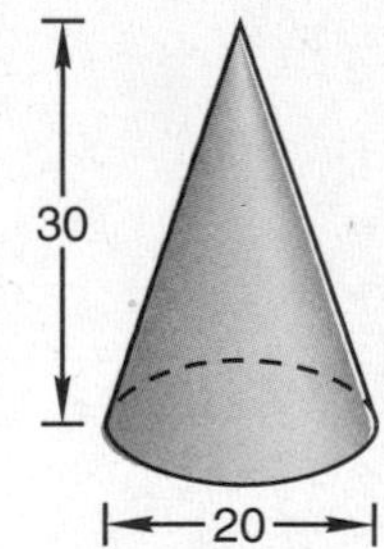

Solution We first find the volume of a cylinder with the same base and height as the cone.

$$\begin{aligned}\text{Volume of cylinder} &= \text{area of circle} \cdot \text{height} \\ &\approx (3.14)(10 \text{ cm})^2 \cdot 30 \text{ cm} \\ &\approx 9420 \text{ cm}^3\end{aligned}$$

Then we find $\frac{1}{3}$ of this volume.

$$\begin{aligned}\text{Volume of cone} &= \frac{1}{3}\,(\text{volume of cylinder}) \\ &\approx \frac{1}{3} \cdot 9420 \text{ cm}^3 \\ &\approx \mathbf{3140\ cm^3}\end{aligned}$$

Volume of spheres

There is a special relationship between the volume of a cylinder, the volume of a cone, and the volume of a sphere. Picture two identical cylinders whose heights are equal to their diameters. In one cylinder is a cone with the same diameter and height as the cylinder. In the other cylinder is a sphere with the same diameter as the cylinder.

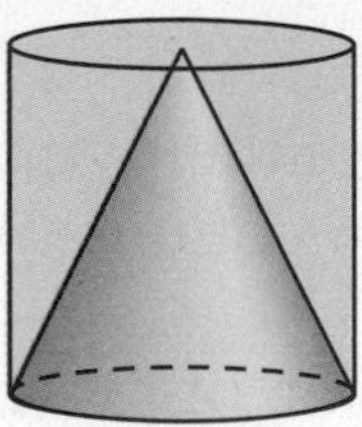

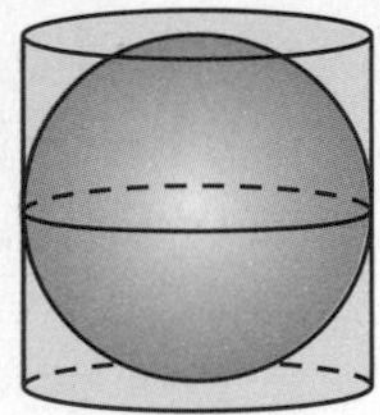

We have learned that the volume of the cone is $\frac{1}{3}$ the volume of the cylinder. Remarkably, the volume of the sphere is twice the volume of the cone, that is, $\frac{2}{3}$ the volume of the cylinder. Here we use a balance scale to provide another view of this relationship.

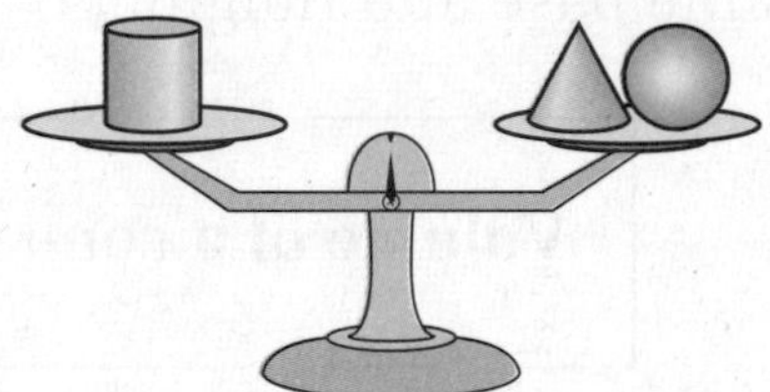

Imagine that the objects on the balance scale are solid and composed of the same material. The diameters of all three solids are equal. The heights of the cylinder and cone equal their diameters. The balance scale shows that the combined masses of the cone and sphere equal the mass of the cylinder. The cone's mass is $\frac{1}{3}$ of the cylinder's mass, and the sphere's mass is $\frac{2}{3}$ of the cylinder's mass.

The formula for the volume of a sphere can be derived from our knowledge of cylinders and their relationship to spheres. First consider the volume of any cylinder whose height is equal to its diameter. In this diagram we have labeled the diameter, *d;* the radius, *r;* and the height, *h.*

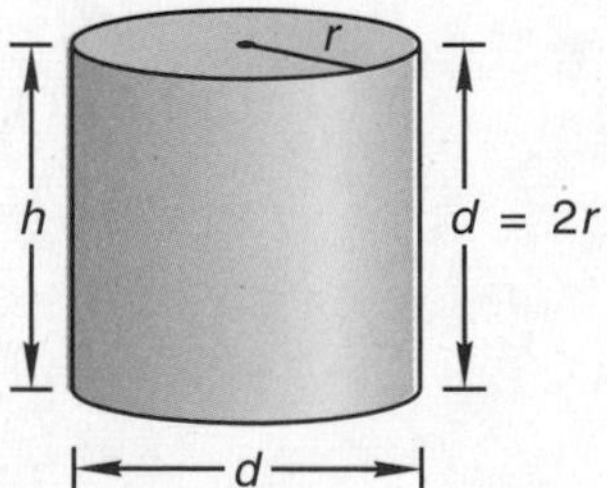

Recall that we can calculate the volume of a cylinder by multiplying the area of its circular base times its height.

$$\text{Volume of cylinder} = \text{area of circle} \cdot \text{height}$$

$$V = \pi r^2 \cdot h$$

For a cylinder whose height is equal to its diameter, we can replace h in the formula with d or with $2r$, since two radii equal the diameter.

$V = \pi r^2 \cdot h$	formula for volume of a cylinder
$V = \pi r^2 \cdot 2r$	replaced h with $2r$, which equals the height of the cylinder
$V = 2\pi r^3$	rearranged factors

This formula, $V = 2\pi r^3$, gives the volume of a cylinder whose height is equal to its diameter. The volume of a sphere is $\frac{2}{3}$ of the volume of a cylinder with the same diameter and height.

$\text{Volume of sphere} = \frac{2}{3} \cdot \text{volume of cylinder}$	
$= \frac{2}{3} \cdot 2\pi r^3$	substituted
$= \frac{4}{3}\pi r^3$	multiplied $\frac{2}{3}$ by 2

We have found the formula for the volume of a sphere.

$$\textbf{Volume of a sphere} = \frac{4}{3}\pi r^3$$

Example 3 A ball with a diameter of 20 cm has a volume of how many cubic centimeters? (Round the answer to the nearest hundred cubic centimeters.)

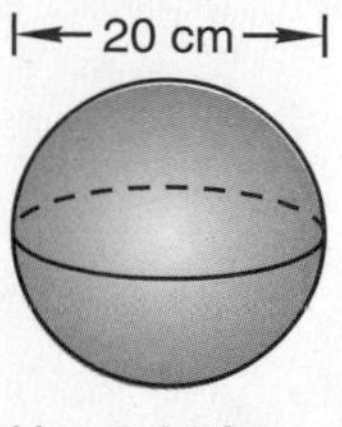

Use 3.14 for π.

Solution The diameter of the sphere is 20 cm, so its radius is 10 cm. We use the formula for the volume of a sphere, substituting 3.14 for π and 10 cm for r.

$V = \frac{4}{3}\pi r^3$	formula
$\approx \frac{4}{3}(3.14)(10 \text{ cm})^3$	substituted 3.14 for π and 10 cm for r
$\approx \frac{4}{3}(3.14)(1000 \text{ cm}^3)$	cubed 10 cm
$\approx \frac{4}{3}(3140 \text{ cm}^3)$	multiplied 3.14 by 1000 cm^3
$\approx 4186\frac{2}{3} \text{ cm}^3$	multiplied $\frac{4}{3}$ by 3140 cm^3
$\approx \mathbf{4200 \text{ cm}^3}$	rounded to nearest hundred cubic centimeters

LESSON PRACTICE

Practice set Pictured are two identical cylinders whose heights are equal to their diameters. In one cylinder is the largest cone it can contain. In the other cylinder is the largest sphere it can contain. Packing material is used to fill all the voids in the cylinders not occupied by the cone or sphere. Use this information to answer problems **a–e.**

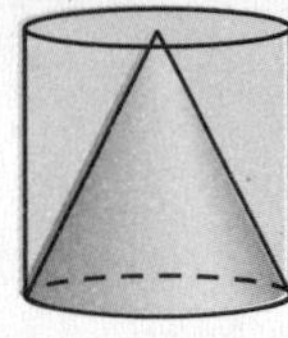
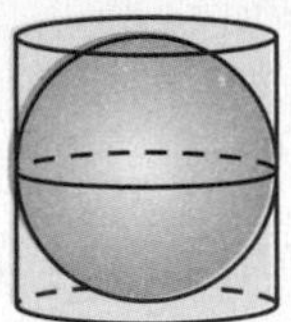

a. What fraction of the cylinder containing the cone is occupied by the cone?

b. What fraction of the cylinder containing the cone is occupied by the packing material?

c. What fraction of the cylinder containing the sphere is occupied by the sphere?

d. What fraction of the cylinder containing the sphere is occupied by the packing material?

e. If the cone and sphere were removed from their boxes and all the packing material from both boxes was put into one box, what portion of the box would be filled with packing material?

f. A pyramid with a height of 12 inches and a base 12 inches square is packed in the smallest cubical box that can contain it. What is the volume of the box? What is the volume of the pyramid?

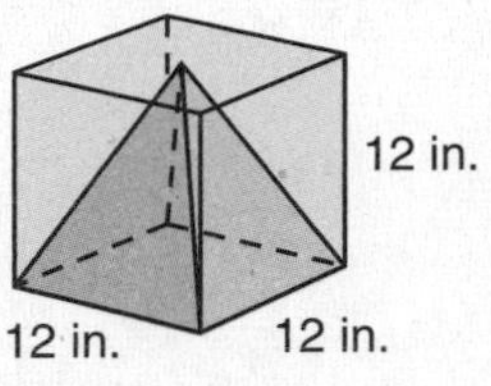

Find the volume of each figure below. For both calculations, leave π as π.

g.
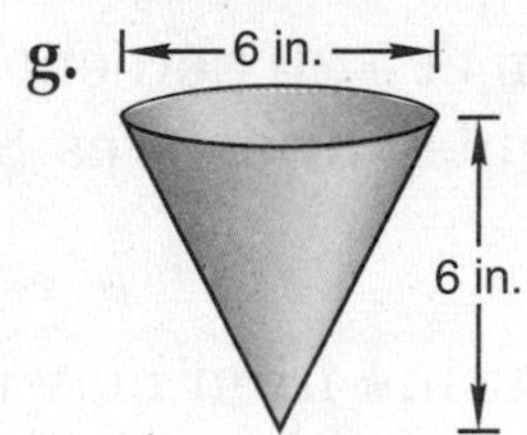

h.
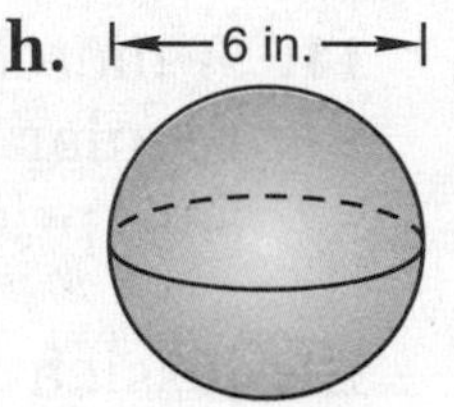

MIXED PRACTICE

Problem set

1. Find the sale price of a $24 item after successive discounts of 25% and 25%. (110)

2. Ten billion is how much greater than nine hundred eighty million? Write the answer in scientific notation. (51)

3. The median of the following numbers is how much less than the mean? (Inv. 4)

   1.4, 0.5, 0.6, 0.75, 5.2

4. Nelda worked 5 hours and earned $24. Christy worked 6 hours and earned $33. (46)

   (a) How much did Nelda earn per hour?

   (b) How much did Christy earn per hour?

   (c) Christy earned how much more per hour than Nelda?

5. Use a ratio box to solve this problem. If 24 kilograms of seed costs $31, what is the cost of 42 kilograms of seed? (72)

6. A kilometer is about $\frac{5}{8}$ of a mile. A mile is 1760 yards. A kilometer is about how many yards? (60)

7. (94) A card is drawn from a deck of 52 playing cards and replaced. Then another card is drawn. What is the probability of drawing a heart both times?

Write equations to solve problems 8 and 9.

8. (77) What percent of \$30 is \$1.50?

9. (77) Fifty percent of what number is $2\frac{1}{2}$?

10. (110) Trinh left \$5000 in an account that paid 8 percent interest compounded annually. How much interest was earned in 3 years?

Use ratio boxes to solve problems 11 and 12.

11. (92) A merchant sold an item at a 20 percent discount from the regular price. If the regular price was \$12, what was the sale price of the item?

12. (98) Jessica sculptured a figurine from clay at $\frac{1}{24}$ of the actual size of the model. If the model was 6 feet tall, how many inches tall was the figurine?

13. (37, 62) The points (0, 4), (–3, 2), and (3, 2) are the vertices of an isosceles triangle. Find the area of the triangle.

14. (112) Use the Pythagorean theorem to find the length of one of the two congruent sides of the triangle in problem 13. (*Hint:* First draw an altitude to form two right triangles.)

15. (113) Roughly estimate the volume of a tennis ball in cubic centimeters by using 6 cm for the diameter and 3 for π. Round the answer to the nearest ten cubic centimeters.

16. (83) Multiply. Write the product in scientific notation.

$$(6.3 \times 10^6)(7 \times 10^{-3})$$

17. (112) Tim can get from point A to point B by staying on the sidewalk and turning left at the corner C, or he can take the shortcut and walk straight from point A to point B. How many yards can Tim save by taking the shortcut? Begin by using the Pythagorean theorem to find the length of the shortcut.

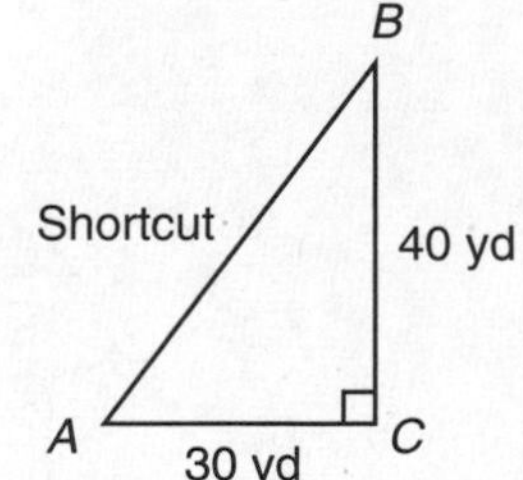

18. (a) Solve for h: $A = \frac{1}{2}bh$
(108)

(b) Use the formula $A = \frac{1}{2}bh$ to find h when $A = 16$ and $b = 8$.

19. Make a table that shows three pairs of numbers for the function $y = -2x + 1$. Then graph the number pairs on a coordinate plane, and draw a line through the points to show other number pairs that satisfy the function. What is the slope of the graphed line?
(Inv. 9, 107)

20. Find the volume of the pyramid shown. Dimensions are in meters.
(113)

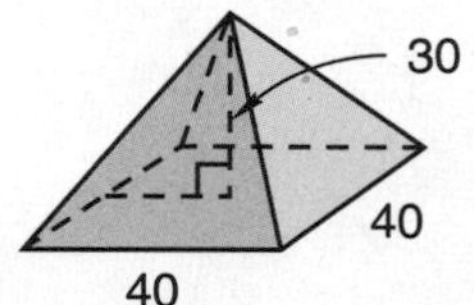

21. Find the volume of the cone at right. Dimensions are in centimeters.
(113)

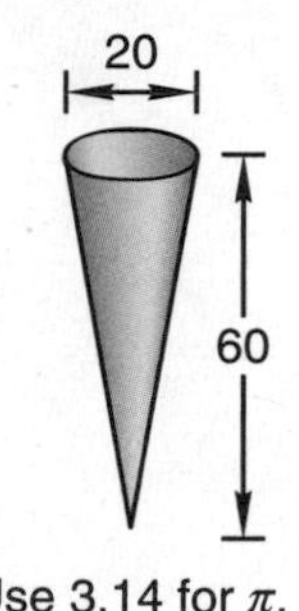

22. Refer to the figure below to find the measures of the following angles. Dimensions are in centimeters.
(40)

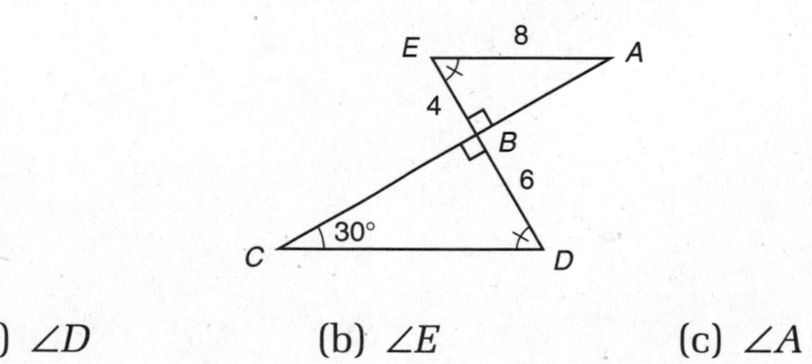

(a) $\angle D$ (b) $\angle E$ (c) $\angle A$

23. In the figure in problem 22, what is the length of $\overline{CD}$?
(97)

Solve:

24. $\frac{7.5}{d} = \frac{25}{16}$
(98)

25. $1\frac{3}{5}w + 17 = 49$
(93)

Simplify:

26. $5^2 - \{4^2 - [3^2 - (2^2 - 1^2)]\}$
(63)

27. $\dfrac{440 \text{ yd}}{1 \text{ min}} \cdot \dfrac{1 \text{ min}}{60 \text{ s}} \cdot \dfrac{3 \text{ ft}}{1 \text{ yd}}$
(88)

28. $1\frac{3}{4} + 2\frac{2}{3} - 3\frac{5}{6}$
(30)

29. $\left(1\frac{3}{4}\right)\left(2\frac{2}{3}\right) \div 3\frac{5}{6}$
(26)

30. $(-7) + |-3| - (2)(-3) + (-4) - (-3)(-2)(-1)$
(85)

LESSON

114 Graphing Linear Inequalities

WARM-UP

Facts Practice: Multiplying and Dividing in Scientific Notation (Test W)

Mental Math:

a. 1001 (base 2)

b. XXIX

c. $(-3)^3 + (-3)^2$

d. $(4 \times 10^8)^2$

e. $10m - m = 9^2$

f. Convert 60°C to degrees Fahrenheit.

g. 150% of \$3000

h. 150% more than \$3000

i. Find 25% of 40, − 1, × 5, − 1, ÷ 2, − 1, ÷ 3, × 10, + 2, ÷ 9, ÷ 2, $\sqrt{\ }$.

Problem Solving:

Four identical blocks marked X, a 250-g mass, and a 500-g mass were balanced on a scale as shown. Write an equation to represent this balanced scale, and find the mass of each block marked X.

NEW CONCEPT

We have graphed inequalities on a number line. We can also graph inequalities on a coordinate plane. For instance, think of two numbers whose sum is less than 10. Possibilities include 2 and 3, 4 and 5, and 0 and 9. You might also think of $1\frac{1}{2}$ and 0.6, $\sqrt{2}$ and $\sqrt{3}$, or −1 and −5. We can show all possible pairs of numbers whose sum is less than 10 on a coordinate plane, as we see in example 1.

Example 1 Graph all pairs of numbers x and y whose sum is less than 10 (that is, $x + y < 10$).

Solution We begin by graphing all pairs of numbers whose sum *is* 10 (that is, $x + y = 10$). This line will serve as a boundary for the inequality.

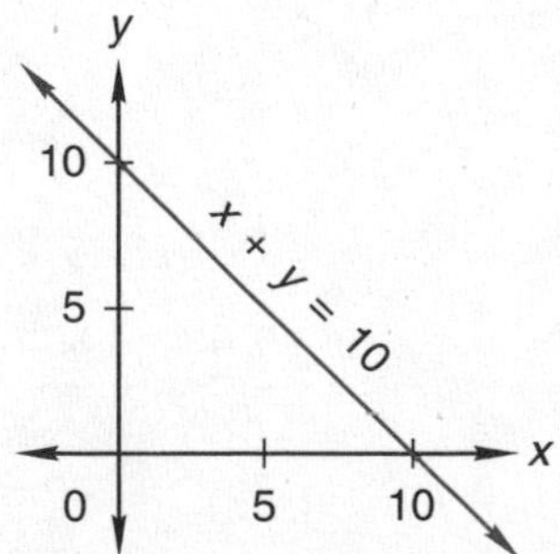

Every point on this line has *x*, *y* coordinates whose sum is 10, such as (10, 0), (6, 4), and (5, 5). Now we will consider the location of the points whose coordinates are less than 10. We will select one point above the line and another point below the line and check the sum of the coordinates.

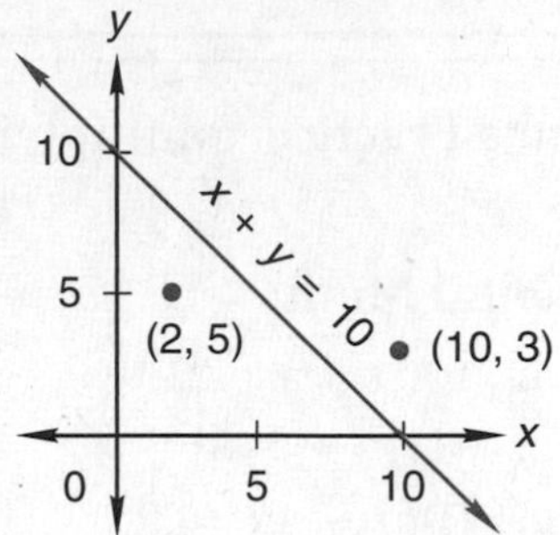

The point (10, 3) is above the line, and the sum of 10 and 3 is greater than 10. The point (2, 5) is below the line, and the sum of 2 and 5 is less than 10. Actually, all the points below the graphed line have coordinates whose sum is less than 10. To show that we are graphing all points below the line, we shade that region of the graph. We assume that the shading continues without end.

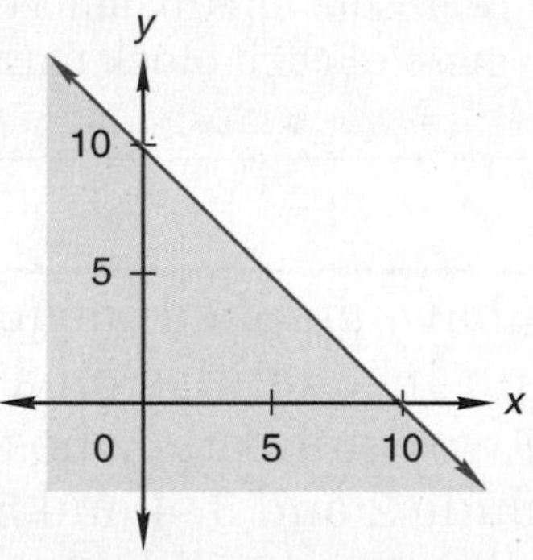

There is one more change to make to this graph. The solid line identifies the *x*, *y* pairs whose sum is 10. So this graph shows the points whose sum is less than *or equal* to 10 (that is, $x + y \leq 10$). However, we are asked to graph only the points whose sum is less than 10. In order to show that the points on the graphed line are excluded, we change the solid line into a dashed line. This completes the graph.

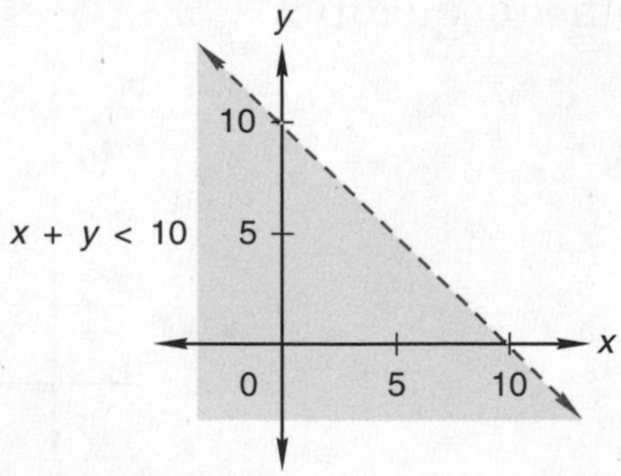

Example 2 Every day Angela runs and/or walks for 15 to 20 minutes. Write two inequalities for Angela's running and walking time, and graph the region that satisfies both inequalities.

Solution We will let x represent the number of minutes Angela walks and y the number of minutes she runs. Since Angela walks or runs for at least 15 minutes, her walk time plus her run time is equal to or greater than 15.

$$\mathbf{x + y \geq 15}$$

However, Angela walks or runs for no more than 20 minutes. So her walk time plus her run time is less than or equal to 20.

$$\mathbf{x + y \leq 20}$$

Now we are ready to graph these two inequalities. We only use the first quadrant because she cannot walk or run negative minutes. We begin by graphing $x + y = 15$ and $x + y = 20$, as shown below.

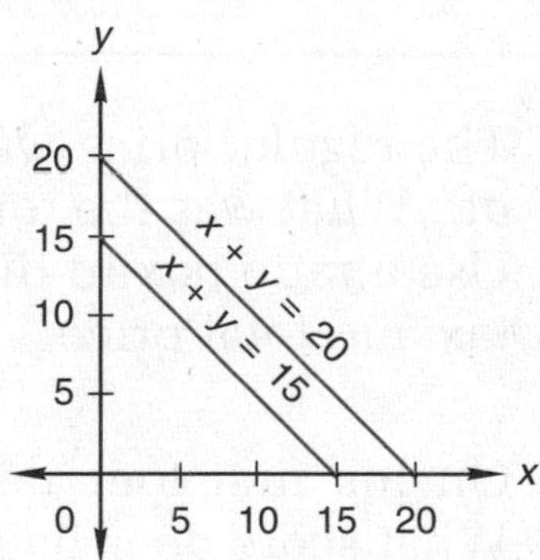

Since Angela runs and/or walks anywhere from 15 to 20 minutes, we shade the region between the two lines. Any point either on or between the graphed lines (including on the axes from 15 to 20) satisfies the conditions of the statement.

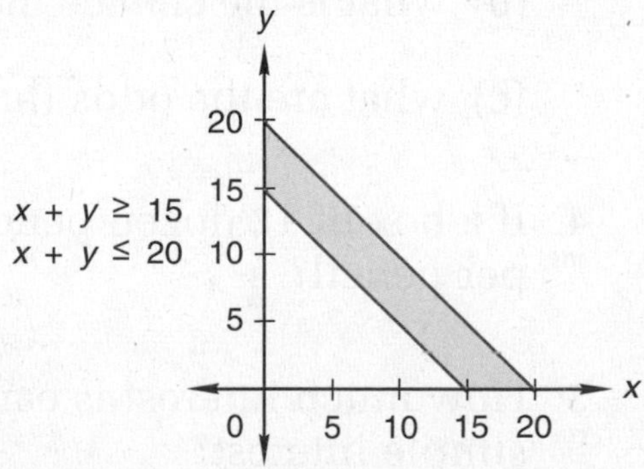

LESSON PRACTICE

Practice set **a.** Jan thought of two numbers whose sum was greater than 5. Write an inequality for the above statement, and graph the inequality on a coordinate plane. (See example 1.)

b. Alberto thought of two numbers. He said that their sum was 8 or less but that neither number was negative. Graph the set of possible solutions on a coordinate plane.

c. Ruthie spends at least two hours reading or studying math every day. However, she spends no more than three hours on these activities. Using x for the number of hours spent reading and y for the number of hours spent studying math, write an inequality for the first sentence and another inequality for the second sentence. Then, in the 1st quadrant, graph the region that satisfies both inequalities.

MIXED PRACTICE

Problem set

1. (46, 92) The regular price was $72.50, but it was on sale for 20% off. What was the total sale price including 7% sales tax? Use a ratio box to find the sale price. Then find the sales tax and total price.

2. (55) On his first four tests, Enrique's average score was 87. What score does he need to average on his next two tests to have a 6-test average of 90?

3. (Inv. 10) In a bag are 27 marbles: 6 red, 9 green, and 12 blue. If one marble is drawn from the bag,

(a) what is the probability that the marble will be blue?

(b) what is the chance that the marble will be green?

(c) what are the odds that the marble will not be red?

4. (46) If a box of 12 dozen pencils costs $10.80, what is the cost per pencil?

5. (110) How much interest is earned in 6 months on $5000 at 8% simple interest?

6. (Inv. 5) One fourth of the trees in the park were oak trees. One third of the trees were pine trees. The rest of the trees were walnut trees.

(a) Draw a circle graph that displays this information.

(b) If six of the trees were oak trees, how many of the trees were walnut trees?

7. (65) The ratio of cars to trucks passing by the checkpoint was 5 to 2. If a total of 3500 cars and trucks passed by the checkpoint, how many were cars?

8. (113) The snowball grew in size as it rolled down the hill. By the time it came to a stop, its diameter was about four feet. Using 3 for π, estimate the number of cubic feet of snow in the snowball.

Write equations to solve problems 9 and 10.

9. (60) What is 120% of \$240?

10. (77) Sixty is what percent of 150?

11. (75, 99) The points (3, 2), (6, –2), (–2, –2), and (–2, 2) are the vertices of a trapezoid.

(a) Find the area of the trapezoid.

(b) Find the perimeter of the trapezoid.

12. (100) (a) Arrange these numbers in order from least to greatest:

$$\sqrt{6}, 6^2, -6, 0.6$$

(b) Which of the numbers in (a) are rational numbers?

13. (48) Complete the table.

FRACTION	DECIMAL	PERCENT
$1\frac{4}{5}$	(a)	(b)

14. (111) Divide. Write each quotient in scientific notation:

(a) $\dfrac{5 \times 10^{-9}}{2 \times 10^{-6}}$ (b) $\dfrac{2 \times 10^{-6}}{5 \times 10^{-9}}$

15. (83) What is the product of answers (a) and (b) in problem 14?

16. (50) Use a unit multiplier to convert 12 inches to centimeters.

17. (108) (a) Solve for d: $C = \pi d$

(b) Use the formula $C = \pi d$ to find d when C is 62.8. (Use 3.14 for π.)

18. (114) James would like to mow the lawn and wash the car but has less than 60 minutes to work. Using x for the number of minutes it will take to mow and y for the number of minutes it will take to wash the car, write an inequality for the first sentence of this problem. Then graph the inequality in the first quadrant.

19. (104) Find the perimeter of the figure at right. Dimensions are in centimeters. (Use 3.14 for π.)

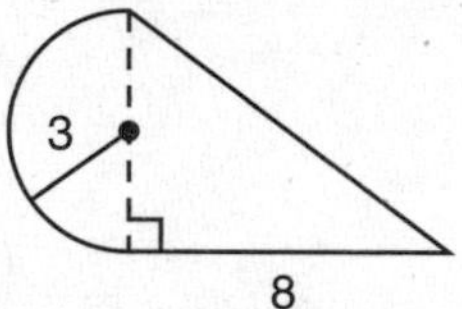

20. (105, 113) (a) Find the surface area of the cube shown. Dimensions are in feet.

(b) If the cube contains the largest pyramid it can hold, what is the volume of the pyramid?

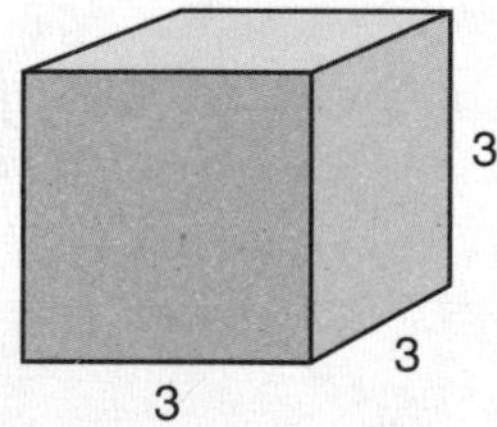

21. (95) Find the volume of this right circular cylinder. Dimensions are in meters. (Use 3.14 for π.)

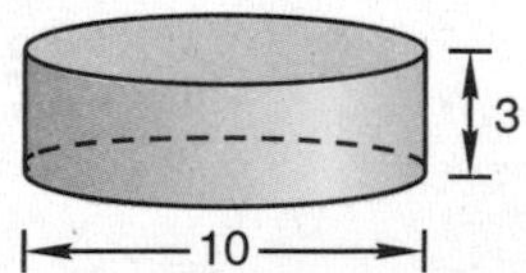

22. (40) Find the measures of the following angles:

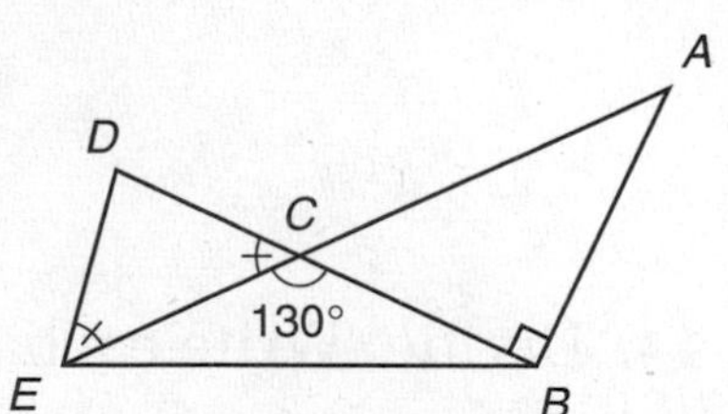

(a) $\angle ACB$ (b) $\angle CAB$ (c) $\angle CDE$

23. An aquarium that is 40 cm long, 10 cm wide, and 20 cm deep is filled with water. Find the volume of the water in the aquarium.
(70)

24. Solve: $0.8m - 1.2 = 6$
(93)

25. Solve this inequality and graph its solution:
(93)

$$3(x - 4) < x - 8$$

Simplify:

26. $4^2 \cdot 2^{-3} \cdot 2^{-1}$
(57)

27. 1 kilogram − 50 grams
(32)

28. $(1.2)\left(3\frac{3}{4}\right) \div 4\frac{1}{2}$
(43)

29. $2\frac{3}{4} - 1.5 - \frac{1}{6}$
(43)

30. $(-3)(-2) - (2)(-3) - (-8) + (-2)(-3) + |-5|$
(85)

LESSON

115 Volume, Capacity, and Mass in the Metric System

WARM-UP

Facts Practice: + – × ÷ Algebraic Terms (Test V)

Mental Math:

a. 10110 (base 2)

b. CLIV

c. 10^{-2}

d. $(4 \times 10^8) \div (4 \times 10^8)$

e. $\frac{1.44}{1.2} = \frac{1.2}{g}$

f. Convert 250 cm to m.

g. $\frac{2}{3}$ of \$1200

h. \$1200 reduced $\frac{1}{3}$

i. A nickel is how many cents less than 3 dimes and 3 quarters?

Problem Solving:

Three tennis balls just fit into a cylindrical container. What fraction of the volume of the container is occupied by the tennis balls?

NEW CONCEPT

Units of volume, capacity, and mass are closely related in the metric system. The relationships between these units are based on the physical characteristics of water under certain standard conditions. We state two commonly used relationships.

One milliliter of water has a volume of 1 cubic centimeter and a mass of 1 gram.

One cubic centimeter can contain 1 milliliter of water, which has a mass of 1 gram.

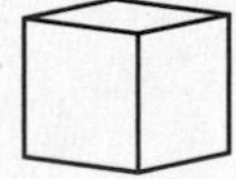

One liter of water has a volume of 1000 cubic centimeters and a mass of 1 kilogram.

One thousand cubic centimeters can contain 1 liter of water, which has a mass of 1 kilogram.

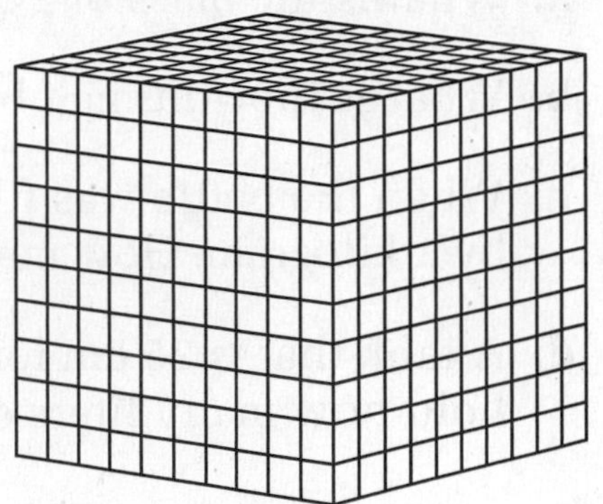

Example 1 Ray has a fish aquarium that is 50 cm long and 20 cm wide. If the aquarium is filled with water to a depth of 30 cm,

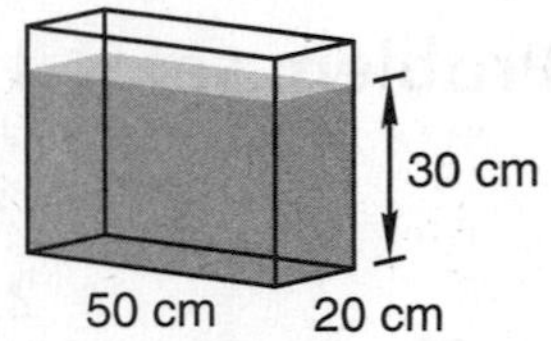

(a) how many liters of water would be in the aquarium?

(b) what would be the mass of the water in the aquarium?

Solution First we find the volume of the water in the aquarium.

$$(50\text{ cm})(20\text{ cm})(30\text{ cm}) = 30{,}000\text{ cm}^3$$

(a) Each cubic centimeter of water is 1 milliliter. Thirty thousand milliliters is **30 liters.**

(b) Each liter of water has a mass of 1 kilogram, so the mass of the water in the aquarium is **30 kilograms.** (Since a 1-kilogram mass weighs about 2.2 pounds on Earth, the water in the aquarium weighs about 66 pounds.)

Example 2 Malaika wanted to find the volume of a vase. She filled a 1-liter beaker with water and then used all but 240 milliliters to fill the vase.

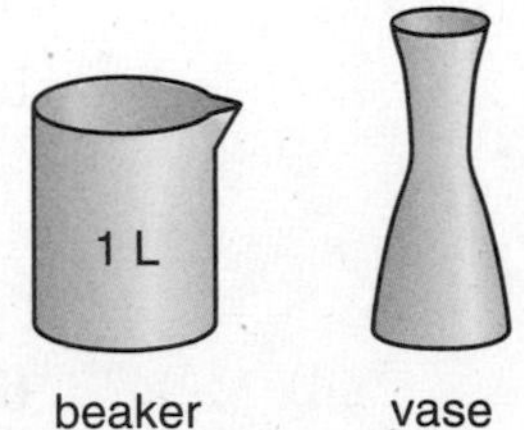

(a) What is the volume of the vase?

(b) If the mass of the vase is 640 grams, what is the mass of the vase filled with water?

Solution (a) The 1-liter beaker contains 1000 mL of water. Since Malaika used 760 mL, (1000 mL – 240 mL), the volume of the inside of the vase is **760 cm³.**

(b) The mass of the water (760 g) plus the mass of the vase (640 g) is **1400 g.**

LESSON PRACTICE

Practice set

a. What is the mass of 2 liters of water?

b. What is the volume of 3 liters of water?

c. When the bottle was filled with water, the mass increased by 1 kilogram. How many milliliters of water were added?

d. A tank that is 25 cm long, 10 cm wide, and 8 cm deep can hold how many liters of water?

MIXED PRACTICE

Problem set

1. (110) How much interest is earned in 9 months on a deposit of \$7000 at 8 percent simple interest?

2. (Inv. 10) With two tosses of a coin,

(a) what is the probability of getting two heads?

(b) what is the chance of getting two tails?

(c) what are the odds of getting heads, then tails?

3. (55) On the first 4 days of their trip, the Schmidts averaged 410 miles per day. On the fifth day they traveled 600 miles. How many miles per day did they average for the first 5 days of their trip?

4. (46) The 18-ounce container costs \$2.16. The 1-quart container costs \$3.36. The smaller container costs how much more per ounce than the larger container?

Use ratio boxes to solve problems 5 and 6.

5. (72) Eve typed 160 words in 5 minutes on her typing test. At that rate, how long would it take her to type an 800-word essay?

6. (65) The ratio of guinea pigs to rats running the maze was 7 to 5. Of the 120 guinea pigs and rats running the maze, how many were guinea pigs?

7. (74) Kelly was thinking of a certain number. If $\frac{3}{4}$ of the number was 48, what was $\frac{5}{8}$ of the number?

8. (92) A used car dealer bought a car for \$1500 and sold the car at a 40% markup. If the purchaser paid a sales tax of 8%, what was the total price of the car including tax?

9. (110) What is the sale price of an $80 skateboard after successive discounts of 25% and 20%?

10. (99) The points (–3, 4), (5, –2), and (–3, –2) are the vertices of a triangle.

(a) Find the area of the triangle.

(b) Find the perimeter of the triangle.

11. (115) A glass aquarium with the dimensions shown has a mass of 5 kg when empty. What is the mass of the aquarium when it is half full of water?

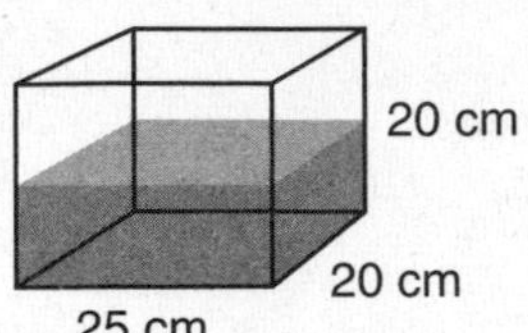

12. (48) Complete the table.

FRACTION	DECIMAL	PERCENT
(a)	0.875	(b)

13. (79) Compare: $a \div b \bigcirc a - b$ if a is positive and b is negative

14. (83, 111) Simplify and express each answer in scientific notation:

(a) $(6.4 \times 10^{6})(8 \times 10^{-8})$

(b) $\dfrac{6.4 \times 10^{6}}{8 \times 10^{-8}}$

15. (50) Use a unit multiplier to convert 36 inches to centimeters.

16. (108) (a) Solve for b: $A = \frac{1}{2}bh$

(b) Use the formula $A = \frac{1}{2}bh$ to find b when A is 24 and h is 6.

17. (Inv. 9, 107) Find three pairs of numbers that satisfy the function $y = -2x$. Then graph the number pairs on a coordinate plane, and draw a line through the points to show other number pairs that satisfy the function. What is the slope of the graphed line?

18. (104) Find the area of this figure. Dimensions are in millimeters. Corners that look square are square. (Use 3.14 for π.)

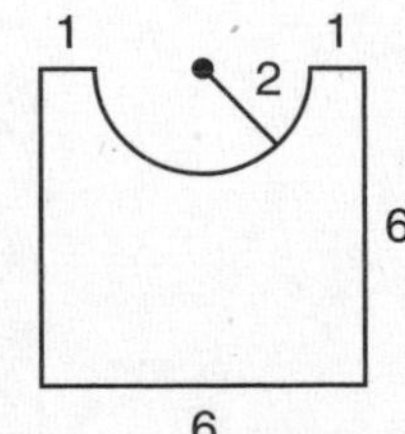

19. (a) Find the surface area of the cube.
(95, 105)
(b) Find the volume of the cube.
(c) How many meters long is each edge of the cube?

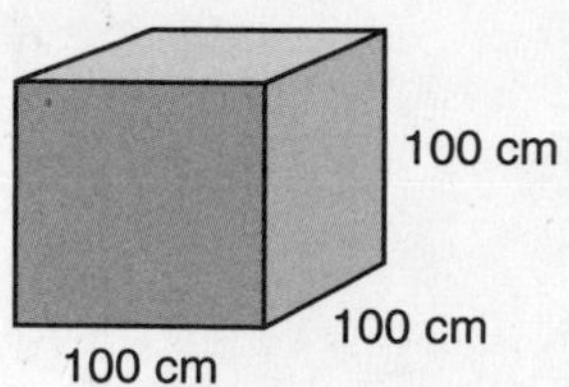

20. (a) Find the volume of the right circular cylinder. Dimensions are in inches.
(95, 113)
(b) If within the cylinder is the largest sphere it can contain, what is the volume of the sphere?

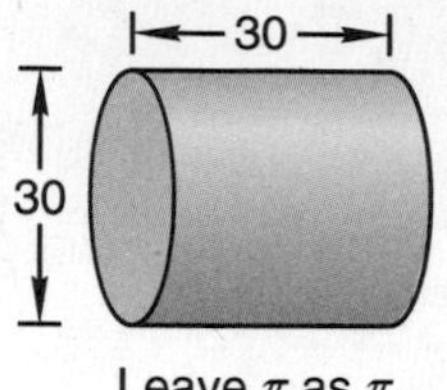

21. Find the measures of the following angles:
(40)

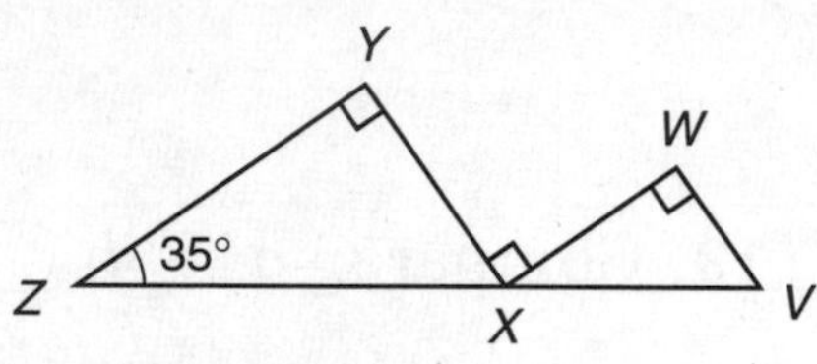

(a) $\angle YXZ$ (b) $\angle WXV$ (c) $\angle WVX$

22. In the figure in problem 21, ZX is 21 cm, YX is 12 cm, and XV is 14 cm. Write a proportion to find WV.
(97)

23. A pyramid is cut out of a plastic cube with dimensions as shown. What is the volume of the pyramid?
(113)

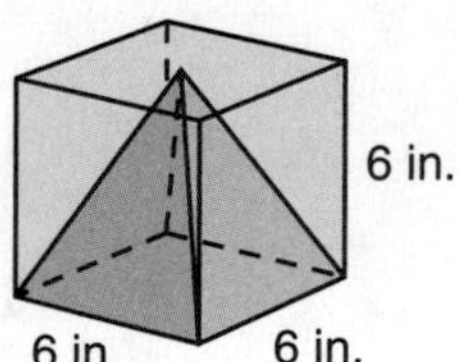

Solve:

24. $0.4n + 5.2 = 12$
(93)

25. $\dfrac{18}{y} = \dfrac{36}{28}$
(98)

Simplify:

26. $\sqrt{5^2 - 3^2} + \sqrt{5^2 - 4^2}$
(20)

27. 3 yd − 2 ft 1 in.
(56)

28. $3.5 \div \left(1\frac{2}{5} \div 3\right)$
(43)

29. $3.5 + 2^{-2} - 2^{-3}$
(57)

30. $\dfrac{(3)(-2)(4)}{(-6)(2)} + (-8) + (-4)(+5) - (2)(-3)$
(85)

LESSON

116 Factoring Algebraic Expressions

WARM-UP

Facts Practice: Multiplying and Dividing in Scientific Notation (Test W)

Mental Math:

a. 101010 (base 2)
b. DCCCXII
c. $(-2)^2 + 2^{-2}$
d. $(5 \times 10^5) \div (2 \times 10^2)$
e. $3x + 1.2 = 2.4$
f. Convert 1 m^2 to cm^2.
g. 125% of \$400
h. \$400 increased 25%
i. Estimate $8\frac{3}{4}$% sales tax on a \$41.19 purchase.

Problem Solving:

Here are the front, top, and side views of an object. Draw a three-dimensional view of the object from the perspective of the upper right front.

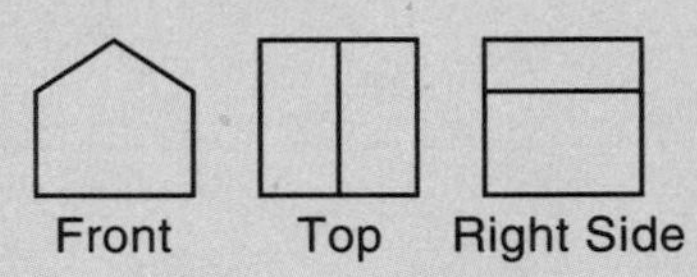

NEW CONCEPT

Algebraic expressions are classified as either **monomials** or **polynomials.** Monomials are single-term expressions such as the following three examples:

$$6x^2y^3 \qquad \frac{5xy}{2w} \qquad -6$$

Polynomials are composed of two or more terms. All of the following algebraic expressions are polynomials:

$$3x^2y + 6xy^2 \qquad x^2 + 2x + 1 \qquad 3a + 4b + 5c + d$$

Polynomials may be further classified by the number of terms they contain. For example, expressions with two terms are called binomials, and expressions with three terms are called trinomials. So $3x^2y + 6xy^2$ is a binomial, and $x^2 + 2x + 1$ is a trinomial.

Recall that to factor a monomial, we express the numerical part of the term as a product of prime factors, and we express the literal (letter) part of the term as a product of factors (instead of using exponents). Here we factor $6x^2y^3$:

$6x^2y^3$	original form
$(2)(3)xxyyy$	factored form

Some polynomials can also be factored. To factor a polynomial we first find the greatest common factor of the terms of the polynomial. Then we use the distributive property to write the expression as a product of the GCF and the remaining polynomial.

To factor $3x^2y + 6xy^2$, we first find the greatest common factor of $3x^2y$ and $6xy^2$. With practice we can find the GCF visually. This time we will factor both terms and circle the common factors.

$$\begin{array}{ccc} 3x^2y & + & 6xy^2 \\ \textcircled{3} \cdot \textcircled{x} \cdot x \cdot \textcircled{y} & + & 2 \cdot \textcircled{3} \cdot \textcircled{x} \cdot y \cdot \textcircled{y} \end{array}$$

We find that the GCF of $3x^2y$ and $6xy^2$ is $3xy$. Notice that removing $3xy$ from $3x^2y$ by division leaves x. Removing $3xy$ from $6xy^2$ by division leaves $2y$.

$\dfrac{3x^2y}{3xy} + \dfrac{6xy^2}{3xy}$	$3xy$ removed by division
$x + 2y$	remaining binomial

We write the factored form of $3x^2y + 6xy^2$ this way:

$3xy(x + 2y)$	factored form

Notice that we began with a binomial and ended with the GCF of its terms times a binomial.

Example 1 Factor the monomial $12a^2b^3c$.

Solution We factor 12 as (2)(2)(3), and we factor a^2b^3c as $aabbbc$.

$$12a^2b^3c = \mathbf{(2)(2)(3)}\boldsymbol{aabbbc}$$

Example 2 Factor the trinomial $6a^2b + 4ab^2 + 2ab$.

Solution First we find the greatest common factor of the three terms. Often we can do this visually. Notice that each term has 2 as a factor, a as a factor, and b as a factor. So the GCF is $2ab$. Next we divide each term of the trinomial by $2ab$ to find what remains of each term after $2ab$ is factored out of the expression.

$$\frac{6a^2b}{2ab} + \frac{4ab^2}{2ab} + \frac{2ab}{2ab} \qquad 2ab \text{ removed by division}$$

$$3a + 2b + 1 \qquad \text{remaining trinomial}$$

Notice that the third term is 1—not zero. This is because we divided $2ab$ by $2ab$; we did not subtract.

Now we write the factored expression in this form:

GCF(remaining polynomial)

The GCF is $2ab$ and the remaining trinomial is $3a + 2b + 1$.

$$\mathbf{2ab(3a + 2b + 1)}$$

LESSON PRACTICE

Practice set Factor each algebraic expression:

a. $8m^2n$

b. $12mn^2$

c. $18x^3y^2$

d. $8m^2n + 12mn^2$

e. $8xy^2 - 4xy$

f. $6a^2b^3 + 9a^3b^2 + 3a^2b^2$

MIXED PRACTICE

Problem set

1. (Inv. 10) If a pair of dot cubes is tossed once,

(a) what is the probability of rolling a total of 5 (expressed as a decimal rounded to two decimal places)?

(b) what is the chance of rolling a total of either 4 or 7?

(c) what are the odds of rolling a total of 12?

2. (20) A kilobyte of memory is 2^{10} bytes. Express the number of bytes in a kilobyte in standard form.

3. (92) Which sign seems to advertise the better sale? Explain your choice.

Sale! **40% off the regular price!**	**Sale!** **40% of the regular price!**

4. (48) Complete the table.

FRACTION	DECIMAL	PERCENT
(a)	(b)	175%
$\frac{1}{12}$	(c)	(d)

5. (80) Triangle *ABC* with vertices *A* (0, 3), *B* (0, 0), and *C* (4, 0) is rotated 180° about the origin to $\Delta A'B'C'$. What are the coordinates of the vertices of $\Delta A'B'C'$?

6. (89) What is the measure of each exterior angle and each interior angle of a regular 20-gon?

7. (92) At a 30%-off sale Rob bought a jacket for $42. How much money did Rob save by buying the jacket on sale instead of paying the regular price?

8. (115) The figure illustrates an aquarium with interior dimensions as shown.

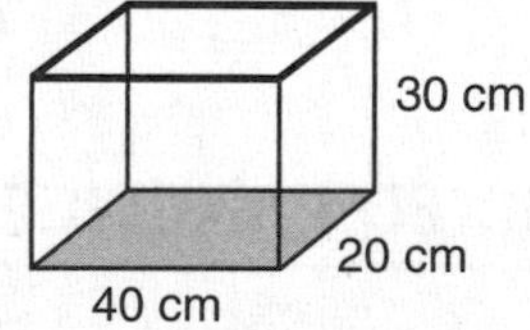

(a) The aquarium has a maximum capacity of how many liters?

(b) If the aquarium is filled with water, what would be the mass of the water in the aquarium?

9. (50) Use a unit multiplier to convert 24 kg to lb. (Use the approximation 1 kg ≈ 2.2 lb.)

10. (101) Write an equation to solve this problem:

Six less than twice what number is 48?

11. Find the measure of the largest angle of the triangle shown.
(101)

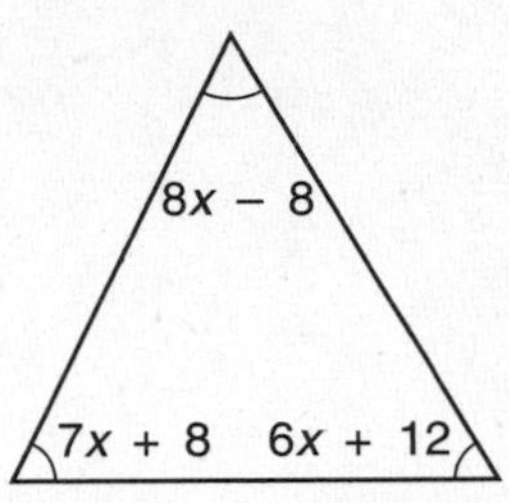

12. Solve for C: $F = 1.8C + 32$
(106)

13. The inside surface of this archway will be covered with a strip of wallpaper. How long must the strip of wallpaper be in order to reach from the floor on one side of the archway around to the floor on the other side of the archway? Use 3.14 for π and round up to the nearest inch.
(104)

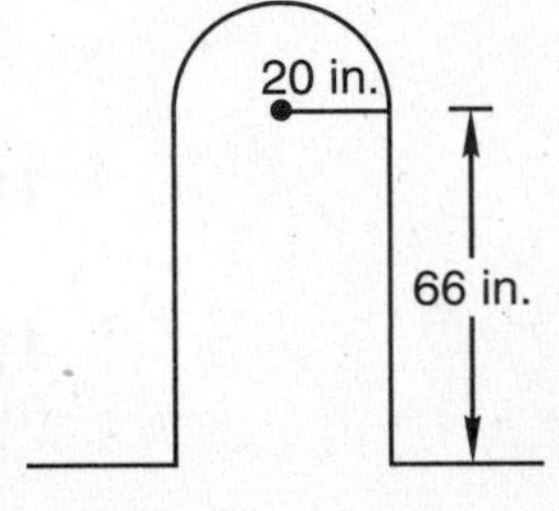

14. What is the total surface area of the right triangular prism below?
(105)

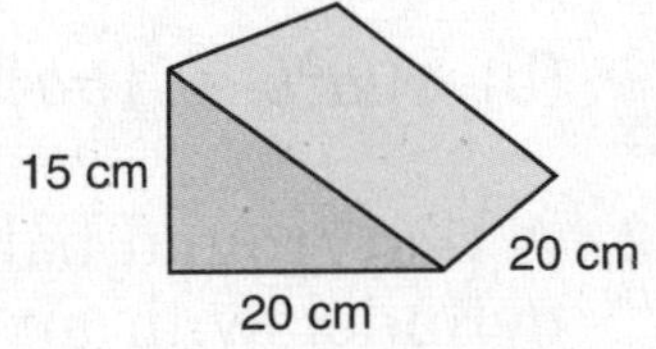

15. What is the volume of the right triangular prism in problem 14?
(95)

16. Find the slope of each line and the point where each line intersects the y-axis:
(107)

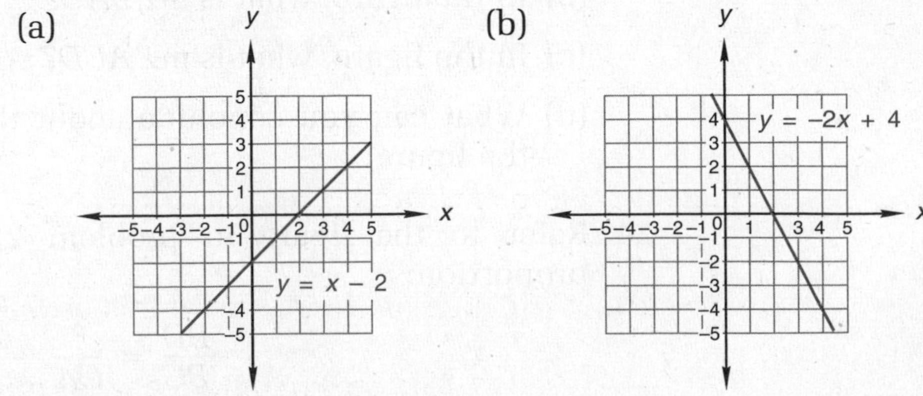

17. (108) The following formula can be used to find the area, A, of a trapezoid. The lengths of the parallel sides are a and b, and the height, h, is the perpendicular distance between the parallel sides.

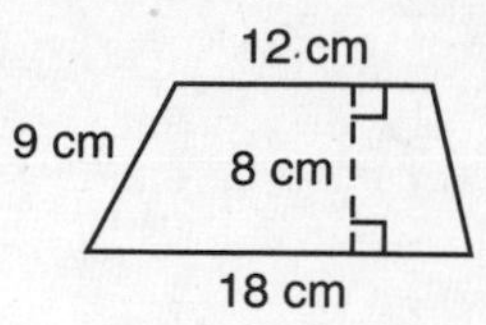

$$A = \frac{1}{2}(a + b)h$$

Use this formula to find the area of the trapezoid shown above.

18. (109) Find two solutions for $3x^2 - 5 = 40$.

19. (111) Express each quotient in scientific notation:

(a) $\dfrac{8 \times 10^{-4}}{4 \times 10^{8}}$ (b) $\dfrac{4 \times 10^{8}}{8 \times 10^{-4}}$

20. (83) What is the product of the two quotients in problem 19? Why?

21. (116) Factor each algebraic expression:

(a) $9x^2y$

(b) $10a^2b + 15a^2b^2 + 20abc$

22. (113) A playground ball just fits inside a cylinder with an interior diameter of 12 in. What is the volume of the ball? Use 3.14 for π and round the answer to the nearest cubic inch.

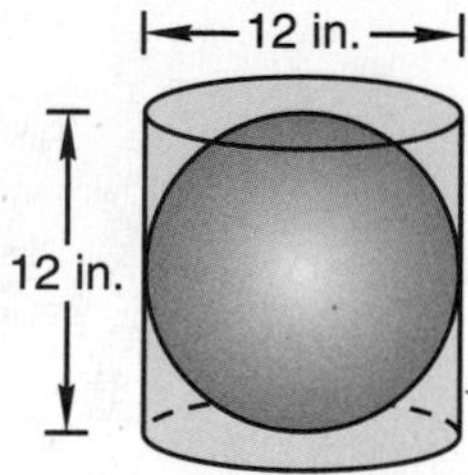

23. (40) (a) In the figure, what is m∠BCD?

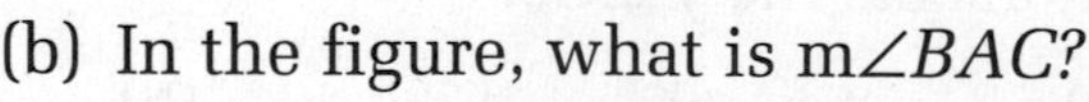
(b) In the figure, what is m∠BAC?

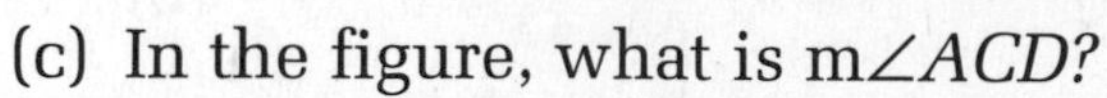
(c) In the figure, what is m∠ACD?

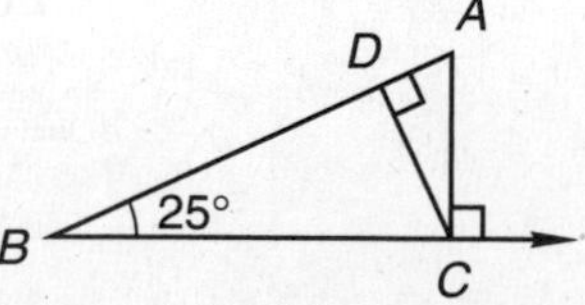

(d) What can you conclude about the three triangles in the figure?

24. (97) Refer to the figure in problem 23 to complete this proportion:

$$\frac{BD}{BC} = \frac{?}{CA}$$

Solve:

25. $x - 15 = x + 2x + 1$
(102)

26. $0.12(m - 5) = 0.96$
(102)

Simplify:

27. $a(b - c) + b(c - a)$
(96)

28. $\dfrac{(8x^2y)(12x^3y^2)}{(4xy)(6y^2)}$
(103)

29. (a) $(-3)^2 + (-2)(-3) - (-2)^3$
(103, 105)
(b) $\sqrt[3]{-8} + \sqrt[3]{8}$

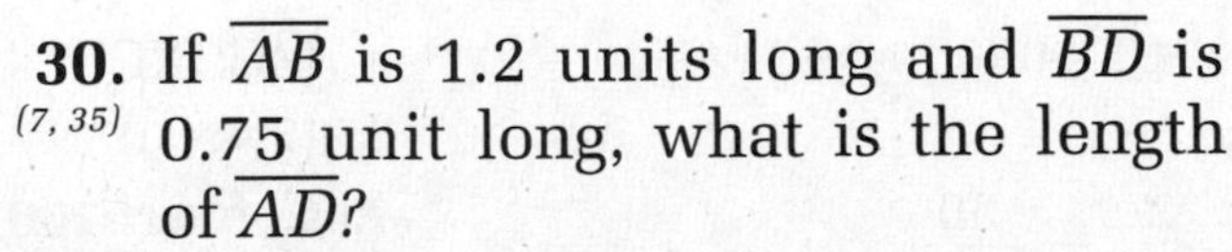

30. If $\overline{AB}$ is 1.2 units long and $\overline{BD}$ is 0.75 unit long, what is the length of $\overline{AD}$?
(7, 35)

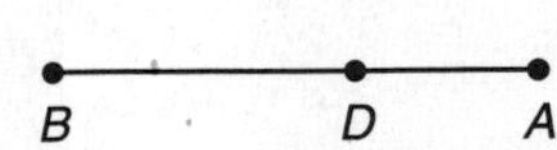

LESSON

117 Slope-Intercept Form of Linear Equations

WARM-UP

Facts Practice: + − × ÷ Algebraic Terms (Test V)

Mental Math:

a. 1000000 (base 2)

b. MCDXCII

c. $\frac{(-9)(-4)}{-6}$

d. $(7 \times 10^{-4}) \div (2 \times 10^{-6})$

e. $2a^2 = 50$

f. Convert 100°C to Fahrenheit.

g. $12\frac{1}{2}\%$ of \$4000

h. $12\frac{1}{2}\%$ less than \$4000

i. Find 10% of 60, + 4, × 8, + 1, $\sqrt{}$, × 3, + 1, ÷ 4, × 5, + 1, $\sqrt{}$, − 7.

Problem Solving:

In this 4-by-4 square we see sixteen 1-by-1 squares, nine 2-by-2 squares, four 3-by-3 squares, and one 4-by-4 square. How many squares of any size are in this 6-by-6 square?

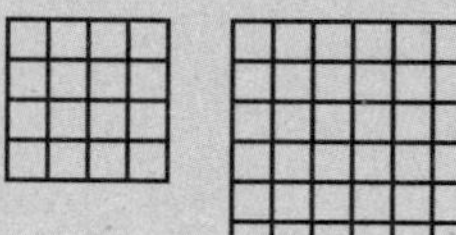

NEW CONCEPT

The three equations below are equivalent equations. Each equation has the same graph.

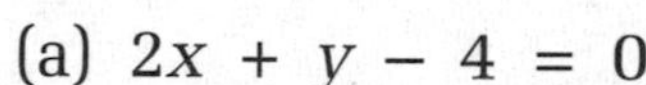

(a) $2x + y - 4 = 0$

(b) $2x + y = 4$

(c) $y = -2x + 4$

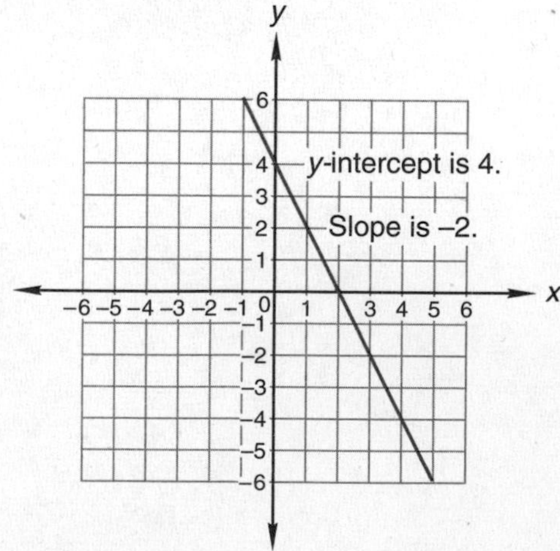

Equation (c) is in a special form called **slope-intercept form.** When an equation is in slope-intercept form, the coefficient of x is the slope of the graph of the equation, and the constant is the ***y*-intercept** (where the graph of the equation intercepts the *y*-axis).

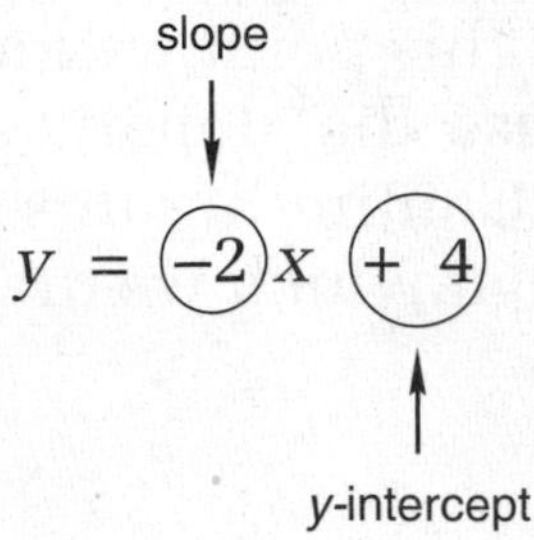

Notice the order of the terms in this equation. The equation is solved for *y*, and *y* is to the left of the equal sign. To the right of the equal sign is the *x*-term and then the constant term. The model for slope-intercept form is written this way:

Slope-Intercept Form
$\boldsymbol{y = mx + b}$

In this model, *m* stands for the slope and *b* for the *y*-intercept.

Example 1 Transform this equation so that it is in slope-intercept form.

$$3x + y = 6$$

Solution We solve the equation for *y* by subtracting $3x$ from both sides of the equation.

$3x + y = 6$	equation
$3x + y - 3x = 6 - 3x$	subtracted $3x$ from both sides
$y = 6 - 3x$	simplified

Next, using the commutative property, we rearrange the terms on the right side of the equal sign so that the *x*-term precedes the constant term.

$y = 6 - 3x$	equation
$\boldsymbol{y = -3x + 6}$	commutative property

Example 2 Graph $y = -3x + 6$ using the slope and y-intercept.

Solution The slope of the graph is the coefficient of x, which is –3, and the y-intercept is +6, which is located at +6 on the y-axis. From this point we move to the right 1 unit and down 3 units because the slope is –3. This gives us another point on the line. Continuing this pattern, we identify a series of points through which we draw the graph of the equation.

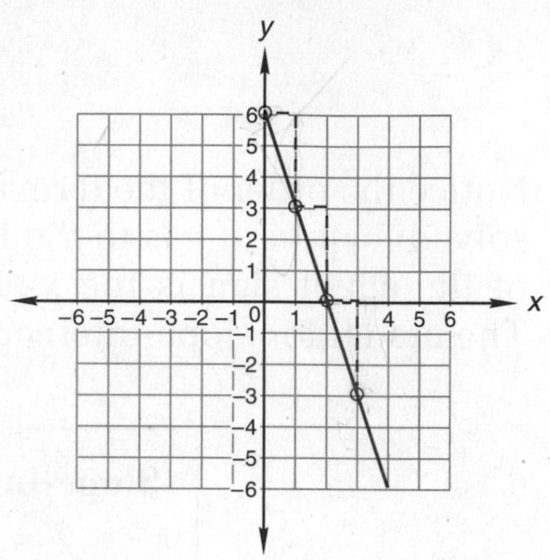

Example 3 Using only slope and y-intercept, graph $y = x - 2$.

Solution The slope is the coefficient of x, which is +1. The y-intercept is –2. We begin at –2 on the y-axis and sketch a line that has a slope of +1.

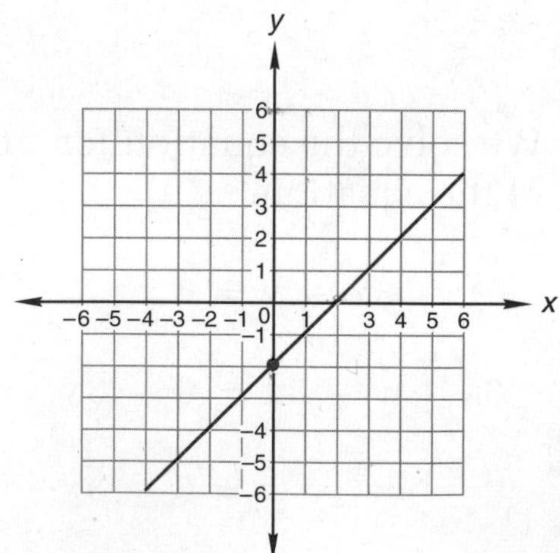

LESSON PRACTICE

Practice set Write each equation below in slope-intercept form:

a. $2x + y = 3$

b. $y - 3 = x$

c. $2x + y - 3 = 0$

d. $x + y = 4 - x$

Using only slope and y-intercept, graph each of these equations:

e. $y = x - 3$ **f.** $y = -2x + 6$

g. $y = \frac{1}{2}x - 2$ **h.** $y = -x + 3$

MIXED PRACTICE

Problem set

1. (110) How much interest is earned in four years on a deposit of \$10,000 if it is allowed to accumulate in an account paying 7% interest compounded annually?

2. (Inv. 10) In 240 at-bats Chester has 60 hits.

(a) What is the statistical probability that Chester will get a hit in his next at-bat?

(b) What are the odds of Chester getting a hit in his next at-bat?

3. (55) On her first four tests Monica's average score was 75%. On her next six tests Monica's average score was 85%. What was Monica's average score on all ten tests?

4. (48) Complete the table.

FRACTION	DECIMAL	PERCENT
(a)	1.4	(b)
$\frac{11}{12}$	(c)	(d)

5. (80) The image of ΔABC reflected in the y-axis is $\Delta A'B'C'$. If the coordinates of vertices A, B, and C are $(-1, 3)$, $(-3, 0)$, and $(0, -2)$, respectively, then what are the coordinates of vertices A', B', and C'?

6. (89) The figure at right shows regular octagon $ABCDEFGH$.

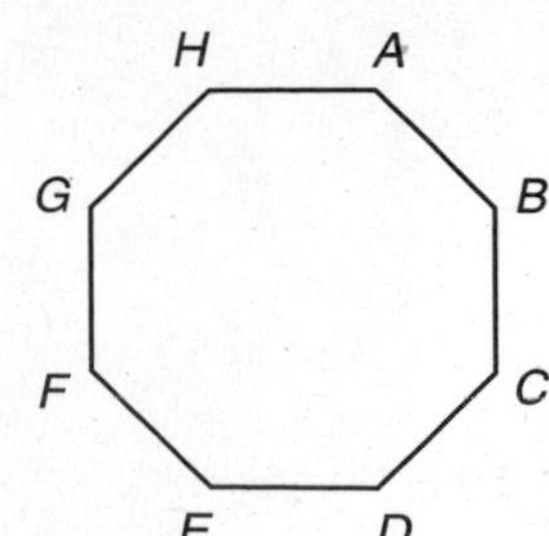

(a) What is the measure of each exterior angle?

(b) What is the measure of each interior angle?

(c) How many diagonals can be drawn from vertex A?

7. (92) In one year the population in the county surged from 1.2 million to 1.5 million. This was an increase of what percent?

8. (115) A beaker is filled with water to the 500 mL level.

(a) What is the volume of the water in cubic centimeters?

(b) What is the mass of the water in kilograms?

9. (88) Use two unit multipliers to convert 540 ft^2 to yd^2.

10. (101) Write an equation to solve this problem:

Six more than three times what number squared is 81?

11. (101) Find the measure of the angle marked y in the figure shown.

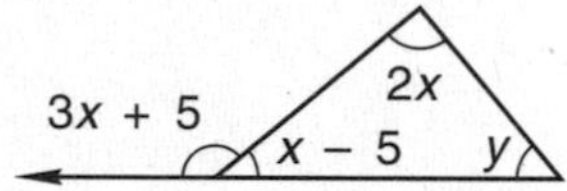

12. (106) Solve for c^2: $c^2 - a^2 = b^2$

13. (Inv. 10) The face of this spinner is divided into four sectors. Sectors B and D are 90° sectors, and sector C is a 120° sector. If the arrow is spun once,

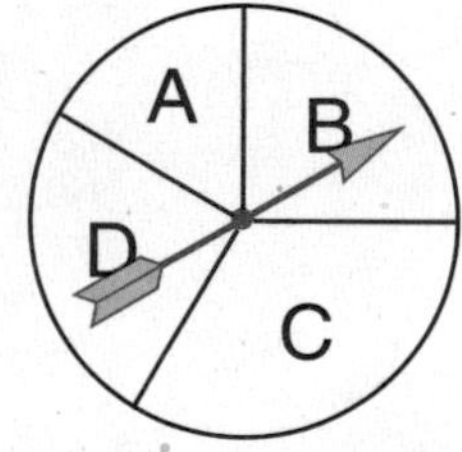

(a) what is the probability (expressed as a decimal) that it will stop in sector B?

(b) what is the chance that it will stop in sector C?

(c) what are the odds that it will stop in sector A?

14. (112) The coordinates of the vertices of a square are (0, 4), (3, 0), (−1, −3), and (−4, 1).

(a) What is the length of each side of the square?

(b) What is the perimeter of the square?

(c) What is the area of the square?

15. (113) A right circular cylinder and a cone have an equal height and an equal diameter as shown.

(a) What is the volume of the cylinder?

(b) What is the volume of the cone?

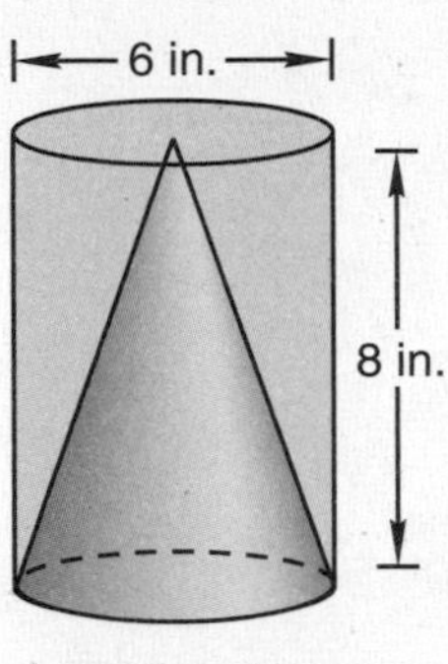

16. (106) The formula for the volume of a rectangular prism is

$$V = lwh$$

(a) Transform this formula to solve for h.

(b) Find h when V is 6000 cm^3, l is 20 cm, and w is 30 cm.

17. (117) Refer to the graph shown below to answer (a)–(c).

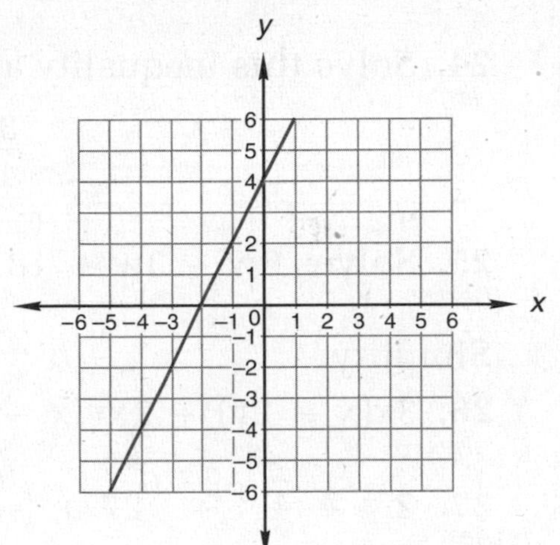

(a) What is the slope of the line?

(b) At what point does the line intersect the y-axis?

(c) What is the equation of the line in slope-intercept form?

18. (117) Write each equation in slope-intercept form:

(a) $y + 5 = x$ (b) $2x + y = 4$

19. (116) Factor each algebraic expression:

(a) $24xy^2$ (b) $3x^2 + 6xy - 9x$

20. (83) Find the area of a square with sides 5×10^3 mm long. Express the area

(a) in scientific notation.

(b) as a standard numeral.

21. (88) Use two unit multipliers to convert the answer to problem 20(b) to square meters.

22. (97) Triangle ABC is a right triangle and is similar to triangles CAD and CBD.

(a) Which side of ΔCBD corresponds to side BC of ΔABC?

(b) Which side of ΔCAD corresponds to side AC of ΔABC?

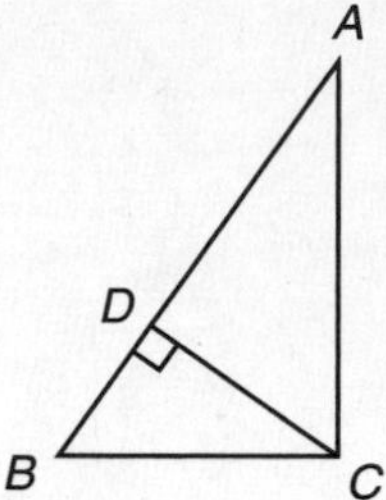

23. (7, 30) Refer to the figure below to find the length of segment BD.

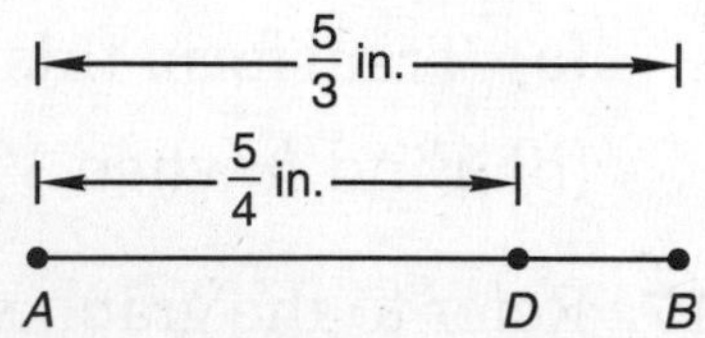

24. (93) Solve this inequality and graph its solution:

$$\frac{3}{4}x + 12 < 15$$

25. (102) Solve: $6w - 3w + 18 = 9(w - 4)$

Simplify:

26. (96) $3x(x - 2y) + 2xy(x + 3)$

27. (57, 103, 105) $2^{-2} + 4^{-1} + \sqrt[3]{27} + (-1)^3$

28. (85) $(-3) + (-2)[(-3)(-2) - (+4)] - (-3)(-4)$

29. (111) $\dfrac{1.2 \times 10^{-6}}{4 \times 10^3}$

30. (103) $\dfrac{36a^2b^3c}{12ab^2c}$

LESSON

118 Copying Angles and Triangles

WARM-UP

Facts Practice: Multiplying and Dividing in Scientific Notation (Test W)

Mental Math:

a. 101011 (base 2)

b. MDCCLXXVI

c. $(-3)^2 + 3^{-2}$

d. $(5 \times 10^{-6})(3 \times 10^{2})$

e. $\frac{k}{33} = \frac{200}{300}$

f. Convert 7500 g to kg.

g. 150% of $4000

h. $4000 increased 150%

i. At an average speed of 30 mph, how long will it take to drive 40 miles?

Problem Solving:

Sylvia wants to pack a 9-by-14-in. rectangular picture frame that is $\frac{1}{2}$-in. thick into a rectangular box. The box has inside dimensions of 12-by-9-by-10 in. Describe why you think the frame will or will not fit into the box.

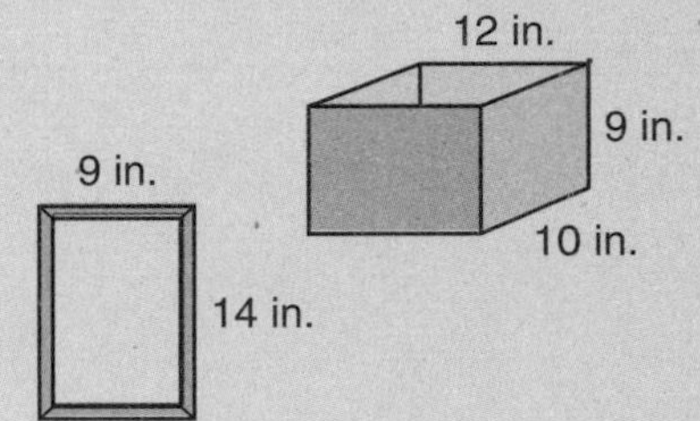

NEW CONCEPT

Recall from Investigations 2 and 8 that we used a compass and straightedge to construct circles, regular polygons, angle bisectors, and perpendicular bisectors of segments. We may also use a compass and straightedge to copy figures. In this lesson we will practice copying angles and triangles.

Suppose we are given this angle to copy:

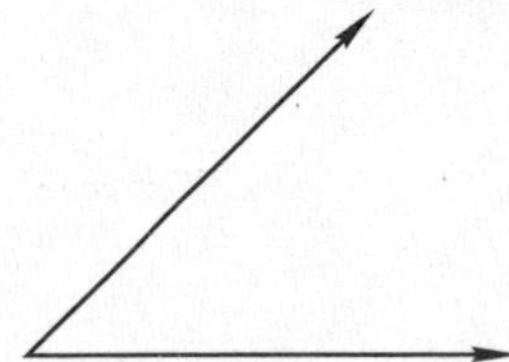

We begin by drawing a ray to form one side of the angle.

Now we need to find a point through which to draw the second ray. We find this point in two steps. First we set the compass and draw an arc across both rays of the original angle from the vertex of the angle. Without resetting the compass, we then draw an arc of the same size from the endpoint of the ray, as we show here.

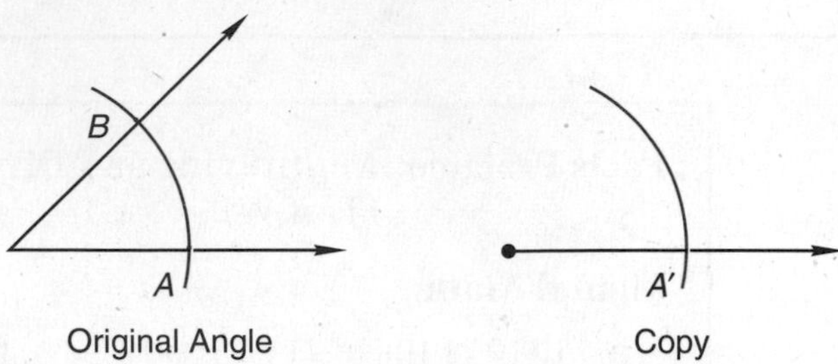

For the second step, we reset the compass to equal the distance from A to B on the original angle. To verify the correct setting, we swing a small arc through point B while the pivot point is on point A. With the compass at this setting, we move the pivot point to point A' of the copy and draw an arc that intersects the first arc we drew on the copy.

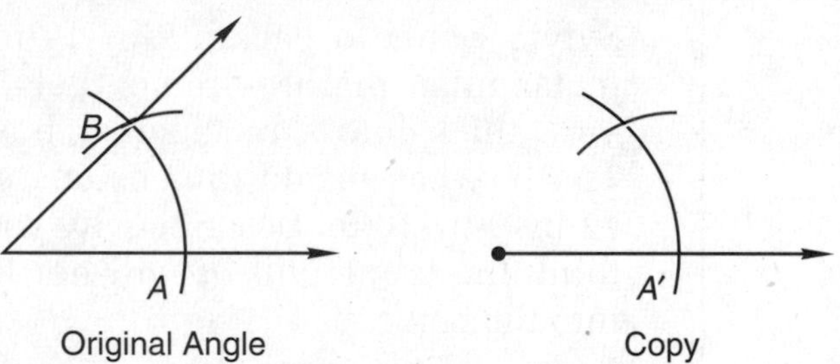

As a final step, we draw the second ray of the copied angle through the point at which the arcs intersect.

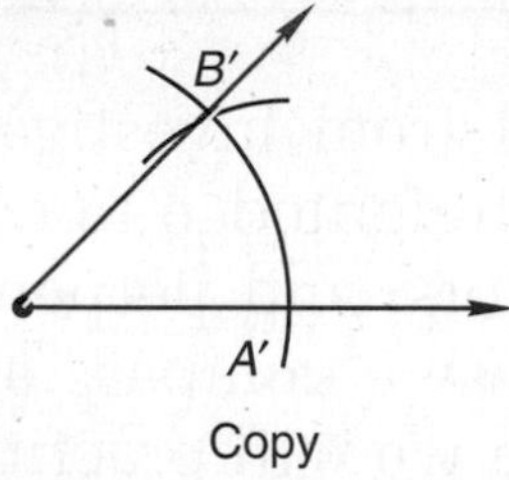

We use a similar method to copy a triangle. Suppose we are asked to copy ΔXYZ.

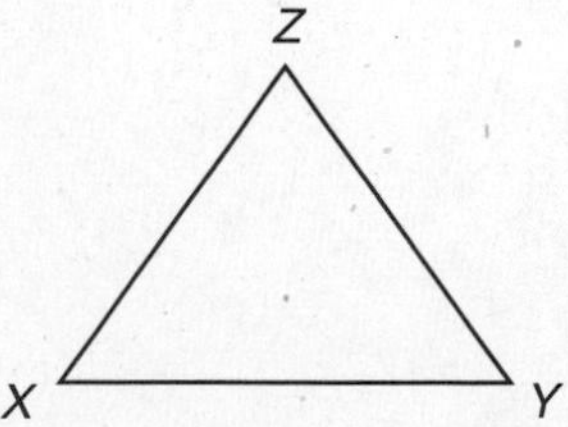

We will begin by drawing a segment equal in length to segment XY. We do this by setting the compass so that the pivot point is on X and the drawing point is on Y. We verify the setting by drawing a small arc through point Y. To copy this segment, we first sketch a ray with endpoint X'. Then we locate Y' by swinging an arc with the preset compass from point X'.

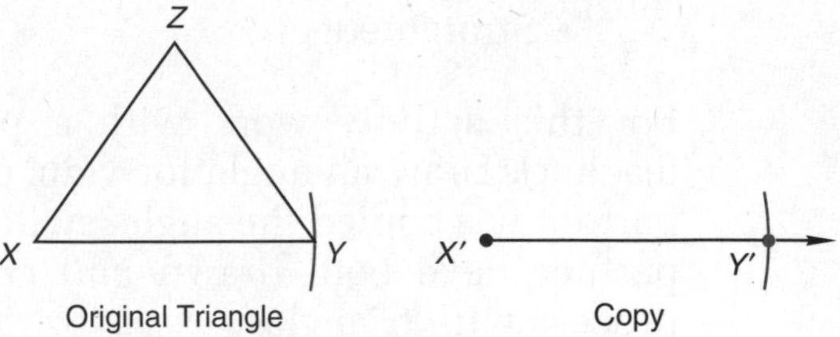

To locate Z' on the copy, we will need to draw two different arcs, one from point X' and one from point Y'. We set the compass on the original triangle so that the distance between its points equals XZ. With the compass at this setting, we draw an arc from X' on the copy.

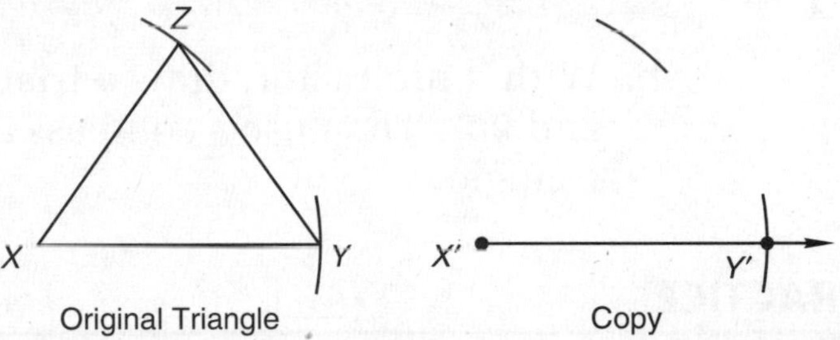

Now we change the setting of the compass to equal YZ on the original. With this setting we draw an arc from Y' that intersects the other arc.

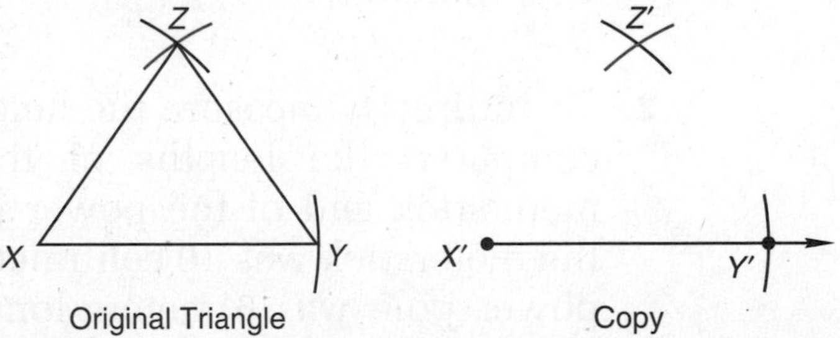

The point where the arcs intersect, which we have labeled Z', corresponds to point Z on the original triangle. To complete the copy, we draw segments $X'Z'$ and $Y'Z'$.

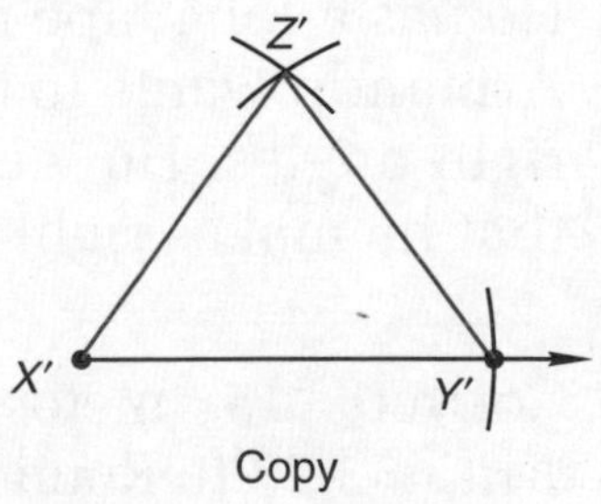

Activity: *Copying Angles and Triangles*

Materials needed:

- Compass
- Straightedge

For this activity work with a partner (who may be your teacher). Draw an angle for your partner to copy. When your partner has copied the angle, switch roles. After you and your partner have both drawn and copied an angle, repeat the process with triangles.

LESSON PRACTICE

Practice set

a. Use a protractor to draw an 80° angle. Then use a compass and straightedge to copy the angle.

b. With a protractor, draw a triangle with angles of 30°, 60°, and 90°. Then use a compass and straightedge to copy the triangle.

MIXED PRACTICE

Problem set

1. (92) The median home price in the county increased from $180,000 to $189,000 in one year. This was an increase of what percent?

2. (97) To indirectly measure the height of a power pole, Teddy compared the lengths of the shadows of a vertical meterstick and of the power pole. When the shadow of the meterstick was 40 centimeters long, the shadow of the power pole was 6 meters long. About how tall was the power pole?

3. (112) Armando is marking off a grass field for a soccer game. He has a long tape measure and chalk for lining the field. Armando wants to be sure that the corners of the field are right angles. How can he use the tape measure to ensure that he makes right angles?

4. (50) Convert 15 meters to feet using the approximation 1 m ≈ 3.28 ft. Round the answer to the nearest foot.

5. (98) The illustration below shows one room of a scale drawing of a house. One inch on the drawing represents a distance of 10 feet. Use a ruler to help calculate the actual area of the room.

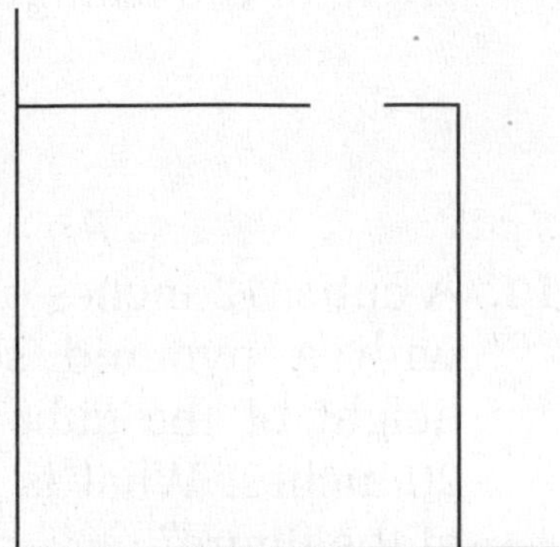

6. (Inv. 10) If a pair of dot cubes is tossed once,

(a) what is the probability of rolling a total of 9?

(b) what is the chance of rolling a total of 10?

(c) what are the odds of rolling a total of 11?

7. (99) Use the Pythagorean theorem to find the distance from (4, 6) to (–1, –6).

8. (115) A two-liter bottle filled with water

(a) contains how many cubic centimeters of water?

(b) has a mass of how many kilograms?

9. (101) Write an equation to solve this problem:

Two thirds less than half of what number is five sixths?

10. (102) In this figure lines *m* and *n* are parallel. If the sum of the measures of angles *a* and *e* is 200°, what is the measure of $\angle g$?

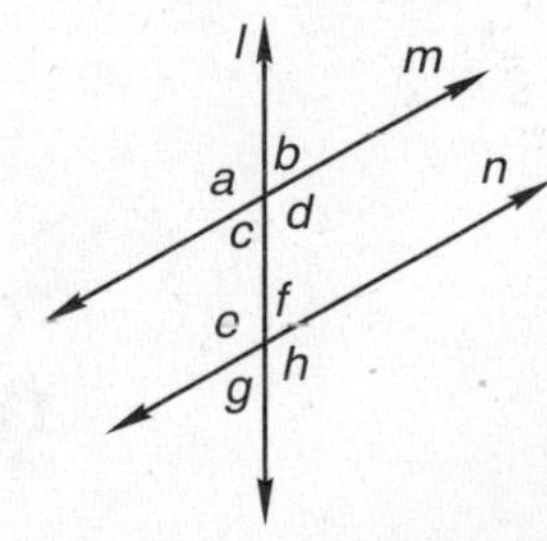

11. (117) Transform the equation $3x + y = 6$ into slope-intercept form. Then graph the equation on a coordinate plane.

12. (101) Find the measure of the smallest angle of the triangle shown.

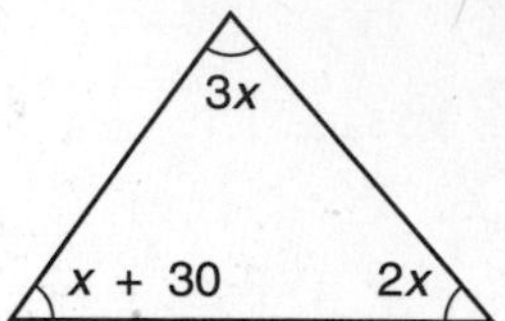

13. (113) A cube, 12 inches on edge, is topped with a pyramid so that the total height of the cube and pyramid is 20 inches. What is the total volume of the figure?

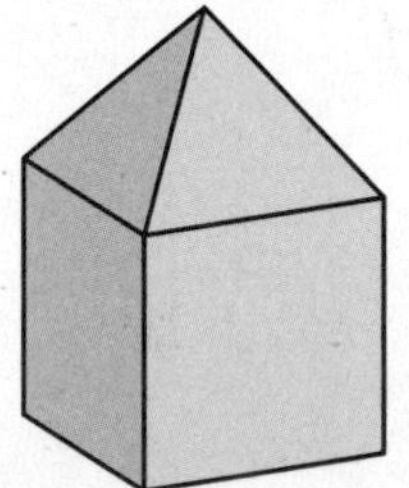

14. (101) The length of segment *BD* is 12. The length of segment *BA* is *c*. Using 12 and *c*, write an expression that indicates the length of segment *AD*.

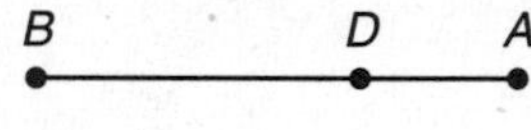

15. (97) The three triangles in the figure shown are similar. The sum of x and y is 25. Use proportions to find x and y.

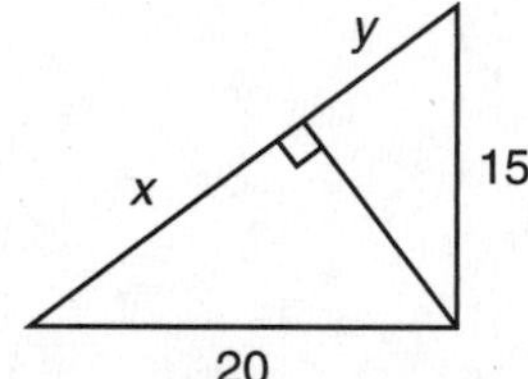

16. (105) Lina cut a grapefruit in half. (The flat surface formed is called a **cross section.**)

Lina knew that the surface area of a sphere is four times the greatest cross-sectional area of the sphere. She estimated that the diameter of the grapefruit was 8 cm, and she used 3 in place of π. Using Lina's numbers, estimate the area of the whole grapefruit peel.

17. (117) Write the equation $y - 2x + 5 = 1$ in slope-intercept form. Then graph the equation.

18. Refer to the graph shown below to answer (a)–(c).
(117)

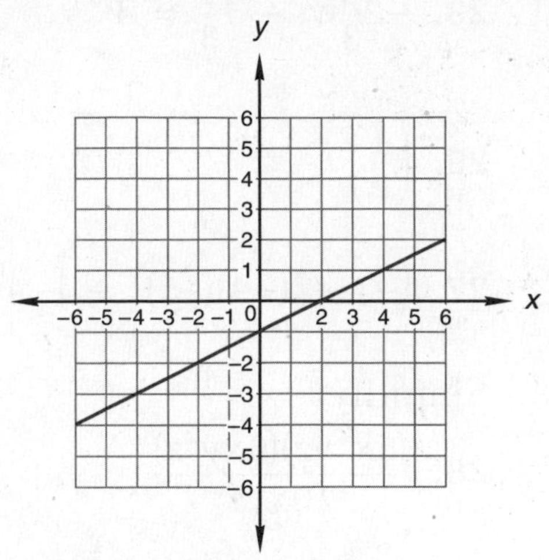

(a) What is the slope of the line?

(b) What is the y-intercept of the line?

(c) What is the equation of the line in slope-intercept form?

19. Draw an estimate of a 60° angle, and check your estimate with a protractor. Then set the protractor aside, and use a compass and straightedge to copy the angle.
(96, 118)

20. A semicircle with a 7-inch diameter was cut from a rectangular half sheet of paper. What is the perimeter of the resulting shape? (Use $\frac{22}{7}$ for π.)
(104)

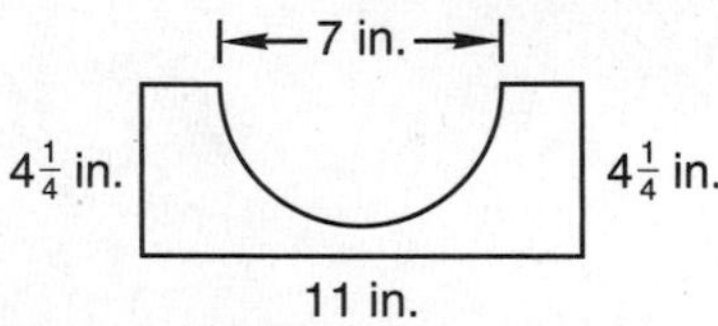

21. A dime is about 1×10^{-3} m thick. A kilometer is 1×10^{3} m. About how many dimes would be needed to make a stack of dimes one kilometer high? Express the answer in scientific notation.
(111)

22. Factor each algebraic expression:
(116)

(a) $x^2 + x$

(b) $12m^2n^3 + 18mn^2 - 24m^2n^2$

Solve:

23. $-2\frac{2}{3}w - 1\frac{1}{3} = 4$ *(93)*

24. $5x^2 + 1 = 81$ *(109)*

25. $\left(\frac{1}{2}\right)^2 - 2^{-2}$ *(57)*

26. $66\frac{2}{3}\%$ of $\frac{5}{6}$ of 0.144 *(48)*

27. $[-3 + (-4)(-5)] - [-4 - (-5)(-2)]$ *(91)*

Simplify:

28. $\frac{(5x^2yz)(6xy^2z)}{10xyz}$ *(103)*

29. $x(x + 2) + 2(x + 2)$ *(96)*

30. *(100)* The length of the hypotenuse of this right triangle is between which two consecutive whole numbers of millimeters?

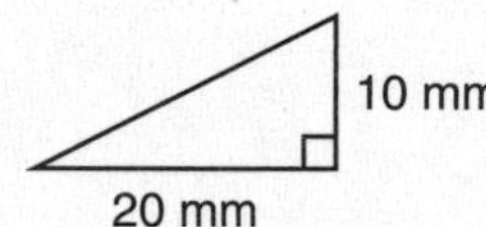

LESSON

119 Division by Zero

WARM-UP

Facts Practice: + − × ÷ Algebraic Terms (Test V)

Mental Math:

a. 11011 (base 2)

b. MLXVI

c. $(2^{-2})(-2)^2$

d. $(1 \times 10^{-8}) \div (1 \times 10^{-4})$

e. $2y + \frac{1}{2} = \frac{1}{2}$

f. Convert 5 cm^2 to mm^2.

g. $66\frac{2}{3}\%$ of \$600

h. \$600 reduced $33\frac{1}{3}\%$

i. What fraction of an hour is 5 minutes less than $\frac{1}{3}$ of an hour?

Problem Solving:

Figure *ABC* is an equilateral triangle whose perimeter is 6 cm. Segment *AD* bisects segment *BC* to form two congruent right triangles. Find the length of segment *AD,* and leave the answer in irrational form.

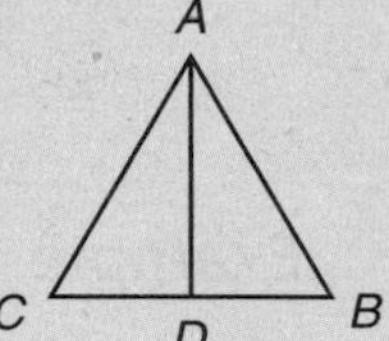

NEW CONCEPT

When performing algebraic operations, it is necessary to guard against dividing by zero. For example, the following expression reduces to 2 only if *x* is not zero:

$$\frac{2x}{x} = 2 \quad \text{if } x \neq 0$$

What is the value of this expression if *x* is zero?

$$\frac{2x}{x} \qquad \text{expression}$$

$$\frac{2 \cdot 0}{0} \qquad \text{substituted 0 for } x$$

$$\frac{0}{0} \qquad \text{multiplied } 2 \cdot 0$$

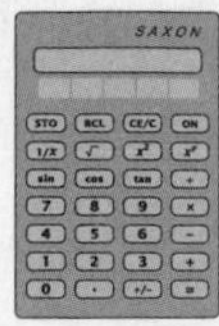

What is the value of $\frac{0}{0}$? How many zeros are in zero? Is the quotient 0? Is the quotient 1? Is the quotient some other number? Try the division with a calculator. What answer does the calculator display? Notice that the calculator displays an error message when division by zero is entered. The display is frozen and other calculations cannot be performed until the erroneous entry is cleared. In this lesson we will consider why division by zero is not possible.

Consider what happens to a quotient when a number is divided by numbers closer and closer to zero. As we know, zero lies on the number line between –1 and 1. Zero is also between –0.1 and 0.1, and between –0.01 and 0.01.

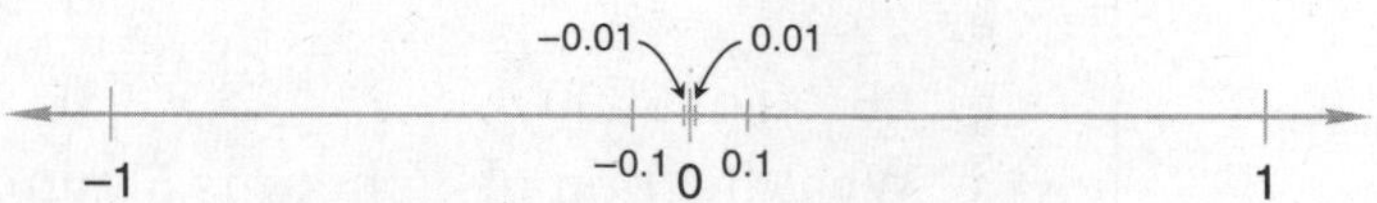

In the following example, notice the quotients we get when we divide a number by numbers closer and closer to zero.

Example 1 Find each set of quotients. As the divisors become closer to zero, do the quotients become closer to zero or farther from zero?

(a) $\frac{10}{1}, \frac{10}{0.1}, \frac{10}{0.01}$

(b) $\frac{10}{-1}, \frac{10}{-0.1}, \frac{10}{-0.01}$

Solution (a) **10, 100, 1000**

(b) **–10, –100, –1000**

As the divisors become closer to zero, the quotients become farther from zero.

Notice from example 1 that as the divisors approach zero from the positive side, the quotients become greater and greater toward positive infinity (+∞). However, as the divisors approach zero from the negative side, the quotients become less and less toward negative infinity (–∞). In other words, as the divisors of a number approach zero from opposite sides of zero, the quotients do not become closer. Rather, the quotients grow infinitely far apart. As the divisor finally reaches zero, we might wonder whether the quotient would equal positive infinity or negative infinity! Considering this growing difference in quotients as divisors approach zero from opposite sides can help us understand why division by zero is not possible.

Another consideration is the relationship between multiplication and division. Recall that multiplication and division are inverse operations. The numbers that form a multiplication fact may be arranged to form two division facts. For the multiplication fact $4 \times 5 = 20$, we may arrange the numbers to form these two division facts:

$$\frac{20}{4} = 5 \quad \text{and} \quad \frac{20}{5} = 4$$

We see that if we divide the product of two factors by either factor, the result is the other factor.

$$\frac{\text{product}}{\text{factor}_1} = \text{factor}_2 \quad \text{and} \quad \frac{\text{product}}{\text{factor}_2} = \text{factor}_1$$

This relationship between multiplication and division breaks down when zero is one of the factors, as we see in example 2.

Example 2 The numbers in the multiplication fact $2 \times 3 = 6$ can be arranged to form two division facts.

$$\frac{6}{3} = 2 \quad \text{and} \quad \frac{6}{2} = 3$$

If we attempt to form two division facts for the multiplication fact $2 \times 0 = 0$, one of the arrangements is not a fact. Which arrangement is not a fact?

Solution The product is 0 and the factors are 2 and 0. So the possible arrangements are these:

$$\frac{0}{2} = 0 \quad \text{and} \quad \frac{0}{0} = 2$$

fact not a fact

The arrangement **$0 \div 0 = 2$ is not a fact.**

The multiplication fact $2 \times 0 = 0$ does not imply $0 \div 0 = 2$ any more than $3 \times 0 = 0$ implies $0 \div 0 = 3$. This breakdown in the inverse relationship between multiplication and division when zero is one of the factors is another indication that division by zero is not possible.

Example 3 If we were asked to graph the following equation, what number could we not use in place of x when generating a table of ordered pairs?

$$y = \frac{12}{3 + x}$$

Solution This equation involves division. Since division by zero is not possible, we need to guard against the divisor, $3 + x$, being zero. When x is 0, the expression $3 + x$ equals 3. So we may use 0 in place of x. However, when x is -3, the expression $3 + x$ equals zero.

$$y = \frac{12}{3 + x} \qquad \text{equation}$$

$$y = \frac{12}{3 + (-3)} \qquad \text{replaced } x \text{ with } -3$$

$$y = \frac{12}{0} \qquad \text{not permitted}$$

Therefore, we may not use -3 in place of x in this equation. We can write our answer this way:

$$\mathbf{x \neq -3}$$

LESSON PRACTICE

Practice set

a. Use a calculator to divide several different numbers of your choosing by zero. Remember to clear the calculator before entering a new problem. What answers are displayed?

b. The numbers in the multiplication fact $7 \times 8 = 56$ can be arranged to form two division facts. If we attempt to form two division facts for the multiplication fact $7 \times 0 = 0$, one of the arrangements is not a fact. Which arrangement is not a fact and why?

For the following expressions, find the number or numbers that may not be used in place of the variable.

c. $\frac{6}{w}$

d. $\frac{3}{x - 1}$

e. $\frac{4}{2w}$

f. $\frac{y + 3}{y - 3}$

g. $\frac{8}{x^2 - 4}$

h. $\frac{3ab}{c}$

MIXED PRACTICE

Problem set

1. (Inv. 10) Robert was asked to select and hold three cards from a normal deck of cards. If the first two cards selected were aces, what is the chance that the third card he selects will be one of the two remaining aces?

2. (92) If Khalid saved \$5 by purchasing an item at a sale price of \$15, then the regular price was reduced by what percent?

3. (78) On a number line graph all real numbers that are both greater than or equal to –3 and less than 2.

4. (89) What is the sum of the measures of the interior angles of any quadrilateral?

5. (48) Complete the table.

FRACTION	DECIMAL	PERCENT
(a)	(b)	0.5%
$\frac{8}{9}$	(c)	(d)

6. (17) (a) Use a centimeter ruler and a protractor to draw a right triangle with legs 10 cm long.

(b) What is the measure of each acute angle?

(c) Measure the length of the hypotenuse to the nearest centimeter.

7. (111) Simplify. Write the answer in scientific notation.

$$\frac{(6 \times 10^5)(2 \times 10^6)}{(3 \times 10^4)}$$

8. (116) Factor each expression:

(a) $2x^2 + x$ (b) $3a^2b - 12a^2 + 9ab^2$

The figure at right was formed by stacking 1-cm cubes. Refer to the figure to answer problems 9 and 10.

9. (70) What is the volume of the figure?

10. (105) What is the surface area of the figure?

11. (108) Transform the formula $A = \frac{1}{2}bh$ to solve for h. Then use the transformed formula to find h when A is 1.44 m^2 and b is 1.6 m.

12. (65) If the ratio of boys to girls on the bus is 3 to 5, then what percent of the children on the bus are boys?

13. (112) If a 10-foot ladder is leaned against a wall so that the foot of the ladder is 6 feet from the base of the wall, how far up the wall will the ladder reach?

The graph below shows line l perpendicular to line m. Refer to the graph to answer problems 14 and 15.

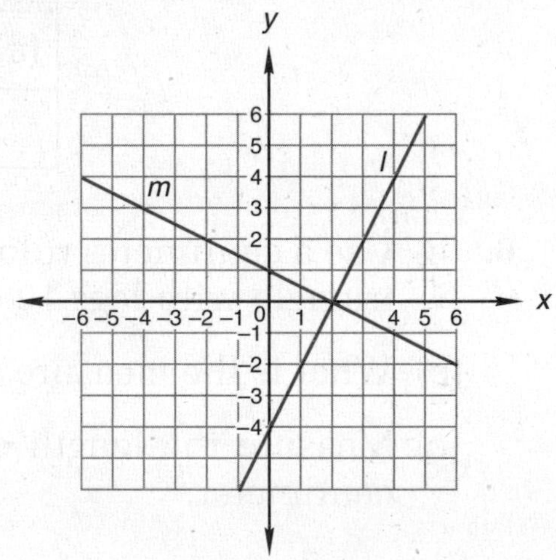

14. (117) (a) What is the equation of line l in slope-intercept form?

(b) What is the equation of line m in slope-intercept form?

15. (107) What is the product of the slopes of line l and line m? Why?

16. (110) If \$8000 is deposited in an account paying 6% interest compounded annually, then what is the total value of the account after four years?

17. (88) The Joneses are planning to carpet their home. The area to be carpeted is 1250 square feet. How many square yards of carpeting must be installed? Round the answer up to the next square yard.

18. (119) In the following expressions, what number may not be used for the variable?

(a) $\frac{12}{3w}$ (b) $\frac{12}{3 + m}$

19. (101) In the figure shown, $\overline{BD}$ is x units long, and $\overline{BA}$ is c units long. Using x and c, write an expression that indicates the length of $\overline{DA}$.

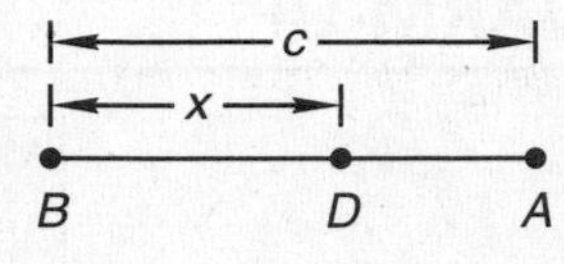

20. (97) In the figure at right, the three triangles are similar. Find the area of the smallest triangle. Dimensions are in inches.

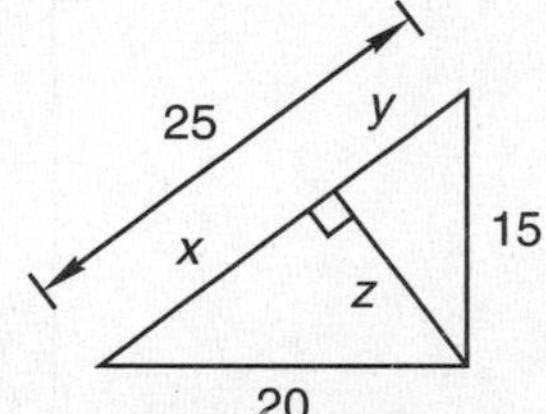

21. (113) A sphere with a diameter of 30 cm has a volume of how many cubic centimeters?

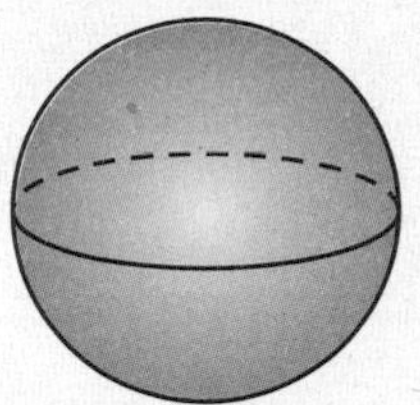

Use 3.14 for π.

22. (96, 118) Draw an estimate of a 45° angle. Then use a compass and straightedge to copy the angle.

Solve:

23. (93) $\frac{2}{3}m + \frac{1}{4} = \frac{7}{12}$

24. (102) $5(3 - x) = 55$

25. (102) $x + x + 12 = 5x$

26. (109) $10x^2 = 100$

Simplify:

27. (20) $\sqrt{90{,}000}$

28. (96) $x(x + 5) - 2(x + 5)$

29. (103) $\frac{(12xy^2z)(9x^2y^2z)}{36xyz^2}$

30. (48) $33\frac{1}{3}\%$ of 0.12 of $3\frac{1}{3}$

LESSON

120 Graphing Nonlinear Equations

WARM-UP

Facts Practice: Multiplying and Dividing in Scientific Notation (Test W)

Mental Math:

a. 1000001 (base 2)

b. MCMLXIX

c. $(10^2)(10^{-2})$

d. $(5 \times 10^{-5})^2$

e. $2x^2 = 32$

f. Convert 0°C to Fahrenheit.

g. 10% of \$250

h. 10% more than \$250

i. 2×12, $+ 1$, $\sqrt{\ }$, $\times 3$, $+ 1$, $\sqrt{\ }$, $\times 2$, $+ 1$, $\sqrt{\ }$, $+ 1$, $\sqrt{\ }$, $- 1$, $\sqrt{\ }$

Problem Solving:

A paper cone is filled with water. Then the water is poured into a cylindrical glass beaker that has the same height and diameter as the paper cone. How many cones of water are needed to fill the beaker?

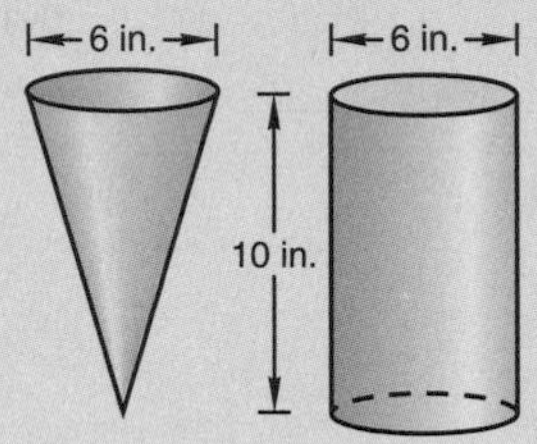

NEW CONCEPT

Equations whose graphs are lines are called **linear equations.** (Notice the word *line* in *linear.*) In this lesson we will graph equations whose graphs are not lines but are curves. These equations are called **nonlinear equations.** To graph each nonlinear equation, we will make a table of ordered pairs and plot enough points to get an idea of the path of the curve.

Example 1 Graph: $y = \dfrac{6}{x}$

Solution We make a table of ordered pairs. For convenience we select x values that are factors of 6. We remember to select negative values as well. Note that we may not select zero for x.

$y = \frac{6}{x}$

x	y	(x, y)
1	6	(1, 6)
2	3	(2, 3)
3	2	(3, 2)
6	1	(6, 1)

x	y	(x, y)
−1	−6	(−1, −6)
−2	−3	(−2, −3)
−3	−2	(−3, −2)
−6	−1	(−6, −1)

On a coordinate plane we graph the x, y pairs we found that satisfy the equation.

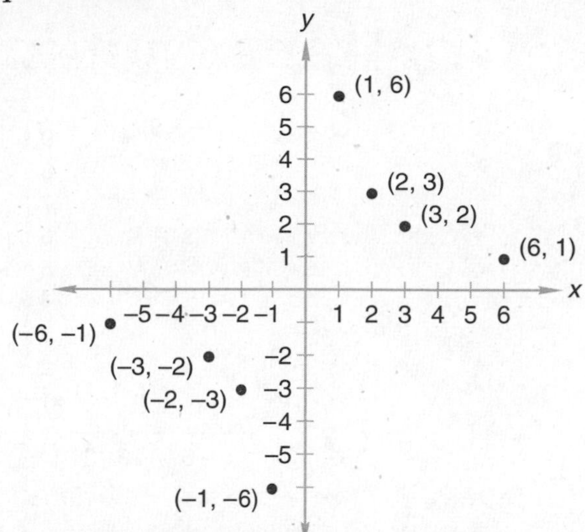

This arrangement of points on the coordinate plane suggests two curves that do not intersect.

We draw two smooth curves through the two sets of points.

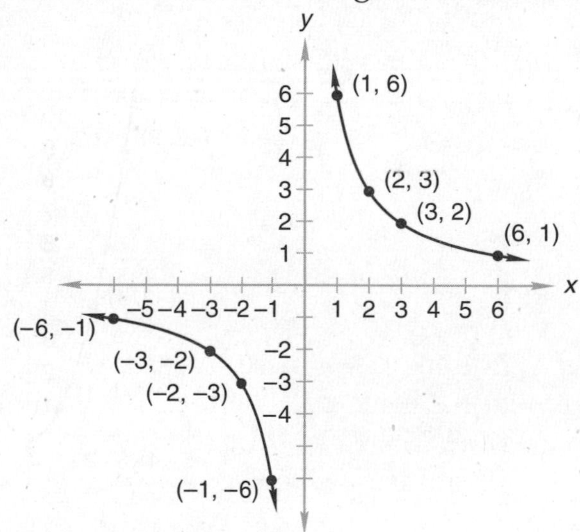

Example 2 Graph: $y = x^2$

Solution We begin by making a table of ordered pairs. We think of numbers for x and then calculate y. We replace x with negative numbers as well. Remember that squaring a negative number results in a positive number.

$$y = x^2$$

x	y	(x, y)
0	0	(0, 0)
1	1	(1, 1)
2	4	(2, 4)
3	9	(3, 9)

x	y	(x, y)
−1	1	(−1, 1)
−2	4	(−2, 4)
−3	9	(−3, 9)

After generating several pairs of coordinates, we graph the points on a coordinate plane.

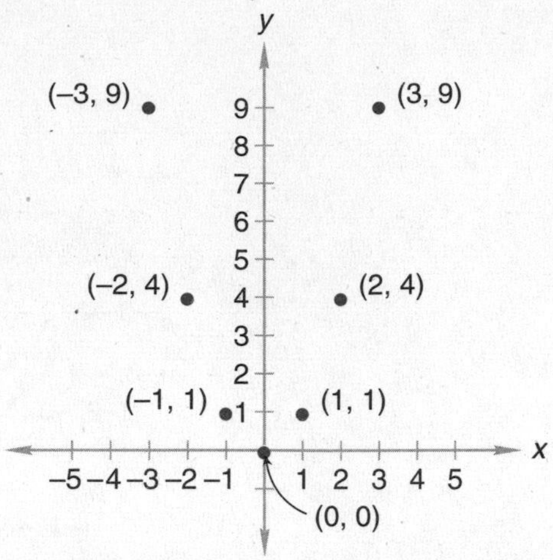

We complete the graph by drawing a smooth curve through the graphed points.

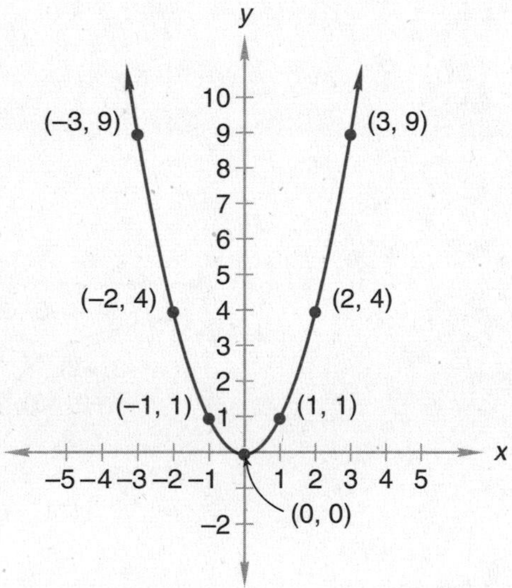

The coordinates of any point on the curve should satisfy the original equation.

LESSON PRACTICE

Practice set

a. Graph $y = \frac{12}{x}$. Begin by creating a table of ordered pairs. Use 6, 4, 3, 2, −2, −3, −4, and −6 in place of x.

b. Graph $y = x^2 - 2$. Compare your graph to the graph in example 2.

c. Graph $y = \frac{10}{x}$. Compare your graph to the graph in example 1.

d. Graph $y = 2x^2$. Compare your graph to the graph in example 2.

MIXED PRACTICE

Problem set

1. Schuster was playing a board game and rolled a 7 with a pair of dot cubes three times in a row. What are the odds of Schuster rolling a 7 with the next roll of the dot cubes?
(Inv. 10)

2. If the total cost of an item including 8% sales tax is \$2.70, then what was the price before tax was added?
(92)

3. Compare: $x^2 \bigcirc y^2$ if $x < y$
(79)

4. If a trapezoid has a line of symmetry and one of its angles measures 100°, what is the measure of each of its other angles?
(58)

5. Complete the table.
(48)

FRACTION	DECIMAL	PERCENT
(a)	(b)	0.1%
$\frac{8}{5}$	(c)	(d)

6. The hypotenuse of this triangle is twice the length of the shorter leg.
(99)

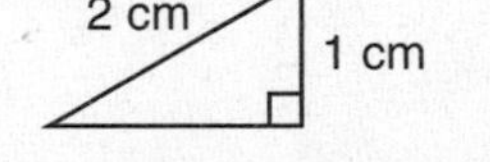

(a) Use the Pythagorean theorem to find the length of the remaining side.

(b) Use a centimeter ruler to find the length of the unmarked side to the nearest tenth of a centimeter.

7. Simplify. Write the answer in scientific notation.
(111)

$$\frac{(4 \times 10^{-5})(6 \times 10^{-4})}{8 \times 10^{3}}$$

8. (116) Factor each expression:

(a) $3y^2 - y$ (b) $6w^2 + 9wx - 12w$

The figure below shows a cylinder and a cone whose heights and diameters are equal. Refer to the figure to answer problems 9 and 10.

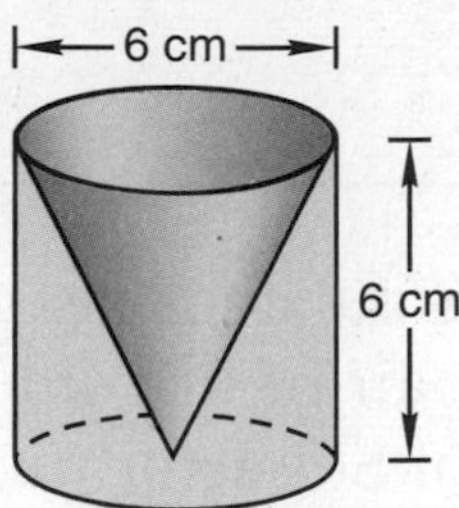

9. (113) What is the ratio of the volume of the cone to the volume of the cylinder?

10. (105) The lateral surface area of a cylinder is the area of the curved side and excludes the areas of the circular ends. What is the lateral surface area of the cylinder rounded to the nearest square centimeter? (Use 3.14 for π.)

11. (106) Transform the formula $E = mc^2$ to solve for m.

12. (54) If 60% of the children at the theater were girls, what was the ratio of boys to girls at the theater?

The graph below shows $m \perp n$. Refer to the graph to answer problems 13 and 14.

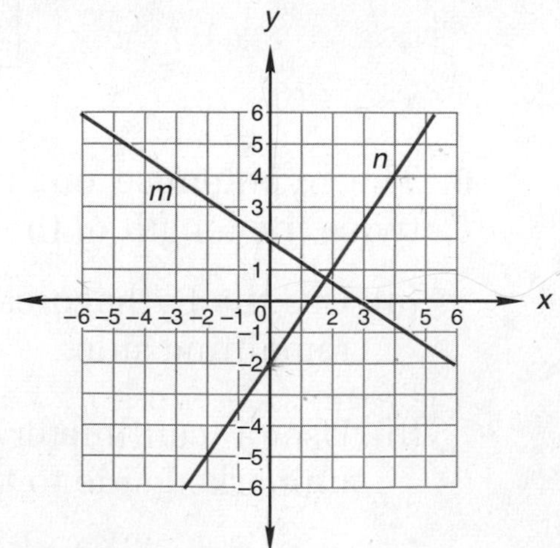

13. (117) What is the equation of each line in slope-intercept form?

14. (107) What is the product of the slopes of lines m and n? Why?

15. (110) If a \$1000 investment earns 20% interest compounded annually, then the investment will double in value in how many years?

16. (112) The stated size of a TV screen or computer monitor is its diagonal measure. A screen that is 17 in. wide and 12 in. tall would be described as what size of screen? Round the answer to the nearest inch.

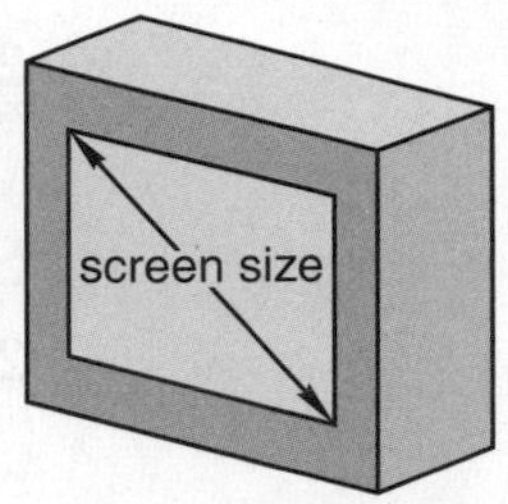

17. (70, 88) Premixed concrete is sold by the cubic yard. The Smiths are pouring a concrete driveway that is 36 feet long, 21 feet wide, and $\frac{1}{2}$ foot thick.

(a) Find the number of cubic feet of concrete needed.

(b) Use three unit multipliers to convert answer (a) to cubic yards.

18. (119) In the following expressions, what number may not be used for the variable?

(a) $\dfrac{12}{4 - 2m}$ (b) $\dfrac{y - 5}{y + 5}$

19. (120) Graph: $y = x^2 - 4$

20. (97) Refer to this drawing of three similar triangles to find the letter that completes the proportion below.

$$\frac{c}{a} = \frac{a}{?}$$

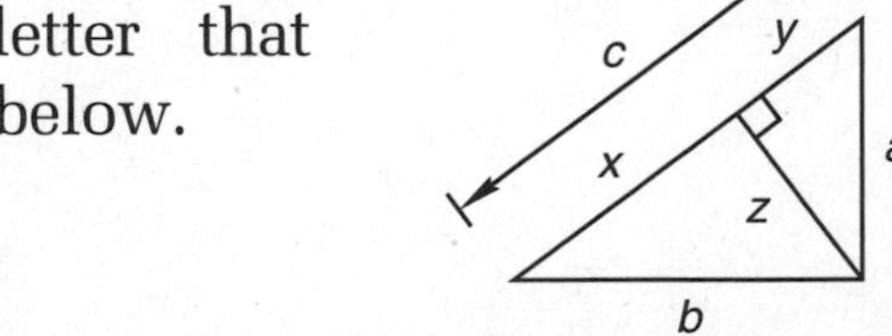

21. (105) Recall that the surface area of a sphere is four times the area of its largest "cross section." What is the approximate surface area of a cantaloupe that is 6 inches in diameter? Use 3.14 for π and round the answer to the nearest square inch.

22. (115) A cup containing 250 cubic centimeters of water holds how many liters of water?

Solve:

23. $15 + x = 3x - 17$
(102)

24. $3\frac{1}{3}x - 16 = 74$
(93)

25. $\frac{m^2}{4} = 9$
(109)

26. $\frac{1.2}{m} = \frac{0.04}{8}$
(98)

Simplify:

27. $x(x - 5) - 2(x - 5)$
(96)

28. $\frac{(3xy)(4x^2y)(5x^2y^2)}{10x^3y^3}$
(103)

29. $|-8| + 3(-7) - [(-4)(-5) - 3(-2)]$
(91)

30. $\frac{7\frac{1}{2} - \frac{2}{3}(0.9)}{0.03}$
(43, 45)

INVESTIGATION 12

Focus on

Proof of the Pythagorean Theorem

When mathematicians want to demonstrate that a particular idea is true, they construct a **proof.** Following logical steps, a proof describes how certain given information leads to a certain conclusion. Indeed, virtually all of the structure of mathematics is built upon conclusions reached by a proof. One extremely important and useful conclusion about our world is that in a right triangle the lengths of the legs (a and b) and the length of the hypotenuse (c) are related in the following way:

$$a^2 + b^2 = c^2$$

Recall that this conclusion is called the Pythagorean theorem. Mathematicians have constructed literally hundreds of proofs of the Pythagorean theorem. In fact, one of America's presidents, James A. Garfield, is credited with providing an original proof of the theorem. In this investigation you will work through one of the many proofs that has been constructed.

The following proof of the Pythagorean theorem is based upon the characteristics of similar triangles. Recall that the corresponding angles of similar triangles are congruent and that the lengths of their corresponding sides are proportional.

1. Begin by sketching a right triangle. For reference, name the vertices A, B, and C, with $\angle C$ being the right angle. This triangle is a generic right triangle, so the measures of the acute angles do not affect the outcome of the proof.

2. It is customary to refer to the lengths of the sides of a right triangle by the lowercase form of the letter of the opposite vertex. So along segment *AB* write a lowercase *c*, along segment *BC* write a lowercase *a*, and along segment *CA* write a lowercase *b*. Remember that these lowercase letters refer to the lengths of the sides. Since this is a generic right triangle, we do not know the lengths of the sides. However, it is not necessary to know the lengths of the sides in order to establish the relationship among the sides.

3. Next draw a segment from vertex *C* across the triangle to side *AB* so that the segment is perpendicular to $\overline{AB}$. Name the point where the segment intersects $\overline{AB}$ point *D*.

4. Point *D* divides segment *AB* into two shorter segments. We have already labeled the distance from *A* to *B* as *c*. Now label the distance from *B* to *D* as *x*. The distance from *D* to *A* is the rest of *c*, which can be found by subtracting *x* from *c*. So label the length of $\overline{AD}$ as $c - x$.

Check your drawing with the following description: If you have performed steps 1–4 correctly, you should have drawn a right triangle *ABC* and divided the right triangle into two smaller right triangles named ΔBCD and ΔACD. The hypotenuse and one leg of ΔBCD should be labeled *a* and *x*. The hypotenuse and one leg of ΔACD should be labeled *b* and $c - x$. The hypotenuse of ΔABC should be labeled *c* and is equal to length *x* plus length $c - x$. (Note that you may need to position *c* on your drawing to indicate that it represents the entire length of $\overline{AB}$.)

We use this figure and what we know about similar triangles to prove the Pythagorean theorem. Before proceeding with the algebraic proof, we need to be convinced that the three triangles in the figure are similar. If their corresponding angles are congruent, then the triangles are similar.

5. What is the sum of the measures of the two acute angles of a right triangle? Why?

6. If $\angle B$ of ΔABC measures m degrees, then how many degrees is the measure of $\angle A$?

7. If the number of degrees in the measure of $\angle B$ is m and the number of degrees in the measure of $\angle A$ is $90° - m$, then how many degrees are in the measure of

   (a) $\angle BCD$? (b) $\angle ACD$?

8. Can we conclude that all three triangles in the figure are similar? Why?

Since similar triangles have sides whose measures are proportional, we can write proportions that relate the lengths of the sides of the triangles. Because we are referring to the lengths of the sides, we will use the lowercase letters on the diagram: a, b, c, x, and $c - x$.

Recall that we began with the largest triangle, ΔABC, and that we divided this triangle into two smaller triangles. We will write two proportions. One proportion will relate the largest triangle to one of the smaller triangles. The second proportion will relate the largest triangle to the other smaller triangle.

9. Write a proportion that relates the hypotenuses of ΔABC and ΔBCD to corresponding legs of ΔABC and ΔBCD.

$$\begin{array}{ccccc} & hyp & & leg \\ \Delta ABC & c & & \square \\ & \overline{} & = & \overline{} \\ \Delta BCD & a & & \square \end{array}$$

10. Write a proportion that relates the hypotenuses of ΔABC and ΔACD to corresponding legs of ΔABC and ΔACD.

$$\begin{array}{ccccc} & hyp & & leg \\ \Delta ABC & c & & \square \\ & \overline{} & = & \overline{} \\ \Delta ACD & b & & \square \end{array}$$

11. Cross multiply the proportions you wrote in problems 9 and 10.

12. If you have finished cross multiplying the proportions from problems 9 and 10, you should see the term cx in both cross products. In the first cross product we see that cx equals a^2. This means that we can replace cx with a^2 in the second cross product. Replace cx with a^2 in the second cross product and write the result.

13. Transform the equation you gave for your answer to problem 12 by solving for c^2. What do we call this equation?

This completes an algebraic proof of the Pythagorean theorem for all right triangles.

Some proofs of the Pythagorean theorem involve dividing up and rearranging areas of squares built on the sides of a right triangle. These proofs are based on the concept that a^2, b^2, and c^2 represent areas of squares, as shown below.

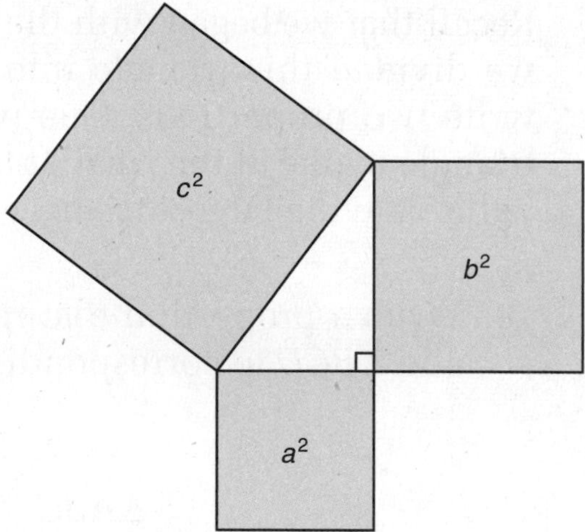

The following activity demonstrates that the combined areas of the squares labeled a^2 and b^2 in the drawing above equal the area of the square labeled c^2.

Activity: *Pythagorean Puzzle*

Materials needed:

- Activity Sheet 12 (available in *Saxon Math 8/7—Homeschool Tests and Worksheets*)
- Scissors

The object of the puzzle is to rearrange the pieces of the squares drawn on the legs of right triangle ABC to form a square on the hypotenuse of the triangle.

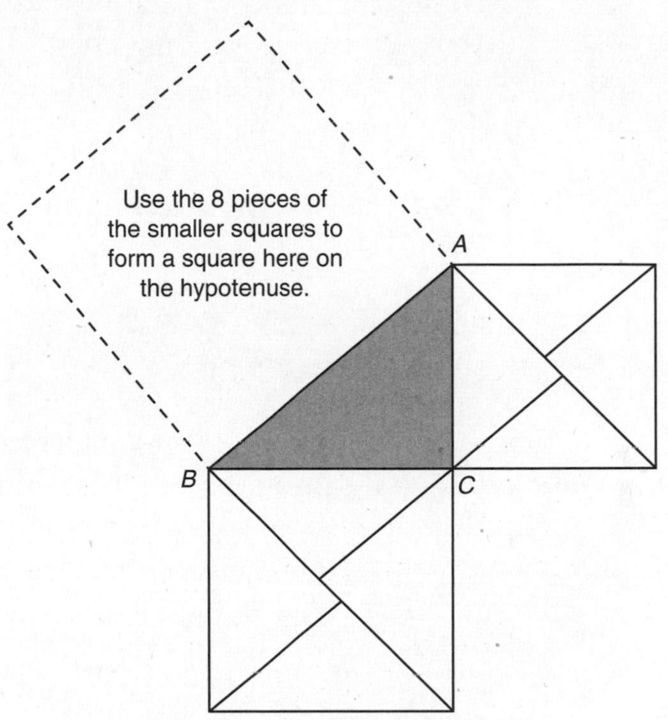

APPENDIX

Additional Topics and Supplemental Practice

TOPIC A

Base 2 • Roman Numerals

NEW CONCEPTS

Base 2 In a world populated with ten-fingered people, we have settled on a base 10 number system for most of our daily counting needs. The structure of a base 10 number system looks like this, with the digits used in each place ranging from 0 to 9:

10^3	10^2	10^1	10^0
___	___	___	___
thousands	hundreds	tens	ones

To find the value of a number, we multiply each digit by its place value. So 1010 equals (1 × 1000) + (1 × 10).

There are other useful number systems. In the world of electronics, in which switches can be either off or on, base 2 serves as a number system. The structure of a base 2 number system looks like this, with the digits used in each place being either 0 or 1:

2^4	2^3	2^2	2^1	2^0
___	___	___	___	___
sixteens	eights	fours	twos	ones

Notice that the value of each place is two times the value of the place to its right. To find the value of a number, we multiply the value of each digit by its place value. So 1010 (base 2) equals (1 × 8) + (1 × 2), which is ten.

1010 (base 2) = 10 (base 10)

Example 1 What base 10 number does 10101 (base 2) represent?

Solution We add the values of the places occupied by 1's.

16's	8's	4's	2's	1's
1	0	1	0	1

We see a 16, a 4, and a 1, which in base 10 totals **21.**

Roman numerals Not all number systems use place value. The value of a Roman numeral is the same whatever its place. Here are the values of the Roman numerals we will consider in this book:

I	1
V	5
X	10
L	50
C	100
D	500
M	1000

To find the value of a Roman numeral, we add the values of the numerals. So MCLXII equals 1000 plus 100 plus 50 plus 10 plus 1 plus 1, which equals 1162. An exception to the rule of adding the values occurs when a numeral of lesser value is to the left of a numeral of greater value. In such a situation we subtract the lesser value from the greater value. The six possible combinations are these:

IV = 4   IX = 9

XL = 40   XC = 90

CD = 400   CM = 900

So, for example, the Roman numeral for 999 is CMXCIX, not IM.

Example 2 Carved into the base of a building was the Roman numeral MCMXXIV, indicating the year in which the building was constructed. In what year was the building constructed?

Solution We will spread out the Roman numeral to show the value of its parts.

M	CM	XX	IV
1000	900	20	4

Adding these values, we find that the building was constructed in **1924.**

LESSON PRACTICE

Practice set Find the base 10 value of these base 2 numbers:

a. 111 (base 2)

b. 1000 (base 2)

c. 1100 (base 2)

d. 10001 (base 2)

Find the value of each Roman numeral:

e. XXXIX

f. LXIV

g. MCMXIX

h. MMII

Supplemental Practice Problems for Selected Lessons

This appendix contains additional practice problems for concepts presented in selected lessons. It is very important that no problems in the regular problem sets be skipped to make room for these problems. Saxon math is designed to produce long-term retention through repeated exposure to concepts in the problem sets. The problem sets provide enough exposure to concepts for most students. However, the problems in this appendix can be completed to provide additional exposure as necessary.

Lesson 3 Find each missing number:

1. $w + 36 = 62$
2. $x - 24 = 42$
3. $5y = 60$
4. $z \div 8 = 16$
5. $18 + m = 72$
6. $24 - n = 6$
7. $6p = 48$
8. $144 \div q = 8$
9. $36 = 4m$
10. $\frac{a}{18} = 3$
11. $\frac{18}{c} = 3$
12. $\begin{array}{r} 84 \\ -\ E \\ \hline 36 \end{array}$
13. $8 + 6 + 5 + x + 4 = 30$
14. $36 + 18 + 27 + w = 90$

Lesson 6 For each number, list the whole-number factors from 1 to 10:

1. 36
2. 3600
3. 350
4. 1326
5. 4320
6. 950
7. 12,000
8. 35,420
9. 36,270
10. 123,450
11. 1,000,000
12. 2520

Lesson 15 Reduce each fraction to lowest terms:

1. $\frac{15}{20}$
2. $\frac{8}{24}$
3. $\frac{9}{24}$
4. $\frac{12}{18}$
5. $\frac{24}{30}$
6. $\frac{16}{32}$
7. $\frac{24}{36}$
8. $\frac{28}{35}$
9. $3\frac{15}{18}$
10. $6\frac{18}{24}$
11. $8\frac{9}{15}$
12. $4\frac{18}{32}$

Lesson 19 Find the perimeter of each polygon. Dimensions are in inches.

1. 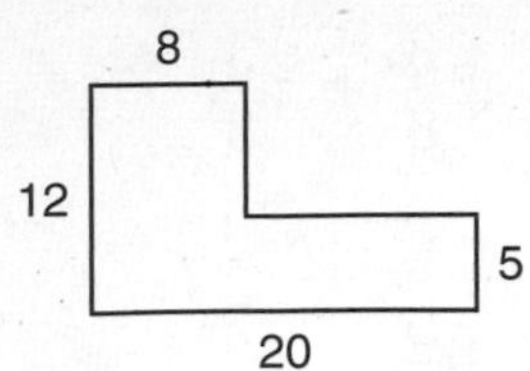

2. 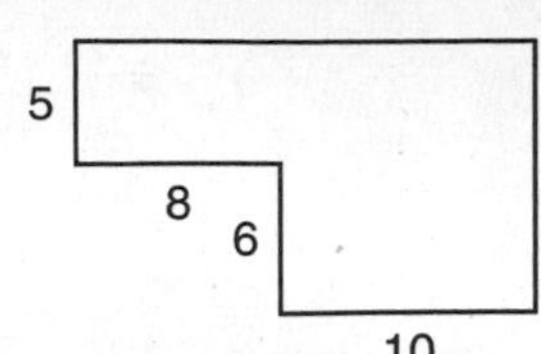

3. 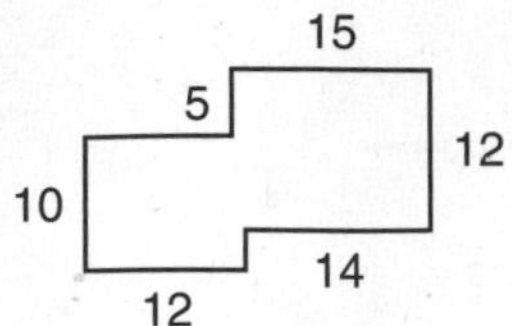

4. 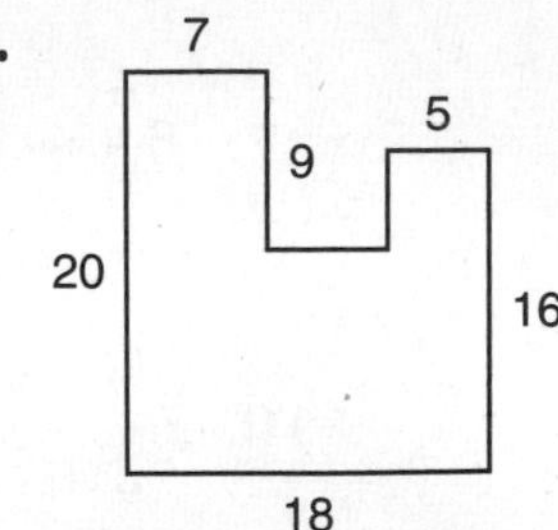

Lesson 20 Simplify:

1. 8^2
2. 2^6
3. 3^3
4. 10^5
5. $3^2 + 2^3$
6. $5^2 - 4^2$
7. 4^3
8. 15^2
9. $\frac{10^4}{10^3}$
10. $\frac{8^2}{2^3}$
11. $5^4 - 5^3$
12. 25^2
13. $\sqrt{81}$
14. $\sqrt{121}$
15. $\sqrt{49}$
16. $\sqrt{144}$
17. $\sqrt{900}$
18. $\sqrt{625}$
19. $\sqrt{196}$
20. $\sqrt{441}$

Lesson 21 Write the prime factorization of each number:

1. 81
2. 300
3. 2000
4. 625
5. 450
6. 1200
7. 440
8. 750
9. 10,000
10. 128
11. 780
12. 1540

Lesson 23 Simplify using regrouping:

1. $5\frac{3}{5} + 2\frac{4}{5}$
2. $7\frac{3}{8} + 1\frac{3}{8}$
3. $2\frac{3}{7} + 3\frac{4}{7}$
4. $5\frac{3}{4} + 3\frac{3}{4}$
5. $6\frac{5}{8} + 5\frac{7}{8}$
6. $8\frac{5}{9} + 2\frac{7}{9}$
7. $6\frac{7}{8} - 2\frac{1}{8}$
8. $5 - 3\frac{1}{4}$
9. $6 - 2\frac{3}{5}$
10. $5\frac{1}{3} - 1\frac{2}{3}$
11. $4\frac{2}{5} - 1\frac{4}{5}$
12. $6\frac{1}{6} - 2\frac{5}{6}$

Lesson 26 Simplify:

1. $3\frac{3}{4} \times \frac{2}{5}$
2. $2\frac{1}{3} \times 3$
3. $1\frac{4}{5} \times 3\frac{1}{3}$
4. $7 \times 2\frac{2}{3}$
5. $\frac{5}{8} \times 3\frac{1}{5}$
6. $2\frac{1}{4} \times 1\frac{3}{5}$
7. $3\frac{1}{2} \div 3$
8. $2\frac{3}{4} \div \frac{3}{4}$
9. $1\frac{1}{2} \div 2\frac{2}{3}$
10. $3\frac{1}{3} \div 1\frac{3}{4}$
11. $6 \div 3\frac{3}{5}$
12. $\frac{5}{8} \div 3\frac{1}{2}$

Lesson 30 Simplify:

1. $\frac{3}{5} + \frac{3}{10}$
2. $\frac{3}{4} + \frac{1}{2} + \frac{3}{8}$
3. $2\frac{5}{6} + 1\frac{1}{2}$
4. $\frac{5}{6} + \frac{3}{4}$
5. $\frac{5}{6} + \frac{3}{8} + \frac{7}{12}$
6. $3\frac{3}{5} + 2\frac{2}{3}$
7. $\frac{5}{8} - \frac{1}{2}$
8. $3\frac{5}{6} - 1\frac{1}{2}$
9. $4\frac{3}{4} - 1\frac{1}{3}$
10. $\frac{8}{12} - \frac{2}{3}$
11. $6\frac{3}{5} - 3\frac{1}{3}$
12. $5\frac{1}{4} - 1\frac{5}{6}$

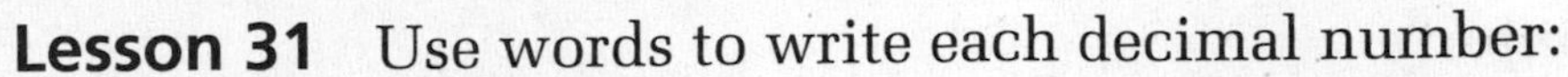

Lesson 31 Use words to write each decimal number:

1. 16.125

2. 5.03

3. 105.105

4. 0.001

5. 160.165

6. 4000.321

Use digits to write each decimal number:

7. one hundred twenty-three thousandths

8. one hundred and twenty-three thousandths

9. one hundred twenty and three thousandths

10. five hundredths

11. twenty and nine hundredths

12. twenty-nine and five tenths

Lesson 33 Round to the nearest whole number:

1. 23.459 2. 164.089 3. 86.6427

Round to two decimal places:

4. 12.83333 5. 6.0166 6. 0.1084

Round to the nearest thousandth:

7. 0.08333 8. 0.45454 9. 3.14159

10. Round 283.567 to the nearest hundred.

11. Round 283.567 to the nearest hundredth.

12. Round 126.59 to the nearest ten.

Lesson 35 Simplify:

1. $45.3 + 2.64 + 3$
2. $0.4 + 0.5 + 0.6 + 0.7$
3. $3.6 + 2.75 + 0.194 + 3$
4. $12.8 + 6.32 + 15$
5. $10 + 1.0 + 0.1 + 0.01$
6. $278.4 + 3.26 + 1.475$
7. $14.327 - 6.5$
8. $10.8 - 9.67$
9. $6.5 - 4.321$
10. $10 - 4.76$
11. $0.1 - 0.019$
12. $5 - 4.937$
13. 0.3×0.12
14. 4.5×5
15. 8×0.012
16. $0.2 \times 0.3 \times 0.4$
17. $1.2 \times 1.2 \times 100$
18. $1.44 \div 12$
19. $0.144 \div 8$
20. $0.144 \div 16$

Lesson 37 Find the area of each triangle. Dimensions are in centimeters.

1.
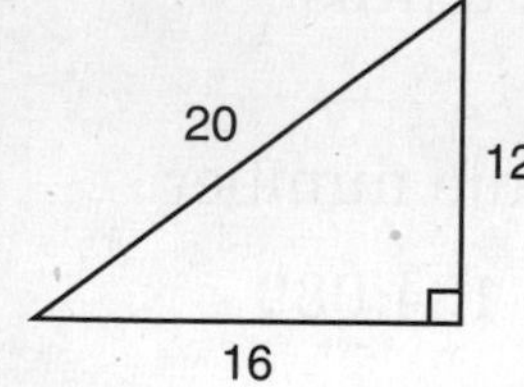

2.
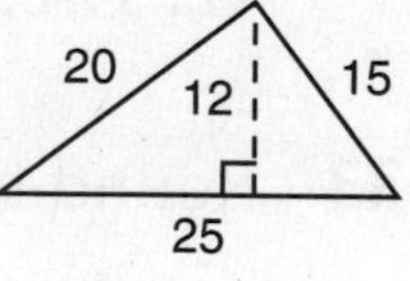

3.
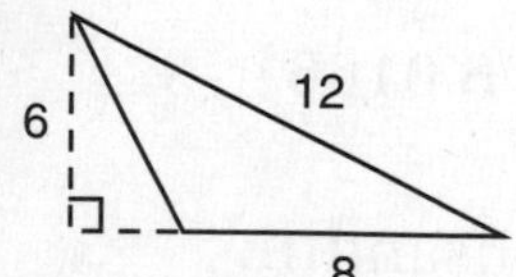

4.
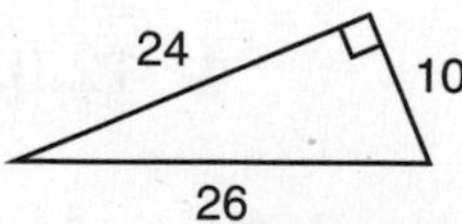

5.
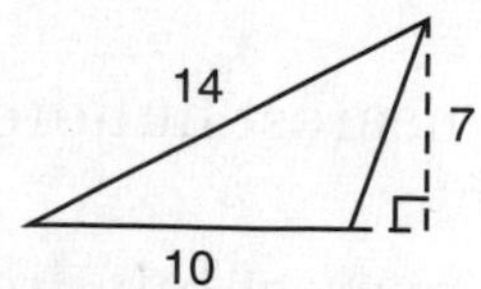

6.
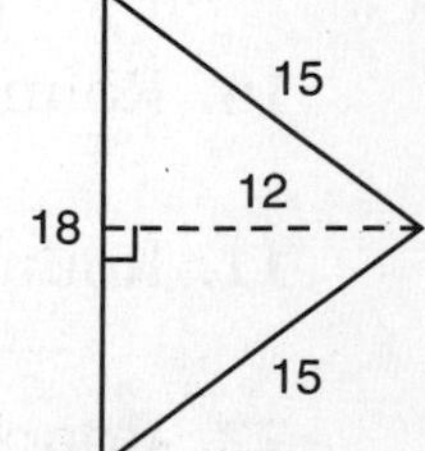

Find the area of each figure. Dimensions are in centimeters.

7.

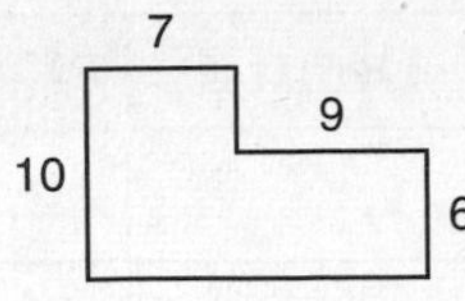

8.

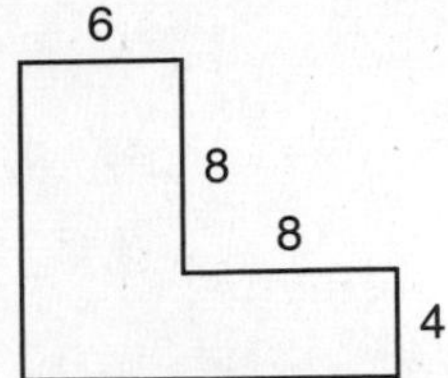

9.

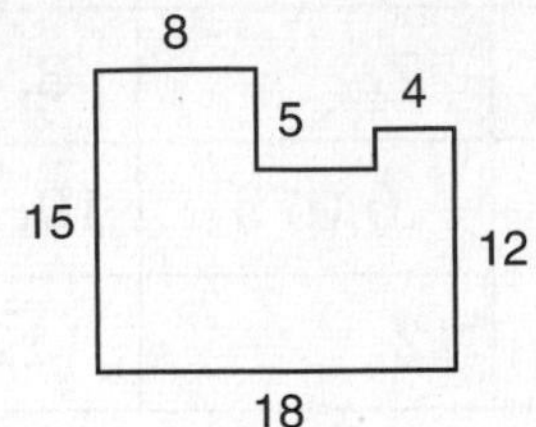

10.

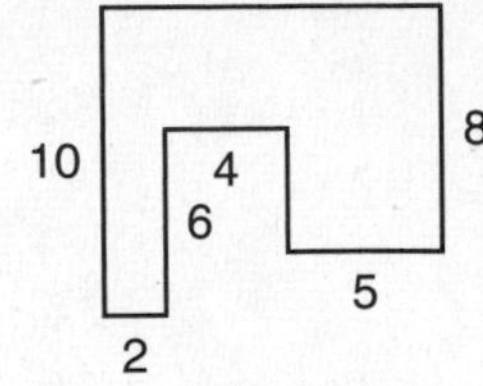

Lesson 43 Change each decimal number to a reduced fraction or mixed number:

1. 0.48 **2.** 3.75 **3.** 0.125

4. 12.6 **5.** 0.025 **6.** 1.08

Change each fraction or mixed number to a decimal number:

7. $\frac{5}{8}$ **8.** $\frac{1}{3}$ **9.** $2\frac{2}{5}$

10. $6\frac{1}{6}$ **11.** $\frac{11}{20}$ **12.** $5\frac{5}{9}$

Lesson 45 Divide:

1. 0.15 ÷ 0.5 **2.** 14.4 ÷ 0.06

3. 18 ÷ 0.4 **4.** 5 ÷ 0.8

5. 12.5 ÷ 0.04 **6.** 288 ÷ 1.2

7. 4.3 ÷ 0.01 **8.** 1.5 ÷ 0.12

9. 9 ÷ 1.8 **10.** 4.5 ÷ 2.5

11. 8 ÷ 0.04 **12.** 12.5 ÷ 0.5

Lesson 48 Complete this table:

FRACTION	DECIMAL	PERCENT
$\frac{5}{6}$	**1.**	**2.**
3.	1.2	**4.**
5.	**6.**	8%
$1\frac{3}{5}$	**7.**	**8.**
9.	0.075	**10.**
11.	**12.**	125%

Lesson 49 Change:

1. 40 inches to feet and inches

2. 200 seconds to minutes and seconds

Simplify:

3. 3 ft 21 in.

4. 2 hr 90 min

Add:

5.
```
  3 yd 2 ft 7 in.
+ 1 yd 1 ft 8 in.
```

6.
```
  5 hr 18 min 23 s
+ 2 hr 45 min 48 s
```

7.
```
  5 lb 10 oz
+ 6 lb  8 oz
```

8.
```
  2 gal 3 qt 1 pt
+ 3 gal 2 qt 1 pt
```

Lesson 50 Use unit multipliers to convert:

1. 24 ft to in.

2. 24 ft to yd

3. 300 min to hr

4. 300 min to s

5. 500 cm to m

6. 500 cm to mm

7. 100 lb to oz

8. 100 pounds to tons

Lesson 52 Simplify:

1. $4 + 4 \times 4 - 4 \div 4$

2. $40 - 20 \div 10 - 5$

3. $5 + 6 \times 7 + 8$

4. $3^2 + 4^2 - 5 \times 2$

5. $\dfrac{10 + 10 \times 10}{10}$

6. $\dfrac{5 + 5 \times 5 \div 5 - 5}{5}$

Evaluate:

7. $ab - bc + abc$ if $a = 5$, $b = 4$, and $c = 2$

8. $xy + \frac{x}{y} - 5$ if $x = 8$ and $y = 4$

9. $abc - ab - \frac{a}{c}$ if $a = 6$, $b = 4$, and $c = 3$

10. $m - mn$ if $m = \frac{3}{4}$ and $n = \frac{1}{2}$

11. $wx + xz - z$ if $w = 1.2$, $x = 0.5$, and $z = 0.1$

12. $ab - ac - \frac{ab}{c}$ if $a = 4$, $b = 3$, and $c = 2$

Lesson 56 Subtract:

1. 5 ft 7 in. − 3 ft 10 in.

2. 10 min 13 s − 3 min 28 s

3. 4 yd 6 in. − 2 ft 8 in.

4. 1 hr 10 min − 24 min 40 s

5. 8 yd 2 ft 4 in.
− 1 yd 2 ft 9 in.

6. 3 hr 17 min 30 s
− 2 hr 48 min 43 s

Lesson 57 Simplify:

1. 4^{-2}

2. 2^{-3}

3. $4^2 \times 4^{-2}$

4. $3^3 \times 3^{-2}$

5. $2^{-2} \times 2^{-3}$

6. $2^{-2} + 2^{-2}$

7. $10^{-2} \times 10$

8. $3(3^{-3})$

9. Write 10^{-3} as a decimal number.

10. What is the reciprocal of 2^{-2}?

Lesson 60 Write an equation to solve each problem. Then solve the equation.

1. What number is $\frac{3}{4}$ of 24?
2. Three fifths of 60 is what number?
3. What number is 0.4 of 80?
4. Six tenths of 60 is what number?
5. What number is 30% of 120?
6. Six percent of 250 is what number?
7. What number is $\frac{5}{6}$ of 300?
8. Two thirds of 90 is what number?
9. What number is 0.5 of 50?
10. Seven tenths of 140 is what number?
11. What number is 75% of 400?
12. Eighty percent of 400 is what number?

Lesson 64 Find each sum:

1. $(-36) + (+54)$
2. $(-15) + (-26)$
3. $(-6) + (-12) + (+15)$
4. $(+4) + (-12) + (+21)$
5. $(-6) + (-8) + (-7) + (-2)$
6. $(-9) + (-15) + (+50)$
7. $(+42) + (-23) + (-19)$
8. $(-54) + (+76) + (-17)$
9. $\left(-3\frac{1}{2}\right) + \left(-2\frac{1}{4}\right)$
10. $\left(-1\frac{1}{3}\right) + \left(+2\frac{5}{6}\right)$
11. $(-1.7) + (-3.2) + (-1.8)$
12. $(-4.3) + (+2.63)$

Lesson 66 Find the circumference of each circle. Dimensions are in centimeters.

1.

40

Use 3.14 for π.

2.

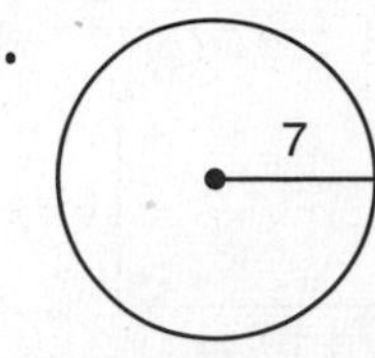

Use $\frac{22}{7}$ for π.

3.

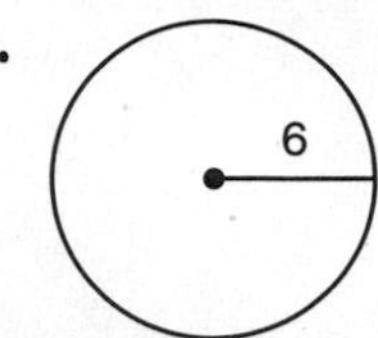

Leave π as π.

4.

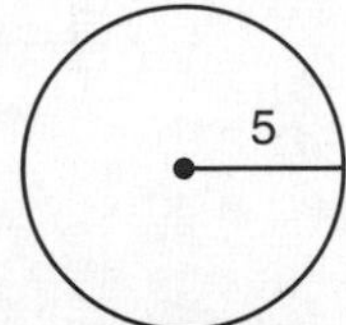

Use 3.14 for π.

5.

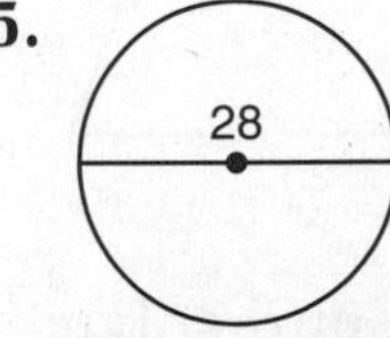

Use $\frac{22}{7}$ for π.

6.

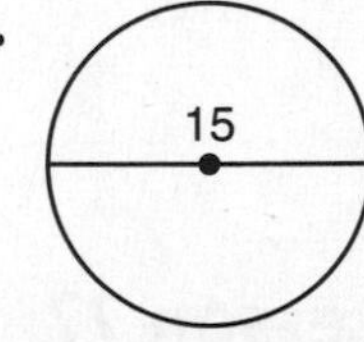

Leave π as π.

Lesson 68 Simplify:

1. $(-3) - (-8)$

2. $(-12) + (+20)$

3. $(+8) - (-15)$

4. $(+6) - (18)$

5. $(-3) + (-4) - (-5)$

6. $(+3) - (-4) - (+5)$

7. $(-2) - (-3) - (-4)$

8. $(+2) - (3) - (-4)$

9. $(-6) - (-7) + (8)$

10. $(+8) - (+9) - (-12)$

11. $(-3) - (-1) - (-8) - (2)$

12. $(-9) - (10) - (-11)$

Lesson 69 Express each number in proper scientific notation:

1. 0.15×10^{7}

2. 48×10^{-8}

3. 20×10^{5}

4. 0.72×10^{-4}

5. 0.125×10^{12}

6. 22.5×10^{-6}

7. 17.5×10^{10}

8. 0.375×10^{-8}

Lesson 75 Find each area. Dimensions are in centimeters.

1.
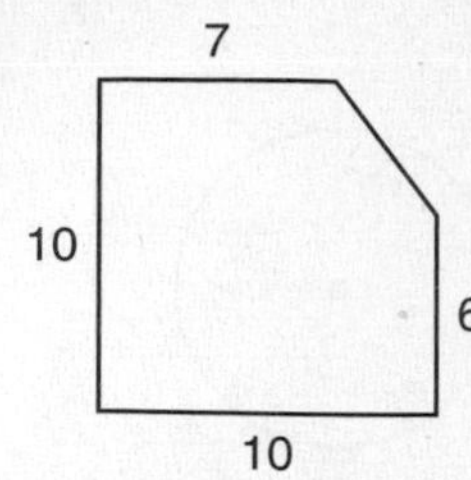

2.
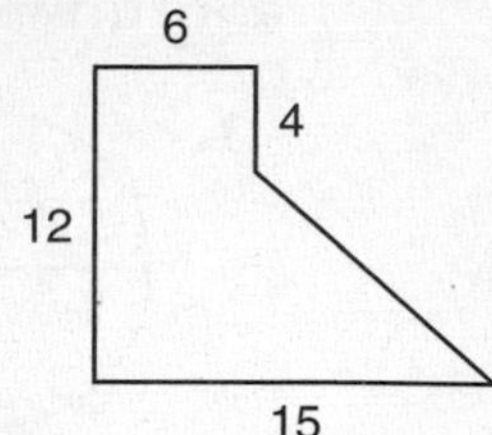

3.
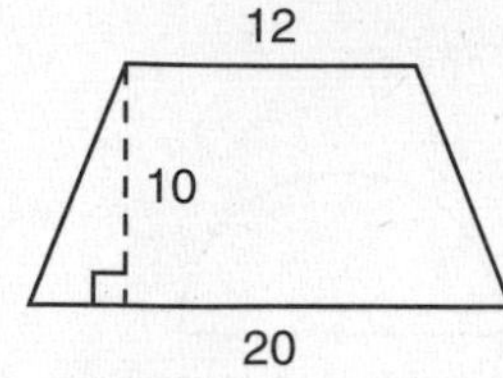

4.
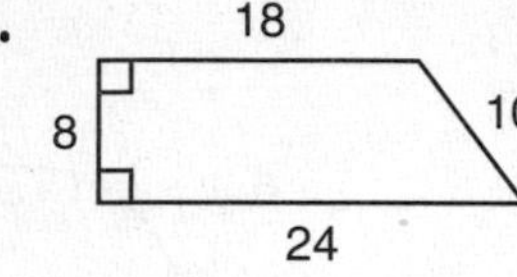

Lesson 77 Translate and solve:

1. What percent of 75 is 60?
2. Sixty is 75% of what number?
3. Thirty is what percent of 90?
4. Thirty is 150% of what number?
5. What percent of 40 is 50?
6. Twenty percent of what number is 50?
7. What percent of \$5.00 is \$3.50?
8. Twelve is $66\frac{2}{3}$% of what number?

Lesson 82 Find the area of each circle. Dimensions are in centimeters.

1.
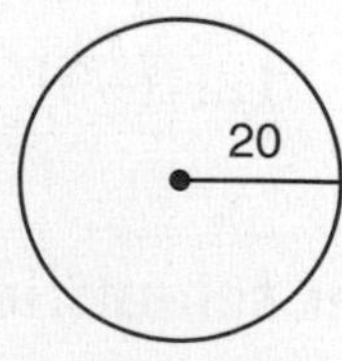

Use 3.14 for π.

2.
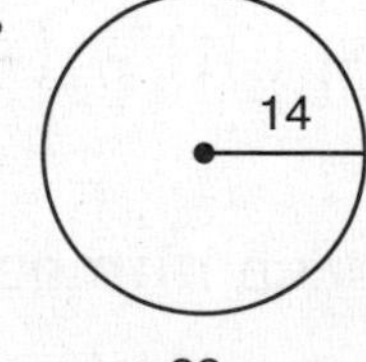

Use $\frac{22}{7}$ for π.

3.
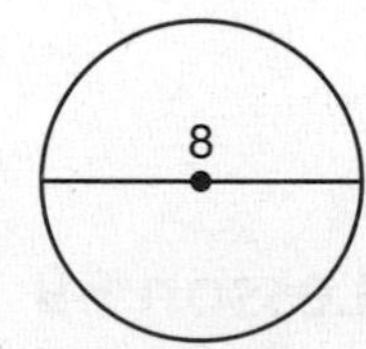

Leave π as π.

4.
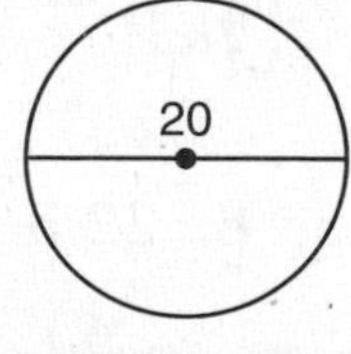

Use 3.14 for π.

5.
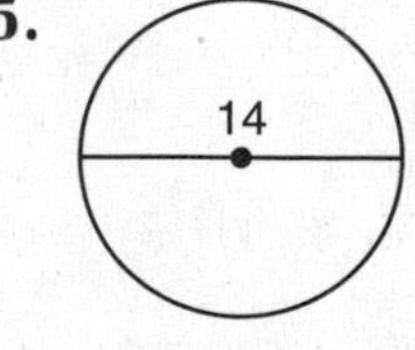

Use $\frac{22}{7}$ for π.

6.
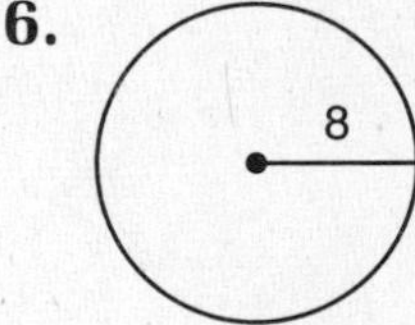

Leave π as π.

Lesson 83 Write each product in scientific notation:

1. $(1.2 \times 10^5)(3 \times 10^6)$
2. $(3 \times 10^6)(6 \times 10^3)$
3. $(4.2 \times 10^8)(2.5 \times 10^{12})$
4. $(2.5 \times 10^5)(4 \times 10^7)$
5. $(4 \times 10^{-3})(2 \times 10^{-8})$
6. $(6 \times 10^{-7})(4 \times 10^{-5})$
7. $(2 \times 10^{-4})(6.5 \times 10^{-8})$
8. $(6 \times 10^{-4})(4 \times 10^8)$
9. $(1.6 \times 10^{-5})(7 \times 10^{-7})$
10. $(7 \times 10^{-9})(3 \times 10^5)$
11. $(1.4 \times 10^7)(8 \times 10^{-5})$
12. $(7.5 \times 10^{-8})(4 \times 10^6)$

Lesson 93 Solve:

1. $3x - 5 = 40$
2. $15 = 2x - 19$
3. $12 + 2x = 60$
4. $80 = 4x - 16$
5. $8x - 16 = 56$
6. $3x + 12 = 54$
7. $0.8x - 1 = 1.4$
8. $0.3w + 1.2 = 3$
9. $\frac{3}{4}w - 12 = 60$
10. $3\frac{1}{3}m + 30 = 120$
11. $-4w + 20 = 8$
12. $-0.2y + 1.4 = 3.2$

Lesson 101 Write and solve an equation for each problem:

1. Six more than twice what number is 72?
2. Five less than the product of 8 and what number is 27?
3. Ten less than half of what number is 50?
4. What number is 12 more than the product of 6 and 4?
5. The sum of what number and 6 is 5 less than 12?
6. Three fourths of what number is 12 less than 60?

Lesson 102 Simplify and solve the following equations:

1. $5m + 6 + m - 18 = 60$
2. $3x + 20 = x + 80$
3. $3(x - 4) = 36$
4. $x + 2(x - 4) = 24 - x$
5. What is the measure of the smallest angle in this figure?

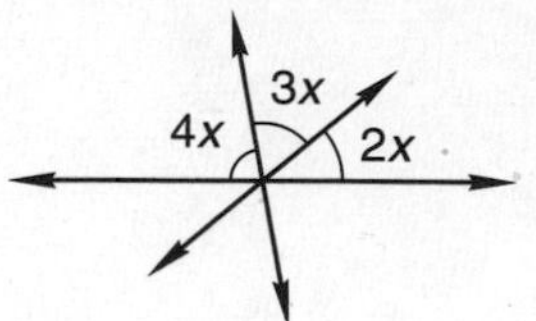

6. Find the measure of the largest angle in this triangle.

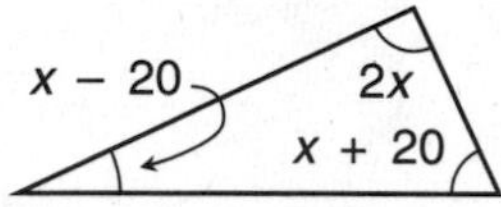

Lesson 111 Write each quotient in scientific notation:

1. $\dfrac{8 \times 10^8}{4 \times 10^4}$
2. $\dfrac{6 \times 10^3}{3 \times 10^6}$
3. $\dfrac{3.6 \times 10^6}{2 \times 10^{12}}$
4. $\dfrac{1.2 \times 10^8}{3 \times 10^4}$
5. $\dfrac{2.4 \times 10^{12}}{8 \times 10^7}$
6. $\dfrac{3 \times 10^7}{4 \times 10^5}$
7. $\dfrac{4.2 \times 10^6}{7 \times 10^9}$
8. $\dfrac{1 \times 10^8}{2 \times 10^{12}}$
9. $\dfrac{1.8 \times 10^7}{6 \times 10^{11}}$
10. $\dfrac{7.5 \times 10^{12}}{5 \times 10^7}$
11. $\dfrac{6.3 \times 10^8}{9 \times 10^4}$
12. $\dfrac{4 \times 10^6}{5 \times 10^{10}}$

GLOSSARY

absolute value The distance from the graph of a number to the number 0 on a number line. The symbol for absolute value is a vertical bar on each side of a numeral or variable, e.g., $|-x|$.

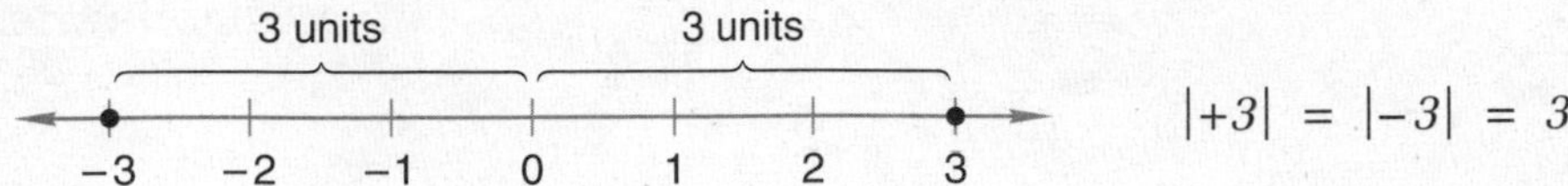

*Since the graphs of −3 and +3 are both 3 units from the number 0, the **absolute value** of both numbers is 3.*

acute angle An angle whose measure is between 0° and 90°.

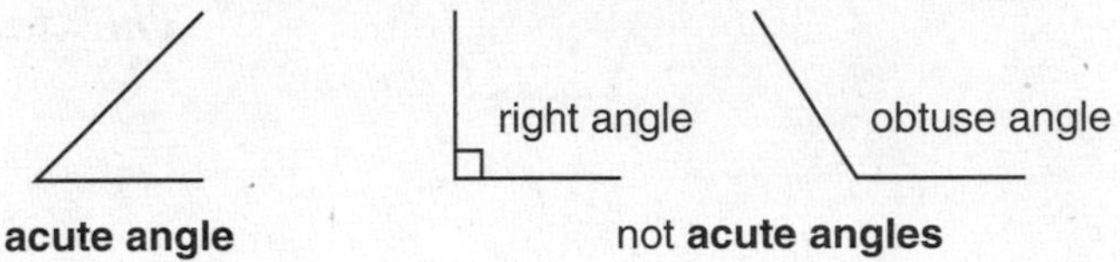

*An **acute angle** is smaller than both a right angle and an obtuse angle.*

acute triangle A triangle whose largest angle measures between 0° and 90°.

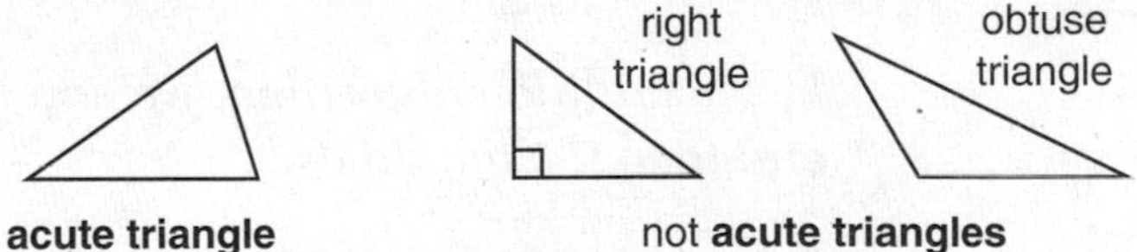

addend One of two or more numbers that are added to find a sum.

7 + 3 = 10 *The **addends** in this problem are 7 and 3.*

additive identity The number 0. *See also* **identity property of addition.**

7 + 0 = 7

↑

additive identity

*We call zero the **additive identity** because adding zero to any number does not change the number.*

adjacent angles Two angles that have a common side and a common vertex. The angles lie on opposite sides of their common side.

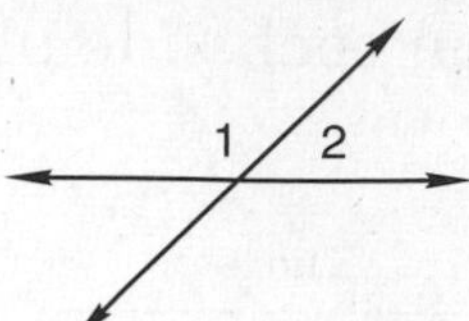

*∠1 and ∠2 are **adjacent angles.** They share a common side and a common vertex.*

adjacent sides In a polygon, two sides that intersect to form a vertex.

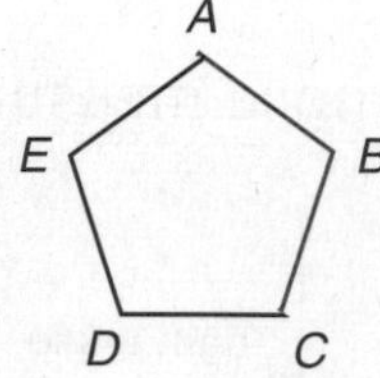

*$\overline{AB}$ and $\overline{BC}$ are **adjacent sides.** They form vertex B.*

algebraic addition The combining of positive and negative numbers to form a sum.

*We use **algebraic addition** to find the sum of −3, +2, and −11:*

$$(-3) + (+2) + (-11) = -12$$

algorithm Any process for solving a mathematical problem.

*In the addition **algorithm,** we add the ones first, then the tens, and then the hundreds.*

alternate exterior angles A special pair of angles formed when a transversal intersects two lines. Alternate exterior angles lie on opposite sides of the transversal and are outside the two intersected lines.

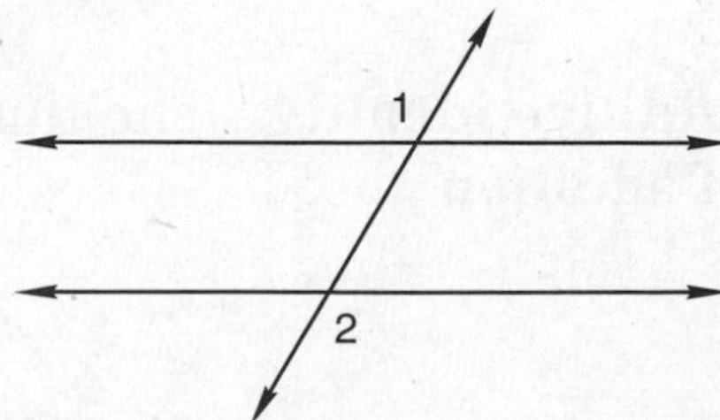

*∠1 and ∠2 are **alternate exterior angles.** When a transversal intersects parallel lines, as in this figure, **alternate exterior angles** have the same measure.*

alternate interior angles A special pair of angles formed when a transversal intersects two lines. Alternate interior angles lie on opposite sides of the transversal and are inside the two intersected lines.

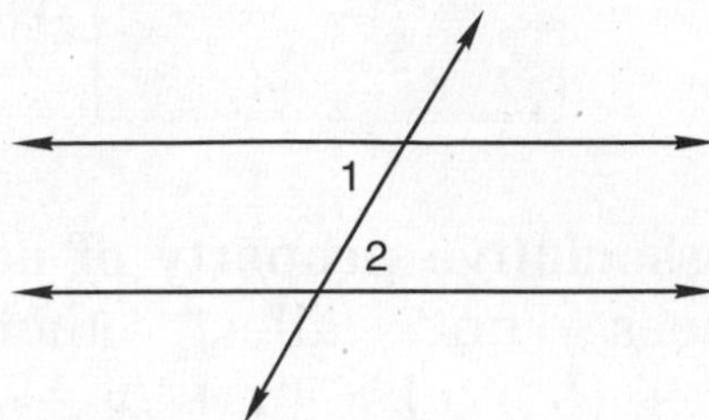

*∠1 and ∠2 are **alternate interior angles.** When a transversal intersects parallel lines, as in this figure, **alternate interior angles** have the same measure.*

altitude The perpendicular distance from the base of a triangle to the opposite vertex; also called *height.*

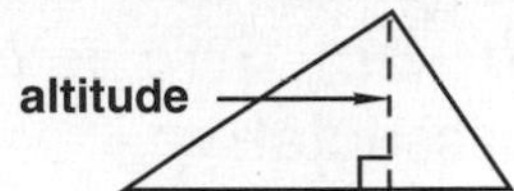

angle The opening that is formed when two lines, rays, or segments intersect.

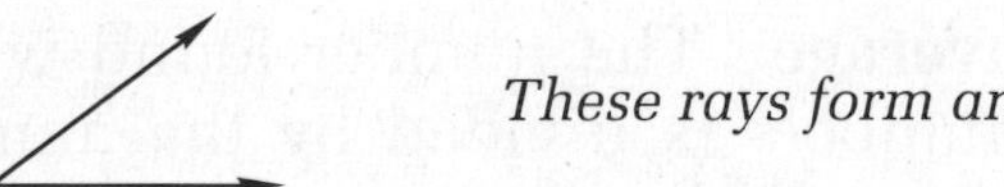

*These rays form an **angle.***

angle bisector A line, ray, or line segment that divides an angle into two equal halves.

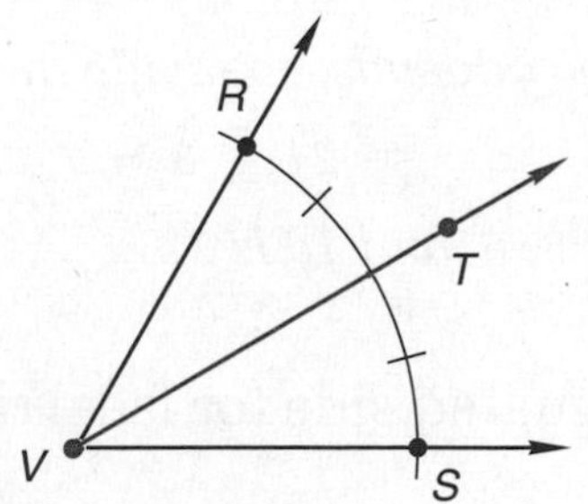

*$\overrightarrow{VT}$ is an **angle bisector.** It divides ∠RVS into two equal halves.*

arc Part of a circle.

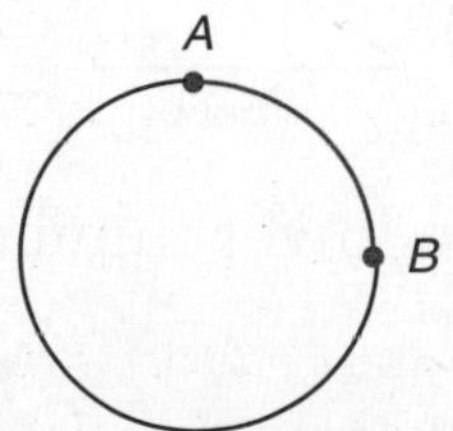

*The portion of the circle between points A and B is **arc** AB.*

area The size of the inside of a flat shape. Area is measured in square units.

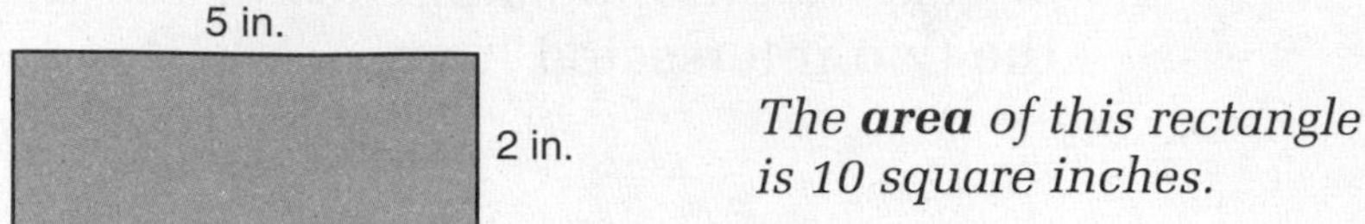

*The **area** of this rectangle is 10 square inches.*

associative property of addition The grouping of addends does not affect their sum. In symbolic form, $a + (b + c) = (a + b) + c$. Unlike addition, subtraction is not associative.

$(8 + 4) + 2 = 8 + (4 + 2)$
*Addition is **associative.***

$(8 - 4) - 2 \neq 8 - (4 - 2)$
*Subtraction is not **associative.***

associative property of multiplication The grouping of factors does not affect their product. In symbolic form, $a \times (b \times c) = (a \times b) \times c$. Unlike multiplication, division is not associative.

$(8 \times 4) \times 2 = 8 \times (4 \times 2)$
*Multiplication is **associative.***

$(8 \div 4) \div 2 \neq 8 \div (4 \div 2)$
*Division is not **associative.***

average The number found when the sum of two or more numbers is divided by the number of addends in the sum; also called *mean.*

*To find the **average** of the numbers 5, 6, and 10, add.*

$$5 + 6 + 10 = 21$$

There were three addends, so divide the sum by 3.

$$21 \div 3 = 7$$

*The **average** of 5, 6, and 10 is 7.*

base (1) A designated side (or face) of a geometric figure.

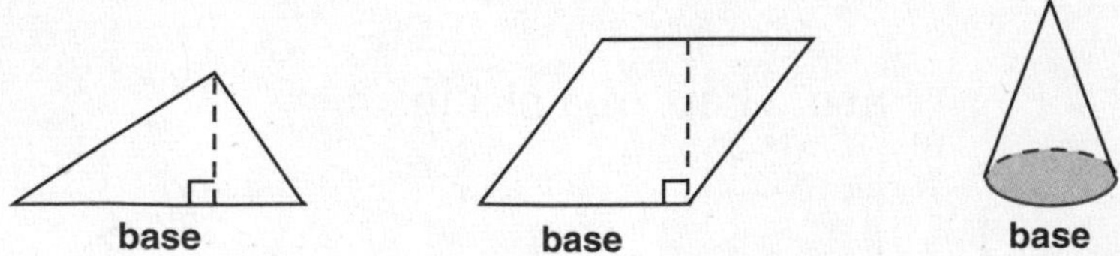

(2) The lower number in an exponential expression.

base → 5^3 ← *exponent*

5^3 means $5 \times 5 \times 5$, and its value is 125.

bisect To divide a segment or angle into two equal halves.

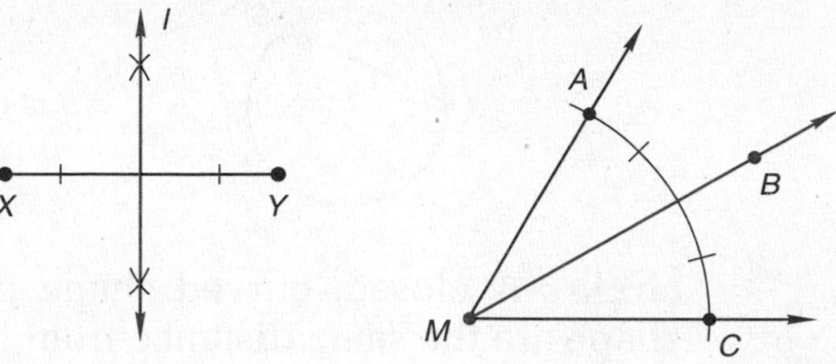

Line l ***bisects*** *$\overline{XY}$.* *Ray MB* ***bisects*** *∠AMC.*

box-and-whisker plot A method of displaying data that involves splitting the numbers into four groups of equal size.

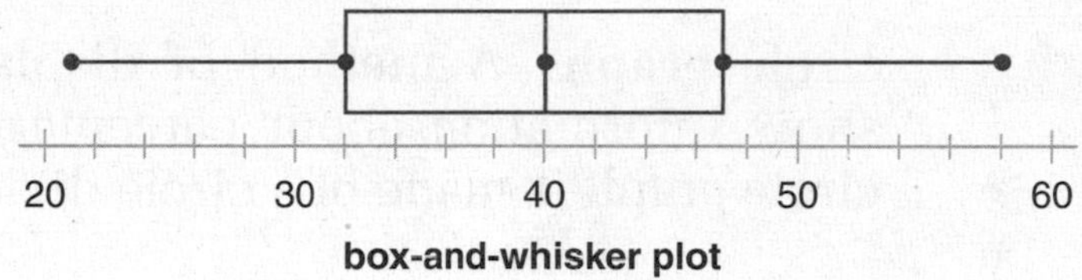

box-and-whisker plot

center The point inside a circle or sphere from which all points on the circle or sphere are equally distant.

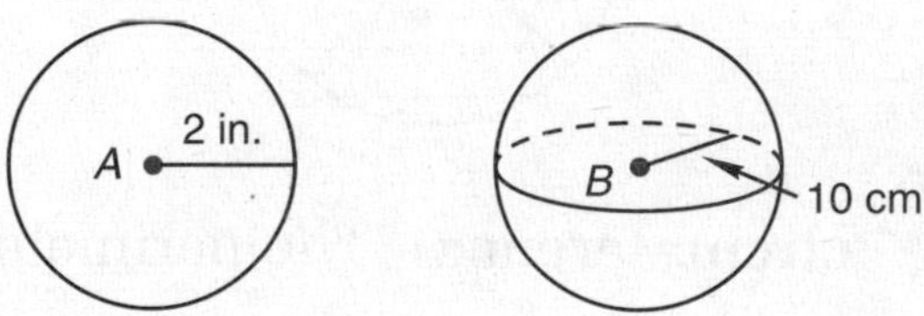

The ***center*** *of circle A is 2 inches from every point on the circle.*

The ***center*** *of sphere B is 10 centimeters from every point on the sphere.*

central angle An angle whose vertex is the center of a circle.

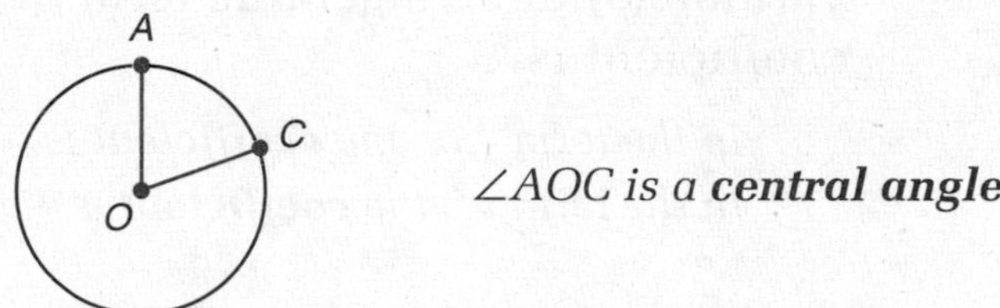

∠AOC is a ***central angle.***

chance A way of expressing the likelihood of an event; the probability of an event expressed as a percent.

The ***chance*** *of snow is 10%. It is not likely to snow.*

There is an 80% ***chance*** *of rain. It is likely to rain.*

chord A segment whose endpoints lie on a circle.

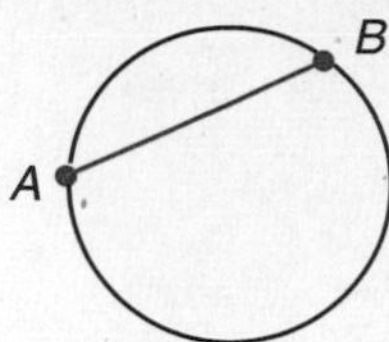

*$\overline{AB}$ is a **chord** of the circle.*

circle A closed, curved shape in which all points on the shape are the same distance from its center.

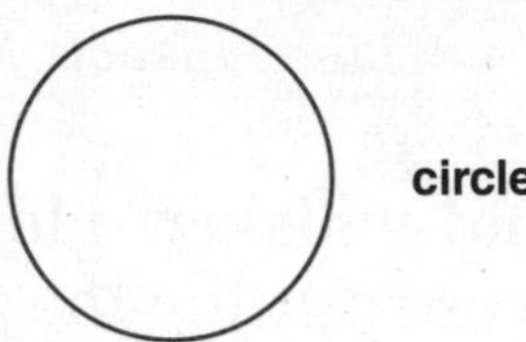

circle graph A method of displaying data, often used to show information about percentages or parts of a whole. A circle graph is made of a circle divided into sectors.

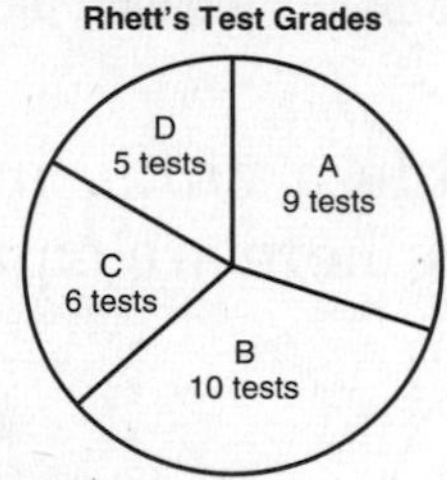

*This **circle graph** shows data for Rhett's test grades.*

circumference The perimeter of a circle.

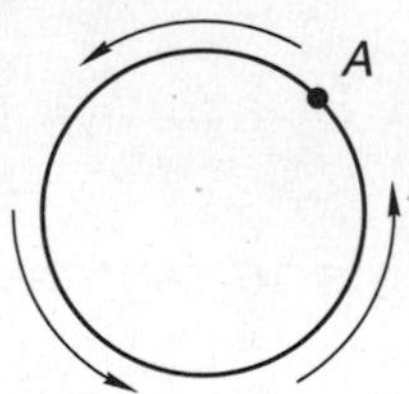

*If the distance from point A around to point A is 3 inches, then the **circumference** of the circle is 3 inches.*

coefficient In common use, the number that multiplies the variable(s) in an algebraic term. If no number is specified, the coefficient is 1.

*In the term $-3x$, the **coefficient** is -3.*
*In the term y^2, the **coefficient** is 1.*

commutative property of addition Changing the order of addends does not change their sum. In symbolic form, $a + b = b + a$. Unlike addition, subtraction is not commutative.

$8 + 2 = 2 + 8$	$8 - 2 \neq 2 - 8$
*Addition is **commutative.***	*Subtraction is not **commutative.***

commutative property of multiplication Changing the order of factors does not change their product. In symbolic form, $a \times b = b \times a$. Unlike multiplication, division is not commutative.

$8 \times 2 = 2 \times 8$	$8 \div 2 \neq 2 \div 8$
*Multiplication is **commutative.***	*Division is not **commutative.***

compass A tool used to draw circles and arcs.

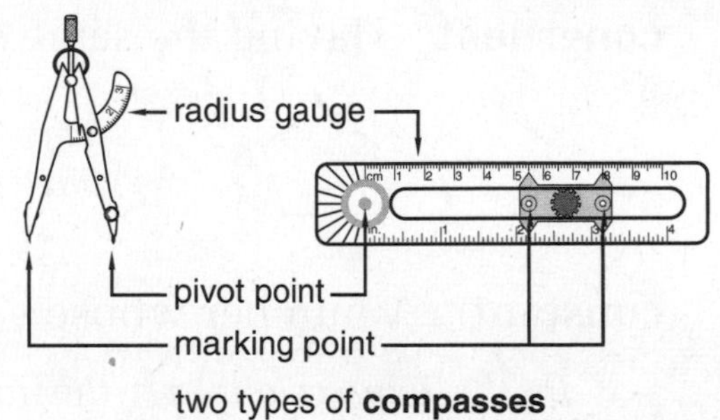

two types of **compasses**

complementary angles Two angles whose sum is 90°.

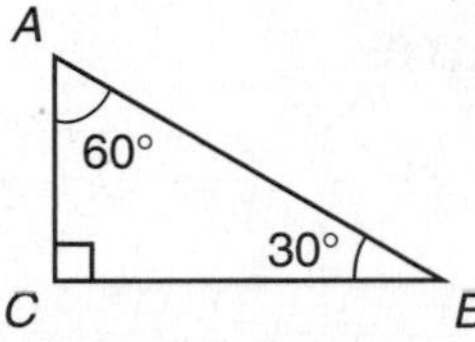

*$\angle A$ and $\angle B$ are **complementary angles.***

complex fraction A fraction that contains one or more fractions in its numerator or denominator.

$\frac{\frac{3}{5}}{\frac{2}{3}}$ $\frac{25\frac{2}{3}}{100}$ $\frac{15}{7\frac{1}{3}}$ $\frac{\frac{a}{b}}{\frac{b}{c}}$ — **complex fractions**

$\frac{1}{2}$ $\frac{12}{101}$ $\frac{xy}{z}$ — not **complex fractions**

composite number A counting number greater than 1 that is divisible by a number other than itself and 1. Every composite number has three or more factors.

*9 is divisible by 1, 3, and 9. It is **composite.***
*11 is divisible by 1 and 11. It is not **composite.***

compound interest Interest that pays on previously earned interest.

Compound Interest

$100.00	principal
+ $6.00	first-year interest (6% of $100.00)
$106.00	total after one year
+ $6.36	second-year interest (6% of $106.00)
$112.36	total after two years

Simple Interest

$100.00	principal
$6.00	first-year interest
+ $6.00	second-year interest
$112.00	total after two years

concentric circles Two or more circles with a common center.

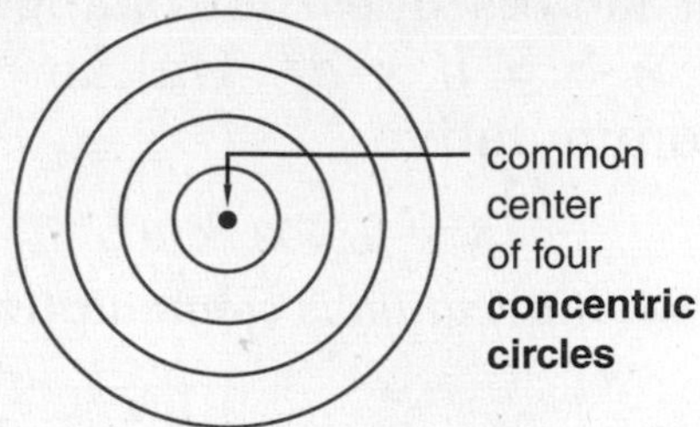

congruent Having the same size and shape.

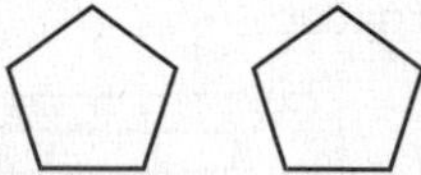

These polygons are ***congruent.*** *They have the same size and shape.*

constant A number whose value does not change.

In the expression $2\pi r$*, the numbers 2 and* π *are* ***constants,*** *while r is a variable.*

construction Using a compass and/or a straightedge to create geometric figures.

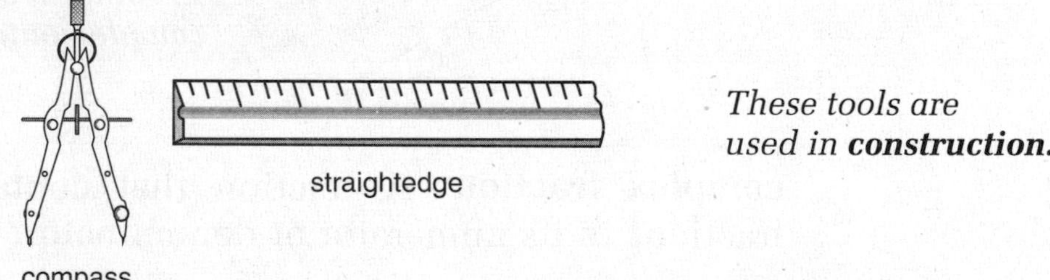

These tools are used in ***construction.***

coordinate(s) (1) A number used to locate a point on a number line.

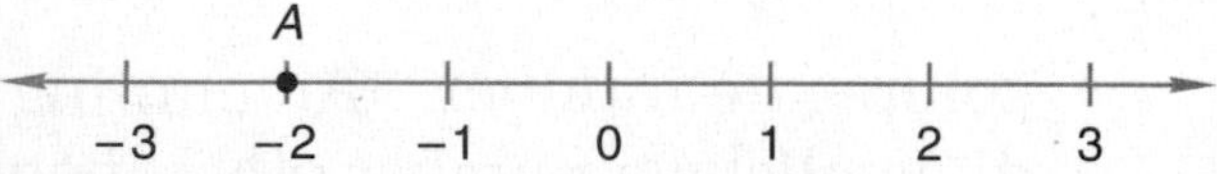

The ***coordinate*** *of point A is –2.*

(2) An ordered pair of numbers used to locate a point in a coordinate plane.

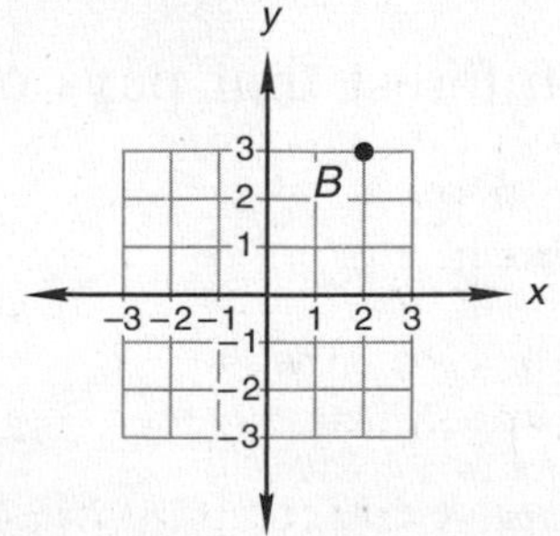

The ***coordinates*** *of point B are (2, 3). The x-coordinate is listed first, the y-coordinate second.*

coordinate plane A grid on which any point can be identified by an ordered pair of numbers.

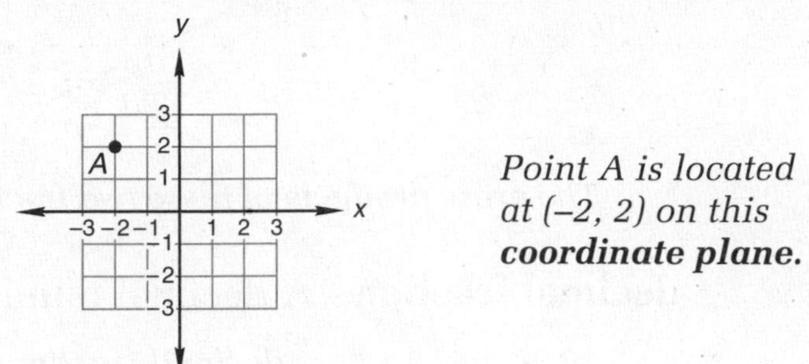

Point A is located at (–2, 2) on this ***coordinate plane.***

corresponding angles A special pair of angles formed when a transversal intersects two lines. Corresponding angles lie on the same side of the transversal and are in the same position relative to the two intersected lines.

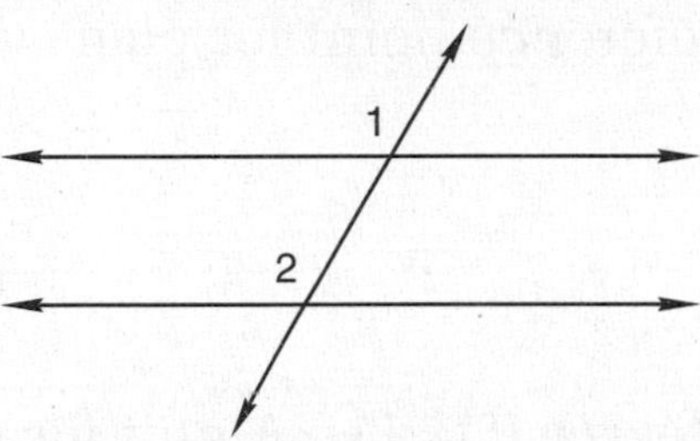

$\angle 1$ and $\angle 2$ are ***corresponding angles.*** *When a transversal intersects parallel lines, as in this figure,* ***corresponding angles*** *have the same measure.*

corresponding parts Sides or angles of similar polygons that occupy the same relative positions.

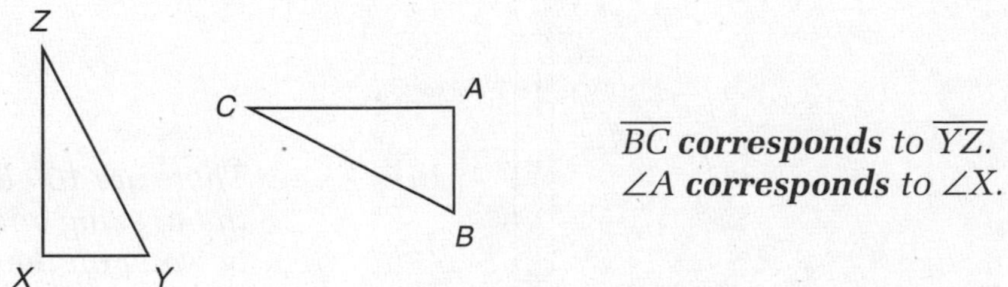

$\overline{BC}$ ***corresponds*** *to $\overline{YZ}$.*
$\angle A$ ***corresponds*** *to $\angle X$.*

counting numbers The numbers used to count; the members of the set {1, 2, 3, 4, 5, …}. Also called *natural numbers*.

1, 24, and 108 are ***counting numbers.***

–2, 3.14, 0, and $2\frac{7}{9}$ are not ***counting numbers.***

cross product The product of the numerator of one fraction and the denominator of another.

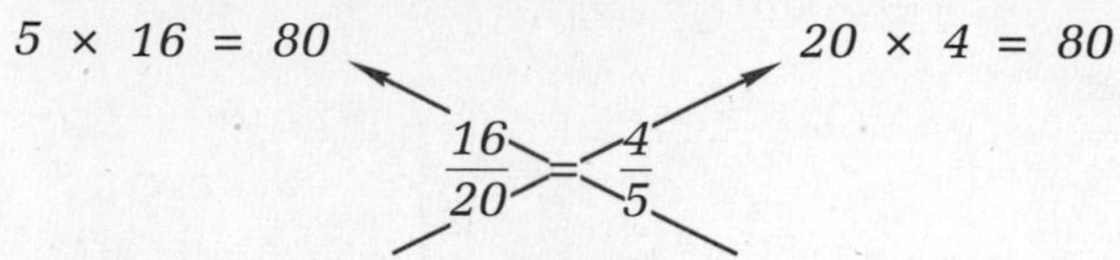

*The **cross products** of these two fractions are equal.*

decimal fraction A decimal number.

*36.28 and 9.12 are **decimal fractions.***
*$3\frac{1}{3}$ and $\frac{-12}{19}$ are not **decimal fractions.***

decimal number A numeral that contains a decimal point.

*23.94 is a **decimal number** because it contains a decimal point.*

decimal point The symbol in a decimal number used as a reference point for place value.

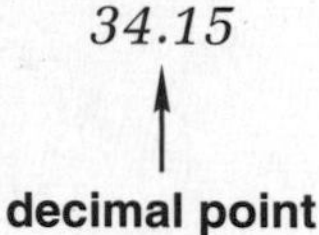

degree (°) (1) A unit for measuring angles.

*There are 90 **degrees** (90°) in a right angle.*

*There are 360 **degrees** (360°) in a circle.*

(2) A unit for measuring temperature.

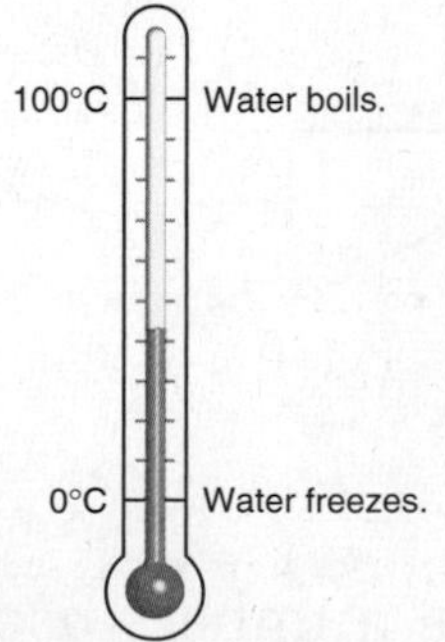

*There are 100 **degrees** between the freezing and boiling points of water on the Celsius scale.*

denominator The bottom term of a fraction.

$\frac{5}{9}$ ← numerator, ← **denominator**

dependent events Two events are *dependent* if the outcome of one event affects the probability that the other event will occur.

Whitney draws a card from a regular deck and does not replace it. Then Chad draws a card from the same deck. The probability of Chad's drawing the queen of hearts is ***dependent*** *on what card Whitney draws. If Whitney draws the queen of hearts, the probability of Chad's drawing it is 0. If Whitney draws a different card, the probability of Chad's drawing the queen of hearts is* $\frac{1}{51}$.

diagonal A line segment, other than a side, that connects two vertices of a polygon.

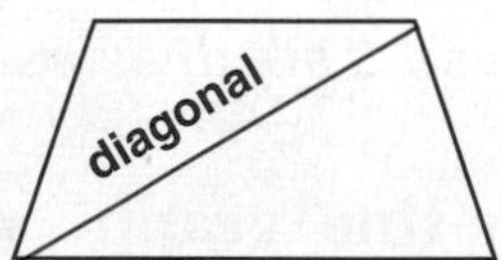

diameter The distance across a circle through its center.

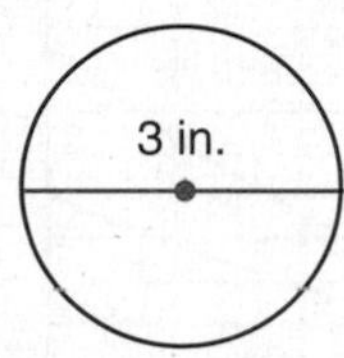

The ***diameter*** *of this circle is 3 inches.*

difference The result of subtraction.

$12 - 8 = 4$ *The* ***difference*** *in this problem is 4.*

digit Any of the symbols used to write numbers: 0, 1, 2, 3, 4, 5, 6, 7, 8, 9.

The last ***digit*** *in the number 7862 is 2.*

directed numbers *See* **signed numbers.**

distributive property A number times the sum of two addends is equal to the sum of that same number times each individual addend: $a \times (b + c) = (a \times b) + (a \times c)$.

$8 \times (2 + 3) = (8 \times 2) + (8 \times 3)$

Multiplication is ***distributive*** *over addition.*

dividend A number that is divided.

$12 \div 3 = 4$ $\quad 3\overline{)12}$ with quotient 4 $\quad \frac{12}{3} = 4$ *The* ***dividend*** *is 12 in each of these problems.*

divisible Able to be divided by a whole number without a remainder.

$4\overline{)20}$ with quotient 5 — *The number 20 is **divisible** by 4, since 20 ÷ 4 has no remainder.*

$3\overline{)20}$ with quotient 6 R 2 — *The number 20 is not **divisible** by 3, since 20 ÷ 3 has a remainder.*

divisor (1) A number by which another number is divided.

$12 \div 3 = 4$ $\quad$ $3\overline{)12}$ with quotient 4 $\quad$ $\frac{12}{3} = 4$ — *The **divisor** is 3 in each of these problems.*

(2) A factor of a number.

*2 and 5 are **divisors** of 10.*

double-line graph A method of displaying a set of data, often used to compare two performances over time.

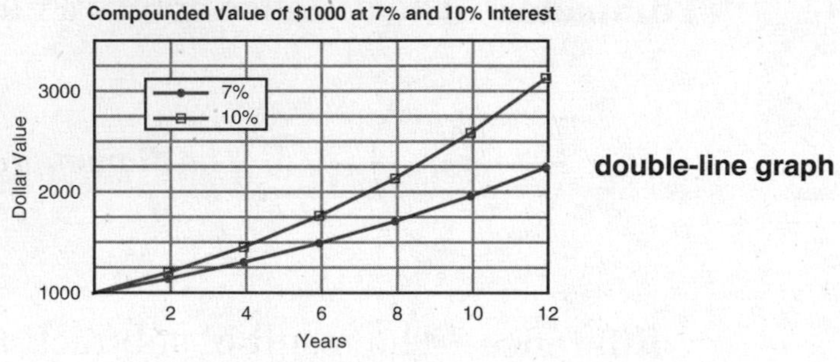

double-line graph

edge A line segment formed where two faces of a polyhedron intersect.

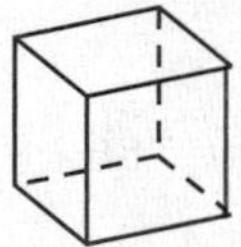

*One **edge** of this cube is in color. A cube has 12 **edges**.*

equation A statement that uses the symbol "=" to show that two quantities are equal.

$x = 3$	$3 + 7 = 10$	$4 + 1$	$x < 7$
equations		not equations	

equilateral triangle A triangle in which all sides are the same length.

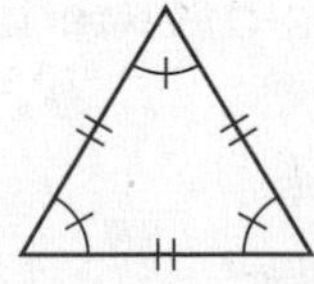

*This is an **equilateral triangle.** All of its sides are the same length.*

equivalent fractions Different fractions that name the same amount.

$\frac{1}{2}$ = $\frac{2}{4}$

*$\frac{1}{2}$ and $\frac{2}{4}$ are **equivalent fractions.***

estimate To determine an approximate value.

*I **estimate** that the sum of 199 and 205 is about 400.*

evaluate To find the value of an expression.

*To **evaluate** $a + b$ for $a = 7$ and $b = 13$, we replace a with 7 and b with 13:*

$$7 + 13 = 20$$

expanded notation A way of writing a number as the sum of the products of the digits and the place values of the digits.

*In **expanded notation** 6753 is written*

$$(6 \times 1000) + (7 \times 100) + (5 \times 10) + (3 \times 1).$$

exponent The upper number in an exponential expression that shows how many times the base is to be used as a factor.

base → 5^3 ← ***exponent***

5^3 means $5 \times 5 \times 5$, and its value is 125.

exponential expression An expression that indicates that the base is to be used as a factor the number of times shown by the exponent.

$$4^3 = 4 \times 4 \times 4 = 64$$

*The **exponential expression** 4^3 is evaluated by using 4 as a factor 3 times. Its value is 64.*

exterior angle In a polygon, the supplementary angle of an interior angle.

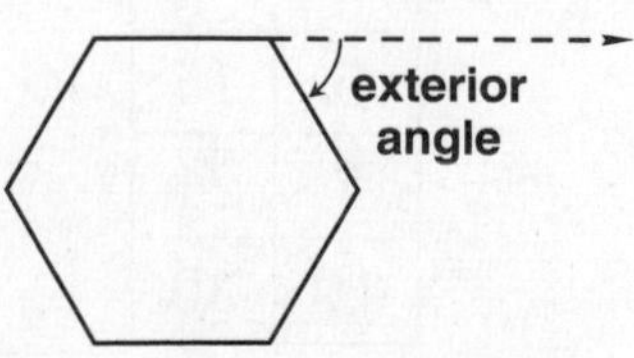

face A flat surface of a geometric solid.

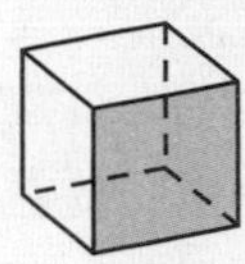

*One **face** of the cube is shaded.*
*A cube has six **faces**.*

fact family A group of three numbers related by addition and subtraction or by multiplication and division. Four mathematical fact statements can be formed using the numbers in a fact family.

*The numbers 3, 4, and 7 are a **fact family**. They make these four facts:*

$3 + 4 = 7 \quad 4 + 3 = 7 \quad 7 - 3 = 4 \quad 7 - 4 = 3$

factor (1) Noun: One of two or more numbers that are multiplied.

$5 \times 6 = 30$ *The **factors** in this problem are 5 and 6.*

(2) Noun: A whole number that divides another whole number without a remainder.

*The numbers 5 and 6 are **factors** of 30.*

(3) Verb: To write as a product of factors.

*We can **factor** the number 6 by writing it as 2×3.*

fraction A number that names part of a whole.

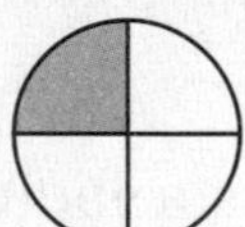

$\frac{1}{4}$ of the circle is shaded.
*$\frac{1}{4}$ is a **fraction**.*

function A rule for using one number (an input) to calculate another number (an output). Each input produces only one output.

$y = 3x$

x	y
3	9
5	15
7	21
10	30

*There is exactly one resulting number for every number we multiply by 3. Thus, $y = 3x$ is a **function**.*

geometric solid A three-dimensional geometric figure.

geometric solids | not **geometric solids**

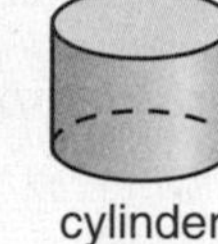

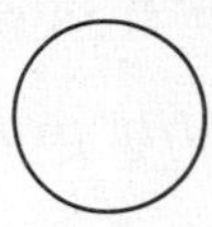

cube | cylinder | circle | rectangle | hexagon

geometry A major branch of mathematics that deals with shapes, sizes, and other properties of figures.

*Some figures we study in **geometry** are angles, circles, and polygons.*

greatest common factor (GCF) The largest whole number that is a factor of two or more indicated numbers.

The factors of 12 are 1, 2, 3, 4, 6, and 12.
The factors of 18 are 1, 2, 3, 6, 9, and 18.
*The **greatest common factor** of 12 and 18 is 6.*

height The perpendicular distance from the base to the opposite side of a parallelogram or trapezoid; from the base to the opposite face of a prism or cylinder; or from the base to the opposite vertex of a triangle, pyramid, or cone. *See also* **altitude.**

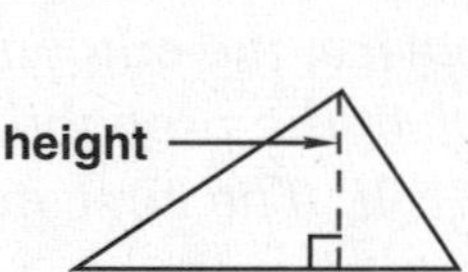

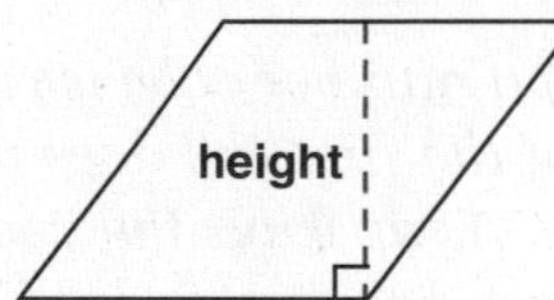

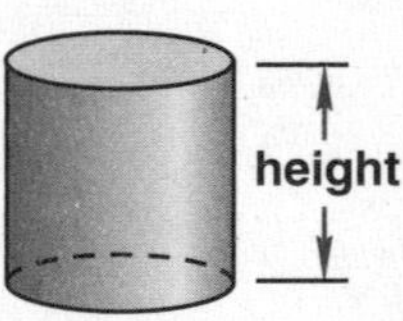

histogram A method of displaying a range of data. A histogram is a special type of bar graph that displays data in intervals of equal size with no space between bars.

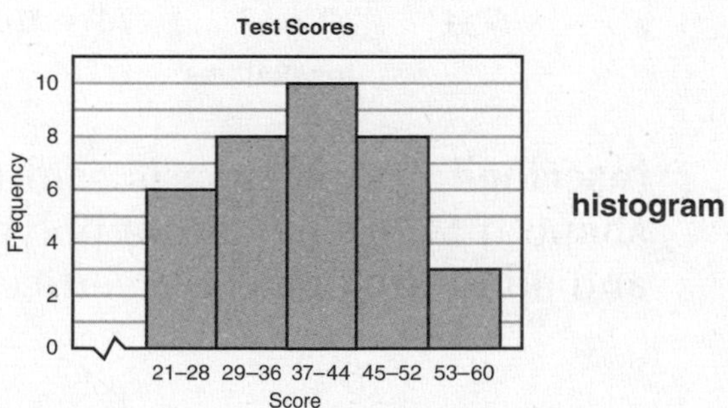

hypotenuse The longest side of a right triangle.

hypotenuse

*The **hypotenuse** of a right triangle is always the side opposite the right angle.*

identity property of addition The sum of any number and 0 is equal to the initial number. In symbolic form, $a + 0 = a$. The number 0 is referred to as the *additive identity.*

*The **identity property of addition** is shown by this statement:*

$$13 + 0 = 13$$

identity property of multiplication The product of any number and 1 is equal to the initial number. In symbolic form, $a \times 1 = a$. The number 1 is referred to as the *multiplicative identity.*

*The **identity property of multiplication** is shown by this statement:*

$$94 \times 1 = 94$$

improper fraction A fraction with a numerator equal to or greater than the denominator.

$\frac{12}{12}$, $\frac{57}{3}$, *and* $2\frac{13}{2}$ *are **improper fractions.***
*All **improper fractions** are greater than or equal to 1.*

independent events Two events are *independent* if the outcome of one event does not affect the probability that the other event will occur.

*If a number cube is rolled twice, the outcome (1, 2, 3, 4, 5, or 6) of the first roll does not affect the probability of getting 1, 2, 3, 4, 5, or 6 on the second roll. The first and second rolls are **independent events.***

inequalities Algebraic statements that have ≠, <, >, ≤, or ≥ as their symbols of comparison.

$x \le 4$	$2 < 7$	$11 \ne 10$	$x = 2$	$9 + 10$
	inequalities		not **inequalities**	

inscribed A polygon is said to be *inscribed* within another shape if all points of the polygon lie within the other shape, and all of the polygon's vertices lie on the other shape.

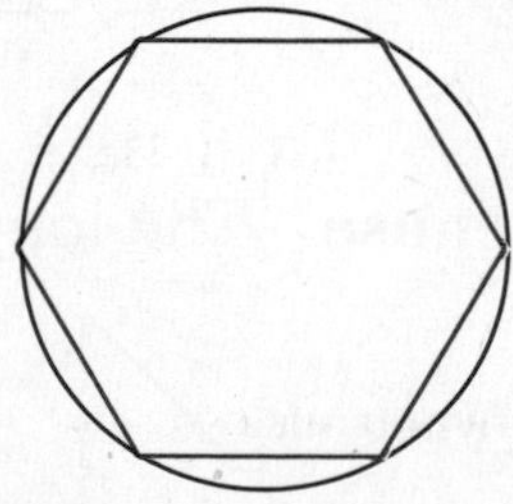

*The hexagon is **inscribed** within the circle.*

inscribed angle An angle that opens to the interior of a circle and has its vertex on the circle.

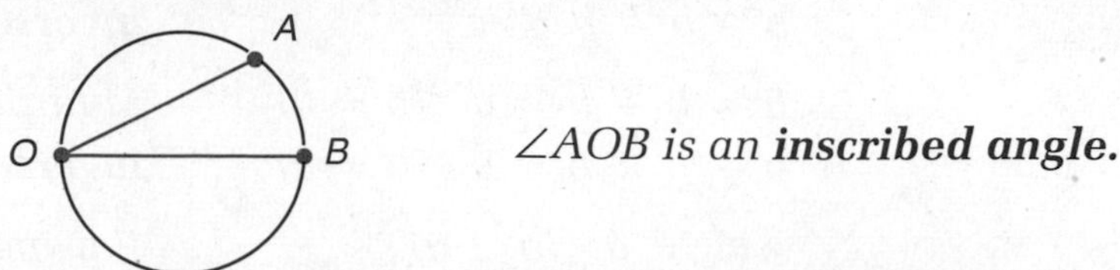

∠AOB is an ***inscribed angle.***

integers The set of counting numbers, their opposites, and zero; the members of the set {..., –2, –1, 0, 1, 2, ...}.

–57 and 4 are ***integers.*** $\frac{15}{8}$ *and –0.98 are not* ***integers.***

interest An additional amount added to a loan, account, or fund, usually based on a percentage of the principal; the difference between the principal and the total amount owed (with loans) or earned (with accounts and funds).

If we borrow \$500.00 from the bank and repay the bank \$575.00 for the loan, the ***interest*** *on the loan is \$575.00 – \$500.00 = \$75.00.*

interest rate A percent that determines the amount of interest paid on a loan over a period of time.

If we borrow \$1000.00 and pay back \$1100.00 after one year, our ***interest rate*** *is*

$$\frac{\$1100.00 - \$1000.00}{\$1000.00} \times 100\% = 10\% \text{ per year}$$

interior angle An angle that opens to the inside of a polygon.

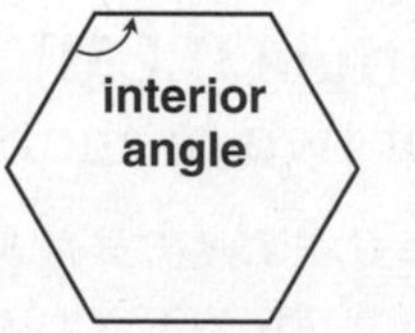

This hexagon has six ***interior angles.***

International System *See* **metric system.**

intersect To share a common point or points.

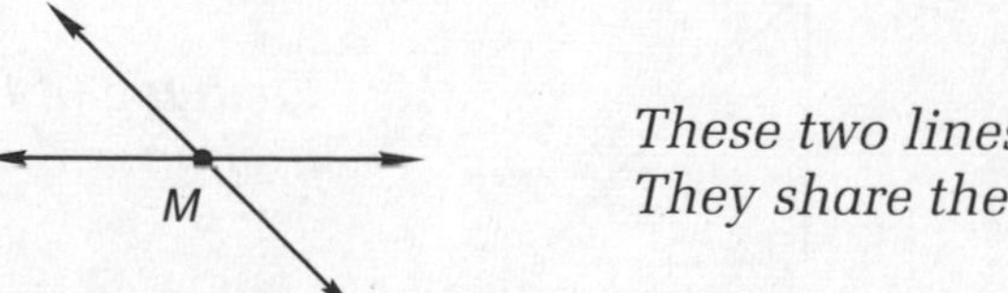

These two lines ***intersect.*** *They share the common point M.*

inverse operations Operations that "undo" one another.

$a + b - b = a$	*Addition and subtraction are*
$a - b + b = a$	***inverse operations.***
$a \times b \div b = a \quad (b \neq 0)$	*Multiplication and division are*
$a \div b \times b = a \quad (b \neq 0)$	***inverse operations.***
$\sqrt{a^2} = a \quad (a \geq 0)$	*Raising to powers and finding*
$(\sqrt{a})^2 = a \quad (a \geq 0)$	*roots are **inverse operations.***

invert To switch the numerator and denominator of a fraction.

*If we **invert** $\frac{7}{8}$, we get $\frac{8}{7}$.*

irrational numbers Numbers that cannot be expressed as a ratio of two integers. Their decimal expansions are nonending and nonrepeating.

*π and $\sqrt{3}$ are **irrational numbers.***

isosceles triangle A triangle with at least two sides of equal length.

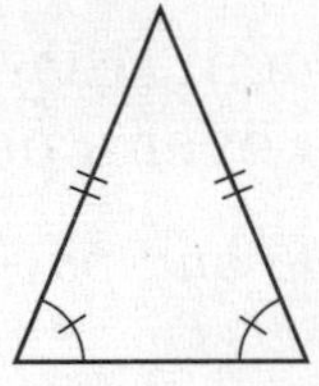

*Two of the sides of this **isosceles triangle** have equal lengths.*

least common denominator (LCD) The least common multiple of the denominators of two or more fractions.

*The **least common denominator** of $\frac{5}{6}$ and $\frac{3}{8}$ is the least common multiple of 6 and 8, which is 24.*

least common multiple (LCM) The smallest whole number that is a multiple of two or more given numbers.

Multiples of 6 are 6, 12, 18, 24, 30, 36, ...
Multiples of 8 are 8, 16, 24, 32, 40, 48, ...
*The **least common multiple** of 6 and 8 is 24.*

legs The two shorter sides of a right triangle that form a 90° angle at their intersection.

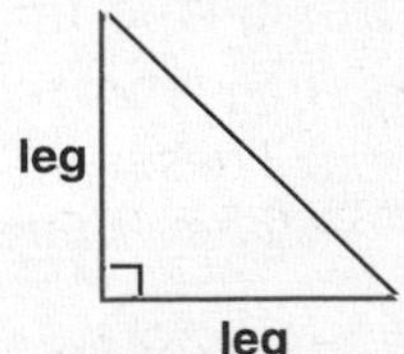

*Each **leg** of this right triangle is shorter than the hypotenuse.*

line A straight collection of points extending in opposite directions without end.

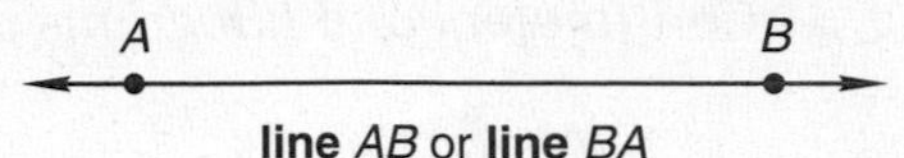

line *AB* or line *BA*

linear equation An equation whose graph is a line.

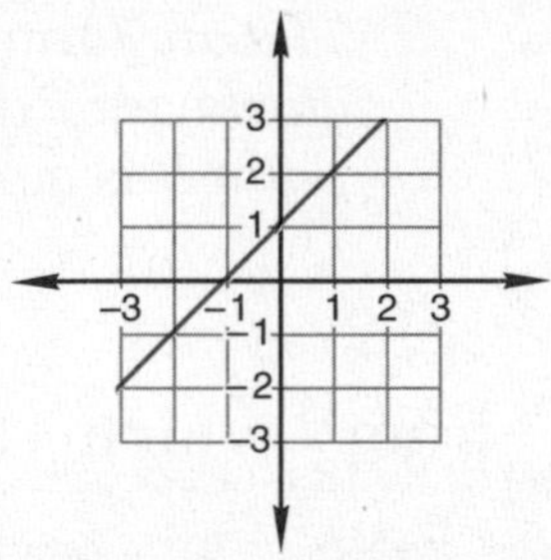

*$y = x + 1$ is a **linear equation** because its graph is a line.*

line of symmetry A line that divides a figure into two halves that are mirror images of each other.

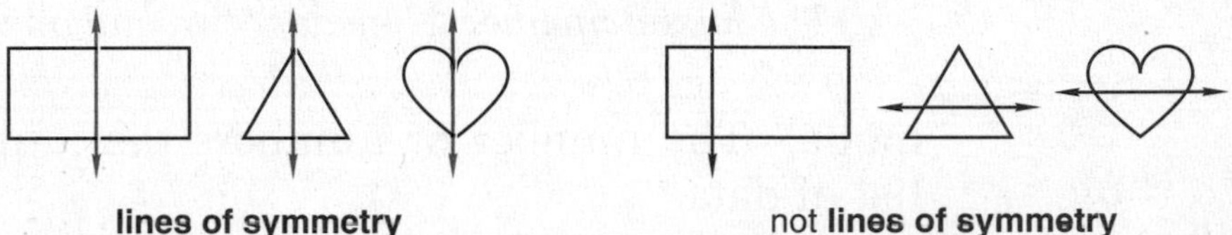

lines of symmetry

not lines of symmetry

lowest terms A fraction is in *lowest terms* if the only common factor of the numerator and the denominator is 1.

*When written in **lowest terms,** the fraction $\frac{8}{16}$ becomes $\frac{1}{2}$.*

major arc An arc whose measure is between 180° and 360°.

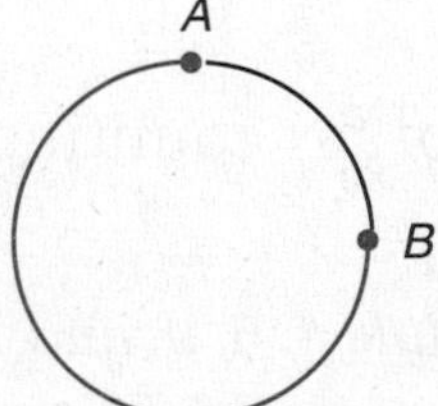

*The arc formed by moving counterclockwise from point A to point B is a **major arc.***

mean *See* **average.**

median The middle number of a list of data when the numbers are arranged in order from the least to the greatest.

1, 1, 2, 5, 6, 7, 9, 15, 24, 36, 44

*In this list of data, 7 is the **median.***

metric system An international system of measurement based on multiples of ten. Also called *International System.*

*Centimeters and kilograms are units in the **metric system.***

minor arc An arc whose measure is between 0° and 180°.

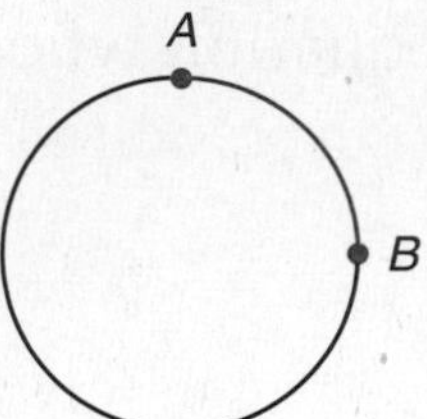

*The arc formed by moving clockwise from point A to point B is a **minor arc.***

minuend A number from which another number is subtracted.

12 − 8 = 4 *The **minuend** in this problem is 12.*

mixed number A whole number and a fraction together.

*The **mixed number** $2\frac{1}{3}$ means "two and one third."*

mode The number or numbers that appear most often in a list of data.

5, 12, 32, 5, 16, 5, 7, 12

*In this list of data, the number 5 is the **mode.***

monomial An algebraic expression that contains only one term.

$3x$	$4ab$	$21mn$	$2 + a$	$x + y + z$	$2r + 3$
	monomials			not monomials	

multiple A product of a counting number and another number.

*The **multiples** of 3 include 3, 6, 9, and 12.*

multiplicative identity The number 1. *See also* **identity property of multiplication.**

$$-2 \times 1 = -2$$

↑ multiplicative identity

*The number 1 is called the **multiplicative identity** because multiplying any number by 1 does not change the number.*

natural numbers *See* **counting numbers.**

negative numbers Numbers less than zero.

–15 and –2.86 are ***negative numbers.***

19 and 0.74 are not ***negative numbers.***

number line A line for representing and graphing numbers. Each point on the line corresponds to a number.

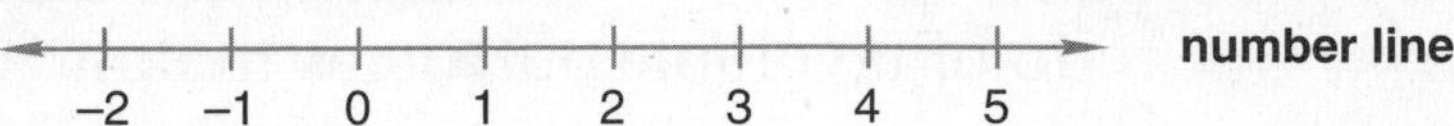

numeral A symbol or group of symbols that represents a number.

4, 72, and $\frac{1}{2}$ are examples of ***numerals.*** *"Four," "seventy-two," and "one-half" are words that name numbers but are not* ***numerals.***

numerator The top term of a fraction.

$\frac{9}{10}$ ← **numerator**
← denominator

oblique line(s) (1) A line that is neither horizontal nor vertical.

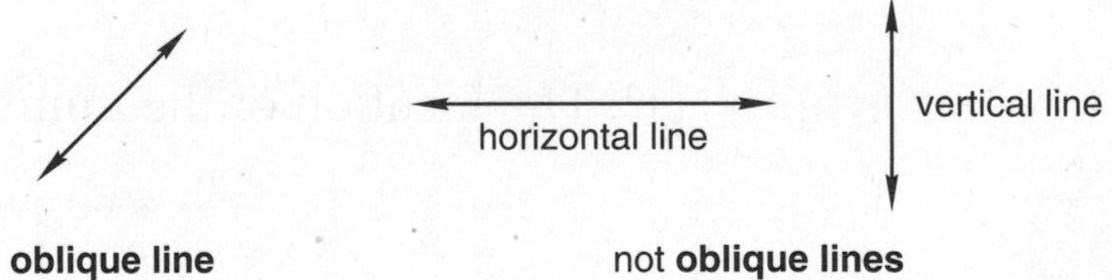

(2) Lines in the same plane that are neither parallel nor perpendicular.

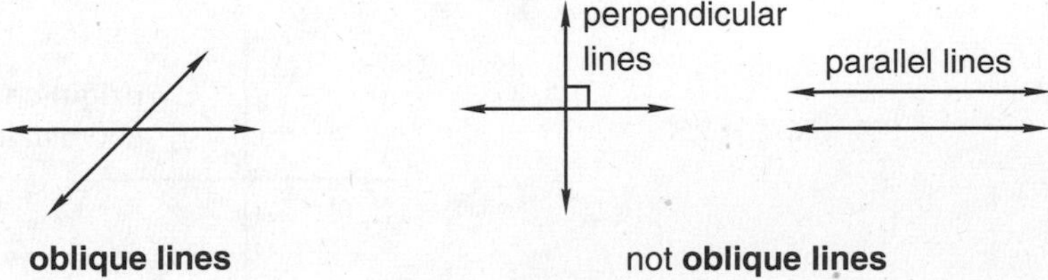

obtuse angle An angle whose measure is between 90° and 180°.

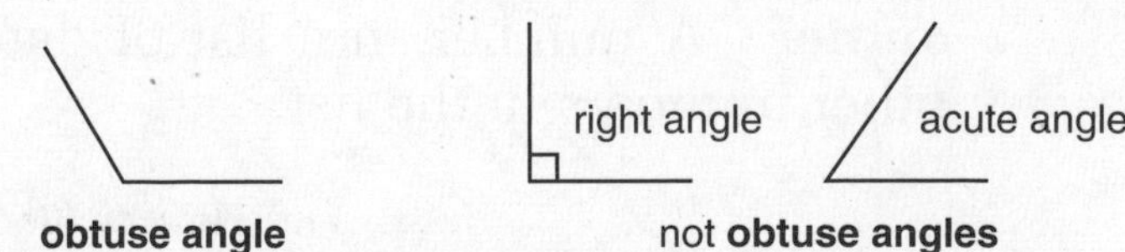

An ***obtuse angle*** *is larger than both a right angle and an acute angle.*

obtuse triangle A triangle whose largest angle measures between 90° and 180°.

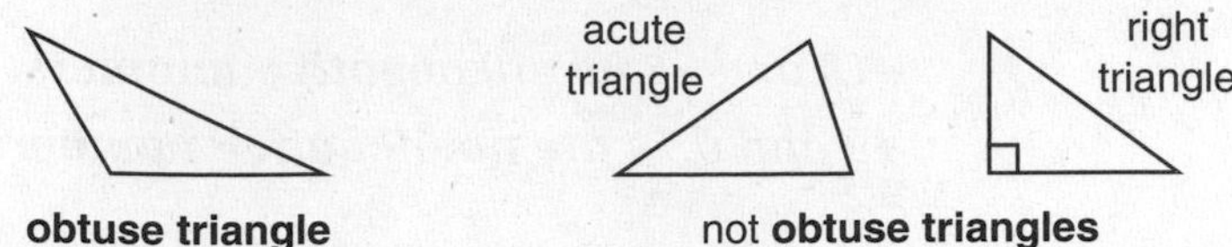

obtuse triangle not **obtuse triangles**

odds A way of describing the likelihood of an event; the ratio of favorable outcomes to unfavorable outcomes.

If you roll a number cube, the ***odds*** *of getting a 3 are 1 to 5.*

operations of arithmetic The four basic mathematical operations: addition, subtraction, multiplication, and division.

$1 + 9 \qquad 21 - 8 \qquad 6 \times 22 \qquad 3 \div 1$

the **operations of arithmetic**

opposites Two numbers whose sum is zero; a positive number and a negative number whose absolute values are equal.

$(-3) + (+3) = 0$

The numbers +3 and –3 are ***opposites.***

origin (1) The location of the number 0 on a number line.

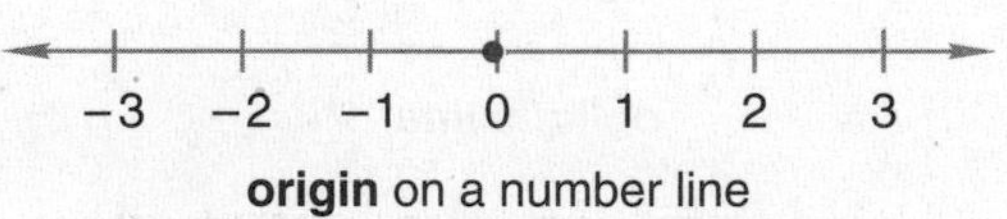

origin on a number line

(2) The point (0, 0) on a coordinate plane.

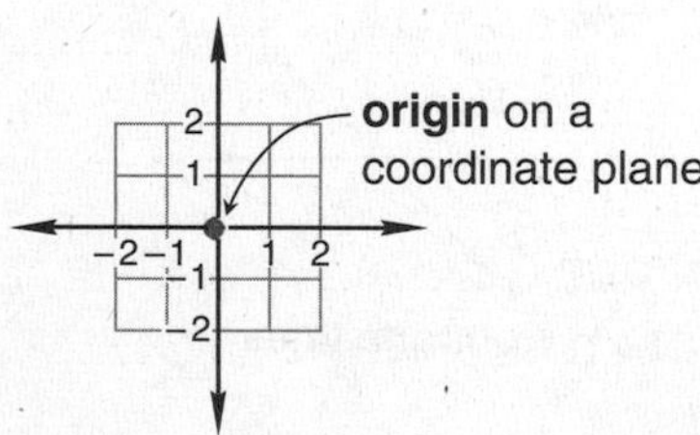

outlier A number in a list of data that is distant from the other numbers in the list.

1, 5, 4, 3, 6, 28, 7, 2

In this list, the number 28 is an ***outlier*** *because it is distant from the other numbers in the list.*

parallel lines Lines in the same plane that do not intersect.

parallelogram A quadrilateral that has two pairs of parallel sides.

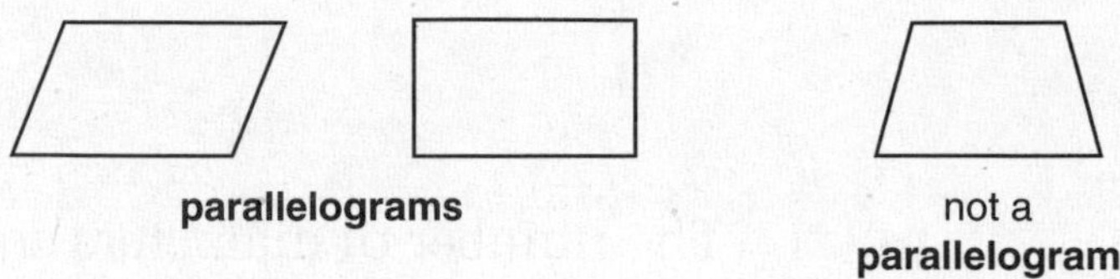

percent A fraction whose denominator of 100 is expressed as a percent sign (%).

$$\frac{99}{100} = 99\% = 99 \textbf{ percent}$$

perfect square The product when a whole number is multiplied by itself.

*The number 9 is a **perfect square** because $9 = 3^2$.*

perimeter The distance around a closed, flat shape.

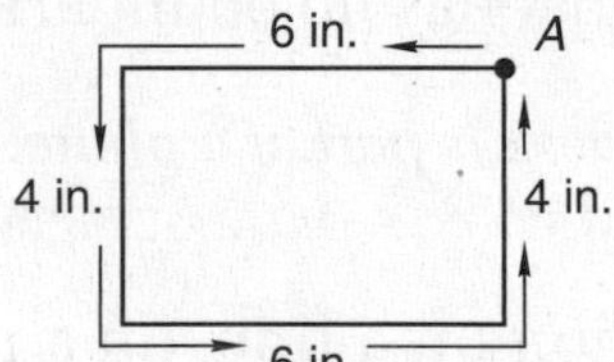

*The **perimeter** of this rectangle (from point A around to point A) is 20 inches.*

permutation One possible arrangement of a set of objects.

2 4 3 1

*The arrangement above is one possible **permutation** of the numbers 1, 2, 3, and 4.*

perpendicular bisector A perpendicular line, ray, or segment that intersects a segment at its midpoint.

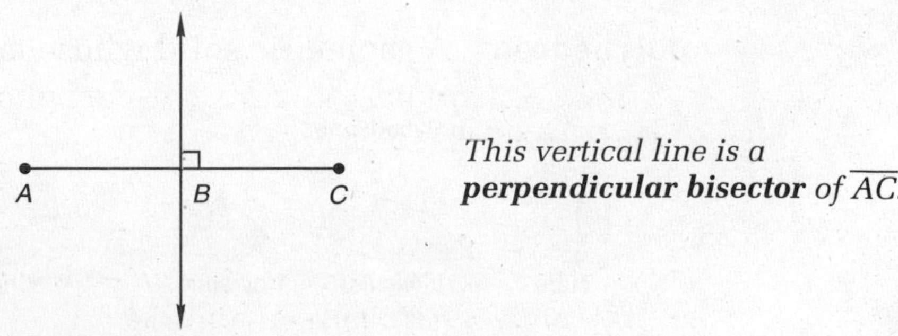

*This vertical line is a **perpendicular bisector** of $\overline{AC}$.*

perpendicular lines Two lines that intersect at right angles.

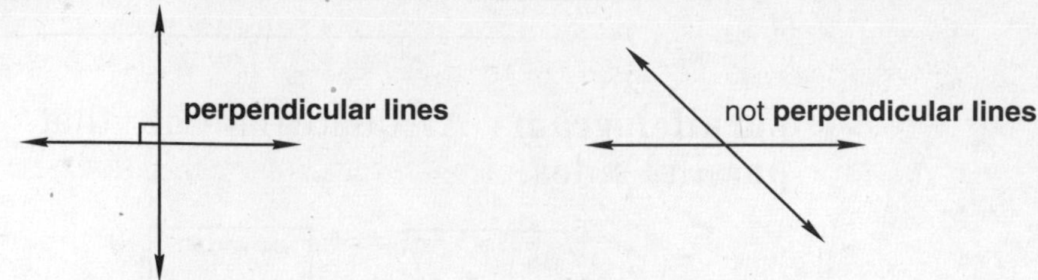

pi (π) The number of diameters equal to the circumference of a circle.

Approximate values of ***pi*** *are 3.14 and $\frac{22}{7}$.*

place value The value of a digit based on its position within a number.

$$\begin{array}{r} 341 \\ 23 \\ +\ \ 7 \\ \hline 371 \end{array}$$

Place value *tells us that the 4 in 341 is worth "4 tens." In addition problems, we align digits with the same* ***place value.***

plane A flat surface that has no boundaries.

The flat surface of a desk is part of a ***plane.***

point An exact position on a line, on a plane, or in space.

• *A* *This dot represents* ***point*** *A.*

polygon A closed, flat shape with straight sides.

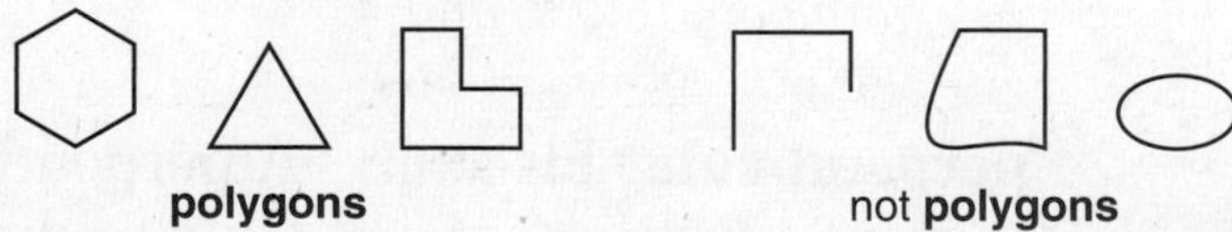

polyhedron A geometric solid whose faces are polygons.

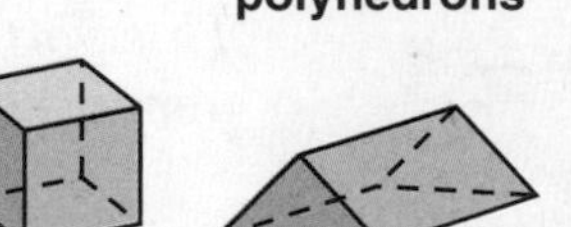
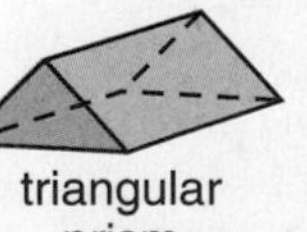
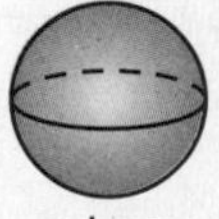
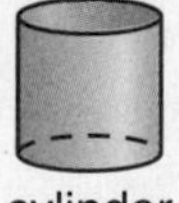

polynomial An algebraic expression that has one or more terms.

The expression $3x^2 + 13x + 12y$ *is a* ***polynomial.***

positive numbers Numbers greater than zero.

0.25 and 157 are ***positive numbers.***

−40 and 0 are not ***positive numbers.***

power (1) The value of an exponential expression.

16 is the fourth ***power*** *of 2 because* $2^4 = 16$.

(2) An exponent.

The expression 2^4 *is read "two to the fourth* ***power.****"*

prime factorization The expression of a composite number as a product of its prime factors.

The ***prime factorization*** *of 60 is* $2 \times 2 \times 3 \times 5$.

prime factors The factors of a number that are prime numbers.

The factors of 45 are 1, 3, 5, 9, 15, and 45. Its ***prime factors*** *are 3 and 5.*

prime number A counting number greater than 1 whose only two factors are the number 1 and itself.

7 is a ***prime number.*** *Its only factors are 1 and 7.*

10 is not a ***prime number.*** *Its factors are 1, 2, 5, and 10.*

principal The amount of money borrowed in a loan, deposited in an account that earns interest, or invested in a fund.

If we borrow $750.00, our ***principal*** *is $750.00.*

prism A polyhedron with two congruent parallel bases.

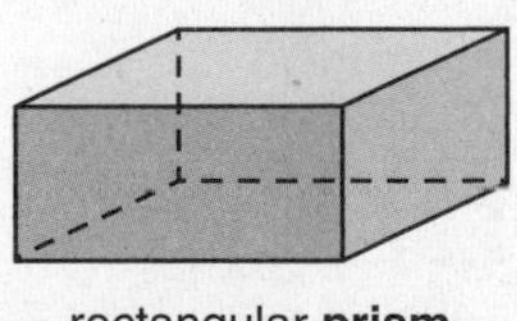

rectangular **prism**

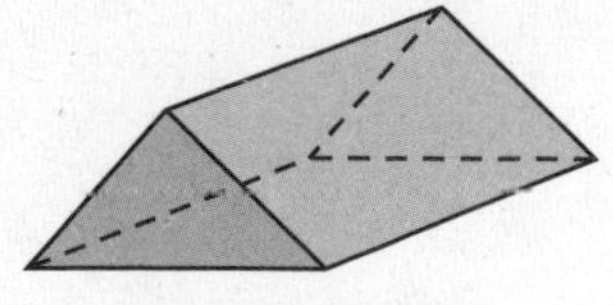

triangular **prism**

probability A way of describing the likelihood of an event; the ratio of favorable outcomes to all possible outcomes.

*The **probability** of rolling a 3 with a standard number cube is* $\frac{1}{6}$.

product The result of multiplication.

$5 \times 4 = 20$ *The **product** of 5 and 4 is 20.*

proof A method that uses logical steps to describe how certain given information can lead to a certain conclusion.

*For an example of a **proof**, please see the **proof** of the Pythagorean theorem in Investigation 12.*

property of zero for multiplication Zero times any number is zero. In symbolic form, $0 \times a = 0$.

*The **property of zero for multiplication** tells us that* $89 \times 0 = 0$.

proportion A statement that shows two ratios are equal.

$\frac{6}{10} = \frac{9}{15}$ *These two ratios are equal, so this is a **proportion**.*

protractor A tool that is used to measure and draw angles.

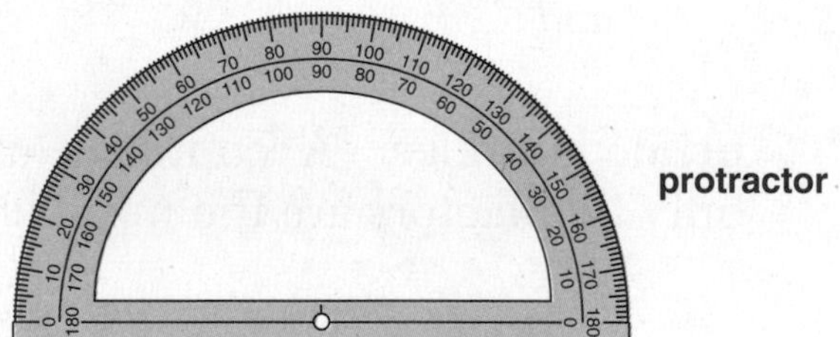

Pythagorean theorem The area of a square constructed on the hypotenuse of a right triangle is equal to the sum of the areas of squares constructed on the legs of the right triangle.

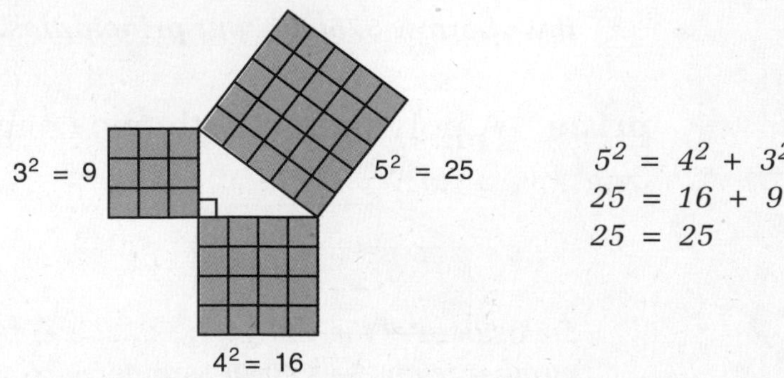

$$5^2 = 4^2 + 3^2$$
$$25 = 16 + 9$$
$$25 = 25$$

quadrant A region of a coordinate plane formed when two perpendicular number lines intersect at their origins.

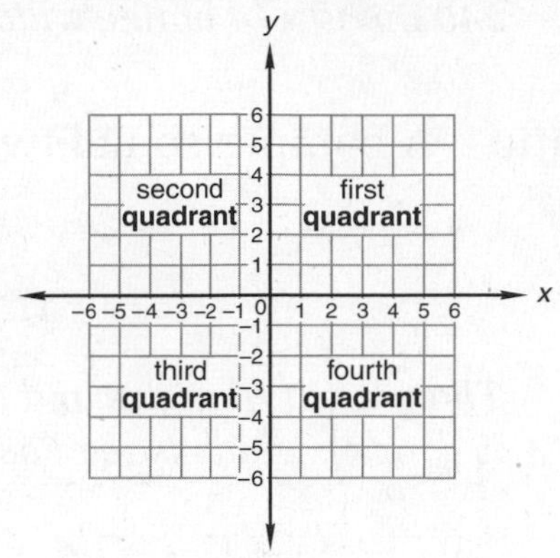

quotient The result of division.

$12 \div 3 = 4$ $\quad 3\overline{)12}$ with quotient 4 $\quad \frac{12}{3} = 4$ *The* ***quotient*** *is 4 in each of these problems.*

radical expression An expression that indicates the root of a number. A radical expression contains a radical sign, $\sqrt{\ }$.

radical expressions		not radical expressions	
$\sqrt{15^2}$	$\sqrt{9}$	$2 + 4$	16
$\sqrt{x}$	$2 + \sqrt{13}$	xy	4133

radius (Plural: *radii*) The distance from the center of a circle or sphere to a point on the circle or sphere.

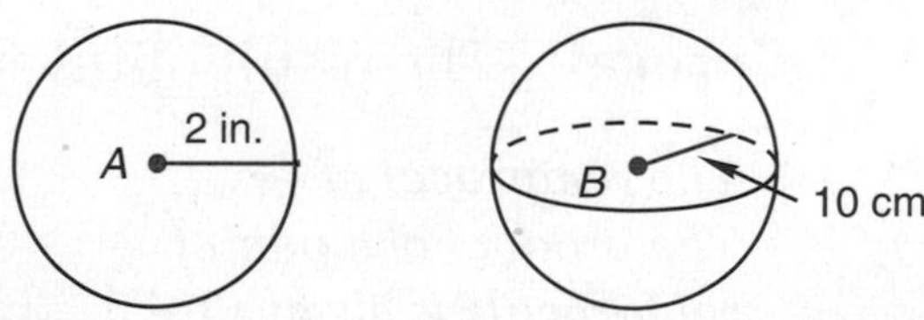

The ***radius*** *of circle A is 2 inches.*
The ***radius*** *of sphere B is 10 centimeters.*

range The difference between the largest number and smallest number in a list.

5, 17, 12, 34, 29, 13

To calculate the ***range*** *of this list, we subtract the smallest number from the largest number. The* ***range*** *of this list is 29.*

rate A ratio of two measures.

If a car travels 240 miles in 4 hours, its average ***rate*** *is 240 miles ÷ 4 hours, which equals 60 miles per hour (mph).*

ratio A comparison of two numbers by division.

△△△

☆☆☆☆☆☆

There are 3 triangles and 6 stars. The ***ratio*** *of triangles to stars is* $\frac{3}{6}$ $\left(\text{or } \frac{1}{2}\right)$*, which is read as "3 to 6" (or "1 to 2").*

rational numbers All numbers that can be written as a ratio of two integers.

$\frac{15}{16}$ *and 37 are* ***rational numbers.***

$\sqrt{2}$ *and* π *are not* ***rational numbers.***

ray A part of a line that begins at a point and continues without end in one direction.

A B

ray *AB*

real numbers All the numbers that can be represented by points on a number line.

The family of ***real numbers*** *is composed of all rational and irrational numbers.*

reciprocal The result of inverting a fraction.

The ***reciprocal*** *of* $\frac{3}{4}$ *is* $\frac{4}{3}$*.*
The product of a pair of ***reciprocals*** *is always 1.*

$$\frac{3}{4} \times \frac{4}{3} = \frac{12}{12} = 1$$

rectangle A quadrilateral that has four right angles.

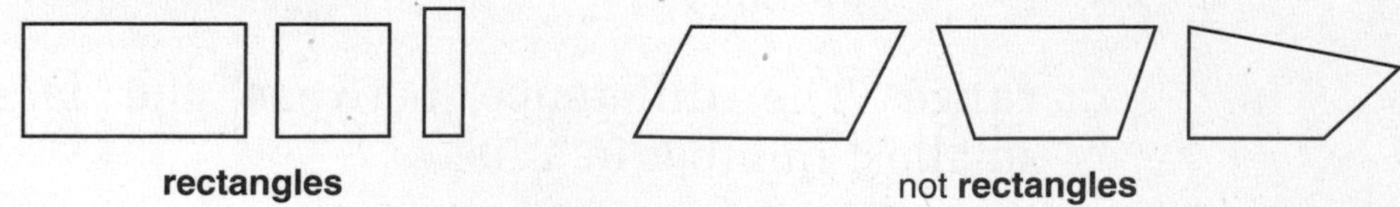

rectangles not **rectangles**

reduce To rewrite a fraction in lowest terms.

If we ***reduce*** *the fraction* $\frac{9}{12}$*, we get* $\frac{3}{4}$*.*

reflection Flipping a figure to produce a mirror image.

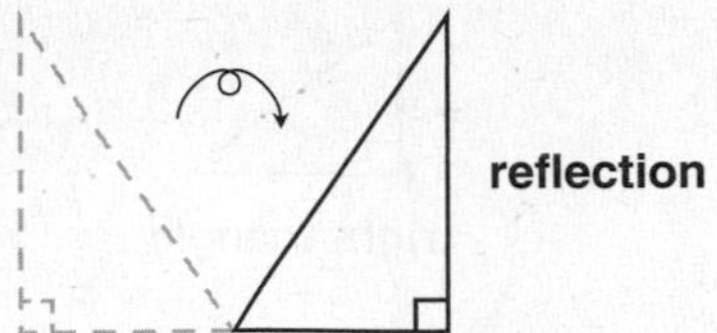

regular polygon A polygon in which all sides have equal lengths and all angles have equal measures.

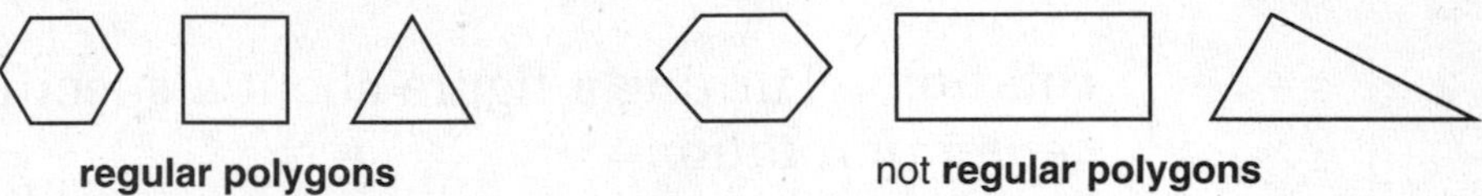

repetend The repeating digits of a decimal number. The symbol for a repetend is an overbar.

$$0.83333333... = 0.8\overline{3}$$

*In the number above, 3 is the **repetend.***

rhombus A parallelogram with all four sides of equal length.

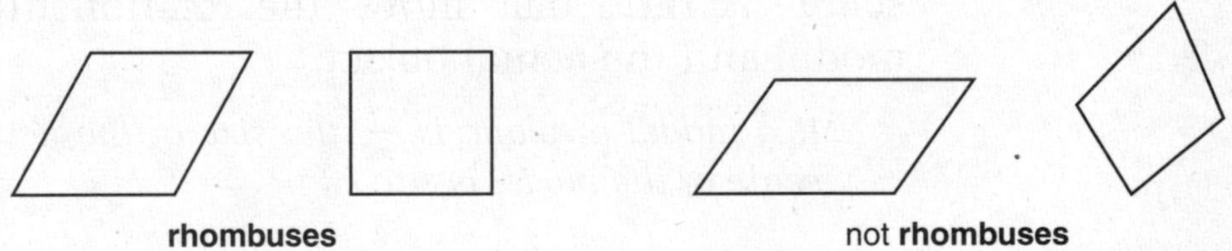

right angle An angle that forms a square corner and measures 90°. It is often marked with a small square.

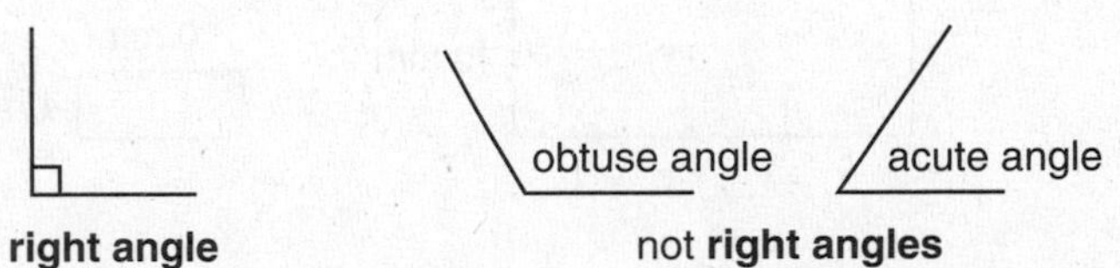

*A **right angle** is larger than an acute angle and smaller than an obtuse angle.*

right solid A geometric solid whose sides are perpendicular to its base.

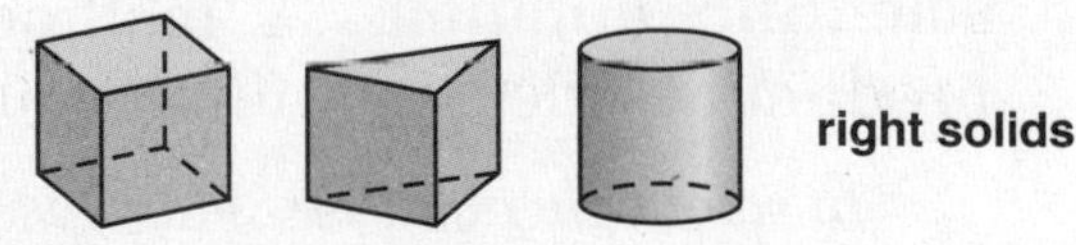

right triangle A triangle whose largest angle measures 90°.

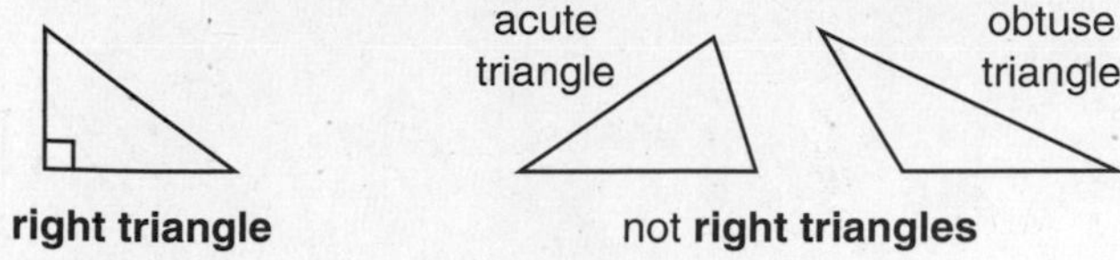

root A value of a radical expression.

$$\sqrt{16} = 4$$

*4 is a **root** of this radical expression.*

rotation Turning a figure about a specified point called the *center of rotation.*

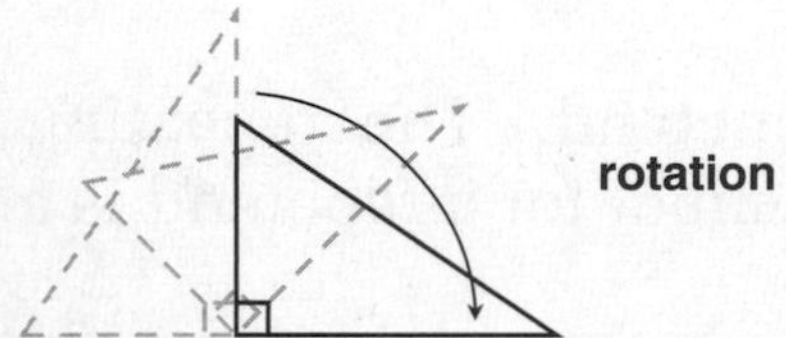

sample space A list of all the possible outcomes of an event.

*The **sample space** of outcomes when flipping a coin consists of* heads *and* tails.

scale A ratio that shows the relationship between a scale model and the actual object.

*If a model airplane is $\frac{1}{24}$ the size of the actual airplane, the **scale** of the model is 1 to 24.*

scale factor The number that relates corresponding sides of similar geometric figures.

25 mm
10 mm
10 mm
4 mm

*The **scale factor** from the smaller rectangle to the larger rectangle is 2.5.*

scalene triangle A triangle with three sides of different lengths.

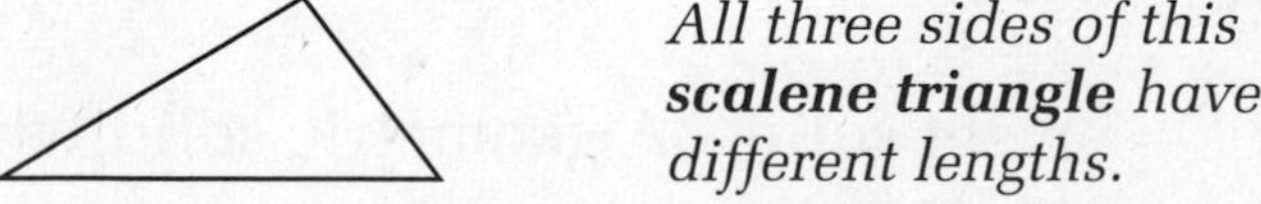

*All three sides of this **scalene triangle** have different lengths.*

scientific notation A method of writing a number as a product of a decimal number and a power of 10.

*In **scientific notation,** 34,000 is written 3.4×10^4.*

sector A region that is bordered by an arc and two radii of a circle.

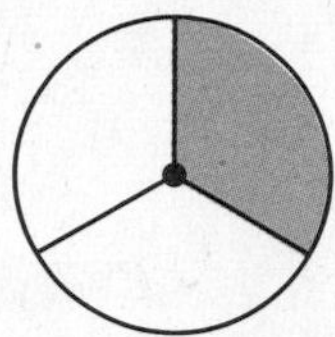

This circle is divided into 3 ***sectors.***

segment A part of a line with two distinct endpoints.

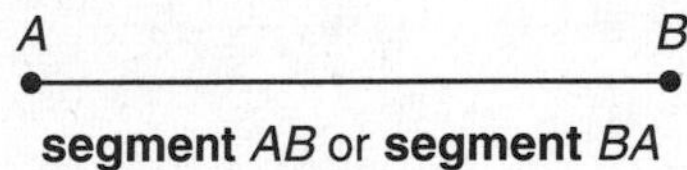

semicircle A half circle.

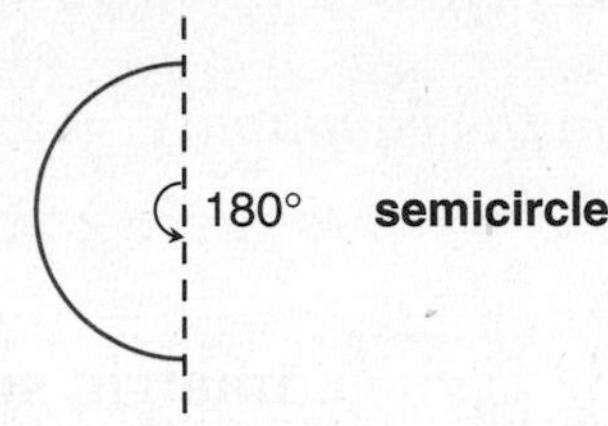

A ***semicircle*** *is an arc whose measure is 180°.*

sequence A list of numbers arranged according to a certain rule.

The numbers 2, 4, 6, 8, ... form a ***sequence.*** *The rule is "count up by twos."*

signed numbers Numbers that are either positive or negative.

–5 and +20 are ***signed numbers.***

0 is the only real number that is not a ***signed number.***

similar Having the same shape but not necessarily the same size. Corresponding parts of similar figures are proportional.

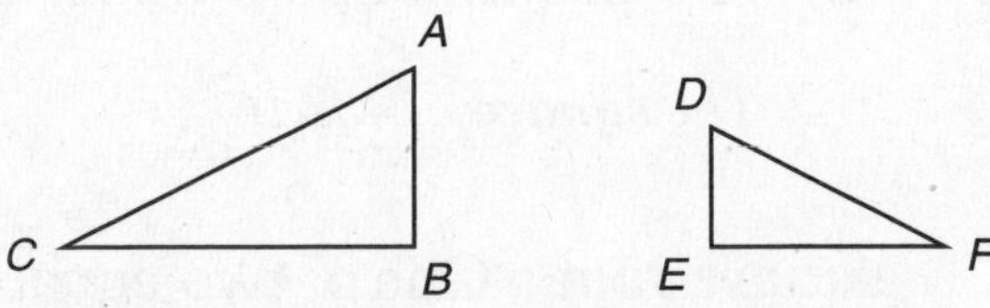

ΔABC and ΔDEF are ***similar.*** *They have the same shape, but not the same size.*

slope The number that represents the slant of the graph of a linear equation.

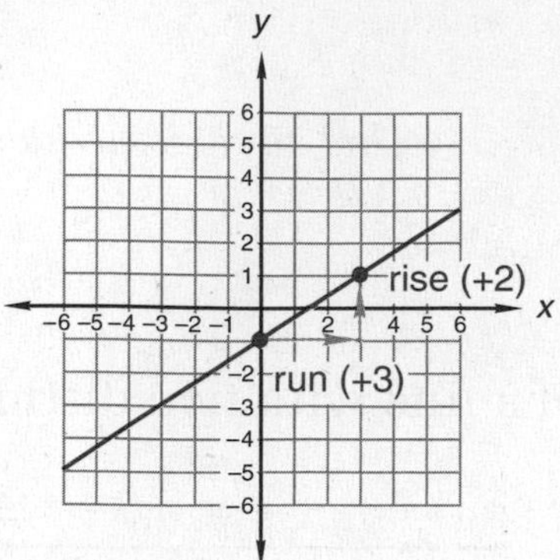

Every vertical increase (rise) of 2 units leads to a horizontal increase (run) of 3 units, so the ***slope*** *of this line is* $\frac{2}{3}$.

slope-intercept form The form $y = mx + b$ for a linear equation. In this form m is the slope of the graph of the equation, and b is the y-intercept.

In the equation $y = 2x + 3$, *the* ***slope*** *is 2 and the* ***y-intercept*** *is 3.*

solid *See* **geometric solid.**

sphere A round geometric surface whose points are an equal distance from its center.

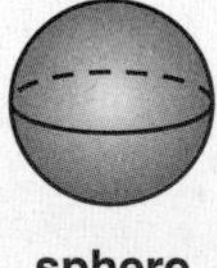

sphere

square (1) A rectangle with all four sides of equal length.

2 in.

2 in. 2 in.

2 in.

All four sides of this ***square*** *are 2 inches long.*

(2) The product of a number and itself.

The ***square*** *of 4 is 16.*

square root One of two equal factors of a number. The symbol for the principal, or positive, square root of a number is $\sqrt{\ }$.

A ***square root*** *of 49 is 7 because* $7 \times 7 = 49$.

stem-and-leaf plot A method of graphing a collection of numbers by placing the "stem" digits (or initial digits) in one column and the "leaf" digits (or remaining digits) out to the right.

Stem	Leaf
2	1 3 5 6 6 8
3	0 0 2 2 4 5 6 6 8 9
4	0 0 1 1 1 2 3 3 5 7 7 8
5	0 1 1 2 3 5 8

*In this **stem-and-leaf plot,** 3|2 represents 32.*

straight angle An angle that measures 180° and thus forms a straight line.

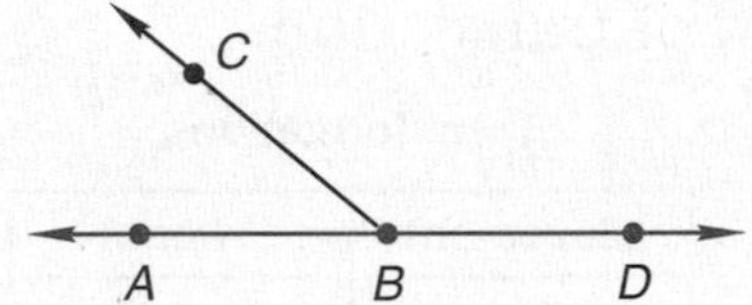

*Angle ABD is a **straight angle.** Angles ABC and CBD are not **straight angles.***

subtrahend A number that is subtracted.

12 − 8 = 4 *The **subtrahend** in this problem is 8.*

sum The result of addition.

7 + 6 = 13 *The **sum** of 7 and 6 is 13.*

supplementary angles Two angles whose sum is 180°.

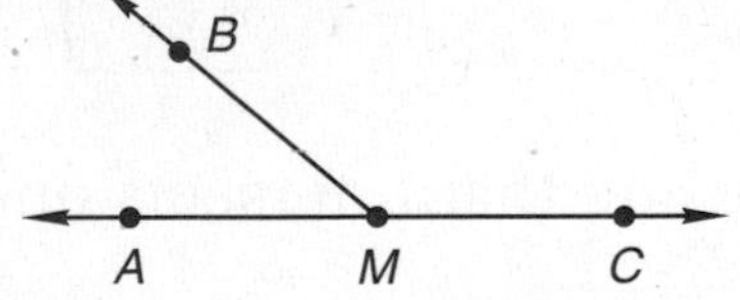

*∠AMB and ∠CMB are **supplementary.***

surface area The total area of the surface of a geometric solid.

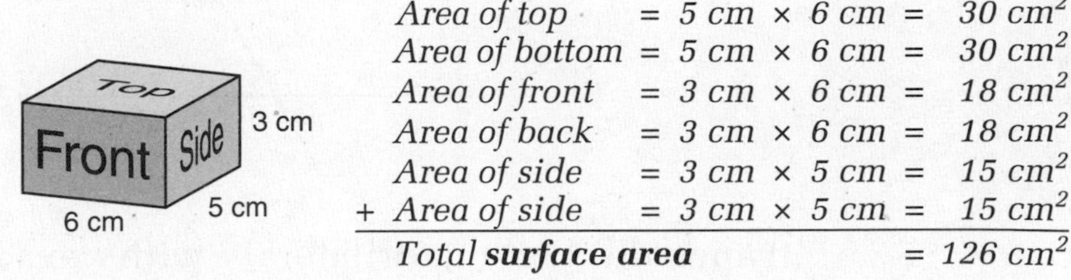

Area of top	= 5 cm × 6 cm =	30 cm^2
Area of bottom	= 5 cm × 6 cm =	30 cm^2
Area of front	= 3 cm × 6 cm =	18 cm^2
Area of back	= 3 cm × 6 cm =	18 cm^2
Area of side	= 3 cm × 5 cm =	15 cm^2
+ *Area of side*	= 3 cm × 5 cm =	15 cm^2
*Total **surface area***		= 126 cm^2

symbols of inclusion Symbols that are used to set apart portions of an expression so that they may be evaluated first: ( ), [ ], { }, and the division bar in a fraction.

*In the statement (8 − 4) ÷ 2, the **symbols of inclusion** indicate that 8 − 4 should be calculated before dividing by 2.*

term (1) A number that serves as a numerator or denominator of a fraction.

$\frac{5}{6}$ ⟩ terms

(2) One of the numbers in a sequence.

1, 3, 5, 7, 9, 11, ...

*Each number in this sequence is a **term.***

(3) A constant or variable expression composed of one or more factors in an algebraic expression.

*The expression $2x + 3xyz$ has two **terms.***

transformation The changing of a figure's position through rotation, reflection, or translation.

Transformations

Movement	Name
flip	reflection
slide	translation
turn	rotation

translation Sliding a figure from one position to another without turning or flipping the figure.

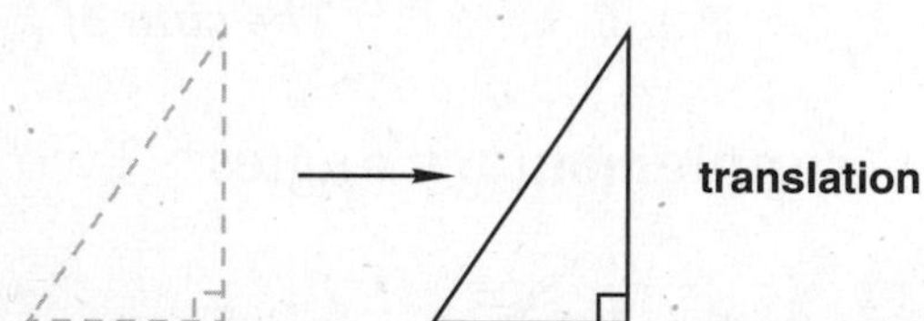

transversal A line that intersects one or more other lines in a plane.

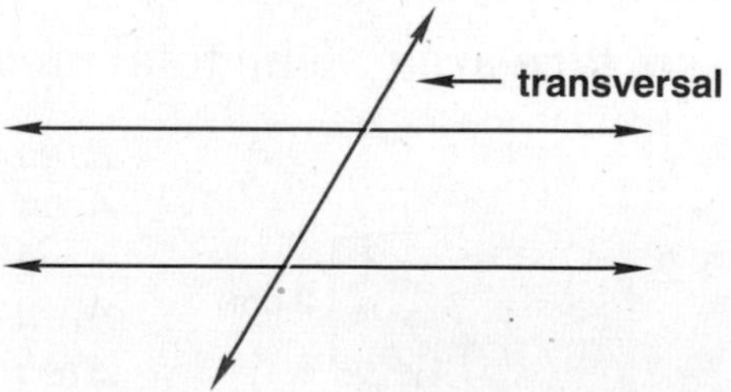

trapezoid A quadrilateral with exactly one pair of parallel sides.

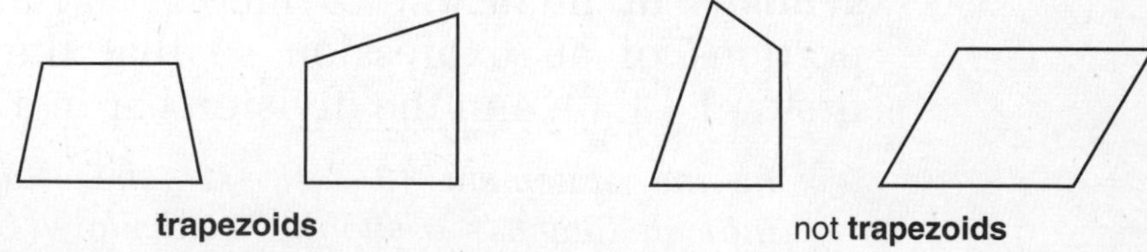

unit conversion The process of changing a measure to an equivalent measure that has different units.

*Through **unit conversion,** we can write 2 feet as 24 inches.*

unit multiplier A ratio equal to 1 that is composed of two equivalent measures.

$$\frac{12 \text{ inches}}{1 \text{ foot}} = 1$$

*We can use this **unit multiplier** to convert feet to inches.*

unit price The price of one unit of measure of a product.

*The **unit price** of bananas is $1.19 per pound.*

U.S. Customary System A system of measurement used almost exclusively in the United States.

*Pounds, quarts, and feet are units in the **U.S. Customary System.***

variable A letter used to represent a number that has not been designated.

*In the statement $x + 7 = y$, the letters x and y are **variables.***

vertex (Plural: *vertices*) A point of an angle, polygon, or polyhedron where two or more lines, rays, or segments meet.

*One **vertex** of this cube is colored. A cube has eight **vertices.***

vertical angles A pair of nonadjacent angles formed by a pair of intersecting lines. Vertical angles have the same measure.

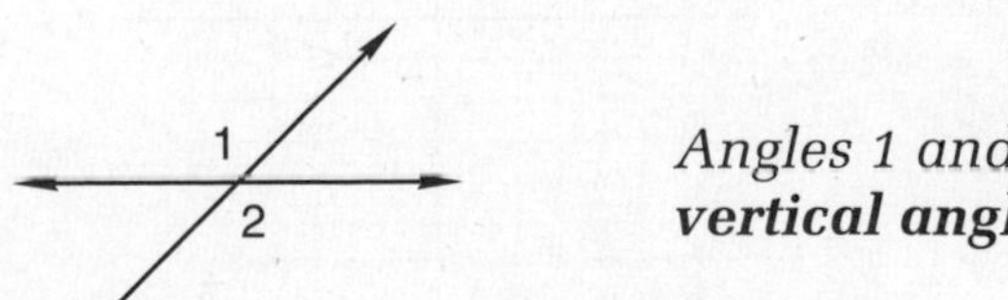

*Angles 1 and 2 are **vertical angles.***

volume The amount of space a solid shape occupies. Volume is measured in cubic units.

*This rectangular prism is 3 units wide, 3 units high, and 4 units deep. Its **volume** is 3 · 3 · 4 = 36 cubic units.*

whole numbers The members of the set {0, 1, 2, 3, 4, ...}.

*0, 25, and 134 are **whole numbers.***

*−3, 0.56, and $100\frac{3}{4}$ are not **whole numbers.***

x-axis The horizontal number line of a coordinate plane.

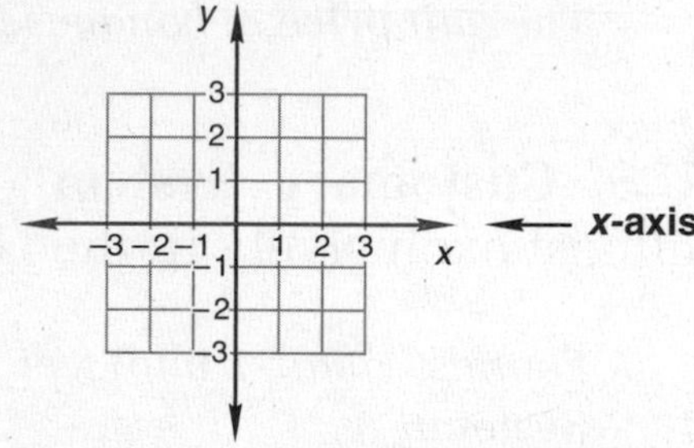

y-axis The vertical number line of a coordinate plane.

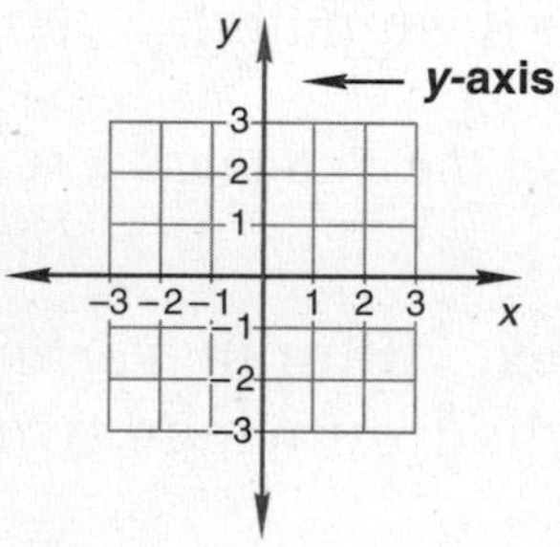

y-intercept The point on a coordinate plane where the graph of an equation intersects the *y*-axis.

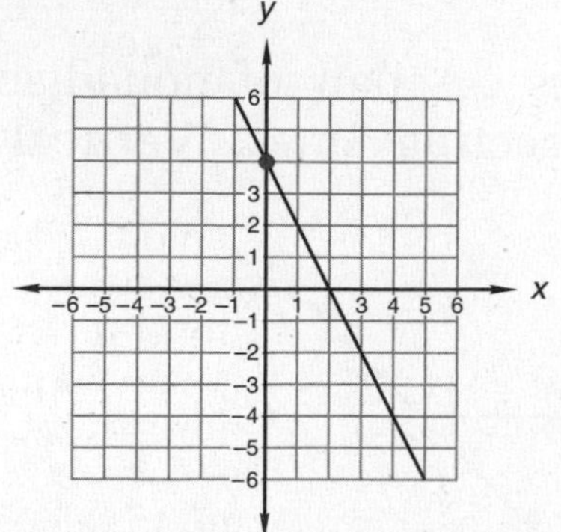

*This line crosses the y-axis at y = 4, so the **y-intercept** of this line is 4.*

INDEX

Note: Asterisks (*) indicate that the cited topic is covered in a lesson Warm-up.

Page locators followed by the letter "n" are references to footnotes on the indicated pages.

D

E

F

G

H

I

K

L

M

N

O

P

Q

R

S

T

U

X

Y

Z